Industrial Chemistry Library, Volume 9

High Pressure Process Technology: Fundamentals and Applications

Industrial Chemistry Library

Advisory Editor: S.T. Sie, *Faculty of Chemical Technology and Materials Science*
Delft University of Technology, Delft, The Netherlands

Industrial Chemistry Library, Volume 9

High Pressure Process Technology: Fundamentals and Applications

Edited by

A. Bertucco
*Università di Padova, DIPIC – Department of Chemical Engineering,
Via F. Marzolo 9, I-35131 Padova PD, Italy*

G. Vetter
*Universität Erlangen-Nürnberg, Department of Process Technology and
Machinery, Cauerstrasse 4, D-91058 Erlangen, Germany*

2001
ELSEVIER
Amsterdam – London – New York – Oxford – Paris – Shannon – Tokyo

ELSEVIER SCIENCE B.V.
Sara Burgerhartstraat 25
P.O. Box 211, 1000 AE Amsterdam, The Netherlands

First edition 2001

Library of Congress Cataloging in Publication Data
A catalog record from the Library of Congress has been applied for.

ISBN: 0-444-50498-2

♾ The paper used in this publication meets the requirements of ANSI/NISO Z39.48-1992 (Permanence of Paper).

Printed in The Netherlands.

v

PREFACE

The application of elevated pressures in the manufacture of high technology products is permanently extending and offering new opportunities. Nowadays this is true not only for reactions and separations during chemical processing, but also for other production activities such as jet-cutting, homogenization, micronization, pressing, plastification, spray-drying and for physico-biological treatments such as pasteurization, sterilization and coagulation. At the dawn of the new century, it is quite evident that high pressure technology is one of the emerging tools and methods for improving product quality, both from the economic and the environmental viewpoints, and for the development of more sustainable processes and products for the future generations.

Although the development of classical high pressure production dates back to the 1920s and 1930s (ammonia, low-density polyethylene, synthetic diamonds, etc.), research in this field has been particularly active in the last twenty-five years, leading to a number of new opportunities expanding to areas such as materials science and microbiology, and to the bulk production of foods, pharmaceuticals, cosmetics, and other products. This is also due to the exploitation of the properties of fluids at the supercritical state, especially supercritical water and supercritical carbon dioxide, which are expected within a few years to offer alternatives to organic solvents in many widespread applications.

On the other hand, high pressure technology is usually regarded as a highly specific field, to which little space is devoted within scientific and technical curricula throughout the world, so that a "high-pressure culture" is not widespread and the related expertise is difficult to find, even among physicists, chemists and chemical engineers. In addition, the fear of dealing with high pressures in production plants always appears as a major issue, and therefore dissemination of the related knowledge and expertise among the manufacturing community deserves maximum attention.

Of course, visions for and problems with the application of high pressure have been discussed and promoted by national and international working groups, both in Europe and overseas, for many years. Within the European Federation of Chemical Engineering the working party on High Pressure Technology, now in its second decade and comprising members from twelve European countries, has developed initiatives for the transfer of scientific and technological knowledge in an outstandingly efficient manner.

In Europe, one important pillar of these activities is represented by the institution of an Intensive Course on High Pressure Technology offered annually to European post-graduate students and funded by the European Union within the framework of the Socrates Programme. The course has now been rotating for several years between major European universities. Most of the working party members as well as other experts have contributed lectures, discussions and class-work problems as well as final examinations. From the beginning of the course programme the firm intention was to publish the high-level teaching and educational documentation as a book for a larger community of users, and we are pleased to present this work now.

A special effort was made to organize and present the matter in such a way that a larger group of readers and experts can take advantage of it. The book is intended to provide a comprehensive approach to the subject, so that it can be interesting not only for specialists, such as mechanical and chemical engineers, but also beginners with high-pressures who would like to apply this kind of technology, but somehow are afraid of dealing with it either on a research- or production scale — biologists, chemists, environmental engineers, food technologists, material scientists, pharmacists, physicists, and others.

The content of the book, structured into nine chapters, each being sub-divided into a number of sections, results from the long-term course presentations and the many connected discussions.

In the *first Chapter* an overview of the general topic is presented. The motivations of using high pressure today are summarized and a number of examples provided which relate to high-pressure production processes applied currently.

Chapter Two deals with the basic concepts of high-pressure thermodynamic and phase equilibrium calculations. Experimental methods and theoretical modelling are described briefly in order to give both a comprehensive view of the problems, and suggestions and references to more detailed treatments.

The problem of the evaluation of kinetic properties is addressed in *Chapter Three,* including both chemical and physical kinetic phenomena.

Then, in the *Fourth Chapter* the design and construction of high pressure equipment is considered, with reference to research and pilot units, and production plants as well. This is a very important part of the book, as it clearly shows that running high pressure apparatus is neither difficult nor hazardous, provided some well established criteria are followed both during design and operation.

Industrial reaction units are discussed in *Chapter Five,* where all the main issues related to catalytic reactors are discussed, and a special emphasis is paid to polymeric reactors.

The problems connected-with separation processes, units, and equipment are treated in the *Sixth Chapter,* focusing the reader's attention on high-pressure distillation and on dense-gas extraction from solids and liquids.

Relevant safety issues arising in the design and operation of high-pressure plants are addressed in *Chapter Seven.* After a general section where testing procedures, safe plant operation, and inspection are summarized, two examples are dealt with in detail: dense-gas extraction units and polymerization reactors.

Chapter Eight is concerned with a major question connected with the development of high pressure technologies in the process and chemical industry, *i.e.*, the economic evaluation of production carried out at high pressures. In this case, also, the matter is discussed in relation to three important examples: dense gas extraction, polymerization and supercritical anti-solvent precipitation processes.

Finally, *Chapter Nine* is a collection of currently used and (mostly) potential applications. Even though it cannot cover all possibilities and ideas put forward continuously by researchers and companies, the proposed examples provide a thorough view of the opportunities offered by the extensive use of high pressure technology in many fields.

The book, written by experts in high pressure technology, is intended to act as a guide for those who are planning, designing, researching, developing, building and operating high pressure processes, plants and components. The large number of references included will support the efficient transfer of the actual state of our knowledge. The examples and problems, which illustrate the numerical application of the formulas and the diagrams, will provide the reader with helpful tools for becoming acquainted with high-pressure technology.

We would like to thank all the contributors for their excellent co-operation, and Elsevier for their support during the editing procedure and for the readiness to publish the book. A special acknowledgement is devoted to Ing. Monica Daminato for her full commitment and precious help during the final editing of the manuscript and preparation of the camera-ready copy.

The editors hope that the book will be well accepted and that it will help to promote the further development of high-pressure technology in the future.

April 2001

Alberto Bertucco   Gerhard Vetter
University of Padova   University of Erlangen-Nürnberg

CONTENTS

LIST OF CONTRIBUTORS

Alberto Bertucco
Dipartimento di Principi e Impianti di Ingegneria Chimica (DIPIC) Università di Padova
Via Marzolo, 9 I-35131 Padova Italy

Maria Josè Cocero
Departamento de Ingenierìa Quìmica, Universidad de Valladolid
Prado de la Madalena SP-47005 Valladolid, Spain

Nicola Elvassore
Dipartimento di Principi e Impianti di Ingegneria Chimica (DIPIC) Università di Padova
Via Marzolo, 9 I-35131 Padova Italy

Theo W. De Loos
Faculty of Applied Science, Department of Chemical Technology
Laboratory of Applied Thermodynamics and Phase Equilibria
Delft University of Technology
Julianalaan 136 NL-2628 BL Delft, The Netherlands

Thomas Gamse
Institut fur Thermische Verfahrenstechnik und Umwelttechnik Erzherzog Johann Universität
Infeldgasse, 25 A-8010 Graz, Austria

Sander van den Hark
Department of Food Science, Chalmers University of Technology
P.O. Box 5401 SE-40229 Göteborg, Sweden

Magnus Härröd
Department of Food Science, Chalmers University of Technology
P.O. Box 5401 SE-40229 Göteborg, Sweden

Željko Knez
Department of Chemical Engineering University of Maribor
P.O. Box 222, Smetanova 17, SI-2000 Maribor, Slovenia

Ireneo Kikic
Dipartimento di Ingegneria Chimica, dell'Ambiente e delle Materie Prime
Universitá degli Studi di Trieste
Piazzale Europa, 1 I-34127 Trieste, Italy

Ludo Kleintjens
DSM Research and Patents
Postbus 18, NL-6160 MD Geleen, The Netherlands

Eduard Lack
NATEX GmbH Prozesstechnologie
Hauptstrasse, 2 A-2630 Ternitz, Austria

André Laurent
Ecole Nationale Supèrieure des Industries Chimiques (ENSIC)
B. P. No. 451, 1. Rue Granville F-54001 Nancy Cedex, France

Gerhard Luft
Department of Chemistry, Darmstadt University of Technology
Petersenstr. 20, D-64287 Darmstadt, Germany

Siegfried Maier
Formerly Research and Development, BASF AG,
D 67056 Ludwigshafen, Germany

Maj-Britt Macher
Department of Food Science, Chalmers University of Technology
P.O. Box 5401 SE-40229 Göteborg, Sweden

Rolf Marr
Institut fur Thermische Verfahrenstechnik und Umwelttechnik Erzherzog Johann Universität
Infeldgasse, 25 A-8010 Graz, Austria

Nicola Meehan
School of Chemistry, University of Nottingham
University Park, Nottingham NG7 2RD England

Poul Møller
Augustenborggade 21B, DK-8000 Aarhus C, Denmark

Paolo Pallado
via M. Ravel, 8   35132 Padova   Italy

Martyn Poliakoff
School of Chemistry, University of Nottingham
University Park, Nottingham NG7 2RD England

Francisco Recasens
Universitat Politécnica de Catalunya
Departamento de Ingenierìa Quìmica
E.T.S.I.I.B. Diagonal, 647 E-08028 Barcelona, Spain

Helmut Seidlitz
NATEX GmbH Prozesstechnologie
Hauptstrasse, 2 A-2630 Ternitz, Austria

Bela Simándi
Budapest University of Technology and Economics, Department of Chemical Engineering
Budapest, XI., Müegyetem rkp. 3. K. ép.mfsz. 56. H-1521, Budapest, Hungary

Sara Spilimbergo
Dipartimento di Principi e Impianti di Ingegneria Chimica (DIPIC) Università di Padova
Via Marzolo, 9 I-35131 Padova Italy

Alberto Striolo
Dipartimento di Principi e Impianti di Ingegneria Chimica (DIPIC) Università di Padova
Via Marzolo, 9 I-35131 Padova Italy

Fakher Trabelsi
Universitat Politécnica de Catalunya
Departamento de Ingenierìa Quìmica
E.T.S.I.I.B. Diagonal, 647   E-08028 Barcelona, Spain

Enrique Velo
Universitat Politécnica de Catalunya
Departamento de Ingenierìa Quìmica
E.T.S.I.I.B. Diagonal, 647   E-08028 Barcelona, Spain

Gerhard Vetter
Department of Process Machinery and Equipment,
University of Erlangen-Nuremberg, Cauerstr. 4, D 91054 Erlangen, Germany

Guenter Weickert
P.O. Box 217 NL-7500 AE Enschede, The Netherlands

Eckhard Weidner
Lehrstuhl für Verfahrenstechnische Transportprozesse, University Bochum
Universitätsstr. 150, 44780 Bochum, Germany

Federico Zanette
Dipartimento di Principi e Impianti di Ingegneria Chimica (DIPIC) Università di Padova
Via Marzolo, 9 I-35131 Padova Italy

ABOUT THE EDITORS

Alberto Bertucco (age 46), Professor of Chemical Engineering at the University of Padova, Italy: Chairman of the Working Party High Pressure Technology of the European Federation of Chemical Engineers, with long-term research activity in the field of Supercritical Fluids Applications.

Gerhard Vetter (age 68), Professor of Chemical Engineering at the University of Erlangen-Nuremberg, Germany: many years of experience in High Pressure Plant Equipment and Process Machinery for Fluids and Bulk Solids.

1

CHAPTER 1

INTRODUCTION

G. Vetter

Department of Process Machinery and Equipment
University Erlangen-Nuremberg, Cauerstr. 4, D-91058 Erlangen, Germany

The definition of high pressure, examples in nature, and the early historical roots of high pressure technology are explained. The motivation of using high pressure today is based on chemical, physico-chemical, physico-bio-chemical, physico-hydrodynamical and physico-hydraulic effects. A survey of today high pressure technology is given demonstrating the large range of applications and comprising many branches and processes of production. A number of examples like the production of polyethylene and fatty alcohols, the decaffeination of coffee beans, the homogenisation of foodstuffs, the water-jet cutting and cleaning, the polymer processing, the ultra-high pressure treatment for the aseptic processing as well as the hydrostatic pressure applications for pressing hydroforming and autofrettage are outlined shortly.

1.1 High Pressure definitions and examples in nature

Within man´s living environment on this planet the pressure ranges from a low vacuum – around 0,25 bar on top of the highest mountain – up to a high pressure of around 1000 bar – on the deepest ocean floor – both exceeding the physiological limits of human beings more or less drastically.

In general living beings on the planet earth are behaving very differently with respect to their compatibility towards pressurized environment. Some species of microbes are able to suffer several thousand bar and there are sea mammals such as whales which dive down to a depth of 1000 m – equal to a pressure difference of 100 bar – within short time intervals, a procedure which would kill human beings immediately.

In the interior of our planet millions of bar are to be expected. On the other hand we are able to develop hundreds of thousands of bar during the technical synthesis of diamonds. Fundamental physical research about the behaviour of matter has now been extended beyond the level of one million bar.

It is a characteristic feature of technical processes with high pressure conditions to exhibit absolutely artificial environments, far beyond those existing in nature. High pressure machinery and containment are required to maintain these, "artificial conditions". With regards to the term "High Pressure" we should not become confused by linguistic terms such as high blood pressure, high-pressure areas in weather forecasts, high political or moral psychological pressure, pressure exerted from above and below, etc.

The "high pressure" this book is focused on represents the physical pressure defined as the force load per the unit of area (Newton/m^2: N/m^2; 10^5 N/m^2 = 1 bar) exhibiting the "normal" atmospheric pressure of our natural environment (ambient or barometric pressure).

1.2 Early historical roots of high pressure technology

The well-known first double-piston pumps of Ktesebios during Archimedes´ time, water supply pipes in the ancient world together with Roman pump developments, as well as Agricola´s (see: Twelve Books of Mining 1596) wooden "high pressure pumps" for the drainage of mines (100 m depth = 10 bar pressure) during the Middle Ages show early applications of high pressure.

James Watt´s steam engine (around 1785) working with several bar steam pressure only, innovated the world´s energy supply and induced an industrial revolution. This steam engine represented one of earliest high pressure processes for power generation.

Starting in the Middle Ages, from the development of firearms and guns based on explosives emerged the problem of designing safe containments (gun barrels) against the high detonation pressure (today, several thousand bar).

As an early milestone of high pressure chemical processing should be mentioned the synthesis of ammonia by Haber and Bosch (Nobel prize 1918). This typical high pressure (300 – 700 bar) process already shows all the characteristics of the similar ones of today. It should be regarded as the initiation of the very successful development of the high pressure chemistry during the last century, including the still up-to-date super-pressure polymerisation of ethylene (3000 bar). Since the mid-20th century diamonds have been synthesized by transforming graphite into diamond at pressures above 120000 bar (3000°C) with a solid-state process and special apparatus.

1.3 High pressure technology today – motivations for using high pressure

High pressure is a proven tool for a number of industrial processes and promising ones in the future. The following effects of high pressure should be distinguished.

The **chemical effect of high pressure** is to stimulate the selectivity and the rate of reaction together with better product properties and quality as well as improved economy. This is based on better physico-chemical and thermodynamic reaction conditions such as density, activation volume, chemical equilibria, concentration and phase situation. Many successful reactions are basically enhanced by catalysis.

The **physico-chemical effect of high pressure,** especially in the supercritical state, to enhance the solubility and phase conditions of the components involved. Supercritical hydrogenation, or enzymatic syntheses are offer new steps with high pressure. Supercritical water oxidation at high pressure represents an efficient method for the decontamination of wastes.

From the application of high pressure liquid or supercritical carbon dioxide as a solvent have emerged a number of promising or successful production processes such as supercritical extraction, fractionation, dyeing, cleaning, degreasing and micronisation (rapid expansion, crystallization, anti-solvent recrystallization). New material properties can be achieved by foam expansion, aerogel drying, polymer processing, impregnation and cell-cracking with high pressure supercritical CO_2 [1, 2].

The **physico-bio-chemical effect of the high pressure treatment** predominantly of foodstuffs and cosmetics, is now emerging. For the sterilization (pasteurisation, pascalisation) high pressure offers an alternative to high temperature. Furthermore, treatment with static high pressure gives a promising improvement of certain organic natural products by advantageous swelling, gelation, coagulation and auto-oxidation effects in combination with fats or proteins. This selection of high pressure effects actually is however only under increasing research however only and successful practical applications have not been achieved yet [3].

The physico-hydrodynamical effect of high pressure is based on the conversion of the potential (pressure) into kinetic energy (high speed fluid jetting: 100 – 1000 m/s). The main applications are the homogenisation of fluid mixtures by expanding them through very narrow clearances, water-jet cutting and water-jet cleaning, and the generation of sprays with fine droplets for efficient combustion or spray-drying of fine particles.

The **physico-hydraulic effect of high pressure** is involved during the conveying of fluids against large differential pressures, for example the filtration of polymer melts, or pipeline transport over long distances. The hydrostatic energy is applied for hydroforming of complex metal parts, isostatic pressing for sintered products, or the autofrettage treatment of high pressure components in order to generate beneficial residual stresses [4, 5].

1.4 High pressure technology today - application survey and examples

Table 1.4-1 gives a summary about high pressure applications, exhibiting the methods, the pressure levels applied, and the products or results of the processes involved. The survey is not complete, as the development is changing and progressing permanently.

It should be pointed out at this stage that the application of high pressure as a beneficial tool for production procedures, from the experience of the past decades, is increasing and decreasing all the time. High pressure equipment and plants are expensive in their development, investment, operation, and safety aspects. So there is the general tendency to reduce the pressures as soon as the process development offers the chances (e.g. by the introduction of new catalysts) to do so.

Table 1.4-1.
Applications of high pressure

Method	Pressure (bar)	Product, application
Solid state reaction	> 125 000	synthetic diamonds
Polymerisation of ethylene	1300 – 3000	low density polyethylene
Synthesis	100 – 700	ammonia
		propionic and acetic acid
		urea (fertilizers)
		butanediol
		methanol
Fischer-Tropsch synthesis		hydrocarbons
Hydrogenation	100 – 300	edible oils
		hydrogasification
		hydrocracking
		desulfurization
		catalytic cracking
		naphtha hydroforming
		coal liquefaction
		fatty alcohols
		1-6-hexanediol
		1-4-butanediol
		hexamethylenediamine
Hydroformylation		C4 to C15 products
Wet (air) oxidation	100 – 400	organic waste elimination
Extraction with supercritical fluids (e.g., CO_2)	80 – 300	decaffeinated coffee (tea)
		spices, hops
		colours
		drugs
		oils, lecithine and fats
		tobacco (nicotine)
		perfumes

Table 1.4-1. Applications of high pressure (continued)

Micronization with supercritical fluids - Crystallization - Rapid expansion - Gas anti-solvent Recrystallization - Precipitation with compressed anti-solvent - Solution-enhanced dispersion - Particles from gas-saturated solutions	80 – 300	fine particles and powders from various products and of designed properties
Dyeing with supercritical fluids Cell structure treatment with supercritical fluids (e.g., CO_2)	80 – 300	dyeing of fabrics tobacco impregnation
Leaching of ores Cryo processing	100 – 300	aluminium (from bauxite) technical gases (N_2, O_2, H_2, He ...) gas liquefaction
Oil/gas production	100 – 400	drying inhibition desulfurization, odorization secondary and tertiary production methods drilling support
Separation of isotopes	300	heavy water
Fluid conveying (transport)	100 – 200	pipeline transport of ores and coal
Polymer processing	100 – 400	polymer spinning polymer filtration polymer extrusion
High performance liquid chromatography	100 – 700	analytical chemistry chemical production
Kinetic fluid (jet) energy with water	up to 4000 up to 2000 up to 600	jet cutting jet cleaning jet treatment of fabrics
Kinetic fluid energy Homogenisation Emulsification Dispersing Cell-cracking	up to 1500	foodstuffs cosmetics pharmaceutical products chemical products bio-products
Potential (pressure) fluid energy	up to 10000 up to 5000 up to 4000	autofrettage (residual stresses) hydroforming isostatic pressing (sintered parts)
Spray drying	50 – 200 (up to 1000)	fine powders of various products

Table 1.4-1.   Applications of high pressure (continued)

Fuel injection	1000 – 2000	diesel motors (improved combustion)
Thermal power generation	100 – 250	steam power plants
Potential (pressure) energy effects on organic products	up to 5000	sterilization pascalisation coagulation gelation of various foodstuffs and other bio-products

The following examples of successful and well developed high pressure processes concentrate mainly on the general aspects and a consideration of the high pressure machinery involved. The explanations will discuss primarily the general aspects and benefits of high pressure as a tool, and will not address details of the methodology.

Example 1: Production of Polyethylene (PE)

The different available high pressure polymerisation processes of polyethylene (PE) yield LDPE (low density PE), LLDPE (linear low density PE) and copolymer features of the same. The various process variations have been developed during recent decades and introduced a number of well developed steps and devices to achieve safe and economical operating conditions at the very high reaction pressures of 1500 to 3000 bar.

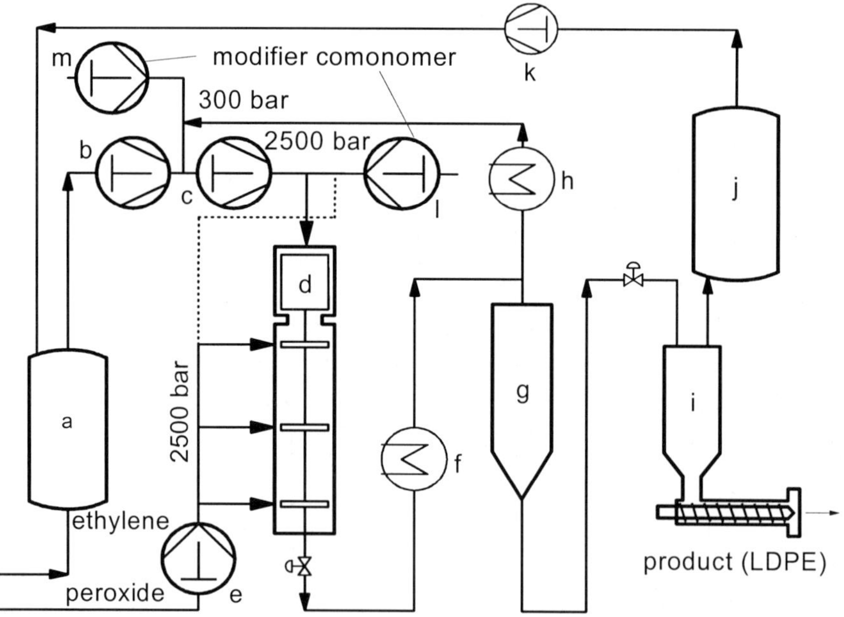

Fig. 1.4-1.   Production of LDPE

There have been ups and downs for this high pressure process, but owing to the increasing need for the types of LPDE, new and scaled-up plants are under erection.

The process (Fig. 1.4-1) makes heavy demands on the pumps, compressors, reactors, piping, fittings and valves, as well as for other devices at the pressure range mentioned. The monomer ethylene (storage tank, a) is compressed by a primary reciprocating compressor with several stages (b), up to around 300 bar, and then by a two-stage "hyper" reciprocating compressor (c) up to around 3000 bar. Between the two piston-type compressors (b and c) is the main location for injecting modifiers, especially co-monomers, in order to achieve certain modifications of the polymer properties. As these additives mainly represent solvents or liquified gases high pressure diaphragm pumps (m) must normally be applied.

The polymerisation reaction takes place in tubular or stirred vessel reactors (d) under careful control of pressure and temperature, enhanced or initiated by the injection of initiator-solvents (e) (as well as co-monomers l) which are frequently based on organic peroxides. The typical injection pumps for this metering problem are of the two-cylinder amplifier types. The further process comprises a number of further steps such as heat exchange (f, h), separation (g, j), gas recycling (k), and polymer discharge (i). The art of producing high pressure PE is based on an excellent understanding of the process and skill in designing and operating the high pressure equipment required.

Example 2: Production of unsaturated fatty alcohols
This hydrogenation process (Fig. 1.4-2) is, among others, the basis for the production of washing detergents.

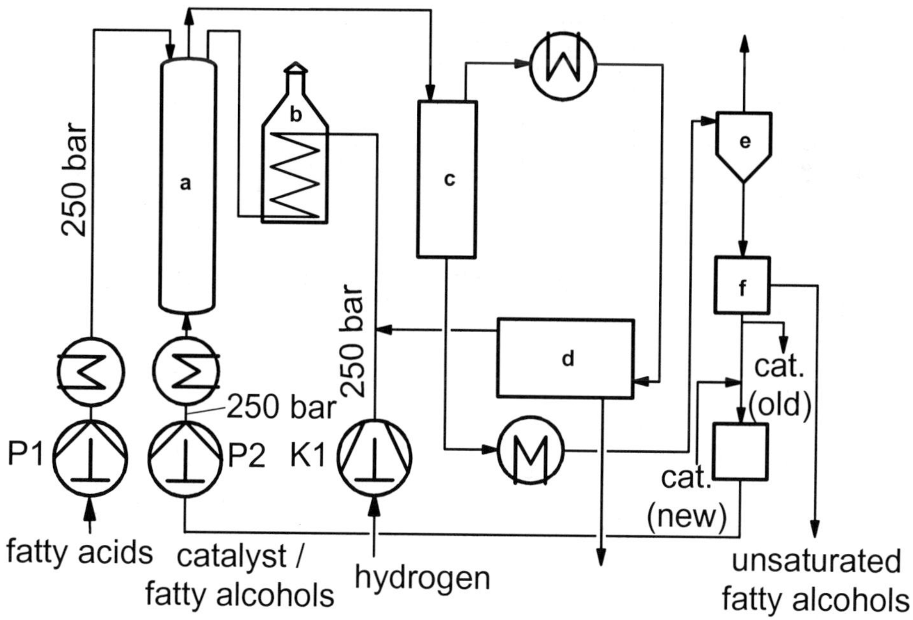

Fig. 1.4-2.   Production of unsaturated fatty alcohols

8

The feed of fatty acids and fatty alcohol with the appropriate catalyst (diaphragm pumps, P1 and P2) reacts with hydrogen (staged dry-running piston compressor, K1) to unsaturated fatty alcohols (reactor, a). Several high pressure steps such as heat exchange, separation, recycling catalyst feed (b to f) together with proper high pressure components, are required. The dry hydrogen compression is avoids any contamination of the product with lubricants. The diaphragm feed pumps offer the best service with respect to endurance and wear protection, with the lowest life-cycle costs.

Example 3: Decaffeination of coffee beans
Of the various extraction processes the decaffeination with supercritical CO_2 exhibits the most commercial advantages for bulk production. The process is a discontinuous one. Fig. 1.4-3 shows a number of serially arranged extractors (5) charged with the supercritical CO_2 feed by the centrifugal circulation pump (1).

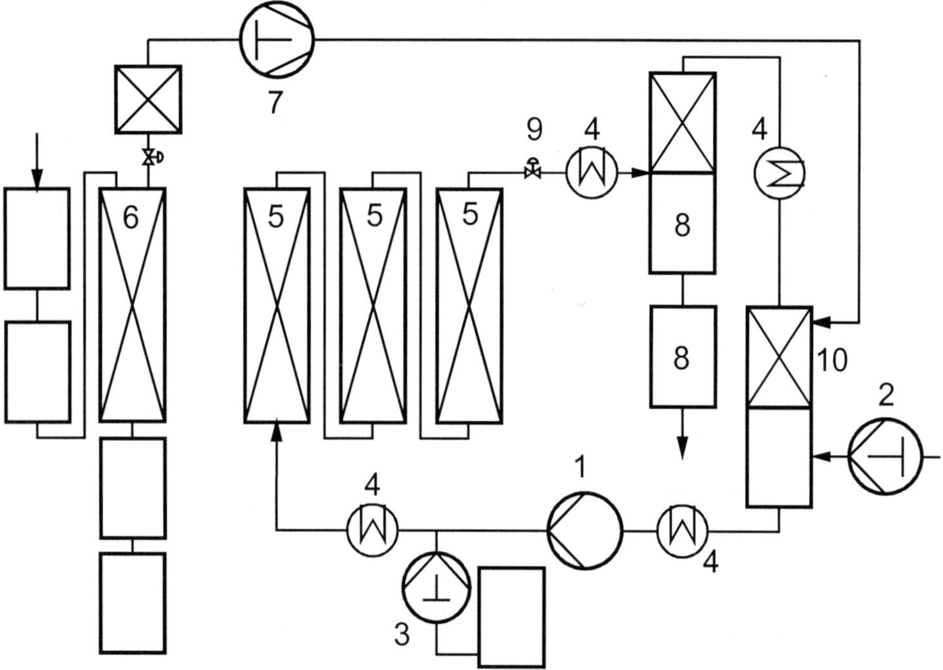

Fig. 1.4-3.   Decaffeination of coffee beans

The supercritical solvent is expanded with the throttling valve (9) in order to remove the caffeine (separator 8) and to bring the solvent back to the liquid state (condenser 10). The gas-recycling (dry running) reciprocating compressor (7), the CO_2 and the co-solvent feed (2, 3; diaphragm pumps) represent variable process components if required. Heat exchangers (4) maintain the suitable thermodynamic conditions.

The coffee bean contents of the extractors are discharged and refilled within certain time-periods (6). The decaffeination process is providing the decaffeinated coffee beans as well as the caffeine as valuable products.

The supercritical extraction of hops, tea, and other foodstuffs can be performed in similar plants. The challenge of the discontinuous extraction of bulk materials is in the design and automatic operation of high pressure extractors which can easily be opened and closed for the filling and discharging procedure.

Example 4: Homogenisation of milk and other foodstuffs

Liquid foodstuffs, for example milk products must be submitted to homogenisation treatment in order to improve their long-term physical stability ("shelf life"). The liquid is pumped at very high pressure by a multiplex reciprocating piston pump through the narrow clearances of a hydraulically controlled homogenisation valve (Fig.1.4-4, C, bottom).

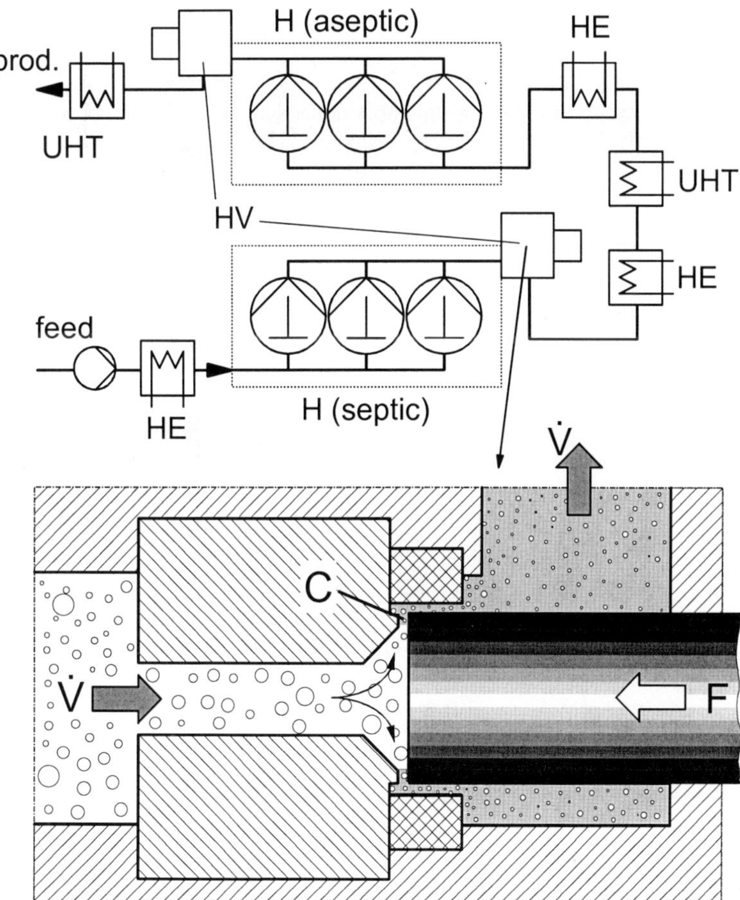

Fig. 1.4-4.   Homogenisation of milk and other foodstuffs

By the action of hydraulic shear forces, cavitation, turbulence and impact owing to the very high flow velocity (several 100 m/s) or high differential pressure (low viscosity liquids, 300 to 400 bar, or more viscous liquids, up to 1500 bar) the liquid is turned into a very fine (homogeneous) dispersion.

The homogenisation process is only one step (or sometimes two stages, see Fig. 1.4-4, top) within the production line. The feed (raw product) is adjusted in temperature by heat exchange (HE), passed through the homogeniser (H septic), then treated by ultra-high-temperature (UHT), homogenized a second time (H aseptic) and UHT-treated, ending with an aseptic final product.

Homogenisation processes now extend up to 1500 bar differential pressures. As the materials to be homogenized exhibit varying properties with respect to viscosity, corrosiveness and abrasiveness the high pressure components, such as homogenising pumps and valves, need very careful design and choice of materials.

Example 5: High speed water-jetting as an efficient tool for production and other treatment steps

The growing demand for fully automated production processes must take benefit of new steps in order to achieve and secure the quality standards requested. During continuous sheet steel production the permanent descaling of the sheet surfaces (Fig. 1.4-5, S) is realized by high speed water-jetting (Fig. 1.4-5, top) at suitable locations in the rolling-mill train (usually 600 bar water supply to the jetting nozzles, N). The high pressure plunger pumps (HP) should provide a smooth volume flow by multiplex design.

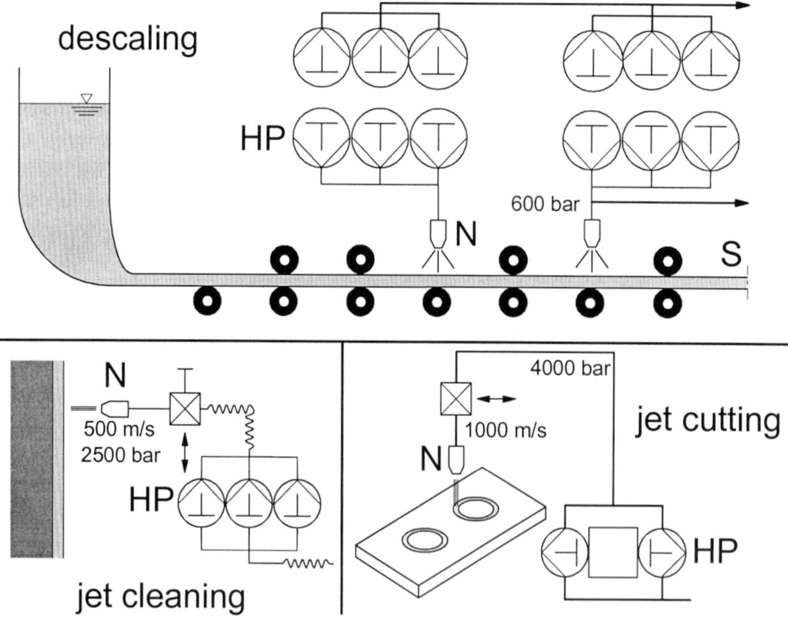

Fig. 1.4-5.   Descaling, cleaning and jet-cutting with high pressure

A very similar process is the high speed water-jet cleaning applied during reconstruction of buildings, cleaning procedures in production processes, for ships, and especially in wastewater systems. Depending on the nature of the surface layers to be removed the required water pressure can approach 2500 bar, and thus make outstanding demands on the high pressure pump design and the installation (Fig. 1.4-5, bottom left side).

The prerequisite of the successful application of water-jet cleaning should be a proper understanding of the parameters involved in the jet-cleaning physics requiring profound case studies.

Super-speed water-jets are further applied increasingly for the production steps requiring the cutting of pieces of material which should be kept at low temperature and which appear soft and restrictive towards mechanical tools. The water-jet as a "hydrodynamic cutter" provides a number of advantages in cases which should be selected by case studies.

Jet-cutting systems need to be compact and suitable for robotic action in automated trains of production. Usually the hyper-pressure plunger pumps for water-jet cutting purposes are based on hydraulic amplifiers, of double-cylinder design, and provide high pressure water of up to 5000 bar.

If very hard materials (e.g., natural stone, or metal sheets) must be cut, the injection of abrasives into the water jet will support and accelerate the cutting procedure (see Fig. 1.4-5, bottom, right side). The water-jet cutting represents a very flexible production method which can be regarded as supplementary to LASER methods if thermal influences on the materials involved cannot be accepted.

Example 6: Polymer processing

During the production of polymers (e.g., polyolefins, polyamide, polystyrene), very viscous (up to $4 \cdot 10^6$ mPas) polymer melts have to be extracted with high pressure gear pumps (PGP) from the reactors (PR) or degasifiers (DG), then transferred through heat exchangers (HE), static mixers (MI), filters (F) and diverters (DI), depending on the process, onto spinning gear (SP) pumps (Fig. 1.4-6).

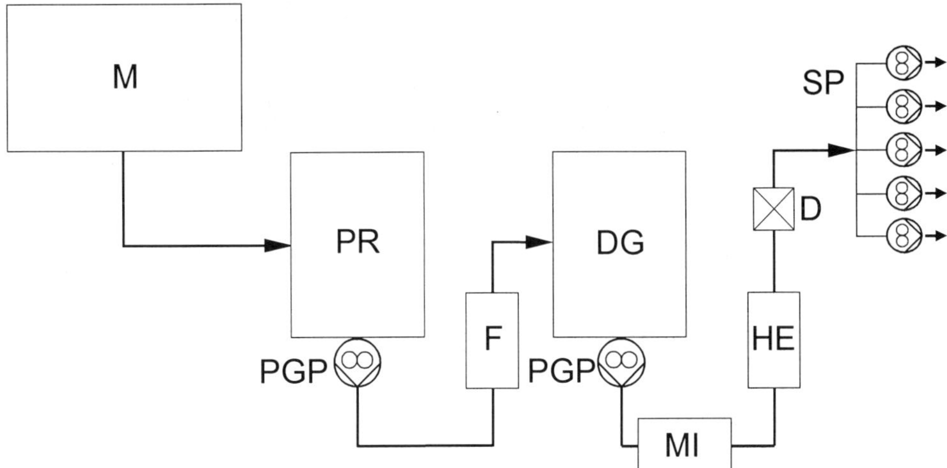

Fig. 1.4-6.   Polymer processing

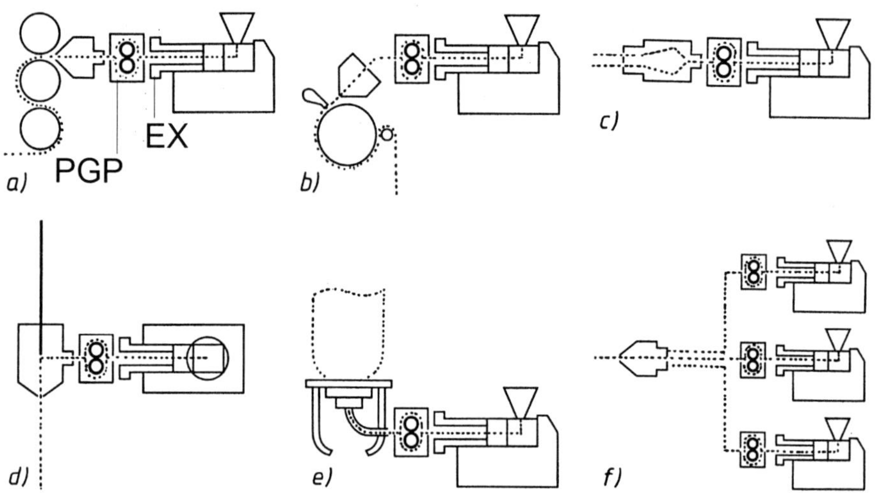

Fig. 1.4-7.   Polymer extrusion
a, b foil extrusion   c bottle extrusion   d cable extrusion   e blow film extrusion
f co – extrusion

The viscosity of the transferred fluids increases from the monomer tank (M) to the polymer reactor (PR) and the degasifier (DG). The highest viscosity (occasionally over 10^6 mPas) is seen in the polymer extraction pump (PGP) behind the vacuum degasifier. As the polymer melt has to pass mixers and filter systems its extreme viscosity requires very high pressures from the polymer gear pumps in order to force the material through the system (up to 400 bar) to the spinning pumps (SP). During extrusion polymer processing the extruder (EX) is responsible for the homogenous melting and the following polymer gear pump (PGP) for generating the high and constant pressure for pressing the material through the extrusion tools (co-extrusion, foil extrusion, cable extrusion etc., Fig. 1.4-7).The gear pumps for extremely viscous polymers must be designed accordingly, with very large inlet nozzles and crescent-shaped clearances in the suction area between the gear wheels and the pump housings.

Example 7: The sterilization of fruit juices with high pressure

This method (ultra-high pressure treatment UHP) for the aseptic processing of food stuffs and other organic products still appears to be some way from extended application.

From a number of pilot applications Fig. 1.4-8 shows the quasi-continuous train for the sterilization of fruit juices with pulp contents. The high pressure sterilization offers valuable advantages with respect to the quality of the final product compared to other sterilization procedures, especially if natural fractions of fruit pulp are desired by the consumers.

The fruit juice enters the autoclaves (5) by the pumping action of the floating pistons (4) involved. The drinking-water supply (vessel 1, low pressure pump 2, high pressure pump 3) is capable of submitting the fruit juice to the high pressure required (around 4000 bar), during a definite time period, through the floating piston. Then the juice is discharged by the water hydraulic-control system. At the same time, other parallel autoclaves perform the same steps with a certain time shift so that quasi-continuous operation of the sterilization process can be achieved.

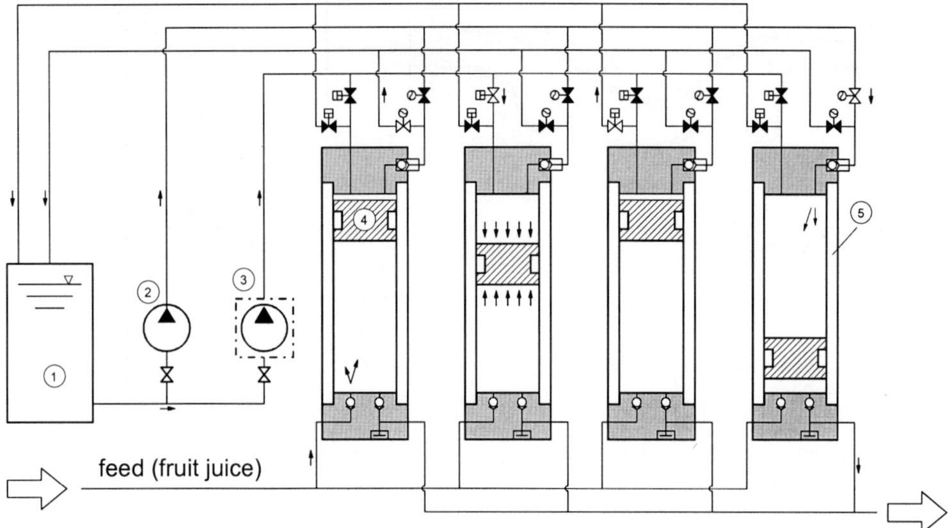

Fig. 1.4-8.   Quasi-continuous sterilization of fruit juice (adopted from Mitsubishi Heavy Ind., Japan 1992)

Example 8: Hydrostatic pressure as an efficient tool for production

The high pressure treatment is growing rapidly for a number of productions steps. Traditional methods such as hot and cold **isostatic pressing** (HIP, CIP) for the production of sintered metallic or ceramic parts have been developed further. They are now also applied as a post-treatment for castings in order to eliminate or heal porosity or internal cracks and to improve the quality. Isostatic pressing is a tool to produce intricately shaped parts demanding high density and homogeneity. The process requires suitable presses to generate pressures of up to 6000 bar at high (up to 2000 °C) or ambient temperature.

Hydroforming is a new method to produce hollow components of intricate shape by internal fluid pressure (Fig. 1.4-9, A). The untreated part may represent for example, a piece of pipe which is fixed with appropriate joints in a swage body and closed at both ends. By admitting an appropriate high internal pressure, various intricate geometrics can be achieved (1000 to 4000 bar). Another similar approach for the production of large flat and curved parts from sheet material is the hydropressing by means of special presses transmitting pressure by appropriate diaphragms.

The **autofrettage treatment** (Fig. 1.4-9, B) is certainly one of the oldest, but still very useful methods to create beneficial residual stresses in thick-walled components (e.g., pipes). The autofrettage pressure must be adjusted to a level so that the material in the thick wall is plastically strained within a certain percentage (e.g., 50 %), the rest staying only elastically strained.

14

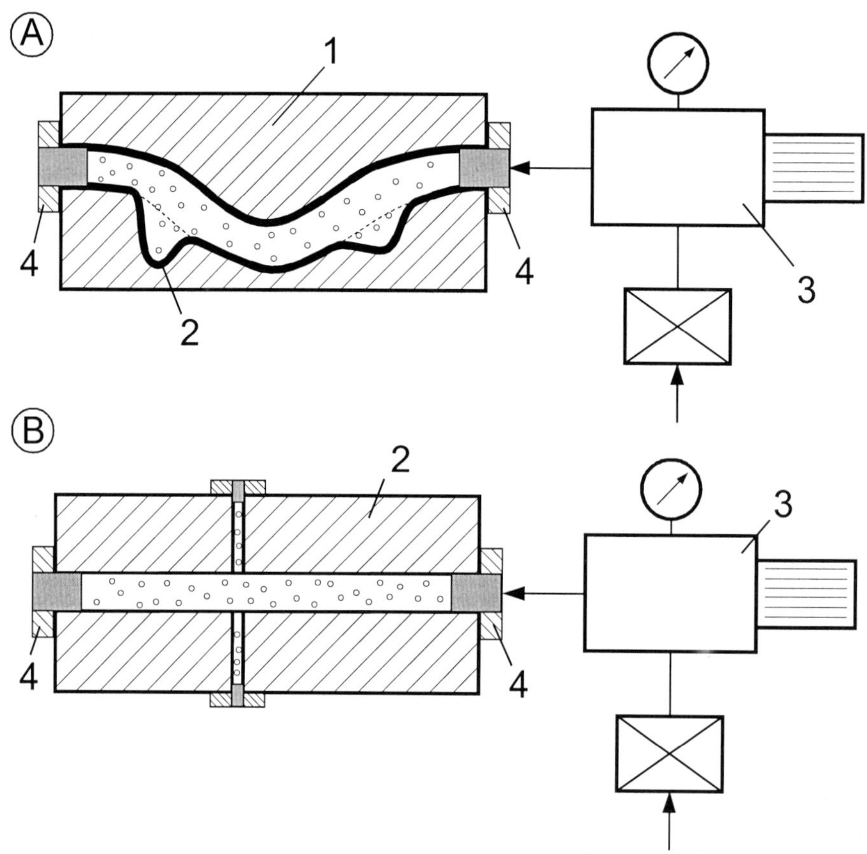

Fig. 1.4-9.   Hydroforming (A), Autofrettage (B)
1 swage   2 treated part   3 pressurization system   4 closures

After the removal of the autofrettage pressure (typically 3000 to 8000 bar) the plastically over-strained region exhibits compressive residual stresses, especially at the internal surface (Fig. 1.4-10). When submitting the thick-walled pipe to the desired operational pressure the compressive internal strains will reduce the operational ones effectively at the inner surface so the same pipe then can carry much more pressure before any failure can occur (compare σ_v and σ_{vA} at the inner diameter). Autofrettage treatment, although first used for the gun-barrel reinforcement hundreds of years ago, is used today for high pressure components in the process industries as well as for appropriate components in common rail diesel injection systems for combustion motors. The autofrettage method can be included in automated manufacturing sequences.

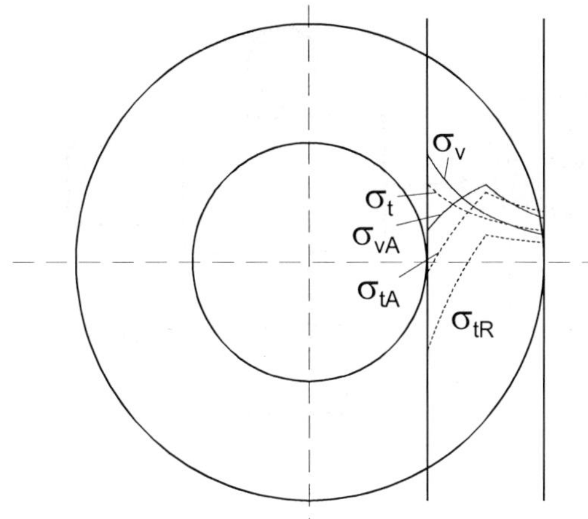

Figure 1.4-10. Stresses in an autofrettaged thick-walled cylinder (plastification 50% through the wall)

σ_t = tangential stress without autofrettage σ_{tA} = tangential stress with autofrettage σ_{tR} = tangential residual stress after autofrettage

σ_v = stress intensity without autofrettage σ_{vA} = stress intensity with autofrettage

References

1. M.B. King, T.R. Bott (editors): Extraction of Natural Products Using Near-Critical Solvents, Blackie Academic & Professional, Glasgow, 1993
2. N.N.: Supercritical Fluids - Material and Natural Products Processing, Nice/France, 1998, Institut National Polytechnique de Lorraine
3. N.N.: International Conference on High Pressure Bioscience & Biotechnology, University of Heidelberg/Germany, Section Physical Chemistry, 1998
4. W.R.D. Manning, S. Labrow: High Pressure Engineering, Leonard Hill, London, 1971
5. I.L. Spain, J. Paauwe (editors): High Pressure Technology, Vol II, Marcel Dekker New York, 1977

CHAPTER 2

THERMODYNAMIC PROPERTIES AT HIGH PRESSURE

I. Kikic[a], T. W. De Loos[b]

[a] Dipartimento di Ingegneria Chimica, dell'Ambiente e delle Materie Prime
Universitá degli Studi di Trieste
Piazzale Europa, 1   I-34127 Trieste, Italy

[b] Faculty of Applied Science, Department of Chemical Technology
Laboratory of Applied Thermodynamics and Phase Equilibria
Delft University of Technology
Julianalaan 136 NL-2628 BL Delft, The Netherlands

In this chapter the basic concepts on high-pressure phase equilibrium are introduced. Phenomenological behaviour, experimental methods and theoretical modelling are briefly described in order to give a general overview of the problematic. References are given to more detailed treatments for the different subjects in order to help the reader to go deeply into them.

2.1   Introduction

For the description of the different processes that take advantage of the use of high-pressure conditions, the knowledge of thermodynamic properties is essential.

Some particular processes can require very high pressures for special applications (i.e. in explosive welding and plating), but pressures between 100 and 1000 bar can be found easily in different industrial processes. Typical examples are the synthesis of ammonia, the synthesis of methanol and the production of low-density polyethylene, but also analytical techniques as high-pressure liquid chromatography. Other important implications are for the storage and transportation of fluids and enhanced oil recovery.

The basic equations needed are essentially the same used in the case of low pressure processes but the difference is given by some peculiar behaviour of fluids when the pressure is relatively high and close to the critical one of some components in the mixture.

In the high-density region the fluids exhibit properties similar to those of liquids, leading to high loading of solutes, high molecular diffusivity and low viscosity. The solubility of a given solute appears to be practically exponential in density: as a consequence very small pressure variations result in large changes of solubility. The enhanced solubility can be explained only by the highly non-ideal thermodynamic behaviour of mixtures. The presence of small amount of a "normal" compound (in the sense that it is far away from the critical point) can also have large effects on the solubility of heavy compounds in the high dense fluid.

Other phenomena can be simply explained by the fact that the critical pressure and temperature for a given mixture is not, as it happens for a pure fluid, the maximum temperature and pressure that allows the coexistence of a vapour and liquid phase in equilibrium. Retrograde condensation phenomena can be easily explained in this way.

In this basic overview of the thermodynamic of fluid mixtures at high-pressure emphasis is given to the behaviour of binary mixtures in order to introduce the reader to the wide variety of phase transitions, which are not observed for pure fluids. As the number of component increases the phase behaviour is mainly determined by interactions between unlike molecules at the expenses of interactions between similar molecules, but fortunately, in most cases, it is only necessary to consider pair interactions and not three or more body interactions. References are given for a more detailed study of the different aspect considered.

2.2   Phase equilibria

2.2.1   Principles of phase equilibria

2.2.1.1   The chemical potential and the phase rule of Gibbs

The equilibrium conditions for phase equilibria can be derived in the simplest way using the Gibbs energy G. According to the second law of thermodynamics, the total Gibbs energy of a closed system at constant temperature and pressure is minimum at equilibrium. If this condition is combined with the condition that the total number of moles of component i is constant in a closed system

$$\sum_{\alpha} n_i^{\alpha} = \text{constant} \tag{2.2-1}$$

where n_i^α is number of moles of component i in phase α, it can easily be derived that for a system of Π phases and N components the equilibrium conditions are [1]:

$$\mu_i^\alpha = \mu_i^\beta = ... = \mu_i^\pi \quad \text{for } i = 1, 2, ..., N \tag{2.2-2}$$

The chemical potential of component i in phase α is defined by

$$\mu_i^\alpha = \left(\frac{\partial(\sum_i n_i^\alpha g^\alpha)}{\partial n_i^\alpha} \right)_{P,T,n_{j\neq i}} \tag{2.2-3}$$

g is the molar Gibbs energy. Since μ_i^α is a function of P,T and $(N-1)$ mole fractions (the additional condition $\sum_i x_i^\alpha = 1$ makes one of the mole fractions a dependent variable), eq.(2.2-2) represents $N(\Pi - 1)$ equations in $2 + \Pi(N-1)$ variables. Therefore the number of degrees of freedom F is

$$F = 2 + \Pi(N-1) - N(\Pi - 1) = 2 - \Pi + N \tag{2.2-4}$$

Eq.(2.2-4) is the *phase rule* of Gibbs. According to this rule a state with Π phases in a system with N components is fully determined (all intensive thermodynamic properties can be calculated) if a number of F of the variables is chosen, provided that g of all phases as function of pressure, temperature and composition is known.

2.2.1.2 Fugacity and activity

In the thermodynamic treatment of phase equilibria, auxiliary thermodynamic functions such as the fugacity coefficient and the activity coefficient are often used. These functions are closely related to the Gibbs energy. The fugacity of component i in a mixture, $\widehat{f}_i$, is defined by

$$d\mu_i \equiv RTd\ln\widehat{f}_i \quad \text{at constant } T \tag{2.2-5a}$$

with

$$\lim_{P \to 0} \frac{\widehat{f}_i}{P_i} = 1 \tag{2.2-5b}$$

According to this definition $\widehat{f}_i$ is equal to the partial pressure P_i in the case of an ideal gas. The *fugacity coefficient* $\widehat{\phi}_i$ is defined by

20

$$\widehat{\phi}_i = \frac{\widehat{f}_i}{P_i} \tag{2.2-6}$$

and is a measure for the deviation from ideal gas behaviour.

The fugacity coefficient can be calculated from an equation of state by one of the following expressions [2]:

$$RT \ln \widehat{\phi}_i = \int_0^P \left(\overline{V}_i - \frac{RT}{P} \right) dP \tag{2.2-7}$$

$$RT \ln \widehat{\phi}_i = -\int_\infty^V \left[\left(\frac{\partial P}{\partial n_i} \right)_{V,T,n_{j\neq i}} - \frac{RT}{V} \right] + RT \ln \left(\frac{\sum_i n_i RT}{PV} \right) \tag{2.2-8}$$

According to eq.(2.2-5) the equilibrium relation eq.(2.2-2) can be replaced by

$$\widehat{f}_i^\alpha = \widehat{f}_i^\beta = \ldots\ldots = \widehat{f}_i^\pi \quad \text{for } i = 1, 2, \ldots, N \tag{2.2-9}$$

The activity a_i is defined as the ratio of $\widehat{f}_i$ and the fucacity of component i in the standard state at the same P and T:

$$a_i \equiv \frac{\widehat{f}_i(P,T,x)}{f_i^0(P,T,x^0)} \tag{2.2-10}$$

An ideal solution is defined by

$$a_i^{id} \equiv x_i \tag{2.2-11}$$

The *activity coefficient* of component i, γ_i, is a measure for the deviation from ideal solution behaviour

$$\gamma_i \equiv \frac{a_i}{a_i^{id}} \tag{2.2-12}$$

so the fugacity of a non-ideal solid or liquid solution can be written as

$$\widehat{f}_i = x_i \gamma_i f_i^0 \tag{2.2-13}$$

The activity coefficient γ_i can be calculated from a model for the molar excess Gibbs energy g^E

$$RT \ln \gamma_i = \left(\frac{\partial \left(\sum_i n_i g^E \right)}{\partial n_i} \right)_{P,T,n_{j \neq i}} \qquad (2.2\text{-}14)$$

In this approach the standard state fugacity of a liquid or solid component is usually the fugacity of the pure solid or liquid component, and is closely related to the sublimation pressure P_i^{sub} or vapour pressure P_i^{sat}, respectively. I.e., on the sublimation curve of a pure component we have

$$f_i^S (P_i^{sub},T) = f_i^V (P_i^{sub},T) = \phi_i^V (P_i^{sub},T) P_i^{sub} \qquad (2.2\text{-}15)$$

According to eq.(2.2-5a)

$$f_i^S (P,T) = f_i^S (P_i^{sub},T) \exp\left(\int_{P_i^{sub}}^P \frac{d\mu_i^S}{RT} \right) = f_i^S (P_i^{sub},T) \exp\left(\int_{P_i^{sub}}^P \frac{v_i^S}{RT} dP \right) \qquad (2.2\text{-}16)$$

v_i^S is the molar volume of pure solid i. Combining eq.(2.2-15) and (2.2-16) we get

$$f_i^S (P,T) = \phi_i^V (P_i^{sub},T) P_i^{sub} \exp\left(\int_{P_i^{sub}}^P \frac{v_i^S}{RT} dP \right) \qquad (2.2\text{-}17)$$

In an analogous way we can derive

$$f_i^L (P,T) = \phi_i^V (P_i^{sat},T) P_i^{sat} \exp\left(\int_{P_i^{sat}}^P \frac{v_i^L}{RT} dP \right) \qquad (2.2\text{-}18)$$

At *low pressure* the fugacity coefficients and the exponential terms are close to 1, so

$$f_i^S \approx P_i^{sub} \quad \text{and} \quad f_i^L \approx P_i^{sat} \qquad (2.2\text{-}19)$$

2.2.1.3 Critical phenomena

Two phases in equilibrium can became identical in a critical point. The calculation of critical points is done by solving the two conditions for a critical point as derived by Gibbs. For a system with N components these conditions are

$$
(D_{sp})_T = \begin{vmatrix}
\dfrac{\partial^2 a}{\partial v^2} & \dfrac{\partial^2 a}{\partial v \partial x_1} & \dfrac{\partial^2 a}{\partial v \partial x_2} & \cdots & \dfrac{\partial^2 a}{\partial v \partial x_{N-1}} \\[2ex]
\dfrac{\partial^2 a}{\partial v \partial x_1} & \dfrac{\partial^2 a}{\partial x_1^2} & \dfrac{\partial^2 a}{\partial x_1 \partial x_2} & \cdots & \dfrac{\partial^2 a}{\partial x_1 \partial x_{N-1}} \\[2ex]
\dfrac{\partial^2 a}{\partial v \partial x_2} & \dfrac{\partial^2 a}{\partial x_1 \partial x_2} & \dfrac{\partial^2 a}{\partial x_2^2} & \cdots & \dfrac{\partial^2 a}{\partial x_2 \partial x_{N-1}} \\[2ex]
\vdots & \vdots & \vdots & \ddots & \vdots \\[2ex]
\dfrac{\partial^2 a}{\partial v \partial x_{N-1}} & \dfrac{\partial^2 a}{\partial x_1 \partial x_{N-1}} & \dfrac{\partial^2 a}{\partial x_2 \partial x_{N-1}} & \cdots & \dfrac{\partial^2 a}{\partial x_{N-1}^2}
\end{vmatrix} = 0 \tag{2.2-20}
$$

and

$$
(D_c)_T = \begin{vmatrix}
\dfrac{\partial D_{sp}}{\partial v} & \dfrac{\partial D_{sp}}{\partial x_1} & \dfrac{\partial D_{sp}}{\partial x_2} & \cdots & \dfrac{\partial D_{sp}}{\partial x_{N-1}} \\[2ex]
\dfrac{\partial^2 a}{\partial v \partial x_1} & \dfrac{\partial^2 a}{\partial x_1^2} & \dfrac{\partial^2 a}{\partial x_1 \partial x_2} & \cdots & \dfrac{\partial^2 a}{\partial x_1 \partial x_{N-1}} \\[2ex]
\dfrac{\partial^2 a}{\partial v \partial x_2} & \dfrac{\partial^2 a}{\partial x_1 \partial x_2} & \dfrac{\partial^2 a}{\partial x_2^2} & \cdots & \dfrac{\partial^2 a}{\partial x_2 \partial x_{N-1}} \\[2ex]
\vdots & \vdots & \vdots & \ddots & \vdots \\[2ex]
\dfrac{\partial^2 a}{\partial v \partial x_{N-1}} & \dfrac{\partial^2 a}{\partial x_1 \partial x_{N-1}} & \dfrac{\partial^2 a}{\partial x_2 \partial x_{N-1}} & \cdots & \dfrac{\partial^2 a}{\partial x_{N-1}^2}
\end{vmatrix} = 0 \tag{2.2-21}
$$

The molar Helmhotz energy a is obtained from an equation of state using the relation $(\partial a / \partial v)_{T,x} = -P$. All derivatives in these two determinants can be obtained from an equation of state. For a pure component only the (1,1) elements of the determinants are left, so we get the well known conditions for a critical point for a pure component:

$$
(\partial P / \partial v)_T = 0 \text{ and } (\partial^2 P / \partial v^2)_T = 0 \tag{2.2-22}
$$

According to eq.(2.2-22) the isothermal compressibility K_T goes to infinity in a critical point because

$$
K_T \equiv -\frac{1}{v}\left(\frac{\partial v}{\partial P}\right)_T \tag{2.2-23}
$$

Experiments and modern physics [3] has shown that the way K_T and other thermodynamic properties diverge when approaching the critical point is described in a fundamentally wrong way by all classical, analytical equations of state like the cubic equations of state and is path dependent. The reason for this is that these equations of state are based on mean field theory, which neglects the strong density fluctuations (and concentration fluctuations in mixtures) close to the critical point. If the distance to the critical point is defined by

$$\Delta P^* \equiv \frac{P - P_c}{P_c}, \quad \Delta T^* = \frac{T - T_c}{T_c}, \quad \Delta \rho^* = \frac{\rho - \rho_c}{\rho_c} \tag{2.2-24}$$

then the way some thermodynamic properties diverge is given by Table 2.2-1.

Table. 2.2-1
Power laws, path, critical exponents and their relations [3]

Property	Power law	Path
isothermal compressibilty	$K_T \propto \left\|\Delta T^*\right\|^{-\gamma}$	critical isochore
isochoric heat capacity	$C_V \propto \left\|\Delta T^*\right\|^{-\alpha}$	critical isochore
coexisting densities	$(\rho^L - \rho^V) \propto \left\|\Delta T^*\right\|^{\beta}$	two-phase
pressure	$\Delta P^* \propto \left\|\Delta \rho^*\right\|^{\delta}$	critical isotherm

The values of the critical exponents α, β, γ and δ are universal and are given in Table 2.2-2

Table 2.2-2
Critical exponent values [3]

Exponent	Classical fluid	Real fluid
α	0	0.110
β	½	0.326
γ	1	1.239
δ	3	4.80

With the relations given in Table 2.2-1 and the critical exponent values given in Table 2.2-2, the thermodynamic behaviour of a pure component close to the critical point can be described exactly, however further away from the critical point also the mean field contributions have to be taken into account. A theory which is in principle capable to describe the thermodynamic properties of a system both far away from the critical point and close to the critical point is the so-called cross-over theory developed by Sengers and co-workers [4]. However this theory is very complicated and can only be applied if the critical coordinates are known.

2.2.2   Classification of phase equilibria

In this section we will discuss the phase behaviour of binary systems. In 2.2.2.1 the classification of fluid phase behaviour according to van Konynenburg and Scott [5] is discussed. The occurrence of solid phases introduces an extra complication in binary phase phase diagrams. This is discussed in 2.2.2.2.

The phase rule for a binary system ($N=2$) is given by

$$F = 4 - \Pi \tag{2.2-25}$$

According to this equation the maximum number of phases that can be in equilibrium in a binary system is $\Pi_{max} = 4$ ($F=0$) and maximum number of degrees of freedom needed to describe the system $F_{max} = 3$ ($\Pi=1$). This means that all phase equilibria can be represented in a three-dimensional P,T,x-space. At equilibrium every phase participating in a phase equilibrium has the same P and T, but in principle a different composition x. This means that a four-phase-equilibrium ($F=0$) is given by four points in P,T,x-space, a three-phase equilibrium ($F=1$) by three curves, a two-phase equilibrium ($F=2$) by two planes and a one phase state ($F=3$) by a region. The critical state and the azeotropic state are represented by one curve.

In a binary system more than two fluid phases are possible. For instance a mixture of pentanol and water can split into two liquid phases with a different composition: a water-rich liquid phase and a pentanol-rich liquid-phase. If these two liquid phases are in equilibrium with a vapour phase we have a three-phase equilibrium. The existence of two pure solid phases is an often occuring case, but it is also possible that solid solutions or mixed crystals are formed and that solids exists in more than one crystal structure.

Often the essentials of phase diagrams in P,T,x-space are represented in a P,T-projection. In this type of diagrams only non-variant ($F=0$) and monovariant ($F=1$) equilibria can be represented. Since pressure and temperature of phases in equilibrium are equal, a four-phase equilibrium is now represented by one point and a three-phase equilibrium with one curve. Also the critical curve and the azeotropic curve are projected as a curve on the P,T-plane. A four-phase point is the point of intersection of four three-phase curves. The point of intersection of a three-phase curve and a critical curve is a so-called critical endpoint. In this intersection point both the three-phase curve and the critical curve terminate.

Another way to represent phase equilibria is in isothermal P,x-sections, isobaric T,x-sections and isoplethic P,T-sections. In Figure 2.2-1 three cases of a binary vapour-liquid equilibrium are shown. The equilibrium composition of the liquid phase and vapour phase are represented by curves, the so-called binodal curves. At a given pressure a mixture with an overall composition in between the two binodal curves will split into two phases with compositions given by the binodal curves. Mixtures outside this region will be in a one-phase state (either liquid or vapour). Two binodals can intersect, which occurs in Figure 2.2-1a at the pure component vapour pressures. They also can be tangent as is the case in Figure 2.2-1b in the azeotropic point or they can merge as is the case in Figure 2.2-1c in the critical point *l=g*. Critical points and azeotropic points are always extrema in pressure or temperature in *P,x*- or *T,x*-sections, respectively.

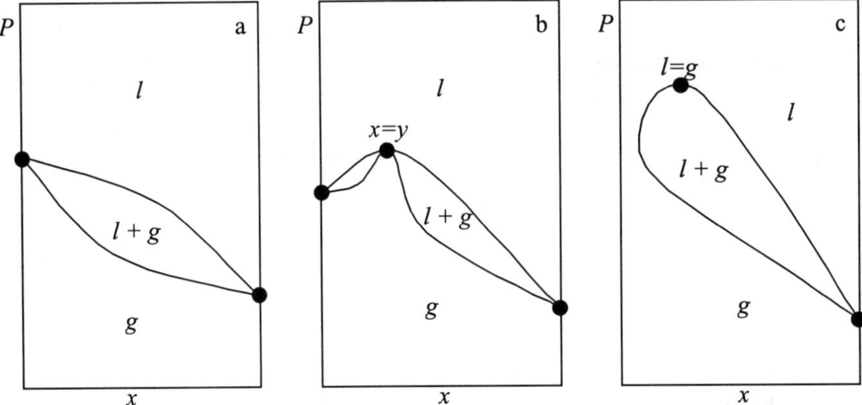

Figure 2.2-1. Two phase equilibria: three cases were the composition of the two phase are equal. a: Pure component boiling point. b: Azeotropic point. c: Critical point.

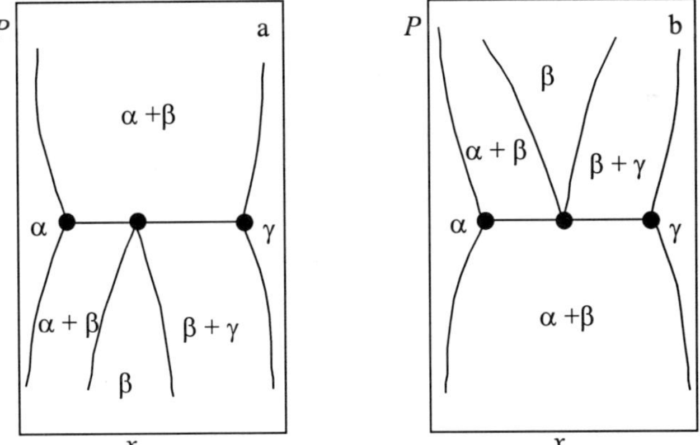

Figure 2.2-2 Possible locations of one- and two-phase equilibria around a three-phase equilibrium in a *P,x*- section.

In a *P,x*- or *T,x*-section a three-phase equilibrium is represented by three points, which give the compositions of the three coexisting phases. All mixtures with composition on the line through these points will split into three phases. Figure 2.2-2 gives a schematic example of a three-phase equilibrium αβγ in a *P,x*-section. Around the three-phase equilibrium three two-phase regions, $\alpha\beta$, $\alpha\gamma$, and $\beta\gamma$ and three one-phase regions, α, β and γ are found. According to the theory of phase transformations the one- and two-phase regions have to be arranged around the three-phase equilibrium, either as is shown in Figure 2.2-2a or as is shown in Figure 2.2-2b. T,x-sections basically look the same.

2.2.2.1   Fluid phase equilibria

According to the classification of van Konynenburg and Scott [5] there are six basic types of fluid phase behaviour. The corresponding P,T-projections are shown in Figure 2.2-3. These P,T-projections can be complicated further by the occurrence of homo- and heteroazeotropes which will not be discussed here.

All six basic types of fluid phase behaviour have been found experimentally. Type I to V can be predicted with simple equations of state like the van der Waals equation, however type VI, which is found in systems in which specific interactions like hydrogen bonding plays a role, is only predicted with more complicated equations of state like the SAFT equation of state [6-8]. In the mean time many more types of fluid phase behaviour have been found computationally [9], but up till now they have not been found in real systems.

In Figure 2.2-3 the curves lg are the vapour pressure curves of the pure components which end in a critical point $l=g$. The curves $l=g$, $l_1=g$ and $l_2=g$ are vapour-liquid critical curves and the curves $l_1=l_2$ are curves on which two liquid phases become critical. The points of intersection of a critical curve with a three-phase curve l_2l_1g is a critical endpoint. Distinction can be made between upper critical endpoints (UCEP) and lower critical endpoints (LCEP). The UCEP is highest temperature of a three-phase curve, the LCEP is the lowest temperature of a three-phase curve. The point of intersection of the l_2l_1g curve with a $l_1=g$ curve is a critical endpoint in which the l_1 liquid phase and the vapour phase are critical in the presence of a non-critical l_2 phase $(l_2+(l_1=g))$ and the point of intersection of the l_2l_1g curve with a $l_2=l_1$ curve is a critical endpoint in which the two liquid phases l_2 and l_1 are critical in the presence of a non-critical vapour phase $((l_2=l_1)+g)$.

In the case of type I phase behaviour there is only one critical curve, the $l=g$ critical curve, which runs continously from the critical point of component A to the critical point of component B. In Figure 2.2-4 some P,x-sections are shown and in Figure 2.2-5 a P,T-section.

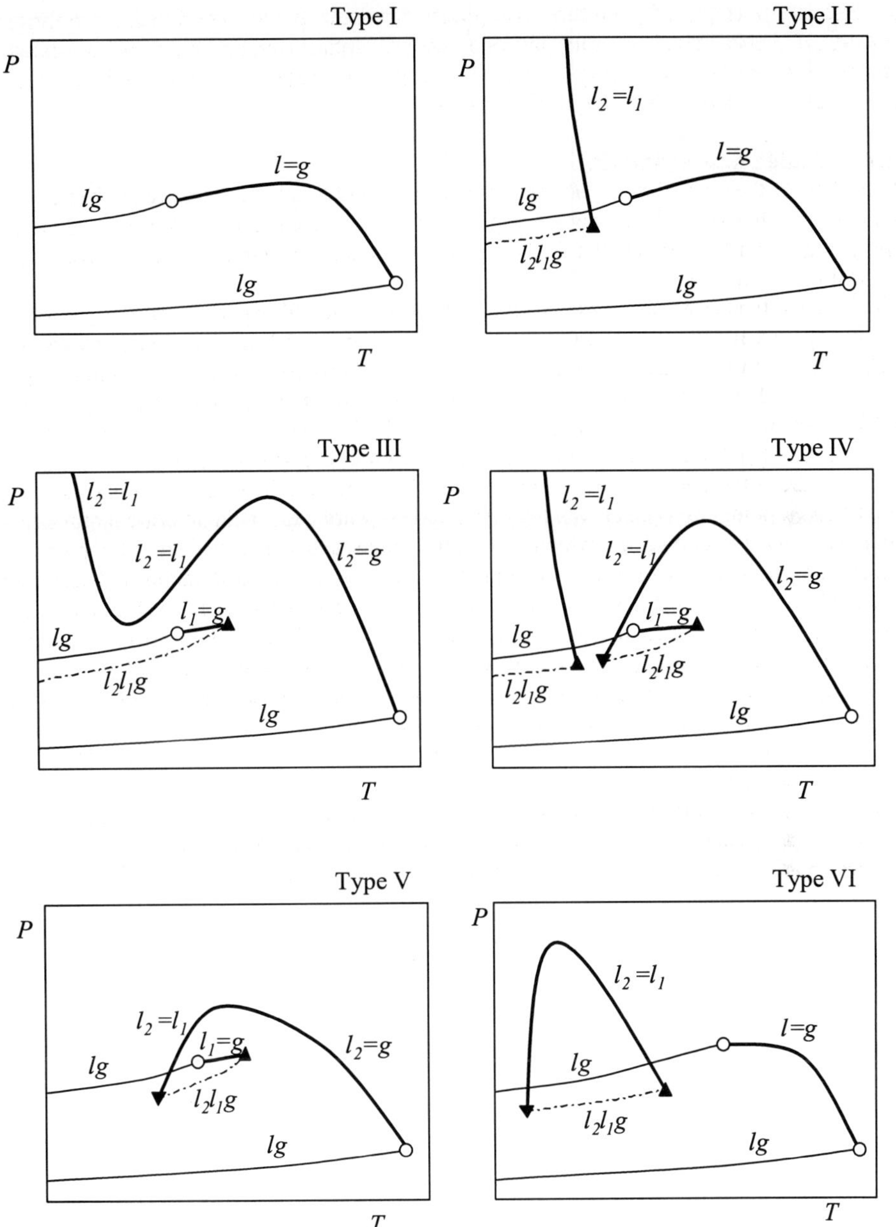

Figure 2.2-3. The six basic types of fluid phase behaviour according to the classification of Van Koynenburg and Scott [5].

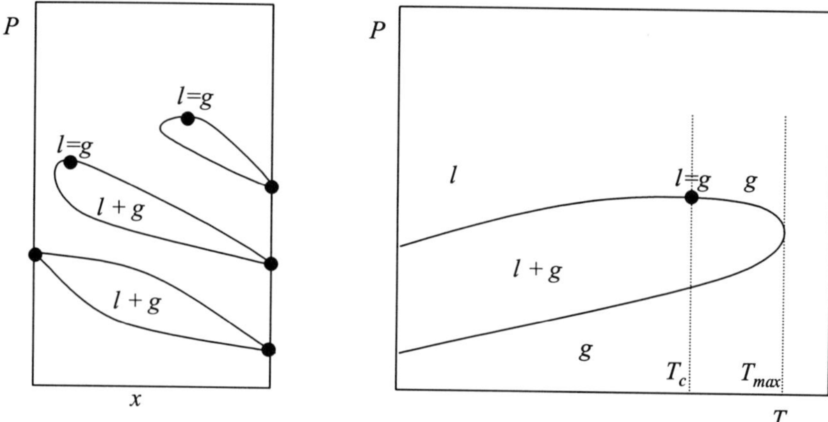

Figure 2.2-4.Type I. *P,x*-sections.   Figure 2.2-5. Type I. *P,T*-section.

In the *P,T*-section the two-phase envelope is tangent to the binary critical curve in the critical point.

Since in the critical point the bubble point curve ($l+g{\rightarrow}l$) and the dew-point curve ($l+g{\rightarrow}g$) merge at temperatures between T_c and T_{max} an isotherm will intersect the dew-point curve twice. If we lower the pressure on this isotherm we will pass the first dew-point and with decreasing pressure the amount of liquid will increase. Then the amount of liquid will reach a maximum and upon a further decrease of the pressure the amount of liquid will decrease until is becomes zero at the second dew-point. The phenomenon is called *retrograde condensation* and is of importance for natural gas pipe lines. In supercritical extraction use is made of the opposite effect. With increasing pressure a non-volatile liquid will dissolve in a dense supercritical gas phase at the first dew point.

In type II fluid phase behaviour next to a continuous *l=g* critical curve at low temperature also a $l_2=l_1$ critical curve and a three-phase curve l_2l_1g is found, which intersect in a UCEP *($l_2=l_1$)+g*. This critical curve runs steeply to high pressure and represents upper critical solution temperatures. In *P,T,x*-space a two phase *(l_2+l_1)* region is found at temperatures lower than the $l_2=l_1$ critical curve and pressures higher than the three-phase curve. Figure 2.2-6 shows three *P,x*-sections for a type II system. Figure 2.2-6a is at $T<T_{UCEP}$, Figure 2.2-6b is at $T=T_{UCEP}$ and Figure 2.2-6c is at $T_{UCEP}<T<T_{c,A}$. In Figure 2.2-6b the bubble-point curve shows a horizontal point of inflection at the $l_2=l_1$ critical point. At higher temperatures the *P,x*-sections are similar to Figure 2.2-4.

Type V fluid phase behaviour shows at temperatures close to $T_{c,A}$ a three-phase curve l_2l_1g which ends at low temperature in a LCEP *($l_2=l_1$)+g* and at high temperature in a UCEP *(l_2+l_1)=g*. The critical curve shows two branches. The branch $l_1=g$ runs from the critical point of pure component A to the UCEP. The second branch starts in the LCEP and ends in the critical point of pure component B. This branch of the critical curve is at low temperature $l_2=l_1$ in nature and at high temperature its character is changed into $l_2=g$. The $l_2=l_1$ curve is a critical curve which represents lower critical solution temperatures. In Figure 2.2-7 four isothermal *P,x*-sections are shown.

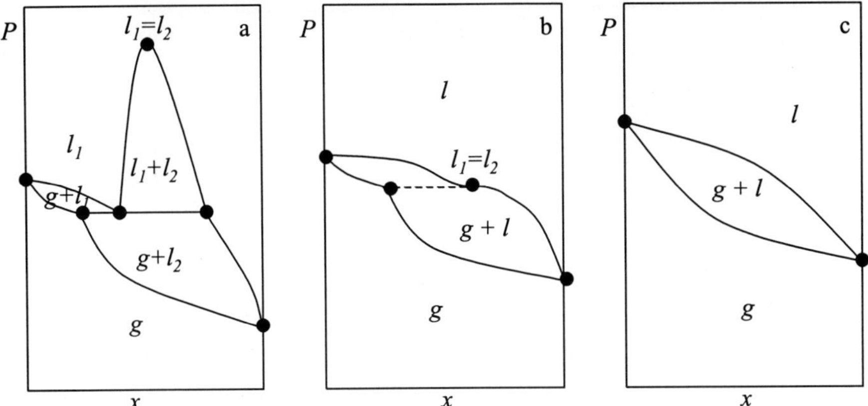

Figure 2.2-6. Type II. P,x-sections. a: $T<T_{UCEP}$. b: $T=T_{UCEP}$. c: $T_{UCEP}<T<T_{c,A}$.

In Figure 2.2-7a the bubble-point curve shows a horizontal point of inflection at the critical point $l_2=l_1$ and in Figure 2.2-7d the binodal shows a horizontal point of inflection at the critical point $l_1=g$. At temperatures lower than T_{LCEP} and temperatures higher than T_{UCEP} the P,x-sections are the same as for type I systems.

Type IV fluid phase behaviour is a combination of type II and type V behaviour. These systems show two branches of the l_2l_1g curve, three branches of the critical curve and three critical endpoints. At low temperature the P,x-sections for this type of systems are similar to Figure 2.2-6, at temperatures close to the critical temperature of the pure component A P,x-sections similar to Figure 2.2-7 are found.

In type III phase behaviour the two branches of the l_2l_1g equilibrium are combined and also the two branches of the $l_2=l_1$ critical curve are united. Only one critical endpoint is left. The $l_2=l_1/l_2=g$ branch of the critical curve can have the shape as is shown in Figure 2.2-3, but it is also possible that this curve goes from the critical point of component B to high pressure via a temperature minimum or that dP/dT is always positive [10].

In type VI phase behaviour a three-phase curve l_2l_1g is found with an LCEP and an UCEP. Both critical endpoints are of the type $(l_2=l_1)+g$ and are connected by a $l_2=l_1$ critical curve which shows a pressure maximum. For this type of phase behaviour at constant pressure closed loop isobaric regions of l_2+l_1 equilibria are found with a lower critical solution temperature and an upper critical solution temperature.

In binary systems of a light gas and members of the n-alkane homologuous series in almost all cases type II phase behaviour is found for the lower members of the n-alkane series. With increasing carbon number type IV phase behaviour is found followed by type III phase behaviour at high carbon numbers. Systems of CO_2 + n-alkanes show type II phase behaviour for n-alkane carbon numbers $n \leq 12$, type IV phase behaviour for $n=13$ and type III phase behaviour for $n \geq 14$ [11, 12]. This transition from type II to type III phase behaviour via type IV phase behaviour seems to be the rule, allthough in most binary families type IV phase behaviour is only found at broken values of n in so-called quasi-binary systems [13, 14].

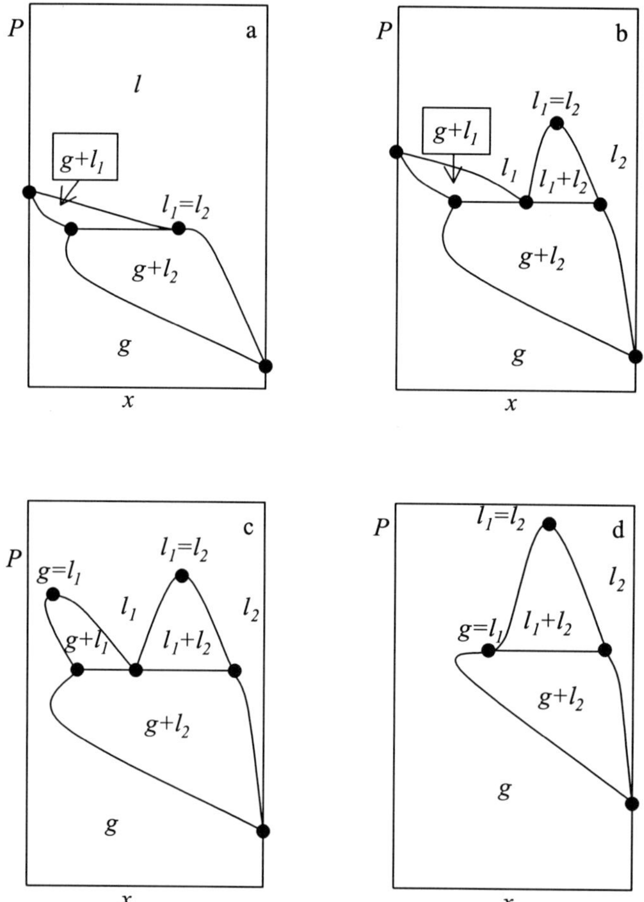

Figure 2.2-7. Type V. P,x-sections. a. $T=T_{LCEP}$. b. $T_{LCEP}<T<T_{c,A}$. c: $T_{c,A}<T<T_{UCEP}$. d: $T=T_{UCEP}$

In Figure 2.2-8 the critical endpoint temperatures for the family of CO_2 + n-alkanes systems are plotted as a function of the carbon number n. If in a particular binary system the three-phase curve l_2l_1g is followed to low temperature then at a certain temperature a solid phase is formed (solid n-alkane or solid CO_2 at low carbon numbers). This occurs at one unique temperature because we now have four phases in equilibrium in a binary system, so according to the phase rule $F=0$. Below this so-called quadruple point temperature the l_2l_1g curve is metastable.

In Figure 2.2-8 also the quadruple point temperatures of the different systems is plotted and the curve through this points intersects the curve through the critical endpoints at a carbon number $23<n<24$. This means that for carbon numbers larger than 23 no stable l_2l_1g and l_2+l_1

equilibria can be found. The consequences of this phenomenon will be discussed in the next section.

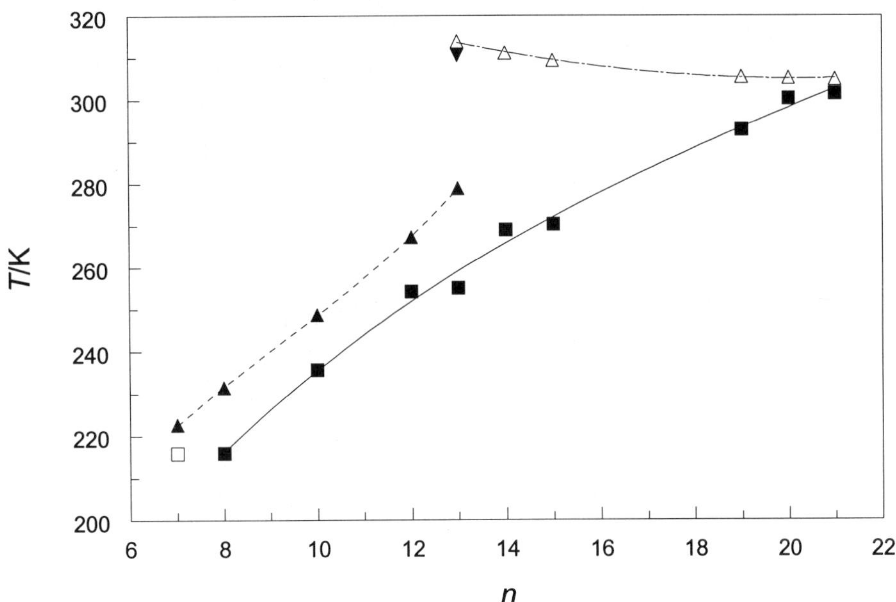

Figure 2.2-8. The family of carbon dioxide with n-alkanes. Critical endpoint temperatures and quadruple point temperatures as a function of carbon number n. $\square$: quadruple point $l_2l_1gs_{CO2}$; $\blacksquare$: quadruple point $s_{\text{n-alkane}}l_2l_1g$; $\blacktriangle$: UCEP $l_2=l_1g$; $\blacktriangledown$: LCEP $l_2=l_1g$; $\triangle$: UCEP $l_2l_1=g$.

2.2.2.2 Phase equilibria with the presence of solid phases

As discussed for CO_2 + n-alkane systems at carbon numbers $n<24$ the three-phase curve l_2l_1g ends a low temperature in a quadruple point $s_2l_2l_1g$. This is shown schematically in Figure 2.2-9a and b. In the quadrupel point three other three-phase curves terminate. The $s_2l_2l_1$ curve runs steeply to high pressure and ends in a critical endpoint where this curve intersects the critical curve. The s_2l_2g curve runs to the triple point of pure component B and the s_1l_1g curve runs to lower temperature and ends at low temperature in a second quadruple point $s_2s_1l_1g$ (not shown).

With increasing carbon number of the n-alkane the melting curve of the n-alkane shifts to higher temperature. As a result also the quadruple point $s_2l_1l_2g$ shifts to higher temperature (Figure 2.2-9b) and eventually coincides with the critical endpoint $l_2+(l_1=g)$ of the l_2l_1g curve

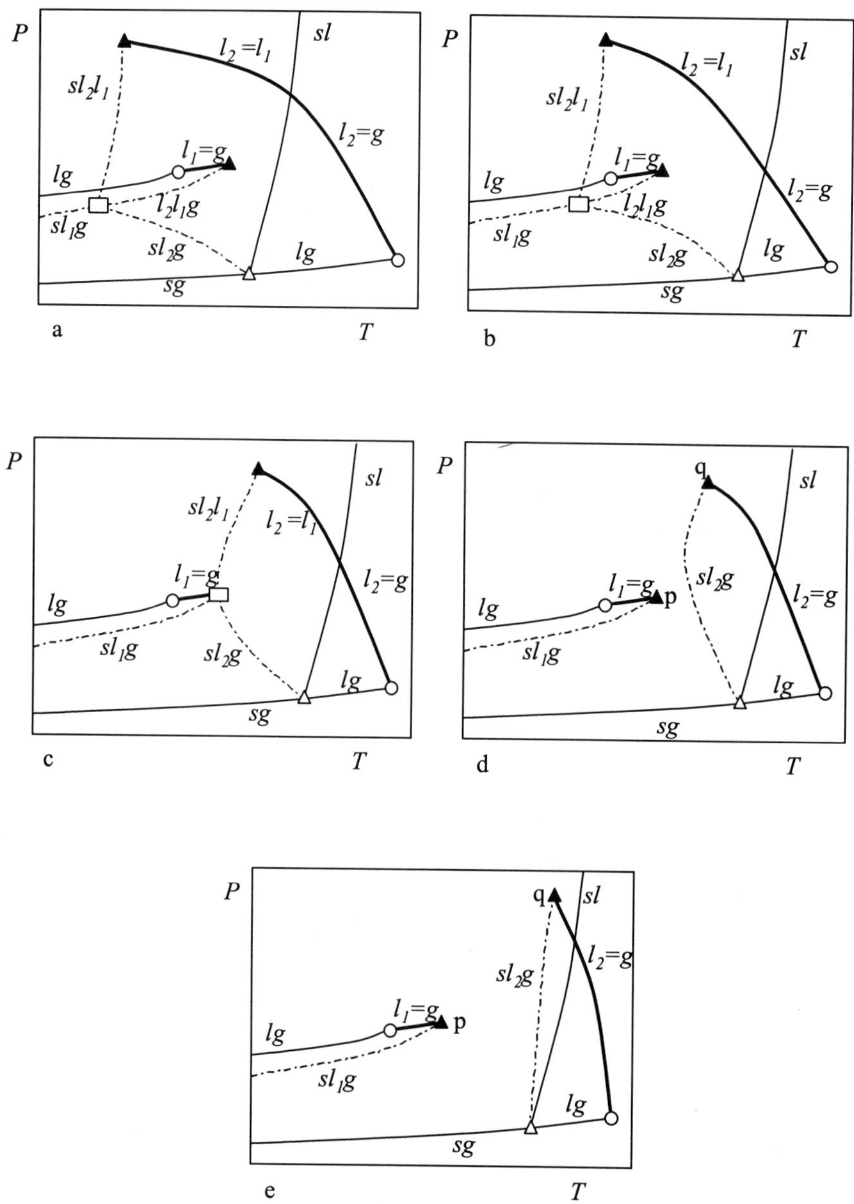

Figure 2.2-9. Type III phase behaviour with interference of the solid phase.

(Figure 2.2-9c). Now the length of the l_2l_1g curve is reduced to zero. At higher carbon numbers the l_2l_1g curve has disappeared and the $s_2l_2l_1$ and the s_2l_2g curve form now one curve

(Figure 2.2-9d). This curve ends at high pressure in a critical endpoint q where the three-phase curve intersects the critical curve. The s_2l_1g curve also shows a critical endpoint p. In both critical endpoints the solid is in equilibrium with a critical fluid phase. In Figure 2.2-9d the three phase curve s_2l_2g shows a temperature minimum which disppears at even higher carbon numbers (Figure 2.2-9e). These p,q-systems are of importance for the extraction of solids with supercritical gases.

In Figure 2.2-10 a number of P,x-sections of a p,q-system at temperatures around the critical temperature of component A are shown. At $T=T_p$ the critical point $l=g$ and the l point and g point of the s_2l_1g equilibrium in Figure 2.2-10b coincide in a horizontal point of inflexion. As a result, at higher temperatures (Figure 2.2-10c) the solubility curve of the solid in the supercritical gas still shows a point of inflexion. This results in a sharp increase of the solubility of the solid in the supercritical gas. This effect plays an important role in supercritical extraction processes.

In Figure 2.2-11 some P,T-sections for a p,q-system at temperature close to T_q are shown. At very low concentration of the heavy component no vapour-liquid equilibria are found (Figure 2.2-11a). At intermediate concentration of the heavy component a part of the three-phase curve s_2l_2g is found and a vapour-liquid region with only dew points. At higher concentration of the heavy component the three-phase curve ends at high pressure in the point q and the vapour liquid region shows a critical point. These diagrams play an important role in the production of low density polyethylene. In this process polyethylene is formed from ethylene at very high pressure.

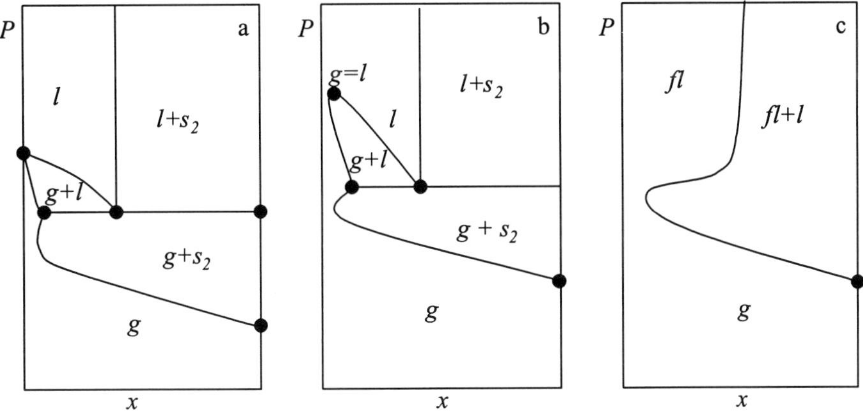

Figure 2.2-10. P,x-section for the phase behaviour as shown in Figure 9d. a: $T<T_{c,A}$. b: $T_{c,A}<T<T_p$. c: $T_p<T<T_q$

The process is run in the one phase fluid region, at pressures higher than the bubble-/dew-point curve in Figure 2.2-11b and 2.2-11c and temperatures higher than the solubility curve in these figures. Depending on the molecular weight of the polyethylene the phase boundary can be found at pressures higher than 200 MPa [15].

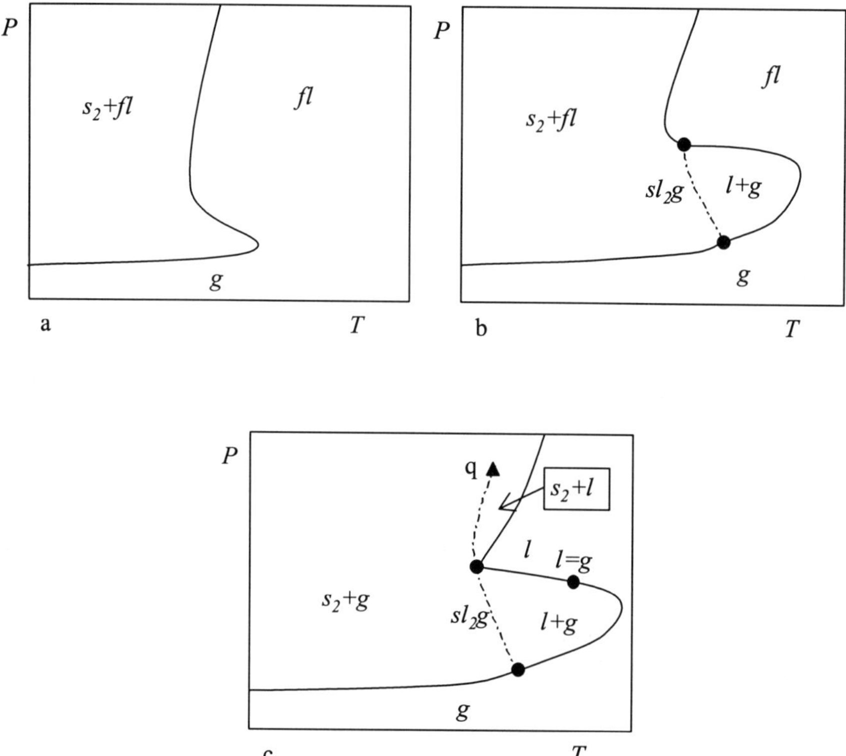

Figure 2.2-11. *P,T*-sections of a p,q-system. See text.

2.3   Calculation of high-pressure phase equilibria

The purpose of phase equilibria calculations is to predict the thermodynamic properties of mixtures, avoiding direct experimental determinations, or to extrapolate the existing data to different temperatures and pressures. The basic requirements for performing any thermodynamic calculation are the choice of the appropriate thermodynamic model and knowledge of the parameters required by the model. In the case of high pressure phase equilibria, the thermodynamic model used is generally an equation of state which is able to describe the properties of both phases.

The general starting equation is the equality of fugacities of each species in each phase:

$$\widehat{f}_i^{\,l}(T,P,x_i) = \widehat{f}_i^{\,v}(T,P,y_i) \tag{2.3-1}$$

or, introducing the fugacity coefficients $\widehat{\phi}_i^{\,v} = \dfrac{\widehat{f}_i^{\,v}}{Py_i}$ and $\widehat{\phi}_i^{\,l} = \dfrac{\widehat{f}_i^{\,l}}{Px_i}$:

$$\hat{\phi}_i^v y_i = \hat{\phi}_i^l x_i \tag{2.3-2}$$

The fugacities are calculated as mentioned above, with an equation of state through the following equation (the equation is written with reference to the vapour phase but a similar equation can be written also for the liquid phase):

$$\ln \frac{\hat{f}_i^v(T,P,y_i)}{y_i P} = \ln \hat{\phi}_i^v = \frac{1}{RT} \int_\infty^V \left[\frac{RT}{V} - \left(\frac{\partial P}{\partial n_i} \right)_{T,V,n_{j \neq i}} \right] dV - \ln \frac{PV}{RT} \tag{2.3-3}$$

where n_i is the number of moles of species i in the mixture and the integral can be solved using an appropriate equation of state (that is an equation of the type $P = P(T,V,n_i,..)$). Note that eq. (2.3-3) can be easily derived from eq. (2.2-8).

The typical problem to solve in considering fluid phases only is to establish the stability of the system or the possibility that the system will split into two or more phases in equilibrium. Many authors [16-18] specifically describe these types of calculations and only the basic outlines will be presented in the following paragraph.

2.3.1 Bubble point-, dew point- and flash calculations

In phase-equilibria calculations different combinations of input variables are possible but, in general, the following choices are the most commonly used in order to solve typical problems.

- For a liquid liquid mixture of given composition and at a given temperature (or pressure) the pressure (or temperature) at which the first bubble of vapour phase appears and its composition is calculated (bubble point calculation).
- For a vapour mixture of given composition and at a given temperature (or pressure) the pressure (or temperature) at which the first drop of liquid phase appears and its composition is calculated (dew point problem).
- For a mixture of given global composition at given conditions of temperature and pressure it can be determined if the mixture is a stable homogeneous liquid or vapour or that the mixture is heterogeneous and splits in a vapour phase and a liquid phase with different composition. If the mixture is heterogeneous the composition and the amount of the vapour phase and of the liquid phase are calculated. This calculation is known as flash calculation. Bubble point calculations and dew point calculations can be considered as particular cases of the general flash calculation.

All these calculations with an equation of state are iterative, and in the following discussion the basic approach for the bubble point pressure and flash calculation will be described briefly.

In the case of the bubble point pressure calculation (see Fig. 2.2-1), the liquid composition, x_i, and the temperature, T, are known and the vapour phase composition, y_i, and the pressure, P, must be calculated. For the iterative calculations an initial guess for the bubble point pressure and the vapour-phase composition is needed. As alternative for the vapour-phase composition an initial guess for the fugacities or the fugacity coefficients in the vapour phase can be used.

An initial guess for the pressure is assumed and the fugacity coefficient of each component in the liquid phase ($\widehat{\phi}_i^l$) can be calculated. An initial guess is also assumed for the fugacity coefficient of each component in the vapour phase ($\widehat{\phi}_i^v$), and consequently a first estimate of the vapour composition is evaluated. With this value of y_i, the fugacity coefficients in the vapour phase are recalculated using the equation of state and a second estimate for y_i is evaluated. This iterative procedure is continued until the difference between two successive values of the composition are below a predetermined error. At this point, the sum of y_i is checked: if the sum is different from unity a new value of the pressure is assumed for a new iteration. The iterative procedure ends when the $\sum y_i$ differs from unity by less then a given value.

With some slight modifications the same scheme can be adopted for dew point calculations.

In the case of the flash calculations, different algorithms and schemes can be adopted: the case of an isothermal, or PT flash will be considered. This term normally refers to any calculation of the amounts and compositions of the vapour and the liquid phase (V, L, y_i, x_i, respectively) making up a two-phase system in equilibrium at known T, P, and overall composition. In this case, one needs to satisfy relation for the equality of fugacities (eq. 2.3-1) and also the mass balance equations (based on 1 mole feed with N components of mole fraction z_i):

$$L + V = 1 \tag{2.3-4}$$

$$z_i = x_i L + y_i V \qquad i = 1,2\ldots,N \tag{2.3-5}$$

Introducing the equilibrium conditions written in the form:

$$y_i = \frac{\widehat{\phi}_i^l x_i}{\widehat{\phi}_i^v} = K_i x_i \tag{2.3-6}$$

$$z_i = \left(L + K_i V\right) x_i \tag{2.3-7}$$

After elimination of V the following equation is obtained for x_i by simple rearrangement:

$$x_i = \frac{z_i}{L + K_i(1 - L)} \tag{2.3-8}$$

or alternatively for y_i:

$$y_i = \frac{K_i z_i}{L + K_i(1-L)}$$

(2.3-9)

These sets of mole fractions must sum to unity:

$$F_x = \sum_{i=1}^{N} x_i = \sum_{i=1}^{N} \frac{(1-K_i)z_i}{L + K_i(1-L)} = 1$$

(2.3-10)

$$F_y = \sum_{i=1}^{N} y_i = \sum_{i=1}^{N} \frac{K_i z_i}{L + K_i(1-L)} = 1$$

(2.3-11)

or equivalently:

$$F = F_x - F_y = \sum_{i=1}^{N} \frac{(1-K_i)z_i}{L + K_i(1-L)} = 0$$

(2.3-12)

The solution of a PT flash problem is accomplished when a value of L is found that makes the function F equal to zero; the advantage of using the F function instead of F_x and F_y is apparent from the fact that the function F vs. L is monotonic and which makes the numerical methods to solve the equations more efficient.

It is not known in advance what the real physical state of a system of given composition at given pressure P and temperature T is. For this reason a bubble pressure (P_{bubl}) calculation of a hypothetical liquid of composition z_i at the temperature T, and a dew pressure (P_{dew}) calculation of a hypothetical vapour of composition z_i at the temperature T, is performed. Only for pressures between P_{dew} and P_{bubl} is the system an equilibrium mixture of vapour and liquid, and flash calulations make sense.

The iterative calculations start by assuming a guess for K_i (it is necessary to assume a value for x_i and y_i, -for example the values previously obtained in the dew- and bubble-points evaluation). With these values of K_i the eq. (2.3-12) is solved iteratively for L. With this value of L, using equations (2.3-8) and (2.3-9) new values of vapour- and liquid-phase compositions are calculated and new values of K_i calculated for the next iteration. The iterations stop when satisfactory convergence is reached.

Various authors have modified this classical procedure by introducing the so-called stability test based on the Gibbs' energy minimization [19-23].

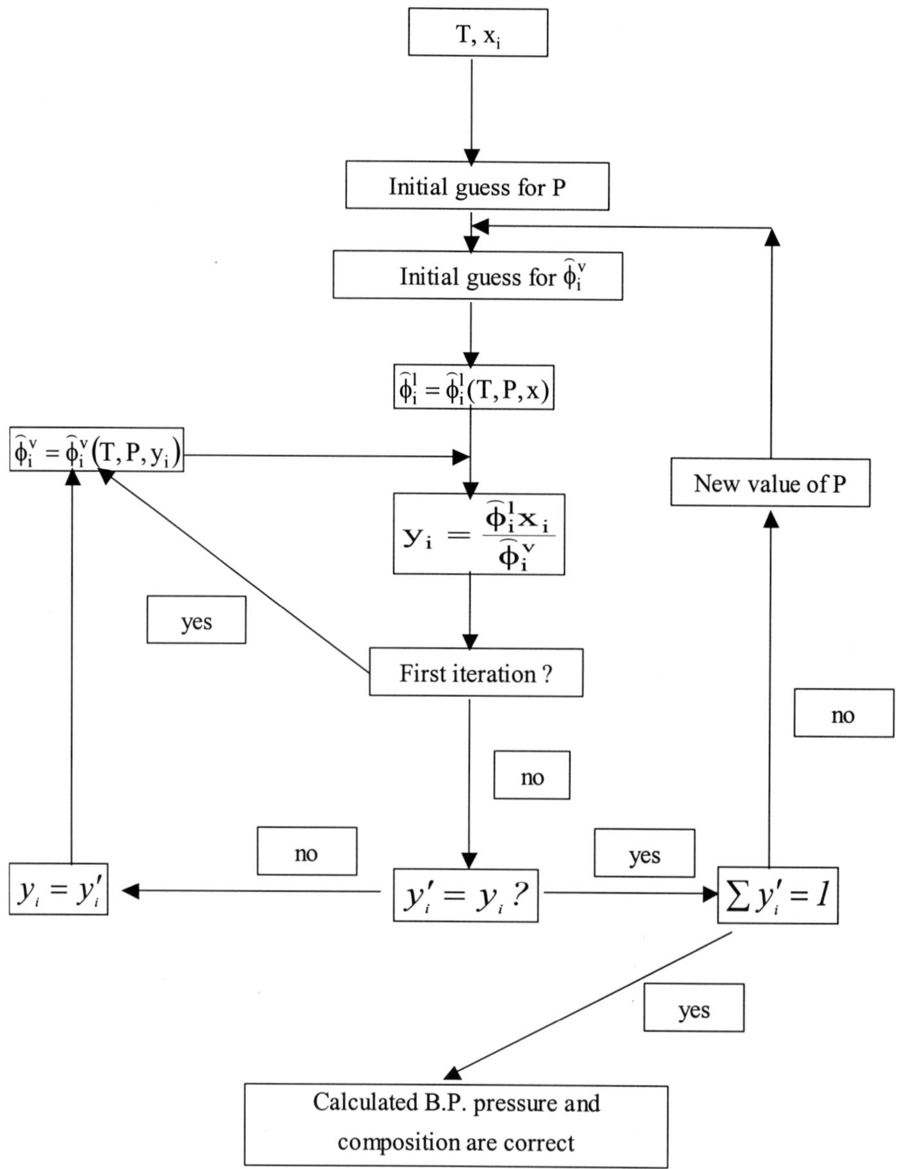

Fig. 2.3-1 Bubble point pressure calculation

2.3.2 Equations of state

The goals of the modelling of phase equilibria are both the correlation of existing data and the prediction of data in regions where experimental results are not available. An ideal model would use easily measured physical properties to predict phase equilibria under all conditions, and would be theoretically based. Unfortunately, no such model exists, and no single model can treat all situations.

Existing correlations of phase equilibrium data contain many regressed parameters, they are often semi-empirical, and they may be successfull in fitting the data in parts of the phase diagram -even with high accuracy. As far as prediction is concerned, models developed for that purpose attempt to justify theoretically a link between the model parameters and real physical phenomena. However, the distinction between these two methods is often lost, since theoretically based models are forced to fit the data better by the introduction of additional adjustable parameters.

Critical reviews of existing methods to model phase behaviour at high pressure using equations of state have been made in recent years, and we refer to these papers for further details [24, 25]. The general conclusion is that modelling is still case-specific. When the critical point is approached, predictions and even correlations of critical curves and solubilities are extremely difficult because of the nonclassical behaviour in this region.

The models used to predict equilibria in the supercritical region can be divided into the following groups [24]:
- the Van der Waals family of cubic equations of state;
- the virial family of equations of state;
- the group contribution equations of state;
- equations of state for associating and polar fluids;
- equations of state from theory and computer simulation.

Table 2.3-1.
Commonly used expressions for the repulsive term in the equation of state

Name	Z
Van der Waals [26]	$\dfrac{1}{1-4\eta}$
Carnahan Starling [27]	$\dfrac{1+\eta+\eta^2-\eta^3}{(1-\eta)^3}$
Hard Convex Body [28]	$\dfrac{1+(3\alpha-2)\eta+(3\alpha^2-3\alpha+1)\eta^2-\alpha^2\eta^3}{(1-\eta)^3}$

Recently [25] a different view was presented and shows clearly the inter-relationships between the different equations of state (E.O.S) presented in the last 50 years. In fact, very often an equation of state is a simple modification of a previously proposed model. A common feature of the equations of state is that they are constructed by combining separate contributions resulting from repulsive and attractive interactions. The classification of different E.O.S. can be based on the degree of the volume-expansion equation, or on the form of the repulsive term. If the first scheme is adopted the equations of state are simply classified into cubic and non-cubic without entering into the details of their form. Instead, following the second approach, we can classify the equations of state on the basis of the representation of the intermolecular interactions.

In order to have an idea of this classification, in Table 2.3-1 some examples of the most used repulsive terms are reported. The table reports the common names that identify the repulsive part of the compressibility factor Z. The equations are expressed using the reduced volume $\eta = \dfrac{b}{4v}$. α is a non-sphericity parameter.

To these repulsive terms, different forms of the attractive terms can be added, generating many equations of state.

At the end, both classifications divide the E.O.S. into empirical (or cubic) and theoretical models. This division does not mean that the cubic E.O.S. are without any theoretical background but simply that in their actual form these models have lost most of the theoretical foundations. In general, all these models are applied to the correlation of phase equilibrium data at low and high pressures. Empirical or cubic equations of state (Van der Waals family) are able to predict qualitatively the main types of I to V of binary phase diagrams of the Van Konynenburg and Scott classification [5], but they are not able to describe type VI fluid phase behaviour, which is associated with the closed loop liquid - liquid immiscibility at high pressure. In the following paragraphs we consider separately the cubic and non-cubic equations of state, both for pure components and mixtures.

2.3.2.1   Cubic equations of state

Pure fluids

The cubic equations of state, which are often also called semi-empirical equations of state are using the repulsive term proposed by Van der Waals (see Table 2.3-1). Various attractive terms have been proposed in the literature and some of them are presented in Table 2.3-2. In this table the different expressions for the attractive term (as $-Z^{att}$) and the name of the resulting equations are reported.

The questions of how to determine the parameters in this equation of state (and others) remains. If the parameters like a and b are independent of temperature, as is the case for the Van der Waals equation, there are two alternatives. The first is to calculate the parameters from the experimental critical point coordinates using the critical conditions given by eq. (2.2-22).

The second way is to fit the parameters to vapour pressures and liquid or vapour densities. With the van der Waals equation, in which the parameters a and b are constants, this can be done only at one temperature. For the equations, for which the parameters are a function of temperature this is done over a range of temperatures, which is discussed below.

The modification of the Van der Waals equation by Redlich and Kwong [29], who introduced a different temperature dependence and a slightly different volume dependency in the attractive term, is very important since it opened the way to a better description of the temperature dependent properties like virial coefficients.

Table 2.3-2.
Attractive terms used in cubic equations of state

Equation	$-Z^{att}$
van der Waals [26]	$\dfrac{a}{RTV}$
Redlich-Kwong [29]	$\dfrac{a}{(V+b)RT^{1.5}}$
Soave [30]	$\dfrac{a(T)}{(V+b)RT}$
Peng-Robinson [31]	$\dfrac{Va(T)}{[V(V+b)+b(V-b)]RT}$
Schmidt-Wenzel [32]	$\dfrac{Va(T)}{[V^2+ubV+wb^2]RT}$
Patel-Teja[33]	$\dfrac{Va(T)}{[V(V+b)+c(V-b)]RT}$

The Redlich-Kwong equation gives a somewhat better critical compressibility ($Z_c = 0.333$ instead of 0.375 as results from the Van der Waals equation), but is still not very accurate for the prediction of vapour pressures and liquid densities.

However, it was Soave's modification [30] of the temperature dependence of the a parameter, which resulted in accurate vapour pressure predictions (especially above 1 bar) for light hydrocarbons, which led to cubic equations of state becoming important tools for the prediction of vapour-liquid equilibria at moderate and high pressures for non-polar fluids.

$$a(T) = 0.4274 \frac{R^2 T_c^2}{P_c} \left[1 + \left(0.48 + 1.57\omega - 0.176\omega^2 \right)\left(1 - \sqrt{T_r} \right) \right]^2 \tag{2.3-13}$$

where ω is the acentric factor defined as [34]

$$\omega = -\log \frac{P(T_r = 0.7)}{P_c} - 1.0 \qquad (2.3\text{-}14)$$

Peng and Robinson [31] used a different volume dependency of the attractive term, which results in slightly improved liquid volumes (because for this E.O.S $Z_c = 0.307$, which is closer to the experimental values) and changed slightly the temperature dependence of a to give accurate vapour pressure predictions for hydrocarbons in the 6- to 10-carbon-number range.

The Peng-Robinson (PR) and the Soave-Redlich-Kwong (SRK) equations are widely used since they require little input information (only critical properties and an acentric factor to calculate the generalized parameters) and require little computer time. However, these equations also have some important shortcomings. In particular, liquid densities are not well predicted, the generalized expressions for the parameters are not accurate for non-hydrocarbons (especially polar and associating fluids), and these equations do not lead to accurate predictions for long-chain molecules. On top of this, these equations are not accurate in the critical region, and vapour pressure predictions are not very accurate below 10 torr.

Peneloux et al. [35] have introduced a clever method of improving the saturated liquid molar volume predictions of a cubic equation of state, by translating the calculated volumes without efffecting the prediction of phase equilibrium. The volume-translation parameter is chosen to give the correct saturated liquid volume at some temperature, usually at a reduced temperature $T_r = T/T_c = 0.7$, which is near the normal boiling point. It is possible to improve the liquid density predictions further by making the translation parameter temperature dependent.

Extension of cubic equations of state to mixtures

The greatest use of cubic equations of state is for phase equilibrium calculations involving mixtures. The assumption inherent in such calculations is that the same equation of state as is used for the pure fluids can be used for mixtures if we have a satisfactory way to obtain the mixtures' parameters. This is most commonly done using the van der Waals one-fluid mixing rules,

$$a = \sum_i \sum_j a_{ij} z_i z_j \qquad (2.3\text{-}15)$$

$$b = \sum \sum b_{ij} z_i z_j \qquad (2.3\text{-}16)$$

In addition, combining rules are needed for the parameters a_{ij} and b_{ij}. The usual combining rules are

$$a_{ij} = \sqrt{a_{ii} a_{jj}} \left(1 - k_{ij}\right) \qquad (2.3\text{-}17)$$

and

$$b_{ij} = \frac{(b_{ii} + b_{jj})}{2}(1 - l_{ij}) \qquad (2.3\text{-}18)$$

where k_{ij} and l_{ij} are the binary interaction parameters obtained by fitting to experimental vapour-liquid equilibrium (VLE) data or VLE and density data. Generally, l_{ij} is set equal to zero, in which case we have the linear mixing rule for the b-parameter $b = \Sigma x_i\, b_{ii}$.

The justification for using the combining rule for the a-parameter is that this parameter is related to the attractive forces, and from intermolecular potential theory the attractive parameter in the intermolecular potential for the interaction between an unlike pair of molecules is given by a relationship similar to eq. (42). Similarly, the excluded volume or repulsive parameter b for an unlike pair would be given by eq. (43) if molecules were hard spheres. Most of the molecules are non-spherical, and do not have only hard-body interactions. Also there is not a one-to-one relationship between the attractive part of the intermolecular potential and a parameter in an equation of state. Consequently, these combining rules do not have a rigorous basis, and others have been proposed.

A shortcoming of the van der Waals classical mixing rules is that they are not applicable to the so-called asymmetric mixtures and to mixtures containing polar compounds. For that reason, different mixing rules have been proposed for the a-parameter, involving essentially a concentration dependence of the k_{ij} [36-39].

Since many mixtures of interest in the chemical industry exhibit much greater degrees of non-ideality, and have been traditionally described by activity coefficient (Gibbs energy) models, Huron and Vidal [40] suggested a method to use excess Gibbs energy models to represent the mixing rule for the a-parameter of the equation of state.

The basic assumptions of the Huron-Vidal method are:

- The excess Gibbs energy G^E calculated from a liquid phase activity coefficient model, and the excess Gibbs energy G^E calculated from the equation-of-state are equal at infinite pressure;
- The co-volume b is equal to the molar volume V at infinite pressure;
- The excess volume at infinite pressure iszero.

By using the linear mixing rule for the volume parameter b , the expression for the parameter a is:

$$a = b\left[\sum_i z_i \frac{a_i}{b_i} - \frac{g_\infty^E}{\Lambda}\right] \qquad (2.3\text{-}19)$$

where Λ is a constant depending from the equation of state used (Λ is ln 2 for the Redlich-Kwong equation) and g_∞^E is the value of the molar excess Gibbs energy at infinite pressure. For g_∞^E it is possible to choose between the different models proposed in the literature [2]: the Wilson, NRTL, or UNIQUAC equations. Huron and Vidal [40] suggested using the NRTL model:

$$g_{\infty}^{E} = \sum_{i} z_{i} \frac{\sum_{j} z_{j} \, exp\left(-\alpha_{ji} \dfrac{C_{ji}}{RT}\right) C_{ji}}{\sum_{k} z_{k} \, exp\left(-\alpha_{ki} \dfrac{C_{ki}}{RT}\right)} \qquad (2.3\text{-}20)$$

In this equation $\alpha_{ij}, C_{ij}, C_{ji}$ are adjustable parameters. When $\alpha_{ij} = 0$ the van der Waals mixing rules are recovered.

This mixing rule when combined with the Wilson or NRTL models, gives excellent results for describing VLE of some highly non-ideal systems. However the Huron-Vidal mixing rule has some theoretical and computational difficulties. The mixing rule may not be successful in describing non-polar hydrocarbon mixtures and this is a problem when a multicomponent mixture contains both polar and non-polar components since all species must be represented by the same mixing rule. Further, it is necessary to draw attention to the difficulties of this mixing rule in correlating low-pressure vapour-liquid equilibrium data.

Some effort has been directed toward relaxing the infinite pressure limit in the Huron-Vidal model [41-43]. The most successful of these is the so-called modified Huron-Vidal first order (MHV1) mixing rule, proposed by Michelsen [42]. In developing the new mixing rule the Soave-Redlich-Kwong equation of state and the Huron-Vidal approach is used but with the equation of state and excess Gibbs energy models matched at liquid density and zero pressure at the temperature of interest:

$$\alpha = \sum_{i} z_{i}\alpha_{i} + \frac{1}{q_{1}}\left[\frac{g_{0}^{E}}{RT} - \sum_{i} z_{i} \, Ln\frac{b_{i}}{b}\right] \qquad (2.3\text{-}21)$$

where $\alpha = \dfrac{b}{RT}$, $\alpha_{i} = \dfrac{a_{i}}{RTb_{i}}$ and for q_{1} a recommended value of -0.593 is suggested.

In addition, an alternative mixing rule (referred as the second order modified Huron-Vidal mixing rule, MHV2) was also derived [43]:

$$q_{1}\left(\alpha - \sum_{i} z_{i}\alpha_{i}\right) + q_{2}\left(\alpha^{2} - \sum_{i} z_{i}\alpha_{i}^{2}\right) = \frac{g_{0}^{E}}{RT} + \sum_{i} z_{i} \, Ln\frac{b}{b_{i}} \qquad (2.3\text{-}22)$$

with $q_{1} = -0.478$, and $q_{2} = -0.0047$

These new mixing rules have the advantage that they allow the use of numerical parameters for the excess-Gibbs-energy models which were obtained by fitting low-pressure vapour-liquid equilibrium data. In particular, the MHV2 mixing rule was used in combination with the UNIFAC group-contribution model, with excellent results [44].

However, these new mixing rules (based both to infinite- or zero pressure limit) give, for the composition dependence of the second virial coefficient, results that are inconsistent with those obtained from statistical mechanics.

Wong and Sandler [45] used the Helmholtz excess energy to develop the following mixing rule, that satisfies the second virial restriction:

$$\frac{a}{b} = \sum_i z_i \frac{a_i}{b_i} + \frac{a_\infty^E}{\Lambda}$$

(2.3-23)

$$b = \frac{\sum_i \sum_j \dfrac{\left(b_{ii} - \dfrac{a_{ii}}{RT}\right) + \left(b_{jj} - \dfrac{a_{jj}}{RT}\right)}{2}(1 - k_{ij})z_i z_j}{1 - \left[\sum_i z_i \dfrac{a_{ii}}{RTb_i} + \dfrac{a_\infty^E}{\Lambda RT}\right]}$$

(2.3-24)

where Λ is a constant dependent on the equation of state selected, (Λ is equal to ln 2 for the Redlich-Kwong equation), and k_{ij} is a binary interaction parameter. A further reformulation of the mixing rule was successively [46] introduced in order to recover smoothly the classical van der Waals one-fluid mixing rule:

$$b = \frac{\sum_i \sum_j \left[\dfrac{(b_i + b_j)}{2} - \dfrac{(1 - k_{ij})\sqrt{a_i a_j}}{RT}\right]z_i z_j}{1 - \left[\sum_i z_i \dfrac{a_{ii}}{RTb_i} + \dfrac{a_\infty^E}{\Lambda RT}\right]}$$

(2.3-25)

The parameter a is again calculated with eq. (2.3-23).

These mixing rules were applied to perform critical-point calculations and the critical behaviour of some highly non-ideal systems [47, 48].

2.3.2.2 Non-cubic Equations of state.

Statistical mechanics and computer simulations have contributed to the development of new generations of equations of state, which in contrast with those discussed so far, have a more sound theoretical basis.

The original van der Waals idea was that pressure in a fluid is the result of both repulsive forces or excluded volume effects, which increase as the molar volume decreases, and attractive forces which reduce the pressure. Since the molecules have a finite size, there would be a limiting molar volume, b, which could be achieved only at infinite pressure. At large intermolcular separations, London dispersion theory establishes that attractive forces increase as r^{-6}, where r is the intermolecular distance. Since volume is proportional to r^3, this provides some explanation also for the attractive term in the van der Waals equation of state.

These arguments provide a plausible explanation for the form of the van der Waals equation, but modern statistical mechanics has shown that neither the repulsive- nor the attractive term in the equation is correct.

The non-cubic equations of state are characterized by the use of a repulsive term that is based on the Carnahan-Starling or on the HCB expressions already reported in Table 3. The attractive part is generally based on that derived from the perturbed hard chain theory (PHCT) [49], or from the statistical associating fluid theory SAFT [50, 51]. These approaches were the precursors of many theoretical attractive terms and consequently of different equations of state. Different authors reported applications of PHCT models to polymer systems [52-54], and to supercritical systems [55, 56]. The SAFT approach was used for the development of the SAFT equation of state [58], and was successively applied to asymmetric [58] and water/hydrocarbon systems [59].

The choice of the equation of state to be used is still now very subjective. The use of cubic equations of state with two binary interaction parameters and with excess Gibbs energy mixing rules allow sometimes also a quantitative description of typical thermodynamic behaviour of binary and ternary mixtures. However it is necessary to have always in mind that cubic equations of state embody a corresponding state theory and this theory assumes similarity of molecular size and force constant. Supercritical fluid solutions are highly asymmetric in force constant and size and as a consequence the cubic equations are forced to describe a situation for which they are not developed. On top of that often the calculation of the critical properties of the pure components, in the case of complicated molecule, is very difficult and ambiguous. Non-cubic equations can represent a more theoretically sound solution but often the gain is only marginal.

2.3.3 Solubility of solids in supercritical fluids

In the following discussion, only the basic equations that allow the calculation of the solubility of a solid material (solute, component 2) in a dense gas will be reported. The equilibrium condition for component 2 is:

$$\hat{f}_2^{\,S} = \hat{f}_2^{\,SF} \tag{2.3-26}$$

where the superscript S indicates the solid phase and the superscript SF the supercritical phase. Since normally the solid phase is considered to be the pure solid solute, according to eq. (2.2-17) the left hand side of this equation is given by:

$$\hat{f}_2^{\,S} = f_2^{\,S} = P_2^{\,sub}\phi_2^{\,V}\,exp\left(\int_{P_2^{sub}}^{P}\frac{v_2^S dP}{RT}\right) \tag{2.3-27}$$

where $P_2^{\,sub}$ is the sublimation pressure of the solid, $\phi_2^{\,V}$ is the fugacity coefficient of the pure component 2 at the pressure $P_2^{\,sub}$ and temperature T in the gas phase, and $v_2^{\,S}$ is the molar volume of the pure solid component, 2, at the temperature T. The fugacity of component 2 in the supercritical phase is given by the well-know relationship:

$$\hat{f}_2^{\,SF} = y_2\hat{\phi}_2^{\,SF}P \tag{2.3-28}$$

and, finally, the solubility is given by the following equation [2]:

$$y_2 = \frac{P_2^{\,sub}}{P}\left[\frac{\phi_2^{\,V}\,exp\left(\int_{P_2^S}^{P}\frac{v_2^S dP}{RT}\right)}{\hat{\phi}_2^{\,SF}}\right] \tag{2.3-29}$$

The quantity contained in square bracketts represents the ratio between the real solubility and the ideal solubility (the supercritical phase is described by the ideal gas law); it is always greater than 1 and it is also called the enhancement factor (*E*). *E* can have values of 10^{+3} or higher.

The enhancement factor contains three terms: $\widehat{\phi}_2^{SF}$, the fugacity in the supercritical phase, ϕ_2^{s}, which takes into account the non-ideal behaviour of the pure component 2 in the vapour phase at the sublimation pressure, and the Poynting factor that describes the influence of the pressure on the fugacity of pure solid 2.

Since the sublimation pressure is normally very low, the fugacity coefficient ϕ_2^{s} is close to 1 and also the Poynting correction is not very different from 1 (usually less than 2). As a consequence, the most important term that contributes to a value of *E* is the fugacity coefficient in the supercritical phase.

The effect of pressure on $\widehat{\phi}_2^{SF}$ is given by the equation:

$$\left(\frac{\partial\, ln\, \widehat{\phi}_2^{\,SF}}{\partial\, P} \right)_{T,y} = \frac{\overline{v}_2^{\,SF}}{RT} - \frac{1}{P} \tag{2.3-30}$$

where $\overline{v}_2^{\,SF}$ is the partial molar volume of component 2 in the supercritical phase. When the solubility y_2 is very low, and the temperature T, and pressure P are close to the critical conditions of the component 1, $\overline{v}_2^{\,SF}$ assumes large and negative values.

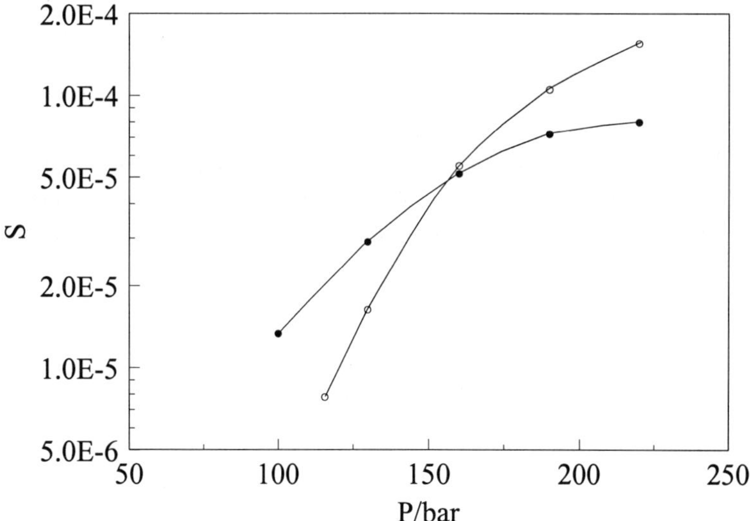

Figure 2.3-2   Solubility S (mole fraction) of a solid compound as a function of pressure at 312.5 K ($\bullet$) and 331.5 K ($\circ$).

As a consequence a small variation of the pressure causes a large variation of the fugacity coefficient and of the solubility.

In Figure 2.3-2, as an example, typical data of the solubility as a function of pressure, at two different temperatures, are reported.

It is necessary to observe, in particular, the effect of temperature on the solubility. At pressures below a given value, which is typical for each binary solute-solvent system, the solubility increases with decreasing temperature. At higher pressures the opposite effect is observed. This characteristic pressure is normally referred to as crossover pressure and it is very important when a process involving solids must be optimized.

According to eq. (2.3-30), the temperature mainly influences the sublimation pressure and the fugacity coefficient in the supercritical phase. The sublimation pressure always increases with increasing temperature, and if this is the main effect the solubility must always increases with temperature. To the contrary, at relatively low pressures (close to the critical pressure of the supercritical fluid) the fugacity coefficient in the supercritical phase plays the most important and preponderant role.

For the evaluation of the solubility it is necessary to know the pure component properties and to use an equation-of-state model for the evaluation of the fugacity coefficients. In general, two problems arise:

- pure-component parameters required by the EOS are lacking for the heavy component;
- the sublimation pressure of the solute is not known.

The critical properties, that are essential basic data if a cubic equation of state is used, can be evaluated using group contribution methods but the numerical values obtained depend on the method used. In particular, this fact represents a problem for multifunctional components that are generally involved in processing natural products and/or pharmaceuticals. As an example, depending on the prediction method used, a critical temperature ranging from 817.8 to 1254.0 K can be obtained for cholesterol [60].

The evaluation of the sublimation pressure is a problem since most of the compounds to be extracted with the supercritical fluids exhibit sublimation pressures of the order of 10^{-14} bar, and as a consequence these data cannot be determined experimentally. The sublimation pressure is thus usually estimated by empirical correlations, which are often developed only for hydrocarbon compounds. In the correlation of solubility data this problem can be solved empirically by considering the pure component parameters as fitting-parameters. Better results are obviously obtained [61], but the physical significance of the numerical values of the parameters obtained is doubtful. For example, different pure component properties can be obtained for the same solute using solubility data for different binary mixtures.

Some of these ambiguities can be partially solved using a simple approach recently proposed by Garnier et al. [62]. The sublimation pressure of a solid can be estimated using experimental fusion properties and the vaporization enthalpy derived from the equation of state. Using the Clapeyron equation P_2^{sub} can be approximated by:

$$ln\left(\frac{P_2^{sub}}{P_1}\right) = -\frac{\Delta H_2^{sub}}{R}\left(\frac{1}{T}-\frac{1}{T_1}\right) \tag{2.3-31}$$

where T_t and P_t are the reference conditions chosen at the triple-point of the pure component 2. The quantity ΔH_2^{sub} is the sublimation enthalpy in the reference state, and is calculated from the fusion and vaporisation enthalpies $\Delta H_2^{sub} = \Delta H_2^{fus} + \Delta H_2^{vap}$.

Using the approximation that for a solid the triple-point temperature T_t is equal to the normal fusion temperature, T_2^{fus}, it is possible to calculate, from the same equation of state that is used for the calculation of $\hat{\phi}_2^{SF}$:

- an estimate the reference pressure P_t at the temperature $T_t \approx T_2^{fus}$;
- an estimate of the vaporisation enthalpy ΔH_2^{vap} at the temperature T_t.

Hence, the estimation of the sublimation pressure requires only the fusion properties, T_2^{fus} and ΔH_2^{fus} which are more easily available in the literature. Using this approach more reliable results are obtained.

The so-called subcooled liquid approach was also suggested in the literature in order to overcome the difficulties connected with the pure component properties of the solid compound. This approach is commonly used for the calculation of the liquid-solid equilibria, and for solubility calculation of solids in supercritical fluids was already suggested [2] and subsequently extensively applied to different supercritical fluid processes [63]. The solubility y_2 is then expressed as:

$$y_2 = \frac{P_0}{P}\frac{\phi_2^L}{\hat{\phi}_2^{SF}}exp\left[\frac{v_2^s(P-P_0)}{RT}\right]exp\left[-\frac{\Delta H_2^{fus}}{RT}\left(1-\frac{T}{T_2^{fus}}\right)\right] \tag{2.3-32}$$

where P_0 is the normal pressure and ϕ_2^L is the fugacity coefficient for the supercooled component 2, considered as a liquid at temperature T and pressure P_0.

Also with this approach the problem of the pure-component properties necessary for the determination of the parameters of the equation of state remains: the advantage lies in the use of experimentally accessible properties of heat-of-fusion and melting point, instead of the sublimation pressure.

2.3.4 Polymer systems

The phase behaviour of many polymer-solvent systems is similar to type IV and type III phase behaviour in the classification of van Konynenburg and Scott [5]. In the first case, the most important feature is the presence of an Upper Critical Solution Temperature (UCST) and a Lower Critical Solution Temperature (LCST). The UCST is the temperature at which two liquid phases become identical (critical) if the temperature is isobarically increased. The LCST is the temperature at which two liquid phases critically merge if the system temperature is isobarically reduced. At temperatures between the UCST and the LCST a single-phase region is found, while at temperatures lower than the UCST and higher than the LCST a liquid-liquid equilibrium occurs. Both the UCST and the LCST loci end in a critical endpoint, the point of intersection of the critical curve and the liquid liquid vapour (l_2l_1g) equilibrium line. In the two intersection points the two liquid phases become critical in the presence of a non-critical vapour phase. In systems of polymers with low-molecular weight solvents the two branches of the three-phase curve l_2l_1g practically coincide with the vapour-pressure curve of the solvent. The high temperature branch of the l_2l_1g curve ends at high temperature (close to the critical temperature of the solvent) in a critical endpoint where the vapour phase and the solvent rich liquid phase become critical in the presence of a the non-critical polymer rich solvent phase.

The reason for the appearance of a UCST curve is usually ascribed to energetic interactions between the two components. While dispersion forces are temperature-independent, polar interactions and specific interactions such as hydrogen bonding are strongly temperature-dependent, and become much more effective when the temperature is lowered. Phase demixing occurs if the polymer and the solvent exhibit stronger polar and specific interactions with themselves than with each other, and when the temperature is low enough to make this kind of interactions more effective than dispersion forces. However also polymer-solvent systems in which no specific interactions occur can show UCST behaviour. In systems without specific interactions the enthalpy of mixing resulting from dispersion forces is small but positive and since the entropy of mixing becomes less important at low temperatures this will result in UCST behaviour at low temperature. Often this behaviour cannot be found experimentally because the polymer crystallises before the onset of liquid-liquid demixing. On the other hand the reason for the presence of a LCST curve is ascribed to entropic effects. When the solution is heated, the expansion of the solvent is higher than that of the polymer, and its free volume becomes larger. To solubilize the polymer, the solvent has to completely surround it. This process causes a penalty in terms of the entropy of mixing. Such a penalty becomes higher when the solvent free volume is higher, i.e., when the solvent is more expanded. Eventually, the unfavourable contribution to the entropy of mixing is so large that the free energy of mixing becomes positive and a phase separation occurs.

In the second case, corresponding to type III fluid phase behaviour, the difference in size and intermolecular potential becomes larger, i.e., when we turn from a liquid organic solvent to a supercritical fluid solvent, the UCST and LCST curves approach each other progressively, until they finally merge. The same can happen if the molecular weight of the polymer is increased. The equilibrium curve can be roughly interpreted as a combination of LCST-type transitions at high temperatures and UCST-type transitions at low temperatures. As suggested by the slope of the critical curve, on the low temperature side, the phase behaviour is strongly temperature-dependent and much less pressure-dependent, while on the high temperature side, both temperature and pressure have a strong effect on the phase behaviour.

Patterson's group [64-66] performed extensive measurements on polymer-organic solvent systems in the late 1960s and early 1970s. In the same period studies on polymer SCF systems were started [67-69], which were continued in the 1980s by the groups of Rätzsch [70-73] and Luft [75-75] and the Delft group [76-77].

In the last ten years, much work in this field has been performed in the groups of Radosz [78-86] and McHugh [87-92]. It is not the purpose of this chapter to review all the results obtained by the quoted researchers. It is important to note that all these results can be successfully placed in the thermodynamic framework described above.

The phase behaviour of binary polymer - supercritical fluid systems can be modelled with an equation of state model. In general, non-cubic equations of state are used, mainly from the PHCT and SAFT families. Lattice-fluid equations of state are also commonly used for the correlation of experimental data for such systems: the reader can refer to [93] for a general review on these equations of state.

2.3.4.1. Glassy Polymers

Supercritical fluids can interact not only with polymers at temperatures higher than softening temperature but also with polymers in the glassy state. Three concomitant effects must be considered:

- The ability of SCFs to dissolve to considerable extent into amorphous glassy polymers (polymer sorption);
- The swelling of the polymer matrix;
- The plasticization of the polymer.

These facts can be used to develop impregnation or devolatilization processes or to promote the compatibility of different polymers [94].

Swelling (i.e., volume increase of the polymer) and sorption (i.e. the weight of the gas dissolved into the polymer) databases are quite sparse. The swelling data are generally determined by measuring the variation of the length of a polymer film resulting from pressurization with the SCF; the volume is consequently calculated, assuming isotropic behaviour.

For the determination of sorption data, gravimetric methods are usually used: the most difficult task is to take into account the buoyancy effect. This effect is negligible at low pressures, but becomes important at SCF densities. Accurate equations of state for pure fluids are used to calculate gas densities, and swelling data are needed to calculate the volume of the polymeric phase. Generally, the amount of dissolved gas increases with increasing pressure until a saturation value is reached.

The so called plasticization effect, i.e., the depression of the glass transition temperature is also an important feature: the sorbed gas acts as a kind of "lubricant", making it easier for chain molecules to slip over one another, and thus causing polymer softening.

Measurements of glass transition temperatures at high pressure were made indirectly using, in particular, creep compliance [95, 96] or directly using differential scanning calorimetric techniques [97, 98]. The measured depression reaches values as high as 60°C for poly(methyl methacrylate) and polystyrene.

Many attempts to model sorption and glass transition temperature depression have been presented in the literature. The possibility of predicting the glass transition shift is also fundamental for choosing the correct thermodynamic approach for the modelling of the sorption. At pressures above the glass transition, the polymer can be treated with the normal equations of state used for fluid-phase thermodynamics. Below the glass transition, the polymer phase is in a "non-equilibrium" condition and the sorption must be described using different approaches.

Regarding the shift of the glass transition temperature, the most important contribution is due to Condo et al. [99]. In this approach the Gibbs Di Marzio criterion is used together with the the Sanchez-Lacombe equation of state. Following this approach, at the glass transition the entropy of the sytem is zero; in this way it was possible to establish the effect of different parameters on the variation of the glass transition. In particular, the possibility of the so-called retrograde vitrification is evidenced as a consequence of the non-monotonic variation of T_g

with pressure. According to this behaviour, when the temperature is reduced at constant pressure it is possible to find a first liquid-glass transition followed by a second transition from glass- to the liquid state again.

The model of Condo permits the determination of the limiting pressure at which the glass state exists, on the basis of a reduced number of parameters (only the glass transition temperature of the pure polymer is needed to calibrate the energy parameter); for this reason, it seems particularly useful for the prediction of the plasticizing effects.

Various authors have considered the glassy state behaviour, and in particular the sorption of gases and supercritical fluids. The main dificulty is to find a thermodynamic model suitable for the description of the properties of the glassy state. Also, the normal equations of state specially developed for polymeric systems are not directly applicable to the non-equilibrium conditions of this state.

Wissinger and Paulaitis [100] developed a model to correlate swelling and sorption in glassy polymers, which is based on the Panayiotou-Vera equation of state (a lattice fluid equation) and on the concept of order parameters. Order parameters are additional variables that describe the thermodynamic state of a system. In the case of a glassy polymer described by a lattice equation two order parameters can be assumed: the fraction of holes in the lattice and the number of nearest-neighbour contacts between polymer segments on the lattice sites. To represent a glassy polymer, the order parameters are frozen at the value they assume at the glass transition temperature. Above the glass transition, the simple minimization of the Gibbs energy of the system (i.e., by imposing isofugacity equations and using the equation of state to calculate fugacity coefficients) is the method used to perform phase-equilibrium calculations. Below the glass transition the system is not at equilibrium, which is accounted for by "freezing in" the order parameters at their glass-transition value before imposing the isofugacity equations. Good correlations of the swelling and sorption data, above and below T_g, are obtained although the model is not predictive, because it uses two binary interaction parameters.

In a further development of the above model, Kalospiros and Paulaitis [101] use the glass-transition temperature of the pure polymer to calibrate the value of the order parameter (fraction of holes in the lattice). This value is kept constant, even if a swelling agent is added to the system, so the glass transition-temperature depression induced by the SCF can be predicted.

The NELF model, recently proposed by Doghieri and Sarti [102], uses the Sanchez Lacombe equation of state for the evaluation of the chemical potential of the supercritical fluid in the gas phase (pure supercritical fluid) and in the polymeric phase. The density of the polymeric phase required for the evaluation of the chemical potential is not calculated through the equation of state, but is taken from experimental swelling data, and is considered to be an order-parameter. This approach can also be modified using a different equation of state for the gas phase, and does not require any binary interaction parameter. In this way, the sorption behaviour can be predicted from the swelling data alone. Since swelling measurements are, in general, more difficult to perform, it was recently [103] proposed to use the model as an empirical method to correlate and predict sorption data. Using one experimental sorption data point, and applying the NELF model, it is possible to predict the sorption at other pressures and temperatures.

2.4 Chemical reaction equilibria

Many industrially important chemical processes are high pressure processes. Examples are the production of ammonia and the production of low density polyethylene. Basically, the pressure affects both the equilibrium yield of a chemical reaction and the reaction rate. Here, only the influence on the equilibrium yield is discussed.

Consider the general chemical reaction

$$aA + bB + cC + dD + \ldots \rightarrow nN + mM + \ldots \ldots \tag{2.4-1}$$

The constituents A, B, C and are called reactants and the constituents N and M are products. This reaction can also be written as

$$nN + mM + \ldots - aA - bB - cC - dD - \ldots \ldots = 0 \tag{2-4-2}$$

or

$$\sum_i \vartheta_i L_i = 0 \tag{2.4-3}$$

The stoichiometric coefficients ϑ_i are positive for products and negative for reactants. The Gibbs energy of the system of products and reactants must be minimal at equilibrium. If this condition is combined with the mass balance that is imposed by eq.(2.4-3) we get the general thermodynamic condition for chemical equilibrium:

$$\sum_i \vartheta_i \mu_i = 0 \tag{2.4-4}$$

The chemical potentials μ_i can be written as

$$\mu_i(P,T,x) = \mu_i^0(P^0,T,x^0) + RT \ln \frac{\widehat{f}_i(P,T,x)}{\widehat{f}_i^0(P^0,T,x^0)} \tag{2.4-5}$$

The superscript ^{0} indicates the standard state, which is taken at equilibrium temperature, standard pressure P^0 and standard composition x^0. The standard pressure is usally taken as 1 bar or 10^{+5} Pa.. For liquid and solid components the standard state is usually the pure liquid or solid component, for gases the standard state is the pure gaseous component in the ideal gas state at P^0 and T. So, for a gaseous component $\widehat{f}_i^0(P^0,T,x^0) = P^0$.

Combining eq.(2.4-4) and eq. (2.4-5) we find:

$$RT \ln \prod_i \left(\frac{\widehat{f}_i(P,T,x)}{\widehat{f}_i^0(P^0,T,x^0)}\right)^{\vartheta_i} = -\sum_i \vartheta_i \mu_i^0(P^0,T,x^0) \tag{2.4-6}$$

which is usually written as:

$$RT \ln K = -\Delta_r G_T^0 \qquad (2.4\text{-}7)$$

K is the thermodynamic equilibrium constant which is only dependent on T, and $\Delta_r G_T^0$ is the standard reaction Gibbs energy of the reaction, which can be calculated with

$$\Delta_r G_T^0 = \Delta_r H_T^0 - T\Delta_r S_T^0 \qquad (2.4\text{-}8)$$

with

$$\Delta_r H_T^0 = \Delta_r H_{298.15}^0 + \int_{298.15}^{T} \Delta_r C_P^0 dT \qquad (2.4\text{-}9)$$

and

$$\Delta_r S_T^0 = \Delta_r S_{298.15}^0 + \int_{298.15}^{T} \frac{\Delta_r C_P^0}{T} dT \qquad (2.4\text{-}10)$$

The standard reaction enthalpy $\Delta_r H_{298.15}^0$ and the standard reaction entropy $\Delta_r S_{298.15}^0$ at 298.15 K and the standard reaction specific heat are calculated from:

$$\Delta_r H_{298.15}^0 = \sum_i \vartheta_i \Delta_f H_{i,298.15}^0 \qquad (2.4\text{-}11)$$

$$\Delta_r S_{298.15}^0 = \sum_i \vartheta_i S_{i,298.15}^0 \qquad (2.4\text{-}12)$$

and

$$\Delta_r C_P^0(T) = \sum_i \vartheta_i C_{i,P}^0(T) \qquad (2.4\text{-}13)$$

The enthalpies of formation $\Delta_f H_{i,298.15}^0$, absolute entropy $S_{i,298.15}^0$ and specific heat values $C_{P,i}^0$ of component i can be found in tables.

After calculation of $\Delta_r G_T^0$, K is calculated and the equilibrium composition is obtained from

$$K = \prod_i \left(\frac{\hat{f}_i(P,T,x)}{f_i^0(P^0,T,x^0)}\right)^{\vartheta_i} \qquad (2.4\text{-}14)$$

For gases, $\hat{f}_i(P,T,x)$ is calculated using an equation of state with eq. (2.2-7) or eq. (2.2-8). For liquid components the same method can be followed or $\hat{f}_i(P,T,x)$ is calculated using an activity coefficient model:

$$\widehat{f}_i(P,T,x) = x_i \gamma_i f_i^0(P,T,x_i = 1) \tag{2.4-15}$$

The standard state fugacicity at pressure P can be calculated from the standard state fugacity at pressure P^0

$$\widehat{f}_i(P,T,x^0) = f_i(P^0,T,x_i = 1)\exp\int_{P^0}^{P} \frac{V_i^L}{RT} dP \tag{2.4-16}$$

The exponential term in eq. (2.4-16), the so-called Poynting-factor, can often be put equal to one,. In that case for a reaction in the liquid phase a simple reation between K and the composition is found:

$$K = \sum_i (x_i \gamma_i)^{\vartheta_i} \tag{2.4-17}$$

For solids no equations of state are available and an activity coefficient approach has to be followed.

2.4.1 Homogeneous gas reactions

As example of the caculation of the equilibrium composition of a homogeneous gas reaction the ammonia synthesis reaction at 450 ^{0}C will be considered:

$$3H_2 + N_2 \rightarrow 2NH_3 \tag{2.4-18}$$

For this reaction eq. (2.4-14) is written as

$$K = \frac{(\widehat{f}_{NH_3}/P^0)^2}{(\widehat{f}_{N_2}/P^0)(\widehat{f}_{H_2}/P^0)^3} = \frac{(y_{NH_3}\widehat{\phi}_{NH_3}P/P^0)^2}{(y_{N_2}\widehat{\phi}_{N_2}P/P^0)(y_{H_2}\widehat{\phi}_{H_2}P/P^0)^3} \tag{2.4-19}$$

The terms in the mole fractions y_i, in the fugacity coefficients $\widehat{\phi}_i$ and in the pressure can be combined

$$K = \frac{y_{NH_3}^2}{y_{N_2}y_{H_2}^3}\frac{\widehat{\phi}_{NH_3}^2}{\widehat{\phi}_{N_2}\widehat{\phi}_{H_2}^3}(P/P^0)^{-2} \tag{2.4-20}$$

If the gas mixture is considered to be an ideal gas mixture then all fugacity coefficients are 1 and since K is a constant, the effect of increasing pressure is an increase of the equilibrium mole fraction of ammonia and a decrease of the mole fractions of nitrogen and hydrogen. However, since the ammonia synthesis is a high pressure process the gas mixture is not an ideal gas and the fugacity coefficients have to be taken into account.

Table 2.4-1
Calculated composition of an equilibrium mixture of ammonia, nitrogen and hydrogen at 450 ^{0}C. The feed is a stoechiometric mixture of nitrogen and hydrogen.

	ideal gas			Redlich-Kwong equation of state		
P/MPa	y_{NH3}	y_{N2}	y_{H2}	y_{NH3}	y_{N2}	y_{H2}
1	0.0023	0.2494	0.7483	0.0023	0.2494	0.7483
10	0.1599	0.2100	0.6301	0.1722	0.2069	0.6208
20	0.2528	0.1868	0.5604	0.2845	01789	0.5367
50	0.4032	0.1492	0.4476	0.4830	0.1292	0.3877
100	0.5206	0.1199	0.3596	0.6412	0.0897	0.2691

At 450 ^{0}C K is calculated from eq. (2.4-7), which results in $K = 0.486 \times 10^{-4}$. In case the reactor is fed with a mixture of hydrogen and nitrogen in the molar ratio 3:1 and the equilibrium conversion is ε, the composition of the equilibrium mixture can be expressed as function of ε: $n_{NH_3} = 2\varepsilon, n_{H_3} = 3 - 3\varepsilon, n_{N_2} = 1 - \varepsilon$. The total number of moles is $n_{tot} = n_{NH_3} + n_{H_2} + n_{N_2} = 4 - 2\varepsilon$, so the equilibrium mole fractions are $y_{NH_3} = 2\varepsilon /(4 - 2\varepsilon), y_{H_3} = (3 - 3\varepsilon)/(4 - 2\varepsilon), y_{N_2} = (1 - \varepsilon)/(4 - 2\varepsilon)$.

The fugacity coefficients are a function of pressure, temperature and the equilibrium mole fractions, so at given pressure and temperature eq. (2.4-20) can be solved for ε and the equilibrium mole fractions can be calculated. Table 2.4-1 gives the calculated equilibrium composition of the reaction mixture at different pressures for an ideal gas mixture and in case the gas is described with the Redlich-Kwong equation of state.

The results of the calculations show that with increasing pressure the equilibrium yield of ammonia is increasing and that the non-ideality of the gas mixtures has in this case a positive effect on the equilibrium conversion.

2.4.2   Heterogeneous reactions

The calculation of the equilibrium conversion of heterogeneous reactions is in most cases much more complicated then in the case of homogeneous reactions, because the calculations involve in general the solution of the conditions for chemical equilibrium and the conditions for phase equilibrium. In the following a relatively simple example is given.

Consider the reaction $C_2H_4(e) + H_2O(w) \rightarrow C_2H_5OH(a)$ at 200 ^{0}C and 34.5 bar. At these conditions the liquid phase is mixture of alcohol and water in which almost no ethene is dissolved and the vapour phase is a mixture of the three reactants. The equilibrium composition can be calculated from the equilibrium condition for the reaction in the gas phase together with the vapour liquid equilibrium conditions:

$$RT \ln K = RT \ln \frac{\widehat{f}_a^V / P^0}{\widehat{f}_e^V / P^0 \times \widehat{f}_w^V / P^0} = -\Delta_r G_T^0 \tag{2.4-21}$$

$$\hat{f}_a^V = \hat{f}_a^L \tag{2.4-22}$$

$$\hat{f}_e^V = \hat{f}_e^L \tag{2.4-23}$$

$$\hat{f}_w^V = \hat{f}_w^L \tag{2.4-24}$$

At given pressure and temperature this set of equations is a set of four equations with four unknowns (x_w, x_e and y_w, y_e) which can be solved. In this case is convenient to describe the gas phase with an equation of state and the liquid phase with an activitivity coefficient model. The equilibrium constant K is calculated in the usual way.

A way to solve this problem is given in [1]. Using eq. (2.4-15) and (2.4-21) K can be written as:

$$K = \frac{x_a \gamma_a \phi_a^{sat} P_a^{sat}}{y_e \hat{\phi}_e P / P^0 \times x_w \gamma_w \phi_w^{sat} P_w^{sat}} \tag{2.4-25}$$

From the vapour-liquid equilibrium conditions we have

$$y_i = \frac{x_i \gamma_i \phi_i^{sat} P_i^{sat}}{\hat{\phi}_i P} \tag{2.4-26}$$

The vapour pressures as function of temperature are known. If we consider the vapour phase as an ideal gas $\hat{\phi}_i = \phi_i$, the fugacity coefficient of pure i at the equilibrium conditions, which can be calculated by an equation of state. y_e is given by

$$y_e = 1 - y_a - y_w = 1 - \frac{x_a \gamma_a \phi_a^{sat} P_a^{sat}}{\phi_a P} - \frac{x_w \gamma_w \phi_w^{sat} P_w^{sat}}{\phi_b P} \tag{2.4-27}$$

And since the liquid phase is a mixture of alcohol and water

$$x_w = 1 - x_a \tag{2.4-28}$$

The activity coefficients are known as function of x_w [104], so x_w, x_a, and y_e can be calculated from eq. (2.4-25), (2.4-27) and (2.4-28). y_a and y_w are then calculated from eq (2.4-26).

2.5   Experimental methods

The wide varity of the properties of chemical compounds does not enable the use of a universal apparatus for the measurement of thermodynamic properties for pure components and mixtures at high pressure. In the case of two-phase equilibria like vapour-liquid equilibria, the typical set of data to be determined is the pressure, the temperature, and the composition of the two phases at equilibrium. Some experimental apparatus also allows the determination of the densities of the coexisting phases. In some types of equipment the pressure, the temperature and the composition of one phase only are determined.

Special classes of apparatus are used for the determination of particular thermodynamic properties, such as activity coefficients at infinite dilution, Henry's constants, or partial molar volumes at infinite dilution [105,106]. These data, together with a thermodynamic model, can be used for the calculation of the compositions of the coexisting phases at equilibrium, and for that reason - in this context - these methods are considered as indirect methods of measurement.

Many different reviews on apparatus used for the experimental determination of phase equilibria at high pressure have been published [107-110], and it will be useful to separate the subject on the basis of the nature of the systems investigated.

2.5.1   Vapour-liquid equilibria

Different types of apparatus have been proposed in the literature, which enable the measurment of the temperature, the pressure, the composition, and the density of coexisting vapour and liquid phases. These types of apparatus can be classified on the basis of the method used [109,110]:

- Static methods
- Recirculating methods

In the first case, the mixture is placed into an evacuated cell, which is placed in a constant temperature bath (usually a liquid or air bath, although also metal-block thermostats have been used). The contents of the cell which are separated into two or more phases of different density, are brought to equilibrium by shaking, stirring, or agitation, and the equilibrium pressure is measured. The sample remains enclosed in the cell, and no phase or part of the system is subject to flow with respect to the others. Because we are dealing with a closed system, this method requires careful evacuation of the cell and tubing and degassing of the materials, in order to avoid errors in the reading of the equilibrium pressure, although this problem is not so crucial as at low pressures.

We generally distinguish between two methods when the determination of the composition of the equilibrium phases is taking place. In the first method, known amounts of the pure substances are introduced into the cell, so that the overall composition of the mixture contained in the cell is known. The compositions of the co-existing equilibrium phases may be recalculated by an iterative procedure from the predetermined overall composition, and equilibrium temperature and pressure data. It is necessary to know the pressure volume temperature (PVT) behaviour, for all the phases present at the experimental conditions, as a function of the composition in the form of a mathematical model (EOS) with a sufficient accuracy. This is very difficult to achieve when dealing with systems at high pressures. Here, the need arises for additional experimentally determined information. One possibility involves the determination of the bubble- or dew point, either optically or by studying the pressure volume relationships of the system. The main problem associated with this method is the preparation of the mixture of known composition in the cell.

The second method is employed more frequently. In this case, samples of the co-existing are withdrawn and analysed. In this method, the initial overall composition of the mixture charged into the cell is established only approximately, just to pre-determine the location of the measured equilibrium data-point in the phase diagram. The major difficulties with this method lie in withdrawing the samples of the equilibrium-phase or -phases from the cell without disturbing the system equilibrium, and with their transport to analytical devices. Sampling is usually accompanied by the depressurization, but any changes in the homogeneity and composition of the phases must be avoided. The advantage of the static method is that only small amounts of substances are needed for the experiments.

The recirculation methods were developed from the static methods. The main part of the recirculation apparatus is a thermostated equilibrium cell. The equilibrium cell has facilities for mechanically driven circulation through external loop(s) of either the lighter (i.e., the top) phase or the heavier (the bottom) phase, or of both phases. Magnetically operated high-pressure pumps usually achieve the circulation.

The purpose of introducing the external circulation of phases is to achieve their more efficient equilibration through stirring and contacting, and thereby to improve the process of equilibration and to reduce the time required. Another reason is that sampling is easier than in the case of a static method. After equilibration has been achieved in a steady-state regime, the phases which circulate through an external loop already represent a separated equilibrium phase. A portion of such a phase may be trapped in a sampling cell, which is removed for analysis, or the phase may be temporarily circulated through an in-line sampling loop such as an injection valve, and then analysed off-line. On-line analysis can also be performed in this case, without disturbing the phase-stream in the circulation loop. Another advantage is that it is also possibile to perform a relatively easy determination of the density of the phases in the recirculation loops, e.g., by using in-line vibrating tube density meters.

These experimental methods, extensively used for the study of binary mixtures, are seldomly used for the determination of phase -quilibria of complex multicomponent mixtures, mainly owing to analytical difficulties.

For details of experimental apparatus the reader can refer to the reviews already cited [109,110].

2.5.2   Equilibria involving solids

Different reviews on the experimental investigation of phase equilibria in systems involving solid components can be found in literature [111,112].

The most popular method used is a dynamic method, the saturation method. In this technique, the non-volatile, heavy solid solute is loaded into a saturator, or a battery of two or more saturators connected in series, and remains there as a stationary phase during the experiment. In most cases the saturator is in the form of a packed column. At constant pressure, a steady stream of supercritical fluid (solvent) passes through a preheater, where it reaches the desired system temperature. Then this fluid is continuously fed to the bottom of the saturator, and the solute is stripped from the stationary heavy phase in the column. The supercritical fluid saturated with the solute leaves the saturator at the top.

Direct sampling and analysis of the effluent stream may be used to determine the solubility of the heavy phase in the volatile component (often a supercritical fluid). Alternatively, the composition can be determined from the total volume of gas (i.e., of the supercritical fluid after expansion) passed through the saturator, and from the known mass of solute extracted during the sample-collecting period. The efflluent stream is expanded to atmospheric pressure via an expansion valve. Then it passes through a cold trap, where the extract is quantitatively precipitated or condensed, and finally proceeds to a dry-test or a wet-test gas meter or other device, where the total amount of the passed gas is measured. The amount of extracted solute is usually determined by weighing the sample collected in the cold trap, or by analysing the solution obtained, by dissolving the collected material in a know amount of liquid solvent.

The method is applicable only to binary two-phase systems. Since only the supercritical-fluid rich effluent phase is sampled, the composition of the equilibrium heavy phase cannot be determined. However, there are some pitfalls in this technique, which the experimentalist should be aware of:

- The danger of unnoticed phase changes occurring in the saturators which can be monitored by using a view cell;
- The necessity of avoiding any entrainment of the particles or drops of the stationary heavy phase with the stream of supercritical-fluid rich phase by using a filter;
- The saturation of the supercritical-fluid phase with the solute should be achieved by adjusting a proper flow rate.
- It is necessary to avoid the precipitation or condensation, and subsequent deposition of the extracted solute, from the effluent stream during the expansion in the expansion valve or in the section of tubing before the stream reaches the collecting trap. Weighing the valve can overcome this problem.

References

1. J.M. Smith, H.C. van Ness and M.M. Abbott, Introduction to Chemical Engineering Thermodynamics, 3rd ed., McGraw-Hill, New York, 1996.
2. J.M. Prausnitz, R.N. Lichtenthaler and E. Gomes de Azevedo, Molecular Thermodynamics of Fluid phase equilibria, 3rd ed., Prentice-Hall inc., Englewood Cliffs, NJ, 1999.
3. J.M.H. Levelt-Sengers in Supercritical Fluids, Fundamentals for Application, E. Kiran and J.M.H. Levelt-Sengers eds., Kluwer Academic Publishers, Dordrecht, The Netherlands, 1994.
4. J.V. Sengers in Supercritical Fluids, Fundamentals for Application, E. Kiran and J.M.H. Levelt-Sengers eds., Kluwer Academic Publishers, Dordrecht, The Netherlands, 1994.
5. P.H. van Konynenburg and R.L. Scott, Phil. Trans. 298A (1980) 495.
6. M.L. Garcia-Lisbona, A. Galindo, G. Jackson, A.N. Burgess, J. Phys. Chem., 100 (1996) 57.
7. M.L. Garcia-Lisbona, A. Galindo, G. Jackson, A.N. Burgess, J. Am. Chem. Soc. 120 (1998) 4191.
8. J.G. Andersen, N. Koak and Th.W. de Loos, Fluid Phase Equil. 163 (1999) 259.
9. A. Bolz, U.K. Deiters, C.J. Peters and Th.W. de Loos, Pure&Appl. Chem. 70 (1998) 2233.
10. Th.W. de Loos in Supercritical Fluids, Fundamentals for Application, E. Kiran and J.M.H. Levelt-Sengers eds., Kluwer Academic Publishers, Dordrecht, The Netherlands, 1994.
11. R. Enick, G.D. holder and B.I. Morsi, Fluid Phase equil. 22 (1981) 3489.
12. D.J. Fall and K.D. Luks, J. Chem. Eng. Data 30 (1985) 276.
13. Th.W. de Loos and W. Poot, Int. Journ. Thermophys. 19 (1998) 637.
14. W. Poot and Th.W. de Loos, Phys. Chem.-Chem. Phys. 18 (1999) 4293.
15. Th.W. de Loos, W. Poot and and G.A.M. Diepen, Macromolecules 16 (1983) 111.

16. L. Asselineau, G. Bogdanic and J. Vidal, Fluid Phase Equil., 3 (1979) 273.
17. M.L. Michelsen, Computers Chem Engng., 17 (1993) 431.
18. N. Orbey and S. L. Sandler, Modeling Vapor-Liquid Equilibria, Cambridge University Press, Cambridge, 1998.
19. M. L. Michelsen, Fluid Phase Equilibria, 9 (1982) 1.
20. M. L. Michelsen, Fluid Phase Equilibria, 9 (1982) 20.
21. M.L. Michelsen, Fluid Phase Equilibria, 33 (1987) 13.
22. Aa. V. Phoenix and R. A. Heidemann, Fluid Phase Equilibria, 150-151 (1998) 255.
23. M.L. Michelsen, Fluid Phase Equilibria, 143 (1998) 1.
24. S. L. Sandler, H. Orbey and B.-I. Lee, Equations of State, chapt. 2 in "Models for Thermodynamic and Phase equilibria Calculations, S. L. Sandler ed., M. Dekker, New York, 1994.
25. Y. S.Wei and R. Sadus, A.I.Ch.E. J.,46 (2000) 169.
26. J. D. Van der Waals, 1873 The equation of state for gases and liquids, in "Nobel lectures in Physics", 1, (1967) 254
27. N.F. Carnahan and K.E. Starling, J. Chem. Phys., 51 (1969) 635.
28. T. Boublik, Ber. Bunsenges. Phys. Chem., 85 81981) 1038.
29. O. Redlich and J. N. S. Kwong, Chem. rev., 44 (1949) 233.
30. G. Soave, Chem. Eng. Science, 27 (1972) 1197.
31. D.-Y. Peng and D. B. Robinson, Ind. Eng. Chem. Fundam., 15 (1976) 59.
32. G. Schmidt and H. Wenzel Chem. Eng. Science, 35 (1980) 1503.
33. N. C. Patel and A. Teja, Chem. Eng. Science, 37 (1982) 463.
34. K. S. Pitzer, D.Z. Lippman, R.F. Curl, Jr., C. M. Huggins and D. E. Petersen, J. Am. Chem. Soc., 77 (1955) 3427.
35. A. Peneloux, E. Rauzy and R. Freze, Fluid Phase Equilibria, 8 (1982) 7.
36. Y. Adachi and H. Sugie, Fluid Phase Equilibria, 23 (1986) 103.
37. A. Z. Panagiotopoulos and R.C. Reid, New Mixing Rules for Cubic Equations of State for Highly Polar, Asymmetric Mixtures, Equation of State - Theories and Applications, K. C. Chao and R.L. Robinson, jr. eds. A.C.S. Symp. Ser., 300 (1986) 571.
38. R. Stryjek and J. Vera, Can. J. Chem. Eng., 64 (1986) 323.
39. J. Schwartzentruber and H. Renon, Ind. Eng. Chem. Res., 28 (1989) 1049.
40. M.J. Huron and J. Vidal, Fluid Phase Equilibria, 3 (1979) 255.
41. J. Mollerup, Fluid Phase Equilibria, 25 (1986) 323.
42. M.L. Michelsen, Fluid Phase equilibria, 60 (1990) 213.
43. S. Dahl and M.L. Michelsen A.I.Ch.E.J., 36 (1990) 1829.
44. S. Dahl, A. Dunalewicz, Aa. Fredenslund and P. Rasmussen, J. Supercritical Fluids, 5 (1992) 42.
45. D.S.H. Wong, S.I. Sandler, A.I.Ch.E.J., 38 (1992) 671.
46. N. Orbey and S. I. Sandler, A.I.Ch.E.J., 41 (1995) 683.
47. M. Castier and S. I. Sandler, Chem. Eng. Science, 52 (1997) 3393.
48. M. Castier and S. I. Sandler, Chem. Eng. Science, 52 (1997) 3579.
49. S. Beret and J.M. Prausnitz, A.I.Ch.E.J., 26 (1975) 1123.
50. W.G. Chapman, G. Jackson and K.E. Gubbins, Mol. Phys., 65 (1988) 1057.
51. W.G. Chapman, K.E. Gubbins, G. Jackson and M. Radosz, Ind. Eng. Chem. Res., 29 (1990) 1709.
52. M.D. Donohue and J.M. Prausnitz, A.I.Ch.E.J., 24 (1978) 849.
53. D.D. Liu and J.M.Prausnitz, J. Appl. Poly. Sci., 24 (1979) 725.

54. D.D. Liu and J.M. Prausnitz, Ind. Eng. Chem. Process des. Dev., 19 (1980) 205.
55. J. Gregorowicz, M. Fermeglia, G. Soave and I. Kikic, Chem.Eng.Science, 46 (1991) 1427.
56. M. Fermeglia and I. Kikic, Chem. Eng. Science, 48 (1993) 3889.
57. S.H. Huang and M. Radosz, Ind. Eng. Chem. Res., 29 (1990) 2284.
58. S.H. Huang and M. Radosz, Ind. Eng. Chem. Res., 30 (1991) 1994.
59. I.G. Economou and C. Tsonopoulos, Chem. Eng. Science, 52 (1997) 511.
60. E. Neau E., S. Garnier, P. Alessi, A. Cortesi and I. Kikic,Proc. 3rd Int. Symposium on High Pressure Technology, Zurich, pp 265- 270, 1996.
61. P. Alessi, A. Cortesi, I. Kikic and N.R. Foster, Chem and Biochem J., 11, (1997) 19.
62. S. Garnier, E. Neau, P. Alessi, A. Cortesi and I. Kikic, Fluid Phase Equilibria, 158-160 (1999) 491.
63. I. Kikic, M. Lora and A. Bertucco, Ind. Eng. Chem. Res., 36 (1997) 5507.
64. L. Zeman, J. Biros, G. Delmas, and D. Patterson, J. Phys. Chem., 76 (1972) 1206.
65. L. Zeman and D. Patterson, J. Phys. Chem., 76 (1972) 1214.
66. K. S. Siow, G. Delmas, and D. Patterson, Macromolecules, 5, (1972) 29.
67. P.I. Freeman and J.S. Rowlinson, Polymer, 1 (1960) 20.
68. P. Ehrlich, J. Polym. Sci., A3 (1965) 131.
69. T. Swelheim, J. de Swaan Arons and G.A.M. Diepen, Recl. Trav. Chim. Pays-Bas, 84 (1965) 261.
70. T. Rätzsch, R. Findeisen and V. S. Sernow, Z. Phys. Chem., 261 (1980) 995.
71. M. T. Rätzsch, P. Wagner, C. Wohlfarth, and D. Heise, Acta Polymerica, 33 (1982) 463.
72. M. T. Rätzsch, P. Wagner, C. Wohlfarth, and S. Gleditzsch, Acta Polymerica, 34 (1983) 340.
73. C. Wohlfarth, P. Wagner, M. T. Rätzsch, and S. Westmeier, Acta Polymerica, 33 81982) 468.
74. G. Luft and N.S. Subramanian, Ind Eng. Chem. Res., 26 (1987) 750.
75. G. Luft and R.W. Wind, Chem. Ing. Tech., 64 (1992) 1114.
76. Th.W. de Loos, W. Poot and G.A.M. Diepen, Macromolecules, 16 (1983), 111.
77. Th.W. de Loos, R.N. Lichtenthaler and G.A.M. Diepen, Macromolecules, 16 (1983) 117.
78. S-J. Chen and M. Radosz, Macromolecules, 25 (1992) 3089.
79. S.-J. Chen, I. G. Economou and M. Radosz, Macromolecules, 25 (1992) 4987.
80. S.-J. Chen, I. G. Economou and M. Radosz, Fluid Phase Equilibria, 83 (1993) 391.
81. C. J. Gregg, S.-J. Chen, F. P. Stein and M. Radosz, Fluid Phase Equil., 83 (1993) 375.
82. C. J. Gregg, S.-J. Chen, F. P. Stein and M. Radosz, Macromolecules, 27 (1994) 4972.
83. C. J. Gregg, S.-J. Chen, F. P. Stein and M. Radosz, Macromolecules, 27 (1994) 4980.
84. C. J. Gregg, F. P. Stein and M. Radosz, J. Phys. Chem., 98 (1994) 10634.
85. B. Folie and M. Radosz, Ind. Eng. Chem Res., 34 (1995) 1501.
86. K.L. Albrecht, F.P. Stein, S.J. Han, C.J. Gregg and M. Radosz, Fluid Phase Equilibria, 117 (1996) 84.
87. M. A. Meilchen, B. M. Hasch and M. A. McHugh, Macromolecules, 24 (1991) 4874.
88. B. M. Hasch, M. A. Meilchen, S.-H. Lee and M. A. McHugh, J. Poly. Sci. Part B: Poly. Phys., 30 (1992) 1365.
89. J. A. Pratt, S.-H. Lee and M. A. McHugh, J. Appl. Polym. Sci., 49 (1993) 953.

90. S.-H. Lee, M. A. LoStracco, B. M. Hasch and M. A. McHugh, J. Phys. Chem., 98 (1994) 4055.

91. B. M. Hasch and M. A. McHugh, J. Poly. Sci. Part B: Polym Phys., 33 (1995) 715.

92. F. Rindfleisch, T.P. DiNoia, M.A. McHugh, J. Phys. Chem., 100 (1996) 15581.

93. I. C. Sanchez and C. G. Panayiotou, Equation of State Thermodynamics of Polymer and Related Solutions, chapt. 3 in "Models for Thermodynamic and Phase equilibria Calculations, S. L. Sandler ed., M. Dekker, New York, 1994.

94. A. R. Berends and G. S. Huvard, in "Supercritical Fluid Science and Technology", K. P. Johnston and J. M. L. Penninger Editors, ACS Symp. Ser. 406 (1988), p. 207.

95. W.-C. V. Wang, E. J. Kramer and W. H. Sachse J. Polym. Sci.: Polym. Phys. Ed., 20 (1982) 1371.

96. R. G. Wissinger and M. E. Paulaitis, Ind. Eng. Chem. Res., 30 (1991) 842.

97. Y.P. Handa, P. Kruus and M. O'Neill, J. Polym. Sci.: Polym. Phys. Ed., 34 (1996) 2635.

98. Z. Zhang and Y. P. Handa, SPE ANTEC Tech. Papers 806 (1997) 2051.

99. P. D. Condo, I. C. Sanchez and C. G. Panayiotou, K. P. Johnston, Macromolecules, 25 (1992) 6119.

100. R. G. Wissinger and M. E. Paulaitis, J. of Polym. Sci.: Part B: Polym. Phys., 25 (1987) 1497.

101. N. S. Kalospiros and M. E. Paulaitis, Chem. Eng. Sci., 49 (1994) 659.

102. F. Doghieri and G. C. Sarti, Macromolecules, 29 (1996) 7885

103. P. Alessi, A. Cortesi, I. Kikic, F. Scotti and F. Vecchione, Proc. 5th Int. Symp. On Supercritical Fluids, Atlanta 2000.

104. H. Otsuki, F.C. Williams Chem. Engr. Progr. Symp. Series No. 6, vol 49,1953, pg. 55

105. U. von Wasen and G. M. Schneider, J. Phys. Chem., 84 (1980) 229.

106. B. Spicka, A. Cortesi, M. Fermeglia, I. Kikic, J. Supercitical Fluids, 7, (1994), 171.

107. G. M. Schneider in "Phase equilibria of liquid and gaseous mixtures at high pressures", B. Le Neindre and B. Vodar eds., Butterworth, London, vol 2., 1975, 787.

108. U. K. Deiters and G. M. Schneider, Fluid Phase Equilibria, 29 (1986) 145.

109. R. Fornari, P. Alessi, and I. Kikic, Fluid Phase Equilibria, 57 (1990) 1.

110. R. Dohrn and G. Brunner, Fluid Phase Equilibria, 106 (1995) 213.

111. P. Alessi and A. Cortesi, "Summer School on Experimental Methods in the Thermodynamics of Fluids", Zakopane (Poland) (1988) 137.

112. T. J. Bruno in "Supercritical fluid technology", T. J. Bruno and J. F. Ely eds., CRC Press, Boca Raton Fl., 1991, 293.

High Pressure Process Technology: Fundamentals and Applications
A. Bertucco and G. Vetter (Editors)
© 2001 Elsevier Science B.V. All rights reserved.

CHAPTER 3

KINETIC PROPERTIES AT HIGH PRESSURE

G. Luft[a] : sections 3.1, 3.2, 3.3

F. Recasens[b] **, E. Velo**[b] : section 3.4

[a] Department of Chemistry, Darmstadt University of Technology
Petersenstr. 20, D-64287 Darmstadt, Germany

[b] Universitat Politécnica de Catalunya
Departamento de Ingenierìa Quìmica
E.T.S.I.I.B. Diagonal, 647   E-08028 Barcelona, Spain

After in the foregoing chapter thermodynamic properties at high pressure were considered, in this chapter other fundamental problems, namely the influence of pressure on the kinetic of chemical reactions and on transport properties, is discussed. For this purpose first the molecular theory of the reaction rate constant is considered. The key parameter is the activation volume $\Delta v^{\#}$ which describes the influence of the pressure on the rate constant. The evaluation of $\Delta v^{\#}$ from measurement of reaction rates is therefor outlined in detail together with theoretical prediction. Typical value of the activation volume of different single reactions, like unimolecular dissociation, Diels-Alder-, rearrangement-, polymerization- and Menshutkin-reactions but also on complex homogeneous and heterogeneous catalytic reactions are presented and discussed.

The second part of the chapter is devoted to the effect of pressure on heat and mass transfer. After a brief survey on fundamentals the estimation of viscosity, diffusivity in dense gases, thermal conductivity and surface tension is explained. The application of these data to calculate heat transfer in different arrangements and external as well as internal mass transfer coefficients is shown. Problems at the end of the two main parts of this chapter illustrate the numerical application of the formulas and the diagrams.

3.1 Interesting features at high pressures[*]

It is well known that the rate of a chemical reaction depends on the pressure. This becomes obvious when we consider the expression for the rate, r, of a simple homogeneous reaction

$$r = k \cdot c_A^{n_A} \cdot c_B^{n_B} \tag{3.1-1}$$

On the one hand, the pressure acts on the concentrations, c_A and c_B, of the reactants, A and B. This pressure effect is steep in gas-phase reactions, and is less steep when the reaction takes place in the fluid phase due to the lower compressibility. The concentrations of gases or of fluids always increase with the pressure, leading to a higher reaction rate.

On the other hand, the pressure influences the rate constant, k. The rate constant may increase or decrease with the presssure, which results in an increase or reduction in the rate. This effect, which is not generally known, can be as steep and as varied as that caused by changes in temperature.

The influence of the pressure is more complex when ions are involved. The tightening or loosening of electrostatic bonds between the reacting species and a solvent leads to a change in volume of the solvent and alters the reaction rate through the rate constant.

In multiple reactions the influence of the pressure on the rate of the various steps is mostly different. This makes the interpretation of the results of kinetic measurements more difficult. On the other hand, by the application of high pressure the ratio of the yield of a desired product to the conversion of initial reactants, the so-called selectivity, can be improved.

The problem is also more complex when heterogeneous catalysed reactions are considered. With porous catalyst pellets, reaction occurs at gas– or liquid–solid interfaces at the outer or inner sphere. When the reactants diffuse only slowly from the bulk phase to the exterior surface of the catalyst, gas or liquid film resistance must be taken into account. Pore diffusion resistance may be involved when the reactants move through the pores into the pellet.

Similarly, the rate of heterogeneous fluid–fluid and gas–liquid reactions depends both on the rate of the chemical reaction and the rate of mass transfer. Since the relative magnitude of the effect of pressure on the two processes can vary greatly we have a large spectrum of possibilities. High pressure can more- or less steeply increase or decrease the overall rate of these reactions.

In order to account for mass-transfer resistance the overall rate equation for heterogeneous reactions contains not only kinetic parameters of the chemical reaction but also mass-transfer coefficients [1,2]:

$$r = \mathrm{f}(\, k, \, c_i, \, D_i, \, k_g, \, k_l, \, \varphi_K, \, Ha, \, E) \tag{3.1-2}$$

These parameters, such as the coefficient of diffusion, D, mass-transfer coefficient in the gas and liquid phase or film, k_g and k_l, Thiele modulus, φ_K, Hatta number, Ha, and enhancement factor, E, are all dependent on the pressure.

In Chapter 3.2 the kinetics of high-pressure chemical reactions are presented. The measurement of kinetic data is then outlined in Chapter 3.3. At last, in Chapter 3.4, the calculation of transport properties at high pressure is considered.

---

[*] References at the end of section 3.3

3.2 Kinetics of high-pressure reactions[*]

In this chapter the influence of high pressure on the rates of different types of reactions is considered. For this purpose, first the molecular theory of reactions at high pressure is briefly presented. The key parameter, the activation volume, is then explained, and its evaluation from experimental data as well as the theoretical prediction are outlined. Examples show the magnitude of the activation volume of some high-pressure reactions of scientific and industrial importance.

3.2.1 Molecular theory of reaction rate constants

The effect of pressure on the reaction rate constant can be interpreted by both the collision-, and the transition state or activated complex theories. However, it has generally been found that the role of pressure can be evaluated more clearly by the transition state approach [3].

The development of transition state theory goes back to Eyring [4], and Evans and Polanyi [5]. The concept assumes that the energy of a bimolecular reaction starting from the initial reactants A and B

$$a \cdot A + b \cdot B \longrightarrow X^{\#} \longrightarrow c \cdot R + d \cdot S \qquad (3.2\text{-}1)$$

proceeds to a state of maximum energy, the so-called transition state, and then decreases as the products R and S separate (Fig. 3.2-1). At the peak, the reactants have been brought to the degree of closeness and distortion such that a small distortion in an appropriate direction will send the system in the direction of products. This critical configuration is called the activated complex, $X^{\#}$.

This complex is in equilibrium with the reactants, A and B. The equilibrium product for the bimolecular reaction under consideration is defined by:

$$K^{\#} = \frac{\left[X^{\#}\right]}{[A]^{a} \cdot [B]^{b}} \qquad (3.2\text{-}2)$$

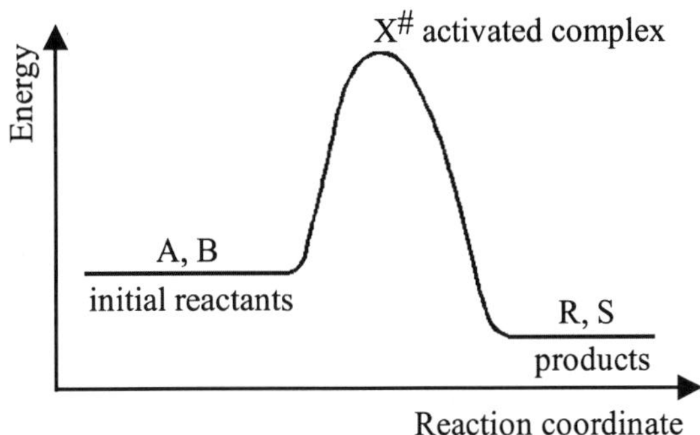

Figure 3.2-1. Reaction profile.

---

[*] References at the end of section 3.3

where [A] and [B] denote the concentrations of the reactants.

The quantity $K^{\#}$ is related to the rate constant, k, through the equation

$$k = \chi \cdot \frac{k_B \cdot T}{h} \cdot K^{\#}$$

(3.2-3)

in which k_B is the Boltzmann's constant, T stands for the absolute temperature, and h is Planck's constant. The product $h/k_B \cdot T$ is the lifetime of the complex in the transition state. The transmission coefficient, χ, is a factor which defines the probability that the activated complex will decompose into the products, R and S, rather than to the original species, A and B. Glasstone, Laidler and Eyring considered that the probability- or transmission coefficient, χ, is close to unity, and is independent of temperature and pressure [6].
We transform eqn. 3.2-3 by logarithm, to

$$\ln k = \ln K^{\#} + \ln T + const.$$

(3.2-4)

and differentiate. At constant temperature we obtain:

$$\frac{\partial \ln k}{\partial p}\bigg|_T = \frac{\partial \ln K^{\#}}{\partial p}\bigg|_T$$

(3.2-5)

The equilibrium product is related to the standard change of free energy, $\Delta G^{\#}$, when the transition state is formed from the reactants.

$$\frac{\partial \ln K^{\#}}{\partial p} = -\frac{1}{R \cdot T}\frac{\partial \Delta G^{\#}}{\partial p} = -\frac{\Delta v^{\#}}{R \cdot T}$$

(3.2-6)

It should be noted that k and $K^{\#}$ must be based on the same concentration scale. If we select the mole-fraction scale, x, we obtain:

$$\frac{\partial \ln k_x}{\partial p}\bigg|_T = -\frac{\Delta v^{\#}}{R \cdot T}$$

(3.2-7)

where

$$\Delta v^{\#} = v_x^{\#} - \left(a \cdot v_A + b \cdot v_B\right)$$

(3.2-8)

Here, $\Delta v^{\#}$ is the activation volume. It is the excess of the partial molar volume of the transition state over the partial molar volume of the initial species, at the composition of the mixture.

Eqn. 3.2-7 is comparable to the well-known Arrhenius law, $\partial \ln k / \partial T = -E_a/RT$, which describes the dependence of the rate constant on the temperature by means of the energy of activation, E_a. For reactions occurring in the liquid phase it is the practice to use molar concentrations instead of mole fractions. This implies some complications. The equilibrium product is now defined by molar concentrations. The quasi-thermodynamic development leads to:

$$K_c^{\#} = K_x^{\#} \cdot v_S^{-(a+b)}$$

(3.2-9)

in which v_S is the molar volume of the pure solvent. The derivative (eqn. 3.2-5) assumes the form:

$$\left. \frac{\partial (R \cdot T \cdot \ln k_c)}{\partial p} \right|_{c=0} = -\left(\Delta v^{\#} \right)^{\infty} + (1 - a - b) \cdot R \cdot T \cdot K_S$$

(3.2-10)

where $(\Delta v^{\#})^{\infty}$ is the activation volume when the solution is nearly infinitely dilute, and

$$K_S = -\left(\frac{\partial \ln v_S}{\partial p} \right)_T$$

(3.2-11)

denotes the compressibility coefficient of the solvent.

Eqn. 3.2-10 does reduce to eqn. 3.2-12

$$\left. \frac{\partial \ln k}{\partial p} \right|_T = -\frac{\Delta v^{\#}}{R \cdot T}$$

(3.2-12)

for first-order reactions when $a = 1$ and $b = 0$. Nevertheless, it is common practice to evaluate the dependence of the rate constant on the pressure by using eqn. 3.2-12 also when molar concentrations are used to express the composition of a solution. This simplification is justified by the fact that the compressibility coefficient of liquids is only small and is in the range of 1 - 4 cm^3/mol.

3.2.2 Activation volume

The activation volume is a composite function to which different effects contribute. It can be split into two major terms:

$\Delta v_R^{\#}$ the change in volume of the reacting molecules of each partial reaction;

$\Delta v_S^{\#}$ the change in volume of solvent, arising from changes in electrostatic forces.

70

3.2.2.1 Terms contributing to $\Delta v_R^{\#}$

The main contribution to $\Delta v_R^{\#}$ results from the formation of new bonds and the stretching of existing bonds in the transition process. Minor contributions are associated with the relaxation of parts of the molecules remote from the reacting centers.

In an unimolecular *dissociation*

$$AB \rightarrow A \cdots B \rightarrow A + B \tag{3.2-13}$$

the covalent bond in the molecule AB is first stretched during the activation process. The activated complex will have a larger volume than the initial state, and therefore the *activation volume is positive* for this type of reaction.

A typical unimolecular reaction is the decomposition of organic peroxides for which always positive activation volumes of up to 15 cm^3/mol have been observed. The decomposition of di(t-butyl)peroxide, an effective initiator for the high pressure polymerisation of ethylene, into two t-butoxyradicals, exhibits a positive activation volume of 13 cm^3/mol (Table 3.2-1, a).

When new bonds are formed as in the *association*

$$A + B \rightarrow A \cdots B \rightarrow AB \tag{3.2-14}$$

the distance between A and B shrinks during the activation process, resulting in a *negative activation volume*.

Large negative values of $\Delta v^{\#}$ (-25 to -50 cm^3/mol) have been found in the investigation of Diels-Alder reactions, such as the dimerization of cyclopentadiene (Table 3.2-1, b). The cyclic transition state has a compact structure similar to the reaction products. $\Delta v^{\#}$ is only a little smaller than the volume change between the initial- and the product-structures, indicating a late transition state.

When *bond-formation and -breaking* take place simultaneously

$$A + BC \rightarrow A \cdots B \cdots C \rightarrow AB + C \tag{3.2-15}$$

the net change of volume is dominated by the contribution of the formation of new bonds. Therefore, the *activation volume is negative*. Examples are rearrangement reactions, such as the Claisen rearrangement of allyl vinyl ether, where $\Delta v^{\#}$ is between 8 and -20 cm^3/mol (Table 3.2-1, c). In these reactions, covalent O-C-bonds are stretched and broken and simultaneously new C-C-bonds are formed. Radical polymerization reactions, such as the polymerization of ethylene, also exhibit negative activation volumes of -10 to -25 cm^3/mol (Table 3.2-1, d). New bonds are formed by the addition of monomer molecules to radicals whereas the double bond of the monomer is stretched in the transition state, resulting in a single bond. By this mechanism the activation volume of the radical polymerization of ethylene, -25.5 cm^3/mol, can be explained.

Table 3.2-1
Typical values of the activation volume of different reactions

	Reaction	Range $\Delta v^{\#}$ [cm³/mol]	Example	$\Delta v^{\#}$ [cm³/mol]
a	Unimolecular dissociation	0 to +15	Rad. decomposition of di(t-butyl)peroxide $(CH_3)_3COOC(CH_3)_3 \longrightarrow 2\,(CH_3)_3CO^{\bullet}$	+13 [7]
b	Diels-Alder reactions	-25 to -50	Dimerisation of cyclopentadiene 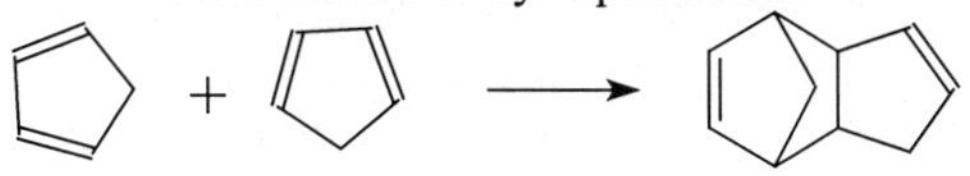	-31 [8]
c	Rearrangement reactions	-8 to -20	Claisen rearrangement of allyl vinyl ether 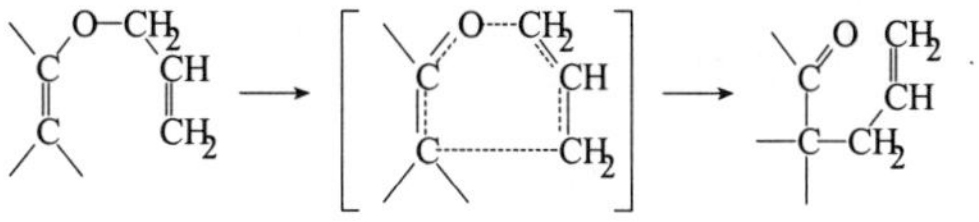	-16 [9]
d	Radical polymerizations	-10 to -25	Radical polymerization of ethylene 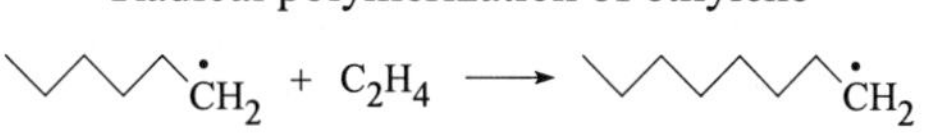 $\sim\!\!\sim\!CH_2 + C_2H_4 \longrightarrow \sim\!\!\sim\!CH_2$	-25.5 [10, 11]
e	Menshutkin reactions	-20 to -40	Reaction between pyridine and ethyl iodide 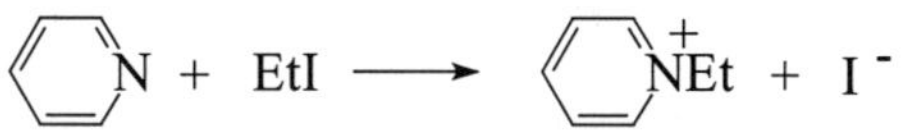 $N + EtI \longrightarrow NEt^{+} + I^{-}$	-30 [1)] -27 [2)] -24 [3)] [12]

1), acetone; 2), cyclohexanone; 3), nitrobenzene.

3.2.2.2 Terms contributing to $\Delta v_S^{\#}$

In the *dissociation of a neutral molecule* AB into two free ions, A^{+} and B^{-},

$$AB \rightarrow X^{\#} \rightarrow A^{+} + B^{-} \tag{3.2-16}$$

electrical charges develop during the transition from AB to $X^{\#}$. The ionic charges exert strong attractive forces on the permanent or induced dipoles of the solvent molecules, causing a contraction. The magnitude of this effect is larger than the increase of volume owing to the stretching of the bond between A and B, leading to a *negative activation volume*.

In the bimolecular *association* to *non-ionic* products

$$A^+ + B^- \rightarrow X^\# \rightarrow AB \qquad (3.2\text{-}17)$$

the reduction in polarity results in an expansion of the solvent. Although there is an opposing contraction owing to bond-forming, the *activation volume* is *positive*. It must be pointed out that the predominance of $\Delta v_S^\#$ over $\Delta v_R^\#$ is often observed but is not a strict rule.

The dependence of $\Delta v^\#$ on the solvent was investigated on Menshutkin reactions. In all solvents the reaction is strongly accelerated by pressure due to high activation volumes of -20 to -40 cm³/mol (Table 3.2-1, e).

3.2.3 Evaluation of the activation volume from experimental data

3.2.3.1 Single homogeneous reactions
The rate, r, of a single homogeneous reaction

$$A \xrightarrow{\;k\;} B \qquad (3.2\text{-}18)$$

can be expressed by the equation:

$$r = \frac{dc_A}{dt} = k \cdot c_A^n \qquad (3.2\text{-}19)$$

In order to determine the volume of activation, first the rate constant, k, is calculated from the measured reaction rate, r, and the concentration, c_A, through eqn. 3.2-19. The exponent n can be obtained from experiments at different concentrations. The value of k is then plotted on a logarithmic scale versus the pressure, and $\Delta v^\#$ (10^{-6} cm³/mol) is evaluated from the slope, α, of the resulting straight line (Fig. 3.2-2). For this purpose α, (MPa⁻¹), is multiplied by the gas constant, R (8.314 J mol⁻¹ K⁻¹), and the temperature, T (K).

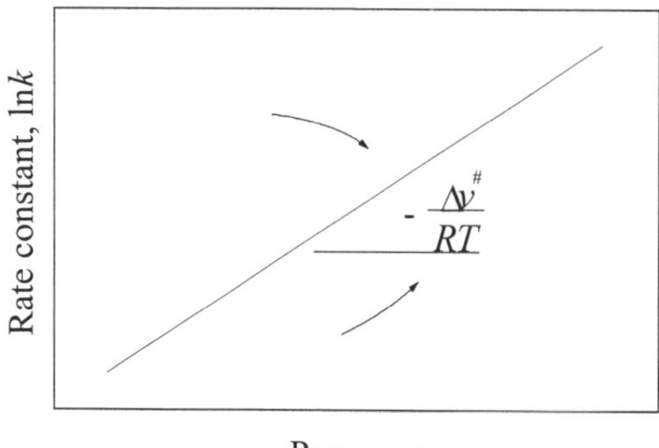

Figure 3.2-2. Determination of the activation volume of single reactions.

The expressions for the rates of multiple reactions are more complex and cannot be easily handled to determine the activation volume. Eqn. 3.2-7 is strictly valid for single reactions.

If a reaction is accompanied by side reactions, which are not accounted for in detail, deviations from the straight line may be experienced in the plot of $\ln k/p$ when the overall rate is considered. This difficulty is well known from the determination of the activation energy for multiple reactions.

3.2.3.2 Parallel reactions

The course of concentration during parallel reactions

$$A \xrightarrow{k_1} R$$
$$A \xrightarrow{k_2} S \qquad\qquad (3.2\text{-}20)$$

is shown in Fig. 3.2-3.

The ratio of the rate constants k_1 and k_2 can be obtained from the ratio of the rates of formation of the products, R and S.

$$\frac{r_1}{r_2} = \frac{dc_R/dt}{dc_S/dt} = \frac{k_1 \cdot c_A^{n_1}}{k_2 \cdot c_A^{n_2}} \qquad\qquad (3.2\text{-}21)$$

If the reactions have the same kinetic order, n, the expression:

$$\frac{r_1}{r_2} = \frac{k_1}{k_2} \qquad\qquad (3.2\text{-}22)$$

results and, using eqn. 3.2-7, the difference of the activation volumes can be determined.

$$\frac{\partial \ln k_1/k_2}{\partial p} = -\frac{\Delta v_1^{\#} - \Delta v_2^{\#}}{R \cdot T} \qquad\qquad (3.2\text{-}23)$$

3.2.3.3 Reactions in series

As an example for reactions in series or consecutive reactions, the simple reaction scheme

$$A \xrightarrow{k_1} R \xrightarrow{k_2} P \qquad\qquad (3.2\text{-}24)$$

should be discussed. The course of the concentrations of the initial A, product R, and consecutive product P, is shown in Fig. 3.2-4. The reduction in the concentration of the initial reactant, A, follows the relationship

$$r_1 = \frac{dc_A}{dt} = k_1 \cdot c_A^{n_1} \qquad\qquad (3.2\text{-}25)$$

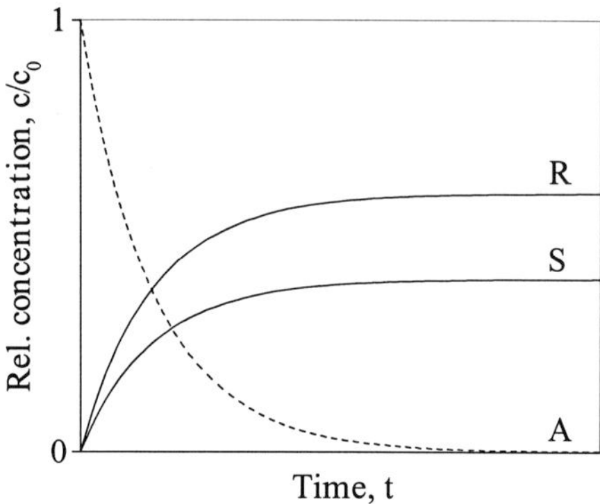

Figure 3.2-3. Change of concentration with time during parallel reactions.

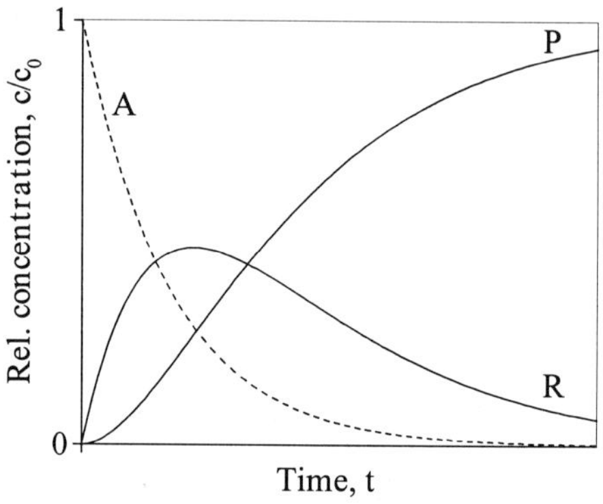

Figure 3.2-4. Change of concentration with time during reactions in series, $k_1 \approx k_2$.

The dependence of k_1 on the pressure is given by eqn. 3.2-7 with the activation volume $\Delta v_1^{\#}$. The change of the concentration of the intermediate product, R, with time is

$$r_2 = \frac{dc_R}{dt} = k_1 \cdot c_A^{n_1} - k_2 \cdot c_R^{n_2} \tag{3.2-26}$$

The influence of pressure is determined by $\Delta v_1^{\#}$ and $\Delta v_2^{\#}$, the activation volumes of the two reactions in series. The determination of $\Delta v_1^{\#}$ and $\Delta v_2^{\#}$ requires one to evaluate first k_1 and k_2.

If n_1 and n_2 are unity, r_2/c_A is plotted versus c_A/c_R. Then k_1 is obtained from the intersection of the resulting straight line and the ordinate, whereas k_2 is its slope. Standard mathematical methods, such as linear- and multiple regression, or search techniques based on least-squares-methods to minimize the deviation of measured and calculated reaction rates, must be applied to determine the rate constants when n_1 and n_2 are different from unity.

3.2.3.4 Chain reactions

An important example of a complex reaction is a chain reaction in which free radicals are involved. These reactions consist of three essential steps: Formation of free radicals or initiation; propagation; and termination.

Radicals, $I^{\bullet}$, can be formed by unimolecular dissociation:

$$I \xrightarrow{k_d} I^{\bullet} \tag{3.2-27}$$

The radicals, $I^{\bullet}$, may react with a further component, M, such as an olefin to form secondary radicals, $R^{\bullet}$.

$$I^{\bullet} + M \xrightarrow{k_i} R^{\bullet} \tag{3.2-28}$$

By successive addition of M to radicals $R_i^{\bullet}$ in the propagation reaction, chain radicals $R_{i+1}^{\bullet}$ are formed

$$R_i^{\bullet} + M \xrightarrow{\;k_p\;} R_{i+1}^{\bullet} \qquad\qquad (3.2\text{-}29)$$

the growth of which is terminated by combination.

$$R_i^{\bullet} + R_j^{\bullet} \xrightarrow{\;k_t\;} P_{i+j} \qquad\qquad (3.2\text{-}30)$$

In the termination reaction, stable products P are obtained.

It will be shown later that the rate of the chain reaction mentioned above can be calculated from the expression:

$$r = \frac{dM}{dt} = k_p \cdot \sqrt{\frac{k_i}{k_t}} [M][I]^{1/2} = k_{br}[M][I]^{1/2} \qquad\qquad (3.2\text{-}31)$$

The dependence of the overall rate constant, k_{br}, on the pressure can be elucidated from eqn. 3.2-7

$$\frac{\partial \ln k_{br}}{\partial p} = -\frac{\Delta v_{br}^{\#}}{R \cdot T} \qquad\qquad (3.2\text{-}32)$$

with

$$\Delta v_{br}^{\#} = \Delta v_p^{\#} + \frac{1}{2}\Delta v_i^{\#} - \frac{1}{2}\Delta v_t^{\#} \qquad\qquad (3.2\text{-}33)$$

where $\Delta v_p^{\#}$, $\Delta v_i^{\#}$, and $\Delta v_t^{\#}$ are the activation volumes of chain propagation, initiation, and termination reaction, respectively.

3.2.3.5 Heterogeneous catalytic reactions

A number of different steps are involved in heterogeneous catalytic processes. Among chemical reaction, adsorption, and desorption, transport processes may influence the overall rate. These steps are illustrated in Fig. 3.2-5.

1. Transport of reactants, A and B, from the bulk phase to the surface of the catalyst pellet.
2. Transport of reactants into the catalyst pores.
3. Adsorption of reactants, A and B, on the catalytic site, L:

$$A + L \rightleftharpoons AL$$
$$B + L \rightleftharpoons BL \qquad\qquad (3.2\text{-}34)$$

4. Surface reaction between the adsorbed species:

$$AL + BL \longrightarrow RL + SL \qquad\qquad (3.2\text{-}35)$$

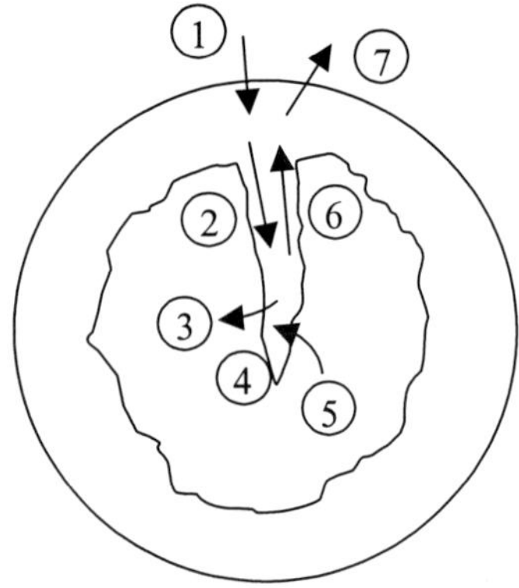

Figure 3.2-5. Steps involved in heterogeneous catalytic reactions.

5. Desorption of the products, R, S:
$$RL \;\rightleftarrows\; R + L$$
$$SL \;\rightleftarrows\; S + L \tag{3.2-36}$$
6. Transport of the products back to the pellet surface.
7. Transport back to the bulk stream.

In the following discussion we will concentrate on the surface reaction, adsorption, and desorption. The complications induced by the transport phenomena will be ignored. In order to develop an expression for the overall rate, the surface concentrations, AL, BL, etc., are related to the concentrations of the reactants in the bulk phase by an "Equilibrium constant". For example:

$$K_A = \frac{[AL]}{[A][L]} \tag{3.2-37}$$

Quite often it is found that one of the above-mentioned steps is much slower than the others and therefore controls the rate of the catalytic reaction. As an example, the rate expression for the bimolecular reaction

$$A + B \;\rightleftarrows\; R + S \tag{3.2-38}$$

is shown below, when the surface reaction of the adsorbed species is rate controlling.

$$r = \frac{k \cdot K_A \cdot K_B \left(p_A \cdot p_B - p_R \cdot p_S / K\right)}{\left(1 + K_A \cdot p_A + K_B \cdot p_B + K_R \cdot p_R + K_S \cdot p_S\right)^2} \tag{3.2-39}$$

Here, p_A, p_B, etc., are the partial pressures of the reactants, $K = \dfrac{p_R \cdot p_S}{p_A \cdot p_B}$ is the equilibrium constant of the reaction, K_A, K_B, etc., are the equilibrium constants of the individual steps, and k is the rate constant of the rate-controlling reaction in which the total concentration of active sites is incorporated.

The volume of activation of the rate-controlling step can be evaluated from the rate constant, k, using eqn. 3.2-7. First, k must be evaluated through eqn. 3.2-39 from the rate of the catalytic process measured at different total pressures. The standard-methods of least squares are applied again to fit values for k and the other parameters, K_A, K_B, etc.

3.2.3.6 Reactions influenced by mass transport

If the rate of a reaction is influenced by mass transport, the effect of the pressure both on the rate of the chemical reaction and on the rate of mass transport must be taken into account. As an example, a heterogeneous catalytic reaction governed by the rate of diffusion within the pores of the catalyst is considered.

The overall rate of reaction, r_{ov}, can be described using the concept of the effectiveness factor, η, by the expression [13]

$$r_{ov} = k \cdot c \cdot \eta \tag{3.2-40}$$

When the pore-diffusion is limiting, the effectiveness factor is $\eta = 1/\varphi$, with the Thiele modulus, $\varphi = r_K/3 \, (k/D_P)^{1/2}$, r_K being the radius of the catalyst particle, and D_P the coefficient of pore-diffusion. The overall rate of the process depends then on the reciprocal modulus, φ:

$$r_{overall} = k \cdot c \cdot \phi^{-1}$$

which means that:

$$r_{overall} \sim (k \cdot D)^{1/2} \quad \text{and} \quad r_{overall} \sim \exp[-(\Delta v^{\#} + \Delta v_D^{\#}) \cdot (p - p_0)/2 \cdot R \cdot T] \tag{3.2-41}$$

Taking into consideration the fact that the activation volume of the diffusion process is small compared to that of the chemical reaction, it follow that:

$$\left. \frac{d \ln r_{ov}}{dp} \right|_T = -\frac{\Delta v^{\#}}{2 \cdot R \cdot T} \quad \text{and} \quad \left. \frac{d \ln r_{ov}}{dp} \right|_T = -\frac{\Delta v_{ov}^{\#}}{R \cdot T} \tag{3.2-42}$$

with $\Delta v_{ov}^{\#} = \Delta v^{\#}/2$. $\tag{3.2-43}$

From eqn. 3.2-42 it can be concluded that the dependence of the overall rate of a heterogeneous catalytic reaction on the pressure is decreased when the pore-diffusion is rate limiting.

3.2.4 Prediction of the activation volume

When covalent bonds break during non-ionic reactions, the contribution to $\Delta v_R^{\#}$ of the elongation, Δl, of the bond whose initial length was l, can be roughly calculated. Hamann [3] has assumed that the stretching occurs along the axis of a cylinder of constant cross-section. For a diatomic molecule, the cross-section can be taken to be the mean of the van der Waals cross-sectional areas of the separating atoms. Using the van der Waals radii, r_A and r_B, it follows that

$$\Delta v_{R,b}^{\#} = \frac{\pi\left(r_A^2 + r_B^2\right)}{2} \cdot \Delta l \tag{3.2-44}$$

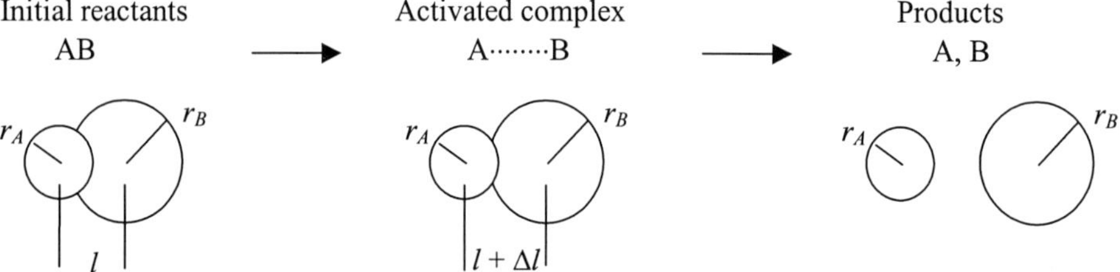

The elongation, Δl, can be evaluated from quantum mechanical calculations. For a rough estimation, a value of

$$\Delta l = (0.10 \text{ to } 0.35) \cdot l \tag{3.2-45}$$

can be assumed. From eqn. 3.2-44 it can be seen that the activation volume is positive when covalent bonds break.

Similarly, when covalent bonds are formed, $\Delta v_{R,f}^{\#}$ can be estimated from the relationship:

$$\Delta v_{R,f}^{\#} = \frac{\pi\left(r_A^2 + r_B^2\right)}{2} \cdot \left[(l + \Delta l) - (r_A + r_B)\right] \tag{3.2-46}$$

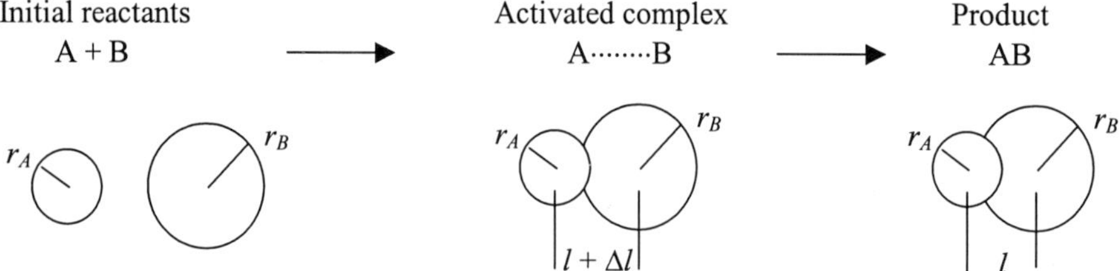

The initial distance of the reactants is taken to be the sum of the van der Waals radii, r_A and r_B. The transition state distance is again $(l + \Delta l)$ and can be roughly estimated from equation

3.2-45. A negative activation volume results when covalent bonds are formed, because $(r_A + r_B) > (l + \Delta l)$. Comparison of equation 3.2-44 and 3.2-46 shows that the value of the activation volume, $\Delta v^{\#}_{R,f}$, for the formation of covalent bonds is invariably greater than $\Delta v^{\#}_{R,b}$ for bond breaking. When bond-formation and the breaking of another bond occur simultaneously, $\Delta v^{\#}_{R,f}$ predominates, and the net activation volume, $\Delta v^{\#}_{R}$, will be negative.

3.2.5 Activation volume as a tool for the elucidation of reaction mechanism

The value of $\Delta v^{\#}$ and its sign can be used to discriminate between possible mechanisms for a reaction. An example taken from the literature [14,15] is the decomposition of diazonium cations in water. The mechanism of this reaction could be bimolecular:

$$RN_2^+ + H_2O \longrightarrow ROH_2^+ + N_2 \qquad (3.2\text{-}47)$$

or unimolecular:

$$RN_2^+ \xrightarrow{\text{slow}} R^+ + N_2 \qquad (3.2\text{-}48)$$

$$R^+ + 2\,H_2O \xrightarrow{\text{fast}} ROH + H_3O^+$$

In the bimolecular reaction (eqn. 3.2-47), bond-breaking and bond-forming take place simultaneously and therefore a negative activation volume would be expected. In the unimolecular mechanism, bond-breaking during the first, rate-determining slow reaction (eqn. 3.2-48) should cause $\Delta v^{\#}$ to be positive. The positive activation volumes evaluated from experiments at high pressures are consistent with the usually accepted unimolecular mechanism.

During the decomposition of peroxyesters, large amounts of CO_2 are formed. The value found for the activation volume can be considered when the mechanism of peroxide decomposition is discussed. The analysis of the gaseous decomposition products by gas chromatography shows large amounts of CO_2, whose formation can take place during the decomposition.

Two possible mechanisms can be discussed:
- sequential decompositions

$$R\overset{\overset{\textstyle O}{\|}}{C}\!-\!O\!-\!O\!-\!R \longrightarrow R\overset{\overset{\textstyle O}{\|}}{C}\!-\!O\cdot + R\!-\!O\cdot \longrightarrow R\cdot + CO_2 + R\!-\!O\cdot \qquad (3.2\text{-}49)$$

- simultaneous decomposition

$$R\overset{\overset{\textstyle O}{\|}}{C}\!-\!O\!-\!O\!-\!R \longrightarrow R\!-\!O\cdot + CO_2 + R\cdot \qquad (3.2\text{-}50)$$

According to eqn. 3.2-49, the decarboxylation occurs in consecutive steps, whereas in the second pathway (eqn. 3.2-50) the fragmentation and the decarboxylation take place simultaneously. The moderately positive activation volumes evaluated from decomposition measurements under high pressures indicate that the sequential decomposition is favoured (eqn. 3.2-49).

Two extreme possibilities can be considered in ligand substitution. In the S_N1 (nucleophilic, monomolecular) mechanism, first the ligand, X^-, to be substituted dissociates in a slow reaction:

$$\left[L_5MX\right]^{n+} \xrightarrow[\text{slow}]{-X^-} \left[L_5M\right]^{(n+1)+} \xrightarrow[\text{fast}]{+Y^-} \left[L_5MY\right]^{n+} \tag{3.2-51}$$

In the bimolecular substitution of the S_N2 type

$$\left[L_5MX\right]^{n+} \xrightarrow[\text{slow}]{+Y^-} \left[L_5M\mathord{<}^X_Y\right]^{(n-1)+} \xrightarrow[\text{fast}]{-X^-} \left[L_5MY\right]^{n+} \tag{3.2-52}$$

the new ligand Y^- attacks the complex. This reaction is also slow and determines the rate of the process. A positive activation volume ($\Delta v^{\#} > 0$) experimentally determined is a hint that the ligand substitution proceeds via the monomolecular mechanism (eqn. 3.2-51) and, $\Delta v^{\#} < 0$ indicates a nucleophilic bimolecular substitution (eqn. 3.2-52).

3.2.6 Change of reaction rate constant with pressure

The sign of the volume of activation, $\Delta v^{\#}$, determines whether the rate constant, k, increases or decreases with the pressure. The value of $\Delta v\#$ is positive when the volume of the activated complex is larger than the volume of the initial reactants. Thus the rate constant, k, decreases when the pressure is increased. When the volume reduces in the transition state, $\Delta v^{\#}$ is negative and k will increase with increasing pressure (Table 3.2-2).

The values for $\Delta v^{\#}$ range between +30 and -60 cm³/mol.

Within a small range of pressure it can be assumed that $\Delta v^{\#}$ does not change with pressure. The integration of eqn. 3.2-7 gives:

$$k_p = k_{p0} \cdot \exp\left[-\Delta v^{\#}(p - p_0)/R \cdot T\right] \tag{3.2-53}$$

Table 3.2-2
How activation volume affects rate constant

Activation volume, $\Delta v^{\#}$	Change of rate constant, k, with pressure, p
positive +	decreases ↓
negative -	increases ↑

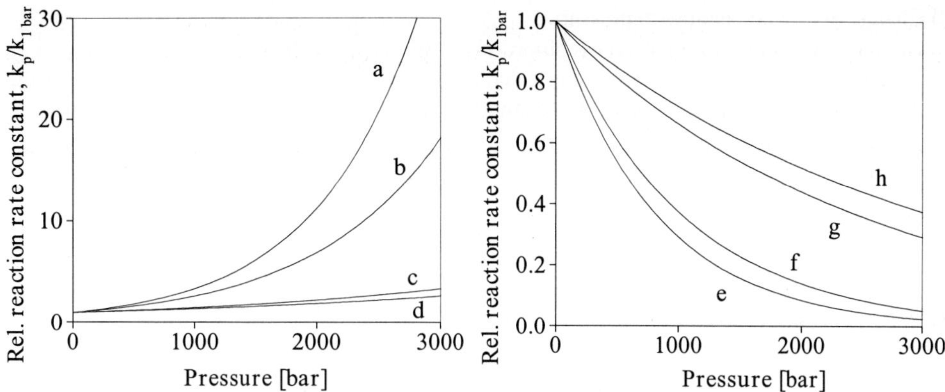

Figure 3.2-6. Relative reaction rate constant $k_p/k_{1\ bar}$. Activation volume: a, -30 cm^3/mol, 298 K; b, -30 cm^3/mol, 373 K; c, -10 cm^3/mol, 298 K; d, -10 cm^3/mol, 373 K; e, +30 cm^3/mol, 298 K; f, +30 cm^3/mol, 373 K; g, +10 cm^3/mol, 298 K; h, +10 cm^3/mol, 373 K.

The change in rate constant at various pressures and temperatures, calculated from eqn. 3.2-53 for several values of $\Delta v^{\#}$ is presented in Fig. 3.2-6. The rate-constant changes exponentially with the pressure. The effect is steeper when the activation volume is large and the temperature is low, and vice versa.

3.2.7 Problems

1) Evaluation of activation volume from experimental data:

The rate constant of a polymerization reaction was determined at 433.8 K and various pressures (Table 3.2-3).

In order to evaluate $\Delta v^{\#}$ the rate constants are plotted on a logarithmic scale versus the pressure. The slope of the resulting straight line is $\alpha = 9 \cdot 10^{-9}$ Pa^{-1}. A value of $\Delta v^{\#} = -32.5 \cdot 10^{-6}$ m^3 mol^{-1} can be obtained from $\Delta v^{\#} = \alpha \cdot R \cdot T$, with $R = 8.314$ J mol^{-1} K^{-1} and $T = 433.8$ K.

Table 3.2-3
Rate constant of a polymerization reaction at various pressures

Pressure, p [MPa]	Relative reaction rate constant, k_p/k_{1bar}
120	0.041
140	0.050
160	0.061
180	0.067

2) Change of rate constant with pressure:

The half-life time, $\tau_{1/2}$, of decomposition of di(t-butyl)peroxide at a temperature of 463 K and ambient pressure is 50 s. One may calculate the rate constant and the half-life time for decomposition at 300 MPa and the same temperature, when the activation volume is $\Delta v^{\#} = +13$ cm^3 mol^{-1}.

$$\text{Rate constant} \quad k = \frac{\ln 2}{\tau_{1/2}} = \frac{\ln 2}{50 \text{ s}} = 0.014 \text{ s}^{-1}$$

$$k_p = k_{(p=300\,\text{MPa})} \cdot e^{-\frac{\Delta v^{\#} \cdot (p - p_0)}{R \cdot T}} = 0.014 \text{ s}^{-1} \cdot e^{-\frac{13 \cdot 10^{-6} \cdot (300 - 0.1) \cdot 10^{-6}}{8{,}314 \cdot 463}} = 5.1 \cdot 10^{-3} \text{ s}^{-1}$$

The half-life time for decomposition at 300 MPa is then: $\tau_{1/2} = \frac{\ln 2}{k} = \frac{\ln 2}{5.1 \cdot 10^{-3} \text{ s}} = 136 \text{ s}^{-1}$

3.3 Measurement of chemical kinetic data at high pressure

The main aim of the methods described in this chapter is to obtain data for the design of chemical reactors, for the simulation of their operation behaviour, and, last but not least, to evaluate the influence of temperature and pressure on reaction rate. For this purpose, the techniques for measuring reaction rates at high pressures are presented. The details of the apparatus are mentioned in Chapter 4.3.4.

3.3.1 Measurement of reaction rates

Generally, the same methods can be used to measure reaction rates at high pressure as at low pressure. Some of them are more suitable than others for use at high pressure. The selection depends whether a homogeneous or a heterogeneous reaction should be investigated, whether it is a gas- or liquid-phase reaction, or a catalyst is used.

The apparatus used are mostly stirred-tank-, tubular-, and differential recycle reactors. Also, optical cells are used for spectroscopic measurements, and differential thermal-analysis apparatus and stopped flow devices are applied at high pressures.

The *stirred-tank reactor* or *stirred autoclave* is continuously operated when rapid changes in concentration do not allow samples to be taken from the reactor for analysis. In the stirred-tank reactor complete uniformity of concentration and temperature throughout the reactor is assumed. The exit stream from the reactor has the same composition as the fluid within the reactor (Fig. 3.3-1). The rate of reaction can be evaluated from the material balance

$$0 = r - \frac{1}{\tau} \cdot (c - c_0) \tag{3.3-1}$$

When the residence time, τ, is known from the reactor volume and the volumetric throughput, $\dot{v}$, and the temperature inside the reactor is measured, the reaction rate can be determined from the concentration, c, of a component in the reactor, and c_0 in the feed.

When a solid catalyst is required the autoclave must be equipped with a rotating basket in which the catalyst is placed.

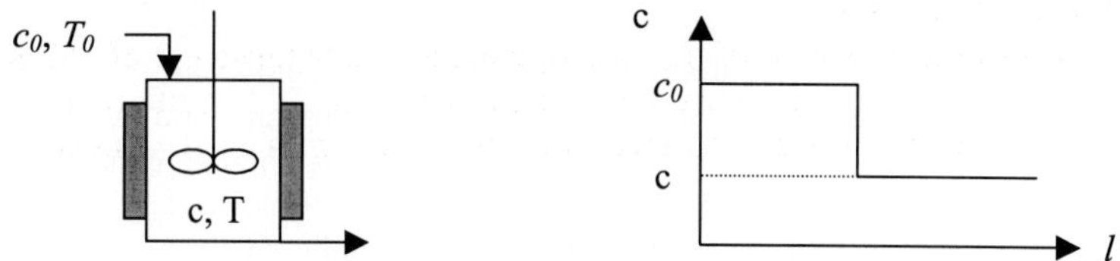

Figure 3.3-1. Continuously operated stirred autoclave.

In a *tubular reactor* the concentration varies from point to point along a flow-path, as shown by the solid curve in Fig. 3.3-2, right. The material balance

$$0 = r - \rho \cdot w \cdot \frac{dc}{dz} + D_{ax} \cdot \frac{d^2c}{dz^2} \tag{3.3-2}$$

is a differential equation, and the concentration of a component must be measured at different locations to determine the reaction rate. A further term, D_{ax} (d^2c/dz^2), must be considered in the material balance if the flow is not plug-flow. D_{ax} is the effective diffusivity in the axial direction, to be determined from measurement of the residence-time distribution. When the tubular reactor could not be run under isothermal conditions, an energy balance

$$0 = r \cdot \left(- \Delta H_R\right) - \rho \cdot c_p \cdot \frac{dT}{dz} + \lambda \cdot \frac{d^2T}{dz^2} - k_w \cdot \frac{F}{V} \cdot \left(T - T_K\right) \tag{3.3-3}$$

must be taken into account and also the course of temperature, T, must be measured together with the temperature, T_K, of the coolant or of the heating fluid. Thus, the rate of reaction cannot be determined directly as a function of concentration and temperature. Material- and energy-balances must be solved by numerical methods and the kinetic parameters must be

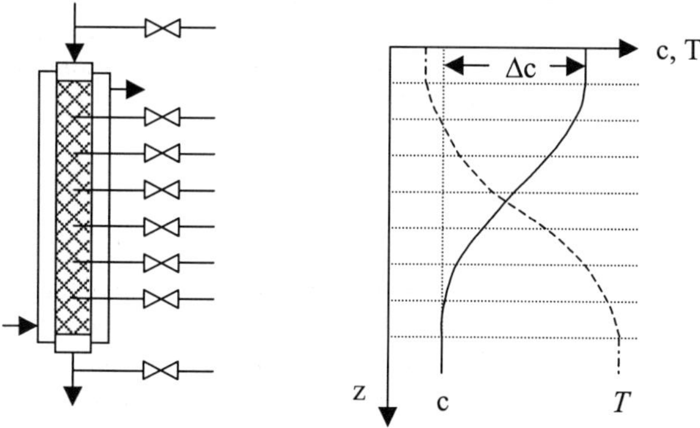

Figure 3.3-2. Tubular reactor.

evaluated by least-squares techniques [16].

On the other hand, the tubular reactor is a simple and inexpensive apparatus. It's small inner diameter requires a low thickness of the tube-wall to resist high pressure, and facilitates the removal or heating of the feed in order to operate the reactor under isothermal conditions. Solid catalyst can easily be placed in the tubular reactor.

The *differential reactor* uses a thin catalyst bed in which only small changes of concentration and temperature occur (Fig. 3.3-3). The rate of reaction, r, can be obtained from the difference in concentration, Δc, over the catalyst bed or its thickness, Δx, the volumetric throughput, $\dot{v}$, or the molar throughput, $\dot{n}_{ges}$, and the quantity of catalyst, W_K, using the material balance:

$$0 = r - \dot{v} \cdot \frac{\Delta c}{W_K} \qquad \text{or} \qquad 0 = r - \dot{n}_{ges} \cdot \frac{\Delta x}{W_K} \tag{3.3-4}$$

The measurement of a small concentration gradient requires more analytical work, and often gives less accurate kinetic data. For this reason, in the *differential recycle reactor* a fraction of the reaction mixture leaving a thin catalyst bed is recycled and added again to the feed (Fig. 3.3-4). This results in a larger difference of concentration, c_o, or mole fraction, x_o, between the feed and c or x at the reactor outlet, which is used to determine the reaction rate from the material balance:

$$0 = r - \frac{1}{\tau} \cdot (c_0 - c) \qquad \text{or} \qquad 0 = r - \frac{\dot{v}_0}{W_K} \cdot (x_0 - x) \tag{3.3-5}$$

Again, W_K is the quantity of catalyst and τ is the average residence time. The recycle ratio, $\eta = \dot{v}_{rec}/\dot{v}_0$, which is determined by the volumetric stream, $\dot{v}_{rec}$, of the recycled reaction mixture and the feed, $\dot{v}_0$, should be large enough that the concentration- and temperature-gradient over the catalyst bed are only small. Criteria for selecting the recycle ratio are presented in References 17 and 18. Appropriate η-values are in the range of 20 - 400,

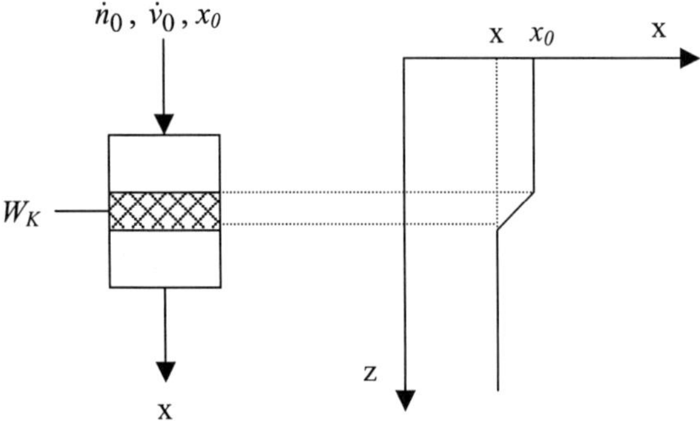

Figure 3.3-3. Differential reactor.

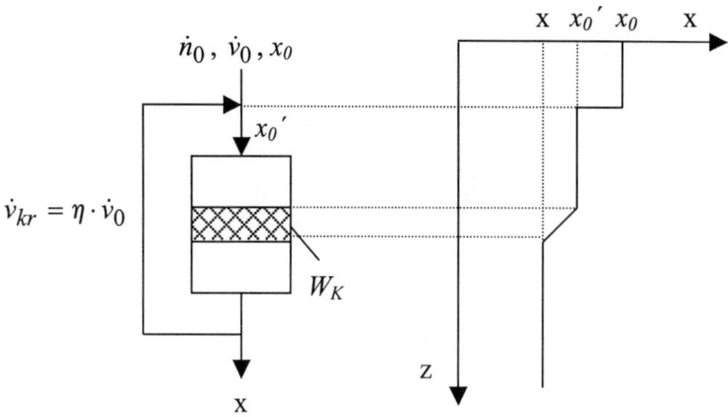

Figure 3.3-4. Differential recycle reactor.

depending on the desired conversion, reaction enthalpy, and heat capacity of the reaction mixture.

On the one hand, the differential reactor with recycle permits kinetic measurements of high accuracy. On the other hand, a transfer equipment is required to recycle a fraction of the reaction mixture. This can be difficult when the pressure is high. For this purpose, a jet loop reactor was developed which is equipped with an ejector to recycle the fluid. The design of the jet loop reactor is described in Chapter 4.3.4.

Problems can arise in the investigation of rapid reactions if the reactants are not heated sufficiently fast to the desired temperature, and if the samples from the reactor are not cooled rapidly to stop the reaction. A more sophisticated approach consists of monitoring the changes in concentration in an optical cell, in situ, by means of spectroscopy. Both infra-red and Raman spectroscopy can be used, depending on the sensitivity of characteristic bonds and the wave-number range of interest.

An *optical cell* for pressures of up to 200 MPa and temperatures to 200°C is presented in Chapter 4.3.4. The cell can be coupled with a commercial Raman spectrometer to measure the course of the intensity of a bond's signal with time. By calibration, the intensity versus time curve can be converted into a concentration versus time curve, from which the rate of reaction and kinetic parameters can be evaluated. The method is explained in Chapter 3.3.2, considering the decomposition of an organic peroxide.

Optical cells can also be used to investigate the kinetics of radical polymerization reactions under high pressure by means of the rotating sector method. Again, the apparatus is presented in Chapter 4.3.4. An example of the method for the evaluation of individual rate constants in radical polymerization of ethylene is given below.

The *stopped-flow method* uses syringe-type pumps, (a), to feed the components, A and B, through a mixing cell, (c), into the reaction cell, (d), which can be an optical cell (Fig. 3.3-5). The pumps, mixing cell, and reactor are well thermostatted. The flow is stopped when the syringe, (e), is loaded and operates a switch, (f), to start the monitoring device. The change in concentration is detected either by spectroscopy or conductivity measurement.

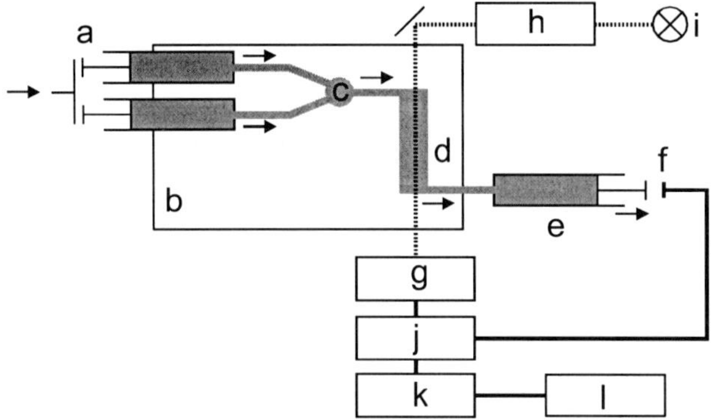

Figure 3.3-5. Stopped flow apparatus [19]. a, Syringe type pump; b, thermostat; c, mixing cell; d, reaction cell; e, stop syringe; f, switch; g, photo multiplier; h, monochromatic filter; i, lamp; j, controller; k, transducer; l, computer.

The heat effects during physical or chemical conversion are used in *differential thermo analysis* (DTA) to evaluate the reaction rate together with kinetic and thermodynamic parameters. The apparatus (Fig. 3.3-6) consists of two identical high-pressure cells which are arranged in an oven, a. In one cell the components to be investigated are placed, the other cell

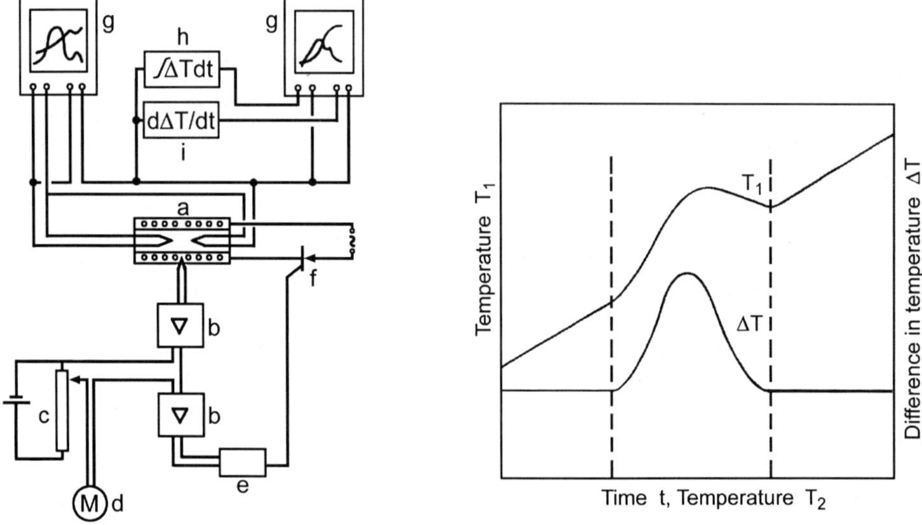

Figure 3.3-6. DTA-device and typical ΔT-curve [20]. a, Oven; b, amplifier; c, potentiometer; d, synchron motor, e, ignition; f, thyristor; g, recorder; h, integrator, i, differentiator; T_1, temperature of the sample cell; ΔT, difference in temperature between sample and reference.

is empty or contains a reference with similar thermodynamic properties as those of the sample. The course of the temperature of the sample cell is measured and compared to the temperature of the reference cell. The course of the temperature of the sample cell in which a physical or chemical conversion takes place is shown in Fig. 3.3-6, right (upper curve),together with the course of the temperature difference between the cells (lower curve). The shape of the ΔT-curve depends on the kinetics and the heat of reaction.

For simple reactions, expressions have been derived to determine the kinetic parameters together with the order of the reaction from the ΔT-curve [21]. A DTA-cell for pressures of up to 300 MPa and temperatures to 300°C is shown in Chapter 4.3.4.

3.3.2 Examples

In this chapter, the application of the apparatus mentioned before to measure reaction rates under high pressure and to evaluate kinetic data, is described. As examples, the decomposition of organic peroxides, the radical polymerization of ethylene, and the synthesis of methanol are selected.

Organic peroxides, which readily decompose into free radicals under the effect of thermal energy, are used under high pressures as initiators for radical polymerizations. The measurement of the influence of pressure on the rate of decomposition gives rise to the determination of the activation volume, which, in turn, allows conclusions to be drawn on the decomposition mechanism and the transition state.

Owing to the very high rate of decomposition, in-situ measurement of concentration by means of Raman spectroscopy was applied. The peroxide used was *t*-butylperoxy pivalate (see Chapter 5.1, Table 5.1-2) dissolved in n-heptane at a concentration of 1 wt.%. In order to observe the change in intensity of absorption of the O-O bond at 861 cm^{-1}, the spectrometer was adjusted to this wave number. The change of intensity is an indication of the reduction in the peroxide concentration, and was recorded as a function of time. The apparatus was calibrated before measuring the intensity of peroxide solutions of different concentrations [22].

In Fig. 3.3-7 the intensity versus time plot is shown for an experiment at 180 MPa and 120°C. It can be seen that the intensity decreases exponentially within 100 s to 58% of its initial value.

By means of the calibration, the absorption intensity curve is converted into a concentration versus time curve. The rate of decomposition is then obtained by differentiating this curve and plotted on a logarithmic scale versus the logarithm of the concentration. From the slope of the resulting straight line, an order of unity for the decomposition is evaluated, and from the intersection at the ordinate the rate-constant is obtained.

In order to determine the activation volume, the rate constants from tests at different pressures are plotted on a logarithmic scale versus the pressure (Fig. 3.3-8). As outlined in Chapter 3.2.3, from the slope of the resulting straight line an activation volume of +7 ml/mol is obtained.

The rotating-sector method was applied to determine the individual rate constants of chain propagation and chain termination of the *radical polymerization of ethylene* [23,24]. The photo-initiator was diphenyldisulfide. First, the overall rate of polymerization was measured under steady illumination at pressures of 50 - 175 MPa and 132 – 199°C (Fig. 3.3-9). It increases first steeply and then less steeply with increasing pressure. At 175 MPa the rate of polymerization is ten times higher than at the low pressure of 50 MPa.

88

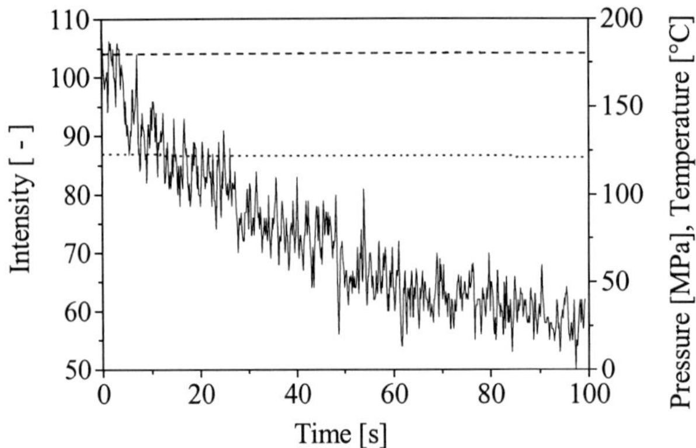

Figure 3.3-7. Peroxide decomposition, change of intensity with time. Solid line, intensity; dashed line, pressure; dotted line, temperature; initial peroxide concentration: 1 wt.%.

The ratio $k_p/k_t^{0.5}$ of the rate constants of chain propagation and termination was determined from the overall rate, r, the known concentration, M, of the monomer and the rate of decomposition, r_i, of the photo-initiator, which was measured separately, using the expression:

$$r = k_p/k_t^{0.5} \cdot [M] \cdot r_i \qquad (3.3\text{-}6)$$

The ratio k_p/k_t can be obtained from the expression:

$$k_p/k_t = 2 \cdot r \cdot [M]^{-1} \cdot \tau \qquad (3.3\text{-}7)$$

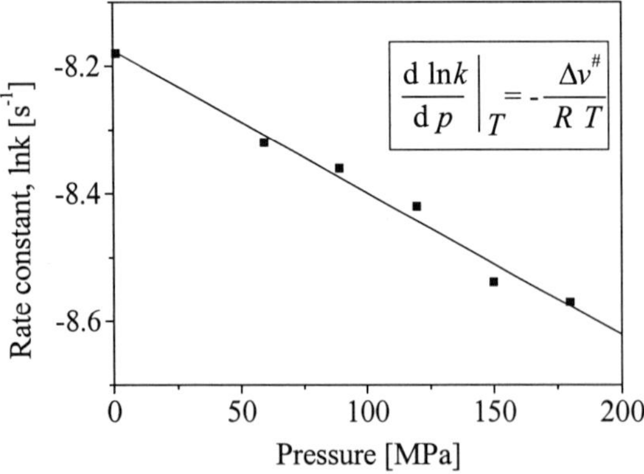

Figure 3.3-8. Determination of the activation volume.

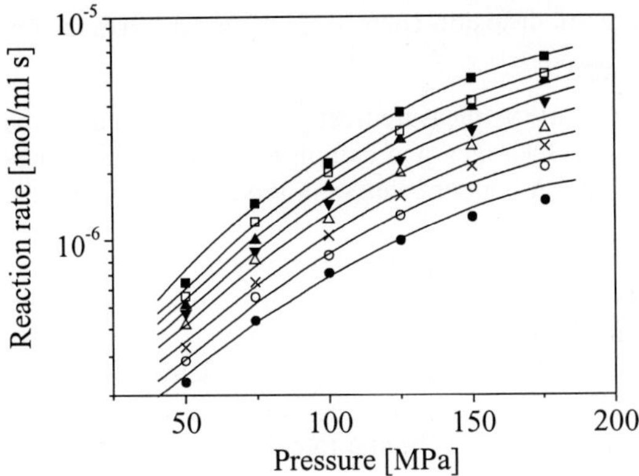

Figure 3.3-9. Overall rate of polymerization. ●, 132°C; O, 144°C; x, 153°C; Δ, 164°C; *, 172°C; σ, 181°C; □, 189°C; v, 199°C.

The life-time, τ, of the radicals can be determined from the ratio of overall rates of polymerization measured at the steady- and unsteady state as a result of intermittent illumination by the rotating sector. In Fig. 3.3-10 the rate constant, k_p, of chain propagation (left) and k_t, that of termination (right), are plotted versus the pressure. Both rate constants increase with increasing temperature. The energy of activation of chain propagation is $E_p = 37$ kJ/mol, and that of chain termination is $E_t = 9.9$ kJ/mol. The influence of pressure is different. k_p increases with increasing pressure. An activation volume of $\Delta v_p^{\#} = -25.5$ ml/mol can be obtained when k_p is plotted on a logarithmic scale versus the pressure. The value of k_t

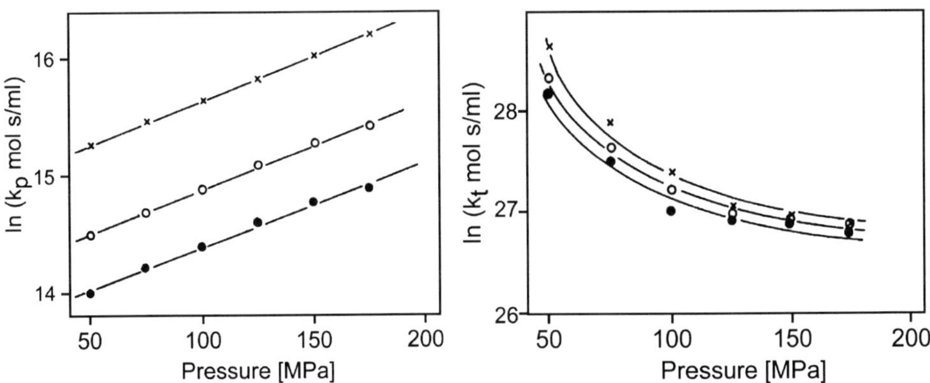

Figure 3.3-10. Rate constants of chain propagation (left) and termination (right) of radical polymerization of ethylene. X, 189°C; O, 153°C; λ, 132°C.

decreases with increasing pressure, because of diffusion-controlled chain termination. On average, a value of $\Delta v_t^{\#} = 7$ cm^3/mol is obtained.

The advantage of the jet-loop reactor for kinetic measurements is demonstrated in the investigation of the synthesis of methanol from hydrogen and carbon monoxide at pressures of 2 - 8 MPa and temperatures of 225 - 265°C. A copper catalyst in the form of cylindrical pellets, with both diameter and length of 5 mm, was used.

The reaction rate was determined from the difference between the methanol concentrations in the reactor outlet and inlet. It increases steeply with increasing pressure and also with increasing temperature, but tends to level off at high temperatures (Fig. 3.3-11).

A rate equation was derived on the assumption that in the first step formaldehyde is formed on the catalyst surface from adsorbed carbon monoxide and hydrogen. The subsequent conversion of formaldehyde to methanol was assumed to be the rate-determining step. The experimental data were best expressed by a rate equation of the Langmuir-Hinshelwood type:

$$r_{CH_3OH} = \frac{\phi_{CO_2} \cdot \phi_{H_2}^2 - \phi_{CH_3OH}/K_{eq}}{\left(A + B \cdot \phi_{CO} + C \cdot \phi_{H_2} + D \cdot \phi_{CH_3OH} + E \cdot \phi_{CO_2}\right)^2} \tag{3.3-8}$$

In this equation, fugacities were used instead of partial pressures, to take into account the non-ideal behaviour of gases at high pressure. The coefficients, A to E, were determined by means of non-linear regression calculation by a method of Marquardt [25]. From the measurements at various temperatures, the frequency factor, k_0, and the activation energy, E, were evaluated. The data are collected in Table 3.3-1.

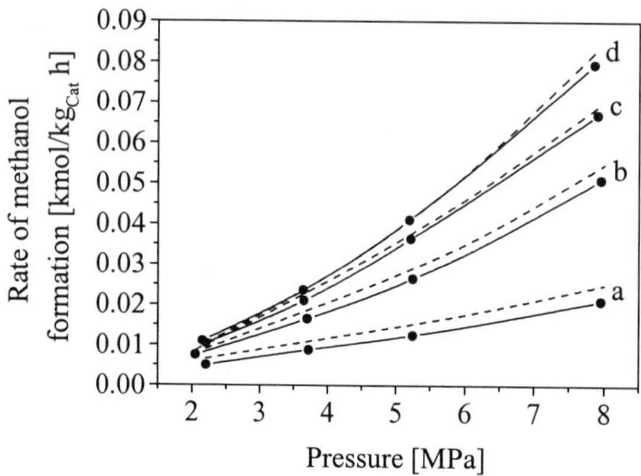

Figure 3.3-11. Rate of methanol formation. Composition of feed: 13 vol.% CO, 82 vol.% H$_2$, 4 vol.% CO$_2$. temperature: a, 225°C; b, 235°C; c, 245°C; d, 255°C; solid lines, experimental; dashed lines, calculated from eqn. 3.3-8.

Table 3.3-1
Coefficients of the rate equation for formation of methanol [26]

	k_0 [bar$^{3/2}$ for A; bar$^{1/2}$ for B to E]	E [kJ/mol]
A	$6.33 \cdot 10^{14}$	128.3
B	$2.28 \cdot 10^{-3}$	-39.4
C	$2.12 \cdot 10^{-6}$	-65.0
D	$8.14 \cdot 10^{0}$	3.9
E	$2.03 \cdot 10^{-11}$	-116.0

References of sections 3.1, 3.2, 3.3

1. M. Baerns, H. Hofmann and A. Renken, Chemische Reaktionstechnik, Bd.1, 100, G.Thieme Verlag, Stuttgart, New York, 1987.
2. O. Levenspiel, Chemical Reaction Engineering, 2nd ed., 349, J.Wiley, New York 1962.
3. S.D. Hamann, Chemical Kinetics, in R.S. Bradley (ed.), High Pressure Physics and Chemistry, Vol. 2, 163, Academic Press, London 1963.
4. H. Eyring, J. Chem. Phys. 3 (1935) 107.
5. M.G. Evans and M. Polanyi, Trans. Faraday Soc. 31 (1935) 875.
6. S. Glasstone, K.J. Laidler and H. Eyring, Theory of Rate Processes, McGraw Hill, New York 1941.
7. G. Luft, P. Mehrling and H. Seidl, Angew. Makromol. Chem. 73, Nr. 1118 (1978) 95.
8. B. Raistrick, R.H. Sapiro and D.M. Newitt, J. Chem. Soc. (1939) 1761.
9. C.K. Ingold, Structure and Mechanism in Organic Chemistry, Bell, London 1953.
10. G. Luft and Y. Ogo, Activation Volumes of Polymerization Reactions, in J. Brandrup and E.H. Immergut (eds.), Polymer Handbook, 3rd ed. J. Wiley, New York 1989.
11. P.-Ch. Lim and G. Luft, Makromol. Chem. 184 (1983) 849.
12. M.G. Gonikberg, Chemical Equilibria and Reaction Rates at High Pressures, Israel Program for Scientific Translations, Jerusalem 1963.
13. G.F. Froment and K.B. Bischoff, Chemical Reactor Analysis and Design, J. Wiley, New York 1979.
14. K.E. Weale, Chemical Reactions at High Pressures, E. and F.N. Spon, London 1967.
15. H.R. Hunt and H. Taube, J. Am. Chem. Soc. 80 (1958) 2642.
16. D.M. Himmelblau, Process Analysis by Statistical Methods, J. Wiley, New York 1969.
17. I.I. Ioffe and L.M. Pissmen, Heterogene Katalyse, Akademie Verlag, Berlin 1975.
18. G. Luft and H.A. Herbertz, Chem.-Ing. Tech. 41, No. 11 (1969) 667.
19. B. Fechner, Doctor Thesis, University of Darmstadt 1995.
20. J. Szabo, G. Luft and R. Steiner, Chem.–Ing.-Tech. 41, No. 18 (1969) 1007.
21. H.J. Borchardt and F. Daniels, J. Am. Chem. Soc. 79 (1957) 41.
22. W. Kessler, G. Luft and W. Zeiß, Ber. Bunsenges. Phys. Chem. 101, No. 4 (1997) 698.
23. G. Luft, P.-Ch. Lim and M. Yokawa, Makromol. Chem. 184 (1983) 207.
24. P.-Ch. Lim and G. Luft, Makromol. Chem. 184 (1983) 849.
25. D.W. Marquardt, J. Soc. Ind. Appl. Math. 11 (1963) 431.
26. O. Schermuly and G. Luft, Ger. Chem. Eng. 1 (1978) 222.

92

3.4 Transport Properties

3.4.1 Fundamentals

In process operations, simultaneous transfer of momentum, heat, and mass occur within the walls of the equipment vessels and exchangers. Transfer processes usually take place with turbulent flow, under forced convection, with or without radiation heat transfer. One of the purposes of engineering science is to provide measurements, interpretations and theories which are useful in the design of equipment and processes, in terms of the residence time required in a given process apparatus. This is why we are concerned here with the coefficients of the governing rate laws that permit such design calculations.

Usually, the values of the transport coefficients for a gas phase are extremely sensitive to pressure, and therefore predictive methods specific for high-pressure work are desired. On the other hand, the transport properties of liquids are relatively insensitive to pressure, and their change can safely be disregarded. The basic laws governing transport phenomena in laminar flow are Newton's law, Fourier's law, and Fick's law. Newton's law relates the shear stress in the y-direction with the velocity gradient at right angles to it, as follows:

$$\tau_{yx} = -\mu \frac{\partial u_x}{\partial y} = -\frac{\mu}{\rho} \frac{\partial (\rho u_x)}{\partial y} = -\nu \frac{\partial (m_x)}{\partial y} \tag{3.4-1}$$

where μ is the dynamic or absolute viscosity, and m_x is the momentum per unit volume in the x-direction. The dimensions of the viscosity are mass/(length x time). The customary unit for viscosity in the c.g.s. system is the Poise (P), but submultiples such as the centipoise (cP) are more often employed. The viscosity of pure water at 20°C is 1 cP [= 1 mPa s = 10^{-3} kg/(m s)]. For Newtonian fluids, the dynamic viscosity depends on the temperature and pressure, but is independent of the velocity gradient.

Analogous equations for the unidirectional transport of heat and mass are the Fourier law and Fick's law, that are written, respectively, as:

$$q_y = -\lambda \frac{\partial T}{\partial y} = -\frac{\lambda}{\rho c_p} \frac{\partial (\rho c_p T)}{\partial y} \tag{3.4-2}$$

$$J_{Ay} = -D \frac{\partial C_A}{\partial y} \tag{3.4-3}$$

For thermal conductivity, the SI units are W/(m K). In laminar flow, the thermal conductivity, λ, and the diffusivity, D, are constant with respect to their respective gradients. Eqn. (3.4-3) indicates that the diffusion flux of solute [mol A/(m^2 s)] is proportional to the transverse concentration gradient, with D as the proportionality constant. The dimensions of D are length^2/ time, and its units are m^2/s in the SI. Eqn. (3.4-2) states that the heat flux [in J/ (m^2.s) = W/m^2] is proportional to the temperature gradient, with a constant $\alpha = \lambda/(\rho c_p)$ that is called the thermal diffusivity. Its dimensions are length^2/time and its SI units are m^2/s. Thus, it is not unexpected that the coefficient $\nu = \mu/\rho$ has the same dimensions and units, m^2/s. The coefficient ν is called the kinematic viscosity, and it clearly has a more fundamental significance than the dynamic viscosity. The usual unit for kinematic viscosity is the Stokes (St) and submultiples such as the centistokes (cSt). In many viscometers, readings are made in cSt from the apparatus constant. The equivalence with the SI units is 1 mm^2/s = 1 cSt = 10^{-6} m^2/s.

In Eqn. (3.4-1), the driving force is the gradient of momentum concentration, which is why v is called the momentum diffusivity. The driving force of the Fourier law, Eqn. (3.4-2), is a gradient of energy concentration, and in Fick's law, Eqn. (3.4-3), is the gradient of the concentration of diffusing species (species A).

It is not known whether high-pressure fluids are Newtonian fluids that behave according to the laws given by Eqns. (3.4-1), (3.4-2), and (3.4-3). With regards to diffusion problems, for example, the Fickian nature of diffusion may be rather the exception than the rule. The diffusivity often depends on solute concentration, not only in extraction with a supercritical gas [1] but also in ordinary low-pressure diffusion in the gas phase and in diffusion in a liquid in multicomponent systems and in porous media.

Analogies between the three transport phenomena are evident and far-reaching [2]. For instance, consider a low pressure, ideal gas, and assume that the kinetic theory of gases holds. It is shown that:

$$\alpha = \frac{\lambda}{\rho\, c_p} = v = \frac{\mu}{\rho} = D = \frac{1}{3}\bar{u}\Lambda \tag{3.4-4}$$

where $\bar{u}$ is the mean particle velocity and Λ the mean free path. It is proved that:

$$Sc = v\,/\,D = 1 \tag{3.4-5a}$$
$$Pr = v\,/\,\alpha = 1 \tag{3.4-5b}$$

These are approximately true for the conditions invoked by the hypotheses of the theory. Thus, for air at atmospheric pressure, Sc = Pr = 0.7. For gaseous carbon dioxide at low pressure, Sc = 0.67 and Pr = 0.81 [3]. The deviation relative to the theory is small. However, for CO_2 at high pressure it has been observed [4] that certain equilibrium properties (c_p, c_v, compressibility), as well as transport properties (Pr, λ) take large values at the critical point. Following the analogy, the Schmidt number should also be very large at the critical point, following the trend of the Pr number, as shown in Fig 3.4-1, where the behaviour of the sound velocity near the critical state can also be seen. As the critical state is approached, the viscosity and the density take finite values. But, in contrast, the diffusivity vanishes as the critical point is approached [5]. Therefore, the values of the Prandtl - and Schmidt numbers can be as high as 2 to 20 for P and T conditions within the region of the near critical fluid. These high values partly explain the good capacity of high-pressure gases as heat- and mass transfer fluids. For liquids at normal pressures, Eqns. (3.4-5) do not hold either. The Prandtl numbers for liquids are around 10 whereas their Schmidt numbers are typically 1000.

Eqns. (3.4-1) to (3.4-3) are practical for describing interfacial transport in combination with transfer coefficients. The use of transfer coefficients is a simplification that involves the linearization of gradients over the boundary layer. Let us assume that a solid dissolves from a non-porous plate into a flowing fluid that sweeps the surface. According to the boundary layer theory, there is a smooth transition between the immobile fluid located on the plane up to the bulk of the fluid that moves with a uniform velocity. In this layer, the three transfer processes

94

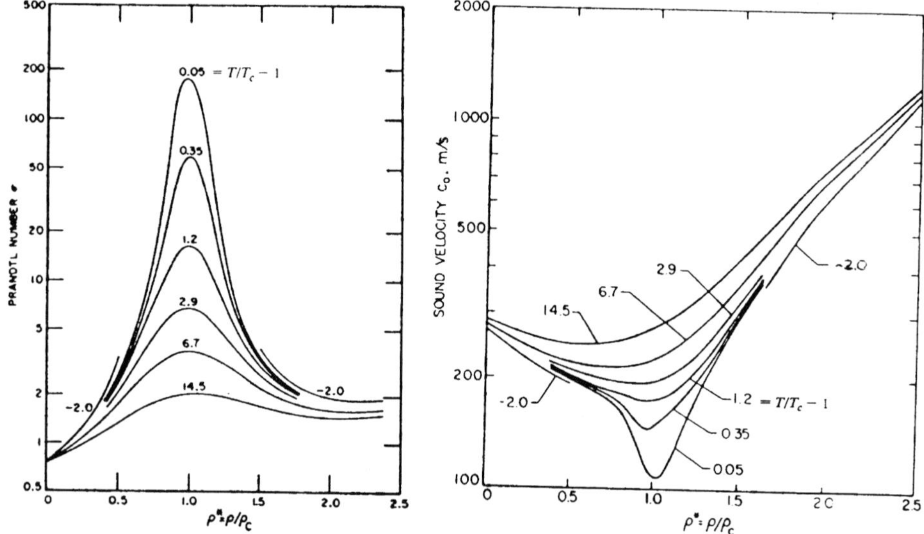

Fig. 3.4-1. Properties of CO_2 near the critical point. Left: Prandtl number. Right: sound velocity from [4] with permission of Butterworths-Heinemann

take place by conduction. For the flat plate geometry, the fluxes in terms of the local coefficients and in terms of the gradients, are:

$$-v\frac{\partial(\rho u_x)}{\partial y}\bigg|_{y=0} = \frac{f}{2}u_0(0-\rho u_0) \tag{3.4-6}$$

$$-\alpha\frac{\partial(\rho c_p T)}{\partial y}\bigg|_{y=0} = h(T_i - T_0) \tag{3.4-7}$$

$$-D\frac{\partial C_A}{\partial y}\bigg|_{y=0} = k_L(C_{Ai} - C_{A0}) \tag{3.4-8}$$

where i represents the conditions at the surface, and 0 that in the bulk fluid. For simple geometries, such as that of the flat plate, the conservation equations for momentum, energy and mass can be solved, and the profiles of u (y, x), T (y, x) and C_A (y, x) can be obtained. Then, the derivatives at the surface, required by Eqns. (3.4-6), (3.4-7) and (3.4-8), can be calculated. Relationships between the transfer coefficients can be written. For the flat plate, two types of relationships are obtained, reflecting the analogy between heat and mass transfer. One relationship is between the values of Re, Sc and Sh, and the other is between Re, Pr, and Nu. The resulting equations in terms of the average values of Nu and Sh, are:

$$\frac{\mathrm{Nu}}{\mathrm{Re}\,\mathrm{Pr}^{1/3}} = \frac{\mathrm{Sh}}{\mathrm{Re}\,\mathrm{Sc}^{1/3}} = 0.664\,\mathrm{Re}^{-1/2} \qquad\qquad (3.4\text{-}9)$$

Where the non-dimensional numbers above are defined as:

$$\mathrm{Re} = \frac{u_0\,L}{v}; \quad \mathrm{Nu} = \frac{h\,L}{\lambda}; \quad \mathrm{Sh} = \frac{k_L\,L}{D} \qquad\qquad (3.4\text{-}10)$$

where L is the distance from the plate's leading edge.

For the case of turbulent flow, the basic transport laws have to be changed to account for turbulence, and many models have been developed since the time of Reynolds. He tried to keep the simple form of Eqns. (3.4-1), (3.4-2), and (3.4-3), but with transport coefficients modified with the so-called eddy diffusivities that are added to the molecular values. So, the net effect of turbulence would be to increase the transfer rates.

The expressions given by Eqns. (3.4-9) correspond to the theory of the boundary layer. Similar expressions are obtained with different theories. In practical work, expressions of the type given below are used for different arrangements. Generally, the exponent of the Reynolds number is less than unity, while the exponent of the Schmidt and Prandtl numbers has been kept as 1/3. The usual expressions for the Nusselt and Sherwood numbers are:

$$\mathrm{Nu} = \mathrm{Nu}_0 + A\,\mathrm{Re}^a\mathrm{Pr}^b \qquad\qquad (3.4\text{-}11a)$$

$$\mathrm{Sh} = \mathrm{Sh}_0 + B\,\mathrm{Re}^c\mathrm{Sc}^d \qquad\qquad (3.4\text{-}11b)$$

There has always been an interest in having correlations for heat- and mass transfer for different geometries in view of the difficulties in: describing turbulence in a meaningful way, and in solving the differential equations of conservation of momentum, energy, and mass transfer for real bodies.

As a result, many correlations are available for heat and mass transfer at moderate pressures that have been developed over time. Perry and Green [6] give a fairly complete amount of data with regards to correlations for different arrangements [7]. On the other hand, very few data and correlations are available in the field of high pressure heat and mass transfer, as will be reviewed later. Correlations are in terms of the individual coefficients, k_L and h, included in dimensionless groups such as those given before in Eqns. (3.4-10).

The correlations exhibit the following types of dependence:

1) Forced convection problems:

for heat transfer, $\mathrm{Nu} = f\,(\mathrm{Re},\,\mathrm{Pr})$

for mass transfer, $\mathrm{Sh} = f\,(\mathrm{Re},\,\mathrm{Sc})$.

2) Free convection problems:

for heat transfer, $\mathrm{Nu} = f\,(\,\mathrm{Gr_H},\,\mathrm{Pr})$

96

for mass transfer, $Sh = f(Gr_D, Sc)$

in which the expressions for the Grashof numbers are:

$$Gr_H = \frac{g\,\beta}{v^2} L^3 |\Delta T| \qquad (3.4\text{-}12a)$$

$$Gr_D = \frac{g\,L^3}{v^2} \frac{|\Delta\rho|}{\rho} \qquad (3.4\text{-}12b)$$

3) For problems of mixed, free and forced convection:

for heat transfer, $Nu_{tot}^3 = Nu_{CN}^3 \pm Nu_{for}^3$ $\qquad (3.4\text{-}13a)$

for mass transfer, $Sh_{tot}^3 = Sh_{CN}^3 \pm Sh_{for}^3$ $\qquad (3.4\text{-}13b)$

in which the plus sign is for assisting flows and the minus sign is for opposing flows [8,9].

For the case of dense-gas heat- and mass transfer, free convective effects are dominant under certain conditions. In engineering work, use is made of these properties.

The solution of the equations of continuity and momentum, along with additional conservation equations for energy and material species, can be found numerically in order to solve problems of non-isothermal flow, mixing, and chemical reaction. The so-called computational fluid dynamics (CFD) constitute a powerful tool for fluid dynamics, both in research and in practice. Commercial CFD software is widely available. Three basic steps are common to all CFD methods. The first step consists in the subdivision of the flow domain into cells (which define a mesh). The second one consists in the discretization of the governing equations (approximating the partial differential equations, by algebraic equations by using finite difference-, finite volume-, or finite element methods). Finally, the solution of the algebraic equations is found using an equation solver. The large quantity of solution data must then be managed by a powerful graphical interface in order to be comprehensible for the user.

CFD methods are used for incompressible- and compressible-, creeping-, laminar- and turbulent-, Newtonian- and non-Newtonian-, and isothermal- and non-isothermal flows. Most commercial CFD codes include the k–ε turbulence model [10]. More accurate models are also becoming available. The accuracy of the solution depends on how the mesh fits the true geometry, on the convergence of the solution algorithm, and also on the model used to describe the turbulent flow [11].

In order to calculate the numerical values of transfer coefficients, values of the molecular properties are required. In the next section, we present estimation methods for viscosity, diffusivity, thermal conductivity and surface tension, for the high-pressure gas.

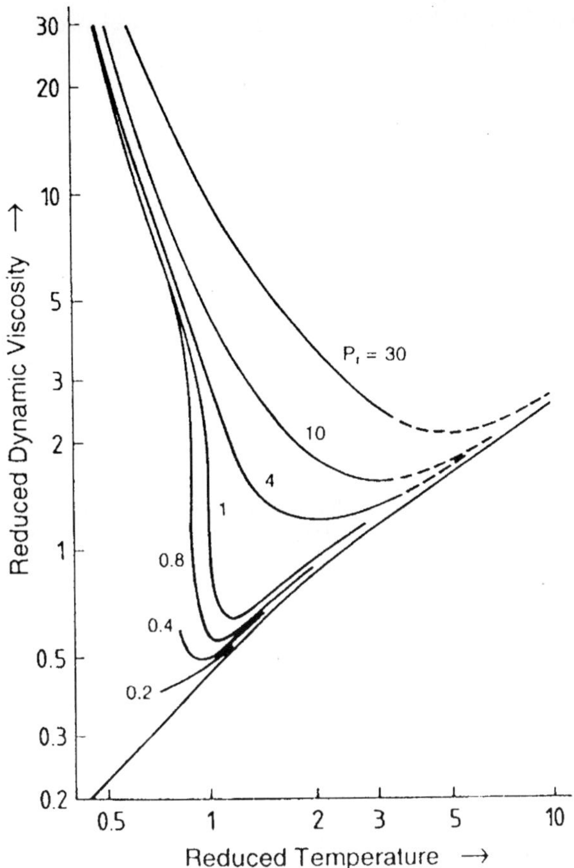

Fig. 3.4-2. Generalized reduced viscosities from [12] with permission of Steinkopff Verlag

3.4.2 Estimation of transport properties

3.4.2.1 Viscosity

We are concerned here with the viscosity of a high-pressure gas phase. Data and correlations for the viscosity of liquids and gases under low-pressure conditions, as a function of temperature, can be found in ref. [13,14].

For a gas, the effect of pressure on the viscosity depends on the region of P and T of interest relative to the critical point. Near the critical state, the change in viscosity with T at constant pressure can be very large. The correlation of Uyehara and Watson [15] is presented for the reduced viscosity estimated from the corresponding-states method. The critical viscosities of a few gases and liquids are available [15]. These are necessary to calculate the

98

reduced viscosity for use with Fig 3.4-2. If they are not available, the critical viscosities can be estimated from the following equation:

$$\mu_c = 7.70 \frac{(M)^{1/2} p_c^{2/3}}{T_c^{1/6}} \tag{3.4-14}$$

$\mu_r = \mu / \mu_c$
M = molecular weight, g/mol
T_c = critical temperature, K
p_c = critical volume, cm^3/mol
where μ_c = critical viscosity, in μP.

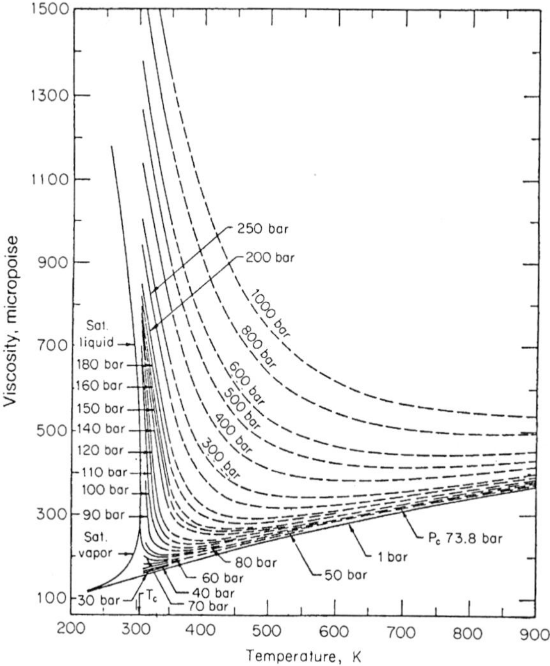

Fig. 3.4-3. Viscosity of CO_2 according to Stephan and Lucas from [13] with permission of McGraw-Hill Book Company.

The results given by Fig. 3.4-2 relative to measured viscosities are about 10%. Fig. 3.4-2 shows that the viscosity tends to a minimum at the critical state. Near the critical point, at constant pressure, there are two branches, one in which viscosity increases with increasing temperature (gas-like behaviour), and the other branch where the viscosity decreases with increasing temperature (liquid-like behaviour).

Fig. 3.4-2 can be applied to mixtures by averaging the component's critical viscosities as:

$$\mu_c^{crit} = \sum_{i=1} y_i \mu_c \qquad \text{(3.4-15)}$$

and then use this value to calculate the reduced viscosity from the Figure. In order to calculate pseudo-critical properties, Kay's rules are recommended. Pseudo-critical values are obtained as mole-fraction averages for the critical temperature, the compressibility factor, and the molar volume. From the latter, the critical pressure, the reduced temperature and the reduced pressure can be calculated for use in the figure. A viscosity correlation of Fig 3.4-2 has been improved by Lucas [7]. In Fig. 3.4-3 the viscosity of pure carbon dioxide is given graphically.

Gaseous non-polar and hydrocarbon mixtures at high pressures

For predicting the vapour viscosity at $T_r > 0.6$, low pressure values, μ_0, are calculated from a low pressure equation and then corrected for high pressure by the method of Dean and Stiel [16] with the following equations:

$$(\mu - \mu_0)\xi = 1.08 \left[\exp(1.439\,\rho_r) - \exp(-1.11\,\rho_r^{1.858}) \right] \qquad \text{(3.4-16a)}$$

$$\xi = \frac{T_c^{1/6}}{M^{1/2} P^{2/3}} \qquad \text{(3.4-16b)}$$

where:

μ_0 = viscosity at low pressure of gas or gas mixture, μP
μ = viscosity of the gas or gaseous mixture at high pressure, μP
ρ_r = reduced viscosity
ρ_c = pseudocritical density of mixture

For mixtures, the simple molar-average pseudo-critical temperature, pressure, and density, and molar-average molecular weight are used as before. For mixtures of polar gases, no appropriate correlation has been given.

In the equations above, the low pressure viscosity appears. This is understood to be at pressures lower than $P_r < 0.6$. A good correlation for this is the Stiel and Thodos [17] equation:

$$\mu_0 = 4.60 \times 10^{-4} \frac{N M^{1/2} P^{2/3}}{T_c^{1/6}} \qquad \text{(3.4-17)}$$

with:

$$N = 0.00034 \times T_r^{0.94} \quad T_r < 1.5 \qquad \text{(3.4-18)}$$

$$N = 0.0001778(4.58 T_r - 1.67)^{0.625} \quad T_r > 1.5 \qquad \text{(3.4-19)}$$

The resultant viscosity is in cP (mPa s), and T_c and P_c are given in K and Pa, respectively. This equation can also be used for light non-hydrocarbon gases except for hydrogen and helium. For hydrocarbons of less than ten carbon atoms, average errors of about 3% can be expected, with errors increasing to 5 – 10% for heavier hydrocarbons.

Pure non-hydrocarbon polar and non-polar gases at high pressure

There is a method of prediction due to Stiel and Thodos that depends on the reduced density as a corrector to the low-pressure gas viscosity (μ_0), and takes various forms for polar gases, according to the reduced density. This is shown in the following equations:

For $\rho_r < 0.1$
$$(\mu - \mu_0)B = 1.656 \times 10^{-7} \rho_r^{1.111} \qquad (3.4\text{-}20a)$$

$0.9 < \rho_r < 2.6$
$$\log\{4 - \log[\mu - \mu_0)10^7 B]\} = 0.6439 - 0.1005\rho_r - d \qquad (3.4\text{-}20b)$$

for $0.9 < \rho_r < 2.2$, $\qquad d = 0 \qquad (3.4\text{-}20c)$

for $2.2 < \rho_r < 2.6$ $\qquad d = 4.74 \times 10^{-4} (\rho_r^3 - 10.65)^2 \qquad (3.4\text{-}20d)$

In the above equations:

$$B = 2173.424 \, T_c^{1/6} \, M^{-1/2} \, P_c^{2/3} \qquad (3.4\text{-}21)$$

In all cases, viscosities are in Pa.s, T_c in K and P_c in Pa. For non-polar gases, the errors are very small, while for polar gases, average errors may reach 10%.

3.4.2.2 Diffusivity in dense gases

The diffusion coefficient as defined by Fick's law, Eqn. (3.4-3), is a molecular parameter and is usually reported as an infinite-dilution, binary-diffusion coefficient. In mass-transfer work, it appears in the Schmidt- and in the Sherwood numbers. These two quantities, Sc and Sh, are strongly affected by pressure and whether the conditions are near the critical state of the solvent or not. As we saw before, the Schmidt and Prandtl numbers theoretically take large values as the critical point of the solvent is approached. Mass-transfer in high-pressure operations is done by extraction or leaching with a dense gas, neat or modified with an entrainer. In dense-gas extraction, the fluid of choice is carbon dioxide, hence many diffusional data relate to carbon dioxide at conditions above its critical point (73.8 bar, 31°C)

In general, the order of magnitude of the diffusivity depends on the type of solvent in which diffusion occurs. Middleman [18] reports some of the following data for diffusion.

3.4.2.3 Binary diffusivity data in different media

Gas *in gas*	10^{-4} m^2/s
Liquid or gas *in liquid*	10^{-9} m^2/s
Gas *in solid*	10^{-10} to 10^{-14} m^2/s
Heavy vapour *in SC fluid*	10^{-8} m^2/s

As can be seen, diffusion in a supercritical fluid is about 10 times larger than in a liquid, and 100 times less than in a gas, and thus between a gas and a liquid. As is known from correlations, the diffusion coefficient is inversely proportional to pressure, so at constant temperature it reduces rapidly with pressure. On the other hand, at constant pressure the gas diffusivity increases with the power 1.5 of the absolute temperature.

Prediction of diffusion coefficient

In high-pressure practice, it is important to have methods for predicting mass transfer of heavy solutes in dense gases, as these solutes or similar ones are extracted with dense gases. The first step is to correlate binary diffusivities of heavy vapours in dense gases. There is now a sufficient database of diffusivities of heavy solutes. Catchpole and King [19] recently collected a large amount of data to propose a fairly simple correlation for estimating the self-diffusion and binary diffusion coefficients over the reduced density range of interest (ρ_r = 1 to 2.5). The method uses the solvent and solute molecular weights, the reduced temperature, the fluid density, and the critical volume of the solute.

The binary diffusivity is calculated with the following correlation (for $1 < \rho_r < 2.5$):

$$D = 5.152 \ D_c T_r (\rho^{-2/3} - 0.4510) R / X \qquad (3.4\text{-}22a)$$

with

$$R = 1.0 \pm 0.1 \qquad (\text{for } 2 < X) \qquad (3.4\text{-}22b)$$

and

$$R = 0.664 \ X^{0.17} \pm 0.1 \ (\text{for } 2 < X < 10) \qquad (3.4\text{-}22c)$$

where X is the size-to-mass parameter which is calculated from:

$$X = \left[1 + (V_{c2} / V_{c1})^{1/3}\right]^2 / (1 + M_1 / M_2)^{1/2} X \qquad (3.4\text{-}23)$$

where 1 is the solvent, and 2 is the solute. In Eqn. (3.4-22), D_c is the self-diffusion coefficient, correlated using critical quantities, as follows:

$$D_c = \frac{4.30 \times 10^{-7} M^{1/2} T_c^{0.75}}{\left(\sum v\right)^{2/3} \rho_c} \qquad (3.4\text{-}24)$$

where

M = molecular weight, g/mol
T_c = in K
ρ_c = in kg/m^3

102

Σv = Sum of atomic diffusion volumes as defined by Fuller *et al.* (see Perry and Green's [6]
       Table 400, p. 2-370)

3.4.2.4 Thermal conductivity

We focus next on the thermal conductivity of gases at high pressure. At ambient pressure
and temperature the thermal conductivity is in the range of 0.01 to 0.025 W/(K m), except for
hydrogen and helium that can have higher values, around 0.18 W/(K m). The change of
thermal conductivity around the critical state is seen on Fig 3.4-4 for carbon dioxide. Near the

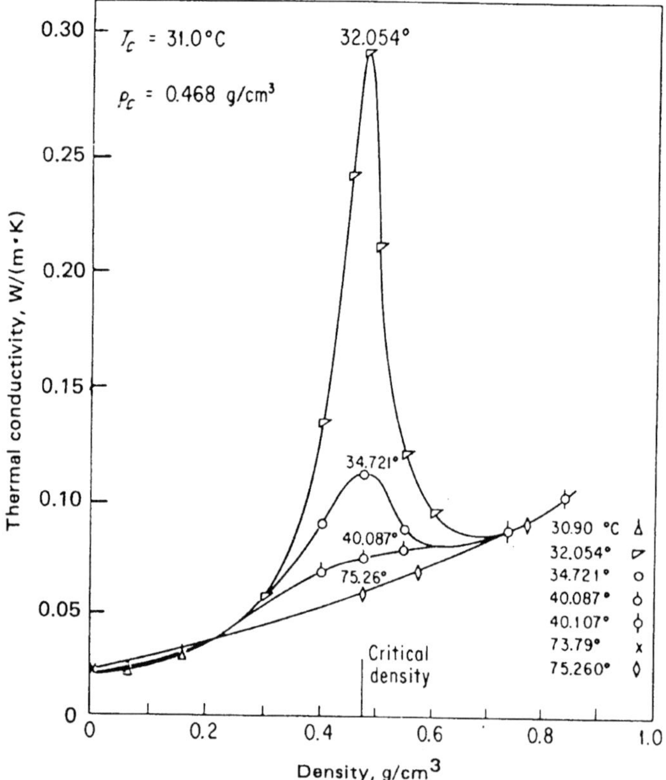

Fig. 3.4-4. Thermal conductivity of CO_2 from [13] with permission of McGraw-Hill Book Co.

critical point large values are reported [0.25 to 0.30 W/(K m)]. This causes the Prandtl
number to reach a high value as well (around 200, which is a liquid-like value), as we have
discussed before (Fig 3.4-1).

Brunner [12] gives an overall picture (Fig. 3.4-5) of the reduced thermal conductivity of
diatomic gases. It has been observed that at constant pressure the thermal conductivity
increases with increasing temperature. This behaviour ceases at high pressures ($Pr \sim 100$).

Under supercritical conditions, the conductivity decreases with increasing temperature, then goes to a minimum, and then increases with temperature. On the other hand, at constant temperature, the thermal conductivity increases with pressure.

Correlating equations for the thermal conductivity. For non-polar gases, the departure of thermal conductivity, λ, from the zero-pressure value, λ_0, at the same temperature is given by the equations of Stiel and Thodos [17] that are valid over different gas-density ranges. These are:

$$(\lambda - \lambda_0)\Gamma Z_c^5 = 1.22 \times 10^{-2}\left[\exp(0.535\rho_r) - 1\right] \qquad \text{for } \rho_r < 0.5 \qquad (3.4\text{-}25a)$$

$$(\lambda - \lambda_0)\Gamma Z_c^5 = 1.14 \times 10^{-2}\left[\exp(0.67\rho_r) - 1.069\right] \qquad \text{for } 0.5 < \rho_r < 2.0 \qquad (3.4\text{-}25b)$$

$$(\lambda - \lambda_0)\Gamma Z_c^5 = 2.6 \times 10^{-3}\left[\exp(1.155\rho_r) + 2.016\right] \qquad \text{for } 2.0 < \rho_r < 2.8 \qquad (3.4\text{-}25c)$$

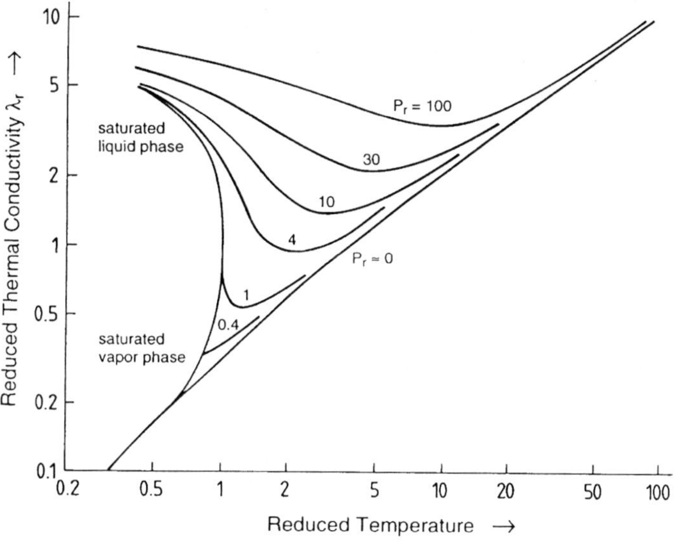

Figure 3.4-5. Reduced thermal conductivity from Brunner [12] with permission of Steinkopff Verlag

The above equations were obtained from twenty non-polar gases including inert gases, hydrocarbons and carbon dioxide (but not hydrogen and helium). Hence, possible errors can be as large as 20%. The maximum pressure corresponds to a reduced density of 2.8. In the above equations, Zc represents the critical compressibility factor. The value of gamma is calculated using Eqn. (3.4-26).

$$\Gamma = \left[\frac{T_c\ M^3 N_0^2}{R^5 P_c^4} \right]^{1/6} \tag{3.4-26}$$

Where SI units are used, $R = 8314$ J/(mol K), $N_0 = $ Avogadro's number $= 6.023 \times 10^{26}$ kmol^{-1}, T_c is in K, M in kg/kmol, and P_c in N/m^2. Γ has the units of K·m/W, and represents the inverse of a thermal conductivity at the critical point. In Eqn. (3.4-25), the low-pressure viscosity (at about 1 bar) can be estimated for linear non-hydrocarbon molecules with the following equation:

$$\lambda_0 = \frac{\mu}{M} \left(1.30\, C_v + 14644.0 - \frac{2928.8}{T_r} \right) \tag{3.4-27}$$

where the viscosity, μ, is expressed in Pa.s, and the heat capacity of the gas, C_v, in J/(kmol.K). This can be obtained from C_p and $R = 8314$ J/(kmol K).

3.4.2.5 Surface tension

Surface tension is not actually a transport property but an equilibrium property related to two-dimensional equilibrium thermodynamics. Surface tension appears in the performance equation of extraction and fractionation columns, where fluid-to-fluid interfaces are present, and represents the imbalance of molecular forces at the interfaces.

Surface tension is defined as the isothermal work required to create a unit surface area. That is:

$$\sigma = \mathrm{d}W/\mathrm{d}A \tag{3.4-28}$$

Therefore the units for σ are J/m^2 = N/m in the SI. Usually, surface tension values are given in the c.g.s. system, hence in dyne/cm (mN/m= mJ/m^2) For liquids, σ is in the range of 20 to 40 mN/m, except for water whose surface tension is larger than these values. It is 72.8 mN/m or mJ/m^2 at 20°C in contact with air.

For pure liquids, an empirical estimation method for surface tension is that of McLeod and Sudgen, based on the parachor (P), which be calculated from the chemical groups of the molecule [13].

$$\sigma^{1/4} = (P)(\rho_L - \rho_v) \tag{3.4-29}$$

In this equation, (P) is the parachor and the ρ values are the liquid- and vapour- molar densities in mol/cm^3. Then σ is obtained in mJ/m^2. For the case of mixtures in the presence of vapour phase, the surface tension of a mixture is calculated in terms of the liquid and vapour composition, as:

$\rho_{iL} = $ density of liquid phase, mol/cm^3
$\rho_{iV} = $ density of liquid phase, mol/cm^3
x_i and $y_i = $ mole fractions of i in the liquid and vapour
(Pi) = parachor for component i (calculated from the structural data)

Surface tension of the mixture is obtained in dyn/cm (or mJ/m^2)

For the case of a high pressure gas in contact with a liquid phase, the dissolution of the gas reduces the surface tension of the liquid. Fig. 3.4-6 shows this effect for the system squalane/CO$_2$ as reported by Brunner [12]. Gases are seen to reduce the surface tension upon dissolution into the liquid at high pressure, so that the surface tension decreases linearly with the mol fraction of the gas dissolved. Apparently the liquid vapour dissolves in the high pressure gas, hence causing a reduction in the surface tension [12].

$$\sigma_m^{1/4} = \sum_{i=1} (P_i)(\rho_{iL} x_i - \rho_{iv} y_i) \tag{3.4-30}$$

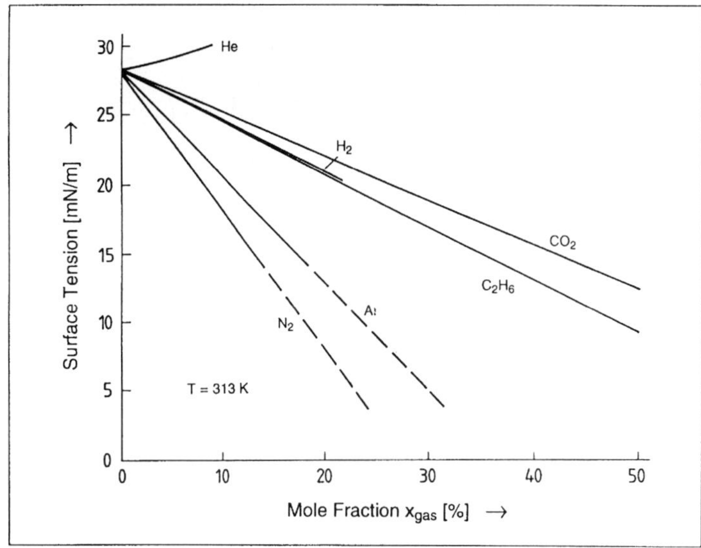

Fig 3.4-6. Effect of pressure on the surface tension of the squalene/CO$_2$ system from [12] with permission of Steinkopff Verlag

106

3.4.3 Heat transfer mechanisms in dense fluids: calculation of heat-transfer coefficients in different arrangements

The interest in heat-transfer for high-pressure systems is related to the extraction of a valuable solute with a compressed gas. The compressed fluid is usually a high-pressure gas-often a supercritical fluid, that is, a gas above its critical state. In this scenario, the prevalent heat-transfer mechanism is convection.

Convection relates to the transfer of heat from a bounding surface to a fluid in motion. The heat-transfer coefficient, h, is defined by the relationship between the heat transferred at the solid-to-fluid interface and the temperature difference.

$$\frac{q}{A} = h\left(T_w - T_f\right) \tag{3.4-31}$$

This equation is known as Newton's law of cooling, and T_w is the surface temperature and T_f is a characteristic fluid temperature. At the wall, the fluid velocity is zero, and the heat-transfer takes place by conduction. Therefore, applying Fourier's law to the fluid at $y = 0$ (where y is the axis normal to the solid surface) and combining it with Newton's law, we have:

$$h = \frac{q\!\big/\!A}{\left(T_w - T_f\right)} = -\frac{\lambda\,(\partial T/\partial y)\big|_{y=0}}{\left(T_w - T_f\right)} \tag{3.4-32}$$

so that the heat-transfer coefficient can be obtained form the temperature gradient, if known. The temperature gradient at the solid-to-fluid interface can be found from the solution of the conservation equations (continuity equation, equation of motion, energy equation, and conservation equation for species). The differential equations describing the problem are too difficult to solve. In laminar flow, the solution of complicated problems is only a matter of having sufficient computer capacity. Heat transfer in turbulent flow is still not well understood theoretically. The use of approximate models by CFD techniques has been remarked upon in Section 3.4.1.

For practical purposes, heat-transfer engineers often use empirical or semi-empirical correlations to predict h values. These formulations are usually based on the dimensionless numbers described before. In this case, the appropriate formulation should be used to prevent significant errors. If dimensionless correlations are applicable under conditions of gas extraction, then heat-transfer coefficients can be determined from these correlations and the influence of parameter variations may be derived also from them.

In extraction processes using supercritical fluids it is of interest to predict the heat-transfer processes that take place in heat exchangers in forced flow.

3.4.3.1 Single phase convective heat transfer

Forced convection in ducts

The transition to turbulent flow in ducts starts at Re = 2300. The flow is fully turbulent at $5 \times 10^4 < \text{Re} < 10^5$, depending on the turbulence of the incoming flow and the shape of the inlet section. Correlations for fully developed turbulent flow in ducts are functions of the type:

$$Nu = A\,Re^m\,Pr^n\,f(D/L) \tag{3.4-33}$$

For smooth straight tubes, $0 < D/L < 1$, $0.6 < Pr < 2000$, and $2300 < Re < 10^6$, Gnielinski proposed the following equation

$$Nu = \frac{(f/8)(Re-1000)\,Pr}{1+12.7\sqrt{f/8}(Pr^{2/3}-1)}\left[1+\left(\frac{D}{L}\right)^{2/3}\right] \tag{3.4-34}$$

where f is the friction factor for turbulent flow. According to Filonenko,

$$f = (1.82\log_{10} Re - 1.64)^{-2} \tag{3.4-35}$$

To take into account the influence of the temperature-dependent physical properties, the Nusselt number can be multiplied by a correction factor of the type

$$\phi = \left(\frac{T_b}{T_w}\right)^n \qquad \text{for gases} \tag{3.4-36a}$$

$$\phi = \left(\frac{Pr}{Pr_w}\right)^{0.11} \qquad \text{for liquids} \tag{3.4-36b}$$

where T_b is the absolute bulk temperature of the gas and T_w is the wall temperature. The exponent, n, is equal to zero for cooled gases and takes different values for each gas for heated gases. For carbon dioxide and steam, $n = 0.15$ in the range $0.5 < T_b/T_w < 1.0$. Pr is the Prandtl number of the liquid at bulk temperature. The range of validity of Eqn. (3.4-36b) is $0.1 < Pr/Pr_w < 10$.

For the prediction of the Nusselt number in ducts of non-circular-cross section (like concentric annular ducts) the same equations can be used for forced convection in the turbulent regime. In this case, the inside diameter should be replaced in evaluating Nu, Re, and (D/L) by the hydraulic diameter defined as,

$$D_h = 4\frac{S}{P} \tag{3.4-37}$$

where S is the cross-sectional area and P is the wetted perimeter of the duct.

Bank of plain tubes

For a bank of plain tubes, the Reynolds number is defined as

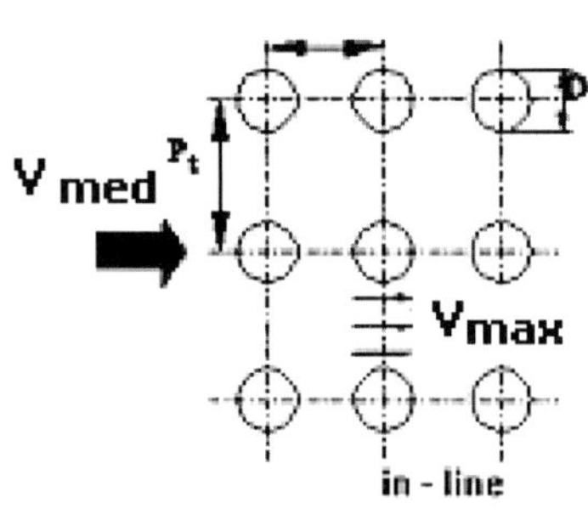
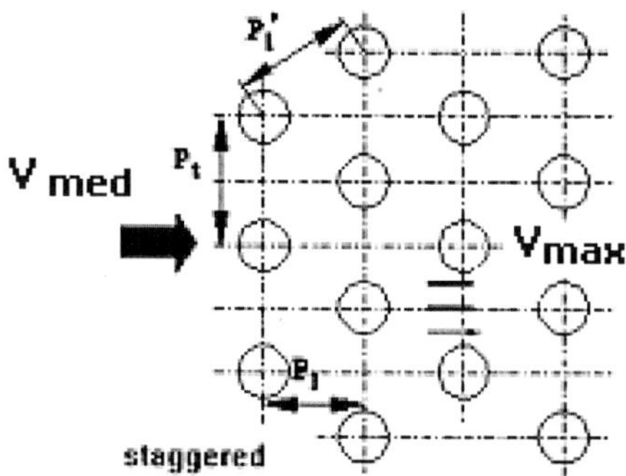

Figure 3.4-7. Tube arrangement

$$\mathrm{Re} = \frac{D\,\mathrm{v}_{max}}{\nu} = \frac{D\,\mathrm{v}_{med}\,\dfrac{a}{a-1}}{\nu} \tag{3.4-38}$$

where (see Fig. 3.4-7)

$$a = \frac{P_t}{D}; \quad and \quad b = \frac{P_l}{D} \tag{3.4-39}$$

The Nusselt number for banks of more than twenty rows can be calculated by the equation proposed by Zukauskas as:

$$\mathrm{Nu}_D = \alpha\,Re_D^{\beta}\,Pr^{0,36}\left(\frac{Pr}{Pr_w}\right)^{0,25} \tag{3.4-40}$$

for liquids, where the properties must be calculated at the bulk mean temperature. Also:

$$\mathrm{Nu}_D = \alpha\,Re_D^{\beta}\,Pr^{0,36} \tag{3.4-41}$$

for gases, where the properties must be calculated at the film temperature (the arithmetic mean between the bulk temperature and the wall temperature).

Free convection
Free convection is fluid flow, induced by density gradients owing, for example, to temperature gradients. In gas extraction the supercritical solvent is subject to density variation with only slight changes in pressure and temperature. Furthermore, flow velocities within the processing equipment are low, so that flow owing to free convection may be important. Therefore, conditions for free convective flow must be considered in such types of systems.
For isothermal vertical plates:

Table 3.4-1
Values of the parameters for the calculation of the Nusselt number of tube banks

Re_D	In - line	Staggered	
$10 << 10^2$	$\alpha=0.9$ $\beta=0.4$	$\alpha=0.9$ $\beta=0.4$	
$10^2 \leq \, \leq 10^3$	$\alpha=0.52$ $\beta=0.50$	$\alpha=0.60$ $\beta=0.50$	
$10^3 \leq \, \leq 2\cdot10^5$	if (a/b)>0.7 $\alpha=0.27$ $\beta=0.63$	if (a/b)<2 $\alpha=0.35\,(a/b)^{0.2}$ $\beta=0.60$	if (a/b)>2 $\alpha=0.40$ $\beta=0.60$
$> 2\cdot10^5$	$\alpha=0.021$ $\beta=0.84$	$\alpha=0.022$ $\beta=0.84$	

$$(Nu_L)_f = \frac{4}{3}(Nu_{x=L})_f = \frac{8}{3}Pr_f^{0,5}\left[\frac{(Gr_L)_f}{240\left(\frac{20}{21}+Pr_f\right)}\right]^{0,25} \tag{3.4-42}$$

for $10^4 < Pr\,Gr_L < 10^9$ and $0.5 < Pr < 10$. All the properties must be calculated at the film temperature.

$$Nu(x)_f = \frac{0,15\,(Gr_x\,Pr)_f^{1/3}}{\left[1+\left(\frac{0,492}{Pr_f}\right)^{9/16}\right]^{16/27}} \tag{3.4-43}$$

for $Pr\,Gr_L > 10^9$ and $0,5 < Pr < 50$ All the properties must be calculated at the film temperature.

For isothermal horizontal plates:

$$Nu_L = C_1\,(Gr_L\,Pr)^n \tag{3.4-44}$$

for $10^4 < Gr_L\,Pr < 10^9$, $n = 0.25$; $C_1 = 0.54$ for the top side of hot plates or the bottom side of cold plates, and $C_1 = 0.27$ for the bottom side of hot plates or the top side of cold plates.
for $10^9 < Gr_L\,Pr < 10^{12}$, $n = 1/3$, and $C_2 = 0.15$ for the top side of hot plates or the bottom side of cold plates.

3.4.3.2 Condensation

Film condensation of pure vapours

In *vertical plates*:

$$h_c = 0.943 \left[\frac{g \, \cos\theta \, \rho_l \, \rho_{lg} \, \lambda_l^3 \, i_{lg}^*}{L \, \mu_l (T_S - T_0)} \right]^{1/4} \tag{3.4-45}$$

where $i^*_{lg} = i_{lg} + (3/8) \, c \, (T_S - T_0)$ is the modified latent heat of vaporization.

This formula is valid for flat plates and cylindrical tubes of diameter larger than around 1/8 inch. In both cases the inclination angle should be lower than 90°. Additionally, $\mathrm{Pr} > 0,5$ and $[c (T_S - T_0) / i_{lg}] < 1$. A better approximation can be obtained when the condensate viscosity is calculated at $[T_0 + 0,31 (T_S - T_0)]$.

For condensation *on horizontal tubes*:

$$h_c = 0,728 \left[\frac{g \, \rho_l \, \rho_{lg} \, \lambda_l^3 \, \Delta i_l^*}{D_0 \, \mu_l (T_S - T_0)} \right]^{1/4} \tag{3.4-46}$$

for $\mathrm{Pr} > 1$ and $[c (T_s - T_0) / i_{lg}] < 1$.

In a *tube bank* the average heat-transfer coefficient for N_f tubes in a vertical bank can be approximated by:

$$h_c = 0,728 \left[1 + 0.2 \, \frac{c \, (T_S - T_0)}{i_{lg}} (N_f - 1) \right] \left[\frac{g \, \rho_l \, \rho_{lg} \, \lambda_l^3 \, i_{lg}^*}{N_f \, D_0 \, \mu_l (T_S - T_0)} \right]^{1/4} \tag{3.4-47}$$

to a good approximation, provided $[c (T_s - T_0) / i_{lg}] < 2$.

For condensation *inside horizontal tubes*, with low condensation rates or short tubes, at low vapour velocities (inlet $\mathrm{Re_V} < 35000$) the liquid condensate runs off in a stratified condition, and:

$$h_c = 0,555 \left[\frac{g \, \rho_l \, \rho_{lg} \, \lambda_l^3 \, i_{lg}^*}{D_i \, \mu_l (T_S - T_0)} \right]^{1/4} \tag{3.4-48}$$

For condensation *inside long tubes*, when the condensation rate or the length are large, the flow regime becomes annular with a vapour core. At even higher vapour velocities, the vapour core may contain a mist of liquid droplets.

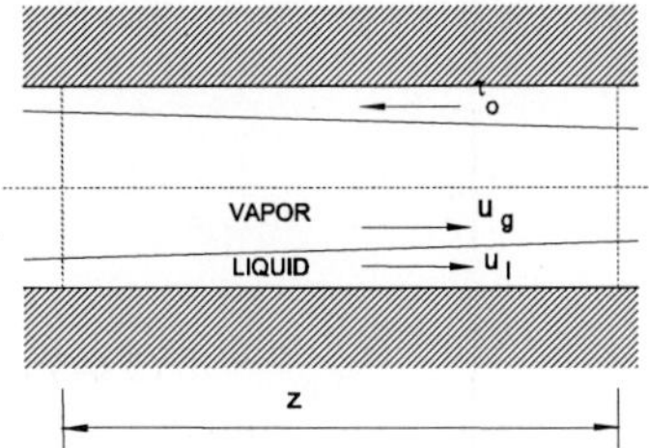

Figure 3.4-8. Film condensation in annular flow

The procedure to size a condenser suggested by Rohsenow [20] is as follows:
Assume magnitudes of the tube-diameter, D, mass velocity, G, and the fluid properties. Divide the length into equal increments of quality change, Δx, and calculate the heat-transfer coefficient, h_z at the mid-quality magnitude of each Δx and assume that h_z is uniform in this quality range. Then the length required to change the quality Δx is

$$\Delta z = \frac{G\, i_{lg}\, D_i\, \Delta x}{4\, h_z\, (T_S - T_0)} \tag{3.4-49}$$

The average heat-transfer coefficient is thereafter calculated from

$$h_c = \frac{\Sigma\, h_z\, \Delta z}{\Sigma\, \Delta z} \tag{3.4-50}$$

The following equations are needed:

$$x = \frac{m_g}{m_g + m_l} = \frac{m_g}{m} \tag{3.4-51}$$

$$h_z = \frac{q/A}{T_S - T_0} \tag{3.4-52}$$

$$\mathrm{Re_l} = \frac{(1 - x)\, G\, D_i}{\mu_l} \tag{3.4-53}$$

$$G = 4\, m/(\pi\, D_i^2) \tag{3.4-54}$$

$$\mathrm{Nu_D} = \frac{h_z\, D_i}{\lambda_l} = \frac{\mathrm{Re}_l^{0.9}\, \mathrm{Pr}_l}{F_2} \left[0.15 \left(\frac{1}{\chi} + \frac{2.85}{\chi^{0.476}} \right) \right]^{1,15} \tag{3.4-55}$$

$$F_2 = 0{,}707\, \mathrm{Pr}_l\, \mathrm{Re}_l^{0.5} \quad \mathrm{Re}_l < 50$$

$$F_2 = 5\, \mathrm{Pr}_l + 5\, \ln[\,1 + \mathrm{Pr}_l\,(0{,}09636\, \mathrm{Re}_l^{0.585} - 1\,)\,] \quad 50 < \mathrm{Re}_l < 1125 \tag{3.4-56}$$

$$F_2 = 5\, \mathrm{Pr}_l + 5\, \ln(\,1 + 5\, \mathrm{Pr}_l\,) + 2{,}5\, \ln(\,0{,}0031\, \mathrm{Re}_l^{0.812}\,) \quad \mathrm{Re}_l > 1125$$

$$\chi = \left(\frac{\mu_l}{\mu_g} \right)^{0,1} \left(\frac{1 - x}{x} \right)^{0,9} \left(\frac{\rho_g}{\rho_l} \right)^{0,5} \tag{3.4-57}$$

112

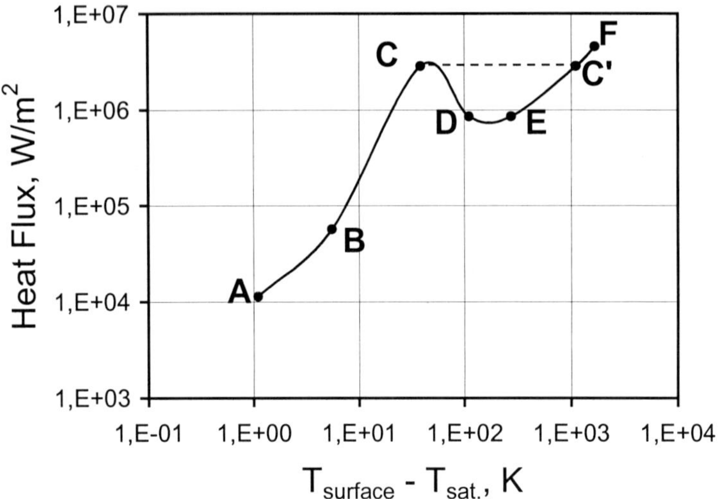

Figure 3.4-9. B–C: Nucleate boiling. D–E–F: Film boiling

3.4.3.3 Boiling

Pool boiling

The relationship between the temperature difference (T_0 - T_S) and the heat flow (q/A) necessary to maintain the boiling, follows the line shown in Fig. 3.4-9 for water heated at atmospheric pressure.

Nucleate pool boiling

The heat-transfer coefficients can be approximated by

$$h_{nb} = 0.18 \, P_c^{0.69} \, P_r^{0.17} \left(q \, / \, A \right)^{0.7} \tag{3.4-58}$$

For a horizontal tube-bank

$$h_{tb} = 1.5 \, h_{nb} + S \tag{3.4-59}$$

where S can be taken as 250 W/(m² K) for hydrocarbons and 1000 W/(m².K) for water.

Since these equations are valid for nucleate boiling, a critical heat-flux must be defined. It can be approximated as

$$(q/A)_{máx} = 253652 \, L \, D_{haz} \, P_c \, (P/P_c)^{0.35} \, (1 - P/P_c)^{0.9} / \, A \tag{3.4-60}$$

Forced convection

Inside a heated duct, the liquid vaporizes in several stages: sub-cooled boiling, saturated or bulk boiling, and the dry wall regime.

Eqn. (3.4-47) gives good results for vapour tittles between 0 and 0.71.

$$h_{TP} = h_l\,F + h_{nb}\,S \tag{3.4-61}$$

where

$$h_l = 0{,}023 \left[\frac{D\,G\,(1-x)}{\mu_l}\right]^{0,8} \left(\frac{c_l\,\mu_l}{\lambda_l}\right)^{0,4} \frac{\lambda_l}{D} \tag{3.4-62}$$

$$h_{bn} = 0{,}00122\,\frac{\lambda_l^{0,79}\,c_l^{0,45}\,\rho_l^{0,49}}{\sigma^{0,5}\,\mu_l^{0,29}\,i_{lg}^{0,24}\,\rho_g^{0,24}}\,\Delta T_S^{0,24}\,\Delta P_S^{0,75} \tag{3.4-63}$$

$$\Delta P_S = \frac{\Delta T_S\,i_{lg}}{T_S\,v_{lg}} \tag{3.4-64}$$

$$F = 1 \qquad\qquad 1/\chi > 0{,}1$$

$$F = 2{,}35 \left(\frac{1}{\chi} + 0{,}213\right)^{0,736} \quad 1/\chi < 0{,}1 \tag{3.4-65}$$

$$S = \left[\,1 + 0{,}12\,Re^{1,14}\right]^{-1} \quad Re < 32{,}5$$

$$S = \left[\,1 + 0{,}42\,Re^{0,78}\right]^{-1} \quad 32{,}5 < Re < 70 \tag{3.4-66}$$

$$S = 0{,}1 \qquad\qquad Re > 70$$

$$Re = \frac{D\,G\,(1-x)}{\mu_l}\,F^{1,25} \times 10^{-4} \tag{3.4-67}$$

3.4.3.4 Overall heat-transfer coefficient for exchangers

Table 3.4-2 summarizes typical values of the overall heat-transfer coefficient for heat exchangers and high-pressure gases.

Table 3.4-2.
Approximate overall heat-transfer coefficients for various types of heat exchangers

Type	Performance	U, W/(m^2 K)
shell and tube heat exchanger	gas (250 bar)–gas (250 bar)	150 to 500
	liquid–gas (250 bar)	200 to 400
Double pipe heat exchanger	gas (250 bar)–gas (250 bar)	150 to 500
	liquid–gas (250 bar)	200 to 600
Helical coil cooler	water–gas (250 bar)	150 to 500

114

3.4.4 Mass transfer mechanisms in dense fluids

3.4.4.1 External mass transfer

The interest in mass transfer in high-pressure systems is related to the extraction of a valuable solute with a compressed gas. This is either a volatile liquid or solid deposited within a porous matrix. The compressed fluid is usually a high-pressure gas, often a supercritical fluid, that is, a gas above its critical state. In this condition the gas density approaches a liquid--like value, so the solubility of the solute in the fluid can be substantially enhanced over its value at low pressure. The retention mechanism of the solute in the solid matrix is only physical (that is, unbound, as with the free moisture), or strongly bound to the solid by some kind of link (as with the so-called bound moisture). Crushed vegetable seeds, for example, have a fraction of free, unbound oil that is readily extracted by the gas, while the rest of the oil is strongly bound to cell walls and structures. This bound solute requires a larger effort to be transferred to the solvent phase.

We first deal with the external convective mass-transfer to a flowing dense gas. Later, we will focus attention on the diffusion of a solute within the solid. In any case, a packed bed of particles (wet or partially dried) is contacted with fresh dense gas, which passes through the bed. When the seed load has been stripped of solute, it is unloaded and a new load of crushed seeds is charged in the extractor. During the fast evaporation stage, the mass transfer rate is independent of the degree of crushing or the grain size.

The usual specific flow-rates for extraction are very small. In terms of space velocities, these are about 5 to 15 kg/h per litre of extractor volume, with superficial velocities in the range of 0.5 to 10 mm/s. With these small velocities, natural convection mass transfer is the favoured mechanism of transport. Gas densities are in the range of 500 to 800 kg/m^3, and viscosities are about 5 x 10^{-7} kg/(m s), thus giving kinematic viscosities of about 10^{-9} m^2/s, which is a very small value for a fluid. For example, the kinematic viscosity of water is 10^{-7} m^2/s and that of ambient air is 2 x 10^{-5} m^2/s. This makes free convection a principal mechanism for mass-transfer in high pressure gases.

For free convection of mass to be significant, the following ratio must be large [21]:

$$\frac{Gr_D}{\mathrm{Re}^2} = \frac{Buoyancy\,forces}{Inertial\,forces} >> 1 \tag{3.4-68}$$

where the Reynolds- and the Grashof numbers are as defined in Section 3.4.1.

In cases where $Gr >> \mathrm{Re}^2$, free convection currents set in, being responsible for the transport processes. In packed beds of seeds, the particle Reynolds number is less than 10 to 50. The Grashof number represents the squared Reynolds number for the velocity of the buoyancy flow [18]. Therefore, the ratio of eqn. (3.4-68) is a comparison of the convective flow owing to buoyancy to the flow owing to circulation in terms of their respective squared Reynolds numbers.

In Fig 3.4-10, a plot similar to the Metais and Eckert [21,22] map is presented for the external mass-transfer in shallow beds of sintered metallic pellets, as developed by Abaroudi [22]. In the map, some recent data have been represented for high pressure work. It is known experimentally that in the region where $Gr.Sc > 10^8$ and $Re < 10^2$, there are considerable effects of free convection on mass-transfer (near the turbulent free convection).

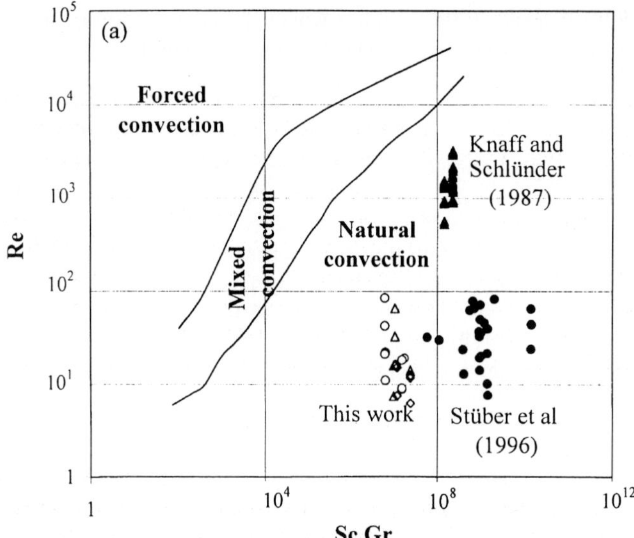

Fig. 3.4-10. Map for the external mass-transfer in shallow beds of sintered metallic pellets [23] with permission of American Chemical Society.

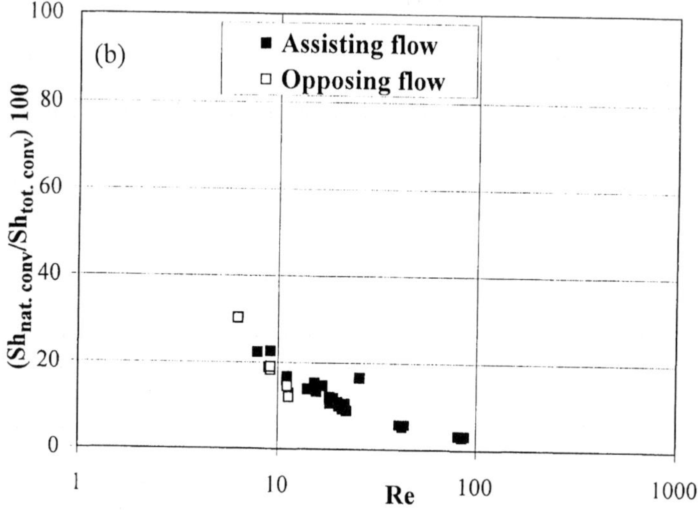

Fig. 3.4-11. Ranges of the Reynolds number where the contribution of the Sh_N over the total Sherwood number is significant [23] with permission of American Chemical Society.

116

Fig 3.4-11 shows the ranges of the Reynolds number where the contribution of the Sh_N over the total Sherwood number is significant [23]. The figures correspond to initial mass-transfer rates from small cylindrical pellets (8x8 mm approx.) impregnated with β-naphthol extracted by dense carbon dioxide.

Free liquid-to-gas mass-transfer in packed beds

The evaporation of a free liquid into a dense fluid is interesting because it represents the fast regime for the extraction of wet crushed seeds. Puiggené *et al.* [24] studied a model system consisting of di-chlorobenzene evaporation from non-porous glass beads. Also, Stüber *et al.* [25] did similar experiments with impregnated sintered (porous) pellets by observing the initial rates of evaporation in shallow beds. They correlated the total Sh number with the forced-
-convection and the natural convection components, since the two mechanisms were present. Their correlation is as follows:

$$Sh_N = Sh_0 + 0.001(GrSc)^{0.33} Sc^{0.244} \qquad (3.4\text{-}69\text{a})$$

$$Sh_0 = 0.3 \qquad (3.4\text{-}69\text{b})$$

$$Sh_F = 0.269 \, Re^{0.88} \, Sc^{0.3} \qquad (3.4\text{-}69\text{c})$$

for the ranges, $8 < Re < 90$, $3 \times 10^7 < Gr\,Sc < 10^9$, $1.5 < Sc < 10$ (Average Absolute Relative Deviation = 18.9%). The resulting total Sh number is obtained by simple addition:

$$Sh_T - Sh_0 = |Sh_F \pm (Sh_N - Sh_0)| \qquad (3.4\text{-}70)$$

where the plus sign is for assisting flows and the minus sign for opposing flows. The authors reasoned that there was not enough precision in the measurements to combine the third powers of Sh as suggested by Churchill [7].

The presence of free convection in extraction shows itself when upflow- and downflow extraction times differ widely (perhaps by 50% or more) for the same apparatus under otherwise similar conditions. For a solute heavier than the gas, convection is assisted by gravity, and therefore extraction is faster in downflow operation. For a solute lighter than the gas, a net buoyancy force lifts the solute over the solvent. In this case, extraction in upflow operation will be faster than in downflow operation. If free convection is strong enough (as in the case of a halogenated solvent in carbon dioxide) the required time for extraction of a bed can be reduced by 50% or more, by operating in assisting-flow mode [25]. Bennecke *et al.* [26] have discussed mixing effects in terms of flow-instability and buoyancy on tracer-gas injection in high-pressure gaseous systems, and how it affects the measured Péclet number. This will be discussed later, in connection with axial dispersion.

Finally, in Fig. 3.4-12 [24], a comparison is given for the overall, gas-based, mass transfer coefficient for several liquid-to-gas and solid-to-gas packed beds and column systems. In Fig. 3.4-12, for a given data point, the vertical distance up to the Tan *et al.* [27] correlation (which is for a solid-to-fluid boundary layer) would provide a measure of the liquid-side mass-transfer resistance associated with the liquid. This is so because amount of the large gas dissolved in the liquid at the high pressure. This markedly reduces the available overall mass-transfer coefficient. Also in Fig. 3.4-12, typical values of low-pressure mass transfer in columns and extractors are given for reference.

Mass-transport at a flat liquid to dense-gas interface

Zehnder and Trepp [23] studied the evaporation of α-tocopherol and two iso-phythols into a dense gas from a flat liquid plate (Re = 100 to 3000) and observed the surface during evaporation, with an optical fibre device. The regressed mass-transfer equation was as follows:

$$Sh \ x = 0.13 \, Re^c \, Sc^{0.63} \, x^{0.63} \left(\frac{d_h}{L} \right)^{0.33}$$ (3.4-71)

(P = 350 bar, T = 353K)

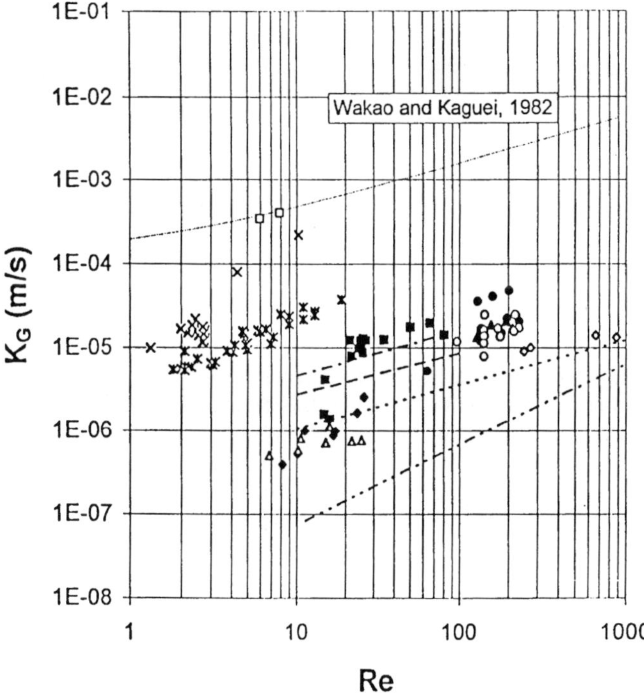

Fig. 3.4-12. Comparison of the overall, gas-based, mass-transfer coefficient for several liquid-to-gas and solid-to-gas packed beds and column systems [24] with permission of Pergamon Press
- Fixed Beds: (□) (◆) (*) (X) (._.) (---)
- SFE Columns: (●) (σ) (μ) (ν) (Γ)
- Low pressure columns: (...) (_.._..)

118

where,

Sh = Sherwood number, with effective diffusivity $D_e = D\,x$, with D evaluated from the Sassiat method [13]

$x =$ equilibrium solubility expressed in mass fraction (transferring solute)

Re = $u\,L/v$

$d_h =$ hydraulic diameter of the passage above liquid

$L =$ length of the passage over the plate

$c =$ exponent, $0.33 < c < 0.5$

Solid-to gas mass-transfer

Knaff and Schlünder [9] studied the evaporation of naphthalene and caffeine from a cylindrical surface (a sintered metallic rod impregnated with the solute) to high-pressure carbon dioxide flowing over an annular space around the rod. They studied the diffusion flux within the bar and in the boundary layer. The mass-transfer coefficient owing to forced convection from cylinder to the gas flowing in the annular duct was correlated, using the standard correlation due to Stephan [7]. For caffeine, it does not require a free-convection correction, as the Reynolds dependence is that expected by a transfer by forced convection. This is

$$Sh_F = Sh_\infty + \left[1 + 0.14\left(\frac{d_i}{d_o}\right)^{0.5}\right] \times \frac{0.19(\mathrm{Re}\ \mathrm{Sc}\ d_h/L)^{0.8}}{1 + 0.117(\mathrm{Re}\ \mathrm{Sc}\ d_h/L)^{0.467}} \tag{3.4-72a}$$

$$Sh_\infty = 3.66 + 1.2\left(d_i/d_o\right)^{-0.8} \tag{3.4-72b}$$

where d_h as the hydraulic diameter of the annular duct and L its length. In the case of naphthalene, the Sherwood numbers for both free- and forced convection are necessary, as the measured Sherwood number is less sensitive to velocity than for caffeine. This is consistent with a larger Gr number for naphthalene than for caffeine. In the naphthalene, Sh numbers were combined using the Churchill equation [8], which is not specific for supercritical fluids.

For the extraction of β-naphthol in packed beds, Tan and co-workers propose a simpler correlation [27], which is useful when the free and forced interactions can be disregarded, or over a narrower range. This is:

$$Sh = 0.38\,\mathrm{Re}^{0.83}\,Sc^{1/3} \tag{3.4-73}$$

with a specific mass-transfer area, $a = 6(1-\varepsilon)/d_p.$

For packed beds of naphthalene and caffeine, Lim et al. [28] took into consideration mass--transfer by both types of convection in opposite flows. Their equation for laminar flow in packed beds (10 to 203 bar and 35 to 45°C) is as follows:

$$\frac{Sh}{(ScGr)^{1/4}} = 0.1813 \left(\frac{\mathrm{Re}^2\,Sc^{1/3}}{Gr}\right)^{1/4} \left(\mathrm{Re}^{1/2}\,Sc^{1/3}\right)^{3/4}$$

$$+1.2149 \left| \left(\frac{\mathrm{Re}^2\,Sc^{1/3}}{Gr}\right)^{3/4} - 0.01649 \right|^{1/3}$$

(3.4-74)

where the Sh appearing in the left-hand side is the total Sherwood number. For a rapid estimate it is better to use the Tan, Liang and Liou equation (eqn. 3.4-71). The equation of Lim *et al.* has recently been updated by Lee and Holder [29] as:

$$\frac{Sh}{(ScGr)^{1/4}} = 0.5265 \frac{\left(\mathrm{Re}^{1/2}\,Sc^{1/3}\right)^{1.6808}}{(ScGr)^{1/4}} +$$

(3.4-75)

$$+2.4800 \left| \left(\frac{\mathrm{Re}^2\,Sc^{1/3}}{Gr}\right)^{0.6439} - 0.8768 \right|^{1/0.6439}$$

over a wide range of Reynolds number 0.3<Re<135, thus including the free- and forced--convection types of flow. In packed beds, the transition of laminar to turbulent flow is around Re = 90 to 120 [28], and therefore the Lee and Holder equation would cover the two regimes.

Axial dispersion in packed beds

When a solute is transferred from a solid into a high-pressure gas, it is then taken downstream in the bulk fluid by convective transport. Depending on turbulence, the solute may travel further by other mass-transport mechanisms such as dispersion. Dispersion spreads the solute axially and radially in a cylindrical stet. Eaton and Akgerman [30] considered both axial and radial effects in a model for the desorption of heavy organics, from carbon, by a dense gas.

It should be pointed out that for a low pressure gas the radial- and axial diffusion coefficients are about the same at low Reynolds numbers (Re<1). Therefore, the two diffusion effects may be important at velocities where the dispersion effects are controlled by molecular diffusion. For Re = 1 to 20, however, the axial diffusivity becomes about five times larger than the radial diffusivity [31]. Therefore, the radial diffusion flux could be neglected relative to the longitudinal flux. If these phenomena were also present in a high-pressure gas, it would be true that radial diffusion could be neglected. In dense- gas extraction, packed beds are operated at Re > 10, so it will be supposed that the Péclet number for axial dispersion only is important ($P\acute{e}_{ax} \ll P\acute{e}_r$).

Neglecting radial effects, the dispersion equation for dispersed tracer with bulk flow in a packed bed is:

120

$$D_{ax}\varepsilon\frac{\partial^2 C}{\partial z^2} - u_0\frac{\partial C}{\partial z} = \varepsilon\frac{\partial C}{\partial t} \tag{3.4-76}$$

where, D_{ax} = axial dispersion coefficient, m^2/s
ε = bed void fraction
C = concentration, kg/m^3
z = axial coordinate, m
u_0 = superficial velocity, m/s

Eqn. (3.4-76) has been solved for a number of initial- and boundary conditions using a variety of techniques [32]. Application to dispersion problems provides information on the axial Péclet number, defined as:

$$P\acute{e} = \frac{u_0 d_p}{D_{ax}\varepsilon} = \frac{Bulk \quad fluid \quad transport}{Dispersion \quad transport} \tag{3.4-77}$$

thus representing a measure of the ratio of the transport owing to bulk flow to the transport owing to diffusion.

Tan and Liou [33] correlated the Péclet number with the product Re·Sc for axial diffusion of methane in high-pressure carbon dioxide in a packed bed of glass spheres. The results are as follows:

$$P\acute{e} = \frac{u_0 d_p}{D_{ax}\varepsilon} = 1.63\,\mathrm{Re}^{0.265}\,Sc^{-0.919} \tag{3.4-78}$$

with an error of about 21%. But the authors found the following regression:

$$D_{ax}\,\varepsilon = 0.085 u^{0.914} d_p^{0.388} \rho_r^{0.725} \mu_r^{0.676} \tag{3.4-79}$$

where the subscript, r, means, "relative to the critical point". The correlation given by eqn. (3.4-79) had a better error performance (8.5%) than eqn. (3.4-68) in fitting the data over the range of the variables ($P < 140$ bar, $T < 328$K, particle-size up to 2 mm diameter). The range of Péclet numbers measured experimentally was from 0.6 to 2.7, which is between those normal for gases and liquids, respectively.

For small particles (<1 mm) diameter, Catchpole et al. [32] developed the following correlation:

$$\frac{1}{P\acute{e}} = \frac{0.018}{\mathrm{Re}} + \frac{10}{1+\dfrac{0.7}{\mathrm{Re}}} \tag{3.4-80}$$

For diameters >1 mm, the usual correlation equations describe dispersion in dense gases, giving values less scattered around Pé = 0.5 to 1, for the supercritical fluid. Catchpole and his co-workers did not find differences in Péclet numbers when up-flow- or down-flow data were compared.

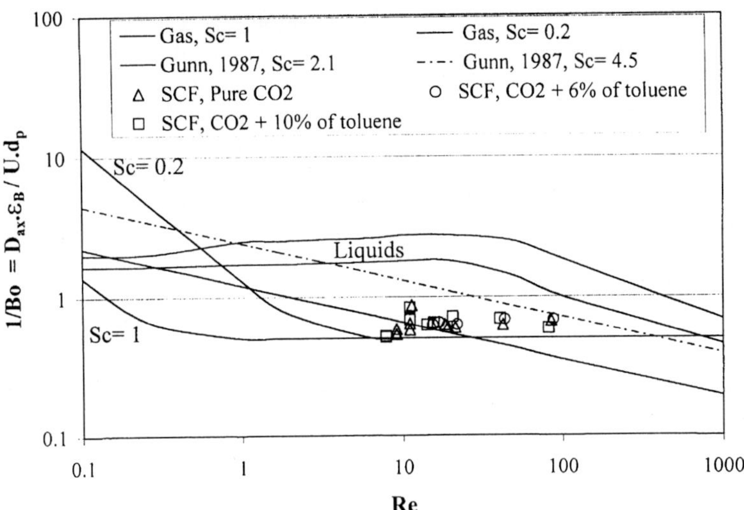

Fig. 3.4-13. Experimental measurements over beds of relatively large particles, after Abaroudi *et al.* [22] with permission of American Chemical Society.

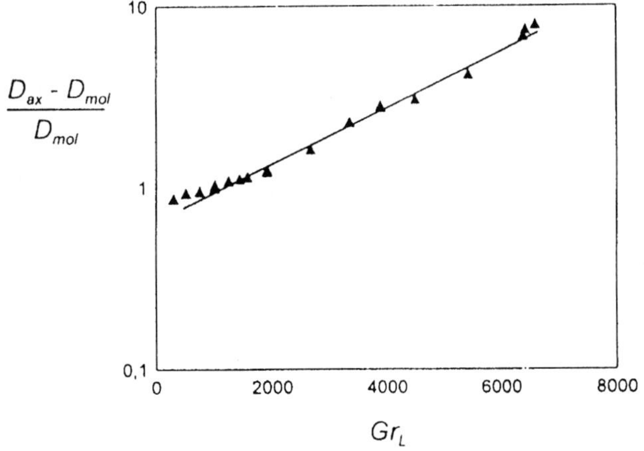

Fig. 3.4-14. Increase of the axial diffusivity as a function of the Grashof number from [26] with permission of the American Institute of Chemical Engineers.

In Fig. 3.4-13, given by Abaroudi *et al.* [22], the measurements over beds of relatively large particles (8 mm cylinders, 5 particles thick) can be observed, including up-flow- and down-flow extractions with dense gas (carbon dioxide, neat and modified with up to 10% toluene). The data points fall nearly on the gap between gas and liquid, near Gunn's correlation [32].

The role of buoyancy on axial dispersion in a high-pressure gas was investigated by Westerterp's group [26] in tracer injection experiments in gas- flows at different densities (at intermediate- to low- pressures, over a range of 17 bar). They define an enhancement factor as the ratio of the axial dispersion coefficient in the presence of buoyant forces and that in their absence. This factor would represent the increase in longitudinal mixing owing to free convection in the case of assisting flow. At low Reynolds numbers, the enhancement factor depends only on the Grashof and Schmidt numbers, since Gr and Sc determine the fluid dynamic stability of the system. At higher Re, the flow pattern depends on the ratio given by eqn. (3.4-68), and therefore, $D_{ax}/(D_{ax})_0 = f(\mathrm{Gr/Re}^2)$, since this variable represents the ratio of the buoyancy forces to the inertial forces. A simple expression for a variable satisfying these two conditions is given in ref. [26]:

$$X = \frac{\mathrm{Gr}}{\left(K/\mathrm{Sc} + \mathrm{Re}^2\right)},$$

(3.4-81)

with K being an scale factor. Hence, it should be possible to correlate the enhancement factor with X. The preliminary plot for the measured enhancement factors are given in ref. [26] as a function of the variable involving Gr, Re and Sc. A result of the theory is given in Fig. 3.4-14 where the increase of the axial diffusivity (relative to molecular diffusivity) is plotted as a function of the modified Grashof number, Gr_L. For large values of the latter, the measured axial diffusion was about ten times larger than in the absence of buoyancy. This situation occurs when the density increases with height, so the axial dispersion is driven by the buoyancy force and by unstable density variations, hence providing an increased axial diffusivity. When the density reduces with the downstream length, axial dispersion is reduced.

Mass-transfer coefficients are also affected by the buoyancy forces in the high Ra = Gr Sc region, as shown in practice by the effect of gravity on the extraction-time for larger Raleigh numbers (Ra = Gr Sc = 10^8), as shown by Stüber *et al.* [25].

Rules-of-thumb to ensure plug-flow conditions in packed beds

In practical cases, a rule-of-thumb is needed to know whether dispersion must be taken into account. One can use the equivalence between the mixing caused by a number of perfect mixers in series, and the degree of axial dispersion that produces the same degree of dispersion. We can use the following equation:

$$N = \frac{1}{2}\frac{u_0 d_p}{D_{ax}\varepsilon}\left(\frac{L\varepsilon}{d_p}\right) \approx 100$$

(3.4-82)

which is true only for a large number of perfect mixers [34].

The Pé is given in Fig. 3.4-13 as a function of ReSc. As is seen, Pé is not far from unity. Assuming that plug-flow is attained for $N \sim 100$ mixers, the Eqn. (3.4-82) provides a value for the necessary ratio, L/d_p. This should be around 200. This is about the same criterion as for liquid flow as expected from Fig. 3.4-13. In other words, a packed bed should be at least 200 diameters long. For a seed of, say, 5 mm size, the bed should be 1m long. Under these design conditions, the equilibrium solubility is possibly approached at the bed exit [35] for the usual linear velocities employed.

3.4.4.2 Internal mass transfer

In Section 3.4.1 we introduced Fick's Law to describe the molecular diffusion flux in a fluid. We are interested here in the flux of a solute diffusing in a porous medium immersed in a high-pressure gas to a dense solvent. Many extractions are governed by the diffusion rate in a porous matrix. Among them are: the high-pressure reaction carried out on a porous catalyst; the extraction of vegetable oil from seeds, during the stage where diffusion in the solid controls the rate; the clean-up of porous solid pellets with high-pressure carbon dioxide; and many catalytic reactions (hydrogenations) with supercritical fluid as solvent. Our goal is to define the catalyst size and shape to avoid internal mass-transfer resistances, as these affect production yields and selectivity.

The rate of diffusion for non-flow conditions of the fluid along the radial direction of a porous sphere is analogous to Fick's law, and is as follows:

$$J_{Ar} = -D_e \frac{\partial C_A}{\partial r} \tag{3.4-83}$$

where r is the radial position from the centre and J_{Ar} is the flux based on the *unit surface of an effective, or pseudo-homogeneous, medium*, whose porosity is the overall medium void-fraction, ε_p, and D_e is the *effective* diffusivity. Physically, the net radial diffusion of a mol of solute will not occur along a straight radial direction from the centre of the sphere to the outer surface, but will follow a complicated path through the pore network, along a tortuous path, until the solute reaches the external surface. Therefore, the expected relationship between the molecular- and the intra-particle, or effective, porosities is usually written as:

$$D_e = \frac{\varepsilon_p D}{\tau} \tag{3.4-84}$$

where τ represents an empirical tortuosity factor to account for the extended path relative to a straight radius. Various authors recommend a value for $\tau \sim 4$ [36, 37]. Only a few authors have measured τ directly in high-pressure, high-density systems [1,22,38]. On the other hand, this has been measured for many catalysts under low-pressure and medium-pressure conditions [39], and is summarized in Table 3.4-3.

If data are available on the catalyst pore- structure, a geometrical model can be applied to calculate the effective diffusivity and the tortuosity factor. Wakao and Smith [36] applied a successful model to calculate the effective diffusivity using the concept of the random pore model. According to this, they established that:

$$D_e = D_M \varepsilon_M^2 + \frac{\varepsilon_\mu^2 (1 + 3\varepsilon_M) D_\mu}{1 - \varepsilon_M} \tag{3.4-85}$$

Table 3.4-3
Tortuosity Factors for Diffusion in Catalysts at 65 atm, as reported by Butt [39]

Catalyst	Internal porosity	Tortuosity factor
Harshaw MeOH synthesis catalyst (pre-reduced)	0.49	6.9
Haldor-Topsøe MeOH synthesis catalyst (pre-reduced)	0.43	3.3
BASF MeOH synthesis catalyst (pre-reduced)	0.50	7.5
Girdler G-52 cat., 33%Ni on refractory oxide support	0.44	4.5
Girdler G-58 catalyst Pd on alumina (6 atm)	0.39	2.8

where D_M and D_μ are the diffusivities in the macropore- and the micropore- regions, respectively. These can be calculated by the following combination of diffusion coefficients,

$$D = \frac{1}{1/D_{AB} + 1/D_K} \qquad (3.4\text{-}86)$$

that applies both in the macropore region and micropore region, to get D_M and D_μ. Here, D_k is the Knudsen diffusivity, but this is not likely to be required in high-pressure work [1].

In order to apply the Wakao and Smith model, the solid porosity and the diffusion coefficients in the two regions are required. Note that Eqn. (3.4-83) does not involve any adjustable parameter, so only structural and molecular transport data are needed.

Most often, only diffusive calculations in the macropore region are necessary for certain large-pore systems. Then, by applying Eqn. (3.4-83) to the macropore region we have:

$$D_e = \varepsilon_p^2 D \qquad (3.4\text{-}87)$$

This, compared with eqn. (3.4-70), provides a value of the empirical tortuosity factor as:

$$\tau = \frac{1}{\varepsilon_p} \qquad (3.4\text{-}88)$$

For a porosity of 35%, τ is close to 3 - 4, the value suggested by Satterfield. On the other hand, for a material with only micropores (silicas, zeolites) a similar equation is obtained in terms of the micropore characteristics, as in eqn. (3.4-75).

Effective diffusion in liquids and dense gases

The role of Knudsen diffusion in dense media seems to be negligible. However, since molecules are closer to each other in a dense medium (liquid or compressed gas), diffusivities are difficult to predict [36]. Direct measurements of D_e values in dense gas have been obtained in situations where the intra-particle rate was controlling. In such cases, the tortuosity factors obtained are very small indeed, as reported by some of the values in Table 3.4-4. Knaff and Schlünder were the first to note an abnormally low ratio of D/D_e (~ 1.5), that was also variable with temperature.

Tortuosity factors less than unity have often been reported for three-phase catalytic reactors where a reactant gas diffuses in a liquid- (or dense) medium. Many examples are available. A discussion of the possible causes of the abnormal tortuosity factors is given by Stüber *et al.*, [38]. These are related to the possible existence of co-operative surface diffusion with bulk pore fluid diffusion. Surface diffusion is not a fully understood phenomenon. It is known, however, that according to one theory [1,36] the resulting effective diffusivity can be written in the following form:

$$D_e = \frac{D\ \varepsilon_p + \rho_p K_A D_s}{\tau} \tag{3.4-89}$$

where K_A is the adsorption equilibrium constant on the solid from the high- pressure fluid, and D_s is the surface-diffusion coefficient on the pore walls. Since these quantities are seldom available, an average or apparent tortuosity factor is employed (see the values in Table 3.4-4. Eqn. (3.4-89) indicated that a change in temperature may largely affect K_A, which may lower the effective diffusivity with increasing temperature, thus showing a reverse- temperature effect.

Table 3.4-4
Tortuosity Factors for High Pressure Extraction of Porous Solids (39)

Authors	Conditions	Solid matrix	Tortuosity factor
Knaff and Schlünder [9]	T=35 to 55°C P= 12-22.6 MPa	Bronze, ε_p=0.3 Pore size 8 to 20 µm	τ independent of d_p and P, τ = 0.4 to 0.54
Recasens et al [41]	Desorption EtAc P <13 MPa, T=300 to 338K (a)	Regeneration of activated C	Linear driving force model $\tau \sim 4$
Madras *et al.* [42]	Naphthalene adsorption, P= 8 to 31 MPa, T=298 to 318 K	Micro-macroporous alumina	τ const. w/P and T τ = 3.49 (av)
Lai & Tan [1]	Toluene adsorption	Micro- and macroporous AC ε_p = 0.45	τ depends on pressure $\tau \sim 0.2$ to 0.6
Stüber *et al.* [38]	diCl-benzene extraction	Bronze, macropores ε_p = 0.20 to 0.25 pore size ~15 µm	τ const. w/P but varies w/T $\tau \sim 0.22$ to 0.62
Lee and Holder [29]	Adsorption of toluene and naphthalene	Silica gel (micro- and macropores)	τ = 3.49 (av)

(a) By modelling Tan and Liou data [33]

3.4.5 Mass-transfer models

Mathematical models derived from mass-conservation equations under unsteady-state conditions allow the calculation of the extracted mass at different bed locations, as a function of time. Semi-batch operation for the high-pressure gas is usually employed, so a fixed bed of solids is bathed with a flow of fluid. Mass-transfer models allow one to predict the effects of the following variables: fluid velocity, pressure, temperature, gravity, particle size, degree of crushing, and bed-length. Therefore, they are extremely useful in simulation and design.

We first give a rather general mass-transfer model, which is useful for most processes of porous-solid extraction with dense gases. Two cases are possible [43] for a single particle loaded with solute. In (a), the solute is adsorbed over the internal surface of the particle, and is desorbed from the sites and diffuses out to the external surface. (b) The solute fills in the pore-cavities completely, and is dissolved from an inner core that moves progressively to the centre of the particle.

Desorption with mass transfer

The rate of desorption can be described by a Langmuir equation, so that the adsorption rate in the general case will be non-linear. We will have:

$$r_d = k_d C_a - k_a C_i (C_m - C_a) \tag{3.4-90}$$

where C_m is the surface concentration for a monolayer coverage, in kg/kg, C_a is the actual occupied coverage, in kg/kg, and C_i is the concentration of solute in the pore volume, in kg/m^3. Here, k_a and k_d are the adsorption- and desorption rate constants, respectively, in consistent units. The conservation equations for the solute in the particle volume and in the flowing fluid are as follows:

$$\varepsilon_p \frac{\partial C_i}{\partial t} = \frac{D_e}{r^2} \left(r^2 \frac{\partial C_i}{\partial r} \right) + \rho_p \, r_d \tag{3.4-91}$$

$$\varepsilon \frac{\partial C}{\partial t} + u_0 \frac{\partial C}{\partial z} = D_{ax} \varepsilon \frac{\partial^2 C}{\partial z^2} + \frac{k_m 3(1-\varepsilon)}{R} \left[C_i(R) - C \right] \tag{3.4-92}$$

With the boundary and initial conditions:

$$-D_e \frac{\partial C_i}{\partial r} \bigg|_R = k_m \left[C_i(R) - C \right] \tag{3.4-93}$$

$$\frac{\partial C}{\partial r} \bigg|_{r=0} = 0 \tag{3.4-94}$$

$$C_i(r, z, t = 0) = C(r, z, t = 0) = C(r, z = 0, t) = 0 \tag{3.4-95}$$

$$C_a(r, z, t = 0) = C_{a0} \tag{3.4-96}$$

where C is the concentration of solute in the fluid, in kg/m^3; Z is the axial bed coordinate; and R is the particle radius, C_i is the pore-fluid concentration, and C_a is the adsorbed solute concentration. In eqn. (3.4-91) it has been assumed that the particle is spherical. Also in eqn. (3.4-92) it has been assumed that there is no radial dispersion and that a uniform velocity profile is present [44]. In any situation, for the general case where the above equations require numerical solution, this may be rather complex and time-consuming, as was shown by Eaton and Akgerman [30] - even for the case of local equilibrium. These authors also showed that not all the steps in the desorption process are significant, while the rates of some steps can be estimated from correlations, without any parameter fitting. In this regard, Tan and Liou [45] overlooked both intra-particle resistance and the reverse-adsorption rate in the previous Langmuir equation. By letting:

$$-\frac{\partial C_a}{\partial t} = k_d C_a \qquad (3.4\text{-}97)$$

an analytical solution is obtained in terms of a single-parameter desorption process constant k_d. This parameter can be correlated with the temperature and fluid density [45].

Two-parameter desorption models

A number of authors [46 to 48] employ the single sphere model in which the packed bed is considered as a set of equal spheres that are under the same state of extraction, and the fluid flowing around them is solute-free. That is, equation (3.4-90) would be valid, but without the generation term [46]. The transport at the solid–fluid interface obeys the boundary condition (Eqn. 3.4-94) with $C = 0$ (fluid-flows at a large velocity). Under these assumptions, there is an analytical solution to the above problem (without axial dispersion) in terms of the Biot number (Bi $= k_m R/D_e$), included in the following equation:

$$\beta_k \cot \beta_k = 1 - Bi \qquad (3.4\text{-}99)$$

as obtained by Reverchon *et al.* [47] and King and Bott [49].

To avoid the difficulties associated with the spherical diffusion equation, a useful hypothesis is the linear-driving-force concept. This arises when a parabolic concentration profile within the spherical particles is supposed – which is a good approximation in cases where there is a Thiele modulus of a maximum volume of 2–5 (that is, with some intra--particle resistance [50]). In these conditions, the volume-averaged intra-particle concentration is defined as:

$$\overline{C}_i(z,t) = \frac{3}{\pi R^3} \int_0^R C_i(r,z,t)\pi r^2 dr \qquad (3.4\text{-}100)$$

so that eqn. (3.4-91) becomes linear [41]. Using a parabolic profile in eqn. (3.4-100), the overall mass-transfer coefficient is obtained as a sum of two resistances in series,

$$\frac{1}{k_p} = \frac{1}{k_m} + \frac{R}{5D_e} \qquad (3.4\text{-}101)$$

128

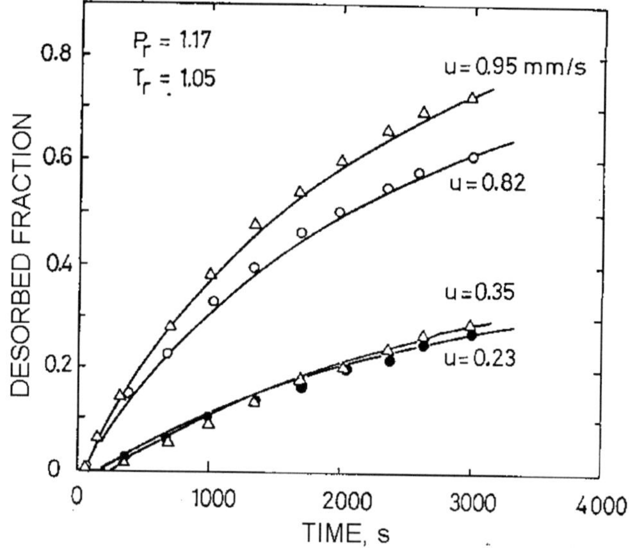

Fig. 3.4-15. Desorption rate for different fluid velocities

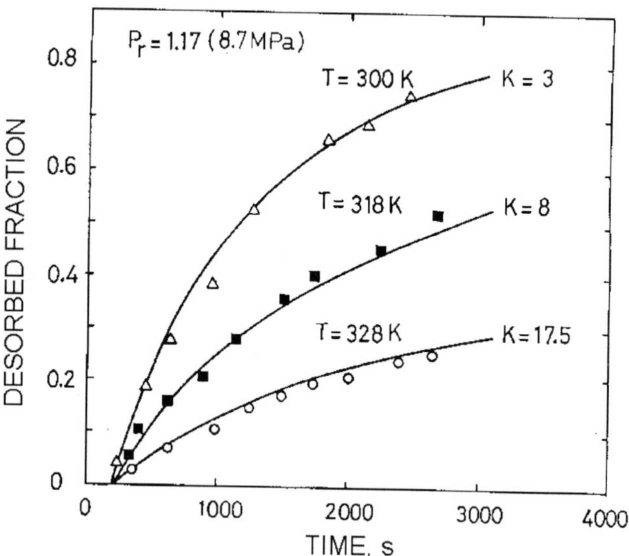

Fig. 3.4-16. Desorption rate at different temperatures

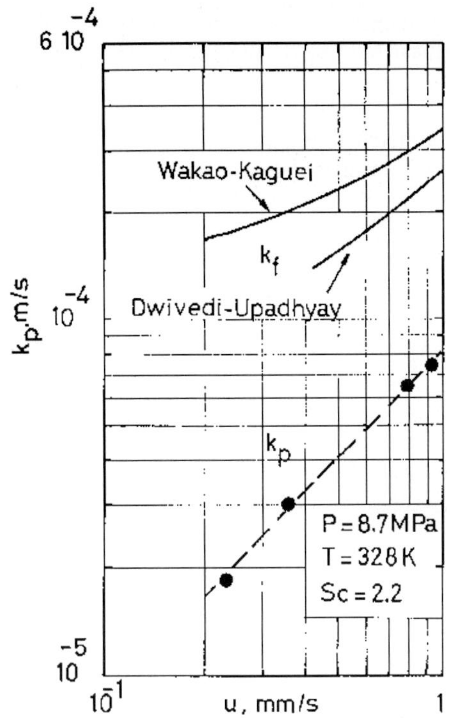

Fig. 3.4-17. Typical values of k_p obtained by fitting the data of Tan and Liou [33] and values estimated from two different correlations.

where the second term represents internal diffusion. With these conditions, the system of equations given before have analytical solutions for cases where the processes are linear [41].

This happens in two cases – when the equilibrium between solid and fluid is linear during desorption, and when there is first-order, irreversible desorption.

In the cases above, a two-parameter model well represents the data. A model with more parameters would be more flexible, but by using a partition constant, K, or a desorption rate constant k_d and k_m, for the mass-transfer coefficients, the data are well described (see Figs. 3.4-15 and 3.4-13). While K would be a value experimentally determined, k_p can be estimated from eqn. (3.4-97) with the external mass-transfer coefficient, k_m, estimated from the correlation of Stüber et al. [25] or from that of Tan et al. [27], and the effective diffusivity from the Wakao–Smith model [36]. Typical values of k_p obtained by fitting the data of Tan and Liou are shown in Fig. 3.4-16. As expected, they are below the usual mass-transfer correlations, because internal resistance diminishes the global mass transfer coefficient. These data correspond to the regeneration of spent activated carbon loaded with ethyl acetate, using high--pressure carbon dioxide, published by Tan and Liou [45].

130

Goto *et al.* [50], also employed the linear-driving-force model for describing the high pressure delignification of wood with supercritical tert.-butyl alcohol, proving that it is a reasonable approximation.

Extraction of solids with high-pressure fluid in mixed-flow
The following treatment applies to the case where the solids are stationary in a shallow packed bed, so that they can be considered to be in well-mixed conditions, and that the solute initially saturates the solid, as in the case of vegetable oil in crushed seeds. For the quasi--steady-state approximation, Brunner [51] derived a practical equation:

$$\frac{dm}{dt} = \dot{m} = m_s \frac{dC_s}{dt} \tag{3.4-102}$$

where m is the total mass present, m_s is the mass of inert solid and C_s is the concentration of solute on the solid. The diffusion rates in the external boundary layer and within the solid matrix are written in terms of transfer coefficients, as:

$$\dot{m} = k_m a(C_M - C) \tag{3.4-103a}$$

$$\dot{m} = k_s a(C_0 - C) \tag{3.4-103b}$$

with a overall mass-transfer coefficient as:

$$\frac{1}{k} = \frac{1}{k_s} + \frac{1}{k_m} \tag{3.4-104}$$

The integration of eqn. (3.4-102) provides:

$$\frac{C_M - C}{C_0 - C} = \exp\left(-\frac{ka}{m_s C_0} t\right) \tag{3.4-105}$$

For the external coefficient, correlations and numerical approximations are available. However, for the internal rate coefficient the experience in using the model can probably dictate some values for the coefficient k_s [51].

Extraction of crushed seeds with fluid in plug flow: the Sovová model
The Sovová model makes use of the rate equations used by Brunner and Lack. During the first stages of the extraction, there is a rapid evaporation, then evaporation slows down. During the extraction, a fraction of the packed bed is still wet, while the rest of the bed is in the dry state. The model of Sovová has three parameters: the two rate-parameters of eqns. (3.4-103a) and (3.4-103b), and x_k the critical extraction yield at which the rate changes from one regime to the other. The basic Sovová model is written in terms of the following equations:

$$-\rho_s(1-\varepsilon)\frac{\partial x}{\partial t} = J(x,y)$$

(3.4-106a)

$$\rho u_0 \frac{\partial y}{\partial z} = J(x,y)$$

(3.4-106b)

with the boundary and initial conditions as follows:

$$x(z, t = 0) = x_0$$

(3.4-107a)

$$y(z = 0, t) = 0$$

(3.4-107b)

The expressions for the mass-transfer rate in the two periods are written in terms of the Brunner coefficients, as:

$$J(x > x_k, y) = k_m a\rho(y_r - y)$$

(3.4-108a)

$$J(x \le x_k, y) = k_s a\rho x\left(1 - \frac{y}{y_r}\right)$$

(3.4-108b)

Analytical solutions for x and y as functions of the bed-length, z, and time, t, are available [45,52]. The expressions are a useful extension of two-phase model applied to plug-flow. These two models are appropriate in describing the extraction of crushed or broken seeds to recover the seed oil, either in shallow beds or in plug flow. As shown by Sovová [52], applying the plug-flow model requires corrections for non-ideal residence-time distribution (non-plug flow) of the fluid in contact with the solid.

The shrinking-core model

The shrinking-core model (SCM) is used in some cases to describe the kinetics of solid and semi-solids-extraction with a supercritical fluid [22,49,53] despite the facts that the seed geometry may be quite irregular, and that internal walls may strongly affect the diffusion. As will be seen with the SCM, the extraction depends on a few parameters. For plug-flow, the transport parameters are the solid-to-fluid mass-transfer coefficient and the intra-particle diffusivity. A third parameter appears when disperse-plug-flow is considered [39,53].

The conservation equations for the SCM are eqn (3.4-92) with $C(r = R)$ at the pore-mouths being slowly variable with time, and with no axial dispersion effects. Then eqn. (3.4-91) (with $r_d = 0$) is also valid for $r_c < r < R$. Actually, equilibrium desorption is assumed, so that at $r = r_c$, we have $Ci = C^*$ and $r_c (t = 0) = R$. Eqns. (3.4-95) and (3.4-96) are also valid. For spherical particles the relationship between the residual loading and the radius of the remaining core is (see Fig. 3.4-18):

$$\frac{C_a}{C_{a0}} = \left(\frac{r_c}{R}\right)^3$$

(3.4-109)

132

King and Catchpole [49] developed a quasi-steady-state solution to the shrinking core extraction model for the case of plug-flow. The near-steady state is obtained if the change in r_c during a residence time is small, and changes from point to point in the extractor at any given time are also small. Then, an analytical expression relating the radius of the core and the extraction time is obtained at $z = L$. That expression is implicit, so it is best worked out numerically. The concentration of the fluid-phase concentration at the bed exit is then obtained, so the breakthrough curve can be calculated.

Goto *et al.* [53] suggest the use of the stoichiometric time to calculate the minimum extraction time for a rapid estimate. This is:

$$t^* = \frac{(1-\varepsilon)LC_{a0}}{u_0 C^*} \tag{3.4-110}$$

where t^* is the time necessary for extraction, in seconds; L is the bed-length, in m; C_{a0} is the initial load of solute, based on particle volume, in kg/m^3; C^* is the solubility of solute in fluid, in kg/m^3. For vegetable oils, the solubility can be estimated by the correlation suggested by King and Catchpole.

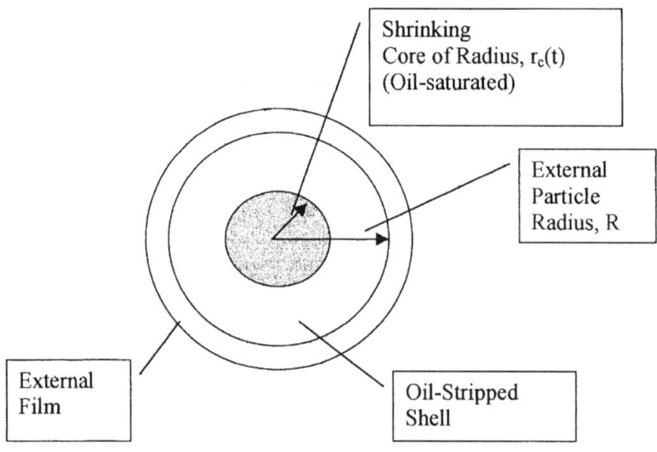

Fig. 3.4-18. Shrinking core for a spherical particle

3.4.6 Numerical examples

Example 1

Give an estimate for the viscosity of toluene vapour at 372°C and 27 atm, using the Uyehara–Watson generalized graph. Critical data for toluene are T_c = 593K and P_c = 41.6 atm.

The reduced pressure and temperature necessary for use in Fig 3.4-2 are:

$$P_r = \frac{P}{P_c} = \frac{27}{41.6} = 0.649$$

$$T_r = \frac{T}{T_c} = \frac{372+273}{593} = 1.09$$

The molecular weight of toluene is 93 kg/kmol, so the critical viscosity estimated from eqn (3.4-14) will be,

$$\mu_c = 7.70 \times \frac{\sqrt{93}}{(593)^{1/6}} (41.6)^{2/3} = 300.5\,\mu P$$

which is close to the given in the Tables, with an error of about 2% (306 μP). With the values of the reduced pressure and temperature (given above) we get from the graph, μ_r ~0.85. So the estimated viscosity will be 255.4 μP.

Example 2

Estimate the thermal conductivity of nitrous oxide (N_2O) at 105 °C and 138 bar. The experimental value at these conditions is 3,9 x 10^{-2} W/(m K) as given by Reid *et al.* [13]. The value of the low pressure thermal conductivity is λ_0 = 2.34 x 10^{-2} W /(m K). The critical and other pertinent data are given as,

T_c = 309,6 K
P_c=72,4 bar
V_c= 97,4 cm^3/mol
Z_c= 0.274
M = 44,01 g/mol

Under the conditions of interest, the reduced values are:

T_r = (105+273)/309.6= 1,22
P_r = 138/72,4=1,90

Interpolation in the corresponding states-compressibility factor gives Z = 0.635. The value of Γ is given by:

$$\Gamma = 210 \left(\frac{T_c M^3}{P_c^4} \right)^{1/6} = 210 \left(\frac{309.6 \times 44^3}{72.4^4} \right)^{1/6} = 209$$

Then the molecular volume and the reduced density required are calculated as:

$$V = \frac{ZRT}{P} = \frac{0.635 \times 8.314 \times 378}{138 \times 10^5} = 144 \times 10^{-6} \, m^3 \, / \, mol$$

$$\rho_r = \frac{V_c}{V} = \frac{97.4}{144} = 0.68$$

We will use eqn. (3.4-25b) to estimate the excess thermal conductivity. We will have:

$$\lambda - \lambda_0 = \frac{1}{\Gamma Z_c^5} \times 1.14 \times 10^{-2} \left[\exp(0.67 \times 0.68) - 1.069 \right] = 1.78 \times 10^{-2}$$

Since λ_0 is given, we will have finally:

$$\lambda = \lambda_0 + 1{,}78 \times 10^{-2} = 2.34 \times 10^{-2} + 1{,}78 \times 10^{-2}$$
$$= 4.12 \times 10^{-2} \, W \, / (mK)$$

Example 3

A cylindrical extractor, 1-m long, is filled with crushed-vegetable-oil seeds. The oil is to be extracted with pumping supercritical carbon dioxide at a density of 500 kg/m^3 through the packed bed. The estimated solubility of the oil in the dense gas at this density is 3.425 kg/m^3. The superficial velocity of the carbon dioxide in the bed will be 1 mm/s. This fluid velocity is sufficiently small for the fluid to become saturated with oil. We are required to estimate the minimum time of operation for complete extraction of the oil from the bed. The initial oil fraction is 12% (wt/wt) based on wet seeds, the void fraction of the bed is 40%, and the density of the particles is 900 kg/m^3.

The minimum time is that required for a stoichiometric amount of fluid to become saturated with oil:

$$t^{sat} = \frac{(1 - \varepsilon_B)LC_{a0}}{u_0 C^*} = \frac{(1 - 0.4) \times 1 \times 0.12 \times 900}{3600 \times 1 \times 10^{-3} \times 3.425} = 5.26h$$

Since this time does not consider the length of the mass-transfer-zone or unsteady-state periods, the real will longer – maybe 10–20 % longer.

Example 4

Assume a condenser built with a single tube if 12.5 mm I.D., which is fed with 150 kg/h of R-22 refrigerant (Du Pont). Calculate the tube length needed, assuming a wall at a constant temperature of 24.4 °C and a saturation temperature of 30°C. The fluid properties are shown in the Table 3.4-5.

Table 3.4-5
Data for example 4

μ_l (Pa.s)	μ_g (Pa.s)	λ_l (W/(K.m))	c (J/(kg.K))	i_{lg} (J/kg)	ρ_l (kg/m³)	ρ_g (kg/m³)	Pr_l
2.303×10^{-4}	1.33×10^{-5}	0.08567	1275	177869	1174	50.65	3.43

Assuming total condensation, and dividing the tube into 20 parts, the tittle change is $\Delta x = 0.05$. The procedure is shown for the tittle changing from; $x_I = 0.725$ to $x_F = 0.675$, with a mean value of $x = 0.7$.

$m = 0.04167$ kg/s
$G = 339.5$ kg/(m² s).

$$Re_l = \frac{(1 - x)\, G\, D_i}{\mu_l} = \frac{(1 - 0.7)(339.5)(0.0125)}{0.0002303} = 5528$$

$Re_L > 1125$, then

$$F_2 = 5\, Pr_l + 5\, \ln(1 + 5\, Pr_l) + 2.5\, \ln(0{,}0031\, Re_l^{0,812}) = 34.69$$

The Martinelli parameter is

$$\chi = \left(\frac{\mu_l}{\mu_g}\right)^{0,1}\left(\frac{1 - x}{x}\right)^{0,9}\left(\frac{\rho_g}{\rho_l}\right)^{0,5} = 0.1289$$

and the Nusselt number

$$N_{u_D} = \frac{h_z\, D_i}{\lambda_l} = \frac{Re_l^{0.9}\, Pr_l}{F_2}\left[0{,}15\left(\frac{1}{\chi} + \frac{2.85}{\chi^{0,476}}\right)\right]^{1,15} = 601$$

the local heat transfer coefficient

$$h_z = \frac{N_{u_D}\, \lambda_l}{D_i} = \frac{(601)(0.0857)}{0.0125} = 4119\,\frac{W}{m^2\,K}$$

$$\frac{q}{A} = h_z(T_S - T_0) = \frac{m\, \Delta x\, i_{lg}}{\pi\, D_i\, \Delta z} = 23.06\,\frac{kW}{m^2}$$

and, finally

$$\Delta z = \frac{G\, i_{lg}\, D_i\, \Delta x}{4\, h_z\, (T_S - T_0)} = 0.409$$

Following this procedure, and neglecting the first and the last Δx's, the values in Table 3.4-6 can be calculated.

Table 3.4-6
Results for example 4

x	Re	h_z	Δz	$h_z * \Delta z$	x	Re	h_z	Δz	$h_z * \Delta z$
0,95	921,4	5248	0,321	1.684,61	0,45	10136	3180	0,530	1.685,40
0,90	1843	4886	0,345	1.685,67	0,40	11057	2968	0,568	1.685,82
0,85	2764	4660	0,362	1.686,92	0,35	11978	2744	0,614	1.684,82
0,80	3686	4471	0,377	1.685,57	0,30	12900	2507	0,672	1.684,70
0,75	4607	4294	0,392	1.683,25	0,25	13821	2252	0,748	1.684,50
0,70	5528	4119	0,409	1.684,67	0,20	14742	1977	0,852	1.684,40
0,65	6450	3943	0,427	1.683,66	0,15	15664	1673	1,001	1.674,67
0,60	7371	3762	0,448	1.685,38	0,10	16585	1324	1,272	1.684,13
0,55	8262	3576	0,471	1.684,30	0,05	17507	895,5	1,882	1.685,33
0,50	9214	3382	0,498	1.684,24					
Σ			4,05	16.848,3				8,14	15.153,8

the overall heat-transfer coefficient

$$h_c = \frac{\Sigma\, h_z\, \Delta z}{\Sigma\, \Delta z} = 2625\ \frac{W}{m^2\,K}$$

Using this value, with $\Delta x = 1$, the length needed to condense completely the saturated vapour is:

$$z = \frac{G\, i_{lg}\, D_i}{4\, h_c\, (T_S - T_0)} = 12.84\ m$$

Example 5

Assume a boiler built of ¾ inch BWG 14 tubes, 4.9 m long. The heat is supplied by saturated steam at 381.7 K. The tube-bank is placed at the bottom of a vertical tank, which produces 3.4563 kg/s of saturated cyclohexane at 356.7 K. How may tubes are needed?

Data:

Cyclohexane

$m_{CH} = 3.4563$ kg/s
$T_{CH} = 356.7$ K
$P = 1.14$ bar (a)
$P_c = 41.16$ bar (a)

$T_c = 553.4$ K

$i_{lg} = 357961$ J/kg

Vapour

$T_V = 381,7$ K

$i_{lg} = 2,232 \times 10^6$ J/kg

Tubes

$D_0 = 0.01905$ m

$D_i = 0.01483$ m

$\lambda_P = 40$ W/(K.m)

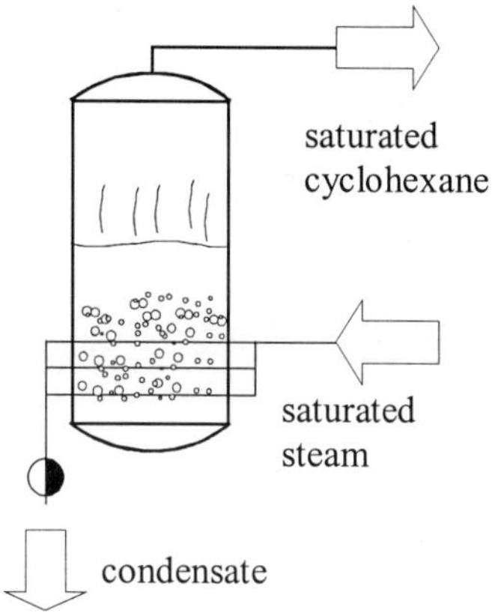

Figure 3.4-19. Schematic of the boiler.

Use as fouling resistance: $r_e = 0.0001$ m^2 K/W for the steam-side; $r'_e = 0.00015$ m^2 K/W for the cyclohexane-side

We can assume as a first approach that the values of h_0, and h_i will have an approximate value of the overall heat transfer coefficient

$$U_0 = \frac{1}{\dfrac{1}{h_0} + \dfrac{D_0}{D_i \, h_i} + \dfrac{D_0 \ln D_0 / D_i}{2 \lambda_p} + r_e \dfrac{D_0}{D_i} + r'_e}$$

The number of tubes needed is then:

$$N_t = \frac{q}{U_0 \, \pi \, D_0 \, L \, (T_V - T_{CH})} = \frac{m_{CH} \, (i_{lg})_{CH}}{U_0 \, \pi \, D_0 \, L \, (T_V - T_{CH})}$$

and the heat transfer area

$$A_0 = \pi \, D_0 \, L \, N_t$$

138

Using this area, we can calculate the heat-transfer coefficients for the cyclohexane boiling (Eqn. 3.4-58), as

$$h_{bn} = 0,18 \; P_c^{0,69} \left(\frac{P}{P_c} \right)^{0,17} \left(\frac{q}{A} \right)^{0,7}$$

for a tube-bank, this value is corrected as follows:

$h_{tb} = 1.5 \; h_{nb} + S$
where $S = 250$ (hydrocarbons). For the steam condensation in horizontal tubes

$$h_i = 0,0105 \; \frac{\lambda_l}{D_i} \; Re_c^{0,8} \; Pr^{0,4} \left(1 + \sqrt{\frac{\rho_l}{\rho_g}} \right)$$

where

$$Re_c = \frac{4 \; m_V \; D_i}{\mu_l \; \pi \; D_i^2 \; N_t}$$

Once the heat-transfer coefficients have been calculated, we can recalculate the overall heat-transfer coefficient and then the heat-transfer area. For this example:

$$q = m_{CH} \; x \; (i_{LG})_{CH} = 1237220 \; W$$

$$m_V = \frac{q}{(i_{LG})_V} = 0,554 \; \text{kg/s}$$

Fluid properties:

    water (T = 381.7 K)

        $\rho_l = 949.79$; $\mu_l = 2.767 \, (10^{-4})$; $\lambda_l = 0.7076$; $Pr_l = 1.655$

    saturated steam

        $P^{set} = 136.37 \; \text{kPa}$; $\rho_g = 0.7742$

As can be seen from Table 3.4-7, about 158 tubes are needed.

Table 3.4-7
Results for example 5

U_0	N_t	A_0	h_0	h_i
1035	164	48.09	2582	5760
1054	161	47.2	2612	5760
...	...	...	...	...
1072	158	46.3	2644	5934

References of section 3.4

1. C.-C. Lai, C.-S. Tan, Ind. Eng. Chem. Res.(1993) 1717.
2. R.B. Bird, W.E. Stewart, E.N. Lightfoot, Transport Phenomena, Wiley, New York, 1960.
3. W. Brötz, Grundiss der Chemischen Reactionstechnik, Verlag Chemie, 1958.
4. S. Walas, Phase equilibria in chemical engineering, Butterworths-Heinemann, Mass., USA 1985.
5. E.L. Cussler, Diffusion, Cambridge University Press, 1984.
6. R.H. Perry, D.W.Green, Chemical Engineers´ Handbook,7th Edtn., McGraw-Hill, New York,1997.
7. VDI-Wärmeatlas, Düsseldorf, 1984.
8. S.W. Churchill, AIChE J. 23 (1977) 10.
9. G. Knaff, E.U. Schlünder, Chem. Eng. Process, 21(1987)151.
10. M. Lesieur, Turbulence in fluids, Kluwer Academic, 1993.
11. J.M. Kay, An introduction to fluid mechanics and heat transfer, Ann Arbor, 1974.
12. G. Brunner, Gas Extraction, Steinkopff Darmstadt-Springer, New York, 1994.
13. R. Reid, J.M. Prausnitz, B.E. Poling, The Properties of Gases and Liquids, 4th Edn, McGraw-Hill, New York, 1987.
14. C.L. Yaws, Chemical Properties Handbook, McGraw-Hill, New York, 1999.
15. O.A. Hougen, K.M. Watson, Chemical Process Principles, Vol III: Kinetics & Catalysis, Wiley, New York, 1947.
16. D.E. Dean, L.I.Stiel, AIChE J. 11(1965) 526.
17. L.I. Stiel, G.Todos, AIChE J., 10 (1964) 26.
18. S. Middleman, An Introduction to Mass and Heat Transfer, J. Wiley & Sons, New York, 1998.
19. O.J. Catchpole, M.B. King, Ind. Eng. Chem. Res. 33(1994) 1828.
20. W.M. Rohsenow, J.P. Hartnett, E.N. Ganiç, Handbook of Heat Transfer Fundamentals, McGraw-Hill, NewYork, 1995.
21. Debenedetti, R.C. Reid, AIChE J. 32 (1986) 2034.
22. K. Abaroudi, F. Trabelsi, B. Calloud-Gabriel. F. Recasens. Ind. Eng. Chem. Res. 38 (1999) 3505.
23. Zehneder, Trepp, J. Supercritical Fluids 6 (1993) 131.
24. J. Puiggené,. M.A. Larayoz, F. Recasens, Chem. Eng. Sci. 52 (1997) 195.

25. F. Stüber, A.M. Vázquez, M.A. Larrayoz, F. Recasens, Ind. Eng. Chem. Res. 35 (1996) 3618.
26. A.H. Bennecke, A.E. Kronberg, K.R. Westerterp, AIChE J. 44 (1998) 263.
27. C.-S. Tan, S.-K. Liang, D-C. Liou, The Chem. Eng. J. 38 (1988) 17.
28. G.-B.Lim, G.D. Holder, Y.T. Sha, J. Supercritical Fluids 3 (1990) 186.
29. C.H. Lee, G.D. Holder, Ind. Eng. Chem. Res. 34 (1995) 906.
30. A.P. Eaton, A. Akgerman, Ind. Eng. Chem. Res. 36 (1997) 923.
31. Coulson and Richardson's Chemical Engineering, 3rd Edtn., Vol. 2, Buckhurst and Harker,1988.
32. O.J. Catchpole , R. Bernig, M.B. King, Ind. Eng. Chem. Res. 35 (1996) 824.
33. C.-S.Tan, C.T. Liou, Ind. Eng. Chem. Res. 27 (1988) 988.
34. O. Levenspiel, Chemical Reaction Engineering, 2nd Edtn., J. Wiley, New York, 1972.
35. K. Abaroudi, F. Trabelsi, F. Recasens, Fluid Phase Equilibria, (in press, to appear 2000)
36. J.M. Smith, Chemical Engineering Kinetics, 3rd Edtn., McGraw-Hill, New York, 1981.
37. C.N. Satterfield, Mass Transfer in Heterogeneous Catalysis, MIT Press, 1970.
38. F. Stüber, S. Julien, F. Recasens, Chem. Eng. Sci. 52 (1997) 3527.
39. J.B. Butt, Reaction Kinetics and Reactor Design, M. Dekker, 1999.
40. P.A. Ramachandran, Chaudhari, Three-Phase Catalytic Reactors, Gordon and Breach Sci. Pub., New York, 1982.
41. F. Recasens, B.J.McCoy, J.M. Smith, AIChE J. 35 (1989) 951.
42. G. Madras, C. Thibaud, C. Erkey, A. Akgerman, AIChE J. 40 (1994) 777.
43. M.C Jones, in T.J. Bruno and J.F. Ely, Supercritical Fluid Technology, CRC Press, Boca Ratón, Fla., 1991.
44. J.J. Lerou, G.F. Froment, Chem. Eng. Sci. 32 (1977) 853.
45. C.-S. Tan, D-C Liou, AIChE J. 35 (1989) 1029.
46. M.M. Esquível, M.G. Bernardo-Gil, M.B. King, J. Supercritical Fluids 16 (1999) 43.
47. E. Reverchon, G. Donsi, L.S. Osséo, Ind. Eng. Chem. Res. 32 (1993) 2721.
48. K. Bartle, A.A. Clifford, S.B. Hawthorne, J.J. Langenfeld, D.J. Miller, R. Robinson, J. Supercrit Fluids 3 (1990) 143.
49. M.B. King and T.R. Bott, Extraction of Natural Products using Near-critical Solvents, Blackie A.&P., Glasgow, 1993.
50. M. Goto, J.M. Smith, B.J. McCoy, Ind. Eng. Chem. Res. 29 (1990) 282.
51. G. Brunner, Ver. Bunsenges. Phys. Chem. 88 (1984) 887.
52. H. Sovová, Chem. Eng. Sci. 49 (1994) 409.
53. M. Goto, B.C. Roy, T. Hirose, J. Supercritical Fluids 9 (1996) 128.

High Pressure Process Technology: Fundamentals and Applications
A. Bertucco and G. Vetter (Editors)
© 2001 Elsevier Science B.V. All rights reserved.

CHAPTER 4

DESIGN AND CONSTRUCTION OF HIGH-PRESSURE EQUIPMENT FOR RESEARCH AND PRODUCTION

G. Vetter[a], **G. Luft**[b] **and S. Maier**[c]

[a] Department of Process Machinery and Equipment,
University of Erlangen-Nuremberg, Cauerstr. 4, D 91054 Erlangen, Germany

[b] Department of Chemistry, University of Technology Darmstadt,
Petersenstr. 20, D 64287 Darmstadt, Germany

[c] Formerly Research and Development, BASF AG,
D 67056 Ludwigshafen, Gemany

Process compressors and pumps, valve with fittings and other piping equipment, and vessels such as reactors and heat exchangers, as well as all the various instrumentation devices for high pressure, demand special design strategies and careful selection of the materials and their treatment. The very high stresses involved, combined with the safety and reliability requirements, have led to typical design features, which are explained together with many examples.

142

4.1 High-pressure machinery[*]

High-pressure processes require suitable liquid pumps, gas compressors, agitators, extruders and other machinery components. High-pressure production methods are generally expensive to develop, to design, to build, to operate, and to maintain. Their application will only be successful if clear advantages can be achieved with respect to the product quality and the costs in general.

Over many years experience, clearly shows that there is a constant challenge to reduce the pressure levels once they have been established owing to new process improvements, for example by using more efficient catalysts.

Because of severe stresses, it is essential to design and fabricate high-pressure machines very carefully.

4.1.1 Requirements and design concepts

Rotary equipment such as pumps and compressors, represent the most demanding group of machinery in high-pressure processes with respect to lifetime, endurance, safety, maintenance and reliability. Their design should therefore be based on the reduction or limitation of the stresses and cope with the strict safety requirements. Because, in many cases the real stress conditions approach the upper limits, and the remaining safety margins are rather low, the design should be well supported by precise stress-analysis and experience from existing high-pressure plants. Some design concepts will be discussed in the following four groups.

Thick-walled internally pressurized components

The hollow parts with predominantly circular cross-section exhibit (see Section 4.3.1) the maximum hoop stresses at the internal surface where the radial stresses effected by the pressure are also superimposed. Especially when they are dynamically loaded, as is frequently the case for machinery components, the design should consider the following aspects: there should be careful stress-analysis by finite-element methods (FEM); reduction of local stress concentrations by avoiding all kinds of notches; surfaces should be smooth, with proper roundings; beneficial compressive residual stresses should be introduced by shot-peening or autofrettage treatment; the components should be designed so that superimposed hoop or bending stresses are avoided, e.g., with vertically split barrel-housings instead of horizontally split flanged ones and stiff design in order to reduce the vibrations.

Material selection

The components submitted to high stresses predominantly consist of high-strength ferritic heat-treated low-alloyed steels of very high purity. Soft martensitic duplex-resp. semiaustenitic or precipitation-hardened, highly alloyed CrNiMo – steels are applied to meet the corrosion resistance required. In general, the materials for high-pressure applications should offer high strength, excellent isotropy, homogeneity, toughness, and well controlled quality. The homogeneity is achieved by the smelting method, and the isotropy by forging the slugs effectively from all sides.

The range of suitable high-strength and corrosion-resistant steels can be extended by alloyed titanium types and others.

The dynamic seals for reciprocating plungers and pistons, as well as rotating shafts, require all the capabilities of modern material developments: high performance plastics, bronzes,

---

[*] References at the end of main section 4.2

sintered carbides, oxide- and silicon ceramics as well as hard- and wear-resistant coatings applied by detonation, flame or plasma-spraying methods, are used.

Dynamic seals

Plunger-, piston- or shaft seals have demanding design problems, especially if large differential pressures have to be sealed. The tribological and tribo-chemical conditions should be studied and understood well, in order to reduce the local friction, temperatures and wear.

Important factors for the sealing of liquids are to reduce the pressure differentials and the compression between the interacting sealing components, to keep particles apart, to avoid dry-running, and enhance lubrication.

Dry-running gas seals for reciprocating pistons require "well-designed" plastic seal rings, and reduction of the pressure in a number of serially arranged stages. The injection of lubricants improves the lifetime greatly but frequently is not acceptable owing to the process contamination involved.

Shaft seals should be balanced by gas- or lubricant pressure in order to reduce the effective pressure differential.

Hermetic sealless pumps

Difficult sealing and leakage problems on shafts and plungers with dangerous liquids during high-pressure service have promoted new concepts which avoid dynamic seals completely by using a hermetic sealless design. In the past two decades reciprocating plunger pumps have been replaced by pumps with hydraulically operated diaphragms. The same has taken place on the centrifugal pump range. The canned-motor types have replaced the centrifugal pumps with mechanical shaft seals for certain applications.

The dry- and sealless gas compression can be achieved in the low capacity range by reciprocating diaphragm compressors which operate up to very high pressures.

The design concepts of the types explained above have to be learned and experienced. There are many detailed problems which cannot yet be calculated or modelled. High-pressure machinery specialists belong to a small international community of engineers. Technical progress is closely coupled to those in material science.

4.1.2 Generation of pressure with pumps and compressors

The thermodynamic state of liquids is a changing polytropic one (close to isentropic) during quick compression (see Fig.4.1-1, for CO_2) and the corresponding temperature rise is low for liquids and large for gases (see AB in Fig. 4.1-1). The final compression temperature limits the compression ratio per stage with gas compressors.

The power required for the compression can be evaluated generally from the difference of the total enthalpies between suction and discharge (for liquids, this term correlates directly with the pressure differential to a good approximation).

The efficiency factors of gas compressors are based on ideal comparative processes such as the isothermal or isentropic ones, and therefore show considerable differences according to the method used for evaluation.

144

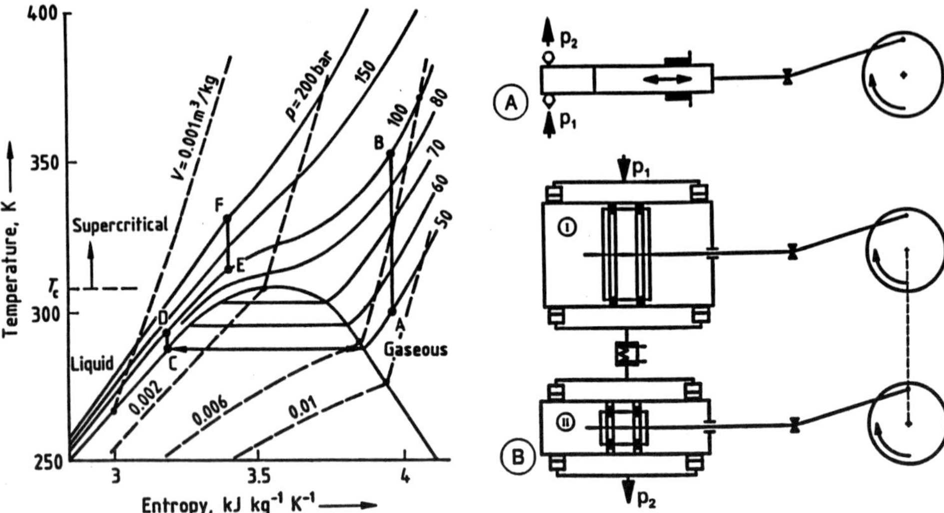

Fig. 4.1-1. Isentropic change of state (CO₂)
AB, Gases; CD, Liquids; EF, Supercritical
fluids.

Fig. 4.1-2. Principles of pressure generation
A, Single-stage plunger pump; B, Two-
stage double-acting piston compressor (intercooling)

For the machinery suitable for generating high-pressures one should distinguish between positive-displacement (hydrostatic), and turbo- (hydrokinetic) types, which are available for liquids and gases.

In general, the positive displacement machines – there are rotating and reciprocating features – show their main application range for high-pressure and lower-capacity conditions. The turbo- (centrifugal) machines, however, are best suited for high capacity but lower pressure conditions. The selection should be based on an analysis of the life-cycle costs (or costs of ownership) [1].

The **reciprocating positive pumps, for relatively incompressible liquids** are capable of discharging against very high pressures (up to 10000 bar) in a single stage (Fig. 4.1-2 A) because the liquid is heated slightly only during compression.

The **reciprocating positive compressors, for gases,** regularly employ a multistage arrangement with intermediate cooling (Fig. 4.1-2 B, shown for two stages) to approximate the energetically favourable isothermal change of state for the most economical performance. The compression ratio per stage varies 2 to 5 depending on whether there is dry- (lower value) or lubricated- (higher value) operation.

The evidence of drastic differences between multistage gas compressors and single stage liquid pumps is shown in Fig. 4.1-3. The five-stage compressor (Fig. 4.1-3, left) performs a gas compression which is, overall, close to isothermal, by intercooling between the stages (1-2-3-4-5-6). The gas-volume is dramatically reduced owing to the higher compressibility of the gas, but the absolute pressure amplitude per stage stays relatively moderate ($\Delta\,p_A$).

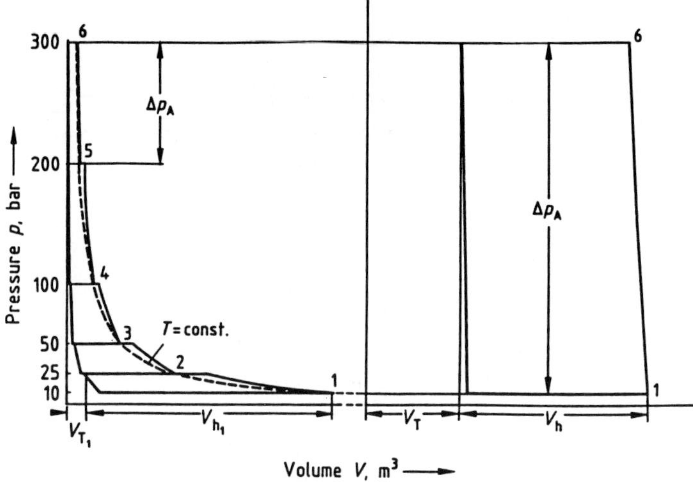

Fig. 4.1-3. PV-diagram for gas- and liquid reciprocating compressors (pumps)
Left: Five-stage gas compressor   Right: Single-stage liquid pump
V_h, Stroke volume; V_T dead space; Δp_A pressure amplitude per stage

In contrast, the single-stage liquid pumps exhibit a very large pressure-amplitude (Δp_A) which must be considered with regards to problems of failure if the components are loaded by pulsating internal pressure (see Section 4.1.4).

It should be noted that owing to the compressibility of the fluid, the dead-space of the working chamber (V_T in Fig. 4.1-3) must be well minimized for gas compressors.
For reciprocating positive pumps and compressors the fluid displacement per stroke depends on the geometry, the fluid compressibility, the polytropic coefficient, and the internal leakages and volumetric losses, but basically not on the speed of machine.

For **hydrodynamic turbomachines** (Fig. 4.1-4), the speed of the fluid flow or the circumferential velocity of the radial impellers involved is of great importance for the pressure increase per stage ($\Delta p \sim u^2 \rho$, where ρ is the density of the fluid, and u is the circumferential velocity). Obviously the high circumferential velocity required for effective pressure generation can be achieved by high angular speed and / or a suitable size of the impeller.

Again, the liquid centrifugal pumps (Figure 4.1-4A) do not require any intercooling. The total differential pressure is achieved by a multistage arrangement as well as by an elevated pump speed. With super-speed, single-stage partial emission pumps it is possible to achieve 100 to 200 bar per stage, with moderate efficiency however (Fig. 4.1-4B). An important limiting factor to increasing the speed is the strength of the impeller.

Radial turbo- compressors provide the best capability for generating high-pressure gas discharges for large capacity. For reasons already explained, intercooling between a number of stages (groups) is necessary (Fig. 4.1-4C). The gas compression requires more stages owing to fluid compressibility, than do liquid pumps for the same total differential pressure. The compressor speed or the circumferential velocity of the impellers is important for achieving the high pressure required. Limiting factors are the strength of the impellers as well as the

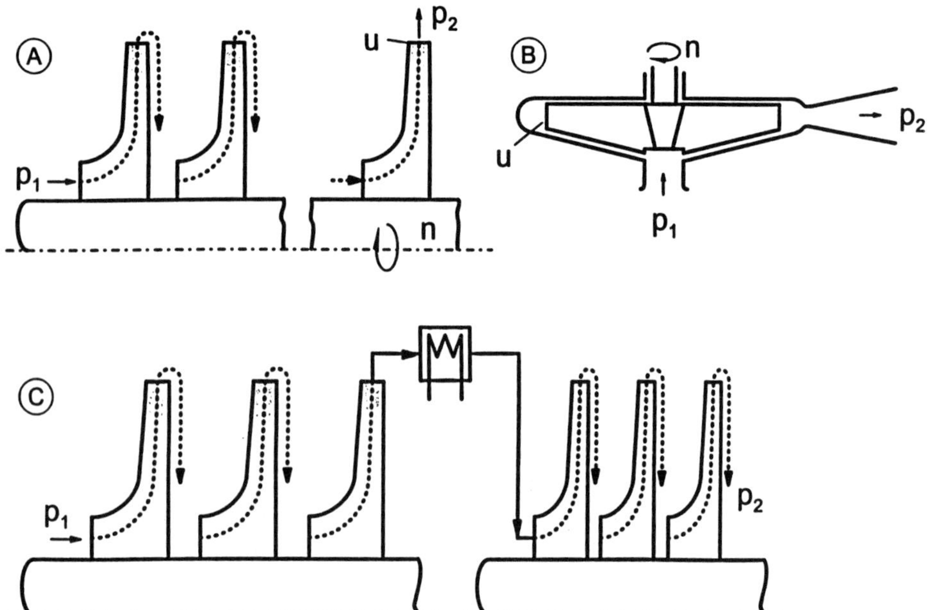

Fig. 4.1-4. Principles of pressure generation
A; Multi-stage centrifugal pump.   B; Single-stage high-speed centrifugal pump.
C; Multi-stage radial compressor (intercooling)

Mach-numbers of the flow; the latter should stay – depending on the gas and the application case-well below 0.7, appropriately below the sonic speed.
Compared to the hydrodynamic turbo-machines, the hydrostatic positive displacement features exhibit a pulsating flow rate (Fig. 4.1-5). Generally turbo-machines operate at comparatively much higher speeds and run more quietly with the advantage of being relatively small in size and requiring less space. Reciprocating positive pumps and compressors exhibit a regularly pulsating flow and need appropriate damping devices in order to keep the installation quiet and safe. The attraction of the reciprocating machines is their lower energy consumption.

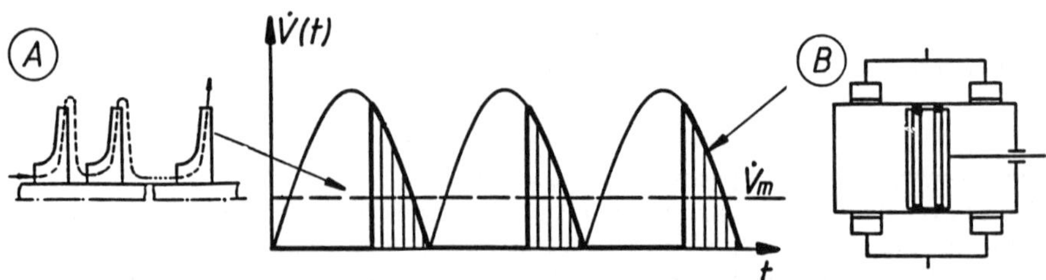

Fig. 4.1-5. Time behaviour of the fluid flow. A; Hydrodynamic pump/compressor. B; Hydrostatic (double-acting) piston machine. V_m Average flow;   V(t) time volume flow

4.1.3 Pumps

In high-pressure systems, pumps are used for the metering, conveying and circulation of all kinds of liquids as well as for general pressurizing. Volume flow-rates range from around 1ml/h to more than 1000 m^3 /h. The pressure range extends up to 300 to 1000 bar and, more rarely, up to 1000 to 4000 (10 000) bar. The range outlined is covered by the various features of the positive displacement and centrifugal type, as shown in Fig. 4.1-6.

The throttling characteristics of pump types differ mainly with regards to their pressure-stiffness (the influence of throttling of pressure on the flow rate, Fig. 4.1-7A). Because of their pressure-elastic curve, centrifugal pumps can also be regulated by throttling (Fig. 4.1-7A: the operating point shifts from A to B). Regulation of the rotational speed is more favourable energetically (Fig. 4.1-7A: the operation point shifts from A to B′). By-pass regulation, in which part of the pump′s flow rate is recycled to suction through a regulation valve, is also customary for large machines and for the case that low energy costs are offered at the location of installation.

Displacement machines have pressure-stiff characteristic curves: they thus impress their flow rate on the pumping system (Fig. 4.1-7A: lines d and e). The control characteristic shows that the flow-rate depends approximately linearly on the regulated variables (rotational speed n, stroke length h_{K0} , Fig. 4.1-7B). With reciprocating displacement pumps, the high differential pressure shifts the characteristic curve from the origin to the right side, owing to elasticity effects of the working space and the fluid [2]. The same applies to rotary displacement pumps

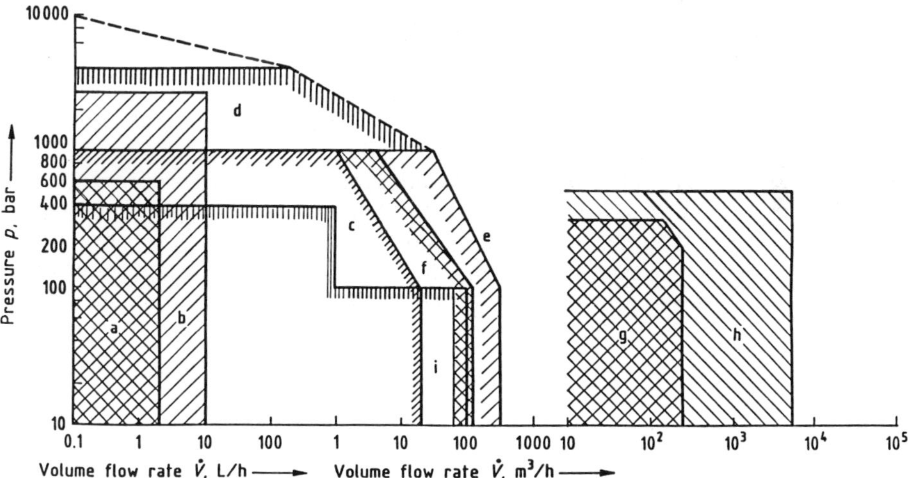

Fig. 4.1-6. Capacity range of various pump types for high-pressure
(a), Micro-metering pumps (piston, diaphragm, gear, spinning pumps); (b), Metering pumps for laboratory and pilot systems (piston, diaphragm/hydraulic); (c), Metering pumps (piston, diaphragm/hydraulic); (d), Ultrahigh-pressure piston pumps; (e), Multicylinder piston pumps; (f,) Multicylinder diaphragm pumps; (g), Centrifugal pumps (high-speed, single- or two-stage); (h), Centrifugal pumps (multistage); (i), Rotating displacement pumps (gear, screw)

148

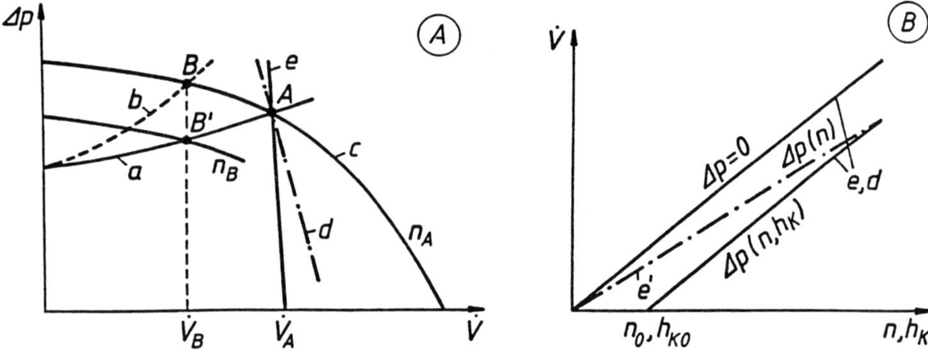

Fig. 4.1-7. Characteristics of various pump types
A; Throttling characteristics of centrifugal (c) -and positive- displacement (e, reciprocating; d, rotary) pumps. B; Control characteristics of positive displacement pumps; a,b, Systems characteristics (b, after speed control); V̇Fluid flow; Δp, Differential pressure; n, Speed; h_k, Stroke length; A, B, B′, Operating points.

at higher pressure differences, owing to interior clearance leakages. With rotational speed n_0 (e.g., gear pumps) and stroke length h_{K0} (e.g., reciprocating metering pumps), the flow-rate is disappearing. Only in the case of reciprocating displacement pumps is there proportionality between the flow rate, V̇, and the regulated variable, n, because the interior leakages tend to vanish (Fig. 4.1-7B, line e′).

4.1.3.1 Reciprocating displacement pumps

This type of pump is typically most suitable for high-pressure (up to 10 000 bar) operated by pneumatic or hydraulic piston, by electrical threaded spindle drives or by a mechanical crank or eccentric gears with electric motor. The capacity is adjustable by the speed or/and the stroke-length [2]. As reciprocating displacement pumps have well sealed internal working chambers (tight check valves and plunger seals), they are well regarded for all kinds of low viscous liquids, including solvents and liquified gases. Reciprocating pumps do not require their plunger seals to come into contact with the liquid to be discharged, if hermetic diaphragm designs are applied [3].

Metering pumps

For very small pulsation-free metering flows (e.g., 10 to 10^3 ml/h), hydraulic and mechanical (spindle) linear drives can be used for pressures up to several thousand bar. By operating two systems in parallel controlled by sensors, these can also be operated without essential interruption (Fig. 4.1-8a and b). By superposing two displacement pistons, or by controlling the angular velocity with a special cam drive, small metering flows with low pulsation are achieved up to a pressure of about 700 bar. This can even be done with leak-free hydraulic

Designation		Functional principle	Displacer kinematics	Adjustment	Displacer system
Spindle drive one- and two-cylinder electro-mechanical hydraulic	a	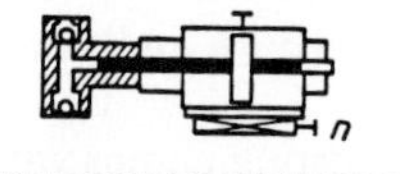	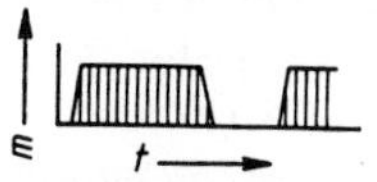	$n(h)$ $n < 30\ \text{min}^{-1}$	P DH
	b	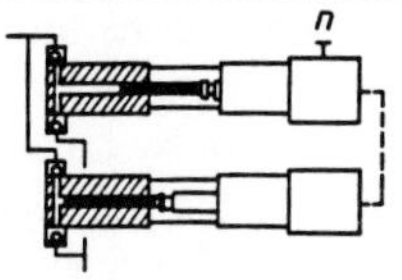	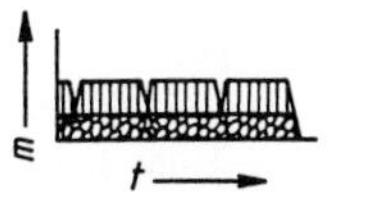		
Cam with pulsating angular velocity	c	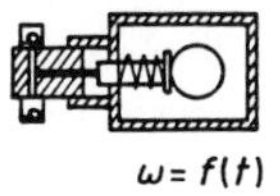 $\omega = f(t)$	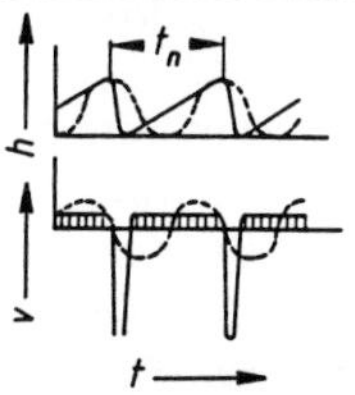	n $n < 60\ \text{min}^{-1}$	P DH
Superposition drive two cylinder with cams	d	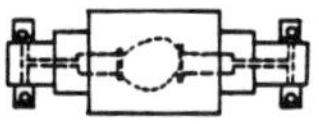	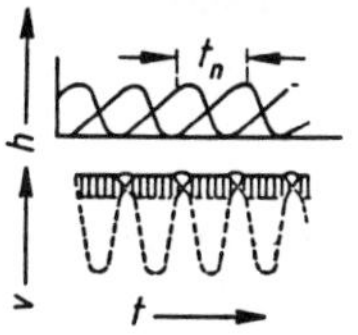	n $n < 100\ \text{min}^{-1}$	P DH
Linear drive single-cylinder magnetic	e	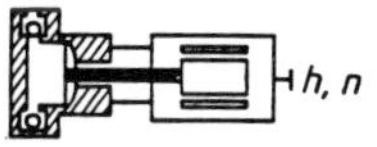	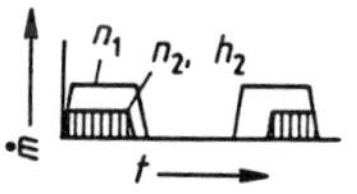	h, n $n < 100\ \text{min}^{-1}$	P DH
Stroke-adjustable quasi-harmonic drive single and multi-cylinder	f	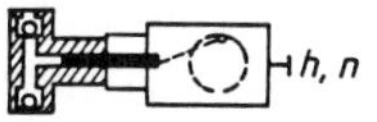	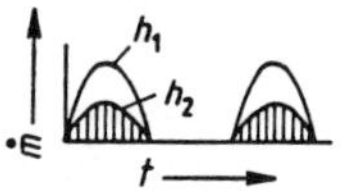	$h(n)$ $n < 350\ \text{min}^{-1}$	P DH
	g	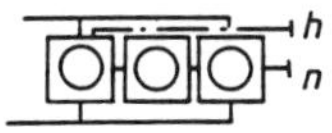	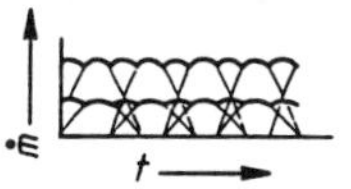		
Stroke-adjustable drive with lost motion drive	h	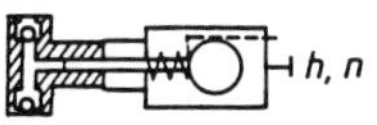	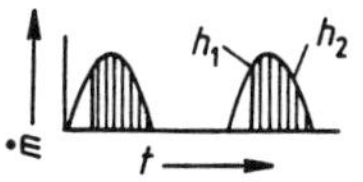	$h(n)$ $n < 150\ \text{min}^{-1}$	P DH
Linear drive super-position, two-cylinder hydraulic	i		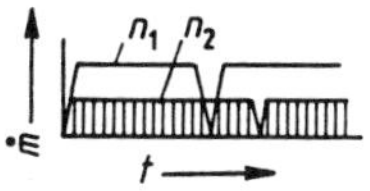	n $n < 100\ \text{min}^{-1}$	P DH

Fig. 4.1-8. Types of reciprocating metering pumps
P, Plunger pump head; DH, Diaphragm pump head (hydraulic)

150

diaphragm pump heads (Fig. 4.1-8c and d). Such pumps are also used in HPLC analysis systems.

Magnetic linear drives (Fig. 4.1-8e) are distinguished by their ease of operation and their small size. They are suitable for flow rates <10 l/h and pressures up to 500 bar with diaphragm pump heads (Fig. 4.1-9).

For larger metering flows, reciprocating metering pumps with an adjustable stroke and mechanical drives have proved suitable. These pumps can be mulitpexed with several cylinders for metering several components or to smooth out pulsations (Fig. 4.1-8f and g) [4]. The automated systems are activated by stroke actuators or by drives with variable speed.

Plunger-type metering pumps are suitable for very high-pressures and not – too - hazardous liquids (Fig. 4.1-10). The plunger seals – packings or lip-rings – as well as the plunger and the connection with the rod of the drive unit, require very careful design and selection of materials.

Diaphragm metering pumps with a hydraulic diaphragm drive are suitable for very high-pressures up to about 500 bar. Their diaphragms consist of polytetrafluorethylen (PTFE): beyond this, up to about 3000 bar, they consist of cold-rolled sheet metal, for example, Cr-Ni-steels an others. The limiting factors for the PTFE diaphragms are the pressure and temperature (actually to be <150°C); their development is continually progressing and focuses on stronger and more temperature-restistant types (e.g., polyetheretherketone, PEEK). Metal diaphragms exhibit the high strength necessary for them to be tightly clamped and sealed, their temperature resistance is superior to that of all suitable polymer types.

With smaller metering flows spring-loaded cam-drive systems predominate because they make it possible to adjust the stroke without backlash, at a low price (Fig. 4.1-8 h). For larger metering flows (max. hydraulic power up to 25 kW per metering element), various types of drive units with adjustable stroke are in use. All of these are actually based on the familiar conventional crank mechanism (Fig. 4.1-11).

The leak-free design, as explained, operates reliably, is easy to maintain, presents no problems with hazardous fluids, and circumvents piston-sealing problems. The lifetime of PTFE diaphragms, which are primarily used up to 500 bar, far exceeds 10 000 h. The rupture of diaphragms can be signalled by sandwich diaphragms (Fig. 4.1-11, items 2, 7), which consist of two diaphragms, the space in between being connected to a pressure-sensor via a check valve. If one of the diaphragms ruptures, the discharge pressure operates a pressure switch. In normal operation, the two diaphragms are connected by an adhesive intermediate fluid (Fig. 4.1-12, item 7) during suction and by direct mechanical support on one another during discharge [5].

Metal diaphragms are more sensitive towards surface damage by particles and often give only a reduced lifetime (< 10 000 h); their deflectability is much less compared to PTFE. This is the reason why diaphragm pumps with metallic diaphragms are more expensive. The motion of the diaphragms is generated by the displacement of the hydraulic oil by the plunger and is subject to the position control (see items 3 to 6, Fig. 4.1-11) by the gate-, snifting-, relief- and venting valves included [2]. Metal diaphragm pumps are predominantly controlled by underpressure with the temporary closure of the snifting (Fig. 4.1-13 a, b; items 1,3,5,6) valve (5) by the plunger moving in the seal bushing (4). Pumps with metal diaphragms for the pressure range between 1000 and 3000 are mainly used for laboratory or pilot plant applications, the diaphragm control being based on underpressure only.

The trend in the development of reciprocating high-pressure metering pumps is still continuing towards the diaphragm technology. Exceptions are the above-mentioned plunger designs for HPLC analysis, where extremely favourable operating conditions prevail.

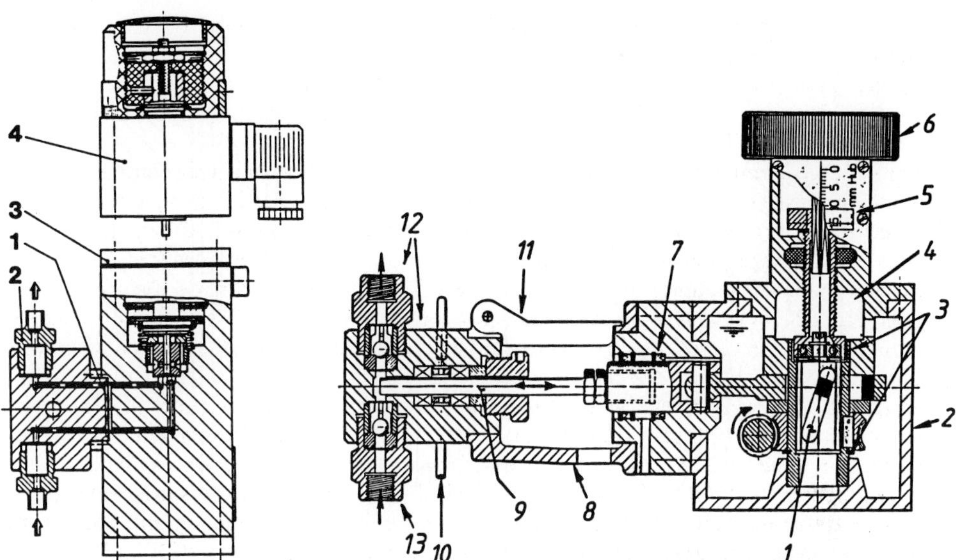

Fig. 4.1-9.Magnetic pump (LEWA).
1/2 Hydraulic diaphragm head; 3/4 Solenoid linear drive.

Fig. 4.1-10. Plunger metering pump (LEWA)
1 to 8, Stroke-adjustable drive unit; 9, Plunger; 10 Seal (flushed); 11, Cover; 12/13, Pump valves.

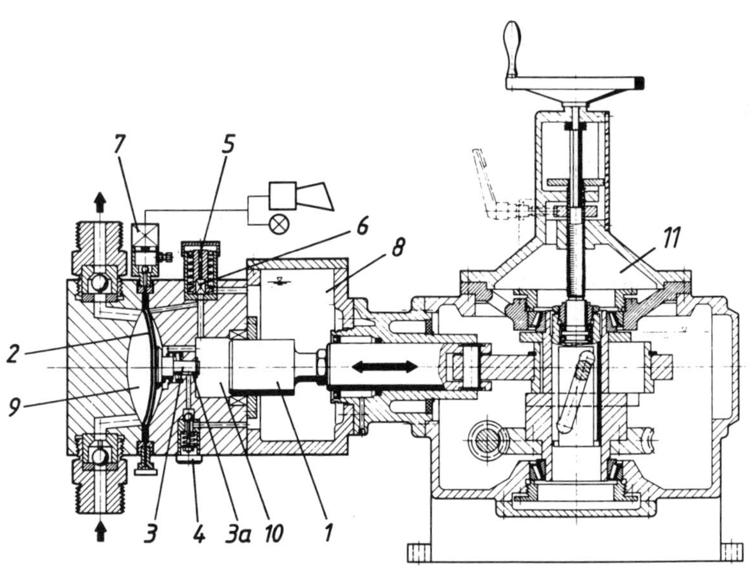

Fig. 4.1-11. Diaphragm pump with hydraulically driven PTFE diaphragm (LEWA)
1, Plunger; 2, Diaphragm; 3/4/5/6, Gate-, snifting-, relief-, and venting valves; 7, Diaphragm-rupture control; 8, Hydraulic system; 9/10, Working spaces; 11,Stroke-adjustable drive unit.

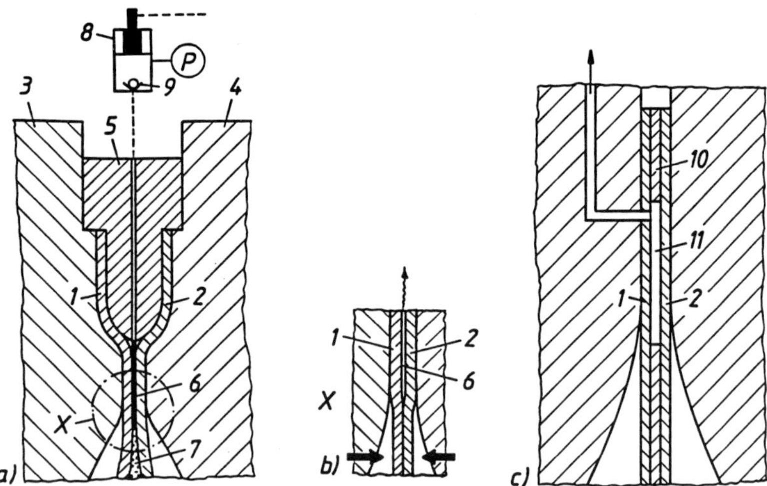

Fig. 4.1-12. Diaphragm-rupture monitoring system.
a), PTFE Sandwich diaphragm (as assembled); b), PTFE Sandwich diaphragm (diaphragms hydraulically coupled, in operation); c) Metal sandwich diaphragm
1/2, Diaphragms; ¾, Covers; 5, Intermediate ring; 6, Intermediate layer; 7, Coupling liquid; 8/9, Rupture sensor; 10, Intermediate diaphragm; 11, Connecting channel.

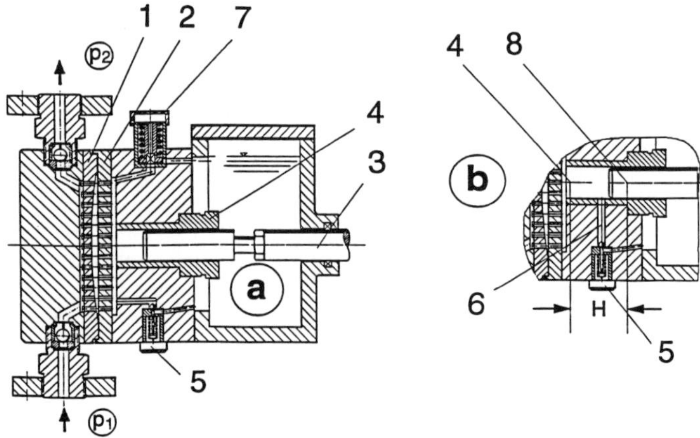

Fig. 4.1-13. Diaphragm pump with metal diaphragm and underpressure control.
1/2, Diaphragm between hole-plates; 3, rod; 4, Seal bushing; 5, Snifting valve; 6, Connecting bore; 7, Venting valve; 8, Plunger.

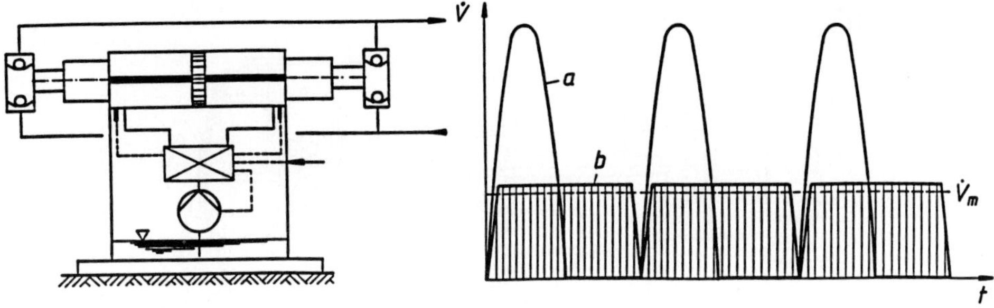

Fig. 4.1-14. Double-head high-pressure intensifier metering pump.
a, Single-head pump with crank drive; b, Double-head intensifier; $\dot{V}$ Volume flow; $\dot{V}_m$, average; t, Time.

For process- and production engineering, plunger metering pumps with a low-frequency, hydraulic long-stroke linear drive, also known as hydraulic intensifiers, are suitable for very high-pressures (1500 – 4000 bar) (Fig. 4.1-8 i). The uniform, slow piston motion leads to an operationally reliable piston seal and, when two pump heads are superposed, to low pulsation of the flow rate [6]. Fig. 4.1-14 shows the obvious advantages of the double-head hydraulic linear drive system compared to a single-head plunger-type pump with quasi-harmonic crank drive: the maximum piston velocity (and frictional energy) are much lower and, the discharge is well smoothed. The hydraulic intensifier pumps can be easily controlled for automation purposes. This type of pump is not only used for the metering of catalysts or initiator solvents during, e.g., LDPE-production, but also for water-jet cutting and the pressurization for hydroforming, hydropressing, hot and cold isostatic pressing, and autofrettage treatment.

Transfer pumps

In contrast to metering pumps, which normally have a single cylinder and an adjustable stroke per metered component, transfer pumps are mainly designed with several cylinders per metered component (e.g., triplex) in a multi-head structure, in order to smooth out pulsations in the flow rate (Fig. 4.1-8 g). Powers over 1 MW are achieved in a vertical or horizontal mode of construction. Drives for multi-head-plunger and diaphragm pumps are very similar (Figs.4.1-15 and 4.1-16) for high-pressure, and when using corrosion-resistant high-strength steels,the monoblock mode of construction (Fig. 4.1-17 A, B) is replaced by a component mode of construction (Fig. 4.1-17 C). The advantage is that the parts which are most subject to fatigue can be replaced more easily. Above 100 – 300 bar, a coaxial valve arrangement (Fig. 4.1-18) should be preferred, because in this way the notch stresses together with the danger of corrosion-fatigue can be reduced by at least a factor of two compared to the customary T-intersection of the pump head bores (see Section 4.1.4). Such designs can deal with differential pressures up to around 2000 bar with long-term operational reliability and endurance. These pressures are presently required, for example, in high-pressure cleaning technology.

154

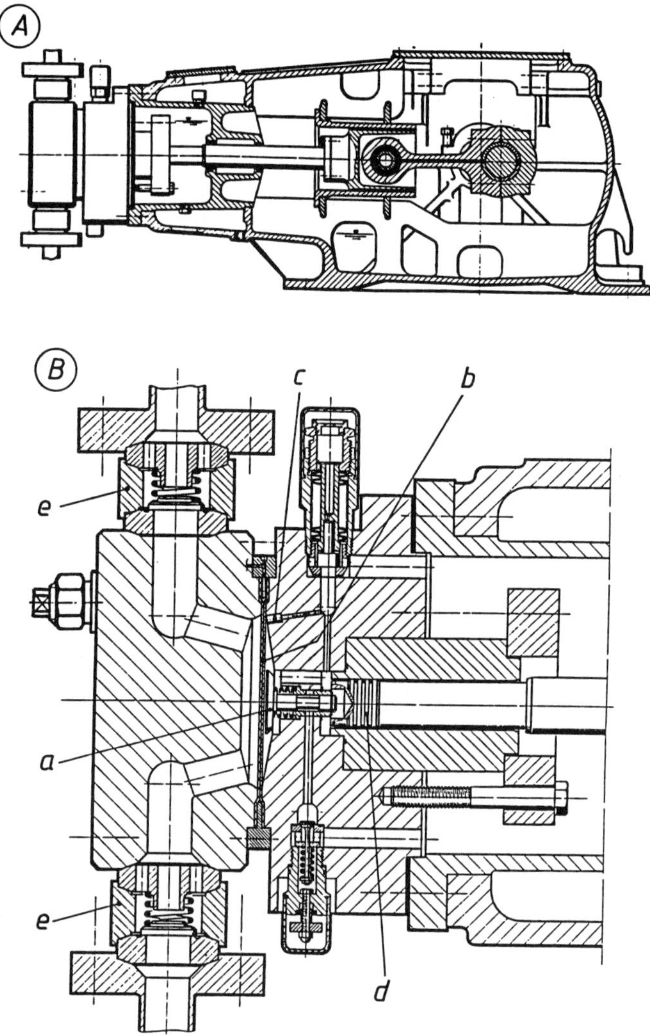

Fig. 4.1-15. Large process diaphragm-pump for high-pressure (LEWA)
A), Drive system with straight thrust crank for multi-head arrangement; B), Diaphragm pump
head for the pressure range up to 500 bar.
a, Sliding gate for controlling the position of the diaphragm; b, Diaphragm (PTFE, sandwich
form); c, Hydraulic cylinder; d, Piston with seal; e, Pump valves.

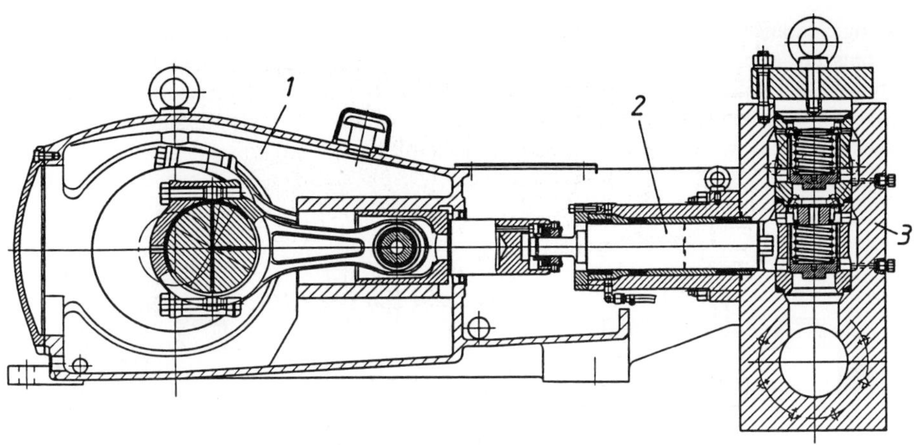

Fig. 4.1-16. Process plunger pump for high-pressure (URACA)
1, Crank drive (multiplex); 2, Plunger/seal; 3, Liquid-end (monoblock).

Diaphragm process-pumps in a multi-head structure (see Fig. 4.1-15), with powers up to several hundred kW, have been built for the pressure range up to 1000 bar. The avoidance of any plunger seals with fluid contact and leakages leads to remarkable operational reliability. For this reason, multi-head diaphragm pumps are frequently more economical for hazardous fluids at high pressures, than are multi-head plunger pumps, despite the higher investment costs.

With the development of pipeline transport of coal or ores, the chemical refining of ores and the coal liquefaction, high-pressure pumps are currently being used for suspensions that are loaded with solids. There are various methods to keep abrasive particles away from the piston seal, e.g., by using hydraulic isolating lines or remote head design, sedimentation spaces, and flushing systems [7]. For predominantly aqueous, abrasive slurries with a high concentration of solids, multi-cylinder diaphragm pumps are suitable up to a pressure of 200 bar and for powers of ca. 1.5 MW. They generally have a four-cylinder arrangement and elastomer diaphragms [8].

A new branch of high-pressure engineering is the conveyance of materials with high consistency, e.g., mortar, concrete, sludge, slurries, over long distances for which pressures of up to 200 bar may be necessary. This can be carried out with slow hydraulic long-stroke piston displacement pumps if the inlet opening of the two-cylinder arrangement is hydraulically controlled and operated with sliding valves [9].

Hydraulically operated linear plunger pumps with automatically controlled valves are also applied for the transport and metering of highly viscous adhesives ($> 10^5$ mPas) against pressures up to several 100 bar [10] for various –partly robotic – production processes.
Cryogenetic piston pumps are applied for the liquefaction of gases and for process applications at high-pressure [11] - and extremely low-temperature conditions.

4.1.3.2 Rotary displacement pumps

Compared with reciprocating pumps which are very tight internally the rotary displacement types have internal working spaces which are sealed by narrow axial-, radial-, and meshing clearances. This is why external-gear and two- or three-screw pumps are suitable for high-pressure applications only for viscous liquids. As the viscosity increases, the internal clearances seal better, and the pump's efficiency and the attainable pressure-difference reach satisfactory values. Gear- and screw the pumps are suitable for high-pressure applications up to around 300 bar, with fluid viscosities > ca. 100 mPa · s.

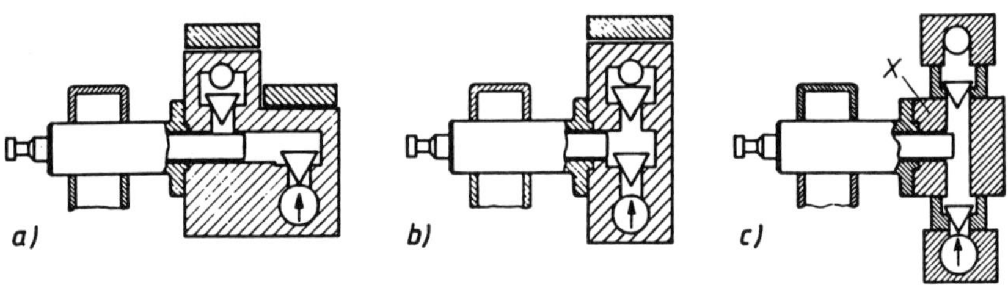

Fig. 4.1-17. Monoblock and component mode of construction
a), Monoblock, valves displaced; b), Monoblock valves in alignment; c), Component mode.

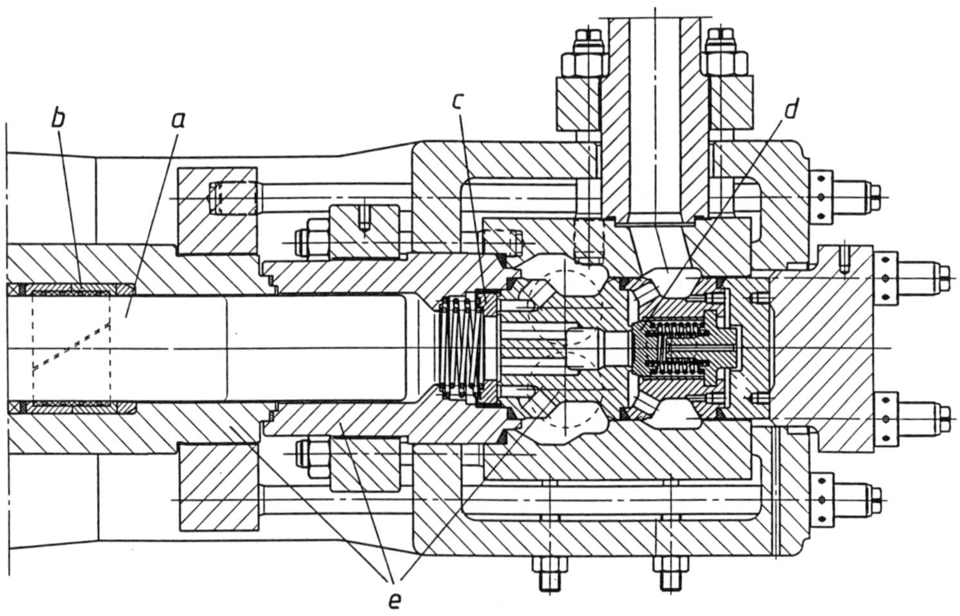

Fig. 4.1-18. Coaxial arrangement of pump valves (URACA). a, Plunger; b, Plunger seal; c, Suction valve; d, Discharge valve; 3, Components with pulsating internal pressure.

One typical application for gear pumps is the injection, conveyance, filtration, and spinning of polymer melts [12] (viscosities > 10^6 mPa·s). Delivery pressures up to 450 bar occur in these applications. In the case of hot- and direct-spinning processes, the heated high-pressure gear pump (Fig. 4.1-19) takes the polymer melt from the reactor and the degassing system which may both be under vacuum, and presses it through filters, heat-exchangers, mixers, and diverters towards the gear spinning pumps, whose flow rate is especially uniform and does not depend greatly on the discharge pressure. Characteristic design criteria for polymer- melt gear pumps include large inlet nozzle, crescent-shaped clearances in the suction region in order to effect hydrodynamically supported filling of the working spaces, and a heating jacketed housing. Gear pumps (Fig. 4.1-20) are used for the extrusion processing of polymers to increase the pressure after the melting extruder (up to 100 to 400 bar) in order to obtain smooth flow.

If the polymer is due to contain fractions of glass fibers or other particles which create abrasiveness [13] the gear rotors are driven by special external timing gears (Fig. 4.1-21) in a manner which is avoid torque transmission between the gear wheels during meshing. This feature is capable of reducing the unfavourable shearing, and the destruction by the adjustment of the clearance dimensions, also.

4.1.3.3 Centrifugal pumps

Centrifugal pumps require appropriately large circumferential speed of the impellers and a number of serially arranged stages, in order to obtain the high-pressure differences. For efficient and economic operation the specific speed, n_q, of the individual pump stages ($n_q = n \cdot \dot{V}^{0,5} / H^{0.75}$; $\dot{V}$ m^3/s; H, m; n, min^{-1}) should stay above 10 to 20. Too small a capacity or hydraulic power transmission will not provide favourable operating conditions and it is then recommended to focus on positive-displacement pumps alternatively.

Multistage centrifugal pumps

High-pressure centrifugal pumps have long been used as feed pumps for steam boilers [14, 15]. For modern multistage centrifugal pumps the power extends to beyond 30 MW, and pressures up to 400 bar. The design of high-pressure centrifugal pumps for process applications varies depending on the pressure level, the temperature conditions (including those during starting), and the directives of the international regulations.

For high-pressure engineering, centrifugal pumps with vertically-split barrel construction have become established starting for powers above around 1 MW in petrochemistry as well as in oil and gas exploration, especially as injection pumps [16, 17]. Injection pumps for surface-water and sea-water impose much more stringent requirement as regards the corrosion- and erosion- resistance of their components (power up to 30 MW, pressure up to 300 bar). The barrel housing design offers several important advantages: the housing is very stiff, and shows favourable stress distribution; the differential pressure at the internal stage-housings (radial or vertical split) is partly negative, and low; the rotating portion and the inner housings can be withdrawn without disassembling the barrel housing, or even removing it out of the piping connections (Fig. 4.1-22).

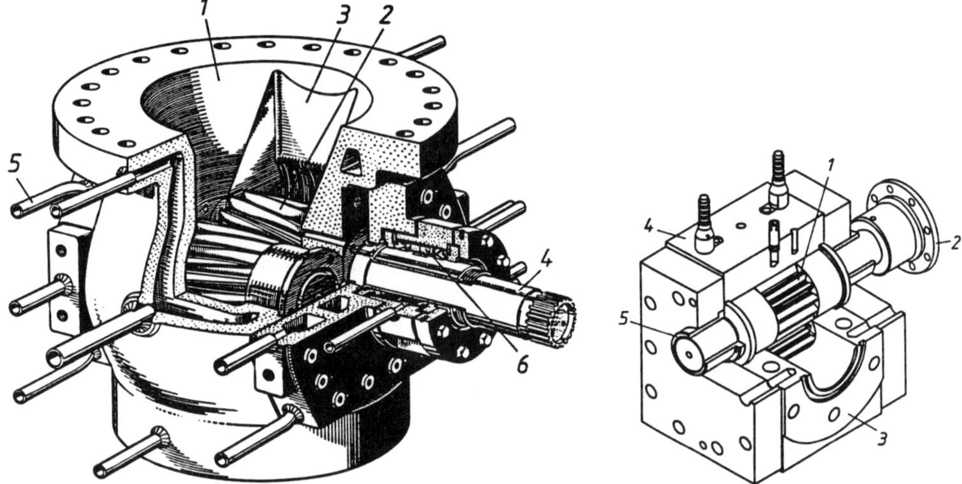

Fig. 4.1-19. High-pressure gear pump for the polymer extraction (adopted from MAAG TEXTRON).
1, Large suction inlet; 2, Gear wheels; 3, Crescent-shaped clearance; 4, Shaft; 5, Heating; 6, Shaft seal.

Fig. 4.1-20. Gear pump as pressure booster behind a melting extruder (adopted from MAAG TEXTRON). 1, Gear wheels; 2, Shaft coupling; 3, Inlet; 4, Outlet; 5, Bearings.

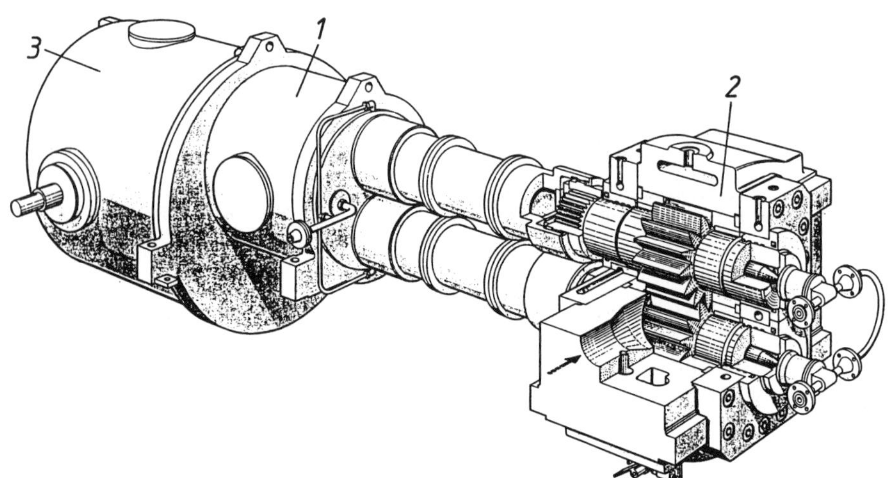

Fig. 4.1-21. Gear pump for sensitive and abrasive polymer melts (adopted from MAAG TEXTRON)
1, External timing gear; 2, Gear pump; 3, Electric motor.

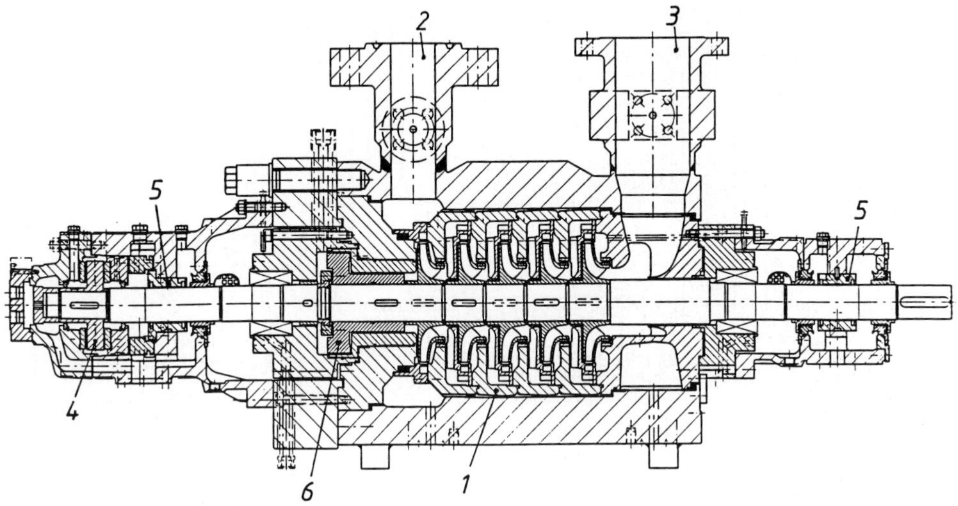

Fig. 4.1-22. Barrel-type five-stage centrifugal pump for petrochemical applications (adopted from KSB)
1, Stage package; 2, Inlet; 3, Outlet; 4, Axial thrust bearing; 5, Radial bearings; 6, Axial thrust compensation "piston".

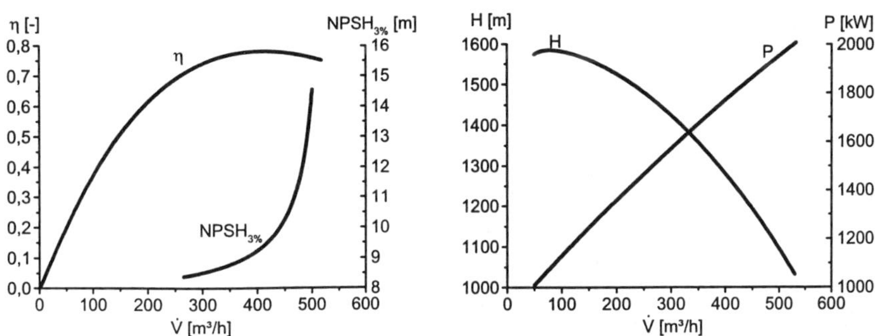

Fig. 4.1-23. Performance diagram of an eight-stage high-pressure centrifugal pump (adopted from KSB)
η, Efficiency; $\dot{V}$, Flow rate; P, Power consumption; NPSHR, net positive suction head required.

The typical performance diagram of an eight-stage centrifugal pump for high-pressure (Fig. 4.1-23) shows that with moderate speed the efficiency is good. The very recently developed new feature of a high-pressure centrifugal pump with barrel housing for ammonia (volume flow around 150 m³/h; head 2500 m NH_3) is characterized (Fig. 4.1-24) by the compensation of the axial thrust at each impeller (clearance seals in front and behind) and double mechanical gas seals flushed by nitrogen [18]. The example of a large six-stage centrifugal with

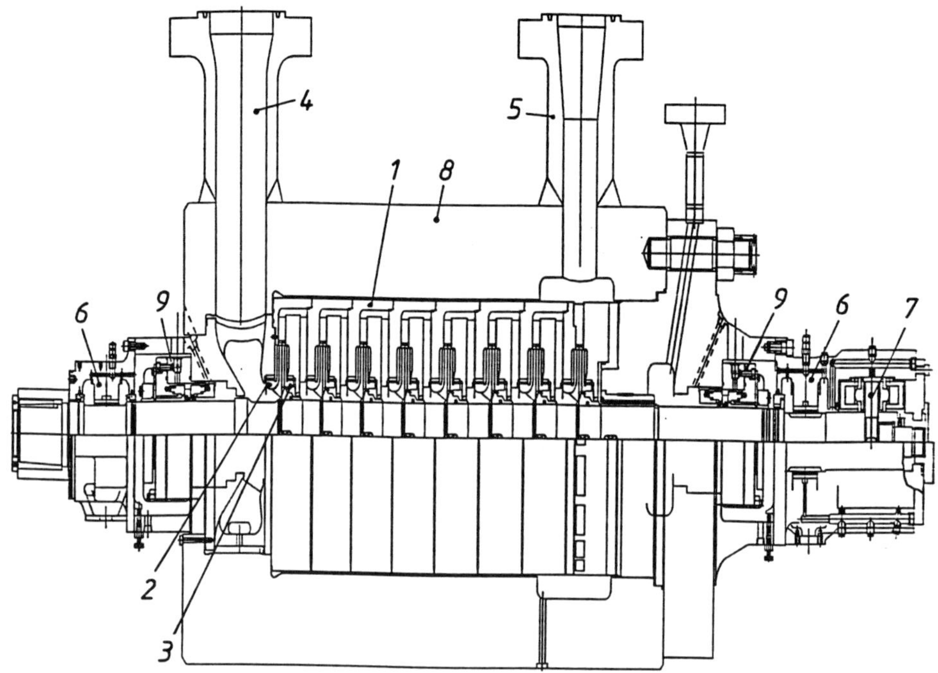

Fig. 4.1-24. Barrel-type eight-stage centrifugal pump for ammonia (adopted from INGER-SOLL DRESSER)
1, Stage package; 2/3, Clearances at the impeller; 4/5, Inlet/outlet; 6, Radial bearings; 7, Axial thrust bearing; 8, Barrel housing 9, Mechanical shaft seals.

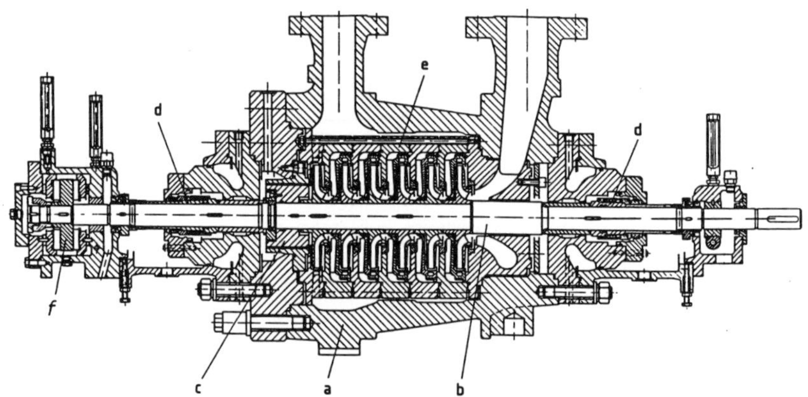

Fig. 4.1-25. Six-stage vertically-split barrel-housing centrifugal pump (SULZER)
a, Barrel housing; b, Shaft; c, Axial thrust compensation; d, Shaft seal; e, Six internal pump stages.

barrel-housing for water injection purposes in oil / gas productions is shown in Fig. 4.1-25. The special operating condition are demanding for axial-thrust compensation (piston c) and the additional external axial-thrust bearing (e).

For pressures below 250 bar, and lower powers, multistage vertical split radial centrifugal pumps are also used with staged housings and modular construction. They generally require lower investment costs. For the pressure range up to 150 bar, a horizontally split housing design (pipeline pump) is suitable for many applications, including boiler feeding, and transferring liquids into chemical processes.

High-speed centrifugal pumps

As the differential pressure per stage depends predominantly on the circumferential velocity of the impeller, super-high-speed single stage pumps (up to 30 000 min^{-1}) offer an alternative to multistaged features. They represent partial emission [19] compared to the full-emission mode of multistage pumps and are a well-developed alternative to multistage centrifugal pumps in the range up to 100 m^3/h. The required high circumferential speed (max. 200 m/s) as generated by the high rotor speed, effects a pressure increase per stage up to around 200 bar. Although the efficiency is generally poorer than in the case of multistage pressure increase, high-speed centrifugal pumps have a compact construction and have lower space requirements. Two-stage gear-centrifugal pumps (Fig. 4.1-26 B) achieve higher pressures than single-stage ones (Fig. 4.1-26 A). However, the shaft seal of the second stage is exposed to higher pressure. The slow-running booster stage that is shown in Fig. 4.1-26 B is used to increase the suction pressure or the head above vapour pressure (NPSHR) to avoid cavitation.

For hydraulic powers below 1 MW (pressures from 100 to 300 bar), only a detailed calculation will generally indicate the most favourable choice between multistage centrifugal pumps, high-speed single- or two-stage centrifugal pumps or reciprocating multi-head-piston or- diaphragm pumps. The decision for the best choice will be provided by analysis of the life-cycle costs.

Hermetic centrifugal pumps

With centrifugal pumps, there is a permanent trend towards avoiding leakages and problems with shaft-seals by means of a leak-free design. Normally, the shaft-seals of centrifugal pumps are exposed only to the intake pressure, and thus can be dealt with relatively easily. With higher intake pressures or system pressures, it is recommended that centrifugal pumps with a canned-motor drive be used, especially in the power range below 500 kW and if the fluids are toxic or hazardous [20, 21]. Experience is available up to system pressures of 1000 bar. The support of the thin-walled containment shells (between the stator and the rotor) of austenitic steel in the grooved stator against the high-pressure is especially important. The rotor bearings, lubricated by the delivery fluid, are now frequently made of silicon carbide.

Fig. 4.1-27 shows a multistage canned-motor centrifugal pump with a separate water-cooling circuit for the motor, such as is used, for example, to circulate carbon dioxide at 300 bar system pressure. The pump rotor is mounted in the motor area; the tandem pump stages operate over-mounted on both sides of the bearings. This arrangement is recommended if the required differ-

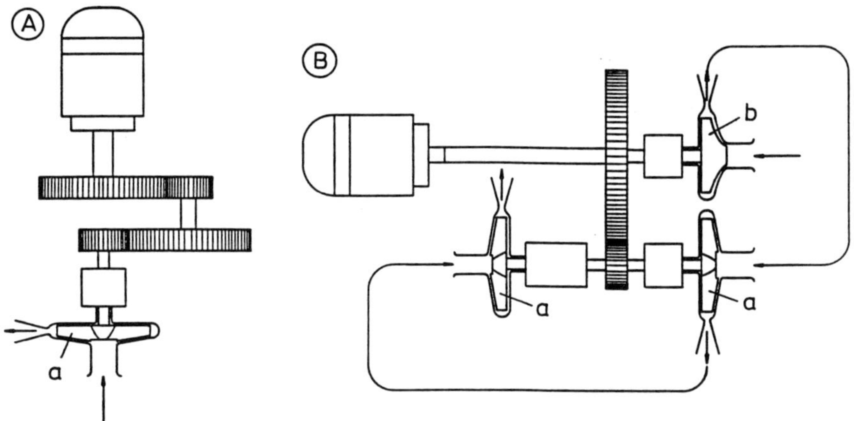

Fig. 4.1-26. High-speed gear centrifugal pump.
A, Single stage; B, Two stage with booster stage for improving the NPSHR; a, Stages; b, Booster stage.

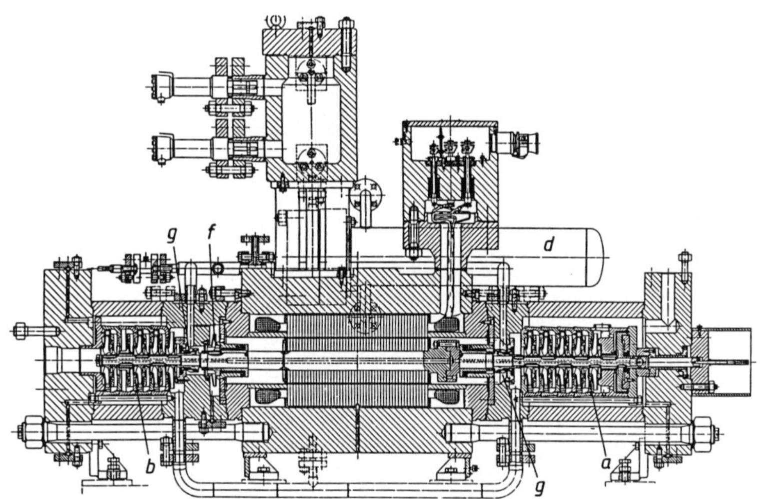

Fig. 4.1-27. Ten-stage canned-motor centrifugal pump (HERMETIC)
a, b, Tandem rotor; c, Canned motor; d, Cooler; e, Terminal box; f, Cooling circuit impeller; g, Mechanical shaft seals.

ential pressure demands for many stages in order to keep the bending deformation of the pump shaft as low as possible.

The internal cooling can be achieved by circulating the discharged fluid with a cooling circuit impeller (f) and an external heat exchanger (cooler d). Semi-hermetic features are offering the possibility of using a liquid for the cooling circuit which is different from the discharged fluid. In this case the mechanical seal (g) serves to separate the two liquid regions (Fig. 4.1-28).

Permanent-magnet drives for pumps and stirrers are increasingly being used in high-pressure technology [22, 23]. The containment shell of the magnet drive, in contrast to canned motors, must take up the full pressure-difference and must have appropriately thick walls.

4.1.4 Compressors

Optimal compressor- types for the various power ranges are plotted in Fig. 4.1-29, which was calculated on the bases of 1 bar intake pressure and an isothermal efficiency of 64% [24]. It can give only approximate reference points for the most favourable area of application, which varies depending on the manufacturer. If the intake pressure is greater than 1 bar, as is the rule, it is necessary to recalculate. For example: the compression of ethylene at a flow-rate of 64000 kg/h ~ 51 130 Nm^3/h (Standard cubic metre per hour) from 231 to 2151 bar, corresponds to a real intake volume flow-rate at 231 bar of 161 m^3/h, and a power consumption of 8200 kW.

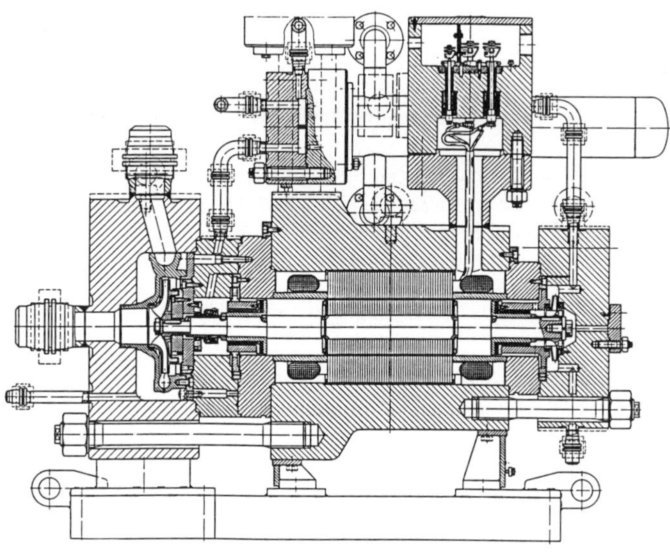

Fig. 4.1-28. Single-stage canned-motor centrifugal pump (HERMETIC)
1, Impeller; 2/3/4, Cooling circuit; 5, Canned motor.

164

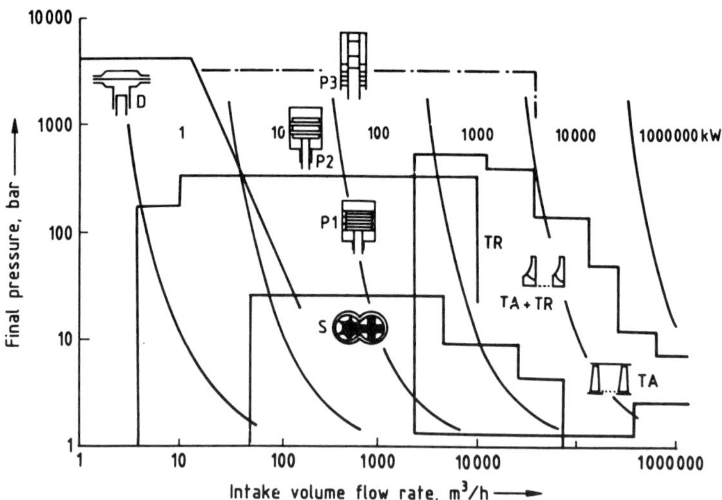

Fig. 4.1-29. Power range of various compressor types for high-pressure [24]. D, Diaphragm compressor; P1, Piston compressor, dry running (piston ring, labyrinth); P2, Piston compressor, lubricated (piston ring); P3, Ultra-high-pressure compressor, lubricated (plunger piston); TR, Radial turbo-compressor; TA, Axial turbo-compressor; S, Screw compressor.

The application boundary between piston- and radial-turbo-compressors lies approximately at 10^4 m³/h intake volume flow rate (power > 1 to 5 MW). In individual cases, the most economical solution must be determined based on the life-cycle costs.

After a lively development in the past two decades during which positive-displacement screw-type and radial-turbo-compressors have replaced piston-type compressors in certain areas. The trend is now towards reciprocating compressors, again owing to their lower power consumption for many applications.

Nevertheless the radial-turbo-compressors demonstrate significant advantages with respect to compactness, space requirements, quieter operation, and lower level of vibrations, as well as giving a smooth non-pulsating flow. Their application for high-pressures in the range of above 5000 N/m³ intake volume, is well established up to pressures of 400 bar and more.

4.1.4.1 Piston compressors

At lower powers (< 2 MW), there is much development of piston compressors in the direction of a higher stroke frequency, a more compact modular construction with a horizontal-, vertical-, or V-shaped arrangement, dry-running compression, sparing lubrication and simpler maintenance. The dry-running range for continuous operation now is extending up to about 300 bar, depending on the type of gas. Labyrinth- and piston-ring-compressors have characteristic fields of application. The advantage of dry-running gas compressors is that process contaminations, and reduction product quality by lubricants, can be avoided completely.

The labyrinth compressor is suitable for gases of higher density (molecular mass > 15 kg / kmol). The labyrinth-seal of the piston (see Section 4.1.5) is non-contacting, operates without

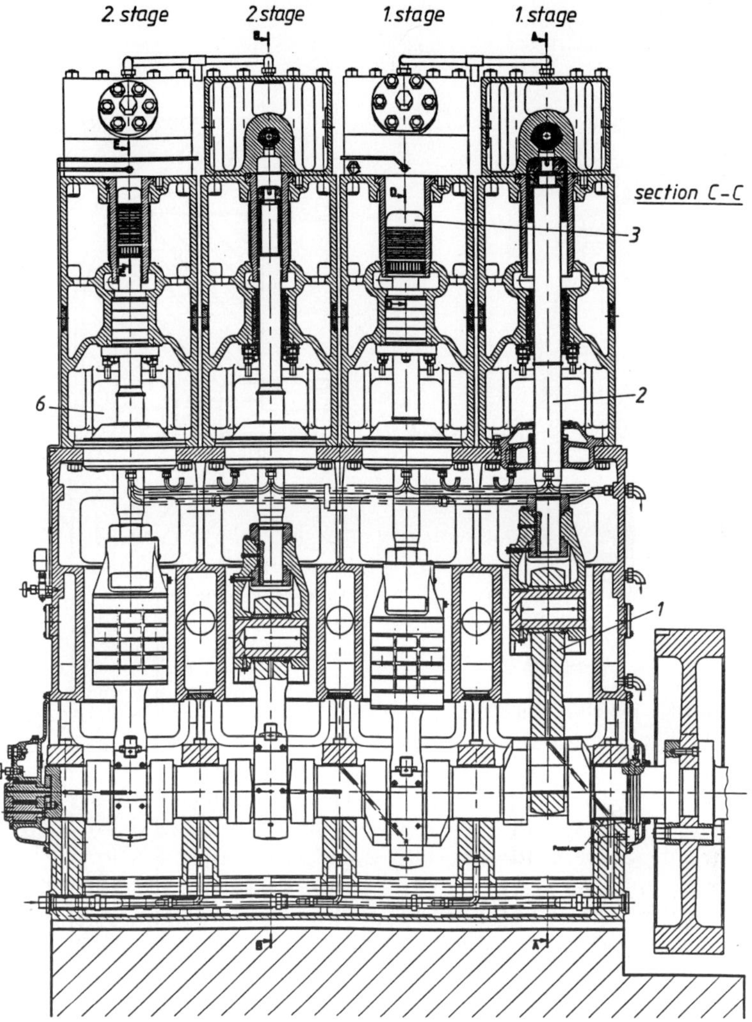

Fig. 4.1-30. Vertical labyrinth piston compressor, longitudinal section (adopted from SUL-ZER)
1, Crank drive; 2, Rod; 3, Piston; 4, Valves; 5, Rod seal; 6, Intermediate "lantern" housing.

friction, requires no maintenance, relatively insensitive towards fine particles and is generally reliable.

Labyrinth compressors are normally built vertical, with a multistage (4 stages up to around 200 bar) arrangement and with separate open- or closed- (flushed) lantern housings, for

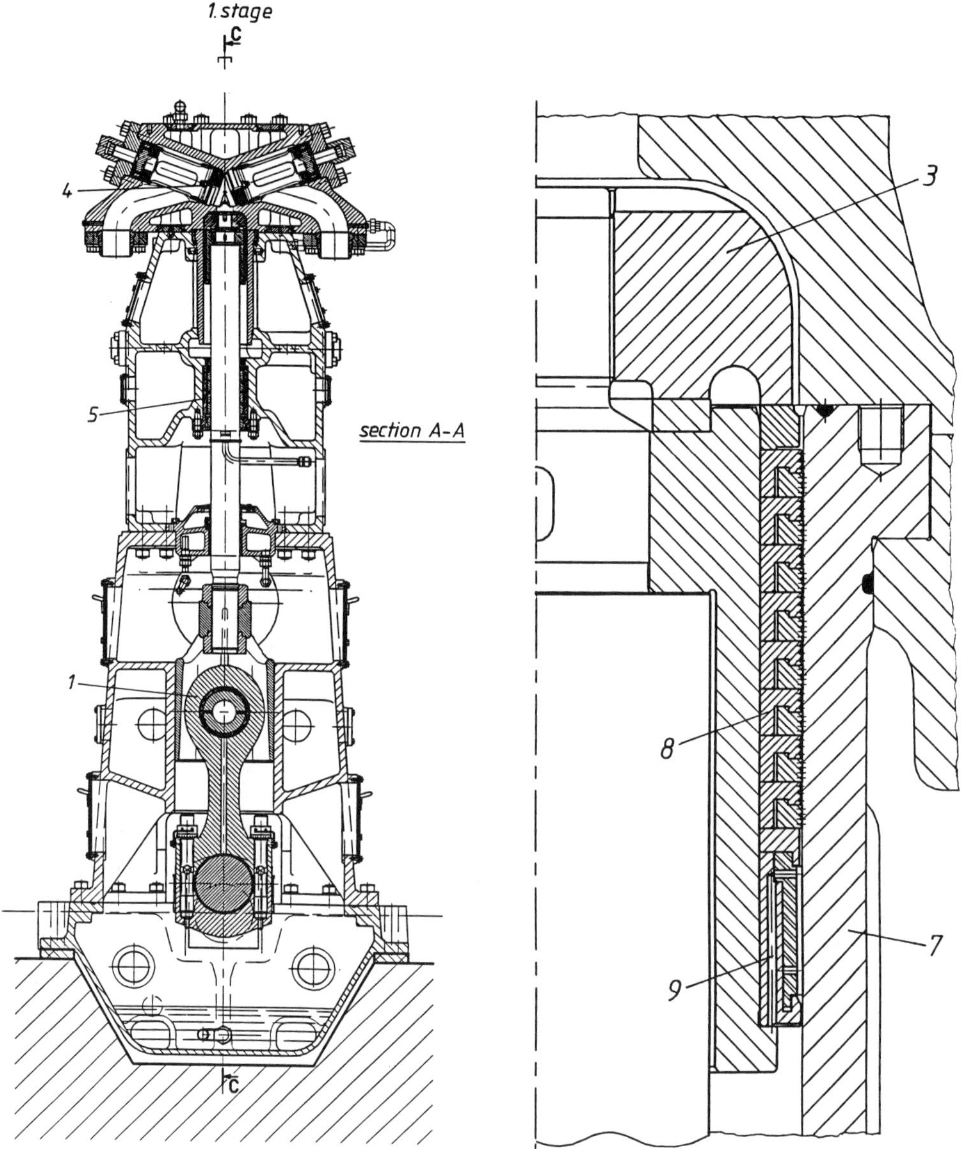

Fig. 4.1-31. Labyrinth compressor as in Fig. 4.1-30.
Left, Cross section; Right, Piston seal; Detail (X); 7, Cylinder bushing; 8, Labyrinth twin-rings; 9, Labyrinth guiding ring.

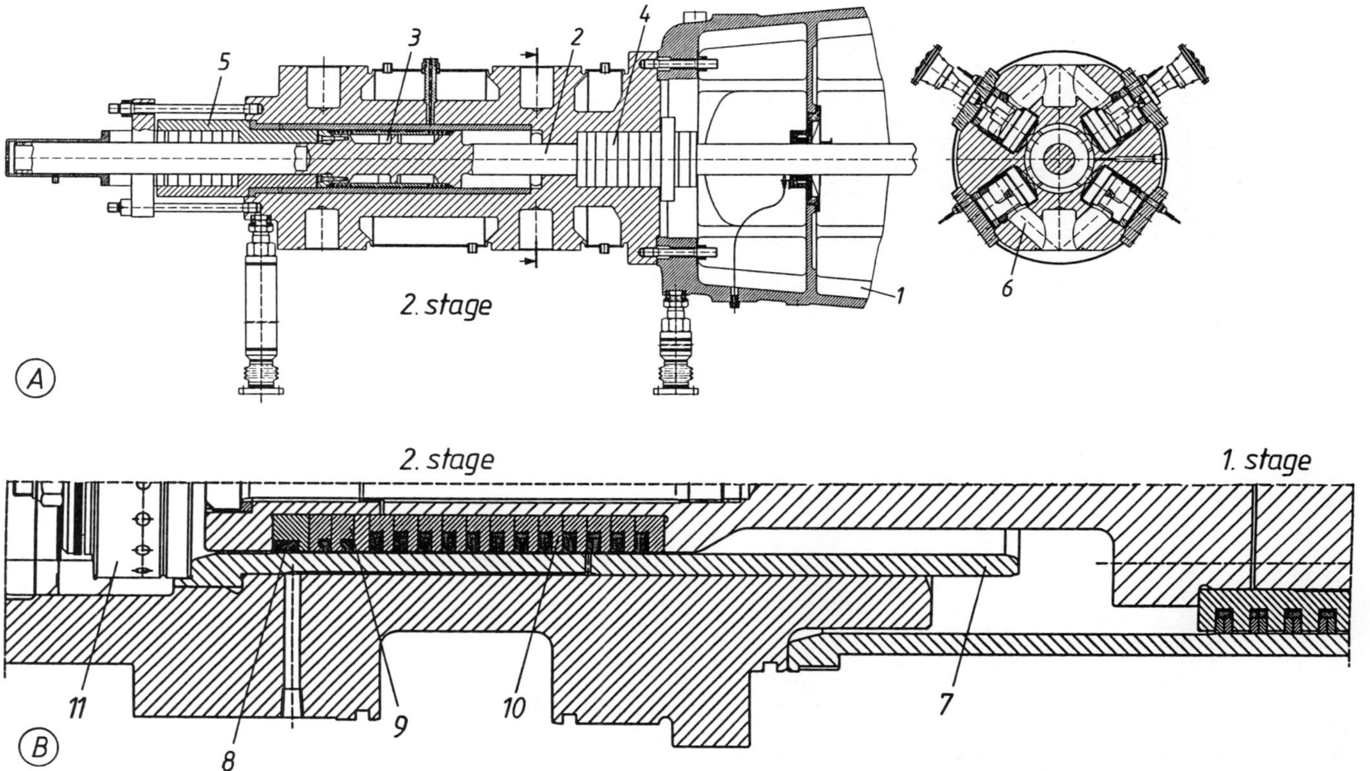

Fig. 4.1-32. Dry-running stages of piston-ring compressors (NEUMAN & ESSER).
A, Second stage of a natural gas compressor (270 bar); B, Two-stage hydrogen compressor (250 bar); 1, Crank drive; 2, Rod; 3, Piston; 4, Rod-seal; 5, Tail-rod-seal; 6, Valve-housing; 7, Cylinder-bushing, 8, Guide-ring; 9, Trapped piston-ring; 10, Twin-piston-rings; 11, Coaxial suction/discharge valve.

hazardous gases, between the compressor head and the crank drive unit. Figs. 4.1.30 and 4.1.31 show a two-stage four-cylinder labyrinth compressor for high pressure. In order to keep the clearances at the piston and rod (labyrinth) seals small, the plunger rod has to be guided precisely and bending forces (vertical) and thermal expansion effects must be reduced as much as possible (by stiff design and by cooling). The tendency towards larger volumetric losses by the labyrinth seal must be balanced by minimal dead-spaces and lower friction of the non-contacting seals. The reliability of labyrinth compressors is unmatched [25] but the application range is restricted to the "heavier" gases only.

Dry-running piston-ring compressors can be used universally, as far as the molecular mass of the gas is concerned. The piston-rings slide and contact the wall of the cylinder under compressive forces. To achieve satisfactory running times (> 10 000 h) improved design and material concepts must be implemented for the piston rings and the compressor valves. The upper pressure limit for reliable dry-running piston-ring seals is steadily increasing, and is now around 300 bar thanks to high performance plastic compound materials [26]. The second stage of a two-stage dry-running piston compressor (natural gas: 30 / 270 bar), shows as important items (Fig. 4.1.32): a steel cylinder with internal bushing; the piston rod with guiding and sealing packing on the crank as well as on the cover side (tail rod), in order to obtain precise piston motion and to avoid lateral forces on the piston rings.

The compression ratio per stage should decrease with the absolute-pressure-differential per stage, in order to reduce the sealing compressions. With growing final pressures (above around 300 bar) the piston compressors with lubricated piston and shaft- seals offer longer lifetimes than those with dry-running features, but the development is continuing.

The so-called hypercompressors for the production of LDPE represent a special case. The ethylene is compressed in a primary piston compressor, with several stages up to around 200 to 300 bar; the hypercompressor (or secondary compressor) brings the gas up from there to 3000 bar. The hypercompressors show pairwise-opposite-cylinders, and are built with up to fourteen cylinders in a multiplex arrangement. The components loaded by an internally pulsating pressure are either shrunk and/or autofrettage-treated in order to implement protective compressive residual stresses (Fig. 4.1-34).

The coaxial arrangement of the compressor valves (plane, spring-loaded, Fig. 4.1-34 B) avoids stress concentrations, by intersections of bores (see Fig. 4.1-18 and Section 4.1-5).

Hypercompressors are working with stationary plunger-seal elements, in general, which is a fundamental difference from standard piston compressors with mobile piston rings. Only with a stationary sealing arrangement does one achieve the required effective high-pressure lubrication, which furthermore demands extremely hard, smooth, and precisely machined plungers. The lubricants injected into the plunger-seal elements with high-pressure pumps must be compatible with the process requirements around 3000 bar in two stages [27,28]. The drive have a robust horizontal cross-head sliding block (Fig. 4.1-33D) in order to provide a precisely guided plunger rod and to keep transverse forces totally away from the stationary seal elements.

Diaphragm compressors, laboratory high-pressure compressors
The diaphragm compressor (Fig. 4.1.35), one of the oldest leak-free process machines, is limited to lower power (< 100 kW) because sensitive metal diaphragms must be used due to the high compression temperatures and pressures. Its applications include laboratories, pilot installations, and special production facilities. An attractive feature is their very high pressure-ratio (up to 20) produced by a single stage as a result of the small dead-space and good cool-

ing. At every stroke, the diaphragm is pressed against the upper contact surface by a hydraulic fluid (smallest clearance volume) [29]. However, for this reason it is sensitive to particles, deposits by gas contamination, liquid droplets, and surface defects, so that its lifetime is frequently less than 5000 h. It is possible to signal diaphragm rupture by means of three-layer sandwich diaphragms. Diaphragm characteristics and rupture-control are similar to those used in diaphragm pumps (Fig. 4.1-36).

Diaphragm compressors are built for pressures up to 4000 bar, at least for short-term operation. They are economical when leak-free operation with no gas contamination is required at

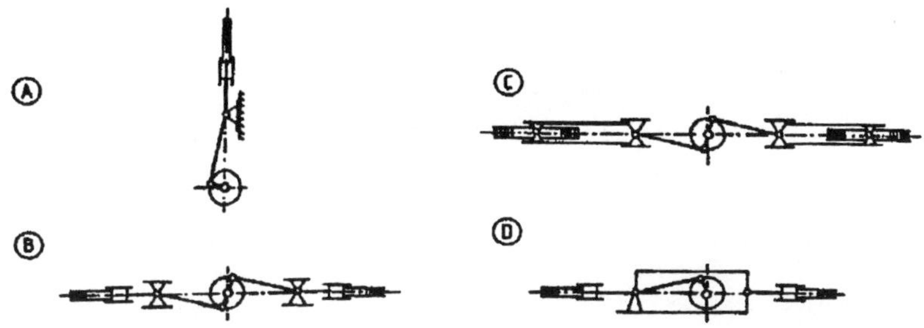

Fig. 4.1-33. Reciprocating drive units for high-pressure piston compressors
A, Vertical crank; B, Horizontal crank C, Horizontal crank (double action); D, Horizontal sliding block.

low-volume flow rates and at high-pressure.

For very small delivery quantities (0.1 to 1 m^3 /h) and pressures up to 2000 bar, air-driven piston- and diaphragm compressors are suitable for transferring and compressing gases. Because of their compact construction, their explosion protection, and their ease of regulation, such machines are attractive for short-term operation in research facilities, especially for booster and filling applications (Fig. 4.1-37).

4.1.4.2 Turbo Compressors

For large flow rates (up to 40 000 m^3/h, 600 bar) various configurations of single shaft radial-turbo-compressors are operating in oil and gas production, pipeline transport and storage of gas, refineries, and petrochemical and chemical industries. A typical feature of the single-shaft turbos is the arrangement of several stages comprising 5 to 8 serially arranged impellers, either straight-through or back-to-back, with intercooling or / and sidestream outlets.

For high pressure the stage housing is of barrel-type (vertically split) on the same reasons as given for centrifugal pumps, above. For hazardous gases this design concepts start already at about 80 bar, owing to more favourable sealing conditions (Fig. 4.1-38). There are good possibilities for compensating the axial thrust by an appropriate arrangement of the impellers, of their clearance, and labyrinth seals between the stages, and of a compensation piston. The higher the pressure the narrower become the flow channels in the impellers, which are consisting of two discs connected by vacuum brazing. The shaft seals must seal against the

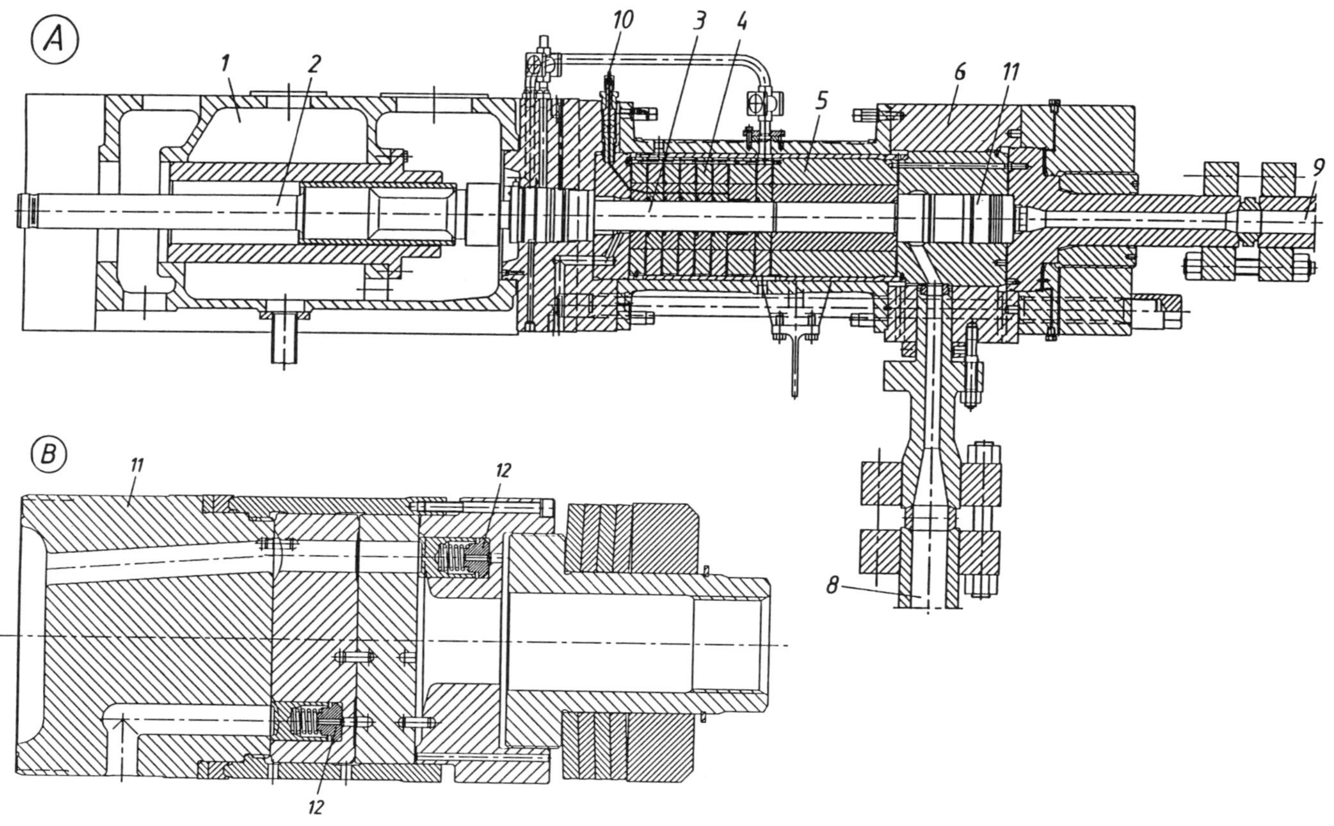

Fig. 4.1-34. Head (A) and Valve package (B) of a hypercompressor (adopted from SULZER BURCKHARDT). 1, Drive unit; 2, Plunger rod; 3, Plunger; 4, Stationary plunger seal-packing; 5, Cylinder; 6, Valve housing; 7, Valve package; 8/9, Inlet/outlet; 10, Lubriction system; 11/12, Valve; 13, Straight smooth bores.

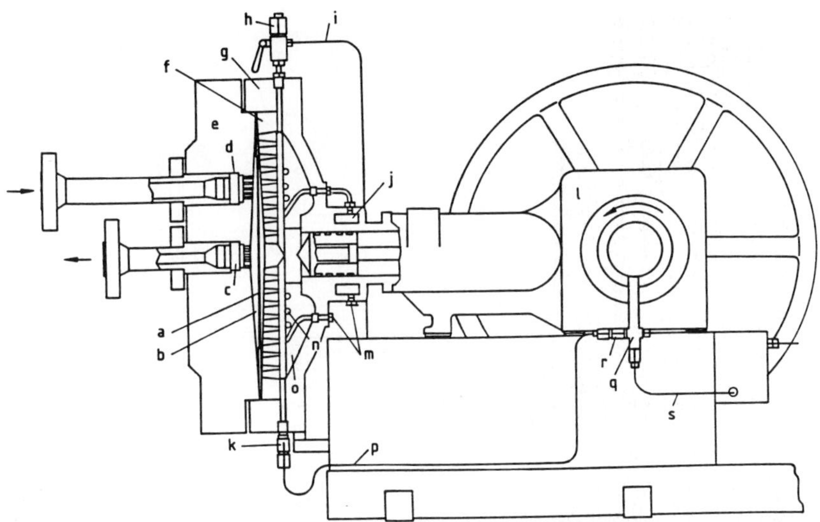

Fig. 4.1-35. Diaphragm compressor (HOFER). a, Diaphragm; b, Gas space; c, Discharge valve; d, Suction valve; e, Diaphragm cover; f, Perforated plate; g, Hydraulic cylinder; h, Oil overflow valve; i, Oil return; j, Cylinder cooling; k, Check valve; l, Crank drive; m, Cooling-water in/out; n, Oil-cooling coil; o, Oil chamber; p, Oil injection (leakage compensation) q, Compensation pump; r, Check valve; s, Oil supply.

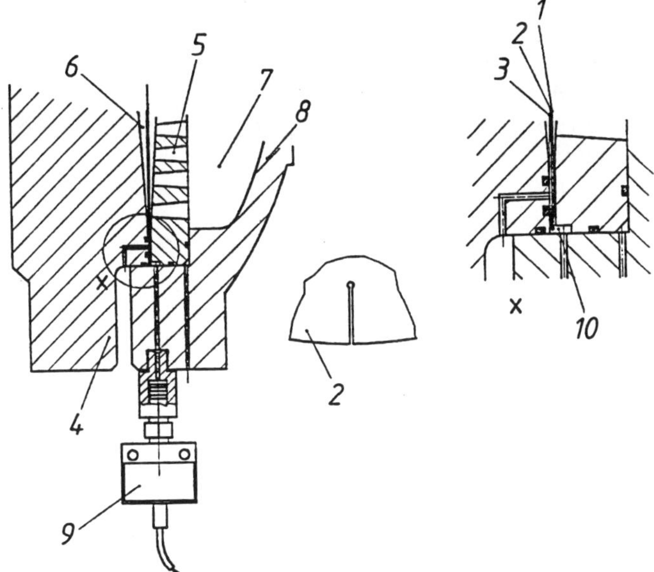

Fig. 4.1-36. Sandwich diaphragm for rupture monitoring (HOFER). 1/2/3, Diaphragm layers; 4, Front cover; 5, Perforated plate; 6, Gas space; 7, Oil space, 8, Cover (housing); 9, Sensor.

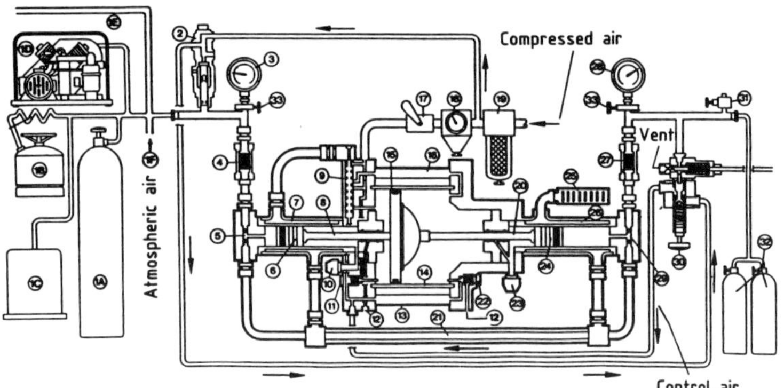

Fig. 4.1-37. Two-stage filling compressor with pneumatic linear drive (HASKEL).

suction-side pressure, and work as labyrinth-, oil-lubricated mechanical contact seals (swimming ring), or mechanical dry-gas seals (see Section 4.1.5).

Radial-turbo-compressors are driven by electric motors, steam- or gas turbines, depending on their size, control strategy, and the most economic energy available.

A series of prototypes was developed to expand the range of application of radial turbo compressors to the level up to 800 bar (Fig. 4.1-39) [30].

The recent development of multistage integrally geared radial compressors [31] is offering a number of advantages: through optimized speed adjustment the stages provide optimal efficiencies; a reduced number of stages through the flexible choice of the impeller-tip velocities; inlet-guide vane control; more compact construction; and lower investment. The schematic arrangement is shown in Fig. 4.1-40 of a 10-stage geared compressor, which bas been built for the compression of 20000 Nm^3/h CO_2 from 1 to 200 bar (Ts-diagram: Fig. 4.1-41). The impeller speed is increasing from around 11000 min^{-1} (first shaft of the gear) up to 50000 min^{-1} (fifth shaft).

4.1.5 Special problems involving high-pressure machinery

4.1.5.1 Strength of the components

In general there is no special problem about dealing with stresses in statically loaded pressure housings especially since a suitable stress-flux can achieve a favourable strain together with the required low deformations, which apply to heat expansion effects, too (e.g., vertically split barrel design) [32]. In contrast, it is much more difficult to dimension components subjected to pulsating internal pressure - as is true for the reciprocating machines. The largest pressure amplitudes regularly have to be faced with reciprocating pumps (see Section 4.1.2).

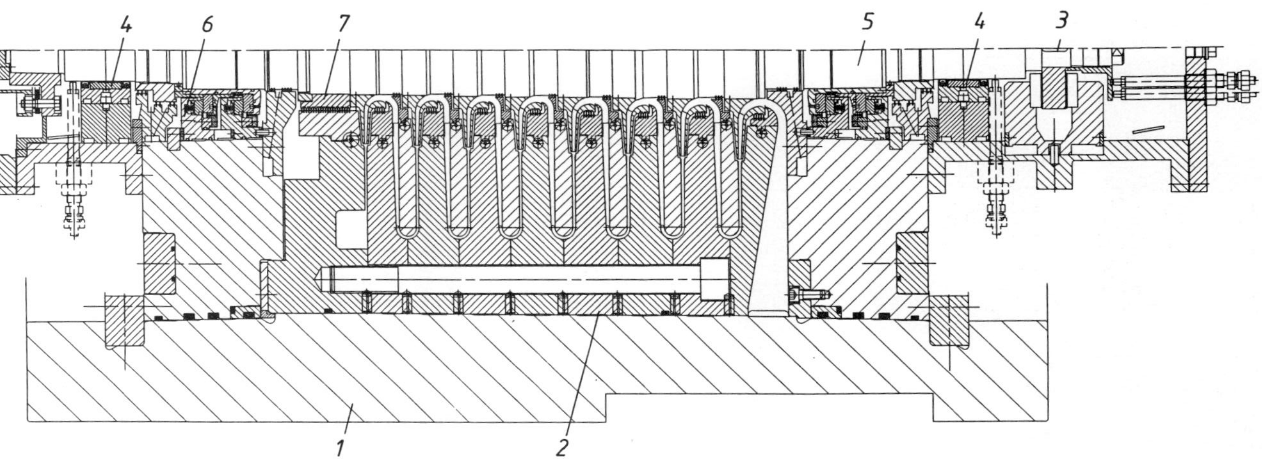

Fig. 4.1-38. Single-shaft eight-stage radial compressor (adopted from MAN GHH BORSIG).
Intake volume, 2100 m^3/h of natural gas; Suction/discharge pressure, 39/186 bar; 1, Barrel housing; 2, Stage housing; 3, Axial thrust bearing; 4, Radial bearing; 5, Shaft; 6, Shaft-seal (gas lubricated); 7, Axial-thrust compensation piston.

174

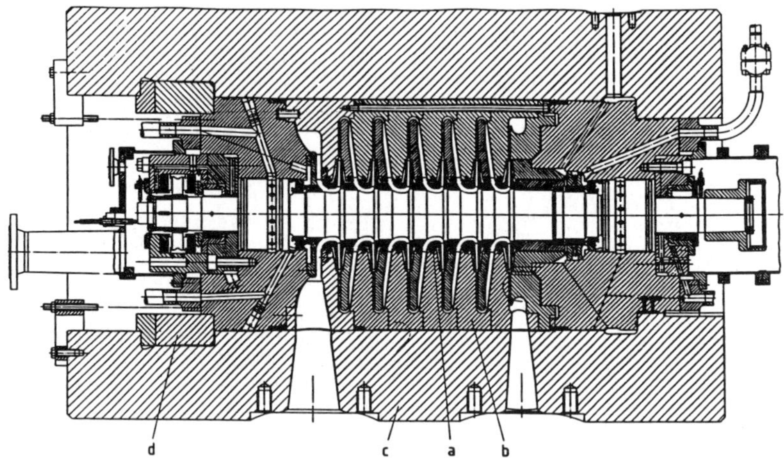

Fig. 4.1-39. Six-stage radial compressor 350/800 bar (MANNESMANN DEMAG).
a, Rotor with compressor stages; b, Stage housing; c, Vertically split barrel housing; d, Closure.

Avoiding local stress-raising effects

Compared to the "smooth" thick-walled pipe, the usually notched parts show notch factors from 1.5 to 3.4 which characterize the local stress-raising effects involved [33 - 37]. The T-intersection of two bores (Fig. 4.1-42a), for example, shows much higher peak stresses (notch-factor around 2) owing to the superimposition of the various stress components involved, than the Y-shaped features (Fig. 4.1-42 b). The best designs with the lowest notch potential have straight "smooth" bores which do not intersect at all (Fig. 4.1-42 d). The high-pressure engineer's answer has already been shown in the Figs. 4.1-18 and 4.1-34 with the design of co-axial valves for pumps and compressors.

The plunger head of an intensifier pump (see Fig. 4.1-14) which is well-established for continuous operation and the metering of initiator solvents in LDPE production (3000 bar) can be seen in Fig. 4.1-43. The components which carry pulsating pressure have either Y-intersected (item 2: autofrettaged) or smooth bores (items 4, 6: shrunk or autofrettaged). Some more design approaches worth understanding can be taken from Fig. 4.1-44. The features in the Figs. 4.1-44 a and b exhibit smooth non-intersecting bores and can carry pulsating pressure amplitudes of well above 3000 bar (if autofrettaged) as required, for example, for jet cutting systems.

Fig. 4.1-40. Ten-stage integrally geared radial compressor (adopted from MAN GHH BOR-SIG)
Intake 20000 Nm^3/h CO_2; Suction/discharge pressure 1/200 bar; Roman Figs: stage numbers; 1, Driving shaft; 2, 3, Gear.

Introduction of beneficial compressive residual stresses

As the radial- and hoop-stress distributions in thick-walled internally pressurized components exhibit a large gradient from the (positive) maximum at the inner- to a minimum at the outer diameter, it is therefore of great benefit to introduce compressive residual stresses at the inner location of the highest operational positive strains. Widely used is the design of thick-walled components of circular shape, radially shrunk with two or more layers (see Section 4.3 and Figs. 4.1-34 and 4.1-43). Thus instead of the uneven stress a more constant stress-distribution and consequently in a total better strength of the containment considered can be achieved (Fig. 4.1-45). Fig. 4.1-44 c shows for the example of the liquid-end (T-intersecting bores) of a triplex plunger pump that, with uni-directional shrinkage it is possible to compensate effectively the local stress-increase by notch effects owing to the hoop stresses superimposed at the corner of the T-intersection.

During shrinkage, the materials involved remain with elastic deformation. The method is well suited for high-strength brittle materials; (e.g., Fig. 4.1-44c: corrosion-resistant high-strength titanium alloys up to 1500 bar for homogenisation pumps).

176

For high-strength steels with excellent toughness (impact strength) methods such as shot-peening and autofrettage, which introduce compressive residual stresses by the local plastification of the material, are well known. Shot-peening, during which particles are blasted against the relevant location, generates local plastification and consequently compressive residual stresses, which reduce appropriately the operational ones as well as stress-raising notch-effects resulting from the geometrics or surface roughnesses.

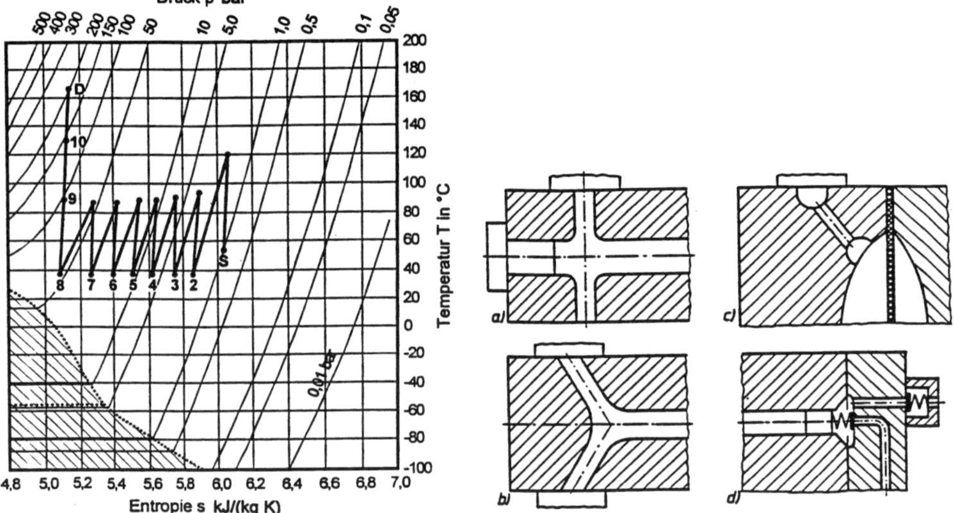

Fig. 4.1-41. Ts-diagram (compressor Fig. 4.1-40).

Fig. 4.1-42. Typical thick-walled components
a), T-intersection; b), Y-intersection; c), Ball/Cylinder-intersection; d), no intersection.

As Fig. 4.1-42 c indicates, the geometrics of the components must cope with the necessary accessibility for the shot-peening tool.

During autofrettage treatment the thick-walled components must be pressurized to such a level as to generate plastic strains through a certain fraction of the wall thickness, the outer region staying elastically strained only.

After the removal of the autofrettage pressure the part shows the compressive residual stresses required to reduce the operational stresses effectively (see Chapter 1: Introduction; Fig. 1.4.10). Successful autofrettage treatment needs tough and sufficiently strong steels which should offer a certain potential for strain-hardening without unacceptable embrittlement.

The finite-element method applied in the elastic-plastic mode gives satisfactory predictions of the achievable improvements from autofrettage. As the residual stresses reduce efficiently the initiation and the growth of cracks, autofrettage is especially well useful for components

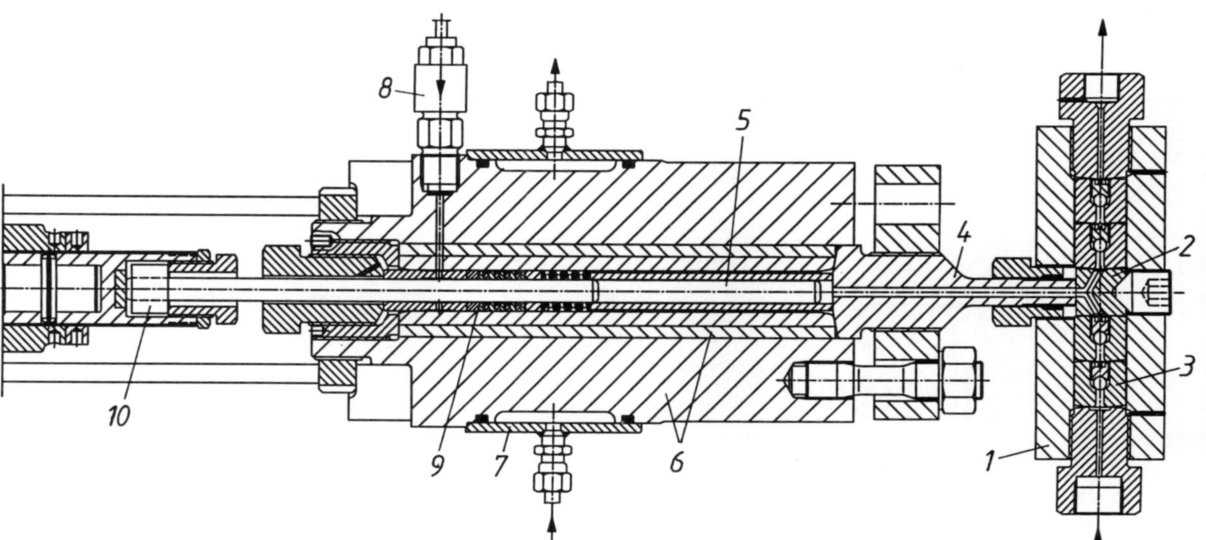

Fig. 4.1-43. Plunger pump head for 3000 bar (KRUPP UHDE).
1, Valve housing; 2, Y-intersection; 3, Valves; 4, Connecting piece; 5, Plunger; 6, Shrunk cylinder; 7, Cooling, 8, Lubrication; 9, Plunger seal (lip-rings); 10, Plunger suspension

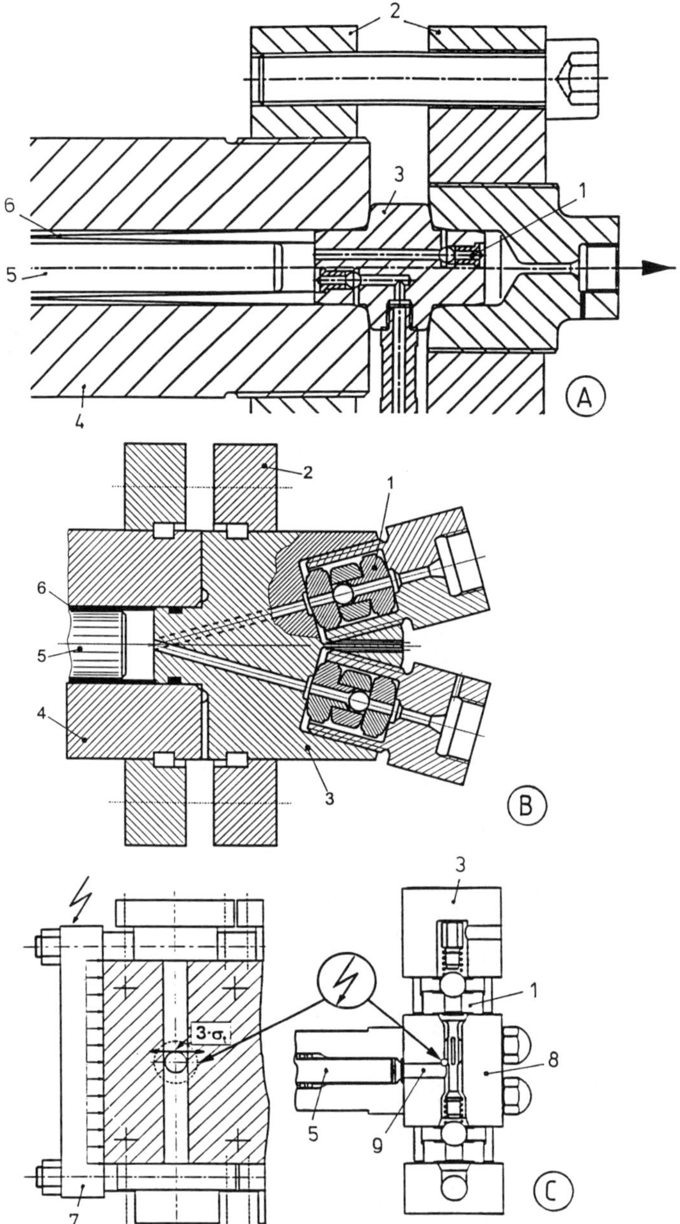

Fig. 4.1-44. Design steps to reduce stress peaks
A, Straight bores, no intersection, in pulsatingly loaded components; coaxial valve(adopted from KRUPP UHDE); B, Offset straight bores, no intersection (adopted from BÖHLER); C, External compressive forces (linear shrinkage).
1, Valve; 2, Flanges; 3, Valve housing; 4, Pump cylinder; 5, Plunger; 6, Bushing; 7, Flanges (linear shrinkage); 8/9, Pump housing/T-intersection.

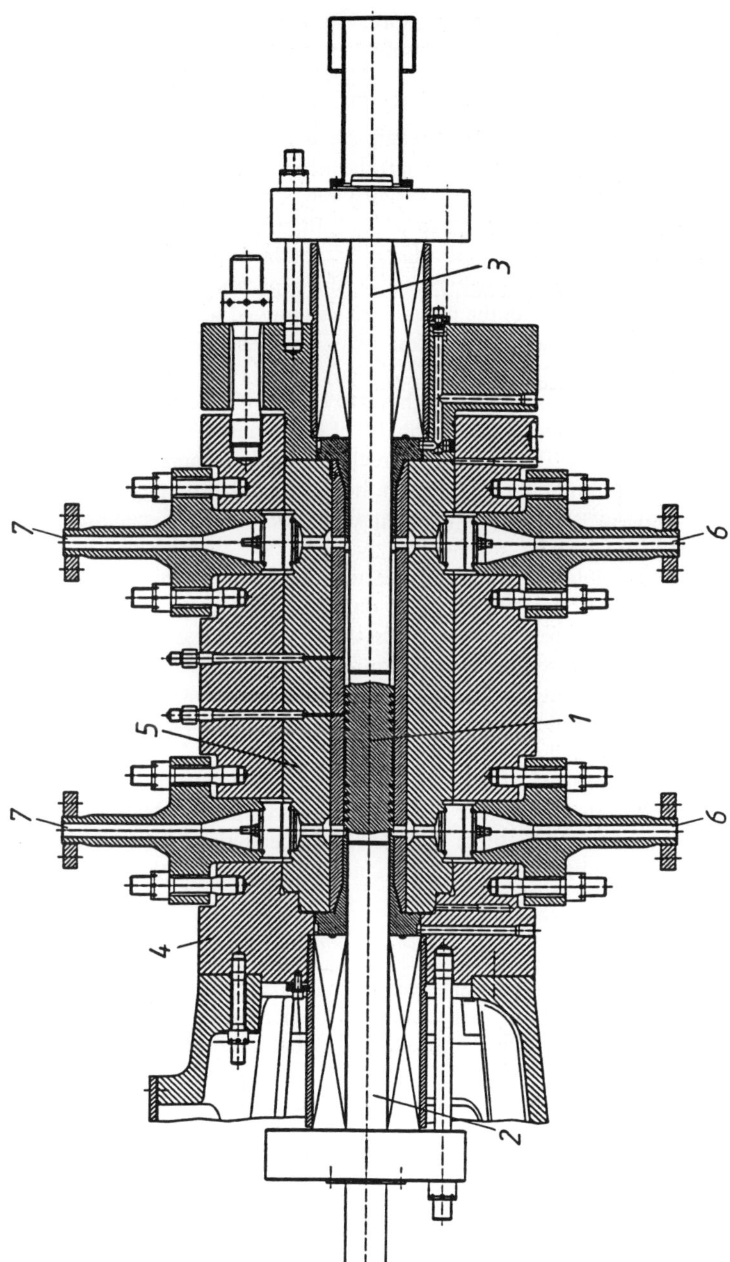

Fig. 4.1-45. Double-acting compressor stage for 700 bar with a shrunk cylinder (NEUMAN & ESSER). 1, Piston; 2, Rod (crank side); 3, Tail rod; 4/5, Shrunk compressor cylinder; 6, Discharge; 7, Suction.

which are submitted to pulsating internal pressure. Fig. 4.1-46 shows the improvement of the endurance limit (admissible amplitude of pulsating pressure) by autofrettage for cross-bored pipe probes from various highly alloyed CrNi-steels (Table 4.1-1) indicating that the notch-effect of the cross-bore can be compensated if the yield-strength of the material is high enough. Further clear evidence is that low-strength steels (see steel No. 1.4306, 20°C) provide only weak improvements with autofrettage, but if the same is applied at extra-low temperature (see steel Nr. 14.306; -90°C) the autofrettage has a very significant positive effect owing to the higher yield-strength at the low temperature [35] [36] [37].

It should be understood that the positive effects of autofrettage are always based on two different sources: compressive residual stresses, which permit a larger amplitude of the operational stress and, depending on the material, the improved local strength of the material owing to strain-hardening by local plastification.

Selection of materials

The steels which are predominantly used have to cope with the requirements of strength and corrosion resistance. The steel types should be of high quality, in general, but especially purity, isotropy and toughness. The production methods should be appropriate with regards to melting, casting, refining, remelting (e.g., ESR), hot- and cold forming, and heat treatment. Table 4.1-1 gives a survey of the steel types applied. All the listed steels are suitable for autofrettage treatment [38].

As an amendment to Table 4.1-1 it should be mentioned that within the range of duplex and superduplex steels there are further types on the market (e.g., X5CrNiMoCu 218 or X5CrNiMoCu 258) which offer similar strength and toughness but partially superior corrosion resistance. In rare cases, other material grades, such like nickel-base and titanium alloys, will be required. As mentioned earlier, brittle materials are not suitable for pulsatingly pressurized components for safety reasons, unless if tensional stresses can be excluded to a very high degree by shrinkage.

It should be remembered [39], that when evaluating data about fatigue limits in the pulsating or alternating mode which have been determined with inert fluid environments, that the corrosion fatigue may not only effect remarkably the reduction of the admissible stresses but also depends on the number of cycles (a continuous small slope of the Wöhler-curve).

4.1.5.2 Seals

In high-pressure technology, metallic seals, such as lenses, ring joints, conical-, flat lapped- or profiled types are used as static seals. Self-sealing elastomer sealing rings, e.g., O-rings, have also proved suitable for easily loosened connections, as long as the pressure, temperature and fluid permit their use.

Where such seals are unfavourable, for example, with horizontally split compressor housings, structural measures, such as the use of vertically split barrel housings, are used to guarantee a reliable seal. The clamping and sealing of diaphragms for diaphragm pumps and compressors requires special static seals [5, 29].

The problem- points in high-pressure machines are certainly the dynamic shaft- and piston- or plunger seals. Seals create special problems when they slide rapidly under high-pressure and if the fluid being sealed has no lubricating actions. Cyclic stress creates additionally fatigue problems in the seal.

Table 4.1-1. Steels for high-pressure application.
F, Ferrite, C, Carbides, A, Austenite; M, Martensite; TS, as tempered; IP, Intermetallic phases; PH, Precipation hardened; H & T, Heat treated and tempered, H, Heat-treated.

Group, structure	No	Name	Yield strength R_{p02} N/mm^2	Impact-strength (ISO-V) J	Corrosion resistance
non-alloyed, heat-treat-able (F,M,TS)	1	34 CrNiMo6 1.6532	600-800	>50	-
	2	30 CrNiMo8 1.6580			
high-alloyed, heat-treat-able, chromium, stainless (3 / 4): M + C, TS	3	X35CrNiMo17 1.4122	500-600 (H&T)	>20	o+
	4	X4CrNi134 1.4313	500-700 (H&T)	>70	++
(5): M + A + F + IP	5	X5CrNiCuNb1744 1.4548	500-700 (PH)	70/40	++
high-alloyed, chromium-nickel, stainless, („Soft-Martensitic")	6	X5CrNiMo165 1.4405	700-900 (H)	>80	+++
high-alloyed chromium-nickel, stainless (A + F) („Duplex", „Superduplex*")	7	X2CrNiMoN2263 1.4462	450 (quenched)	>100	++++
	8	X2CrNiMoN2654 1.4467	690 (quenched)	>65	++++
	9	X2CrNiMoCu2674*[)] (similar to 1.4501)	550 (quenched)	>70	+++++
high-alloyed, chromium-nickel, stainless (A, „Superaustenitic**")	10	X1NiCrMoCu25205**[)] 1.4539	230 (quenched)	>100	++++++
	11	X6CrNiMoTi17122 1.4571	240 (quenched)	>100	+++++
	12	X2CrNi1911 1.4306	240 (quenched)	>100	+++++
	13	X5CrNi1810 1.4301	700 (cold-rolled)	>50	+++++

Non-contact seals with defined narrow clearances, such as labyrinth seals in turbo-machines (Fig. 4.1-47 A), are less problematic owing to the relatively low differential pressures; however, the fluid should not have an erosive effect. Differential pressures up to several hundred bar occur at the axial-thrust compensation pistons of centrifugal pumps and compressors, but the fluid expansion is distributed over the many labyrinth stages.

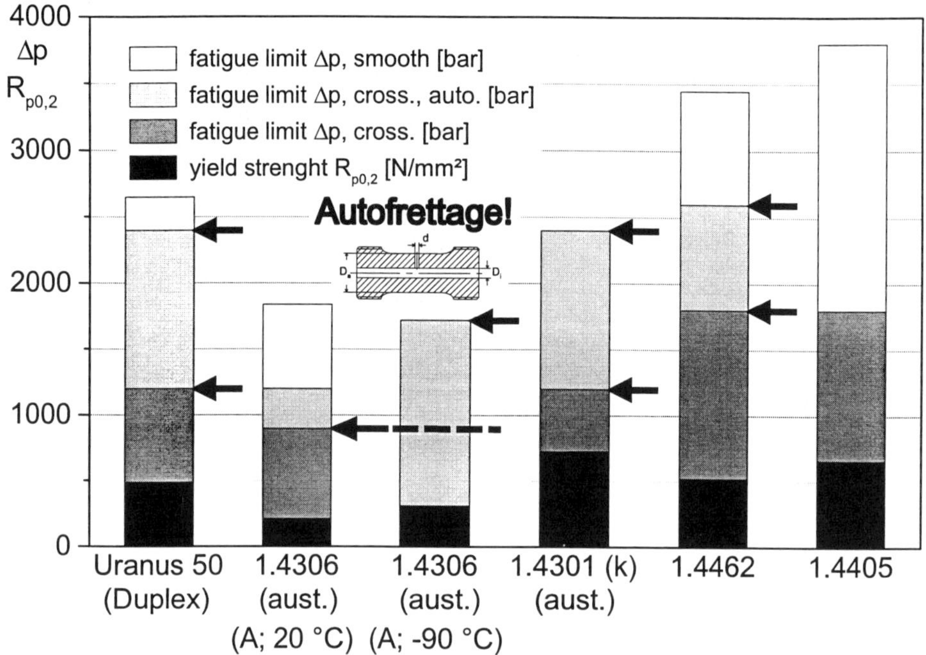

Fig. 4.1-46. Endurance limit of cross-bored pipe probes.
Da = 30 mm; Di = 10 mm; d = 3 mm; full autofrettage (100 % through the wall); specification of the materials see Table 4.1-1.

Labyrinth seals on reciprocating pistons require very precise alignment, and are limited to gases that do not have too low a molecular mass, as well as to pressure differences below around 100 bar (Fig. 4.1-47 B). The materials of the cylinder bushing and the labyrinth-rings (see Fig. 4.1-31, right) should allow run-in contacts without fretting and sparking especially for the compression of O_2. The twin-ring design for labyrinth-pistons offers flexible application advantages.

The leakage-flow of non-contact gap and labyrinth seals can be considerable, and must be recycled internally. It should be noted that labyrinth compressors use this type of seal on the rods, also.

The technology for shaft-seals of high-pressure turbo-compressors has seen further development recently [40]. With the floating ring-seal, a typical shaft-seal for high-pressure turbo-compressors, the high-pressure sealing oil system maintains a slight overpressure compared to the gas pressure at the floating ring, so that the sealing problem is reduced to the gap-seal for the oil [30] and is suitable for several 100 bar. In this case the sealing liquid is the oil, so an appropriate oil supply system must be installed and a very small amount of oil will enter the gas system. Principally, the floating ring seal represents a mechanical seal blocked by high-pressure oil (Fig. 4.1-48).

Tandem mechanical seals, "lubricated" and blocked by a seal gas with slightly higher pressure than the gas discharged by the compressor offer major advantages. There is no contami

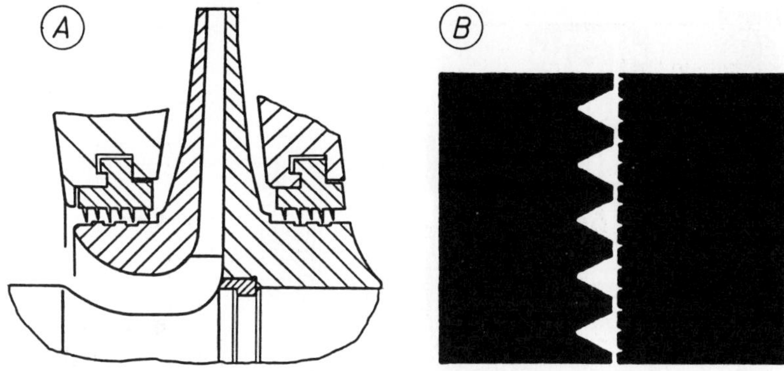

Fig. 4.1-47. Labyrinth seals: A, Radial impeller, B, Labyrinth piston.

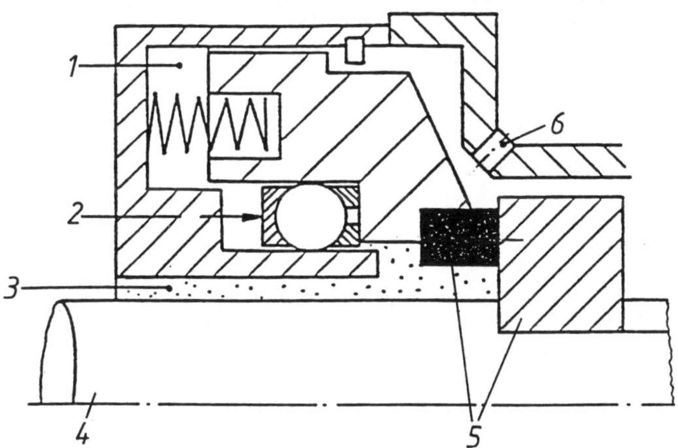

Fig. 4.1-48. Floating ring shaft-seal (MANNESMANN DEMAG).
1/6, Oil supply; 2, O-ring; 3, Gas space; 4, Shaft; 5, Seal rings.

nation of the process with oil. The very low friction is achieved by the generation of a stable gas film due to the profiled seal rings. A large circumferential velocity is admissible. Pressure differences of around 250 bar can be sealed. The sealing fluid is the seal gas and the seal rings are predominantly made from silicon carbide. Dry gas mechanical seals can be flexibly designed for varying operating conditions by flushing with clean gas before the shaft seal, adequate choice of the seal gas, and by flushing and blocking any leakages so that the sealing systems do not pollute the environment.

The shaft-seal development for high-pressure centrifugal pumps is quite similar to that explained for compressors. The traditional mechanical seals lubricated by liquids are now being partly replaced, for obvious reasons, by gas-lubricated ones.

184

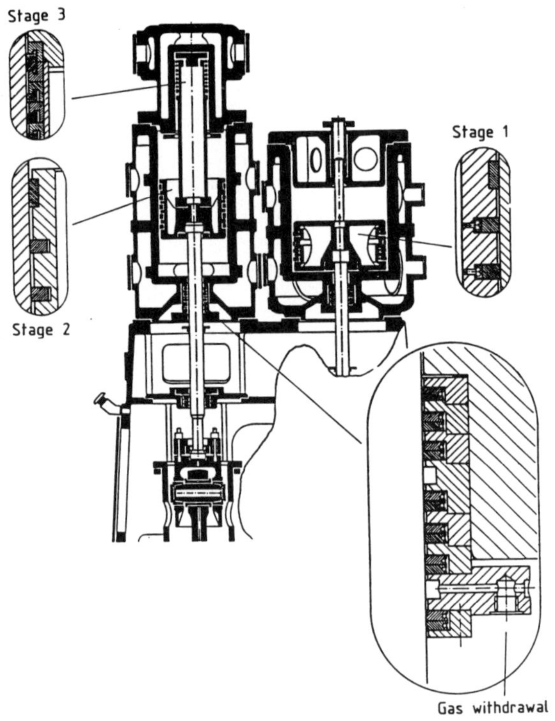

Fig. 4.1-49. Various designs of piston and rod seals (NEUMAN & ESSER).

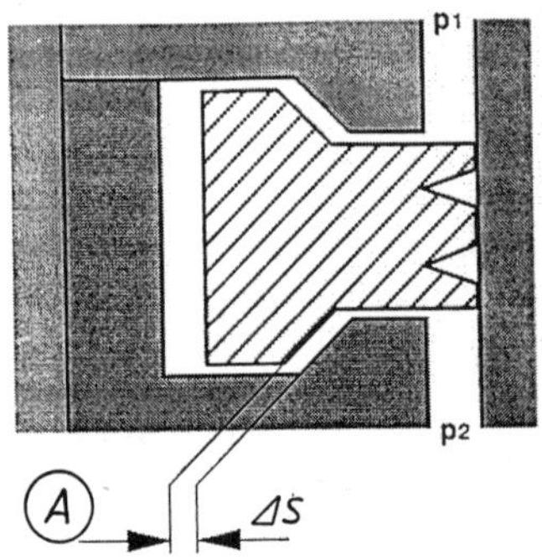

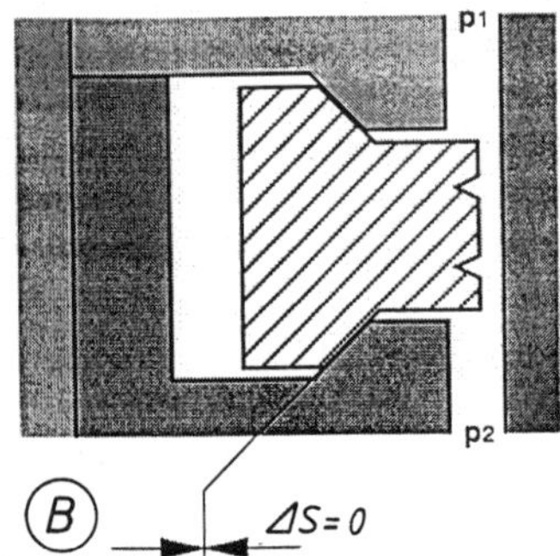

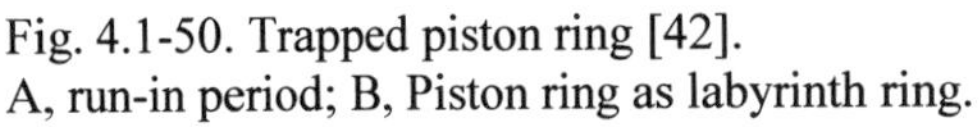

Fig. 4.1-50. Trapped piston ring [42].
A, run-in period; B, Piston ring as labyrinth ring.

For centrifugal pumps the pressures to be sealed are regularly much lower than with compressors owing to the pressure relief by labyrinth seals down to the suction level [18].
Conventional mechanical seals for liquids are limited to differential pressures below 100 bar. The limits are determined by the product of the sliding speed and the differential pressure based on the friction power arising in the gap. With pressures above 100 bar, pressure blocking systems are used to reduce the differential pressure, to avoid evaporation, to improve lubrication, and to provide cooling. High-pressure mechanical seals are needed for centrifugal pumps only if the system pressure or vapour pressure is high. For lower powers, leak-free canned-motor pumps are generally used.

It is especially difficult to seal pistons with reciprocating motion. Piston rings are suitable as piston seals in piston compressors up to 1500 bar. Above 200 bar, lubrication is generally required.
With dry-running, or sparsely lubricated piston compressors, the use of piston rings made of PTFE composites (Fig. 4.1-49) has proved to be effective as an alternative to non-contact labyrinth seals, but without any limitations with regard to the gas properties.
Continued research [41, 42] on the tribological and tribo-chemical effects [43] involved, and improved design features are extending the lifetime of dry-running piston-ring seals towards 15 000 h and more [27]. New and more temperature-resistant plastic materials (polymer blends of PTFE, PEEK, PI or PPS) and additives such as glass- or graphite fibres (or powders), bronze- or other metal powders and MoS_2 to improve the strength, thermal conductivity, and the sliding properties have contributed much to major progress in dry-running gas compressors. The design of piston-rings depends on the pressure level and the differential pressure to be sealed.
The rectangular piston-ring (also with internal spring elements, Fig. 4.1-49, stages 1 and 2) is common in the pressure range below 50 bar. At higher pressure the "trapped" piston ring (see Fig. 4.1-49, stage 3) is especially interesting. It is supported after a run-in period and thus makes contact at low compression or acts as a labyrinth ring with extremely small clearance (Fig. 4.1-50). Typical high-pressure dry-running compressors (Fig. 4.1-32 B) show the piston ring package consisting of guiding rings (item 8), trapped piston rings (item 9) and twin-rings (item 10) sliding in a well designed (frequently hard-coated) cylinder bushing (item 7). The twin-design of piston rings has proven advantages with respect to the pressure expansion along the sealing gap of the individual ring, the tightness and a more flexible choice of the geometry and materials (Fig. 4.1-51).
In contrast to the piston rings, which are radially expanded and pressed to the cylinder, the rod seals are radially contracted. The rod seal usually consists of a serially arranged number of segmented seal rings (Fig. 4.1-52). It is of great importance to have precisely machined smooth and hard rod surfaces (e.g. detonation coated) and to take care that there is proper cooling in order to remove the heat from the friction. The rod and piston seals demonstrate no essential differences with regards to the materials and tribological conditions, however.

Plunger seals of ethylene secondary compressors for the polyethylene production are a special case (Fig. 4.1-53). Supercritical ethylene is compressed from around 300 bar, in two stages, to 3000 bar (specific volume at 4500 bar: 1.5 dm^3 / kg). Stationary bronze seals which run against hard-metal (e.g. sintered tungsten carbide) pistons have proved suitable. The sealing elements consist of a sequence of radially and tangentially slotted rings, generally preceded by a throttle ring. Lubricating fluid is injected between the individual sealing elements, against the high-pressure produced by the compressor [44]. Similar sealing elements (high performance plastic composites) are also used successfully for sealing piston rods at pressures up to 500 bar when taking advantage of lubrication.

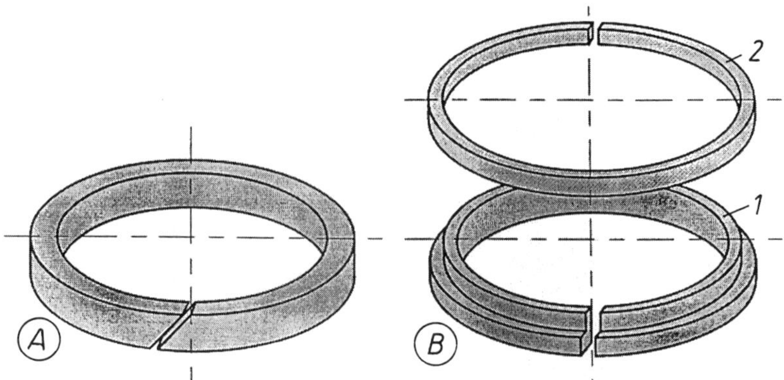

Fig. 4.1-51. Piston ring design [42].
A, slit rectangular rind; B, Slit twin-ring (clearance covered); 1, Filling ring; 2, L-ring.

Piston seals for reciprocating displacement pumps must cope with the highest dynamic pressure differences, because compression is always performed in a single stage, as well as with a wide spectrum of fluid properties (e.g., toxicity, corrosiveness, abrasiveness, low vapour pressure). Lubrication is often ineffective because the pumped fluid acts as a solvent. The seals vary widely in terms of shape and material. Solid or braided PTFE and graphite composites together with other corrosion- and temperature-resistant plastics are often used. The plunger should be hard, smooth, geometrically precise, guided centrally and linked flexibly in the rod of the drive unit so that it runs in the seal "without force". The sealing elements should press just as hard as is required for sealing. Dry running must be avoided, possibly by lubrication or flushing.

The plunger-packing of a process plunger pump as shown in Fig. 4.1-54 demonstrates the essential design steps: the plunger (1) is well guided (guiding rings 2) and flexibly linked to the rod (3). The packing rings are tensioned by the adjusting nuts (4,5) separately, and lube- and flushing connections permit the establishment of favourable operating conditions according to the process requirements. Packing-ring seals have proved to be suitable for the pressure range up to 300 bar and above. If the fluid contains abrasives, or tends to crystallize, there is a possibility of injecting a suitable clear liquid from an external injection pump during the suction into the working space before the packing in order keep abrasives away from the plunger seal.

The maintenance and installation of piston seals requires a great deal of experience; the necessary flushing- and lubricating systems are sometimes quite complicated. Frequently, the optimal design (3000 to 8000 h lifetime) must be determined empirically.

At very high-pressures (> 1000 bar), special measures are needed, using plungers and guide bushings of sintered hard metals; extremely precise, smooth machining and seals made of polymer- blends compounded with graphite. To this is added an external lubrication, which supports the discharge fluid lubrication, as well as sealing elements which operate as automatically as possible under pressure (Fig. 4.1-43: spring-loaded lip-rings). A low piston speed reduces the friction power and increases the lifetime of piston sealing elements. In this respect, hydraulic linear drives (Fig. 4.1-14) are more favorable.

It should be noted that shaft- and plunger-seal problems can be avoided by using hermetic (seal-less) pump types.

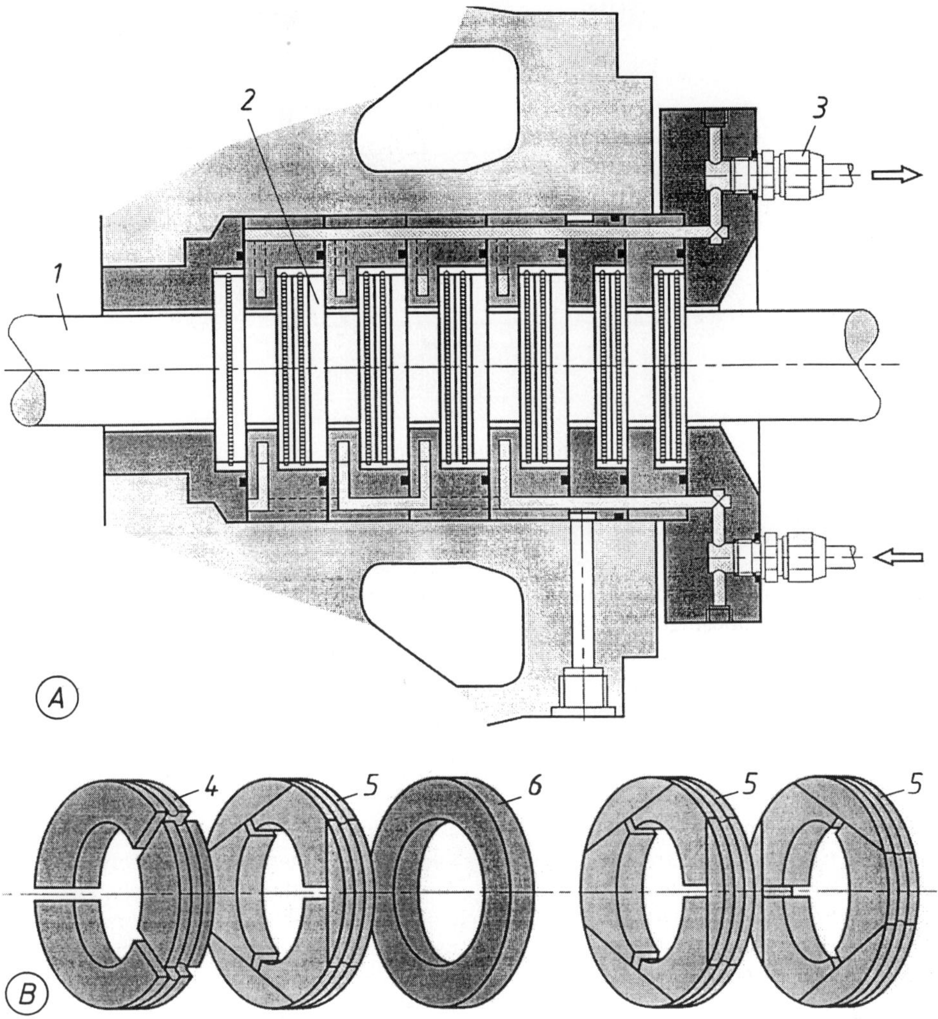

Fig. 4.1-52. Shaft seal [42].
1, Shaft; 2, Seal element; 3, Cooling; 4, Three-piece segmented ring; 5, Six-piece semented ring; 6, Support ring.

188

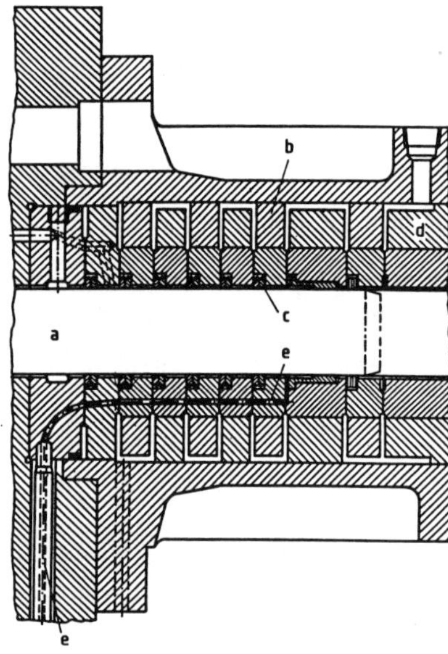

Fig. 4.1-53. Plunger seal of an ethylene hypercompressor 3000 bar (NUOVO PIGNONE).
a, Plunger; b, Seal housing; c, Sealing rings; d, Compressor cylinder; e, Lubrication oil.

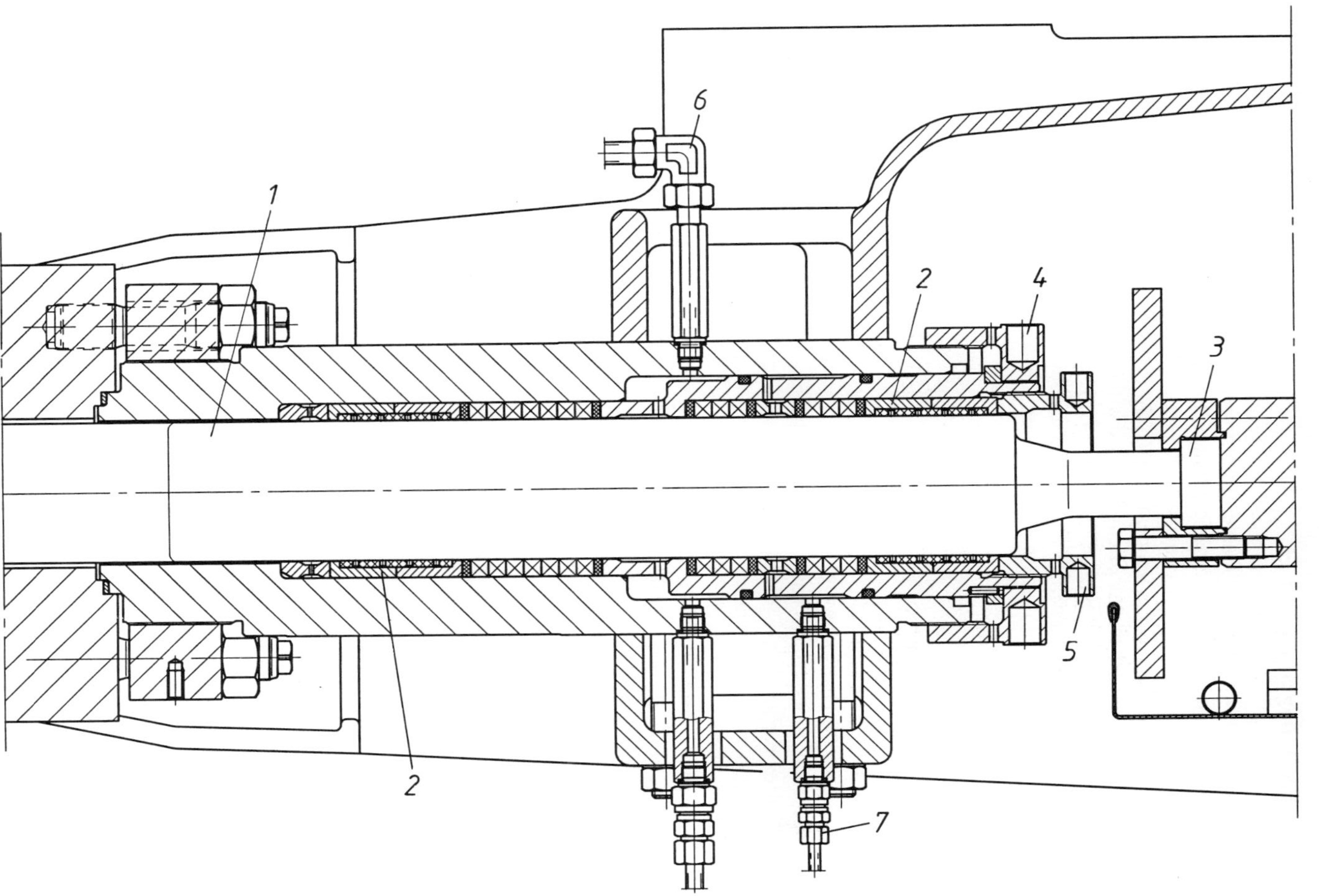

Fig. 4.1-54. Plunger seal with packing rings (URACA).

1, Plunger; 2, Guide bushings; 3, Plunger joint; 4/5, Adjustment nuts; 6, Lube oil; 7, Flushing liquid.

190

4.2 High-pressure piping equipment

High-pressure processes require suitable fittings, seals, tubing and valves. The design concepts, the material selection and the basic attempts to secure safety and tightness are very similar to those for high-pressure machinery (see Section 4.1). As the piping equipment is usually more exposed to the process temperatures (from cryogenic up to more than 500° C) this aspect should be focused on the strength and the thermal expansions involved. In the special case of tubular reactors as applied, e.g., for LDPE-production (up to 3000 bar) or for bauxite-leaching (up to 200 bar) a part of the tubing (several 100 metres in length) is included within the reaction system of the process together with the heating and cooling devices (see Sections 4.3 and 5.1.2).

4.2.1 Tubing and fittings

The Standard DIN EN or ANSI ASME components are generally only used for the pressure range below 200 bar in the process industry and will not be discussed further here.
There are historical roots which have generated special high-pressure standards for the chemical industries earlier than was the case for other industrial applications, e.g., for the power engineering sector. The technical reasons for this fact are numerous and are based on the special requirements in the process industries:
-   The large pressure-range, which is demonstrated by the standardized nominal pressures (PN according to the "IG-Standard") PN 325 DN 6-330, PN 700 DN 6-100, PN 1600 DN 6-60, PN 3000 DN 6-65 (DN nominal diameter);
-   the various liquids involved, including corrosive, abrasive, and the appropriate materials;
-   the large temperature range from cryogenic up to 700°C;
-   the robust operating conditions;
-   the high safety requirements;
-   the need for easy maintenance and repair;
-   the modularity.
One of the most applied high-pressure standards for the piping components is based on the so--called IG-Standard, which was established by an association of the German Chemical Industry beginning with 1930. Similar developments took place in the US and elsewhere. The pipes applied are of seamless design. The joints and seals which have been approved for high-pressure are shown in Fig. 4.2-1.
The lens ring seals (Fig. 4.2-1 A) together with (threaded) flanges are used in Europe for pressures up to around 4000 bar. The shape of the lens is crowned (spherical), the counter-piece conical and so that the sealing compression takes place along a narrow sealing line and the two joint pieces can be adjusted flexibly to a certain extent. The conical ring seal (Fig. 4.2-1 B) is most widely used in the US for pressures up to 2000 bar. The Grayloc® Connector Seals (other similar features are on the market) is characterized by a clamp, first (Fig. 4.2-1 C), and by a sealing ring with a seal lip, second. It is used up to the range of more than 1000 bar (middle size DN). The ease of the clamp adjustment and the pressure-energized seal lip action have proved to be very suitable. The ring joint seal (Fig. 4.2-1 D, e.g. ANSI ASME) gives good experience for the pressure range up to 200 bar only, but for temperatures up to 700°C. Flat seals are rarely used in high-pressure plants.
The design concept with totally metallic, such as lens seal rings, is found to be safer and more practicable. The conical joint with cap nut (Fig. 4.2-1 E) can be used for small nominal pipe

sizes and pressures up to around 10000 bar (DN 25/1400 bar, DN 8 / 10000). The sealing action is enhanced by the fact that the male cone has a slightly smaller included angle (58°) compared with the female (60°). The cap nut transmits the sealing compression during its adjustment through the collar threaded on the pipe (Fig. 4.2-2). The conical engagement is sealed by the adjustment force induced by the cap-nut and by an additional pressure-energized "lip seal" effect of the male cone. If a leakage should happen the „weep-hole prevents pressure imposition on the cap nut and facilitates leak detection. The coned and- threaded connection as in Fig. 4.2-2 is superior with regard to the admissible pressure and safety compared with features that employ tube-gripping or friction elements, which are limited to a pressure level of below 300 to 700 bar and only offer reduced reliability and safety.

Fittings for larger nominal diameters, e.g., pipe bends, are produced from pipes which are bent when cold or hot, or from hollow cylinders which are bent in a forging die. Tighter bending radii can be achieved with bending in a die (regard reduction of the wall–thickness!). T-pieces are produced similarly by forging procedures (Fig. 4.2-3).

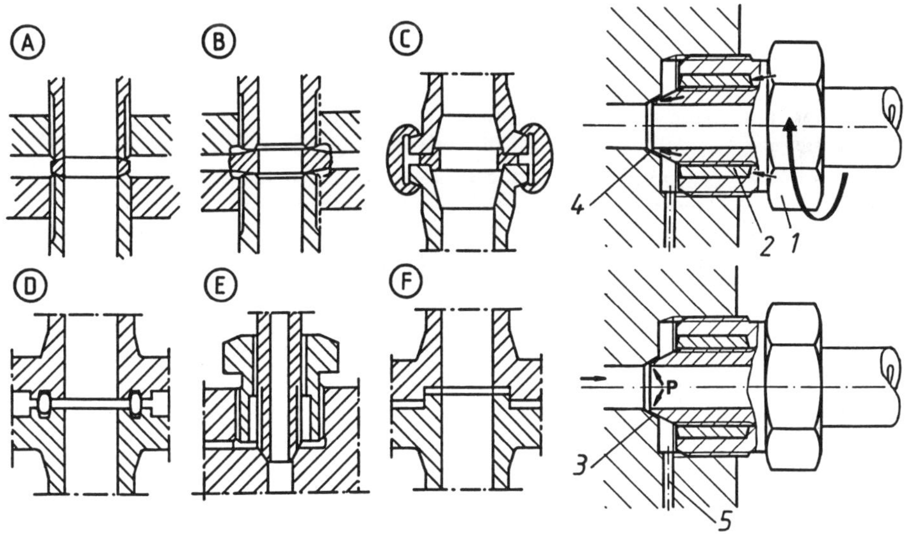

Fig. 4.2-1 (left). Connections and seals for high-pressure piping (see text for explanations).

Fig. 4.2-2 (right). Coned and – threaded connection (adopted from SITEC).
1, Cup nut; 2, Threaded collar; 3, Male cone; 4, Female cone; 5, Weep-hole.

192

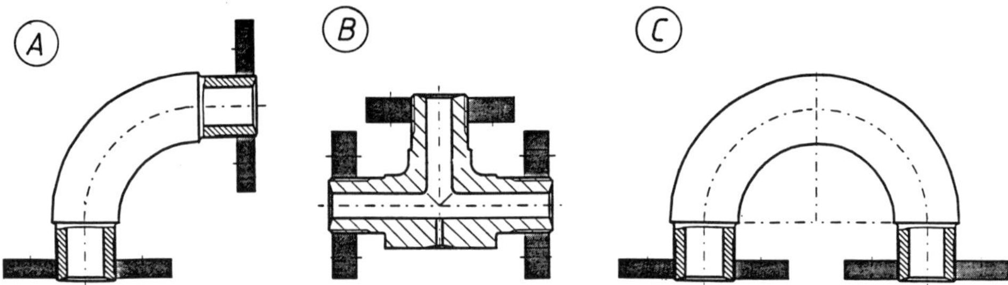

Fig. 4.2-3. High-pressure fittings for larger nominal diameters (up to PN 3000, adopted from BASF)

Connections by welding are used if the materials applied can be welded safely, and only up to the pressure level of 700 bar. The usual range of materials comprises low-alloyed- and higher-alloyed ferritic steels as well as austenitic types.

For the smaller sizes in nominal pipe diameters with coned -and- threaded connections it is usual to design the pipe-work with standard fittings such as crosses, unions, bulk-head couplings, reducers, adapters, elbows and Tees (for some examples see Fig. 4.2-4). These fittings are normally in cold-worked austenitic steel (e.g., 1.4571 or AISI 316Ti), but features with inserts of other more corrosion-resistant materials are available.

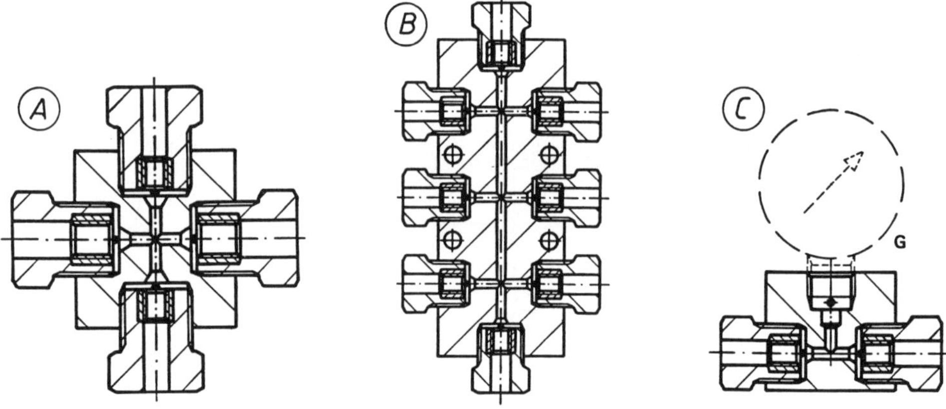

Fig. 4.2-4. High-pressure fittings for smaller nominal diameters (up to 7000 bar, adopted from SITEC).
a,Tee (with gauge); b, Cross; c, Multiconnector.

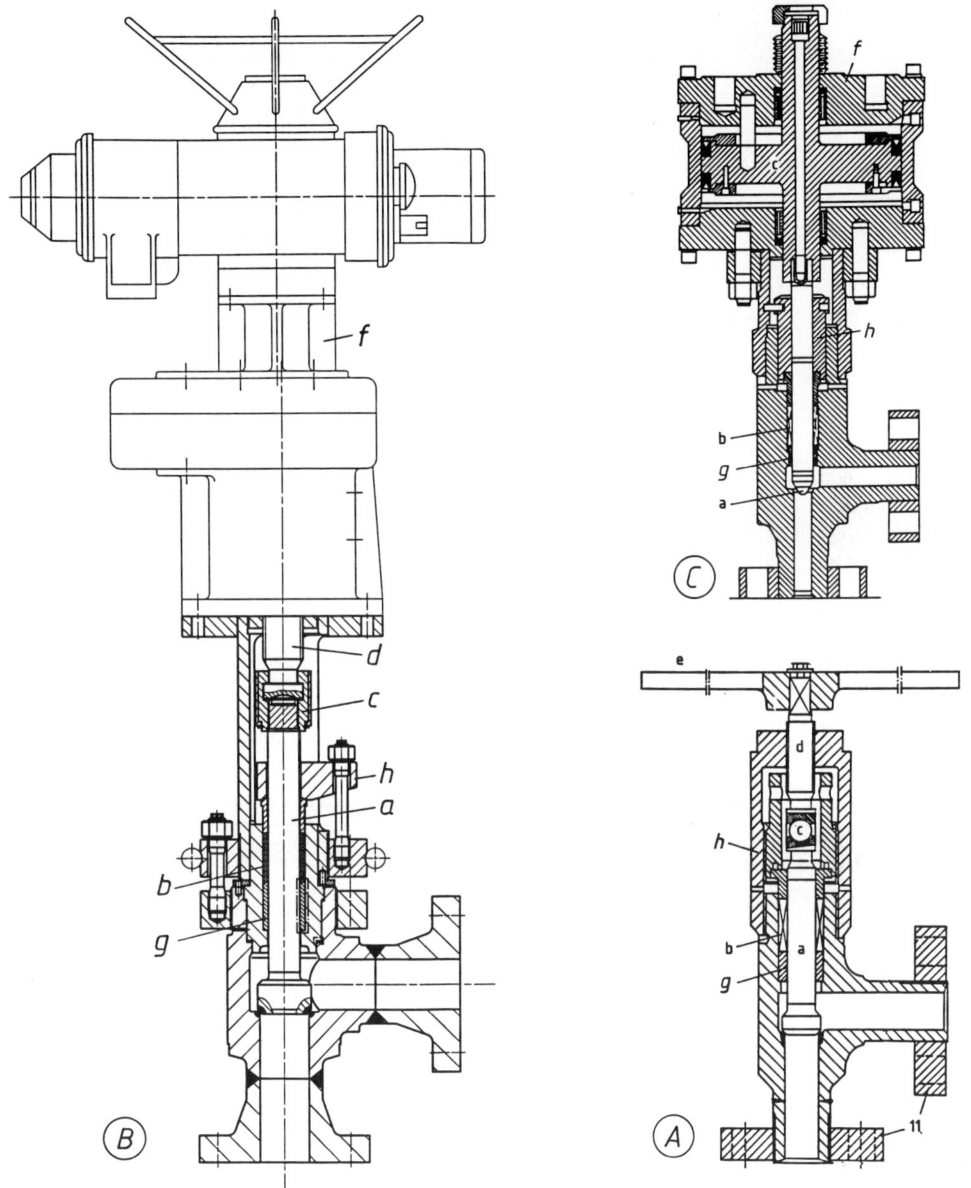

Fig. 4.2-5. High-pressure angle valves.
A, manually operated (325 bar, adopted from BASF); B, electrically operated (375 bar, adopted from BÖHLER);
C, hydraulically operated (3600 bar, adopted from BASF), a Lower stem; b, Gland; c, Flexible coupling; d, Upper stem; e, Hand wheel; f, Actuator; g, Bushing; h, Adjustment of the gland.

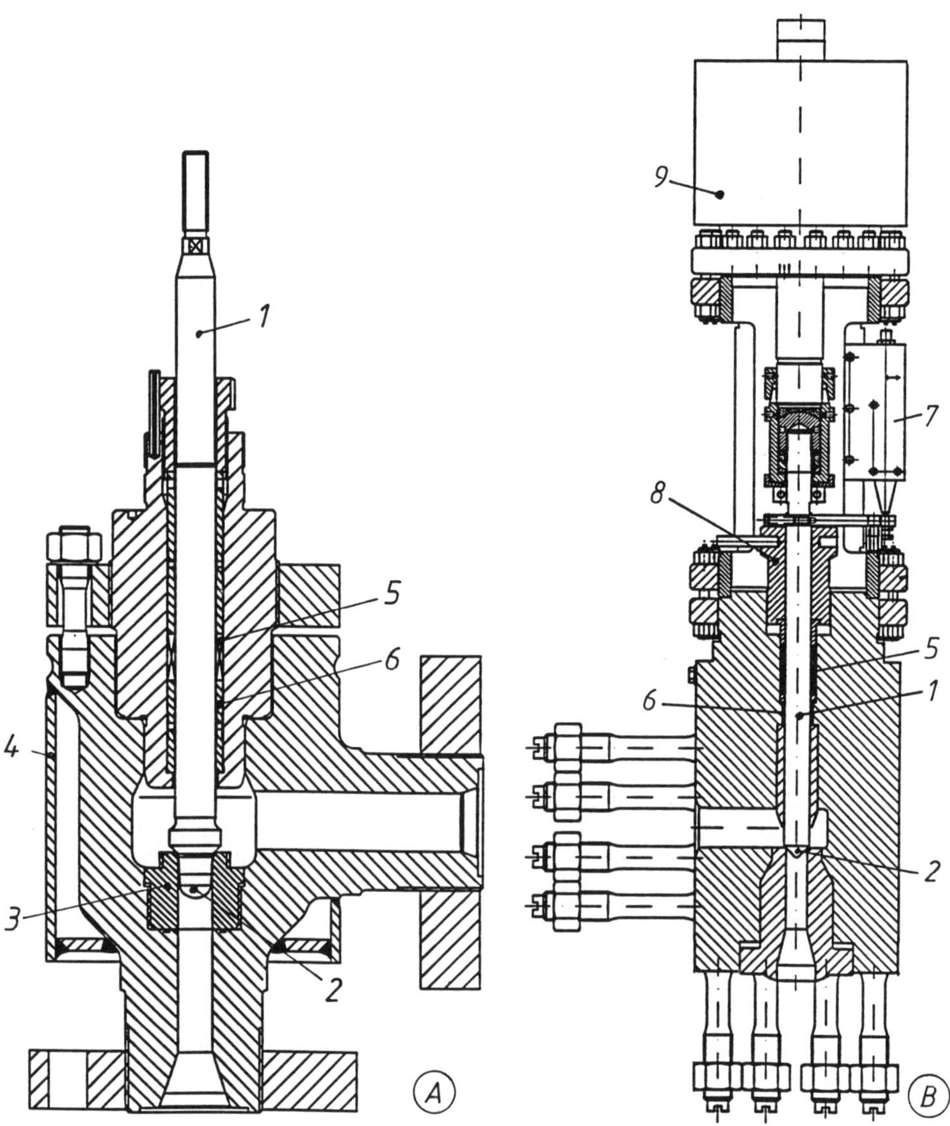

Fig. 4.2-6. Control valves
A, PN 385 bar (adopted from UHDE); B, Let-down valve, PN 3600 / DN 76mm (adopted from BASF); 1, Stem; 2, Cone; 3, Insert (seat); 4, Heating jacket; 5, Gland; 6, Bushing; 7, Position monitoring; 8, Adjustment of the gland; 9, Actuator.

4.2.2 Isolation and control valves

Valves are used for shut-off (isolation), control and safety or relief operations. For the pressure level of 300 bar and above stem valves are typically applied; up to around 200 bar, ball-valves are well regarded also.
For easy production and manipulation, shut-off valves are designed predominantly as angle valves. Fig. 4.2-5 shows three valves for the pressure levels of 325 and 3600 bar.

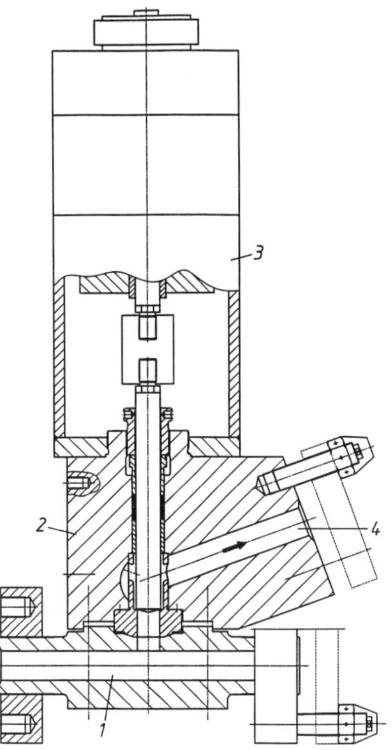

Fig. 4.2-7. In-line safety- and side-stream valve (PN 2500, DN 60/32, adopted from BÖHLER). 1, Tubular reactor train; 2, In-line valve; 3, Hydraulic actuator; 4, Side outlet (inlet).

The manually operated angle-valve (Fig. 4.2-5 A) has a coupling (c) between the upper- (d), and the lower- (e), valve spindle which eliminates the need for the spindle to rotate in the seal during opening and closing. This design feature improves the tightness and the endurance of the valve´s function. The same design concept is seen in the angle-valve Fig. 4.2-5 B which is operated by an electrical actuator, (f). The use of V-shaped rings (Compounds of PTFE) or rectangular packing rings (various braided types of PTFE, graphite, and other high performance plastics) are suitable for the gland (b) of the stem at temperatures up to 500°C. The angle-valve, Fig. 4.2-5 C, has a hydraulic actuator, (f).

The design of the closing section of the stem requires special attention to the material choice and shape. Usually the stem, as well as the seat, is locally heat-hardened or coated for better wear resistance.

If the valve is applied for control purposes the closing section is usually cone-shaped in order to obtain the appropriate control characteristics. One is recommended to install easily replaceable wear-resistant inserts as valve seats (Fig. 4.2-6 A).

The LDPE production with tubular reactors (see Section 5.1) requires some sophisticated control valves [45]. The let-down valve (Fig. 4.2-6 B) controls the polymerization reaction via the pressure and temperature by a high-speed hydraulic actuator (9) together with an electronic hydraulic transducer. The position of the valve relative to the stem is determined by a high-resolution electronic positioner (7). The cone-shaped end of the valve stem (2), as well as the shrunk valve seat (3) are made from wear-resistant materials (e.g., sintered tungsten carbide) in order to tolerate the high differential pressure of around 3000 bar during the expansion of the polymer at that location.

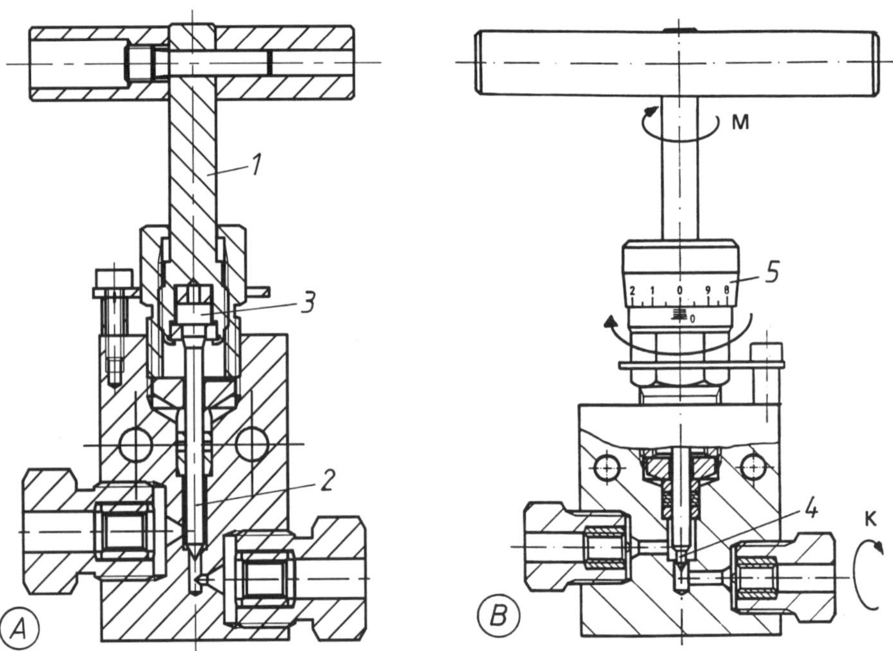

Fig. 4.2-8. High-pressure valves for smaller nominal diameters (adopted from SITEC). A, Shut-off valve; B, Micrometering valve; 1, Upper stem; 2, Lower stem; 3, Flexible coupling; 4, Cone (needle); 5, Micrometer scale.

The in-line safety valve [45] as shown in Fig. 4.2-7 is used in LDPE plants during emergency situations in order to depressurize the system by the outlet (2). The valve is directly included in the tubular reactor train (1) in a manner to avoid dead spaces where harmful decomposition of the polymer could take place. The control of the safety valve is effected again by a special hydraulic actuator (3). The valve shown is also used to inject quench gas if required; measures must be taken to avoid excessive heating-up of the ethylene in the side nozzle when the valve is closed to avoid undesired reactions.

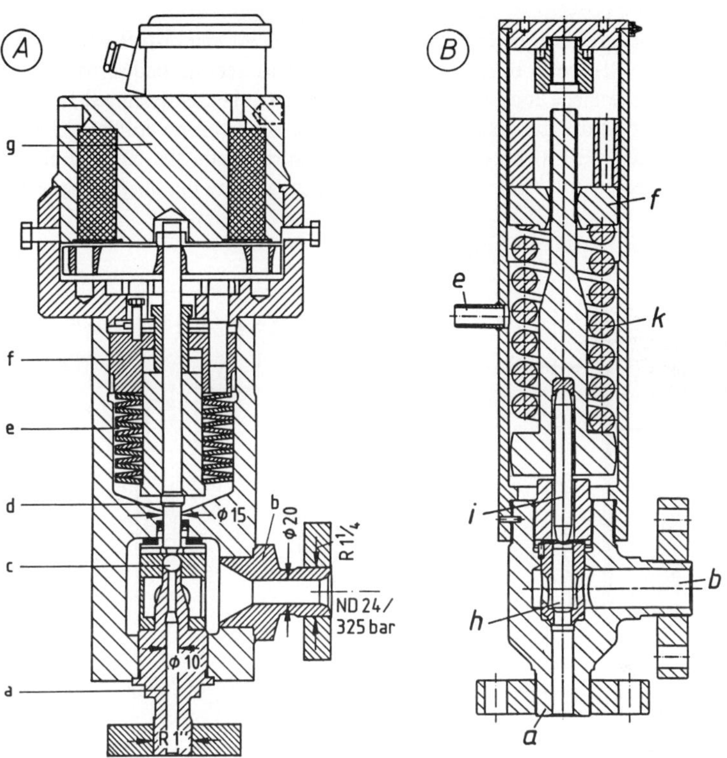

Fig. 4.2-9. Full-stroke relief valves
A, PN 3600/325, DN 10/25 [3]
B, PN 1600/500, DN 24/45 (BÖHLER)
A, Inlet; b, Outlet; c, Ball; d, Stem; e,, Disk-spring package; g, Magnetic lifting; h, Stem with cone; i, Intermediate stem; k, Spring; L, Venting.

As explained already in Section 4.2.1 the design of isolation- or control valves for small sizes of nominal pipe diameters is based on the coned-and-threaded connections. The valve stem is prevented from rotating during closing or opening by an appropriate coupling (3) between the upper (1) and the lower (2) stem section (Fig. 4.2-8 A). Basically, the same valve module can be used for micro-metering purposes when equipped with a precisely fitted metering taper, (4), and a micrometer scale, (5): (Fig. 4.2-8 B). The valves are available in austenitic cold-worked steels as the piping and fittings, but also in other special materials. There are features with pneumatic actuators, for extra-low and-high temperatures, and with valve-seats inserted into the housing for better wear- protection and maintenance.

4.2.3 Safety valves and other devices

Safety - or relief - valves are designed as spring-loaded proportional or full-stroke valves. The Fig. 4.2-9 A shows a full-stroke relief valve for 3600 bar, spring-loaded by a disc spring package (e), owing to the large load required. The disc chamber is designed for an internal pressure of 325 bar so that the gas (ethylene) to be discharged can be collected in a pressure vessel. The stainless steel ball, (c), effects the tight closure. Both, the ball (c) and the seat-insert (a) can be replaced easily. The relief valve can be opened by the magnet (g) at any

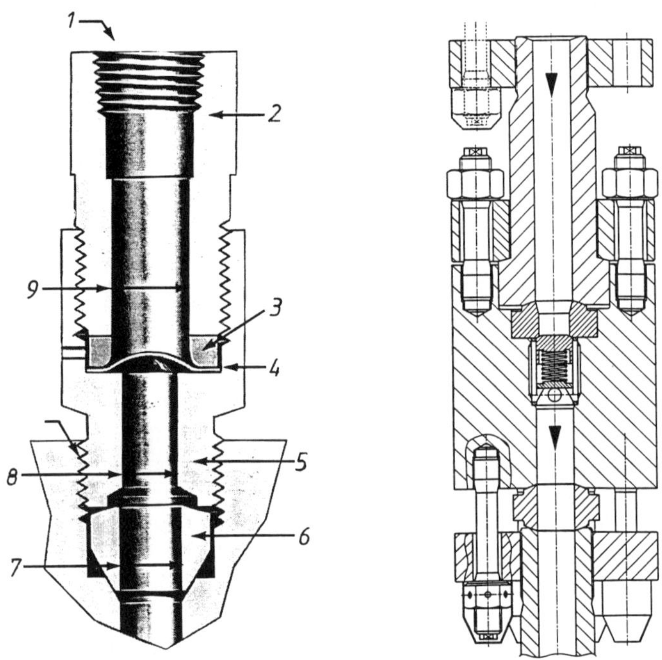

Fig. 4.2-10 (left). Rupture disc (adopted from Autoclave).
Fig. 4.2-11 (right). Non-return valve (PN 3200, DN 32, adopted from BÖHLER).

operating pressure. Spring-loaded safety valves are available for all nominal sizes required (Fig. 4.2-9 B). For smaller nominal diameters rupture discs are also used for pressures up to 4000 bar (Fig. 4.2-10). As they are one-way acting only, their application is designated for ultimate safety measures.
Non-return ball valves are manufactured up to a nominal width of around 30 mm: for larger sizes spring-loaded spherically shaped mobile valve parts are used (Fig. 4.2-11).

Other high-pressure fittings and piping devices to be mentioned, which are available up to 4000 bar are filters, injection nozzles, and adapters for instrumentation purposes.

References

Sections 4.1 and 4.2

[1]   F. Geitner, D. Galster: Chemical Engineering 2000, 80 – 86
[2]   G. Vetter: Dosing Handbook, ELSEVIER Advanced Technology Oxford 1997
[3]   G. Vetter, E. Karl: High Pressure Technology
       Ullmans´s Encycl. Ind. Chemistry, Volume B4, 1992
[4]   G. Vetter: Leakfree Pumps and Compressors, ELSEVIER Advanced Technology Oxford 1995
[5]   G. Vetter, E. Schlücker, W. Horn, E.W. Neiss: Proc 12th Int. Pump Users Symp. Houston, 1995, 207-220
[6]   I.P. Körner, P.S. Crofton I_a: Chem. Tech. (Heidelberg) 8, 1900 No. 11, 561-565
[7]   W. Dettinger: Chem. Ing. Techn. 54 (1982) 500 – 501
[8]   G. Vetter: in G. Vetter (ed.) Leckfreie Pumpen, Verdichter und Vakuumpumpen, Vulkan Verlag Essen 1992, 251-260
[9]   U. Schuster: in G. Vetter (ed.) Pumpen 2. Ausgabe, Vulkan-Verlag Essen 1992, p. 128 -134
[10]   E. Schlücker: in G. Vetter (ed.) Dosing Handbook s. Ref. [2], 699 – 720
[11]   E. Turnwald: Linde Berichte aus Technik und Wissenschaft 45 (1979) p. 104 – 107
[12]   T. Bartilla: Der Extruder im Extrusionsprozeß VDI-Verlag Düsseldorf 1989, 201 – 240
[13]   J.M. Mc. Kelvey, W.T. Rice; Chem. Engng. 1983, 89 – 94
[14]   Ch. Laux et al: ASME Paper 94-IPGC-PWR-47, 1994
[15]   G. Feldle: Neue Baureihe Hochdruck-Kreisel-Pumpen, KSB Technische Berichte 1986, 50 – 55
[16]   J.F. Gülich: Kreiselpumpen Springer Berlin Heidelberg, 1999
[17]   G. Füssle: Techn. Rundschau Sulzer 1985, 12–17
[18]   U. Capponi: High Pressure Pumps for Ammonia and Carbamate, Krupp Uhde Fertilizer Symposium Dortmund 1998
[19]   V.S. Lobanoff, R.R. Ross: Centrifugal Pumps-Design and Application, Gulf Publ. Co., Houston 1985
[20]   R. Neumaier: Hermetische Kreiselpumpen für extreme Förderparameter, Verlag und Bildarchiv W.H. Faragallah, 1999
[21]   R.Krämer, R. Neumaier: in G. Vetter (Hrsg.) Pumpen 2. Ausgabe, Vulkan-Verlag Essen 1992, 293 – 296
[22]   M. Knorr: in G. Vetter (Hrsg.) Leckfreie Pumpen, Vulkan-Verlag Essen 1990
[23]   J.P. Körner: 2nd Int. Symp. High Pressure Chem. Eng., Erlangen 1990 Abstract Handbook, 489 - 496
[24]   H.R. Kläy: in G. Vetter (ed.), Handbuch Verdichter 1st ed. Vulkan-Verlag Essen, 1990 1–5
[25]   H.R. Kläy: in G. Vetter (ed.) Handbuch Verdichter 1st ed. Vulkan-Verlag Essen, 1990, 207 – 212
[26]   J. Nickol: Proc 1st EFRC-Conf. Dresden 1999, 127 – 135
[27]   W. Bammerlin: Sulzer Tech. Rev. 4/1997

[28] E. Giacomelli, Ch. Xiaowen, K. Yaokun: Quadernipignone 66, 1999, 4 – 9
[29] M. Dehnen: in G. Vetter (ed.) Leckfreie Pumpen, Verdichter und Vakuumpumpen, Vulkan-Verlag Essen 1999, 273 – 291
[30] NN.: Mannesmann-Demag Druckschrift MA 25.69.dt/11.81, 1981
[31] M. Stark, M. Schölch: Konstruktion 50 Nr. 1 / 2 (1998) p. 23 – 27
[32] K. H. Hoff: Konstruktion Nr. 1 / 2 (2000) p. 51 – 55
[33] G. Mischorr: Diss. Univ. Erlangen-Nürnberg 1990
[34] G. Vetter et al: Chem. Engg. Technol. 15 (1992) p. 1 – 13
[35] B. Donth: Doct. Thesis Universität Erlangen-Nürnberg 1996
[36] H. Mughrabi, B. Donth, H. Feng, G. Vetter: Fatigue Fract. Eng. Struct. 20 (1997) 555 - 604
[37] H. Feng, H. Mughrabi, B. Donth: Modell Simul. Mater. Sci. Eng. 6 (1998) 51 – 69
[38] H. Mughrabi, B. Donth, H. Feng, G. Vetter: Proc. Fifth Int. Conf. On Residual Stresses (ICRS-5) Linköping 1997
[39] G. Vetter, L. Depmeier: in High Pressure Chemical Engineering (ed. RV. Rohr, Ch. Trepp) ELSEVIER 1996, p. 633 – 638
[40] NN.: Burgmann Gasgeschmierte Gleitringdichtungen 1997 (Edition GSBD/D1/3000/06.96/1.11.)
[41] G. Vetter: 1st EFRC-Conference the Recip.-a State of the Art Compressor 1999 Dresden
[42] N. Feistel: Sulzer Tech. Review 4/96, 42 – 45
[43] G. Vetter, U. Tomschi: Pumpen + Kompressoren 1, (1995) 66 - 76
[44] K. Scheuber: Ölhydraulik und Pneumatik 25 (1981), 581 – 586
[45] J.P. Körner, M. Köpl: Equipment for high pressure processes in the Chemical Industry, UHDE Doc. Nr. 924.3, 1993

4.3 High pressure vessels and other components[*]

High pressure vessels are used to realize the physical and chemical process conditions with respect of pressure and temperature to successfully producing chemicals and products for a wide range of applications. In this chapter will be reported about the central aspects like the type of vessels and components, their design, construction and strength calculation for static and dynamic plant conditions and about materials. Informations about the operability of such installations in chemical plants are given. Both aspects of industrial and laboratory applications are discussed. The safe operation of all parts is a fundamental demand when using high pressures and temperatures.

Authorities responsible for national pressure vessel and piping codes recommend detailed design procedures. Furthermore a short introduction to national regulations is given.
Specific design aspects for specialities and for superpressure facilities e.g. for the diamond synthesis and others are not addressed.

This chapter will give the reader a condensed presentation about the present knowledge in this field only.

4.3.1 Calculation of vessels and components

Before design and strength calculations for any part of a high pressure plant are performed, a number of points should be considered e.g.:
-   The maximum pressure and temperature and their variation during operation to which the parts will be subjected.
-   The volume of the component.
-   The shape of the apparatus.
-   The kind of closure.
-   Sealing.
-   Dynamic aspects during operation e.g., pressure cycling and its influence on fatigue of materials.
-   Knowledge of corrosion attack on the material.
-   Notches and side-holes, and the induced stresses.
-   Special requirements such as fast - opening and closing for a quick loading of the vessel and for cleaning.

The safety aspects are of extreme importance as the regularly large pressurized fluid - volumes are representing a great danger if failures resp. rupturing should occur. Each high pressure component e.g. a vessel, must therefore obey the codes adopted by the individual regulatory national bodies. Table 4.3-1 gives a survey about the most common codes, which impose conservative design criteria. However they do not cover all working conditions or materials. In these cases appropriate solutions must be found.

A well know design strategy is summerized by the remarkable sentence: "Leakage Before Rupture", which means that tough materials must be selected and local embrittle-ments due to the three-dimensional stress-distributions involved should be avoided for high pressure applications.

---

[*] References at the end of section 4.4

202

Table 4.3-1
Pressure Vessel Codes For Non-Creep Conditions.

Country	National Pressure Vessel Code	Body responsible for Code	Legal standing	Design stress
France	CODAP	Service de Mines	Enforced	(0.2%-proof stress at working temperature/ 1.5) ;(ultimate tensile strength /2.4)
Germany	AD – Merkblatt	Technischer Überwachungs-verein	Enforced	0.2%-proof stress at working temperature/1.5
Japan	Unified Pressure Vessel Code	Japanese Industrial Standards Commitee	Enforced	0.2%-proof stress at working temperature/1.7
United States of America	A.S.M.E. – Code Section VIII, Div.2 and Div. 3	American Society of Mechanical Engineers	Enforced in some States	Div. 2: Defined strength data as listed in this code, Safety-factors are included in the defined strength data. Div. 3: 0.2%-proof stress at working temperature / 1.5-2 correlated with data for material used.
	Div.: 3 , for > 700 bar.			

In former times, many severe failures of high pressure components evolved have led to a high standard in the knowledge of calculating and designing such parts. The most important aspects for understanding the design and strength calculations of a modern, safe plant components are given in this chapter.

It has previously been difficult to obtain an analytical solution, which includes all the aspects mentioned, for calculating the dimensions of high pressure components. Since the 1970s major progress has been made, using the "Finite Element Method", to analyse the stresses and the stress distributions in high pressure components including special stress rising effects, e.g. holes or the influence of cracks in the structure. In the following section the relevant equations to calculate the strength of hollow cylinders regarding the resistance of the materials under elastic, semiplastic and plastic behaviour will be discussed. Useful design criteria for both thin- and thick - walled hollow bodies are explained for cylindrical components of great length. The radial-, axial and tangential stresses in the wall of such a component are then only functions of the diameter ratio. The derivation of the various equations will be neglected, but the references included will give the reader the necessary scientific links.

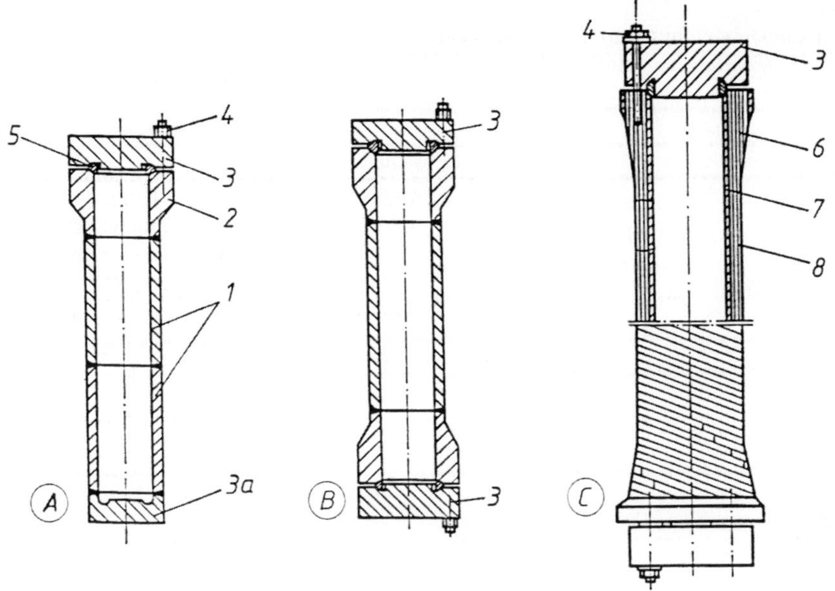

Fig. 4.3-1. Basic features of the thick-walled vessels (adopted from Lewin et. al. [1]).
A   welded, with cover and closed-end
B   welded, with two covers
C   multilayer with core-tube and two covers
1, Thickwalled cylindrical pieces; 2, Flange piece; 3; 3a, Cover; 4, Bolts; 5, Seal;
6, Multilayer flange section; 7, Multilayer cylinder section; 8, Core tube.

4.3.1.1 The hollow cylinder under static loading

Most of the high pressure vessels as used for reactors, scrubbers, heat exchangers, buffer vessels or pulsation dampers, are mainly consisting of hollow cylinders.

The vessels are used as monobloc resp. solid-wall or multilayer features with appropriate connections, mainly consisting of endpieces and covers (Fig. 4.3-1).

The relative thickness of the wall (thick- or thin-walled) is depending on the operational pressure.

The calculation of the strength resp. the admissible internal pressure varies with the wall-thickness: thick-walled hollow cylinders are calculated by neglecting the radial stress (equal to the pressure) which is small compared to the tangential. On the other side the thick-walled hollow cylinders are calculated with the Lamé equations (1833).

Table 4.3-2 shows the adequate formulae for closed and long enough cylinders during internal pressure p, the external pressure staying ambient. In this case the internal pressure is inducing an axial stress by acting on the covers resp. the end-pieces.

Table 4.3-2

Stresses in thick- and thin-walled hollow cylinders: p internal pressure N/mm^2; σ stress N/mm^2; $u = D_o/D_i$ diameter ratio

Stress	Inside dia.		Outside dia.	
	Thick-walled hollow cylinders			
tangential	$\sigma_{ti} = p \cdot (u^2 + 1) / (u^2 - 1)$	(4.3-1)	$\sigma_{to} = p \cdot 2 / (u^2 - 1)$	(4.3-1a)
axial	$\sigma_a = p \cdot / (u^2 - 1)$	(4.3-2)	$\sigma_a = p / (u^2 - 1)$	(4.3-2a)
radial	$\sigma_{ri} = -p$	(4.3-3)	$\sigma_{ro} = 0$	(4.3-3a)
equivalent GE hyp.	$\sigma_{vGEi} = p \cdot \sqrt{3} \cdot u^2 / (u^2 - 1)$	(4.3-4)	$\sigma_{vGEo} = p \cdot \sqrt{3} / (u^2 - 1)$	(4.3-4a)
equivalent Shear hyp.	$\sigma_{vShi} = p \cdot 2u^2 / (u^2 - 1)$	(4.3-5)	$\sigma_{vsho} = p \cdot 2 / (u^2 - 1)$	(4.3-5a)
	Thin-walled hollow cylinders			
tangential	$\sigma_{tm} = p / (u - 1)$			(4.3-6)
axial	$\sigma_a = p / (u^2 - 1)$			(4.3-7)
equivalent Shear hyp.	$\sigma_{vSh} = p / (u - 1)$			(4.4-8)

Equation (4.3-6) can be put in the form, $\sigma_{tm} = Di/2s$ (σ_{tm} mean tangential stress, s wall thickness).

The stress distribution exhibits well uniform for thin-walled, and extremely gradient afflicted (Fig. 4.3-2) for thick-walled hollow cylinders, the latter offering a potential to exploit their strength to a larger extent. A closer analysis reveals that calculations should be performed according to the "thick-walled mode" starting with Do/Di > 1.1 or 1.2 in order to avoid too large erroneous predictions. It should be noted that the calculations according to Table 4.3-2 have been well experimentally verified [1][2].

This is also true for the calculation of the internal pressure p_{el} at which the elastic behavior ends and the plastic deformation (overstrain) begins, based on the GE (v. Mises) hypothesis a ($\sigma_{0,2} = 0.2\%$ proof stress in N/mm^2, p_{el} = internal pressure in N/mm^2, u diameter ratio)

$$p_{el} = \sigma_{0,2} \cdot (u^2 - 1) / (u^2 \cdot \sqrt{3}) \qquad (4.3\text{-}9)$$

The same calculation based on the shear stress hypothesis is yielding a slightly more conservative prediction.

Reviewing the calculation methods as implemented in the various pressure vessel codes (see Table 4.3-1) shows that these are based on the "scientific" formulae as per Table 4.3-2 but partly appear in a different shape due to approximations, simplifications, additional safety margins and other factors taking care of manufacturing tolerances, weakening by corrosion and welding seams [3][4][5] involved.

A new ASME code for calculating high pressure vessels (Sect VIII Div. 3) is based on the formulae to determine the internal pressure $p_{compl\cdot pl}$ for complete plastic yielding through the full wall with some assumptions, e.g. perfectly elastic-plastic material behavior and the GE-hypothesis [2].

$$p_{compl\cdot pl.} = \sigma_{0,2} \cdot (2 / \sqrt{3}) \cdot \ln u \qquad (4.3\text{-}10)$$

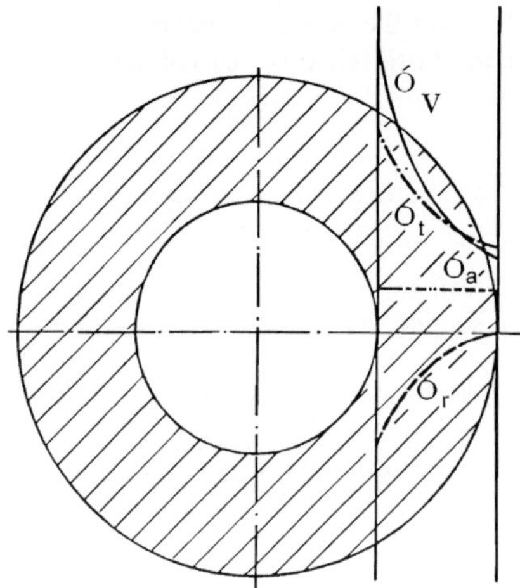

Fig. 4.3-2. Stress distribution in a thick-walled cylinder.
σ_r radial stress, σ_t tangential stress, σ_a axial stress; σ_v equivalent stress (GE hypoth.)

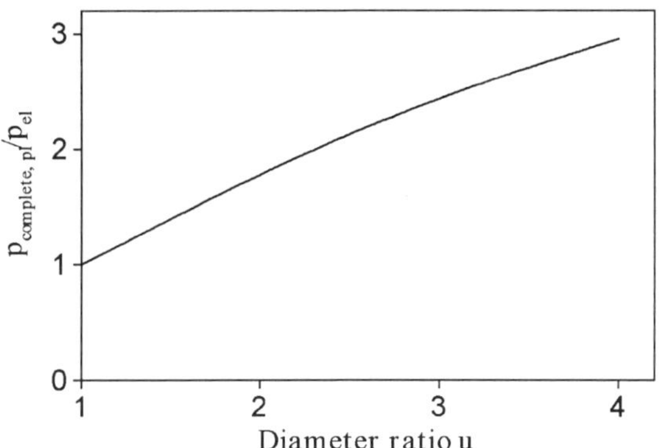

Figure 4.3-3. Ratio of complete plastic yielding to elastic deformation as a function of the diameter ratio.

For strain-hardening materials $p_{compl.pl}$ would be higher (correction algorithm see [4][7]). The ratio $p_{compl.pl}/p_{el}$ depending on the diameter ratio (Figure 4.3-3)gives clear evidence about the potential of improved exploitation of the strength of thick-walled hollow cylinders against internal pressure.

206

The above mentioned ASME code [3] uses the type of calculation presented above and selects a safety factor of S=2 to complete plastic yielding according to the GE hypothesis or S=1.732 with respect to complete plastic yielding according to the shear hypothesis Hence the permissible internal pressure is:

$$p_{ASME} = \sigma_{0,2} \cdot (1 \, / \, 1.732) \cdot \ln u \tag{4.3-11}$$

The following conditions should be satisfied: stress must not exceed the tensile yield strength at any point; fatigue conditions should be considered and brittle fractures avoided.

4.3.1.2 Strengthening the thick-walled hollow cylinder under static loading
The stress distribution within the thick-walled structure can be turned to better eveness by introducing appropriate benefitial pre-stresses with multilayer shrink fitting or autofrettage.

Shrunk compound cylinders
The tangential stresses at the inner side of a cylindrical vessel can be substantially reduced if two or more cylinders are shrunk one onto the other. Shrunk vessels are manufactured by heating the outer cylinder before combining the both parts. The maximum heating temperature is around 500°C. In special cases it is usual to cool additionally the inner cylinder. Another method is to press two or more conical parts mechanically together. All parts used in shrunk constructions must be exactly fabricated and controlled.

The shrink parameters must be selected so that at the later operating pressure the inner and the outer layer of the vessel work under nearly the same strain-conditions.
Fig. 4.3-4 (ABC) gives the superimposed stress distribution in the walls of a two-layered vessel under internal pressure. It can be clearly recognized that the compressive tangential pre-stresses by shrink-fitting (Fig. 4.3- 4B) are decreased at the inner layer and increased at the outer layer towards a more even stress distribution (Fig. 4.3- 4 C) compared to that for a mon-obloc cylinder (Fig. 4.3- 4A). The theoretical fundamentals for the dimensioning of shrink-fit multilayer cylinders can be taken from [2][8][9].
The optimum admissible internal pressure for a two-layer shrunk cylinder yields

$$p_{shrink}= 2 \, \sigma_{0,2} \cdot (u - 1) \, / \, (\sqrt{3} \cdot u) \tag{4.3-12}$$

with the definition that the maximum equivalent stress according to the GE-hypothesis as well as the individual diameter ratios of the two shrunk layers are the same. About limitations and the calculation of shrunk layer compounds see [2]. Shrink-fitting of more than two thick-walled layers are rarely applied as the manufacture is demanding and expensive.
The generation of benefitial residual prestresses for larger thick-walled cylinders can be achieved easier by wrapping with welding, shrink-fitting or coiling many thin layers or winding strips onto a core tube, which can be corrosion resistant whereas the other layers stay totally apart from any contact with the fluids involved.
The different methods to produce mulitlayer shrink-joints are based on the thermal contraction of welding seams, as well as the mechanical tension and thermal contraction during strip- wire- or coil-winding after cooling down. Fig. 4.3-4 D shows the ideal distribution of residual

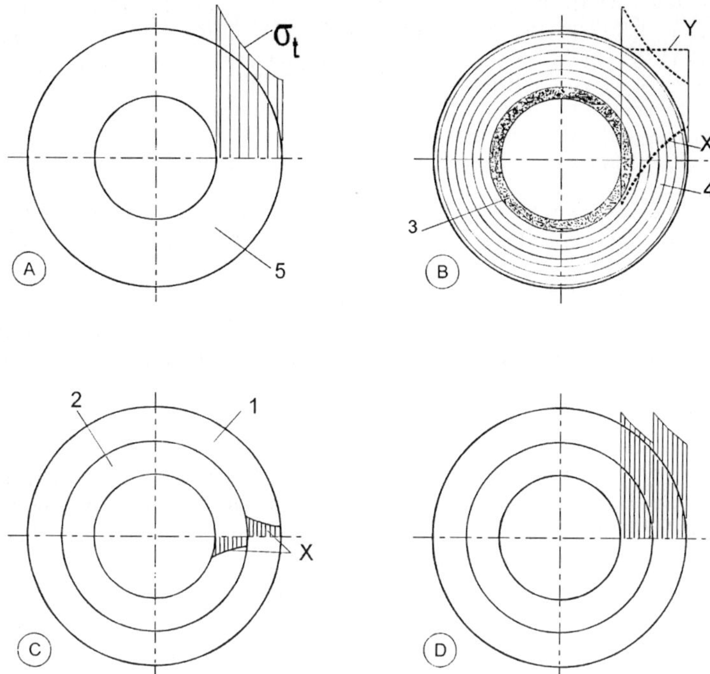

Fig. 4.3-4. Stresses in shrunk thick-walled cylinders (schematically).
A, Monobloc; B, Multilayer structure (ideal);
C, Shrink fit: residual σ_t ;
D, Shrink fit: σ_t when pressurized as A.
σ_t tangential stress; X residual stress; Y even stress distribution;
1,2, Shrunk thick-walled layers; 3 Core tube; 4 Multilayer package; 5 Monobloc (single wall)

stresses (X) and their influence on the achievement of a constant stress distribution (Y) during internal pressure loading. This has the consequence of a remarkable reduction of the peak stresses which are usual in monobloc walls.

The calculation of multilayer thick-walled cylinders is still a matter of experience and the control of the production method [2][3][5][8][9] applied.

Autofrettaged cylinders

If the pressure in a thick-walled cylinder is raised beyond the yield pressure p_{el} according to the equation (9), the yield will spread through the wall until it reaches the outer diameter [10]. For a perfectly elastic-plastic material the ultimate pressure for complete plastic deformation of the thick wall $p_{compl\cdot pl}$, also called collapse pressure, can be calculated by equation (4.3-10).

As the ductile materials used for high pressure equipment generally demonstrate strain hardening, the real burst pressure will be higher than the collapse pressure (empirical formulae and theory see [10]).

The autofrettage treatment practically is extended to a certain fraction of the thick wall only where overstrain occurs during the application of the autofrettage pressure. When the pressure is removed from a partially or totally yielded cylinder a part of the internal wall region will be forced into a state of compressive residual stresses. As long as the material which is imposed to the highest residual stresses is not forced into reversed yield, it will now behave nearly completely elastic when being respressurized.

The strengthening effect of a monobloc cylinder due to autofrettage practically is achieved by two effects: First the introduction of the compressive (tangential) residual stresses which extend the elastically admissible internal pressure and second the increased available material strength by strain hardening. The maximum admissible pressure for optimum autofrettaged cylinders based on perfectly elastic-plastic materials, completely elastic stress (plus/minus) conditions at the inner bore diameter and the assumption of the GE-hypothesis can be calculated as [11].

$$p_{aut.} = 2\,\sigma_{0,2} \cdot (u^2 - 1)\,/\,(\sqrt{3} \cdot u^2) \tag{4.3-13}$$

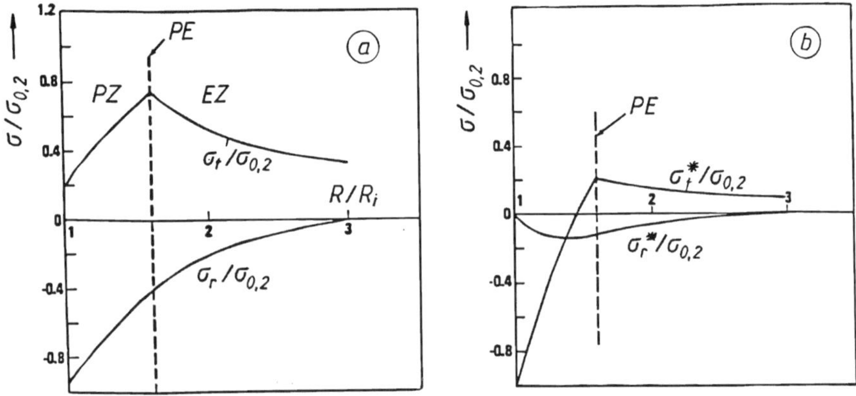

Fig. 4.3-5. Stress distributions during (a) and after the release (b) of the autofrettage pressure (adopted from Spain et. al. [11]).
D_i/D_o inner / outer diameter, σ_t tangential stress, σ_t^* residual tangential stress; σ_r, σ_r^* radial stress, $\sigma / \sigma_{0,2}$ stress ratio; R, R_i radius, inner; $\sigma_{0,2}$ yield stress
$u = D_o/D_i = 3$; $D_i/D_{pl} = 1.6$; $p/\sigma_{0,2} = 0.956$; $p_{autpr}/p_{el} = 1.86$; PE plastic/elastic transition; PZ, EZ plastic, elastic zone

Autofrettage is a useful measure for statically as well as dynamically pressurized vessels (see section 4.1.5.1). There is also a benifitial impact on the decrease of the rate of crack growth and the increase of the critical crack depth.

The residual stresses can be calculated either by approximation on analytical base with the assumption of ideal elastic-plastic material behavior or numerically by FEM for arbitrary material properties.

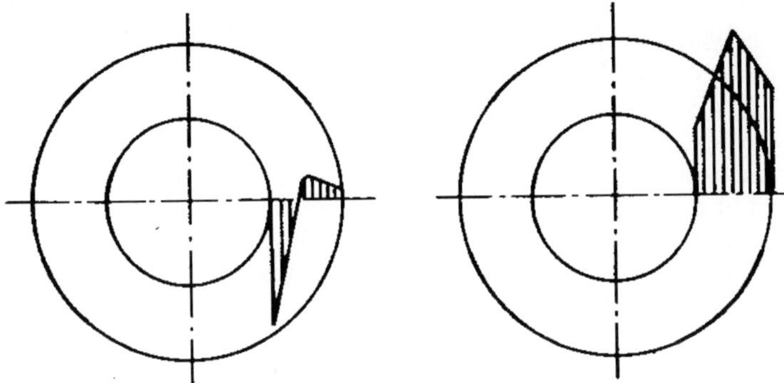

Fig. 4.3-6. Reduction of the stresses at the inner bore diameter by autofrettage (schematical) left: residual stress; right operational stresses after pressurization.

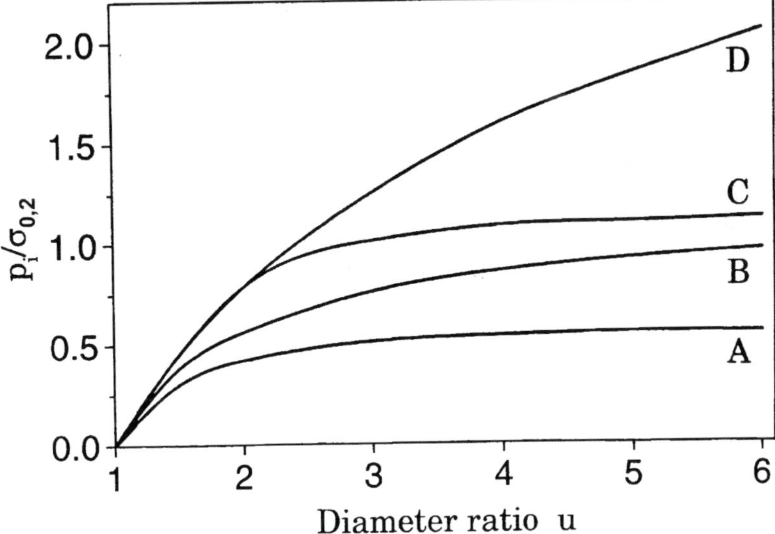

Fig. 4.3-7. Comparison of different design strategies of high pressure cylindrical vessels A, Monobloc (eq. 4.3-9); B, Two piece shrink fit $R_m = \sqrt{R_i \cdot R_o}$ (eq. 4.3-12); C, Partially autofrettaged cylinder (eq. 4.3-13); D, collapse pressure ($p_{compl.\ pl}$) of a Monobloc (eq. 4.3-10); $\sigma_{0,2}$ yield strength; p imposed internal pressure, R_m radius of the shrink fit.

The ASME Section VIII Div. 3 recommends to determine the residual stresses after autofrettage by the measurement of the mean tangential expension at the outside or the mean residual expension at the inside of the thick wall.

In order to understand the stress distributions involved during the autofrettage pressurization and after its release Fig. 4.3-5 presents an example adopted from [11] for a partially autofrettaged thick-walled cylinder. Fig. 4.3-5a shows the stress ratios $\sigma_t/\sigma_{0,2}$ and $\sigma_r/\sigma_{0,2}$ through the wall during the application of the autofrettage pressure p_{autpr}.. After the release of the autofrettage pressure (Fig. 4.3-5b) the compressive tangential stresses (see ratio $\sigma_t*/\sigma_{0,2}$) appear, which evidently reduce the tangential as well as the equivalent stresses when the operational pressure is imposed as shown schematically in Fig. 4.3-6 (see also Chapter 1, Fig. 1.4-10).

As a final result of the explanations about strengthening measures the admissible static internal pressure for thick-walled cylinders is compared in Fig. 4.3-7 for different design strategies according to the equations (4.3-9), (4.3-10), (4.3-12) and (4.3-13) and the explained assumptions and optimisations. In the case of the monobloc (A), the two-piece shrink fit and the autofrettaged cylinders the maximum stress at the inner diameter stays within the elastic limit ($\sigma_{0,2}$). Comparatively much larger is the admissible pressure when complete plastic yielding occurs as shown for the collapse pressure ($p_{compl\cdot pl.} = p_{coll}$; D) .

The strengthening effect due to autofrettage of a monobloc cylinder is impressive, the two-piece shrinkage yielding a result in between. The application of larger diameter ratio demonstrates not very efficient but the increased safety against collapsing (bursting) should be considered.

4.3.1.3 Influence of temperature gradients on design

When the wall of a cylindrical pressure vessel is subjected to a temperature gradient, every part expands depending on the coefficient of linear thermal expansion of the steel. The parts at lower temperature impede the expansion of those parts with higher temperature, and induce additional thermal stresses. To estimate the transient thermal stresses which regularly appear e.g. during start-up or shut-down of process components or as well as a result of process interruptions and in the case of pulsating temperature conditions during operation. Informations about the temperature distribution across the vessel wall as a function of radius and time [12] are required:

$$\sigma_r = -(\beta / 2) \{ (u^2 / u^2 - 1) (1 - (R_i / R)^2) - \ln (R / R_i) / \ln u \} \tag{4.3-14}$$

$$\sigma_t = -(\beta / 2) \{ (u^2 / u^2 - 1) (1 - (R_i / R)^2) - \ln (R / R_i) / \ln u - 1 / \ln u \} \tag{4.3-15}$$

$$\sigma_r = -\beta \{ (u^2 / u^2 - 1) - (\ln (R / R_i) / \ln u) - 1 / 2 \ln u \} \tag{4.3-16}$$

where $u = D_o/D_i$; D_i, r_i inner diameter, radius; D_o, r_o outer diameter radius, and with

$$\beta = \alpha \cdot E \cdot \Delta T / (1-v) \tag{4.3-17}$$

$\Delta T = T_i - T_o$ is the steady-state temperature difference between the internal radius R_i and the external radius R_o. When the above three equations are derived it is assumed that the cylinder is made of an isotropic and initially stress free material and the temperature does not vary along the cylinder axis. Further the effect of temperature on α, the coefficient of thermal expansion, and E, Young's modulus, may be neglected within the considered range. Furthermore it is assumed that the temperatures everywhere in the cylinder are low enough so that no relaxation of the stresses as a result of creep could happen.

The thermal stresses as per the equations ((4.3-14) – (4.3-17)) must be superimposed to those generated by the internal pressure as explained above (sections 4.3.1.1 – 4.3.1.4).

Creeping of the material under pressure and temperature must be additionally taken into account. The time shield limit of the material must be known for a correct stress calculation. The results influence the life span of the high pressure vessels and other components. Time shield values are listed in the pressure vessel codes.

Thermal stresses must be taken into account in general if the heating occurs from outside through the wall or when the vessel wall is cooled by the contents. It should be noted with regards to the heat transfer through thick-walled high pressure vessels that multilayer features exhibit much lesser heat conductivity in general growing with increasing internal pressure [13] however.

4.3.1.4 End pieces, side-holes and surface influences

The end pieces are normally forged hemisperical or flat plates, either welded on or joint to the cylindrical vessel by means of threaded bolts for all dimensions required. The hemisperical end pieces require less wall thickness (around 60 %) of the cylindrical wall and offer a good distribution of the stresses from the hemispherical to the cylindrical shape [14].

Flat end pieces (plates) are frequently chosen for production reasons. For monobloc forgings of high pressure vessels or welded on end pieces it is important to avoid stress risers at the transition locations. For conical transition the stresses yield only slightly larger than those in the cylindrical wall (Fig. 4.3-8).

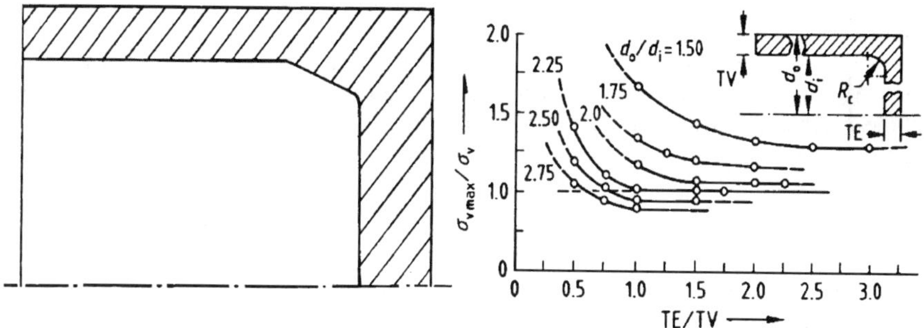

Fig. 4.3-8. Favorably shaped conical transition from cylinder to end plate of a pressure vessel.

Fig. 4.3-9. Stress ratio σ_{vmax}/σ_v at the radius of transition according to [15] (for $R_C/TV = 0.25$ only).

The shape factor with flat end pieces and radius transition are shown in Fig. 4.3-9 ($R_c/TV = 0{,}25$). Larger radii R_c will not reduce the shape factors involved much more [15].

The stress rising effect of orthogonal side-holes in high pressure vessels and pipings has been investigated by a number of authors [15][16][17][18][19] showing that shape factors of around three must be expected depending on the thickness ratio of the wall, the ratio between

212

the diameter of the side-hole and the internal diameter of the vessel and (slightly) on the rounding radius at the intersection between. As the side-holes are representing a location of a high failure potential for static and dynamic load they should be avoided whenever possible.

It is an experience that in the different constructions of pressure vessels side-holes, cracks as well as rough surfaces influence often dramatically the stability of a pressure vessel during pulsating internal pressure. In a lot of experiments experiences were gained how to handle those questions. The next figures present experimental results of the influence of side-holes (Fig. 4.3-10, [20]) as well as that of the quality of a wall surface (Fig. 4.3-11, [12]). Cracks and holes at the surface of a vessel are often the starting point for fatigue failures. The question of so-called critical crack dimensions and their behaviour under dynamic working conditions must be considered. The consequences could be cracks and sometimes also leacks. The risk of fatigue failures can be reduced by rounding off and polishing the edges and the surfaces of high pressure vessels. Dynamic and pulsating conditions additionally influence constructive details in a wide range and must be considered. Dynamic conditions influence the choice of the material and reduce the maximum allowable shear stress values. That means that the horizontal part of the so-called "Wöhler-curve" [21] is reduced to lower allowable stress values for a long time use [22].

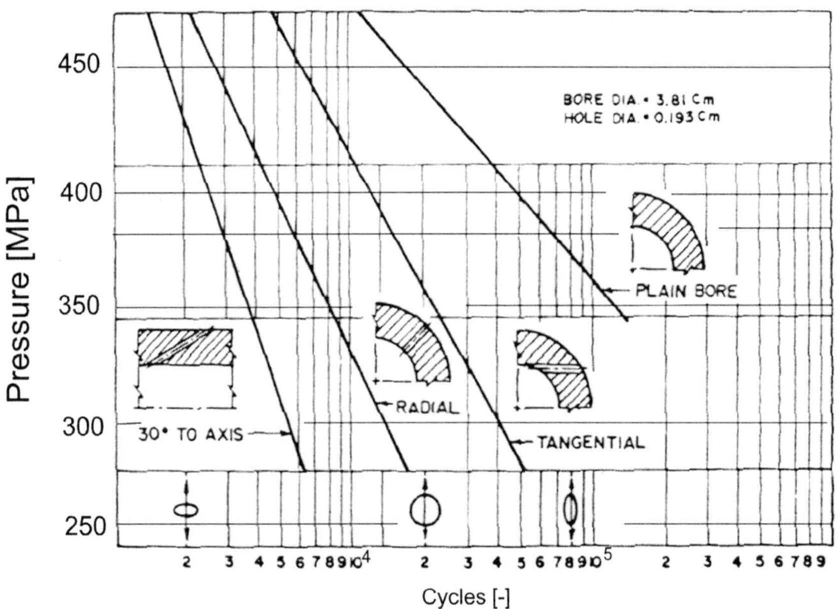

Fig. 4.3-10. Influence of side holes intersecting a vessel wall [20].

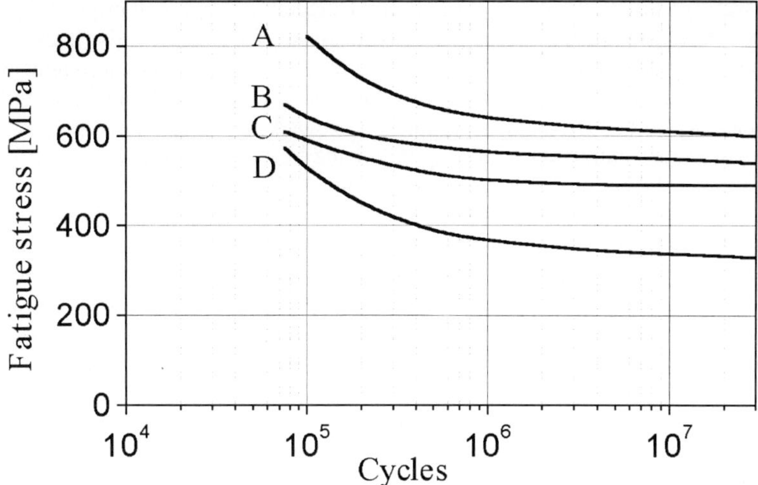

Fig. 4.3-11. Influence of surface quality on the allowable fatigue strength rate for high pressure pipes made of a special material (30CrMoV9). The results were measured in a high frequency pulsator. [12].
A, Autofrettaged; B, Nitrided; C, Electropolished; D, Untreated.

4.3.2 Materials

In the following section, typical materials for use in high pressure equipment are mentioned. During the selection process, both technical and economic aspects must be considered. Attempts must be made to optimize the weight of a vessel, the material input and - last but not least - its price. A sufficient knowledge about the material properties, as well as the working conditions (pressure, temperature, dynamic operation conditions and corrosion attack) should be available. This chapter can give only a very brief summary of some of the requirements for selecting appropriare materials.

4.3.2.1 Typical materials for apparatus and other equipment

Low-alloy steels (as specified, e.g., in the German Standard: DIN 10083 [23]) are generally used for the load-bearing walls of pressure vessels. For relatively thick walls, good annealing characteristics and good toughness are important. Modern established manufacturing processes lead to homogenous and uniform properties over the cross -section of the materials. For this reason, monobloc reactors are preferred to multi-wall constructions. This development is highly influenced by the testing possibilities for the completed apparatus.

To test multi-wall constructions readily is very difficult, or in some cases, impossible. Mono-bloc vessels could be tested much better. Increasingly the operation of such facilities is only allowed when a adequate acceptance-test by officials at the completed apparatus is possible. Furthermore, the use of modern materials is also an essential economic factor of the manufacturing process.

All multi-wall constructions are considered for a temperature limit of approx. 300°C. Above this temperature the stresses from the shrinking process are disintegrated. For higher temperatures, mono-bloc vessels are used. On the other hand, it may necessary to use multi-wall vessels when te inner wall must be cooled, for example as in ammonia reactors, where the cold inlet gas is fed along the inner wall to keep the temperatures at allowable levels.

Increase of the wall temperatures above 350°C may produce temperature embrittlement. This can be reduced by limiting the content of impurities (S, P, Zn, As). Hydrogen-resistant materials are necessary at temperatures above 200 °C with hydrogen-containing media. It is generally necessary to use high-strength steels at operating pressures above 1000 bar. Since elongation at rupture, and toughness reduce with increasing strength, good toughness of the selected material is important. Mechanical investigations fracture are advisable if these steels are used for wall-thicknesses exceeding 150 mm. Some typical steels for high-pressure vessels and piping are listed in Table 4.3-3.

Table 4.3-3
Some common steels used for high-pressure applications. [14]

German Standard specification	German Standard Material No.	American Standard AISI No./ UNS No.	AFNOR No. or AFNOR NF No.	Application for	Additional notes
21CrMoV5 7	1.7709			Forgings	
20CrMoV13 5	1.7779				
30 CrNiMo8	1.6580		30CND8		Parts >1000 bar
30CrMoV9	1.7707				
X20CrMoV12 1	1.4922				
Ck 35	1.1181	1035/1038	XC32,XC32H1	Covers,	
24CrMo5	1.7258			flanges,	
21CrMoV9	1.7709			bolts, etc.	
30CrNiMo8	1.6580		30CrNiMo8	Pipes	
30CrMoV9	1.7707				For H_2, > 450°C
X20CrMoV12 1	1.4922				PE-pipes
24CrMo10	1.7273			Apparatus	For H_2, > 200°C
16CrMo9 3	1.7281		20CD8		
X6CrNiMoTi 17-12-2	1.4571	316Ti	X6CrNiMoTi 17-12-2		

4.3.2.2 Corrosion-resisting materials

Common materials as austenites, titanium, Hastelloys, tantalum, nickel, and silver are used to avoid corrosion attack. These materials are used as lining materials in vessels or as explosion-bonded steel sheets. The sheets are formed to parts of an internal liner of a vessel, and these parts are welded, using special welding procedures, to a complete liner.

For more information about corrosion attack see Ref. [24].

In this part, only brief details are given about corrosive attack, especially that caused or intensified by high pressure.

4.3.2.3 H_2 – Attack at Elevated Temperatures: Nelson Diagram

The attack of molecular hydrogen on the construction materials at high pressure and elevated temperatures must be considered in a number of high-pressure processes, e.g., hydrogenation and ammonia production. At elevated temperatures (> 200 °C) and high pressure, hydrogen tends to decarburize carbon steels, which causes embrittlement and deterioration of the mechanical properties. Hydrogen attack can be prevented by using
carbon-binding alloying elements such as chromium, molybdenum, vanadium, and tungsten. The hydrogen-resistant steels contain 2 to 6% Cr, 0.2 to 0.6% Mo, and a few types are additionally alloyed with up to 0.86% V and 0.45% W. Fig. 4.3-12 shows the resistance limits for different steels. The so-called "Nelson Diagram" [25] is permanently updated by the American Petroleum Institute [26].

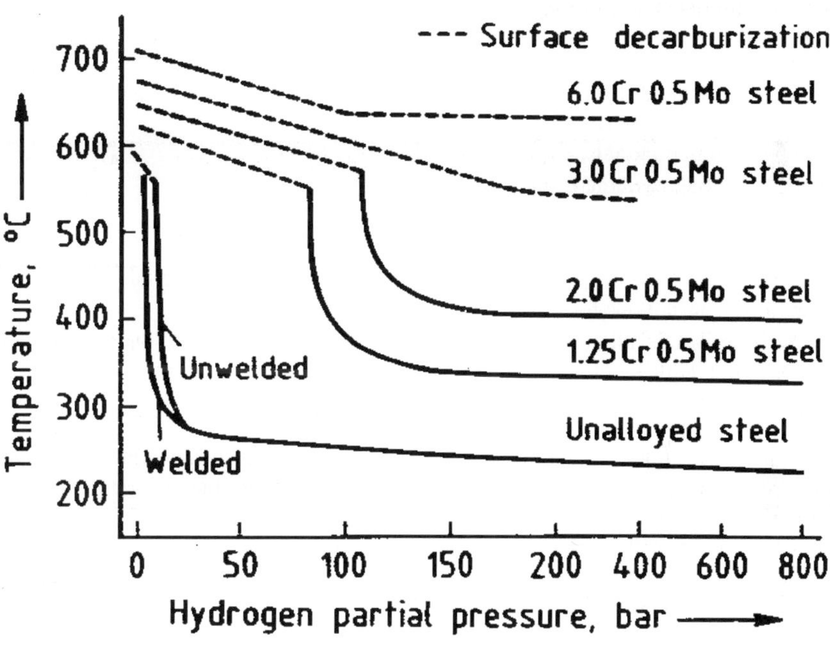

Fig. 4.3-12 . Resistance limits for the attack of pressurized hydrogen on steels (Nelson Diagram).

4.3.2.4 Corrosion by carbon monoxide

Carbon monoxide under pressure attacks unalloyed and low-alloyed steels at 130 to 140°C with formation of ironpentacarbonyl. Attack virtually ceases above 350°C. Where high-alloy chromium and chromium nickel steels are used the danger of damage is significantly reduced. Chromium steels with 30% Cr, and austenitic steels with 25% Cr and 20% Ni are completely resistant. If iron pentacarbonyl is entrained in a synthesis cycle, it may decompose exothermically at a higher temperature to form active, pyrophoric iron.

4.3.2.5 Nitriding by Ammonia

At temperatures above 250°C, ammonia forms a nitride layer on steels and embrittles the material. Welding is then difficult owing to the nitrogen enrichment. These parts are often irreparably damaged after long operating times. Apart from Armco iron, austenitic steels are resistant to this kind of attack. A nitride layer does form with austenitic steels at temperatures above 350°C, which protects the material. The layer must be ground off before repair-welding, but the welding properties of the base material are not affected.

High temperature materials are not described here. For more information see ref. [27,28].

4.3.3 Vessels and other apparatus

The bodies of thick-walled pressure-vessels with an outer- to inner-diameter ratio $d_a/d_i > 1.2$ can either be manufacured as monobloc (solid-wall vessels) or be constructed of several layers (multi-wall vessels). During the process of design and fabrication of such vessels some significant and common rules must be observed:

- Selection of tough materials, with high ductility,
- Approved materials suitable for welding,
- Construction with smooth changes in wall thickness,
- Walls with minimized notches and side holes,
- Specific material constants, such as the thermal expansion must be considered,
- Appropriate constructions for testing procedures,
- Different materials for all non-positive load transmission between the single components.

Companies manufacturing large and heavy components for the process industries must possess all the facilities for the different steps of the manufacturing process. The largest parts which can be handled have weights of approx. 450 tons. For transport on streets and on railways the diameter is limited to approx. 4.5 m. Heavier and larger components could only be transported with ships. Sometimes it is more economical to erect the apparatus direct at the plant site.

Since the beginning of the manufacture of high-pressure vessels in the early 1910's their weight have increased continuously up to the beginning of the 1970's. As an example, the development of ammonia-reactors is shown in Table 4.3-4 [29].

4.3.3.1 Thick-walled vessels

The term **solid-wall or monobloc vessel** is applied to all components where the cylindrical wall consists of a single layer. Solid-wall vessels are suitable for all types of pressure vessels, in particular for those operated under high temperatures. Thermal stresses arising during heating or cooling are smaller than in multilayer vessels because of the good thermal conduction across the wall. Therefore solid-wall vessels are especially suitable for batchwise operation.

As decribed in Chapter 4.3.2, it is increasingly common to design and manufacture such apparatus as monobloc reactors. Furthermore with modern testing facilities with, for example, ultrasonic methods it is possible to check the finished apparatus up to wall-thicknesses of 350 mm. These developments are an excellent contribution for the safe operation of high-pressure vessels.

Table 4.3-4
Development of size and weight of ammonia reactors since 1910 [29].

Year of fabrication	1910	1912	1913	1915	1956	1963	1969	1972	1982 [1]
Inner diameter [mm]	90	160	300	800	1200	2900	2000	2400	3350
Height [m]	1,8	4	8	12	12	22	22	34	9
Weight [tonnes]			3,5	65	105	270	386		190
Volume of catalyst [m³]			0,09	1,1	4,85	64	36	56	53
Operating pressure [bar]			200	300	300	150	350	210	210
Capacity [tonnes/day]			3,6-4,8	85	195	910	1200	1580	1800

[1] Two reactors are installed.

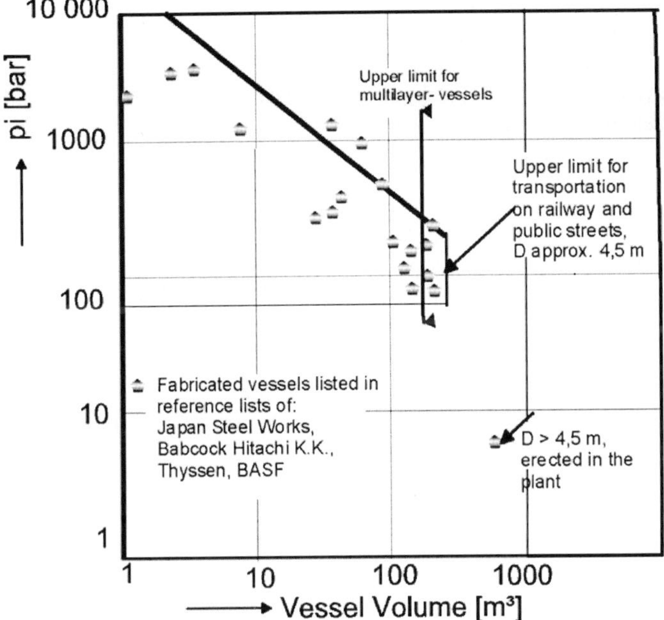

Fig. 4.3-13. Limits for the manufacturing of large high-pressure vessels
as a function of the internal volume and the pressure.

Solid-wall vessels can be manufacured by different methods:
- A one-piece hollow body is forged without welding seams,
- Forged rings are joined to form the vessel by circular welding seams,
- Thick sheets are bent around and welded into sections by one or more longitudinal seams.
  The sections are then joined by circular welding seams to form the vessel,

- Pressure vessels are also produced by small-gap submerged-arc welding. The vessel wall is joined together with a flange and bottom by deposit-welding onto a thinner internal tube. Vessels of any shape can be produced using this technique.
- The pressure vessel is fabricated by non-cutting shaping layers positioned one above the other. The vessel is formed by welding the layers starting on the inner liner. At first, such vessels were fabricated for testing purposeses, but the development was stopped because the energy costs of fabrication had become too high.[30]

4.3.3.2 Multiwall vessels

Multiwall vessels are suitable for all types of pressure vessels, in particular for those with large diameters, -lengths, and -wall thicknesses. Thermal conduction through the wall is lower than in solid-wall vessels, so the lower heating and cooling rates must be taken into account. The type of joint between the wall elements is used as the distinguishing feature. Multiwall vessels can be produced by shrinking one or more seamlessly forged cylinders onto a core shell. The difference required between the outer diameter of the smaller cylinder and the inner diameter of the larger cylinder must be accurately calculated. Before shrinking, the larger cylinder is heated and the smaller may be cooled. During the shrinking process, compressive stresses are produced in the smaller cylinder, and tensile stresses in the larger cylinder, both in the circumferential and longitudinal directions. Today, this process is applied only in special cases for compressor cylinders or for vessels having volumes below 1.5 m^3 and pressures above 2000 bar used in batch processing Fig. 4.3-14 shows an example [32].

Another well-known methods for making a pressure vessel is the use of thin squared sheets, strips or wires to build up the wall in single layers as shown in Fig. 4.3-15 [31].

Fig. 4.3-16 shows a pressure vessel fabricated by the BASF Schierenbeck strip-winding process [35]. In this process, a profiled steel strip is gas- or electrically heated to between 650 and 950°C and helically wound onto a core, often made of an corrosion-resistent material. The strip is cooled, annealed by blowing air on it, and shrunk under high tension onto the underlying strip layers. The strip layers are joined one to the other. The first layer is fixed directly on the specially grooved inner liner. The inner liner possesses a helical channel which during operation could be connected to a leak–detecting system. The system of the shrunk grooved strips is combined so that, apart from the circumferential stresses, they are also able to absorb longitudinal stresses generated by the internal pressure. The end pieces are produced by forging ingots and are joined to the cylindrical shell by overwinding. The flanges are wound from the same profiled strip, and the threaded bores for the cover bolts can then be drilled directly into the wound flange. Further fabrication processes are known where the pressure vessel is formed of steel sheets. By this technique the pressure- vessel wall is build up of different shaped sheets which are welded together to form a shell. Fig. 4.3-15 shows widely used manufacturing processes for multi-layer-wall vessels [31].

Cylindrical shaped shells having different diameters are produced from 25 up to 50-mm thick sheets welded with longitudinal seams. These shells are shrunk, one to another, without prior machining, to produce individual shell sections. These are joined to the forged end-pieces and flanges by circular welded seams.

In the multilayer design, half-shells, formed from sheets 6 up to 22 mm-thick sheets, are applied to the core tube, which is seamless or welded by a longitudinal seam. The shells are joined one to the other, and to the underlying layer, by two longitudinal seams (three-sheet seams). The half-shells are tangentially pre-stressed by the welding contraction. Uniform dis

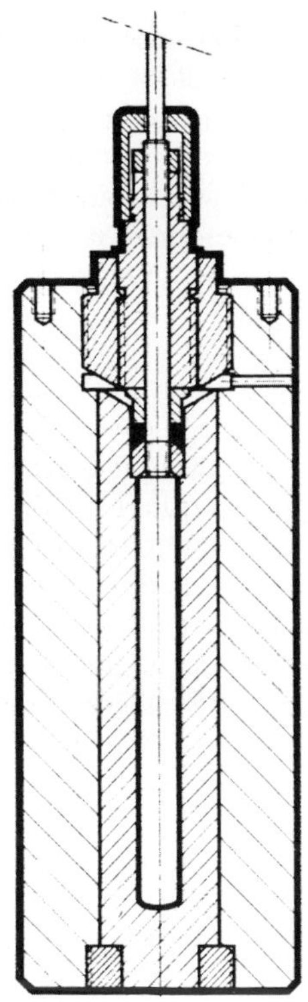

Fig. 4.3-14. Shrunk two-layer pressure vessel for 4000 bar and 600°C.

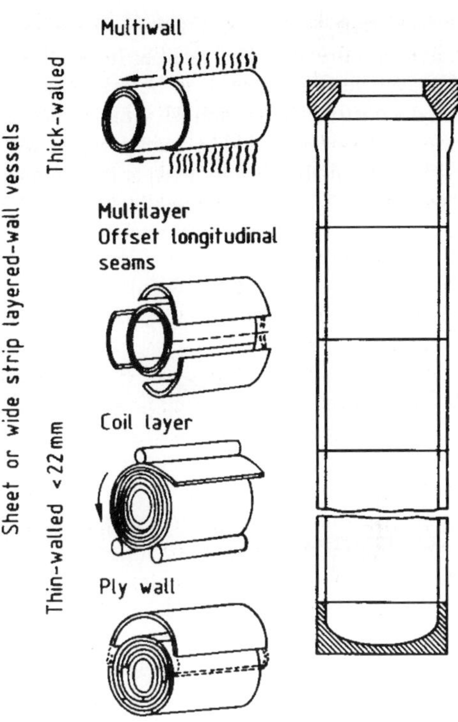

Fig. 4.3-15. Typical examples for the design of multi-wall vessels [31].

tribution of the welds around the circumference is achieved by offsetting the longitudinal seams from layer to layer. When larger diameters are required, three or four sheet segments are arranged on the circumference instead of half-shells. The individual sections are joined by circular welded seams.

Three further processes to manufacture multi-layer vessels are applied:
- The Coil Layer Process. The shell is produced by winding a 3 up to 5-mm-thick sheet strip onto the core liner. The outer shell is applied after completion of the winding. The beginning and end of the winding require wedges. Longitudinal seam welds are made at the beginning and end, or when ever a sufficiently long strip is not available.
- The PIy Wall Process. This design is a combination of the coil-layer and multilayer designs the spiral construction of the shell section is combined with three-sheet longitudinally welded seams.

- Another strategy is to manufacture the body of a pressure vessel from wires which are wound on a prefabricated internal liner. This technique is similar to that for strip-wound multilayer vessels, mentioned before. The wires made of a high strength material. The special construction combined with a surrounding frame was proposed by the ASEA company, Sweden [33]. The main problem with these constructions is to rule-out the axial strains by a suitable design. Further design material is presented in the American Pressure Vessel Code [3]. Such constructions are used in isostatic pressing. A sketch of the principle of a wire wound pressure vessel is given in Fig. 4.3-17.

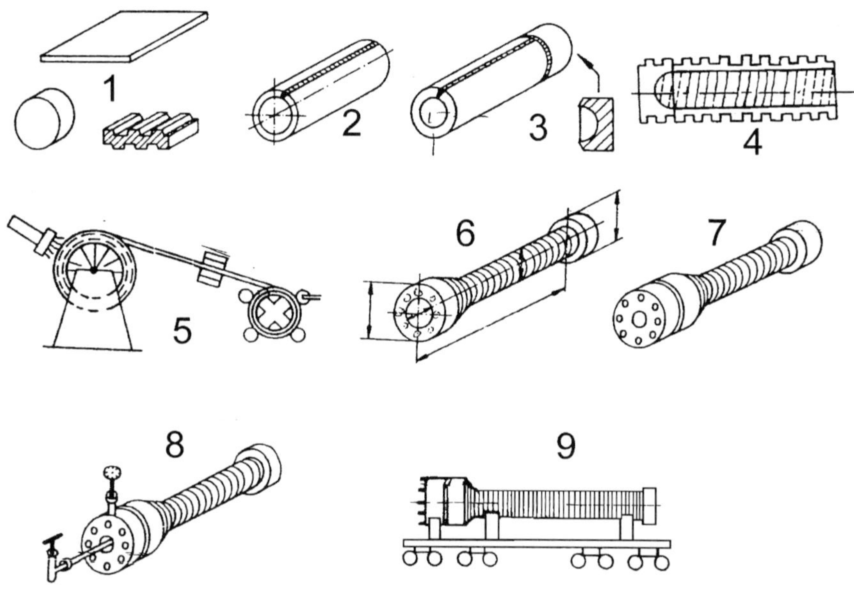

Fig. 4.3-16. Procedure for fabricating a strip-wound pressure vessel, as first proposed by Schierenbeck, BASF, D-67056 Ludwigshafen, Germany [36].
1, Raw materials; 2, Inner shell; 3, Inner shell with welded bottom; 4, Spiral groove inner shell; 5, Wrapping bands onto inner shell; 6, Dimensions inspection; 7, Assemble closures; 8, Hydrotest; 9, Transport to destination.

A further step in the fabricating process, after having built up the mantle of the pressure vessel, is the design of the flanges. Several constructions are technically well tested, and the four mainly used are presented in Fig. 4.3-19 [26]. At the left side, a forged flange ring shrunk onto the multilayer cylindrical body of the pressure vessel is shown. Fig. 4.3-19 (1) The bolts of the closure are fixed into this forged ring. In Fig. 4.3-19 (2) a smaller forged flange ring is arranged on the cylindrical vessel body. In this example, the bolts are fixed in the ring and in the multilayer part of the body. The third construction shows a flange made of the same strips as used for the body. The bolts are positioned in the middle of the multilayer head of the reactor Fig. 4.3-19 (3). The fourth example shows the connection of a forged flange-ring directly with the internal liner and the cylindrical body Fig. 4.3-19 (4).

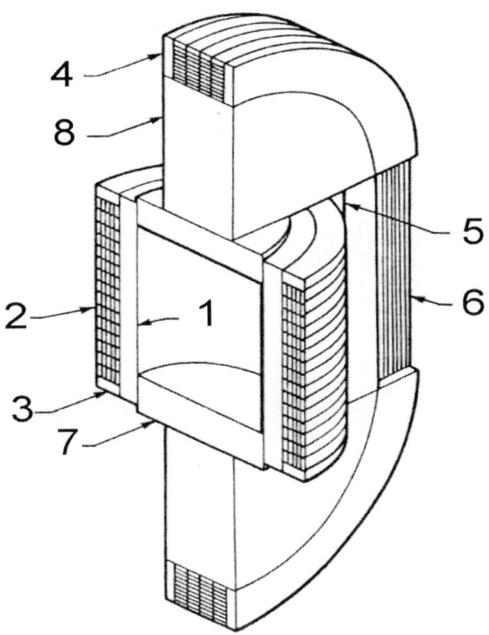

Fig. 4.3-17. Wire wound vessel and frame construction
1, cylinder; 2, cylinder winding; 3, cylinder flange; 4, frame flange; 5, column; 6, frame winding; 7, closure; 8, yoke.

4.3.3.3 Closures and sealings

The most commonly used types of closures are shown in Fig. 4.3-18 [14]. The cover seals (Fig. 4.3-18 A - E) are attached by means of threaded bolts and are used for all diameters and pressure ranges. The bolts are generally tightened hydraulically or pneumatically. It is, however, also possible to pre-stress the bolts longitudinally and then to tighten the nuts.
In the so-called Bredtschneider seal (Fig. 4.3-18, F), the force acting on the cover owing to the internal pressure is led via the sealing ring (a) onto the split ring (b), which is supported in the groove of the shell. This design requires only small bolts which bring the cover and sealing ring together in the pressureless state. The clamp seal (Fig. 4.3-18, G) permits rapid opening and closing of the pressure vessel. The two- or three-part clamp is pushed over the collar on the cover of the vessel and is fixed. The elastic seal is pressed onto the sealing surface by the internal pressure. Because the vessel can be automatically closed this joint is frequently used in batch processes, for example in extraction with supercritical gases. Screwed-in covers (Fig. 4.3-18, H) are used for smaller inner diameters. The Hahn and Clay finger-pin

222

seal (Fig. 4.3-18, I) is suitable for large diameters. An elastomeric seal is necessary . Opening and closing can easily be fully mechanized.

Metal seals are usually necessary as cover seals in chemical high-pressure plants. The single cone (Fig. 4.3-18, A) with an angle of 10° is preferentially used for cover diameters of up to 800 mm and pressures up to 400 bar. The double cone ring (Fig. 4.3-18, B) with an angle of 30° is more complicated than the single cone, and is used for covers having diameters above 500 mm and pressures up to 700 bar. The delta ring (Fig. 4.3-18, C) and the ring-

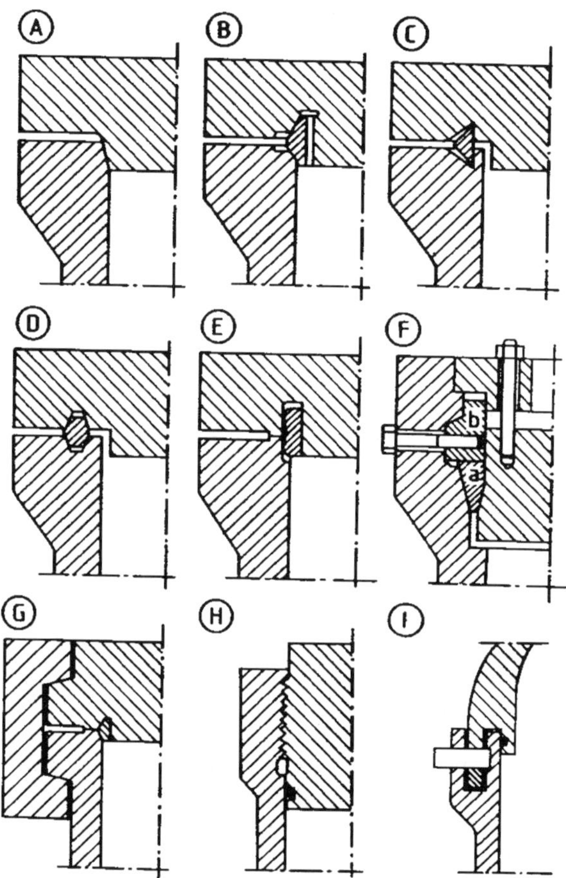

Fig. 4.3-18. Constructive details for covers and seals of high pressure vessels. (Explanations are given in the text).

joint-seal-ring (Fig. 4.3-18, D) are smaller versions of the double cone ring. The wave-ring seal (Fig. 4.3-18, E) is pressed by the internal pressure against the cylinder sealing surface and is plastically deformed in the process. This is a special seal for stirred-tank polyethylene reactors run at temperatures up to 300°C and 2500 bar. Flat sealing rings and metallic O-rings are seldom used in chemical plants. All contact surfaces must have a high quality surface free from any cracks in the region of the seal.

4.3.3.4 Design details - corrosion-protecting of inner surfaces

Only vessels for small units are completley made of corrosion- resistant material. This protection method is to expensive for large vessels. They can be more economically protected by a liner of corrosion-resistant material containing Mo, Cr, Ni and V. Multilayer vessels require corrosion-resistant steel only for the innner layer. They are often equipped with venting channels, so-called „Bosch-holes" to ensure that for example hydrogen diffusing through the inner layer is drained and could be analysed.

Cladding

Abrasion can be prevented by rollbonding cladding or weld-overlay cladding. However, this type of protection is often not possible with the steels and wall thicknesses used for high-pressure vessels. An advantage of this method is that the cladding materials (for example, austenitic steel) often have a thermal expansion different fromthe vessel wall. This leads to thermal stresses in the cladding during heating and cooling. If the cladding is joined
solidly to the vessel wall, the expansion of the base material is imposed on every part of the cladding. The thermal stresses are, therefore, uniformly distributed in all parts of the cladding. Disadvantages of this method include the facts that: -
- The base material is attacked in the cladding and is damaged at distinct points; any start - ing damage often cannot be detected in an early state.
- In weld-overlay cladding, the solidifying melt forms columnar crystals perpendicular to the wall surface. Along the grain boundary of these crystals, the corrosion resistance is lower than in the rolled material. In the case of heavy corrosion, at least a two-layer weld-overlay cladding should therefore be selected.
When hydrogen diffuses through the cladding, the cladding does not protect the base material against hydrogen attack.

Lining

With loose linings, the tightness can be monitored during operation. For this purpose, the space between the lining and the pressure-bearing wall is monitored by means of venting holes. The lining must be attached to the pressure-bearing wall in such a way that different
thermal expansions cannot produce a displacement between the lining and the vessel wall. An advantage over cladding is that the lining can be checked for leaks at any time. Disadvantages of this method compared to the cladding method include:
- When large differences in thermal expansion exist, or the liner is unsuitably attached to the pressure-bearing wall, the expansion can concentrate on the weakest point and cause damage to the liner.
- The joining of a nozzle lining to the lining of the vessel can be a preferred place of damage. Fig. 4.3-20 shows the attachment of a loose titanium lining by means of mortised titanium strips. The venting hole through the vessel wall is used to introduce the protective gas [29].
- Floating Bladders. The "floating bladder" principle is used under extreme corrosion conditions. The balloon is made of an alloy which is resistant to the extremely aggressive medium and is surrounded by pressurized water so that the internal pressure of the bal loon

224

and the pressure of the surrounding water are virtually identical (see Fig. 4.3-21). An advantage of this method is that the high pressure shell and the high-pressure seals are not

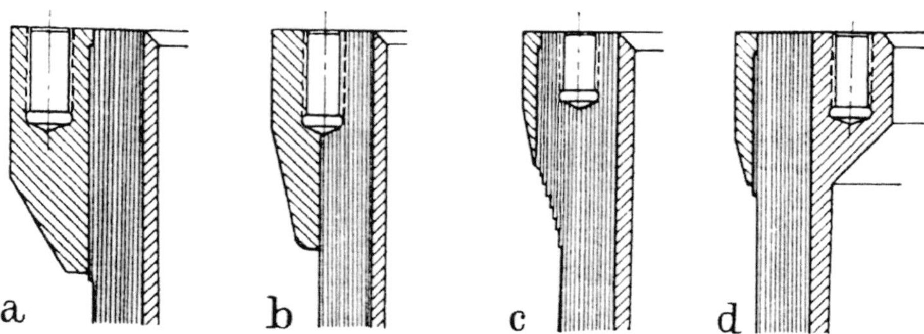

Fig. 4.3-19. Different kinds of flanges of multilayer high pressure vessels [34].
a, forged flange shrunk onto the multilayer vessel body; b, forged thinner flange ring shrunk onto the multilayer vessel body; c, flange made out of the same material as vessel body; d, forged flange ring welded directly to the inner liner of the multilayer vessel.

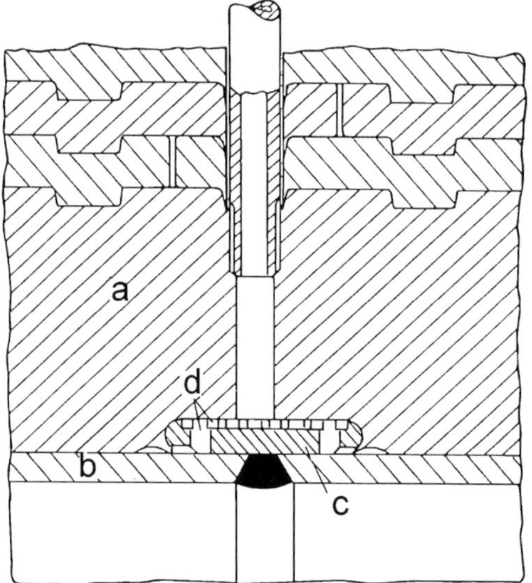

Fig. 4.3-20. Titanium clad pressure vessel attached by a mortised corrosion-resistant titanium strip [34].
a, core liner of a multilayer vessel; b, titanium sheet; c, grooved titanium strip; d, hole for venting blanketing gas (so-called "Bosch-hole") and connecting device for an analyser.

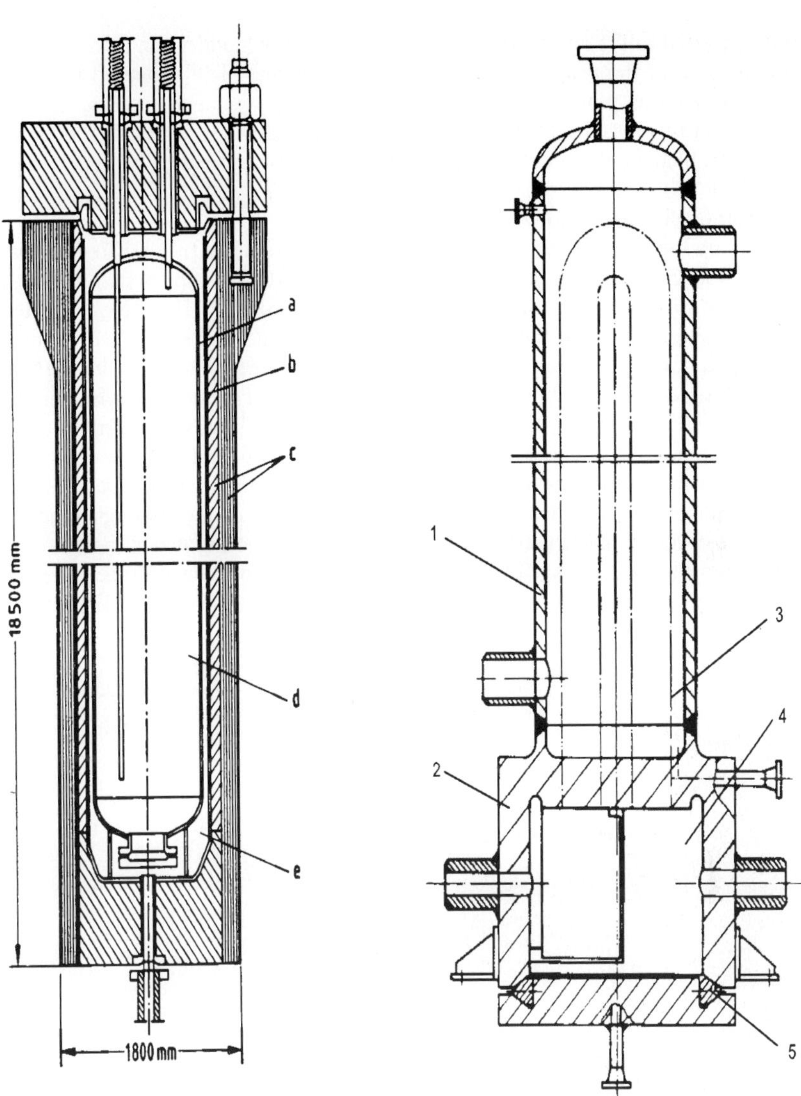

Fig. 4.3-21, Pressure vessel with corrosion resistent internal bladder [34]
a, bladder; b, austenitic steel liner;
c, BASF-Schierenbeck multilayer-vessel;
d, corrosive medium;e, intermediate phase of water.

Fig. 4.3-23. Constructive details of a hairpin heat-exchanger, for conditions similiar to whose in Fig. 4.3-22 (UHDE-Hochdrucktechnik) [37].
1, low-pressure mantle; 2, high-pressure mantle; 3, high pressure hair-pin tubes;
4, pre-chamber of heat-exchanger;
5, double cone seal ring.

directly exposed to the aggressive medium. Leaks in the bladder can be detected by chemical monitoring of the pressurized water. A disadvantage of this method is that additional equipment, such as piping, measuring instruments, and balancing tanks, is necessary to ensure the same pressure of the ballon and pressurized water. The difference between the inside of the bladder and outside is approx. 2 bar, and is automatically regulated [29].

4.3.3.5 Heat exchangers and others

High pressure heat exchangers, with working conditions up to more than 325 bar and temperatures of 400°C and higher, are normally designed as double-pipe systems or as shell- and tube-apparatus provided with normal tubes or with so-called UHDE Hochdrucktechnik fin-tubes to enlarge the square of the exchanger unit. The shell- and tube-construction is used as a plate-to-plate construction, as well as with a high-pressure plate and with welded-in high-pressure hair-pin tubes. Both types of apparatus can be built to meet individual requirements. With both types of apparatus, the heat-exchange tasks can s be solved safely and economically. Heat exchangers for higher operating conditions must be individually designed.

Double-tube heat exchangers often consist of an inside high-pressure tube with a thin-vessel mantle. Seldom is the mantle also operated at high pressure. At the cooling- or heating-side of the apparatus different media are used. They range from cooling water, thermostatted water or another fluid such as oil from an oil circuit, up to steam, to solve the special heat-exchanger problems. A well known example is the arrangement used as reactor in a PE-plant.

Fig. 4.3-22. Shell- and tube-heat-exchanger for 325 bar and 250°C operating conditions (UHDE-Hochdrucktechnik) [32].

Another type is the shell- and tube-arrangement. In Fig. 4.3-23 an example with a combined inlet- and outlet chamber is shown, which is designed as a high-pressure vessel. One end of this vessel is the tube plate into which the hair-pin tubes are welded. The lower-pressure side could be connected to a low-pressure heating or cooling system. A further apparatus is the multi-tube hair-pin heat exchanger presented in Fig. 4.3-24.

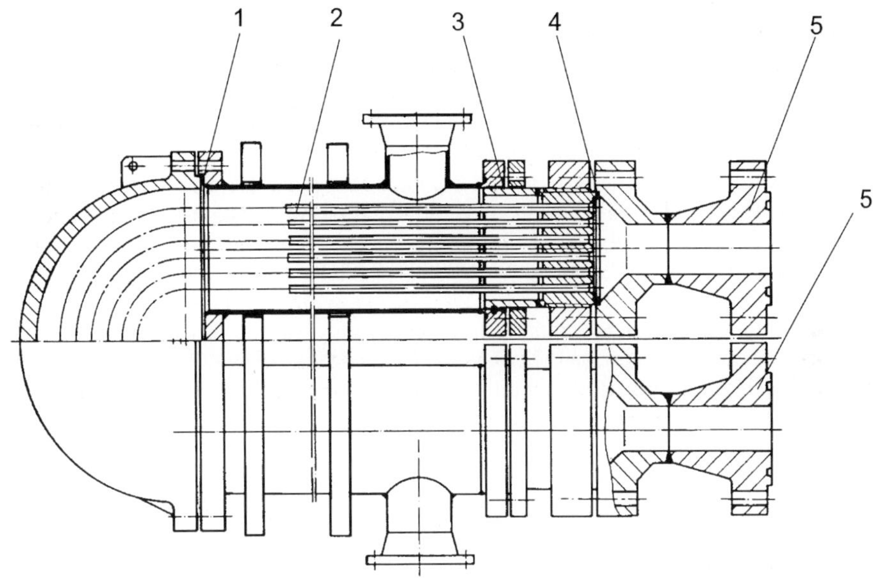

Fig. 4.3-24. A hair-pin multi-tube high-pressure heat exchanger for 325 bar and temperatures above 300°C [37].
1, low-pressure seal; 2, high-pressure hair-pin tubes; 3, low-pressure seal-ring; 4, high-pressure seal; 5, high-pressure flange.

The hair-pin tubes are welded into different plates. The low-pressure side is closed with an end-cove [37]. This construction exhibits a good high-pressure performance together with an effective heat exchange under such extreme conditions and is an economical solution for such a task. If the chemical process needs high pressure also at the mantle-site of the heat-exchanger it is usual to design this part of the apparatus also as a high-pressure vessel.

An other possibility of a modular standardized heat exchanger program for the high pressure region, is the simple kind to enlarge needed heat exchanger surface for p.e., a debottlenecking in a chemical plant, is the use of finned-tubes enlarging the surface of the single tubes. Modular units up to 400 m² are possible. Modernly designed heat exchangers include the necessity of efficient cleaning in the plant.

For such constructions the common high-pressure materials reported in Chapter 4.3.2 should be chosen.

An often-used design is the high-pressure tube coil submerged in low-pressure vessels. Coils of high pressure tubes are easy to manufacture.

In addition to apparatus mentioned apparatus many special constructions, up to large sizes are described in the literature.

4.3.4 Laboratory-Scale Units

In research and development, facilities for investigating chemical reactions are required. Besides screening tests, reaction-rate and kinetic parameters must be determined. For accurate and safe scale-up of the apparatus of an industrial chemical plant, furthermore, thermodynamic- and transport-data must be measured. For this purpose, a variety of apparatus is described in the open literature. In this chapter, some typical devices for use at high pressure are presented.

4.3.4.1 Reactors

The first example for small-scale reactors is a stirred vessel for a maximum pressure of 32.5 MPa and 350°C (Fig. 4.3-25). It has a volume of 0.4 l and can be used batchwise or in continuous operation, preferably for gas-liquid reactions, without- or with soluble or suspended catalysts.

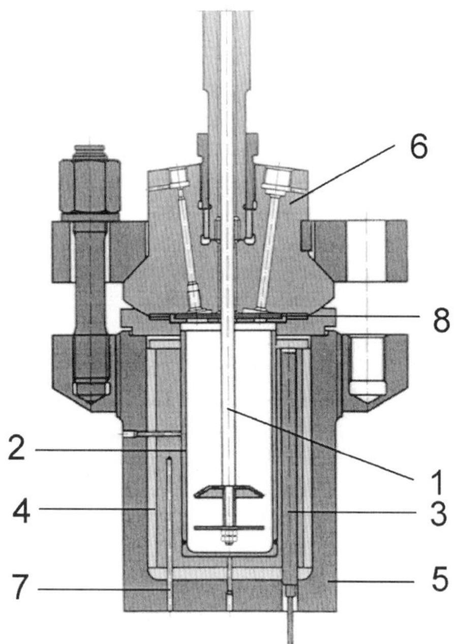

Fig. 4.3-25. Stirred pressure vessel with internally mounted electrical heat pipes [38];
0.4 l internal volume, 32.5 MPa, 350°C, heating rate 60°C/min.
1, stirrer; 2, internal liner; 3, heating rod; 4, internal insulation; 5, high pressure mantle;
6, head; 7, thermocouple.

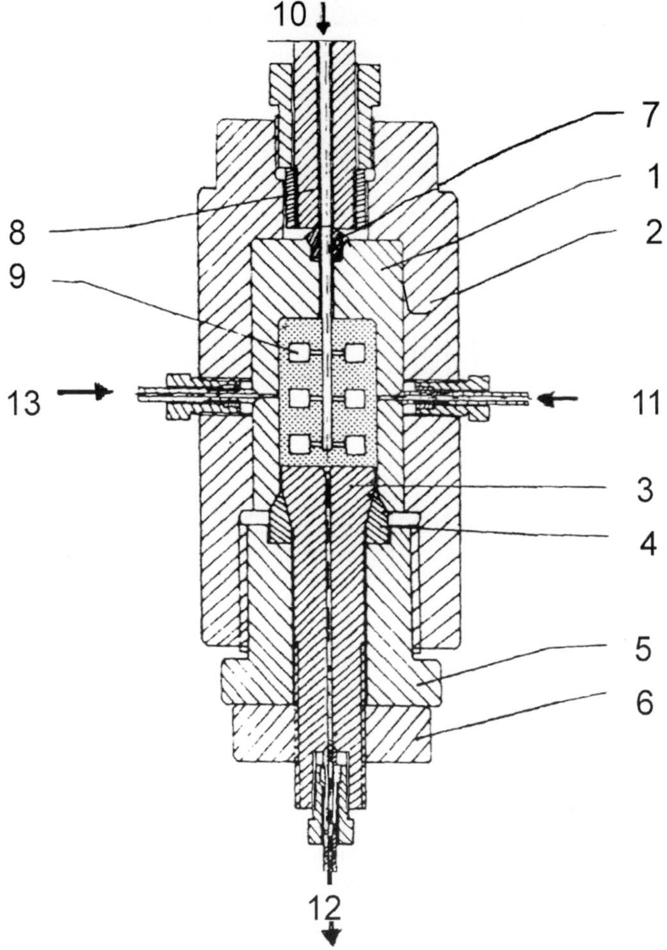

Fig. 4.3-26. Stirred pressure vessel, 250MPa, internal volume 100 ml [39].
1, internal tube; 2 liner; 3, pressure tube; 4, pressure collar; 5, locking screw; 6, lock nut;
7, sealing; 8, shaft of magnetic stirrer; 9, stirrer; 10, inlet; 11, catalyst port; 12, outlet;
13, temperature- and pressure measurement device.

A small reactor for the much higher pressure of 250 MPa and up to 300°C is shown in Fig.
4.3-26. The capacity of the shrinked vessel is only 100 ml. It is equipped with a fast -running
stirrer, a magnetic drive, and an electric heater. The reactor is suitable also for polymerization
reactions [39].

The third example (Fig. 4.3-27) is a loop reactor with internal recycle, developed by G.
Luft. This reactor can advantageously be used to study kinetics of heterogenous catalytic reactions at pressures up to 40 MPa and temperatures to 500°C. The internal recycle

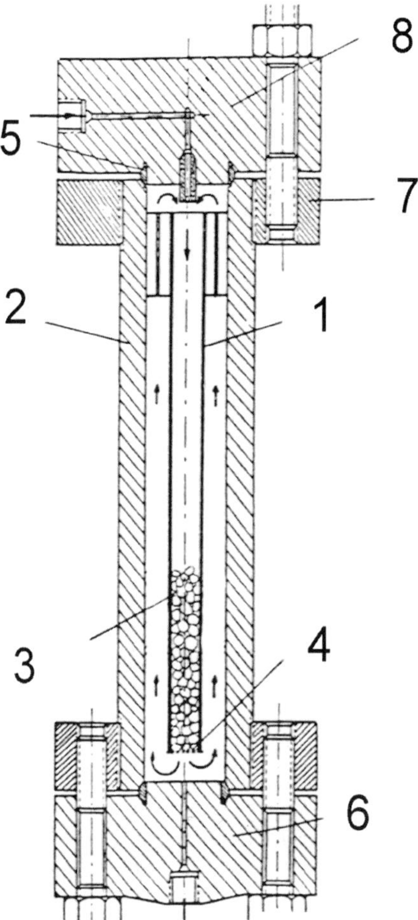

Fig. 4.3-27. Jet loop reactor, 40 MPa [40]
1, inside tube; 2, mantle; 3, catalyst; 4, sieve plate; 5, seal; 6,bottom closure; 7, flange;
8, head closure

of the reactants is generated by a jet from a nozzle in the flange. This system avoid problems
arising from the use of blowers at high pressures. The recycle-ratio can be adjusted, varying
reaction conditions, by using nozzles of different diameter and different pressure of the feed
[40].

4.3.4.2. Optical cells

Optical cells are used to investigate phase equilibria at high pressures as well as for analytical purposes. A more sophisticated approach consists of monitoring the changes in concentration to determine the rate of chemical reactions.

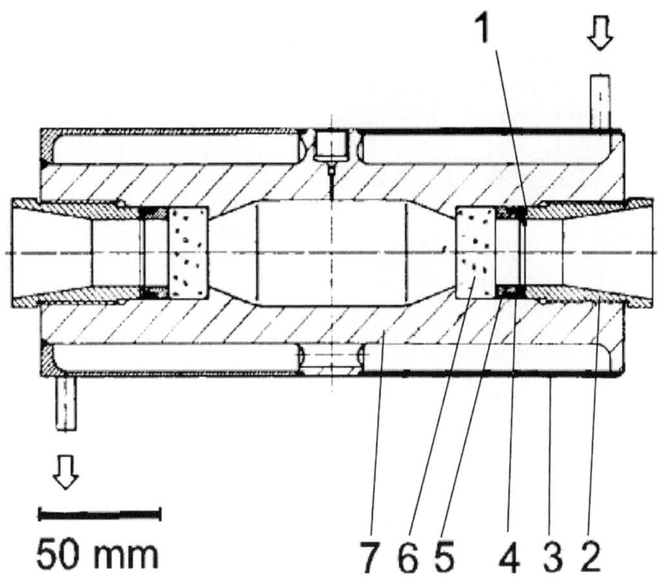

Fig. 4.3-28. Optical cell [41]: 200 MPa, 350°C, 110 ml internal volume.
1, pressure distribution ring; 2, pressure screw; 3, heating jacket; 4, seal ring; 5, pressure ring; 6, window; 7, pressure vessel.

According to the specific task, different materials are used for the windows. Typical materials are sapphire, calcium fluoride, zinc selenide, diamond, and normal quartz. The selection of the materials depends on the pressure and, for spectroscpic investigations, also on the wavelength corresponding to the bonds of the species to be analysed.

The windows are mostly sealed with thin films of PTFE, gold, indium and other corrosion-resistant easily deformable materials.

Optical cells can be designed for very high pressure of 1000 MPa and, more and temperatures from –273 to 800°C.

First, a simple cell for 200 MPa and 250 °C, with two sapphire windows, is shown. The volume of the cell is approx. 110 cm³, the free diameter of the windows is 16 mm. This cell is suitable for spectroscopic measurements, as well as for phase observation (Fig. 4.3-28) [41].

An optical cell developed for Raman spectroscopy is presented in Fig. 4.3-29 [42].

Four sapphire windows are arranged in two planes displaced by 90° from each other. The temperature is measured in the reaction chamber by a Ni-Cr/Ni thermocouple. The cell can be electrically heated by four heating rods placed in the jacket. Through the left-hand window, the light of a laser is introduced. Behind the lower and the right-hand window there are mirrors which reflect the light into the cell. Through the upper window, the light passes into the spectrometer detector system.

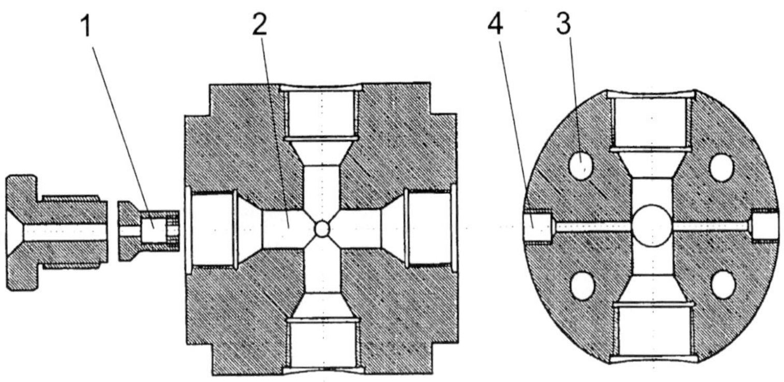

Fig. 4.3-29. Optical cell for spectroscopic measurements under high pressure
Max. pressure, 200 Mpa; max temperature, 200°C.
1, window; 2, chamber; 3, heating rod; 4, medium inlet.

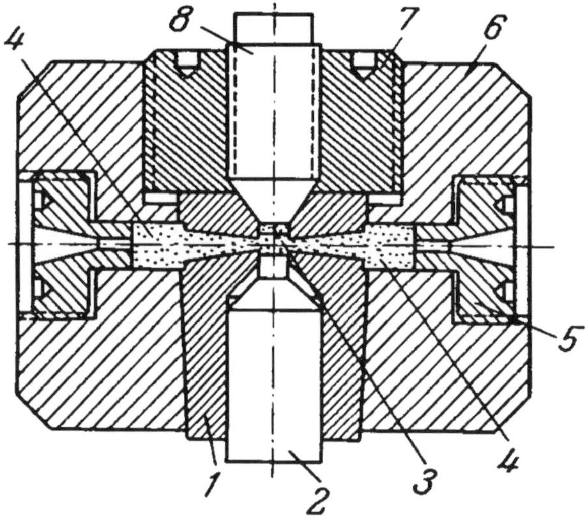

Fig. 4.3-30. Chamber for optical measurements up to approx. 3000 MPa.
1, die; 2, 8 piston; 3, steel packing; 4, windows; 5, obturaters; 6, mount ; 7, bracing nut.

The next Fig. 4.3-30 presents a cell for very high pressure with specially shaped windows.
This shape is suitable for supporting the windows against the extreme pressure of 3000 MPa
[36].

4.3.4.3. Other devices

An apparatus for differential thermal analysis (DTA), which can be used up to 250 MPa and 300°C is shown in Fig. 31 [37]. It consists of a cylindrical body (1) with 6 mm inside-, 13 mm outside-, diameter and 0.55 ml volume and is equipped with a thermocouple and

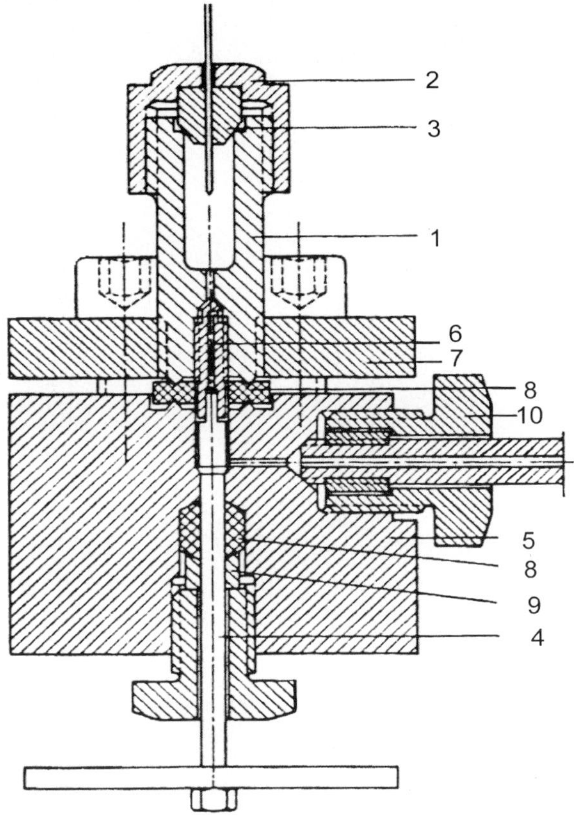

Fig. 4.3-31. DTA-device.
1, body; 2, cup; 3, seal; 4, needle valve; 5, valve body; 6, stem; 7, flange; 8, seal; 9, collar; 10, pressure screw.

closed by a cap through which solid samples can be placed. Gaseous samples can be fed through a valve at the bottom. For this purpose the cell is joined to a block equipped with needle valves. Also, apparatus for differential thermal gravimetry (DTG) at pressures up to 100 MPa is available [45].

The rolling-ball viscometer shown in Fig. 4.3-32 is designed for pressures up to 200 MPa and temperatures of 250°C. It consists of a standard viscometer made of glass (1), which is arranged inside a high pressure autoclave (2). The autoclave is heated by silicone oil in a jacket (3). When the steel ball (4) passes the measuring coils (6) a timer is started. From the rolling time between the coils the viscosity can be evaluated [46].

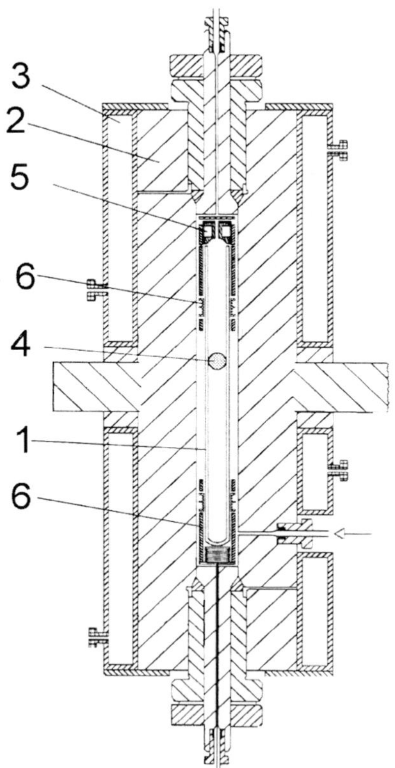

Fig. 4.3-32: Rolling ball viscosimeter [44].
1, glass tube; 2, high-pressure body; 3, heating jacket; 4, steel ball; 5, electromagnet;
6, measuring coil.

Also, rotating viscometers for pressures up to 200 MPa are available [47, 48, 49]. Further
pressure applications using small-scale vessels can be found in the literature.

4.3.4.4 Small-scale high-pressure plants

High-pressure pilot plants are used for scaling-up the plants for production of chemicals and
for separation processes. As an example, a small scale plant for supercritical extraction is
shown in Fig. 4.3-33. The pilot plant manufactured by the SITEC-company [50] contains all
the components of a large-scale plant to gain, for example, concentrate of hops from the natural product, and is fully equipped with controll- and measuring devices. On the left -hand side,
the extractor can be seen, to which the separator is joined (right hand side). The pilot plant is
designed for pressures up to 300 bar and temperatures of 250°C. It is movable, and can be
applied to separate different products from sundry natural and other materials.

Similar plants are available for further extraction processes, as well as for testing chemical
reactions or for the manufacturing of small lots of special products. Units for pressures up to
1000 bar, combined with additional separation steps, have been constructed.

Fig. 4.3-33. SITEC super-critical fluid extraction small-scale plant [50].

4.4 Instrumentation of High-Pressure Facilities

To operate high-pressure plants, adequate control- and measuring devices are required. High-pressure instrumentation is available for most purposes. In the following, a short selection will be presented.

4.4.1 Pressure measurement

Devices for measurement of high pressures are divided into primary- and secondary-gauges.

The dead-weight pressure-balance apparatus belong to the first group. They are mostly equipped with a vertical plunger in a close-fitting cylinder which is loaded with calibrated weights. Such devices originally designed by Michels [51], and Bridgman [52], are used to calibrate instruments for pressure measurement. All other instruments, beside dead-weight pressure balances, belong to the second group.

All over the world instruments of the "Bourdon gauge type" are used, and many manufacturers make such devices. Some of the best instruments of the Bourdon-type are from [53] with standard ranges up to approx. 7000 bar with a tolerance of only 0.01 % of span, provided with up to 16 inch diameter dials.

Another well proved pressure gauge is equipped with a metallic membrane cell [54]. The principle of this cell is the inclination of the membrane which causes the pressure-proportional variation of the electrical resistance of the system. The pressure range of those systems reaches up to 400 bar. The allowable temperatures of the metering system are near

Table 4.4-1
Secondary Pressure Measurement methods; their range and accuracy [11].

Method	Pressure range kbar	Possible accuracy	Temperature control for accuracy of ± 1 bar	Comments
Elastic strain element, e.g. Bourdon tube [2])	1 - 15	~ 1 in 10^3	~ 0.5 for typical gauge	Mostly used type of gauge
Special Bourdon tube device	0 – 7	<0.01% of span	Up to ~ 50°C zero, system temperature compensated	Proved standard device of [53]
Special strain -gauges	0 – 12		±0.01% of span	Proved standard device [55]
Change of resistance Special sensor [3])	0 - 8	Max. ± 0.3 % of span	< ± 0.5%	Proved standard method, Types of devices: PMP731, PMD235 [3]) [54], type HP-1 [58]
Resistance element, e.g. manganin [2])	0 – 30	~ 1 in 10^3	~ 0.1 (depends on pressure)	Probably the mostly used secondary gauge for the range 10 – 30 kbar
Measurement of ultrasonic velocity [2])	0 – 20	~± 2 bar for p<20 kbar	~ 0.5	Gauge has also been developed for use up to ~ 50 kbar
Melting line of mercury [1])	0 – 15	~ 1 bar	< 0.01	Provisional pressure scale 0 – 15 kbar

[1]) The limiting factor in the temperature effect is the slope of the melting line.
[2]) The limiting factor in the temperature effect is the thermal influence on the special sensor.
[3]) Maximum pressure 400 bar, maximum allowable temperature, 100°C

100°C. For higher temperatures, a capillary system can be used as a pressure transmitter and the sensitive cell is mounted in the cold part of an application.

Further devices use strain-gauge transducers as so-called dead-end instruments or sensors mounted on a pipe with internal flow. The principle of this arrangement is the elastic deformation of a metallic cylinder, measured with the strain gauges. Pressures up to approx. 15 kbar can be measured [11].

Well-proven devices using strain-gauges for pressure measurement are presented by [55]. The precision pressure-transducers of this type are of sturdy and compact construction with a very small dead volume. Their range is 0–1000 bar. The measuring element of a precision pressure-transducer consists in a diaphragm, to whose reverse side is applied a strain-gauge rosette, which is an assembly of four active-resistance strain-gauges arranged in a bridge circuit. The space behind the diaphragm is under atmospheric pressure, which means that the measurment of pressure is performed against atmospheric pressure. The medium to be measured is led via the pressure port onto the diaphragm. As a result of the pressure acting on the diaphragm it is deformed and consequently the resistance of the strain-gauges is changed. By applying a voltage to the strain-gauge bridge, the resistance change is transformed into an output voltage which is directly proportional to the pressure.

The results are influenced by the change of the physical properties of the wires which are used. Above the technical range used for super-pressures, fixed points such as the melting point of a substance, or the change of a spectral line, can be used.

The application of so-called intensifiers is another possibility for pressure measurement. A good survey of other possibilities for pressure measurement is given by [11].

4.4.2 Temperature measurement

Two main principles of temperature measurement use thermocouples and the so-called resistance thermometer. In chemical plants both methods were applied because they are easy to fit and to maintain.The accuracy of the measurement is influenced by, for example, radiation, which must be taken into account. Thermocouples can be inserted into the pressure system using special sealing techniques, or they may be mounted within a protective tube which is introduced into the pressurized volume. Thermocouple-wires are usually protected with an isulating input in closed-end capillaries with outer diameters of at least 0.5 mm. Thermocouples are technically well tested for pressures up to 6 kbar and temperatures to approx. 800°C. Above these ranges the exact measurement is negatively influenced by several parameters, and the deviations must be taken into account. The accuracy of the temperature measurement devices is normally better than 1°C.

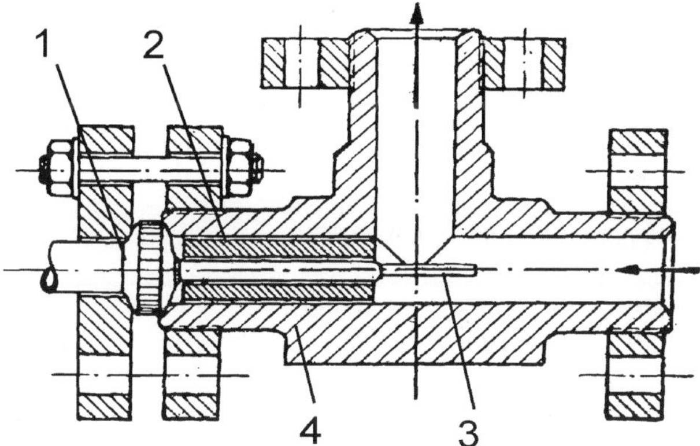

Fig. 4.4-1. High-pressure temperature measurement device [9]
1, high pressure lens; 2, protection ring; 3, resistance thermometer; 4, body.

The same featuresas mentioned before can be discussed for the resistance thermometers. The outer minimum diameter of these sensors is approx. 4 mm [9]. A typical high-pressure temperature measurement device is presented in Fig. 4.4-1.

Both methods cover a wide range (thermocouples, 250°C-1000°C; and resistance thermometers, 230°C-2800°C, with an accuracy of better than 0.2%). The response-time of such devices is short, and the small dimensions are favourable for use under high pressure.

The temperature measurement devices which do not contact the hot surfaces, for example, optical -, radiation pyrometers, and infrared techniques, are not typical for high-pressure application.

4.4.3 Flow measurement

Flow measurement will be performed in most cases at the low-pressure side of a process step. For this reason, such devices are only available for moderate pressures. In research, the use of flow measurement devices in the high-pressure loop of a process step in small-scale plants is necessary to check the mass balance. These applications normally need the measurement of small mass-flow rates. Therefore the development of these devices for higher pressures ends normally in the range up to approx. 400 bar. Different kinds of measuring principles are in applied. The different methods of flow measurement are presented in Table 4.4-2, together with the range of pressure.

Concentric-orifice devices can be easily installed in high-pressure lenses. The high-end- and the low-end connection of such devices could be coupled with differential pressure transmitters, for example, with the before-mentioned devices of [54]. The devices are mounted in different ways. Mostly, the open pipe technique is used. Furthermore, both connected sides of the transmitter are fed with an inert gas, for example, nitrogen. For cases where the systems must separated, membrane devices can be arranged between them. All the mentioned orifice devices are technically proved and are applied in the high-pressure area.

Magnetic flow meters are seldom used, and often special constructions are required.

The turbine flow-meters are suitable for use at high pressure up to 4000 bar. The measuring device could be easily mounted into high-pressure tubes. For this purpose, small instruments have been developed [56]. The use of turbines is limited to the measurement of flow of extremely clean components. Fig.4.4-2 shows an example of such a device. Other devices are so-called helical- and gear flow meters which are in use up to pressures of approx. 400 bar [56].

Table 4.4.2
Flow-measurement devices for use under high pressure.

Measurement method	Pressure range bar	Allowable flow-range	Accuracy	Additional notes
Concentric orifice	1000	3 : 1	< 1 % of span	Viscosity dependent
Magnetic flow meter	400	30 : 1	~ 1% of span	No pressure loss in the device, suitable for slurries
Turbine flow meter	Up to 4000	15 : 1	~ 0.5% of span	Vicosity dependent, only for pure fluids or gases [56]
Coriolis flow meter	400	10 : 1 100 : 1	~ 0.15 % ~ 1.5 % of span	For small flow rates [57]
Thermal /caloric flow meter	400 liquids 700 gases	50 : 1	0.5-0,8% 0.1-0,2% of span	Only for very small flow rates [59,60]

Devices using the Coriolis effect are available for moderate pressures, up to 400 bar. For high flow-rates the the maximum pressure is approx. 250 bar [11]. The advantage of these instruments is their wide range of flow.

For small flow-rates devices using the heat transfer rate of a fluid are approved. These devices cover a pressure range of up to 1000 bar. A main field of application is in small scale plants. The minimum flow-rates are in the range of only some millilitres per hour.

Fig. 4.4-2. Turbine flow meter for a maximum pressure of 400 bar [56].

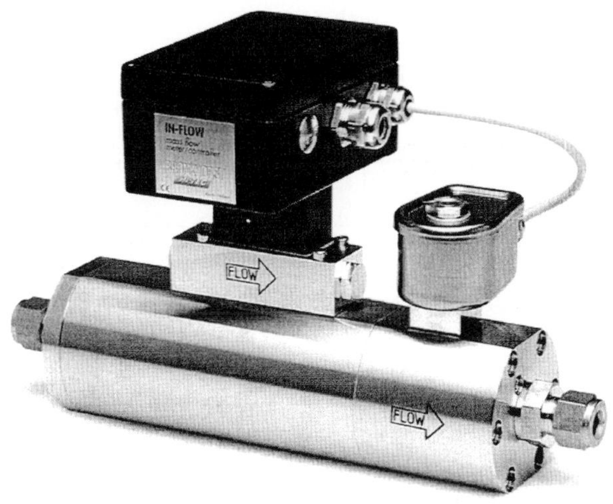

Fig. 4.4-3 Thermal conductivity flow-meter for low flow-rates and a maximum pressure of 700 bar [60].

Well-known makers are described in the papers [59] and [60].

A typical example of such devices for the flow measurement of gases is shown in Fig. 4.4-3 [60] .

4.4.4   Level measurement

The last kind of devices used in high-pressure service are level indicators. The knowledge of the level of fluids in different apparatus is important for safe and continuous operation of chemical plants. Levels must be measured under high pressure, methods using various physical properties have been developed for this purpose. A selection of methods used in high-pressure technology is presented in Table 4.4-3.

Sight-glasses are available for pressures up to 700 bar. They can be easily installed, but it is sometimes difficult to seal the system tightly.

The differential-pressure transmitters are only available for moderate pressures, up to 400 bar. Membrane systems give the possibility of choosing corrosion-resistant materials for the parts of a device (wet system), or to protect the inside of the device by using an additional membrane which divides the instrument side from corrosive media (dry system).

A further very simple method for level-measurement is a technique of feeding an inert gas or fluid into the upper and lower pipes of the measuring system which is connected with an apparatus. The differential pressure is a measure of the height of a fluid in an apparatus, for example a settler, or the level in a reactor of a chemical plant.

The use of floaders is also well proved. For higher pressures, often open floaders - even fabricated in glass- are in use. The open systems are necessary to reduce the weight of the floader. The metering signals are magnetic or electrical values which are transmitted through the wall of the system.

Table 4.4-3.
Level measurement devices for use under high pressure.

Method	Pressure Range/ bar	Level range/ m	Accuracy	Additional notes
Sight glass systems	700	Short range only, < 0.5 m	< 1 cm	Very sensitive systems
Differential pressure transmitters used in: • Membrane systems dry- and wet systems	Up to 400	Several metres	< 1 cm	Wet system: device is in contact with the medium Dry system: device is protected against the medium through an additional membrane
• Gas-bubble systems	1000	Several metres	< 3 cm	
Level float sensors	4000	Several metres	< 5 cm	A floader is mounted at the inside of a tube
Ultra-sonic devices	1000	Several metres	< 5 cm	Special devices
Nuclear radiation devices	4000	Several metres	< 10 cm	

Special constructions use also ultrasonic signal devices to measure values for the height of fluids or solids in high-pressure apparatus.

An expensive method is the use of nuclear radiation to obtain information on the level in an apparatus. The nuclear sensor is mounted at one side, and at the other side a scintillation counter is fixed near the surface of the apparatus. Both systems are sheathed with lead-screen shields to give protection from nuclear radiation. A continous level indicator using nuclear radiation is very complicated and therfeore seldom applied.

Because of the limited space available in such volumes, no further aspects of process control units instrumentation operated at high pressure can be discussed [61].

References of sections 4.3 and 4.4

1.   G. Lewin, G. Lässig, N. Woywode: Apparate und Behälter - Grundlagen der Festigkeitsberechnung, VEB Verlag Technik, Berlin 1990

2.   W.R.D. Manning, S.Labrow: High Pressure Engineering, Leonard Hill, London 1971

3.   ASME Boiler and Pressure Vessel Code Sect. VIII Div. 3, 1997

4.   E. Klapp: Festigkeit im Apparate- u. Anlagenbau, Werner Verlag, Düsseldorf 1970

5.   AD Merkblätter: Heymanns Verlag, Köln 1990

6.   A. Nadai: Der bildsame Zustand der Werkstoffe, Springer Verlag, Berlin 1927

7.   J.H. Faupel: Yield and Bursting Characteristics of Heavy Wall Cylinders, ASME, Paper No. 55, H. 79, August 1962

8.   D.M. Fryer, J.F. Harvey: High Pressure Vessels, Chapman & Hall, Int. Thomson Publishing, New York 1998

9.   H. H. Buchter: Apparate und Armaturen der chemischen Hochdrucktechnik, Springer Verlag, Berlin 1967

10.   R.W. Nichols (ed.): Development In Pressure Vessel Technology – 4, Applied Science Publishers, Chapter 2 by B. Crossland, London 1983

11.   I.L. Spain, I. Paauwe: High Pressure Technology, Vol I, Marcel Dekker Inc., New York 1977

12.   M. Spähn, I. Class: Zeitschrift für Werkstofftechnik, 4 (1973) 31-39

13.   R. Eggers in: Extraction of Natural Products Using Near-critical Solvents (ed. M.B. King, T.R. Bott) Chapter 8, Blackie Academic & Professional, Glasgow UK, 1993

14.   E. Karl in: Ullmann´s Encycl. of Ind. Chemistry, Volume B4, Verlag Chemie, Weinheim 1992

15.   A. Chabaan, PhD Thesis, University of Waterloo, Canada 1985

16.   H. Fessler, B.H. Lewin: Brit. J. Appl. Physics, Vol 7 (1956) 76-79

17.   I. C. Gardeen: Trans. ASME I. of Eng. for Industry, 1972, 815-824

18.   R.E. Little: Machine Design, Vol. 37, 1965, No. 30, 133-135

19.   B.N. Cole, G. Craggs, I. Ficenec: J. Mech. Eng. Science, Vol. 18

20.   T.E.Davidson, B.B.Brown, D.P.Kendall : Journal Mech.Engineering (1977) 63-71

21.   W.Herold: Die Wechselfestigkeit metallischer Werkstoffe, Wien, 1934

22.   D.I.Burns, E.Karl, J.Liljeblad: ASME, PVP Vol.98-8 (1985) 213-223

23.   DIN/EN 10083 Part 1.and 2: Vergütungsstähle–technische Lieferbedigungen Oct. 1996

24.   Dechema Werkstofftabellen, DECHEMA, D-60061 Frankfurt/Main, Germany Dec. 1999

25.   G. H.Nelson: Journal Petroleum Refiner 29, No.9 (1967) 140

26.   API Refining Department, API Publication, (May 1983) 941

27. J. W.Freeman, H.Vorhees: ASTM STP 187, ASTM, Philadelphia, Pa. , (1956) 7

28. Metals Handbook, Vol.1, 10th ed., ASWM, Materials Park, Ohio, (1990) 631, 934, 976.

29. M.Appl BASF Ludwigshafen, Germany: Indian Institute of Chemical Engineers, Hyderabad, India, Dec.19., 1986

30. Thyssen Krupp AG, 40211 Düsseldorf, Germany 1984

31. R. Tschirsch: Stahlbau 4 (1976) 108-119

32. UHDE Hochdrucktechnik, D-58187 Hagen, Germany 2000

33. ASEA AB, 722 13 Vasteras, Kingdom of Sweden

34. H.H.Buchter: Apparate unf Armaturen der chemischen Hochdrucktechnik, Springer Verlag , Berlin 1967

35. W. Witschakowski: Capital and Process Engineering 1 (1968) 63-66

36. J. Schierenbeck: Brennstoff-Chemie (1959) 375-381

37. J.P.Körner, P.Hoffmann, Chemie Technik Heft 4 (1988)

38. S.Maier, F.J.Müller: VDI Berichte 607, VDI-Verlag (1985) 869-892

39. G. Luft: Private communication, March 2000

40. G.Luft, O.Schermuly: DOS 2808 366,27,2.78

41. E.Brunner, S.Maier, K.Windhaber: Journal of Phys. E:Scientific Instrumentation, Vol.17 (1984) 44-48

42. W.Kessler, G.Luft, W.Zeiß: Ber.Bunsenges.Phys.Chem. 101 (1997) 698

43. Ym.A.Klyev:Dokl. Akad. SSSR, 144 (1962) 538

44. J.Szabo, G.Luft, R.Steiner: Chemie-Ing.-Technik 41 No.18 (1969) 1007

45. Netzsch GmbH, D-95140 Selb

46. U.Stanislawski, G.Luft: Ber. Bunsenges. Phys.Chem. 101 (1997) 698

47. H.Gerißen, F.Gernandt, B.A.Wolf, H.Lentz: Makromol.Chem. 192 (1991) 165

48. K.H.Dudziak,E.U.Franck: Ber. Bunsenges. Physik. Chemie 70 (1966) 1120

49. M. Kinzl, G.Luft, R.Horst, B.A.Wolf: Journal of Rheology, submitted

50. R. Sieber: SITEC-Sieber Engineering AG, CH-8124 Maur/Zürich, Switzerland 2000

51. A. M.J.F.Michels: Ann. Phys. 72 (1923), 285

52. P. W.Bridgman: The Physics of High Pressure, 1st. Edition, Bell, London (1931) 65

53. Dressel Industries, Instrument Division, Stratford, Connecticut , USA 2000

54. Endress+Hauser Messtechnik GmbH+Co., D-79574 Weil am Rhein, Germany 2000

55. Burster Präzissionsmesstechnik GmbH+Co.,KG: D-76587 Gernsbach,Germany 2000

56. Küppers Elektromechanik GmbH, D-85757 Karlsfeld, Germany 1999

57. Micro Motion, Fisher Rosemount GmbH&Co, D-64625 Bensheim, Germany 2000

58. WIKA Alexander Wiegand GmbH & Co, D-63911 Klingenberg, Germany 2000

59. Brooks Instruments 407 West Vine Street Hatfield, PA 19440-0903, USA 2000

60. Bronkhorst HI-TEC 7261 AK Ruurlo, Niederlande represented by Wagner Mess+Regeltechnik GmbH, D-63067 Offenbach, Germany 2000

61. G.W.Schanz: Sensortechnik aktuell-Ausgabe 2000, Oldenburg Verlag, München Wien 1999

243

CHAPTER 5

INDUSTRIAL REACTION UNITS

G. Luft[a] : section 5.1

A. Laurent[b] : section 5.2

F. Recasens[c], **F. Trabelsi**[c]: section 5.3

G. Weickert[d] : section 5.4

[a] Department of Chemistry, Darmstadt University of Technology
Petersenstr. 20, D-64287 Darmstadt, Germany

[b] Ecole Nationale Supèrieure des Industries Chimiques (ENSIC)
B. P. No. 451, 1. Rue Granville F-54001 Nancy Cedex, France

[c] Universitat Politécnica de Catalunya
Departamento de Ingenierìa Quìmica
E.T.S.I.I.B. Diagonal, 647   E-08028 Barcelona, Spain

[d] P.O. Box 217 NL-7500 AE Enschede, The Netherlands

The first section of this chapter describes the most important high pressure process run under homogeneous conditions to manufacture Low Density PolyEthylene (LDPE). The radical polymerization of ethylene to LDPE is carried out in tubular reactors or in stirred autoclaves. Tubular reactors exhibit higher capacities than stirred autoclaves. The latter are preferred to produce ethylene copolymers having a higher comonomer content.

The second section presents a review of studies concerning counter-currently and co-currently down-flow conditions in fixed bed gas-liquid-solid reactors operating at elevated pressures. The various consequences induced by the presence of elevated pressures are detailed for Trickle Bed Reactors (TBR). Hydrodynamic parameters including flow regimes, two-phase pressure drop and liquid hold-up are examined. The scarce mass transfer data such gas-liquid interfacial area, liquid-side and gas-side mass transfer coefficients are reported.

In the third section an extensive writing on two types of slurry catalytic reactors is proposed: Bubble Slurry Column Reactors (BSCR) and Mechanically Stirred Slurry Reactors (MSSR). All the variables relevant in the design and for the scale-up and the scale-down of slurry catalytic reactors are discussed particularly from the point of view of hydrodynamics and mass transfer. Two examples of application are included at the end of the section.

Finally, in the fourth section the fundamentals of the modelling concerning two basic olefin polymerization processes are examined: heterogeneous slurry polymerization and gas-phase polymerization. The SPERIPOL process for making High Impact PolyPropylene (HIPP) is then described as an illustrative example for combining fundamentals and elements of product and technology development.

244

5.1 Reactors for homogeneous reactions

5.1.1 Polymerization of ethylene

The most important high-pressure process run under homogeneous conditions on an industrial scale is the radical polymerization of ethylene to manufacture low-density polyethylene (LDPE). It exhibits the typical features of a high-pressure process and is therefore described as an example.

LDPE was occasionally found in 1933 by R.O. Gibson and E.W. Fawcett, when they tried to perform reactions with ethylene [1]. Based on their invention, Imperial Chemicals Ltd (ICI), Great Britain, developed a process with a stirred autoclave in which ethylene was radically polymerized under high pressure [2]. Later, BASF AG in Germany designed a tubular reactor to produce LDPE under similar high-pressure conditions [3].

The production capacity of LDPE increased steeply and amounted in 1997 to 4.8 million tonnes per annum in Western Europe. The rapid development was made possible by the progress achieved in high-pressure engineering to construct large reactors and compressors, in control engineering and, last but not least, in improved process technology.

A simplified flow sheet of the industrial process is shown in Fig. 5.1-1. In the first section fresh ethylene is mixed with the low-pressure recycle at 5 MPa and is compressed to 15 – 35 MPa by means of a five-stage piston compressor. Fresh ethylene should have a high purity of above 99.9 vol.%. Further specifications of polymerization-grade ethylene are given in Table 5.1-1.

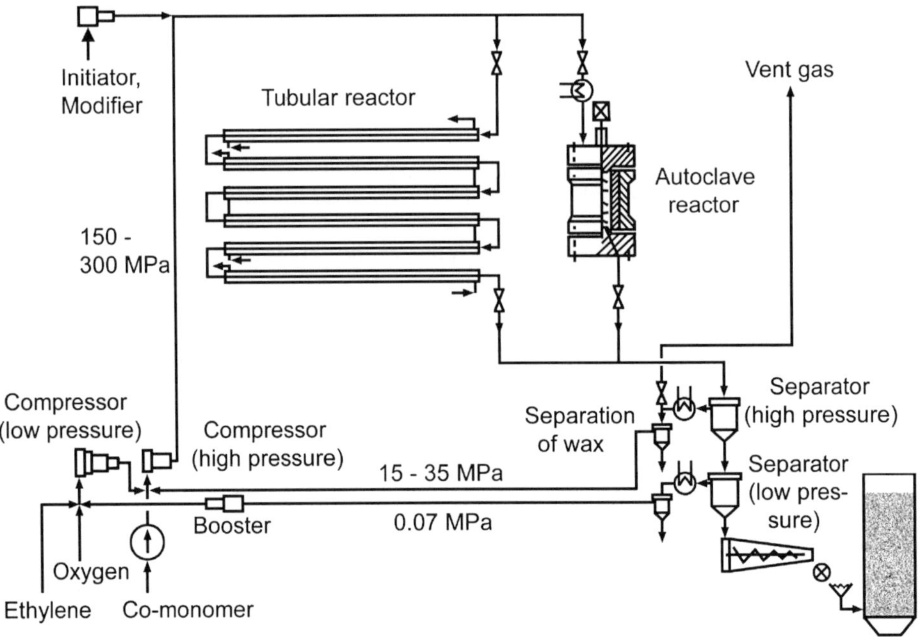

Figure 5.1-1. Flow sheet of the industrial high-pressure polymerization of ethylene.

Table 5.1-1
Specifications for polymerization-grade ethylene

Impurity	Maximum limits [vol. ppm]	Typical values [vol. ppm]
Ethane	500	340
Methane	1000	50
Propene	10	*
Acetylene	5	0.2
Hydrogen	10	*
Carbon monoxide	2	*
Carbon dioxide	5	1
Oxygen	50	0.1
Nitrogen	100	*
Water	10	0.2
Methanol	10	*
Hydrogen sulfide	2 [wt. ppm]	*
2-Propanol	10	*
1-Propanol	10	*

*, below detection limits.

After the primary compressor the ethylene is mixed with the medium-pressure recycle and further compressed by a secondary, two-stage compressor to pressures between 150 and 350 MPa to be fed into the reactor.

Using piston pumps or other conveying equipment the diluted organic peroxide initiators are either mixed with the compressed ethylene or metered directly into the reactor. Here, 0.5 - 1.5 kg initiator per ton polyethylene are used [4]. Organic peroxides frequently used in the high-pressure polymerization of ethylene are listed in Table 5.1-2, together with the temperature for a half-life of 1 minute. This temperature is a measure of their activity and is used to select appropriate peroxides. Oxygen, which is only used as an initiator in the tubular reactor process, is usually added before the compression stage in a quantity giving a concentration of 10 - 80 ppm by weight at the reactor inlet.

A similar procedure is used for the molecular weight modifiers or chain-transfer agents. They are required in quantities of approximately 50 kg per ton PE to control the molecular weight of the polymer. Typical modifiers are isobutane, propylene, isobutylene, 1-butene, or aldehydes.

The co-monomers such as vinyl acetate, acrylate esters, or carbon monoxide are fed together with ethylene, or introduced by liquid pumps, into the suction of the secondary compressor. The concentration in the feed of the co-monomer which is required to achieve a certain level of the co-monomer in the resulting polymer depends on the reactivity ratios, r_1 and r_2, which are the ratios of rate constants of chain-propagation reactions [5]. The values for the co-monomers used in the high-pressure process are presented in Table 5.1-3. In the case of vinyl acetate, both reactivity ratios are identical and therefore the composition of the copolymer is the same as that of the feed. The concentration of vinyl acetate, for example, in

Table 5.1-2
Organic peroxide initiators

Peroxide	Formula	M [g/mol]	Form	$T_{1/2}$ [°C][1]
Dicyclohexyl peroxy dicarbonate		286.3	Powder	90
t-Butylperoxy neodecanoate		244.4	Solution in aliphatics	100
t-Butylperoxy pivalate		174.2	Solution in aliphatics	110
t-Butylperoxy 2-ethylhexanoate		216.3	Solution in aliphatics	130
t-Butylperoxy benzoate		194.2	Liquid	165
Di(t-butyl)peroxide		146.2	Liquid	190

1), Temperature for a half-life of 1 minute; $R_1 + R_2 = C_7H_{16}$.

the feed is between 5 and 30 wt.% and remains constant in the reactor because the two monomers are consumed at the same rate. Most of the other co-monomers are incorporated into the copolymer at a higher rate than ethylene, as can be seen from r_2.

On the industrial scale, high-pressure polymerization is carried out in tubular reactors or in stirred autoclaves, each of which has particular advantages and disadvantages, and requires a certain know-how because of their characteristic operating behaviour. In the tubular reactor, higher pressures and slightly higher temperatures can be used than in the autoclave. In general, pressures of 130 - 330 MPa are applied. The temperatures used are in the range of 130 - 350°C. This wide temperature range is due in part of the heat of polymerization generated, which is largely absorbed by the reaction mixture. However, the variation in temperature is necessary to obtain polymers of differing properties. For economic reasons not given here, the temperature difference between the reactants inside the reactor and those entering should be as large as possible.

Table 5.1-3
Reactivity ratios

Co-monomer	r_1	r_2	Conditions	Ref.
Vinyl acetate	0.88	1.03		[5]
	1.08	1.07	90°C, 101 MPa	[6]
	1.00	1.00	200 - 240°C, 110 - 190 MPa	[7]
Methyl acrylate	0.05	2.40	180°C, 190 MPa	[8]
	0.05	3.00	180°C, 110 MPa	[8]
	0.09	2.10	230°C, 110 MPa	[8]
Butyl acrylate	0.01	13.94		[5]
Carbon monoxide	0.50	0		[5]
	0.19	0	195°C, 180 MPa	[9]

Monomer: 1, ethylene; 2, co-monomer.

With a single passage, approximately 20 - 40% of the ethylene introduced is converted into polyethylene in the reactor. On the basis of current reactor dimensions, this corresponds to a LDPE production of 100,000 - 300,000 t/a per reactor.

The pressure in the reactor is controlled by an automatic outlet valve at the reactor outlet. To separate off the polyethylene, the pressure must be reduced below the pressure for homogeneity given by the thick line in Fig. 5.1-2, the so-called phase boundary- or cloud-point curve. At pressures above the curve, a mixture of ethylene and LDPE is homogeneous, whereas at pressures below, the polymer separates off. The cloud-point pressure increases with increasing concentration of polymer in the mixture, achieves a maximum, and reduces again. It is higher when the temperature is low and when the polymer exhibits a high molecular weight. Most co-monomers reduce the cloud-point pressure when they are added to the mixture or when they are incorporated into the polymer [11].

The thin lines, the so-called co-existence curves, give the composition of the ethylene-rich phase (Fig. 5.1-2, left) and that of the polymer-rich phase (Fig. 5.1-2, right) for different total concentrations of polymer. With decreasing pressure the concentration of polymer in the ethylene-rich phase decreases, and increases in the polymer-rich phase.

In the industrial process pressures of 15 - 35 MPa are chosen to separate off most of the polymer in the high-pressure separator. This range of pressure is a compromise between the separation efficiency and compression energy savings. At low pressures the separation efficiency is higher, but so also is the energy which is required for the compression of the unreacted ethylene.

The polymer collected in the large-volume high-pressure separator is fed to a low-pressure separator for further degassing and is then pelletized in a melt-extruder by a die-face cutter. The pellets are dried and conveyed to silos.

The ethylene from the high-pressure separator is recycled into the secondary compressor through a system of coolers and separators in which the entrained low-molecular waxes are removed. The ethylene from the low-pressure separator is recycled to the primary compressor by means of a booster compressor.

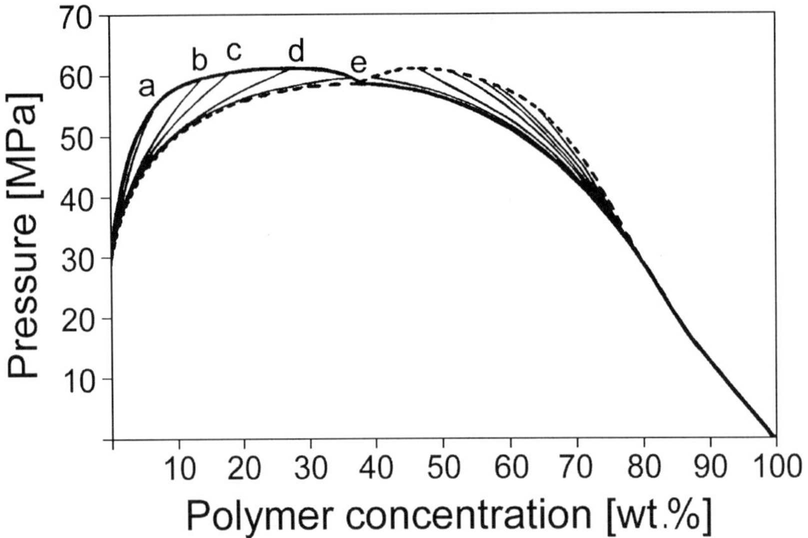

Figure 5.1-2. Phase behaviour of ethylene–LDPE mixtures [10]. Thick line, cloud-point curve; dotted line, shadow curve; thin lines, co-existence curves; total polymer concentration: a, 6.1 wt.%; b, 11.4 wt.%; c, 18.6 wt.%; d, 28.0 wt.%; e, 36.5 wt.%.

5.1.2 Tubular reactor

Tubular reactors are operated under pressures of 250 - 330 MPa measured at the reactor inlet. When a pressure drop of 0.01 - 0.03 MPa/m is taken into account the pressure at the reactor outlet is lower by 30 - 60 MPa.

The reactor consists of high-pressure tubes, 10 - 15 m long and with inner diameters of 34 - 68 mm and outer diameters of 71 - 150 mm. These are flanged together to straight sections which are connected by 180° bends to form a helical tube of 500 - 2,000 m length. The tubes are made of steel alloys of high yield strength, e.g., German DIN standard no. 1.7707 (designation 30 CrMoV 9) or 1.6580 (30 CrNiMo 8). In order to resist the high pressure the tubes were subjected to autofrettage (see also Chapter 4.3). The high pressure tubes are equipped with a jacket for cooling or heating by means of water or steam (Fig. 5.1-3).

Through the lens rings of pipe junctions thermocouples are introduced to measure the temperature of the reaction mixture (Fig. 5.1-4). Furthermore, peroxide initiator can be metered at these locations.

Ethylene decomposition can start from hot spots, dead zones at high temperature, because of too high an initiator concentration, or when decomposition-sensitizing compounds accumulate. Decomposition propagates with a high rate of pressure- (100 - 1,000 MPa/s) and temperature rise [12]. In order to prevent destruction of the high-pressure tubes when a decomposition occurs, safety valves having a large inside diameter are arranged at the reactor inlet and outlet.

Figure 5.1-3. Tubular reactor.

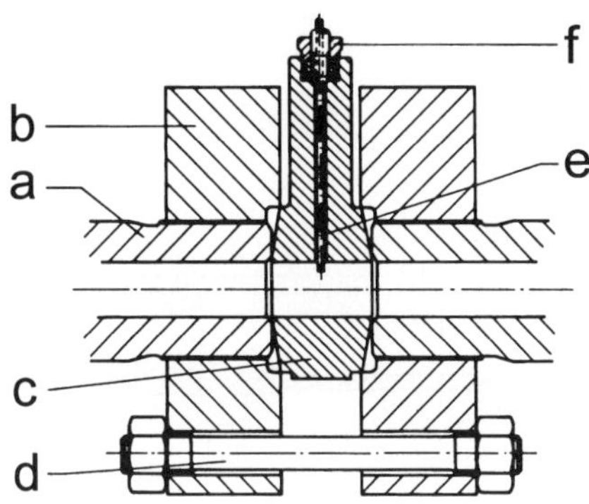

Figure 5.1-4. Arrangement of thermocouples and peroxide injection. a, High pressure tube; b, flange; c, lens-ring; d, bolt; e, thermocouple; f, gland nut and sleeve.

In the first zone of the tubular reactor the ethylene is heated to the decomposition temperature of the initiator. This is a temperature of 130°C when an active organic peroxide such as dicyclohexyl peroxy dicarbonate (Table 5.1-2) is used, or 160 - 170°C when oxygen is the initiator. In the second zone, the first actual polymerization zone, the temperature increases substantially to 270 - 330°C. The high rates of conversion can be achieved by the so-called cold-gas addition or multiple-feed method. In the course of this process only approximately 25 - 65% of fresh ethylene is heated in the heating zone. The remaining quantity is fed, at the compressor discharge temperature or after being cooled to lower temperatures, directly into the reactor at various locations (Fig. 5.1-5). In this case, the initiator is either added to the cold gas, which is preferable in the case of oxygen, or metered in at suitable points downstream of the site of cold gas introduction. In this way, the temperature is repeatedly reduced to 180 - 220°C and polymerization is re-initiated. In the last zone, the reaction mixture is cooled to prevent its temperature exceeding the decomposition temperature of ethylene during pressure release, as a result of the negative Joule–Thomson effect. Also, the cooling system consists of a number of sections run at different temperatures in the range of 160 - 200°C. The coolant flows in co- and counter-current senses.

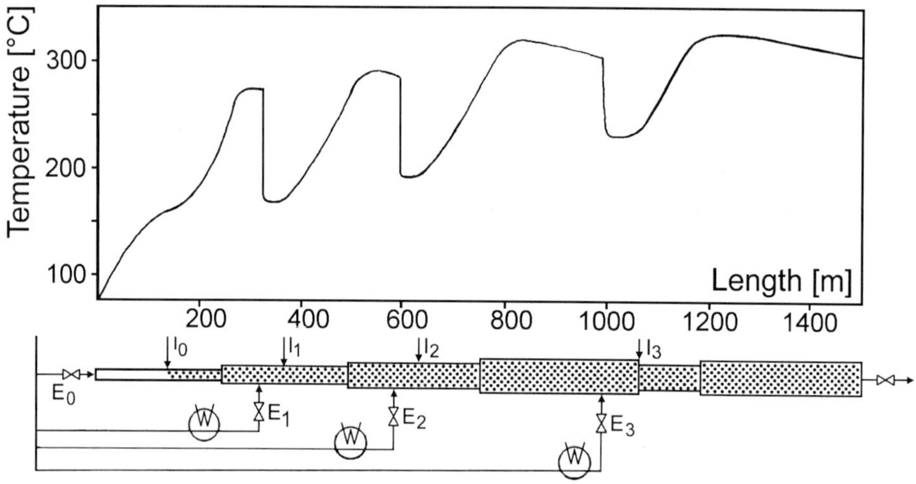

Figure 5.1-5. Tubular reactor with multiple feed. E, fresh ethylene; I, initiator.

In the multiple-feed process an initiator mix is used to obtain a smooth temperature profile and low initiator consumption. The initiator mix consists of three or four organic peroxides from Table 5.1-2 differing in activity. To cover the range of lower temperatures, dicyclohexyl peroxy dicarbonate, t-butylperoxy neodecanoate, or t-butylperoxy pivalate are used. The peroxide for moderately high temperatures is frequently t-butylperoxy 2-ethylhexanoate or t-butylperoxy benzoate. The high-temperature initiator in the mix is di(t-butyl)peroxide.

Downstream, the inner diameter of the high-pressure tubes increases because of the increased throughput by side-streams of fresh ethylene. In between, tubes having a lowerinner diameter are arranged to increase the turbulence and to improve heat transfer. The Reynold numbers are around 10^6. The heat-transfer parameters are typically in the range of 800 - 1200 $W/m^2/K$. They are steeply influenced by polymer deposits onto the inside wall of the high-pressure tube. In some processes the polymer film is removed by a steady or temporary high velocity of the reaction mixture. The latter is generated by a pressure oscillation effected by an intermittently operating outlet valve.

5.1.3 Autoclave reactors

The autoclave reactors used today in the high-pressure polymerization of ethylene are single stirred-tank reactors, cascades of stirred autoclaves, and multi-chamber autoclaves.

Single stirred-tank reactors were run in the first industrial scale processes to manufacture LDPE. Today they are used only for plants having lower capacities. The design of a single autoclave is shown in Fig. 5.1-6. It consists of a thick-walled forged-, or two-layer shrunk mantle. The ratio of inside length to inner diameter is typically in the range of one to two, and the volume is 1 - 2 m^3.

Ethylene and initiator are fed through the top closure which also contains the bursting-disc ports. The autoclave is equipped with a stirrer driven by an external or internal electric motor. The external motor can be arranged at the top or at the bottom closure, sealed with a number

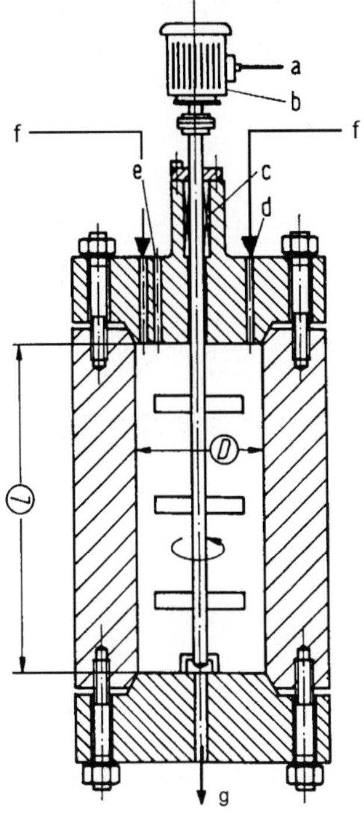

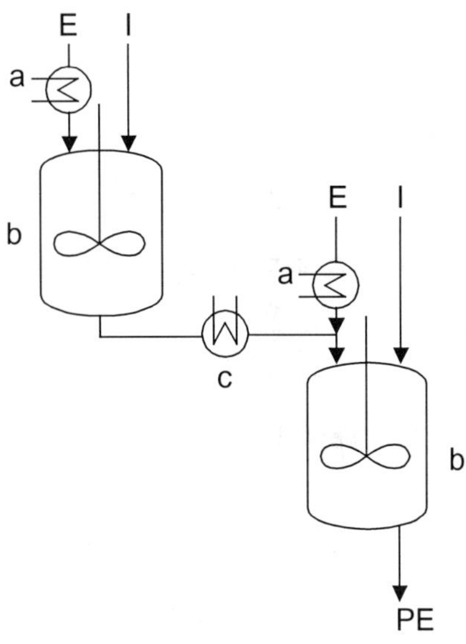

Figure 5.1-6. Design of single autoclave [13].
a, Connection; b, motor; c, packing elements;
d, thermocouple; e, bursting disc port; f,
ethylene and initiator feed; g, product outlet.

Figure 5.1-7. Cascade of stirred autoclaves.
E, Ethylene; I, initiator; PE, product;
a, cooler; b, reactor; c, inter cooler.

of packing elements and pressurized barrier-liquid. The design of the stirrer is proprietary knowledge of the LDPE manufacturers. The stirrer should provide good mixing of initiator, fresh ethylene, and polymer and high turbulence near the inside wall to avoid polymer deposition. Generally, the single stirred autoclave having a low L/d_i-ratio gives good mixing and temperature distribution.

The residence time is typically in the range of 15 - 60 s. The temperature and pressure are a little lower then in a tubular reactor. Pressures are in the range of 130 - 220 MPa, and the temperatures mostly do not exceed 260°C. When ethylene–vinyl acetate resins are produced, single autoclaves are run at temperatures which are 30 - 50°C lower then in the production of homopolyethylene.

Autoclave reactors are operated adiabatically, which means that the heat of reaction must be removed by the fresh ethylene entering the reactor. The conversion is related, therefore, to the difference in temperature between the feed and the reactor temperature. This limits the conversion to 15 - 20%. Taking into account the fact that the percentage conversion, Δx, can be approximated by using the average specific heat, c_p, the enthalpy of the polymerization,

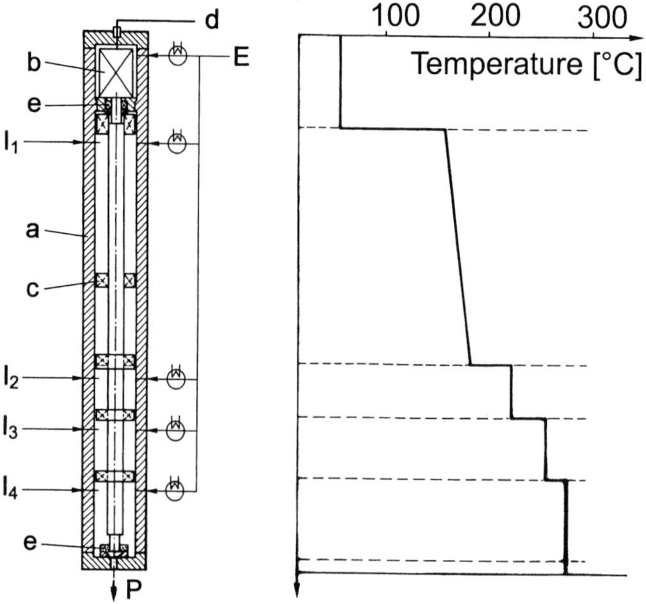

Figure 5.1-8. Multi-zone autoclave reactor. E, Ethylene; I, initiator; P, product; a, reactor; b, motor; c, stirrer; d, connecting cable; e, bearing.

ΔH_R, and the difference in temperature, ΔT, between the feed and the reactor temperature to give $\Delta x = 100 \cdot (c_p/\Delta H_R) \cdot \Delta T \approx \Delta T/13$, the feed temperature should be in the range of 65 - 0°C when the reactor temperature is 260°C.

The main feature of the resins from the single-autoclave process is a broader molecular-weight distribution, which gives excellent properties for shrink-films and heavy duty packaging.

The autoclave reactor process was improved when the reactors were arranged in series (Fig. 5.1-7). By feeding fresh ethylene into the line between the reactors and/or removing heat by means of coolers [14,15] the conversion could be increased above 20%. When peroxides of different activity are used the reactors can be run at different temperatures. Furthermore, the split of the feed of fresh ethylene can be varied.

Compared to the single-autoclave process, the cascade arrangement gives more flexibility to adjust the properties of LDPE resins.

High conversion, similar to that in a tubular reactor with multiple feed, is obtained in a multi-chamber or multi-zone autoclave reactor. This is an elongated cylindrical vessel with a ratio of length to inner diameter of 10 - 12. It is subdivided by baffles and the stirrer into two to five chambers (Fig. 5.1-8, left).

The chambers have different volumes: mostly, the top chamber is the largest one, having 30 - 50% of the total volume. Each chamber is equipped with thermocouples and feed lines for fresh ethylene and initiator.

Table 5.1-4
Properties of LDPE

Density [g/ml]	0.915 - 0.934
Melt flow index [g/10 min]	0.15 - 8
Crystallinity [%]	40 - 50
Melting point [°C]	90 - 120
Tensile strength at yield [MPa]	9 -15
Tensile strength [MPa]	10 - 29
Elongation at break [%]	50 - 70
Haze (at 50 μm films) [%]	< 6
Gloss (at 50 μm films) [%]	> 90

The multi-zone autoclave is mostly run with a temperature profile. The temperature in the top chamber is moderately high, and is typically in the range of 165 - 200°C, whereas the temperature in the bottom zone is 280 - 290°C to adjust the density of the polymer (Fig. 5.1-8, right).

Through the temperature profile, which is controlled by the initiator feed and the split of fresh ethylene, the properties of the polymers can be varied in a wide range. Different peroxides, e.g., *t*-butylperoxy neodecanoate together with *t*-butylperoxy 2-ethylhexanoate, or *t*-butylperoxy perbenzoate, or di(*t*-butyl)peroxide are required when the zones are run at different temperatures.

5.1.4 Conclusions

In summarizing, it can be concluded that in all processes for manufacturing LDPE, high pressure gives a high rate of polymerization and allows one conduct the reaction in a single phase where the polymer is completely dissolved in compressed ethylene. Only by radical polymerization under high pressure can polyethylene of typical structure – namely, molecules with a large number of short side-chains and a lower number of long side-chains – be obtained. Properties of LDPE are listed in Table 5.1-4.

Today, modern tubular reactor processes exhibit higher capacities for manufacturing commodities than do autoclave processes. The latter are preferred for producing ethylene co-polymers having a higher co-monomer content.

References

1. R.O. Gibson, The discovery of polyethylene, Lecture series, Vol 1., Royal Institute of Chemistry 1964.
2. M.W. Perrin, Research 6 (1953) 111.
3. R.G. Klimesch, The future of the high pressure process, Lecture at the conference "The 1990's Polyethylene" of the Plastic and Rubber Institute, London 1992.
4. H. Seidel and G. Luft, J. Macromol. Sci.-Chem., A15(1) (1981) 1.
5. J. Brandrup, E.H Immergut and E.A. Grulke (eds.) Polymer Handbook, 4th Ed., II/210, J. Wiley & Sons, New York 1999.

6. R.D. Burkhardt and N.L. Zutty, J. Polym. Sci. A1 (1963) 1145.
7. G. Luft and H. Bitsch, Angew. Makromol. Chem. 32 (1973) 17.
8. F. Stein, Ph.D. Thesis, Technical University Darmstadt 1991.
9. G. Luft and D. Freitag, Ph.D. Thesis, Technical University Darmstadt 2000.
10. R. Spahl and G. Luft, Ber. Bunsenges. Phys. Chem. 85 (1981) 379.
11. F. Folie, C. Gregg, G. Luft and M. Radosz, Fluid Phase Equilibria 120 (1996) 11.
12. J. Albert and G. Luft, AIChE Journal, Vol. 45, No. 10 (1999) 2214.
13. Ullmanns Encyclopädie der Technischen Chemie, 4. Aufl., Vol.19, 172, Verlag Chemie, Weinheim 1980.
14. Ch.R. Donaldson and C.J. Stiles, Dual Reactor Apparatus for Polymerizing Ethylene, US Patent No. 4,229,416 (1980), National Distillers Corporation, USA.
15. S. Kita, F.Hiki, M. Shimizu and A. Kondou, Process for the High Pressure Production of Polyethylene, US 4,123,600 (1978), Sumitomo Chemical Company, Japan.

5.2 Hydrodynamics and mass transfer in fixed-bed gas-liquid-solid reactors operating at high pressure

Multiphase chemical reactors are used for reactions involving at least two reactant species. Normally each of the reactants enters the reactor in a different phase. Typically one reactant is introduced in the liquid phase and the other in the gas phase. More than one phase is present in the reactor. Any phase may participate in the reaction as a reactant or as a product; however one phase, generally a solid phase, may act as a catalyst.

A fundamental division of multiphase reactors may be made, depending on whether the solid phase is present as a moving-or as a fixed bed. In principle, one gas-liquid-solid reactor with the fixed bed of solids can be operated in three ways, depending upon the relative orientation of the superficial gas-mass G and superficial liquid-mass L flow-rates (see Figure 5.2–1).

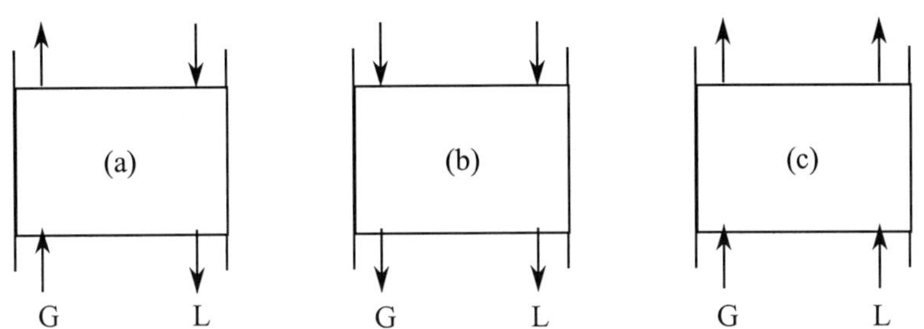

Figure 5.2–1. Types of gas-liquid-solid fixed bed reactors. (a), countercurrent flow; (b), cocurrent downflow; (c), cocurrent upflow.

The gas and liquid can each flow countercurrently, cocurrently downwards or cocurrently upwards. The hydrodynamics and the mass-transfer conditions are different in each of these flow conditions.

The following section presents a review of the reported studies concerning countercurrently and cocurrently downwards-flow conditions in fixed-bed gas-liquid-solid reactors operating at elevated pressure.

5.2.1 Countercurrent gas-liquid flow in solid fixed-bed columns

For the normal arrangement in a fixed-bed column operating in countercurrent flow, the liquid is evenly distributed over the top of the packing, runs down over it in thin layers, films, rivulets, etc., and leaves at the bottom. The gas enters as a continuous phase at the bottom and leaves at the top.

The countercurrent-flow fixed-bed operation is often used for physical absorption or for gas-liquid reactions rather than gas-liquid-solid processes. Shah [1] gives a comparison between a gas-liquid-solid (catalytic) fixed bed reactor and a gas-liquid-solid (inert) fixed-bed reactor. The major difference between these two types of reactors are the nature and the size of the packing used and the conditions of gas and liquid flow-rates.

The packed-bed gas-liquid reactors use non-porous, large-size packings, so that they can be operated at high gas and liquid flow-rates without excessive pressure drop. The shape of the

256

packing is also designed to give larger gas-liquid interfacial areas. The packed-bed gas-liquid reactors are often operated near flooding conditions.

The conditions in countercurrent fixed beds have been investigated for many years in order to improve the understanding of two-phase flow and to develop reliable design methods. However the proposed correlations available for fluid dynamics and mass transfer are practically all based on experimental data obtained at atmospheric pressure. Extrapolation of the results to high pressure is questionable and not recommended. Moreover the results of systematic investigations in the high-pressure range are scarce in the open literature.

5.2.1.1 Hydrodynamics in countercurrent fixed beds

Mozenski and Kucharski [2] examined the pressure-drop, overload limit, and flooding limit of a column (0.5 m diameter) packed with Pall rings (35 mm diameter) and Bialecki rings (35 mm and 50 mm diameter) sprayed with propylene carbonate up to 15 bar. Some specific correlations have been proposed and compared with literature data for atmospheric pressure, particularly with the use of the Sherwood diagram for loading-and flooding capacities.

Krehenwinkel and Knapp [3, 4] investigated several types of industrially important packings (Berl saddles, Raschig rings, Interpak, Pall rings, Novalox saddles) to measure pressure-drop in dry and irrigated columns as a function of gas-and liquid loads. Flooding points were also determined using several liquids of varying viscosity and density over a pressure range of 5-100 bar. The evaluation of the results shows that the method for calculating pressure-drops in dry packing by the representation of the friction-factor as a function of the gas Reynolds number remains valid with the pressure. However, in order to estimate two-phase pressure-drop in irrigated beds, and to determine the upper capacity limit, appropriate corrections have to be made.

Benadda *et al.* [5] showed the influence of the total pressure (1-15 bar) to characterize the flow behaviour and measured the hold-up and the axial dispersion coefficients of the fluid phases. The gas-flow diverges from plug-flow when the pressure increases. The pressure has no effect on the gas hold-up. No influence of pressure was found on the liquid-flow or liquid hold-up.

Stockfleth and Brunner [6] reported the hydrodynamic behaviour of aqueous systems (water+CO_2, and water+surfactant+CO_2) between 80 and 300 bar with Sulzer gauze packing operating in the countercurrent mode (pressure drop, hold-up, and flooding). A distinctive change in the flooding mechanisms was observed as a function of the flow parameter. The foamability of the surfactant solution decreased significantly with increasing pressure, but its influence on the flooding behaviour could not be proved.

Simoes *et al.* [7] studied the hydrodynamics of a Sulzer structured-gauze packing column operating in a countercurrent manner under supercritical conditions (110-240 bar) with two different systems: olive oil+SCCO$_2$, and squalene+oleic acid+SCCO$_2$. Flooding points for the two systems were measured with different liquid phase compositions and density. The density difference between the two contacting phases appears to be an important factor, as far as flooding is concerned. In a loading diagram giving a capacity-factor as a function of the flow-parameter, the authors demonstrated as preliminary results, the relative concordance between their experimental values and the few literature data.

A major conclusion from these relatively few studies is, that it is necessary to promote the investigation of the hydrodynamics of countercurrent fixed beds and the physical properties of the liquid and gaseous phases at elevated pressure. At present, the relatively unstudied subject of the hydrodynamics of packed towers under high-pressure conditions should be improved owing to the importance of establishing the pressure drop, the hold-up and the ultimate capacity of the column.

5.2.1.2 Mass transfer in countercurrent fixed beds

The influence of pressure on the mass transfer in a countercurrent packed column has been scarcely investigated to date. The only systematic experimental work has been made by the Research Group of the INSA Lyon (F) with Professor M. Otterbein *et al.* These authors [8, 9] studied the influence of the total pressure (up to 15 bar) on the gas-liquid interfacial area, a, and on the volumetric mass-transfer coefficient in the liquid phase, k_La, in a countercurrent packed column. The method of gas-liquid absorption with chemical reaction was applied with different chemical systems. The results showed the increase of the interfacial area with increasing pressure, at constant gas-and liquid velocities. The same trend was observed for the variation of the volumetric liquid mass-transfer coefficient. The effect of pressure on k_La was probably due to the influence of pressure on the interfacial area, a. In fact, by observing the ratio, k_La/a, it can be seen that the liquid-side mass-transfer coefficient, k_L, is independent of pressure.

The lack of gas-side mass-transfer data at elevated pressure for countercurrent fixed beds packed with large and small solid particles should be underlined.

5.2.2 Cocurrent gas-liquid downflow fixed-bed reactors: trickle-bed reactors (TBR)

The cocurrent gas-liquid downflow fixed-bed reactors well known as trickle-bed reactors (TBR) are among the most widely used three-phase reactors.

The TBR consist of a column that may be very high (above 10-30 m) equipped with a fixed bed of solid catalyst, through which a gas-liquid cocurrent downward flow occurs. In this cocurrent downflow configuration, the liquid flows mainly through the catalyst particles in the form of rivulets, films and droplets.

TBR are better suited for long on-stream lives, or operations in which a high production rate is required. TBR, devised originally for the petroleum refining and petrochemical industries, have been used increasingly for chemical processes. Nevertheless, they are still mainly employed for hydrodesulfurization, demetallization, denitrogenation of petroleum fractions, hydrocracking of heavy oils, and hydrofinishing of lubricating oils.

Table 5.2–1 lists some examples of reactions carried out in downflow TBR.

Examples of three-phase hydrogenations that are, or can be, performed in TBR are shown in Table 5.2–2, with the indication of the required high pressure.

Table 5.2–3 shows the values of the characteristic parameters in the most widely used TBR.

Table 5.2–4 lists evaluations of the real and potential advantages and disadvantages of TBR.

Generally speaking, the design of TBR requires knowledge of hydrodynamics and flow regimes, pressure-drop, hold-ups of the phases, interfacial areas and mass-transfer resistances, heat transfer, dispersion and back-mixing, residence time distribution, and segregation of the phases.

Several authors have published a considerable amount of information concerning TBR, mostly limited to atmospheric pressure.

Table 5.2–1
Examples of reactions worked in downflow TBR (after Gianetto *et al.* [10]).

Hydrodesulfurization, demetallization, denitrification.
Catalytic hydrocracking.
Catalytic hydrotreating.
Catalytic hydrofinishing.
Other catalytic hydrogenation:
- selective hydrogenation of acetylene to separate this compound from C_4 fractions in the presence of butadiene.
- hydrogenation of various petroleum fractions, nitrocompounds, carbonyl compounds, and carboxylic acids to alcohols (adipic acid to 1,6-hexanediol).
- hydrogenation of benzene to cyclohexane.
- hydrogenation of α-methylstyrene to cumene.
- hydrogenation of alkyl anthraquinones to alkylhydroquinones.
- hydrogenation of aniline to cyclohexylaniline.
- hydrogenation of glucose to sorbitol.
- hydrogenation of coal liquefaction extracts.
- hydrogenation of benzoic acid to hydrobenzoic acid.
- hydrogenation of caprolactone to hexanediol.
- hydrogenation of organic acid esters to alcohols.

Production of calcium acid sulfite (Jenssen tower process).
Synthesis of 1,4-butynediol (acetylene and formaldehyde).
Oxidation of formic acid in water, and of lean SO_2 to SO_3 on active carbon.
Oxidation of organic matter in waste-waters with Pd catalyst or fixed micro-organisms (biological filters).

Table 5.2–2
Some examples of three-phase hydrogenation reactions in TBR with the operating conditions of temperature and pressure (after L'Homme [11]).

Feedstock	Product	Catalyst	Temperature (°C)	Pressure (bar)
Acetone	Methylisobutyl-ketone	Pd	120-160	20-50
Adiponitrile	Hexamethylene diamine	Co-Cu-Fe-Ni Ni-Fe Ni-Cr	75-180	30-650
Benzoic acid	Cyclohexane-carboxylic acid	Pd-C	160	10-17
Feedstock	Product	Catalyst	Temperature (°C)	Pressure (bar)
Benzene	Cyclohexane	Raney Ni- Pt-Li-Ni	200-300	30-50
2-butyne-1,4-diol	2-butene-1,4-diol	Ni-Fe-Pd-Zn		
2-butyne-1,- diol	1,4-butanediol	Ni	80-160	40-300
γ-butyrolactone	1,4-butanediol	Ni-Co-ThO$_2$	250	100

Table 5.2–2 (continued)
Some examples of three phase hydrogenation reactions in TBR with the operating conditions of temperature and pressure (after L'Homme [11]).

Caprolactone or Hydroxy caproic acid or adipic acid	1,6-hexanediol	Cu-Co-Mn Copper chromite	170-250	150-300
Cyclododeca-1-3-5-triene	Cyclodecane	Ni	200	10-15
Dinitrotoluene	Diaminotoluene	Ni or Pd-C	100	> 50
Dimethyl-Terephtalate	Cyclohexane-1-4-dicarboxylic acid dimethyl-ester	Pd	160-180	300-400
Cyclohexane-1,-4-dicarboxylic acid dimethyl-ester	1-4-dimethyl-olcyclohexane	Copper chromite		
Fatty esters Fatty acids	Fatty alcohol	Copper chromite	200-300	250-300
Furfural	Furfuryl alcohol	Copper chromite	200	50
Glucose	Sorbitol	Ni Raney	120-150	30-50
Maleic anhydride	Tetrahydrofuran	Ni-Re		
Mesityl oxide	Methyl isobutyl ketone or Carbinol	Ni-Cu	150-200	3-10

Table 5.2–3
Typical values of the parameters used in TBR (after Gianetto et al. [10]).

Catalyst loading (% by volume)	$\simeq 0.5$
Liquid holdup	0.05-0.2
Gas holdup	0.45-0.30
Particle size (mm)	1-5
Catalyst external area (m^{-1})	1000
Gas-liquid interfacial area (m^{-1})	200-600
Power consumption ($Watt.m^{-3}$)	100-1000
Maximum reactor volume (m^3)	200
Maximum operating pressure (bar)	> 80 bar possible

Table 5.2–4
Advantages and disadvantages of TBR (after Gianetto and Specchia [12]).

Advantages	Disadvantages
1) Liquid flow approaches piston flow (no mixing) leading to higher conversions for most reactions.	1) Lower catalyst effectiveness and selectivity owing to the large catalyst particle size used in this reactor type.
2) Low catalyst loss, permitting the use of costly catalysts such as Pt, Pd or Ru.	2) Limitations on the use of viscous or foaming liquids in the reactor.
3) No moving parts and therefore lower maintenance cost and fewer leaks.	3) Risk of increasing pressure drop or obstructing catalyst pores when side reactions lead to fouling or solid products.
4) Possibility of operating at higher pressure and temperature because of the absence of seals.	4) Strong influence of hydrodynamic regimes on reactor performance, making the reactor type very sensitive to flow-rates.
5) Larger reactor sizes (cheaper construction).	5) Relationship between liquid residence time and the downward velocity of the trickling liquid reduces the range of possible gas and liquid flow-rates.
6) Low liquid-solid volume ratio and hence less occurrence of homogeneous side reactions.	6) Incomplete and/or ineffective wetting of the catalyst with low liquid flow–rates and low column diameter/particle size ratio (< 15/20); possibility of liquid by-passing along the reactor wall.
7) Lower investment and operating costs.	7) Sensitivity to thermal effects and problems with temperature control, although these drawbacks can be avoided by recycling part of the discharge liquid after cooling, or by injecting cooled gas or liquid between the sections of the reactor.
8) Low pressure-drop and no flooding (at least for cocurrent, downflow reactors).	8) Difficulties in recovery of the heat of reaction.
9) Possibility of operating partially or wholly in the vapour phase by varying the liquid flowrate according to catalyst wetting, heat of vaporization, and mass-transfer resistances in the liquid phase.	

Notwithstanding the fact that some commercial TBR are operated at high pressure (up to 200 bar, occasionally 300 bar) and at high temperature (up to 400°C), only few papers concerning TBR with downward flow up to a maximum pressure of about 100 bar, but at low temperature (20-50°C), have been published.

In the past ten years, systematic experimental studies have been made, mainly by the Research Group of the University of Twente (NL) with Professor K.R. Westerterp *et al.*, by the Research Group of the ENSIC Nancy (F) with Professor A. Laurent *et al.*, by the Research Group of the Washington University Saint Louis (USA) with Professor M. Dudukovic, and by the Research Group of the University Laval (Canada) with Professor F. Larachi.

One of the main effects of the high pressure on TBR hydrodynamics and performance is produced by the increased gas-phase density. We will now examine the various consequences of the presence of elevated pressures.

5.2.2.1 Flow regimes

Various flow regimes may exist in cocurrent gas-liquid downflow TBR. Owing to the vast difference in the hydrodynamic characteristics of the encountered regimes, the respective heat–and mass-transfer rates, hold-up and pressure-drop differ in a significant way. For this reason it is imperative to have a good understanding of the hydrodynamics and to predict which flow-regime to expect for a given reactor system and a specified set of operating conditions.

The four basic flow-regimes in TBR are: trickling, pulsing, spray, and bubble flows.

a - Trickling flow regime. In trickling flow the liquid flows down the column on the surface of the packings, while the gas-phase travels as a continuous phase in the remaining void space.

At low gas-and liquid flow-rates, the liquid flow is laminar and a fraction of the packing remains unwetted. If the liquid rate is raised, the partial wetting regime changes to the complete wetting trickling regime in which the packing is totally covered by a liquid film.

b - Pulsing flow regime. The pulsing flow regime refers to gas-and liquid slugs traversing the column alternately at high gas-and liquid flow-rates.

This begins when the flow channels between packings are plugged by a slug of liquid followed by blowing-off of the slug by the gas flow. It should be noted that the gas pulse is not completely devoid of liquid. A thin film of liquid always exists on the surface of the packings within the gas pulse. Similarly, the liquid pulse also contains some small gas bubbles, particularly at the front end of the liquid pulse.

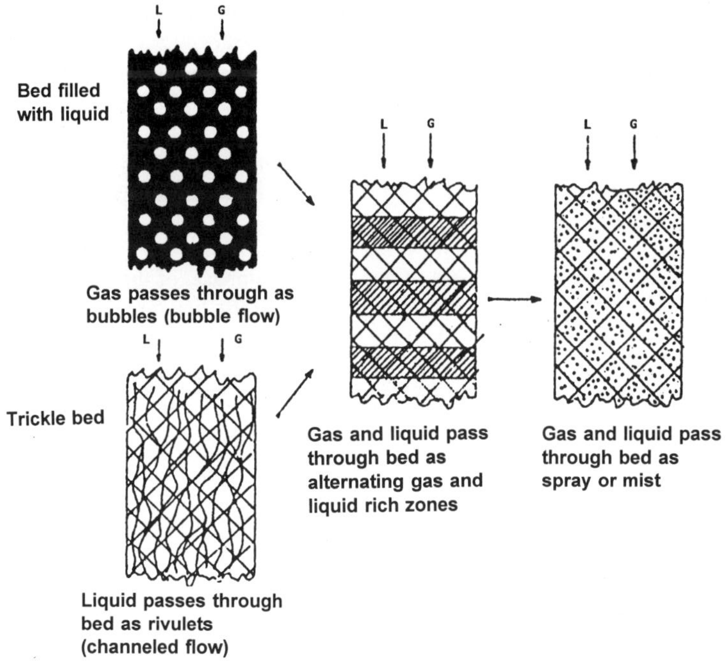

Figure 5.2–2. Patterns in gas-liquid downwards flow in fixed beds (after Charpentier [13]).

c - Spray flow regime. Transition to the spray flow regime occurs when the gas flow-rate is high while the liquid flow-rate is kept at a low value. The liquid phase travels through the column in the form of very small droplets entrained by the continuous gas phase.

d - Bubble flow regime. The bubble-flow regime appears at high liquid flow-rates and low gas flow-rates. The whole column is filled with the liquid (liquid continuous phase) and the gas phase is dispersed in the form of slightly elongated bubbles, highly irregular in shape.

Note that if the liquid mass flow-rate is kept constant and the gas mass flow-rate is started, the different flow patterns are observed as the gas mass flow-rate increases (Fig. 5.2–2).

When the liquid phase foams during gas-flow, two new regimes appear between trickling– and pulsing-flow: foaming-and foaming-pulsing-flow regimes.

In industry, TBR operate mainly in the trickling and the pulsing flow regimes.

The flow phenomena in TBR are not easy to predict, because of the large number of variables such a bed porosity, size and shape of the catalyst, viscosity, density, interfacial tension, flowrates, and reactor dimensions.

5.2.2.2 Flow charts

In the literature one finds a wealth of flow-regime diagrams essentially obtained from pilot–scale columns of 0.30 m or less in diameter and under moderate operating conditions.

For example, typical for the empirical flow charts is the flow-regime map of Charpentier and Favier [14], whose use has been recommended by Tosun [15]. It uses the coordinates proposed by Baker [16] for two-phase flow in horizontal tubes. The abscissa is the superficial mass flow-rate of the gas; the ordinate is the ratio of the liquid-to the gas-mass flow-rates. The properties of the gas and the liquid are taken into account by the parameters:

$$\lambda = \left(\frac{\rho_G}{\rho_{AIR}} \frac{\rho_L}{\rho_W} \right)^{0.5} \tag{5.2–1}$$

$$\Psi = \frac{\sigma_W}{\sigma_L} \left[\frac{\mu_L}{\mu_W} \left(\frac{\rho_W}{\rho_L} \right)^2 \right]^{0.33} \tag{5.2–2}$$

where ρ_{AIR}, ρ_G, ρ_L and ρ_W are respectively the density of air, of the gas-phase, of the liquid-phase, and of water, μ_L and μ_W are respectively the dynamic viscosity of the liquid phase and of water, σ_L and σ_W are respectively the surface tension of the liquid phase, and of water.

As an example, patterns and transitions are indicated in Figure 5.2–3 for cylindrical (1×5 mm) and spherical (2.4 and 3 mm) alumina catalyst pellets with nitrogen and different foaming and non-foaming hydrocarbon and viscous organic liquids. This diagram covers the fluid properties range:

$770 < \rho_L < 1200$ kg.m^{-3}
$0.15 < \rho_G < 2$ kg.m^{-3}
$0.3 < \mu_L < 67$ mPa.s
$19 < \sigma_L < 75$ mN.m^{-1}

The lines on this diagram which separate the flow regimes are actually transition regimes rather than points of abrupt change from one flow type to another. It is also worth noting that

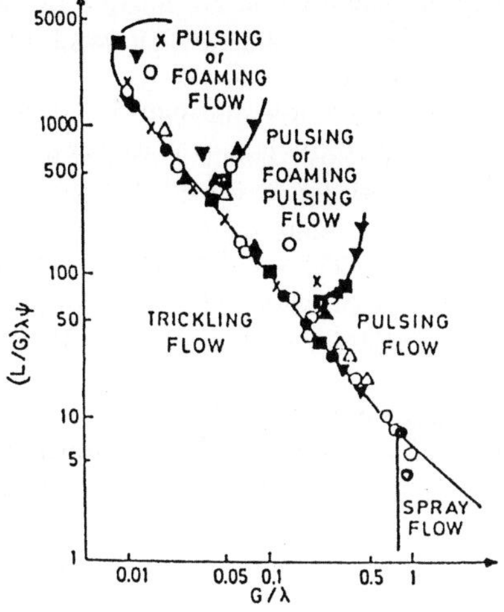

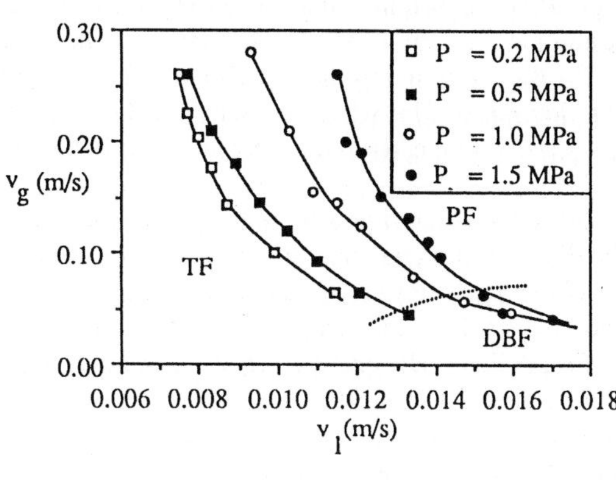

Figure 5.2–3. Flow regimes in TBR (Charpentier diagram after Charpentier and Favier [14])

Figure 5.2–4. Effect of the total reactor pressure on the transition between trickle-and pulsing flow for the N_2-H_2O system (after Wammes [17])

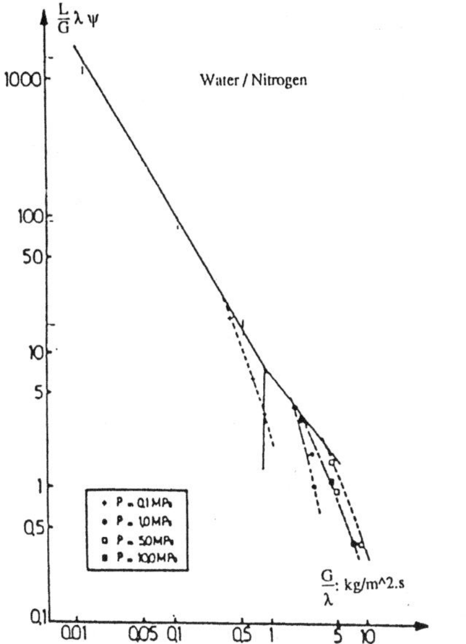

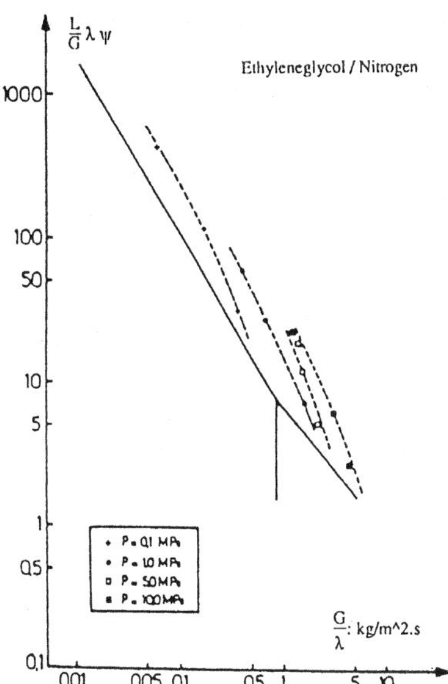

Figure 5.2–5. Influence of the operating pressure on the transition between trickling and pulsing regimes for two gas-liquid systems (Charpentier diagram) (after Hasseni *et al.* [18]).

264

the line that separates trickling flow from the other flow patterns is considered as the transition line between the low-gas-interaction regime (trickling flow) and the high interaction regime.

At atmospheric pressure, in our experience, this diagram works reasonably well.

The transition between the trickle-flow regime and the pulse-flow regime is plotted in the Figure 5.2–4 as a function of the superficial gas velocity, u_G, the liquid velocity, u_L, and the total reactor pressure, P, for the water-nitrogen system [17]. This figure shows clearly that the transition depends strongly on the pressure in the reactor.

We consider now the experimental data obtained under high pressure, using Charpentier's diagram. Here the parameter, λ, does not represent in a satisfactory manner the influence of the total pressure on the flow transition between the low-and high-interaction regime (Figure 5.2–5). At elevated pressure, the transition zone shifts to higher G/λ values [18].

Other flow charts did not lead to a better agreement with experimental data at different pressures. Larachi *et al.* [19] have suggested the use of a modified Charpentier's diagram. Based on experimental data available to date on the high-pressure trickle-pulsed transition, the extended diagram is proposed to quantify directly the effect of pressure in non-foaming systems.

Figure 5.2–6 shows the boundary between low-and high interaction regimes versus the classical Baker's coordinates and the function, Φ. The influence of the total pressure is included in the empirical function, Φ, defined by the relationship :

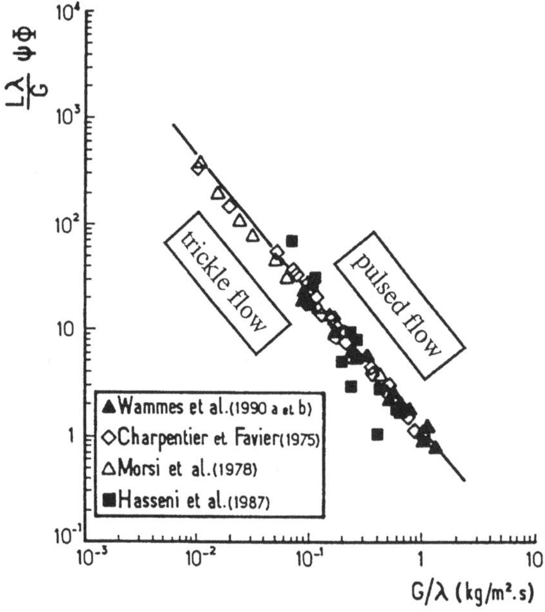

Figure 5.2–6. Extended Charpentier diagram including the influence of the total pressure via parameter Φ (after Larachi *et al.* [19]).

$$\Phi = \frac{1}{4.76 + 0.5 \dfrac{\rho_G}{\rho_{AIR}}}$$ (5.2–3)

One must be careful in using this extended regime map: all the data used were obtained in small-scale columns and at ambient temperature. Experiments in large reactors are needed to confirm the validity of this proposition.

At present, to the best of our knowledge, no single approach can be recommended to propose a universal flow chart or correlation. Each of them can only be employed in an interpolative manner for conditions falling within those that served for its establishment.

5.2.2.3 Models for the hydrodynamics of TBR

In the literature, approaches to modelling the hydrodynamics can be divided into two categories:
- a microscopic approach which examines the flows at the pore level;
- a macroscopic approach (volume-averaged) which captures the gross flow characteristics.

A - Microscopic approach

An example of a *microscopic* approach has been proposed by Ng [20]. Let us consider the transition from trickling-to pulsing flow as described by Ng's model. This model tries to represent what is happening locally at the place where pulsing is likely to be initiated.

As illustrated by Figure 5.2–7, the place is just above the constrictions of the porous medium. The gas velocity here is the highest. The gas flow tends to induce the formation of a bridge, while the surface tension tends to keep the films apart. Ng applies Bernoulli's law between the points A and C. When the pulse is about to be initiated and the film breaks down, the liquid flow stops momentarily and the pressure difference between B and A is given by the Young-Laplace equation. By writing that the total gas flow-rate is the same at C and A, one obtains the following relationship at the flow transition from trickling-to pulsing flow:

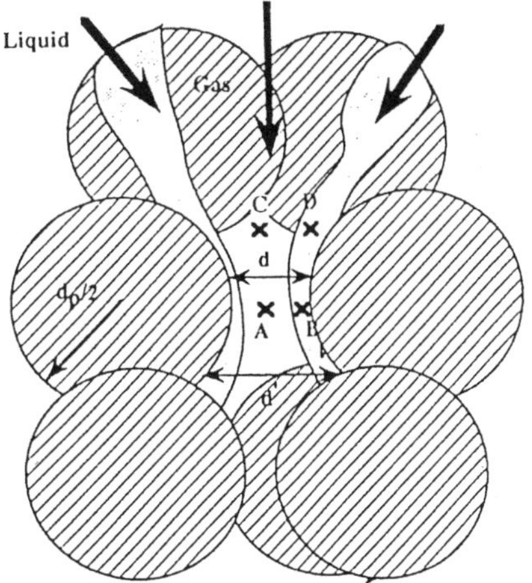

Figure 5.2–7. Mechanism of formation of a liquid pulse (after Ng [20]).

266

$$\frac{1}{2} \rho_G u_G^2 \left[1 - \left(\frac{d}{d'} \right)^4 \right] = \frac{4\sigma_L}{d_p} - \rho_L g \frac{d_p}{2} \qquad (5.2\text{--}4)$$

where d, d' and d_p are respectively the diameter of the pore, the diameter of the constriction and the diameter of the particle.

Here u_G, the interstitial velocity of the gas, is related to the superficial mass flow-rate, G, by the following equation:

$$G = \varepsilon \, u_G \, (1 - \alpha) \, \rho_G \qquad (5.2\text{--}5)$$

where

$$\alpha = 4 \left[\sqrt{1 - \beta} - (1 - \beta) \right] \qquad (5.2\text{--}6)$$

and ε is the porosity of the packed bed.

The liquid saturation, β, is estimated by the correlation proposed by Wijffels et $al.$ [21]:

$$\beta = \left[\left(\frac{200}{Re_L} + 1.75 \right) \frac{u_L^2}{g d_p} \frac{1 - \varepsilon}{\varepsilon^3} \right]^{1/4} \qquad (5.2\text{--}7)$$

Figure 5.2–8 shows a comparison between predictions of Ng's model with measurements with water, nitrogen and 3 mm glass beads at different pressures (up to 101 bar) made by Hasseni et $al.$ [18]. The agreement is acceptable.

If one considers results obtained with ethyleneglycol, the liquid flow-rate at the trickling/pulsing boundary predicted by Ng's model is one order of magnitude smaller than the value measured by Hasseni et $al.$ [18] (see Figure 5.2–9). It seems, therefore, that this

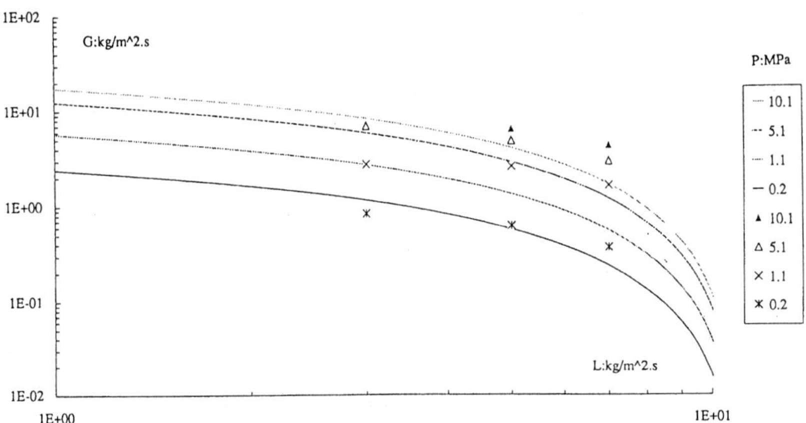

Figure 5.2–8. Comparison of the observed trickling-pulsing transition with Ng's model at different pressures for the N_2-H_2O system (after Wild et $al.$ [28]).

model does not represent well the influence of the liquid viscosity under pressure on the trickling/pulsing flow transition.

Some comments can be made on this model:

• Regardless of the viscosity, trickling flow cannot exist as soon as the right side of Equation (5.2–4) becomes zero or negative. There is a maximum value of the particle diameter, d_{pMAX}, for which trickling flow may exist:

$$d_{pMAX} = \left(\frac{8\sigma_L}{\rho_L g}\right)^{0.5}$$

(5.2–8)

With water, this limit is 8 mm. However with the usual organic liquids at ambient temperature, it would be about 4 mm for $\sigma_L = 20$ mN/m, and with higher temperature organic liquids, 3 mm or less for $\sigma_L = 10$ mN/m.

• The influence of the liquid velocity and of the liquid viscosity are taken into account via the liquid saturation β. Unfortunately, very few reliable experimental investigations on the influence of viscosity on the liquid hold-up are to be found in the literature. The correlation proposed by Wijffels *et al.* [21] is based mainly on data obtained with water.

• We have some misgivings about the basic assumptions of the model which leads to Equation (5.2–4). We fear that these assumptions oversimplify a very complex problem.

B - Macroscopic approach

A large number of models consider the macroscopic approach for the prediction of the hydrodynamic behaviour of TBR. In this subsection some of the reported models are presented. First, an example of the macroscopic approach developed by the team of Sundaresan [22-24] is detailed. Next, a rapid review of the other models is given.

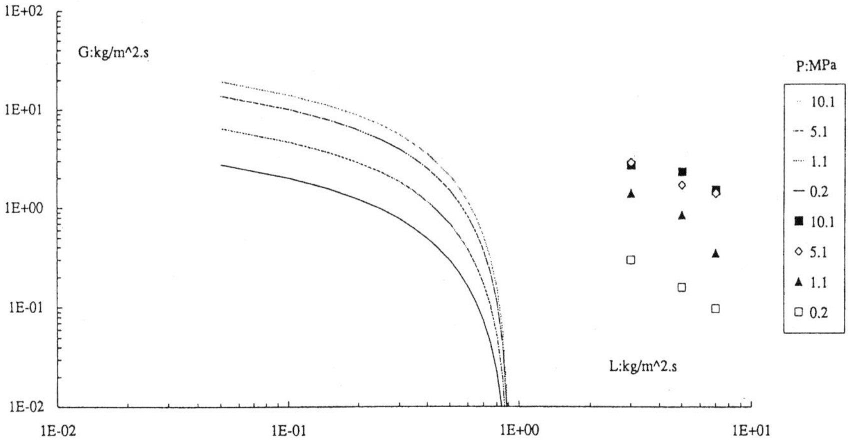

Figure 5.2–9. Comparison of the observed trickling-pulsing transition with Ng's model at different pressures for the N$_2$-Ethyleneglycol system (after Wild *et al.* [28]).

268

a - Sundarasan approach. The authors propose a macroscopic volume averaged model based on the loss of stability (criterion 1) and/or the loss of existence (criterion 2) of a steady-state solution of the flow-momentum-and continuity equations. The theoretical formulation of their equations was extended from that developed for single-phase fluid motion through porous media [25].

The model assumptions are the following:

- As long as the flow regime is trickling, the gas phase is incompressible, which means that neither the velocities nor the holdups depend on the axial position (no axial gradients of these variables).

- Wall effects are neglected and the bed is supposed to be uniformly packed.

- Phase holdups and local averaged velocities are uniform in the lateral directions.

- The boundary-condition problem is reduced to that of an infinite medium in all directions.

The authors write the equations of motion for the trickling flow (space conservation, mass– and momentum conservation of both phases). The latter take the following form:

$$\rho_L\,\varepsilon_L\left(\frac{\partial u_L}{\partial t} + u_L\,\frac{\partial u_L}{\partial z}\right) + \varepsilon_L\,\frac{\partial P_L}{\partial z} - g\varepsilon_L\rho_L - F_L(z) + \frac{\partial}{\partial z}\left(\varepsilon_L\mu_L^*\,\frac{\partial u_L}{\partial z}\right) = 0 \qquad (5.2\text{–}9)$$

$$\rho_G\,\varepsilon_G\left(\frac{\partial u_G}{\partial t} + u_G\,\frac{\partial u_G}{\partial z}\right) + \varepsilon_G\,\frac{\partial P_G}{\partial z} - g\varepsilon_G\rho_G - F_G(z) + \frac{\partial}{\partial z}\left(\varepsilon_G\mu_G^*\,\frac{\partial u_G}{\partial z}\right) = 0 \qquad (5.2\text{–}10)$$

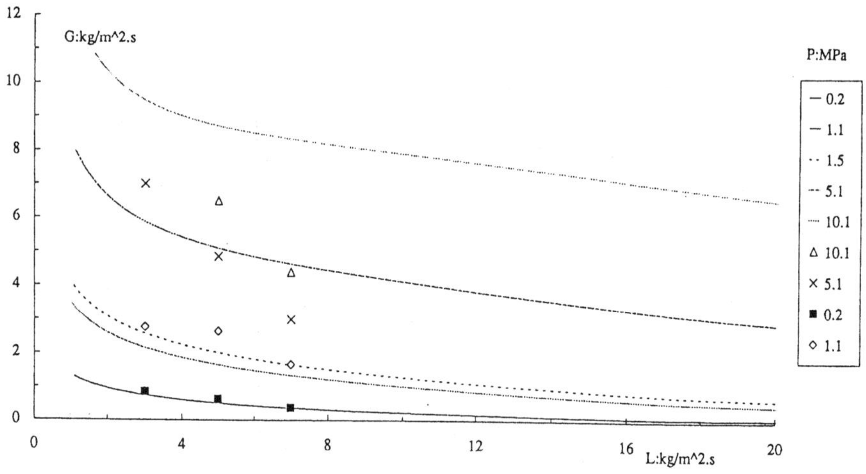

Figure 5.2–10. Comparison of the observed trickling-pulsing transition (Hasseni *et al.* [18]) with the model of Dankworth *et al.* [23] at different pressures for the N_2-H_2O system (after Wild *et al.* [28]).

The last terms in each of Eqn. (5.2–9) and Eqn. (5.2–10) represent the divergence of the deviatoric stress including viscosity and pseudo-turbulence. The quantities μ_L^* and μ_G^* are effective viscosities of each phase including bulk and shear viscosities. F_L and F_G represent the volume-averaged forces exerted on the liquid-and gas-phase (respectively) by the other phases across the common interfaces.

In addition to these equations, one can set up a constitutive relationship between P_G and P_L ($P_G-P_L = P_c$, capillary pressure) by means of the so-called Leverett's J function:

$$P_c = \sqrt{\frac{\varepsilon}{k}}\, \sigma_L\, J\,(\varepsilon_L) \tag{5.2–11}$$

$$J(\varepsilon_L) = 0.48 + 0.036 \ln\left(\frac{\varepsilon - \varepsilon_L}{\varepsilon_L}\right) \tag{5.2–12}$$

Here k is the permeability of the dry medium and $J(\varepsilon_L)$ characterizes hysteresis behaviour in the trickling regime. It should be noted at this level that Eqn. (5.2–12) was derived by these authors from available data on two-phase imbibition and drainage curves, implicitly identified to the trickling flow regime in trickle-beds. The J function may be multi-valued and depends on the history of the flow, however Grosser *et al.* [22] as well as Dankworth *et al.* [23] assume it to be single-valued.

Two additional constitutive equations have to be defined for the total drag-force per unit bed volume experience by each phase:

$$F_G(z) = -\left(\frac{A\,\mu_G(1-\varepsilon)^2\,\varepsilon^{1.8}}{d_p^2\,\varepsilon_G^{2.8}} + \frac{B\,\rho_G(1-\varepsilon)\varepsilon^{1.8}}{d_p\varepsilon_G^{1.8}}\,|u_G-u_L|\right)(u_G-u_L) \tag{5.2–13}$$

$$F_L(z) = -\,u_L\left(\frac{A\,\mu_L(1-\varepsilon)^2\,\varepsilon_L^2}{d_p^2\,\varepsilon^3} + \frac{B\,\rho_L(1-\varepsilon)\varepsilon_L^3}{d_p\varepsilon^3}\,u_L\right)\left(\frac{\varepsilon-\varepsilon_L^r}{\varepsilon_L-\varepsilon_L^r}\right)^{2.43} - F_G(z) \tag{5.2–14}$$

Such relationships have been inspired from the work of Saez and Carbonell [26] to correlate macroscopically the pressure-drop data corresponding to the low-interaction regime in trickle-bed reactors.

The residual (or capillary) liquid hold-up was correlated also by Saez and Carbonell [26] as:

$$\varepsilon_L^r = \frac{1}{20 + 0.9\,E\ddot{o}} \tag{5.2–15}$$

where $E\ddot{o}$ is the Eotvos number.

If the aim is to determine the stability limit of the trickling flow (that is, the limit of existence of a truly stationary flow), one may neglect the effective viscosity terms in Eqn. (5.2–9) and Eqn. (5.2–10). Furthermore, in trickling flow, all derivatives with respect to time or height (z) are equal to zero; a stability analysis of the solution of the system of equations against small variations of t or z yields an algebraic equation which permits to calculation of the limit of liquid mass-flow-rate for a given gas flow rate.

270

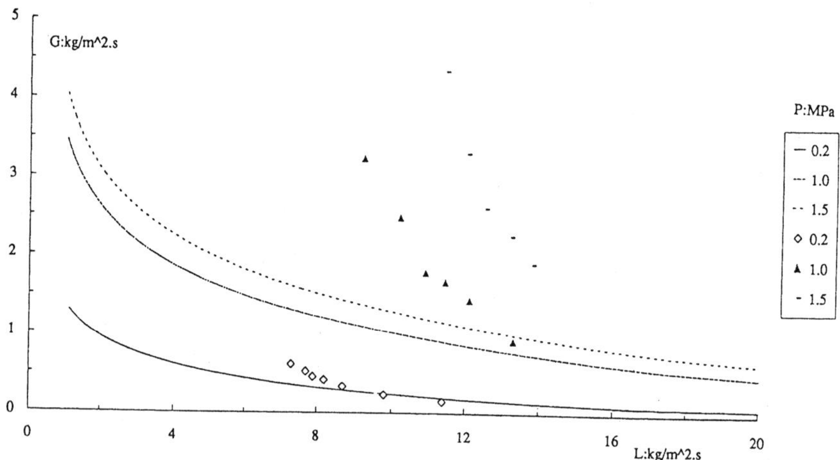

Figure 5.2–11. Comparison of the observed trickling-pulsing transition (Wammes *et al.* [27]) with the model of Dankworth *et al.* [23] at different pressures for the N_2-H_2O system (after Wild *et al.* [28]).

The mathematics are not very complicated but quite cumbersome and are not be presented here. This technique seems quite elegant. However, it cannot be better than the equations used: the Equations (5.2–13) and (5.2–14) contain some constants whose values are empirical (A, B, the exponents 2.8, 2.43...); the permeability is not always well known, and it is not certain that Leverett's approach remains valid in the case of cocurrent flow.

Figures 5.2–10 and 5.2–11 show some results obtained by Hasseni *et al.* [18] and by Wammes *et al.* [27] with water, 3 mm glass beads, and nitrogen at different pressures compared with the predictions of the model of Dankworth *et al.* At moderate pressure (20 bar) the model agrees fairly well with measurements made by both teams. At higher pressures, Dankworth's model overestimates slightly the gas mass flow-rate at the trickling/pulsing boundary measured by our team. On the other hand, it underestimates grossly the gas flow-rate measured by Wammes *et al.* for the same transition. The reason for this difference is not easily to explain. We suspect that it is a matter of definition of the experimental transition from trickle to pulse flow [28].

The same observation can be made for more viscous liquids (ethyleneglycol, in the work by Hasseni *et al.*, Figure 5.2–12, and an aqueous solution of ethyleneglycol, in the work of Wammes *et al.*, Figure 5.2–13).

In brief, it is obvious with this presentation that the data at elevated pressures do not fit into

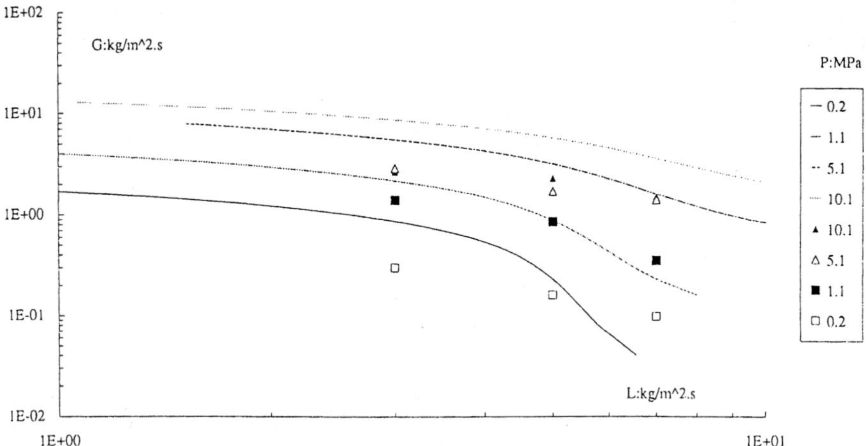

Figure 5.2–12. Comparison of the observed trickling-pulsing transition (Hasseni *et al.* [18]) with the model of Dankworth *et al.* [23] at different pressures for the N_2-Ethyleneglycol system (after Wild *et al.* [28]).

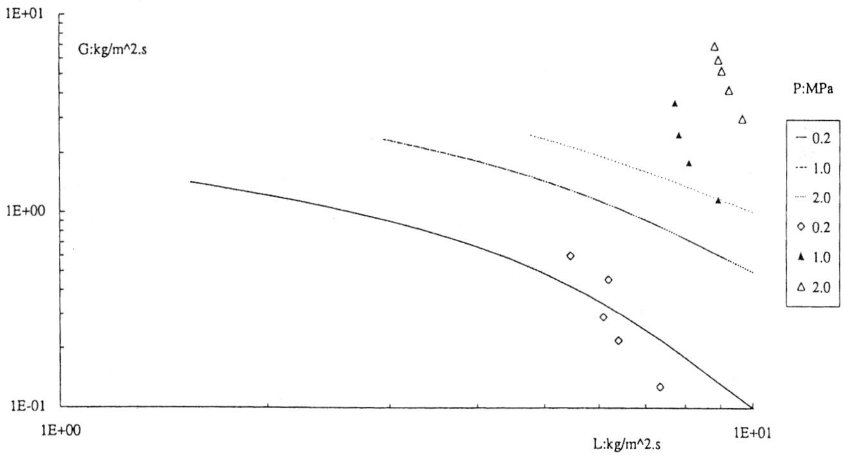

Figure 5.2–13. Comparison of the observed trickling-pulsing transition (Wammes *et al.* [27]) with the model of Dankworth *et al.* [23] at different pressures for the N_2-Ethyleneglycol system (after Wild et al. [28]).

the proposed model equations or flow charts to give a good prediction for the transition of the trickle-flow regime to the pulse-flow regime in TBR.

b - Slit models. Holub [29] developed a mechanistic pore-scale 1–D two-fluid segregated flow model in the form of two-phase flow Ergun like momentum equation inside an inclined slit. The model sketches the two-phase flow structure as a gas-free liquid flowing film totally covering the slit walls (complete wetting) and a gas flow in the central core of the slit. One of the models intrinsic assumptions is that full wetting predominates, regardless of the operating conditions.

In the original model, the gas-liquid interface was assumed hermetic to momentum transfer (shear-free boundary) and the interfacial gas velocity was zero [30]. This assumption implies that the gas flow does not influence the liquid flow. However, the experimental studies have shown that the gas flow has a considerable influence on the hydrodynamics of TBR, especially at high operating pressure [31-41].

An extension, by accounting for gas-liquid interfacial interactions via velocity-and shear–slip factors was then proposed by Al-Dahhan *et al.* [42] to lift the model disparities observed for conditions of high gas throughputs and elevated pressures. Later, Iliuta *et al.* [43] derived more general slip-corrective correlations.

Recently Iliuta and Larachi [44] developed a generalized slit model for the prediction of frictional two-phase pressure-drop, liquid hold-up, and wetting efficiency in TBR operated under partially-and fully wetted conditions. This proposed model mimicked the actual bed void by two geometrically identical inclined slits, a wet slit and a dry slit (see Figure 5.2–14).

In the first slit, the liquid wets the wall with a film of uniform thickness; the gas being in the central core (wet slit). The second slit is visited exclusively by the gas (dry slit). The high–pressure-and high-temperature-wetting efficiency, liquid hold-up and pressure-drop data reported in the literature for TBR in the trickle-flow regime were successfully forecasted by the model.

The authors indicated that the comparison with other alternative models or with widely used empirical correlations, showed that this current mechanistic model performed better.

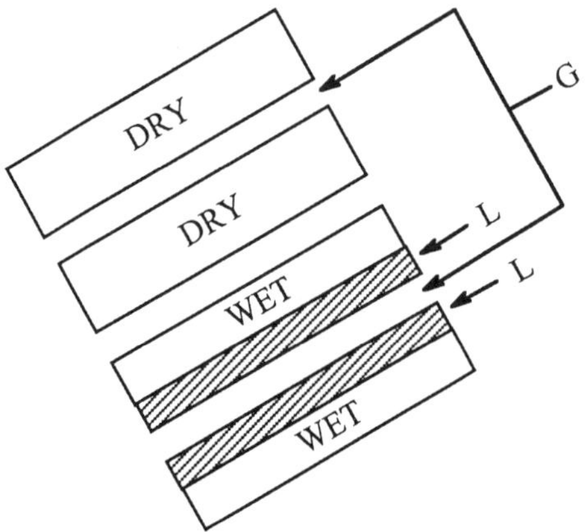

Figure 5.2–14. Double slit model representation.

Nevertheless, the generalized slit model needs a preliminary determination of the two Ergun constants from single-phase flow experiments, which is not easily to perform in the industrial practice, and of the two slip factors, which is unfortunately more difficult.

c - 1-D CFD Attou's model. Attou *et al.* [45,46] developed a physical one-dimensional model to describe the hydrodynamics of a steady-state cocurrent gas-liquid trickle-flow regime through a TBR. The trickle-flow regime is idealized by a flow in which the gas and liquid phases are completely separated by a smooth and stable interface. As a consequence, each fluid behaves as a continuous medium for which the macroscopic balance laws can be applied in the Eulerian formalism. The formulation of the trickle-flow model involves the global-mass-and momentum-balance equations applied to each fluid across the interstitial void volume. The closure equations describing the various interactions between the phases are formulated on the basis of theoretical considerations, by taking into account the assumed idealized flow pattern. The authors indicated that the resulting model has the fundamental characteristic of involving no parameter from fitting single-phase or two-phase flow data. It must be said, nevertheless, that the application of the 1-D CFD model implies knowledge of the two momentum-transfer coefficients between phases and the mean fraction of the interstitial void volume occupied by the gas phase. A good agreement is generally observed between the predictions of pressure-gradient and liquid-saturation from the 1-D CFD model and the respective experimental results obtained over a large range of operating pressure (1-100 bar). The proposed model underestimates the pressure gradient in the conditions of high operating pressures and superficial gas velocities owing to the appearance of the phenomena of roll waves pattern of the gas-liquid interface and the entrainment of droplets.

c - Recommendations for the choice of a mode or correlation. Iliuta *et al.* [47] recently published an overview of hydrodynamics and mass-transfer in TBR based on extensive historic experimental flow data (22,000 experiments) from the literature. The state-of-the art of trickle-bed fluid dynamics was presented, and a set of unified and updated estimation methods relying on neural work, dimensional analysis, and phenomenological hybrid approaches was discussed. Updates and re-tuned correlations are available at the web address:

http://www.gch.ulaval.ca/~larachi

In the same way, Larachi *et al.* [48] evaluated with an important trickle-flow-regime database (4,000 experiments) different phenomenological models for liquid holdup and two-phase pressure drop in TBR. Table 5.2–5 summarizes the respective scatters (mean relative error and deviation) between the experimental values of pressure drop, $\Delta P/Z$, and liquid holdup, β_t, and their predictions by the different models.

The following main remarks emerge from the analysis of the Table 5.2–5:

- the slit model predicted well the liquid hold-up but its performance at forecasting pressure-drop was as weak as that from the two empirical correlations;

- the Attou's 1-D CFD model, and the extended slit model constitutive shear and slip correlations did not improve much the pressure-drop prediction;

- the Attou's 1-D CFD model was tested with different simplifying assumptions, which do not provide significant gains. It is suggested that one should use this model in its simplest form;

- the extended slit model of Iliuta *et al.* [43], and the double slit model, outperformed all the available models in term of the pressure-drop prediction;

- all six models performed almost equally well, and can be recommended equally in liquid–hold-up prediction;

- the double slit model was the only model able to predict the wetting efficiency.

Table 5.2–5
Statistical evaluation of models for two-phase pressure drop and liquid holdup in trickle flow regime (after Larachi *et al.* [48]).

Trickle flow	All data				Partial wetting data			
Statistical parameter (%)	Mean relative error		Deviation		Mean relative error		Deviation	
Hydrodynamic parameter	$\Delta P/Z$	β_t	$\Delta P/Z$	β_t	$\Delta P/Z$	β_t	$\Delta P/Z$	β_t
MODEL								
Permeability [26]	53	19	56	48	46	26	26	82
Slit [30] (*)	74	15	19	14	69	21	25	19
Slit [30] (**)	70	18	22	15	-	-	-	-
Extended slit [42]	68	13	22	13	-	-	-	-
Extended slit [43]	35	12	26	11	41	25	22	16
Double slit [44]	-	-	-	-	32	24	23	29
	62	£20	48	17	57	25	46	31
1-D-CFD Attou [45]	61	§20	45	18	58	23	48	17
	65 ◊	19	45	18	-	-	-	-
CORRELATION								
Elmann *et al.* [49-50]	70	25	27	30	-	-	-	-
Larachi *et al.* [37]	72	20	33	22	-	-	-	-

(*) Ergun constants determined from single-phase flow experiments
(**) Fixed values of Ergun constants (180 and 1.8)
£ Original model
§ Simplified model (neglects hold-up axial variation)
◊ Original model (with Ergun constants correlation of [43]).

5.2.2.4 Two-phase pressure drop

a - Some experimental results. The most important experimental investigation on two-phase pressure drop has been carried out by Larachi *et al.* [37]. They studied several aqueous and organic liquids (foaming and non-foaming; more and less viscous) with different gas phases.

Two sizes of glass beads (1.4 and 2 mm diameter) were used, both in the trickle-and pulsing-flow regime, as well as in the transition zone (see Table 5.2–6) at total pressures between 2 and 81 bar.

As typical results, Figure 5.2–15 shows the two-phase pressure drop, $\Delta P/Z$, as a function of the gas-and the liquid-mass flow-rates for the system ethyleneglycol-nitrogen at 51 bar. At a given pressure, $\Delta P/Z$ increases with both mass flow-rates. Such trends are in agreement with the reported observations made at atmospheric pressure.

In the same way, Figure 5.2–16 shows the two-phase pressure drop at a given liquid mass flow-rate for different values of the total pressure. Similar results are presented in Figure 5.2–17, where the density of the gas was varied by changing the nature of the gas (He-N$_2$-Ar) and maintaining the total pressure constant at 51 bar. In both cases, at given mass flowrates, $\Delta P/Z$ decreases when the gas-density increases via the operating pressure or the gas molar weight.

In brief, increasing the gas-and liquid velocities or mass fluxes, the liquid viscosity, or gas density, or reducing the particle diameters, increases the pressure drop.

Table 5.2–6
Experimental conditions investigated for $\Delta P/Z$ (after Larachi *et al.* [37]).

System	Behaviour	Packing glass beads d_p (mm)	ε (%)	Pressure MPa	Liquid flow-rate kg.m^{-2}.s^{-1}	Gas flow-rate kg.m^{-2}.s^{-1}
Water-nitrogen	Aqueous, non-foaming	2	38	0.2-8.1	1.8-24.5	0.003-3
		1.4	35	0.3-5.1	13.2	0.009-1.9
Water-helium	Aqueous, non-foaming	1.4	35	0.3-5.1	13.2	0.005-0.37
Water-argon	Aqueous, non-foaming	1.4	35	0.3-5.1	13.2	0.05-2.2
Ethanol-nitrogen	Organic, non-foaming	2	38	0.3-5.1	6.6 and 13.2	0.009-1.9
Propylene carbonate-helium	Organic, moderately viscous weakly foaming	2	38	0.3-5.1	6.6. and 13.2	0.005-0.4
		1.4	35	1.1 and 2.1	13.2	0.005-0.45
Propylene carbonate-nitrogen	Organic, moderately viscous weakly foaming	2	38	0.3-5.1	6.6 and 13.2	0.009-1.8
		1.4	35	0.3-5.1	13.2	0.009-1.9
Ethylene glycol-nitrogen	Organic, viscous	2	38	0.3-5.1	2-13.2	0.009-1.8
Water + 40 % sucrose-nitrogen	Aqueous, moderately viscous	2	38	0.3-5.1	3.3-13.2	0.009-1.9
Water + 1 % ethanol-nitrogen	Aqueous, foaming	2	38	0.3-5.1	4.2-13.2	0.009-1.9

b - Correlations. The major problem in the representation of the two-phase pressure drop by a dimensionless correlation is the description of the geometry of the bed. It is well known that the single-phase pressure drop is a very rigid function of the porosity and of the bed structure. Three types of pressure drop correlations are available in the literature [49]:

• *Dimensional, entirely empirical correlations* whose use is limited to the range of variables of the experimental work it is based upon.

• *Dimensionless correlations based on momentum-or energy balances, and using the Lockhart and Martinelli parameter, χ_L, that is the ratio of the single-phase pressure drops at the same velocities:*

$$\chi_L = \sqrt{\left(\frac{\Delta P_L}{\Delta P_G}\right)} \qquad (5.2\text{--}16)$$

Our team proposed a number of such correlations, following the advances in our work in this field; Eqn. (5.2–17) proposed by Tosun [15] for the low interaction regime is of this type:

276

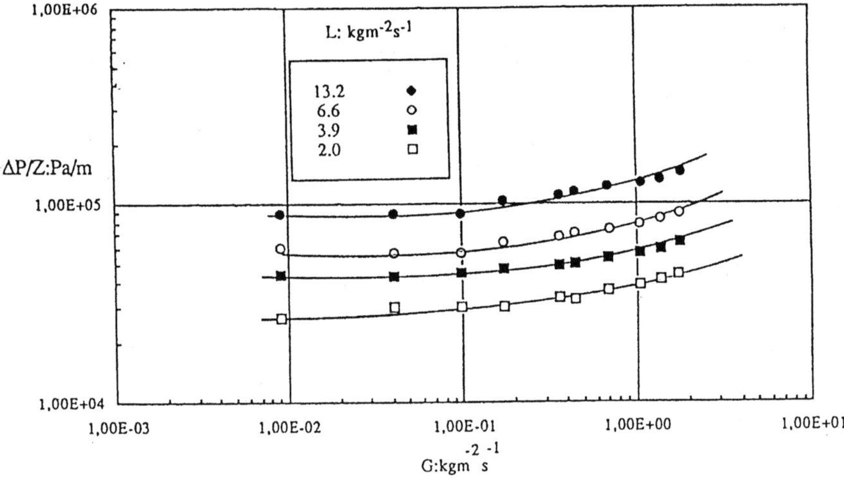

Figure 5.2–15. Influence of the gas-and liquid flow-rates on the two-phase pressure drop for the N_2-Ethyleneglycol system at 51 bar (after Larachi *et al.* [37]).

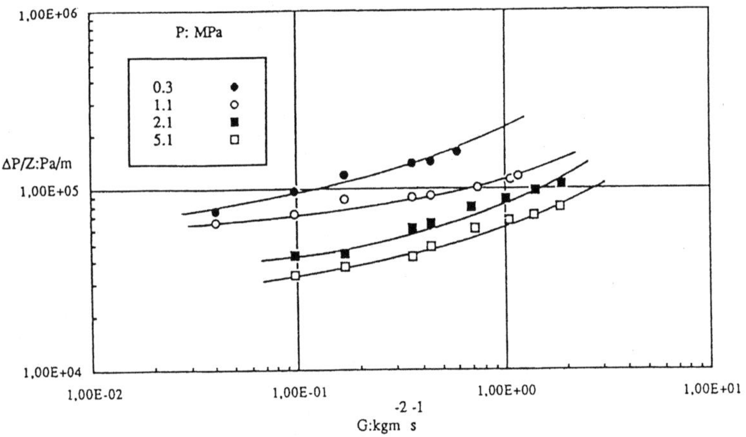

Figure 5.2–16. The effect of the total pressure on two-phase pressure drop for the system N_2-Propylene carbonate ($L = 13.2$ kg.m^{-2}.s^{-1}) (after Larachi *et al.* [37]).

$$\Phi_L = \sqrt{\frac{\Delta P_{LG}}{\Delta P_L}} = 1 + \frac{1}{\chi_L} + \frac{1.424}{\chi_L^{0.576}} \qquad (5.2-17)$$

This way of correlating the pressure drop has the advantage that the single-phase pressure drop is a convenient means of combining geometric properties of the packing.

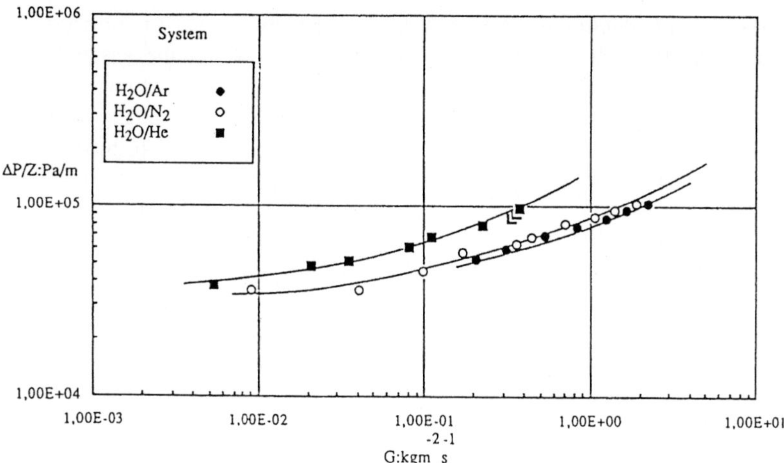

Figure 5.2–17. Influence of the gas density on the pressure drop for He, Ar and N_2-water

The disadvantage is that there is no reliable correlation of the single-phase pressure drop: in almost all cases it is possible to use an equation of Ergun's type, but the empirical coefficients of this correlation have to be determined experimentally: *a priori* calculation can be very inaccurate.

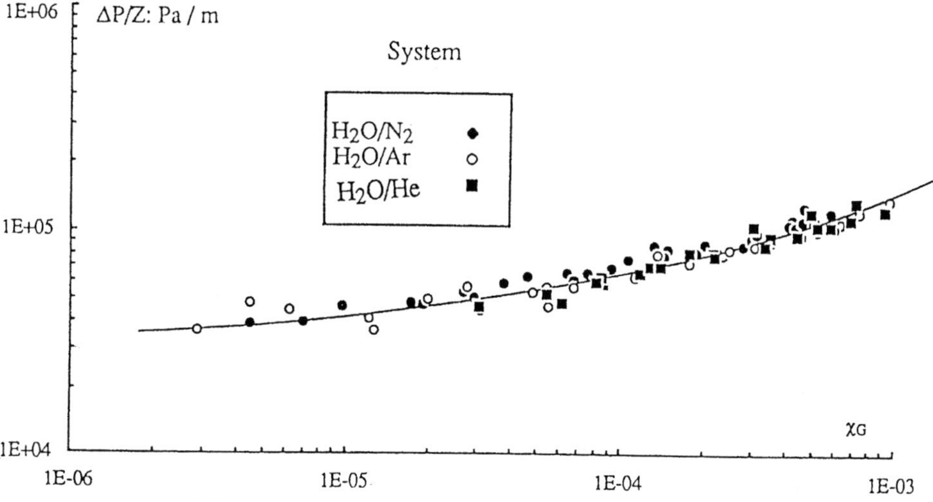

Figure 5.2–18. Two-phase pressure drop versus parameter χ_G for various gases (after Wild *et al.* [28]).

• *Dimensionless, entirely empirical correlations:* The problem is, that the pressure drop depends on many variables: the gas-and liquid velocities, u_L and u_G, the gas-and liquid densities, ρ_L and ρ_G, the particle-diameter, d_p, and eventually the diameter and shape distribution, the surface tension, σ_L, the viscosity, μ_L, of the liquid and eventually of the gas, μ_G, the bed porosity, ε, and the column diameter and height, and the type of distributor.

It is not always easy to decide which variables should be taken into account in order to obtain significant correlations, and one must be extremely careful not to use such correlations outside the range of variables of the data on which the correlation is based.

In TBR operating at elevated pressures, the influence of the relevant variables can be seen for example in the following Figures.

In Figure 5.2–18, the pressure drop is represented as a function of the simplified Lockhart-Martinelli parameter, χ_G, $(\chi_G = (G/L)\,(\rho_L/\rho_G)^{0.5})$. This figure shows that, at a given mass flow rate of the liquid, the pressure drop is a function of the ratio of $G/\sqrt{\rho_G}$, regardless of the nature of the gas.

Figures 5.2–19 and 5.2–20 stress the influence of the liquid mass flow-rate and of the liquid viscosity on the two-phase pressure drop. These figures confirm that, at a given liquid flow-rate, the pressure drop depends on the ratio $G/\sqrt{\rho_G}$ (that is the square root of the gas momentum flow-rate, $\rho_G\,u_G^2$). It is therefore possible to predict the pressure drop by replacing one gas by another one, provided the gas momentum flow-rate is kept constant.

When the main variables have been determined, one has to use a database as large as possible to obtain a correlation.

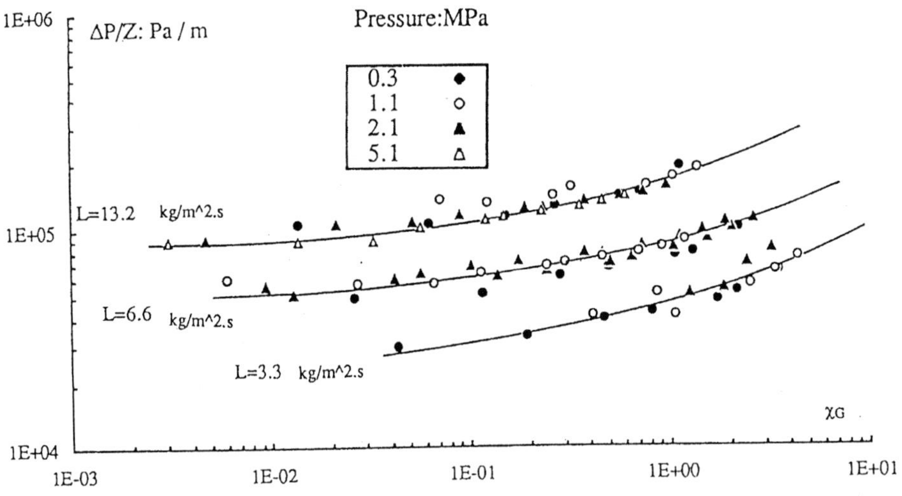

Figure 5.2–19. Effect of the liquid flowrate on two-phase pressure drop (after Wild *et al.* [28]).

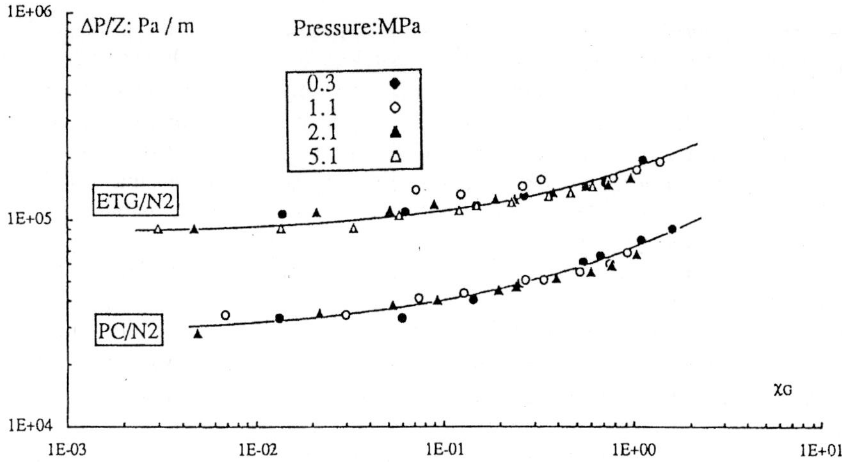

Figure 5.2–20. Liquid viscosity effect on the two-phase pressure drop (after Wild *et al.* [28]).

Larachi *et al.* [37] presented a simplified version of Ellman's correlation. A friction factor, f_{LGG}, is represented as a function of dimensionless groups which takes inertia, viscosity and surface-tension effects into account by using, respectively, χ_G, Re_L, and We_L:

$$f_{LGG} = \frac{1}{K^{1.5}}\left(A + \frac{B}{K^{0.5}} \right) \tag{5.2–18}$$

where

$$K = \chi_G \left(Re_L\, We_L \right)^{0.25} \tag{5.2–19}$$

$$f_{LGG} = \frac{(\Delta P/Z)\, d_H\, \rho_G}{2\, G^2} \tag{5.2–20}$$

with $A = 31.3 \pm 3.9$; $B = 17.3 \pm 0.6$ (97.5 % confidence).

This correlation is valid for non-foaming liquids.

Considering the diversity of the gas-liquid systems used, the fit is quite satisfactory (Figure 5.2–21).

Wammes *et al.* [33] also proposed a theoretical correlation to estimate the two-phase pressure drop. The following assumptions were made.
- the TBR operates under stationary and isothermal conditions,
- the gas density is constant,
- the gas-liquid surface tension does not play a role.

280

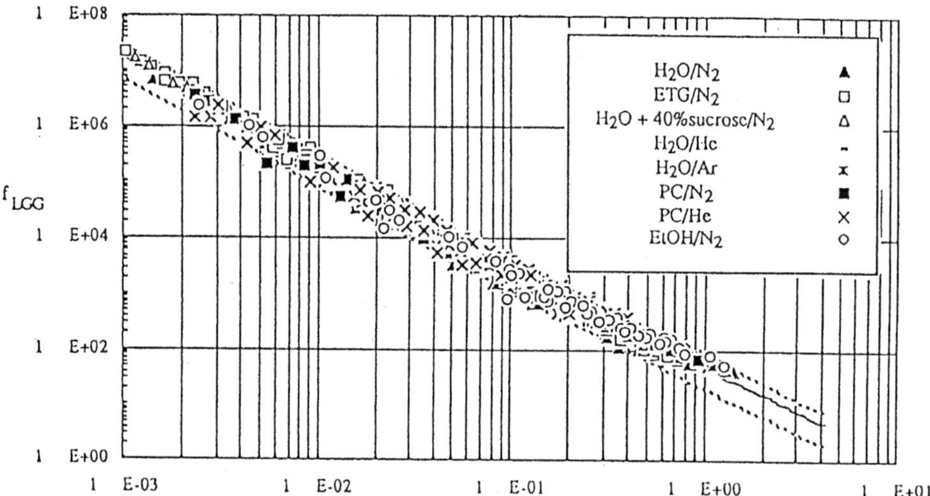

Figure 5.2–21. Correlation of the two-phase pressure drop with its ± 50 % limits and 97.5 % confidence (after Larachi [36] and Larachi *et al.* [37]).

The force balance over the gas phase then can be described by:

$$\varepsilon(1-\text{\ss}) \, \Omega Z \rho_G g + \Delta P \, \varepsilon(1-\text{\ss})\Omega = \tau_{LG} a_{LG} \Omega Z \qquad (5.2\text{–}21)$$

The authors assumed that the shear stress, τ_{LG}, at the right-hand side can be described by $\frac{1}{2} f_{LG} \rho_G u_G^2 \, a_{LG}$, in which f_{LG} is an empirical two-phase friction factor.

In principle, the difference between the gas-and liquid velocity should be used instead of the mean interstitial gas velocity, but in their operating conditions, the mean interstitial liquid velocity is small compared to the mean interstitial gas velocity. To a first approximation, a_{LG} was considered to be proportional to the external geometric area, a_c, of the packing:

$$a_{LG} \simeq a_c = \frac{6(1-\varepsilon)}{d_p} \qquad (5.2\text{–}22)$$

The expression for the pressure gradient is then:

$$\frac{\Delta P}{Z} = \frac{1}{2} f_{LG} \rho_G \, u_G^2 \, \frac{(1-\varepsilon) \, 6}{d_p \, \varepsilon(1-\text{\ss})} \qquad (5.2\text{–}23)$$

The effects of many flow variables on the phenomena at the gas-liquid interface, and on the pressure drop, are all included in the empirical two-phase friction factor, f_{LG}.

An expression for f_{LG} has to be derived from experimental data, by means of a power-product equation containing the gas density, the interstitial mean gas velocity, and a film-liquid Reynolds number. They found a value of -0.37 for both the exponent of ρ_G and of u_G, indicating that f_{LG} depends on a gas-phase Reynolds number. For the exponent of the liquid film Reynolds number the value was about zero.

In Figure 5.2–22, the pressure-gradient, $\varepsilon(1-\beta)(\Delta P/Z)$, is plotted versus the interstitial mean gas velocity, u_G, at constant gas densities. Along each line for a given density, the liquid flow-rates vary considerably, but the relationship between the pressure gradient and u_G is unique, and independent of the liquid mean velocity.

On the basis of the TBR experiments of Wammes et al. [33] the following relationship is proposed for the pressure gradient; this is valid in the range of gas-phase Reynolds numbers $200 < \mathrm{Re}_G < 5000$:

$$\frac{\Delta P}{0.5\,\rho_G\,u_G^2}\left(\frac{d_p}{Z}\right) = 155\left(\frac{\rho_G\,u_G\,d_p}{\mu_G\,(1-\varepsilon)}\right)^{-0.37}\left(\frac{1-\varepsilon}{\varepsilon(1-\beta)}\right) \qquad (5.2\text{–}24)$$

A comparison between the experimental data and the predicted values is given in Figure 5.2–23. However the validity of Equation (5.2–24) is more limited than that of Equation (5.2–18), as it is based on a narrower range of operating conditions. Moreover, Equation (5.2–24) needs prior evaluation of the liquid hold-up, β.

Other correlations for two-phase pressure drops in TBR have been summarized by different authors [40, 41, 51], the majority of them being mostly empirical and restricted to their specific narrow ranges of process conditions.

At the present time, the best new correlation for two-phase pressure drop, both for low and high interaction regimes, is that proposed by Iliuta et al. [47], and derived using a neural

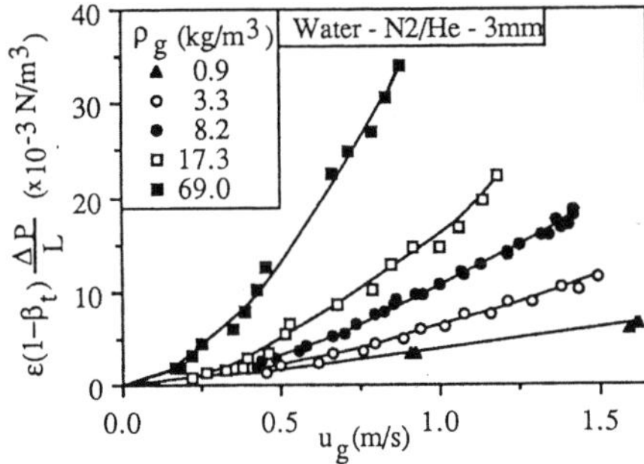

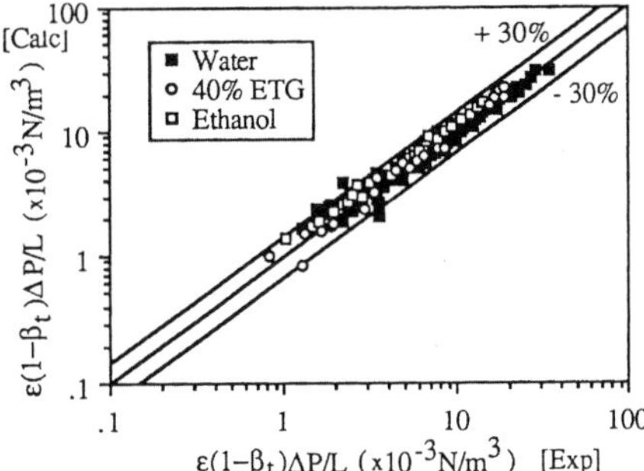

Figure 5.2–22. Effect of gas velocity on pressure drop for different gas densities (after Wammes et al. [33]).

Figure 5.2–23. Predicted values of the pressure drop (Eqn. 5.2–24) versus the experimental pressure drop data (after Wammes et al. [33]).

network dimensionless-group-approach along with a flow database of 17,800 measurements on 74 gas-liquid systems, 29 column diameters, and 47 packing materials. The detailed normalized input and output variables of the neural network model are presented in two tables with the neural network weights for the low and high interaction correlations [47]. These two correlations outperform all existing correlating tools of pressure drops, respectively in the low interaction range with an average absolute relative error of 16.1 %, and in the high interaction range with 23.5 %.

5.2.2.5 Liquid hold-up

The total hold-up of the liquid in a TBR consists of two parts:
• the static (or capillary) liquid hold-up is the internal liquid hold-up held inside the pores of the catalyst by the capillary forces;
• the dynamic liquid hold-up is the external liquid hold-up partially occupying the void volume of the packed bed.

For the TBR design the dynamic liquid hold-up is a basic parameter because it is related to other important hydrodynamic parameters (including the pressure drop, wetting, and mean-residence-time of liquid).

The total external liquid hold-up is equal to the sum of the static-and dynamic-hold-ups.

The static liquid hold-up can be determined from the difference with the weight of the TBR after a minimum time of draining with the same TBR dry packing.

Two different methods are generally applied to determine the liquid hold-up.

The first method for determining the dynamic liquid hold-up uses weighing experiments. After the TBR has reached its desired operating point, the gas-and liquid inlets are closed simultaneously. The amount of liquid trickling out of the column is called the dynamic hold-up.

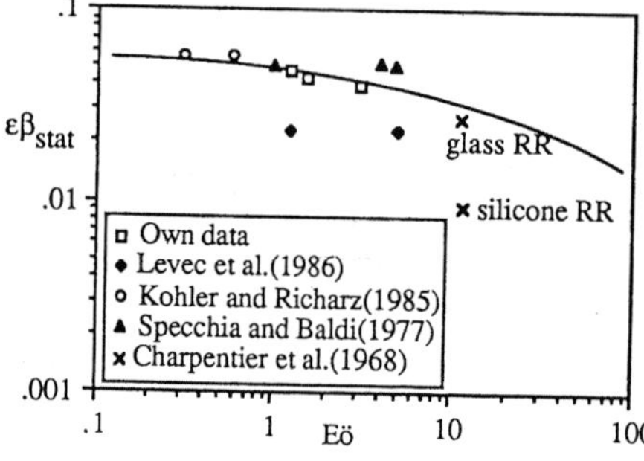

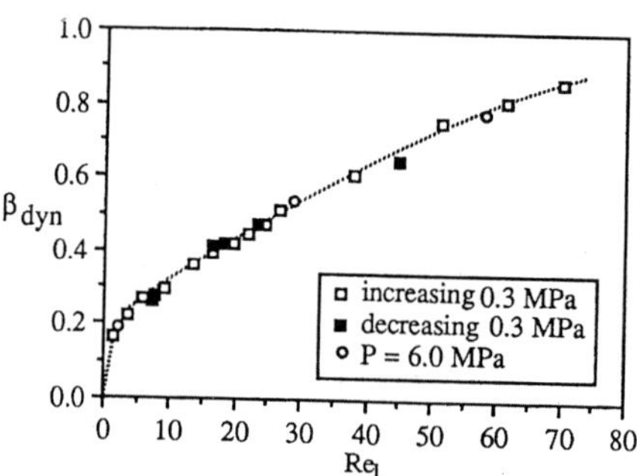

Figure 5.2–24. Experimental static liquid hold-up as a function of the Eötvos number (after Wammes *et al.* [34]).

Figure 5.2–25. Influence of the total reactor pressure and of the liquid flow-rate on the dynamic liquid holdup for the single water operation (after Wammes *et al.* [34]).

The second method is an indirect method, based on the liquid's average residence time evaluated with the tracer injection technique. From the first moment of the RTD curve the total external liquid hold-up can be calculated.

a - Static liquid hold-up. The static hold-up, β_{stat}, situated around the contacting points of the particles, results from the balance between the capillary and gravitational forces and is independent of the gas-flow, liquid-flow, and liquid viscosity.

Van Swaaij [52] and Charpentier *et al.* [53] proposed a relationship between the static liquid hold-up, β_{stat}, and the dimensionless Eötvos number, Eö. At high Eötvos numbers the static hold-up is inversely proportional to Eö, whereas at low Eötvos numbers, the static hold-up reaches a maximum value.

This correlation gives, for perfectly wettable solids, fairly good estimates of the static hold-up for different particle-geometries and sizes. Saez and Carbonnel [26] used the hydraulic diameter, instead of the nominal particle diameter, as the characteristic length in the Eötvos number, to include the influence of the particle geometry on the static hold-up. However, no improvement could be obtained in correlating the data with this new representation.

Wammes *et al.* [34], by employing three different liquids (water - ethanol and 40 % ethyleneglycol aqueous solution) with 3 mm glass spheres, obtained experimentally determined static hold-up data. Figure 5.2–24 shows the values of the static holdup as a function of the Eötvos number together with data of other authors. Wammes *et al.* [34] concluded that the static liquid hold-up is not affected by the total reactor pressure.

b - Dynamic liquid hold-up. Wammes *et al.* [33, 34] determined the dynamic hold-up, β_{dyn}, by using the direct method of the drainage technique. Figure 5.2–25 shows an example of the dynamic liquid hold-up experiments with single liquid flow (water) as a function of the liquid Reynolds number, Re_L. It can be concluded that in the single liquid flow TBR, the dynamic

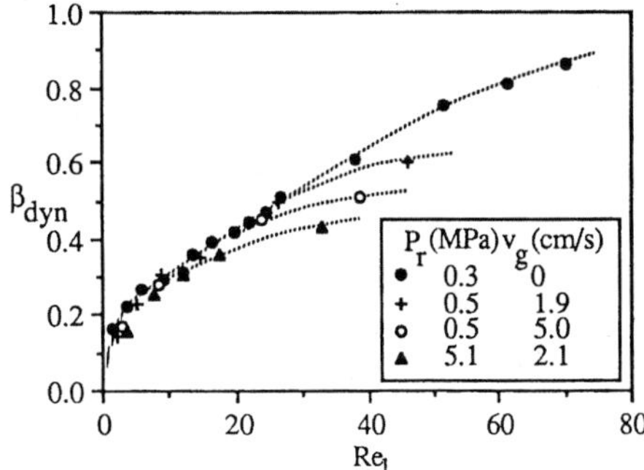

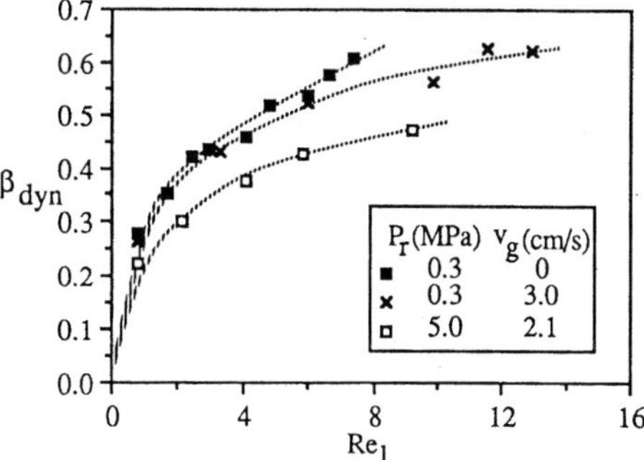

Figure 5.2–26. Dynamic liquid hold-up versus liquid Reynolds number at different gas velocities and total pressures for the N_2–H_2O system (after Wammes [17]).

Figure 5.2–27. Dynamic liquid hold-up versus liquid Reynolds number at various gas velo cities and total pressures for the N_2-40 % Ethyleneglycol-H_2O system (after Wammes [17]).

liquid hold-up is pressure-independent and shows no hysteresis.

Figures 5.2–26 and 5.2–27 show, respectively, the influence of the total reactor pressure and of the superficial gas velocity on the dynamic liquid hold-up with water-nitrogen and aqueous 40 % ethyleneglycol-nitrogen. Similar trends are observed for the two systems. In the trickle-flow regime, the total operating pressure has no influence on the dynamic liquid hold-up at low liquid flow-rates and at low gas velocity (lower than few mm/s). Note that the influence is however, more noticeable for the viscous system.

Then, in contrast to the operation at atmospheric pressure a small increase in the superficial gas velocity reduces considerably the dynamic liquid hold-up. This effect is more pronounced at higher liquid flow-rate values and total reactor pressures. Because of this high influence of the gas flow on the hydrodynamics, Wammes *et al.* [34] recommend avoidance of the use of the term, "low interaction regime", for the trickle-flow regime at high pressure.

In processing the quantitative effect of the liquid flow-rate, the superficial gas velocity, and the total reactor pressure on the dynamic liquid hold-up, good results have been obtained by using:

• the liquid-phase Reynolds number:

$$\mathrm{Re_L} = \frac{\rho_L \, u_L \, d_p}{\mu_L} \tag{5.2-25}$$

• the modified Galileo number, which takes into account the ratio between the pressure-gradient and gravitational force:

$$\mathrm{Ga_L^*} = \mathrm{Ga_L} \left(1 + \frac{\Delta P/Z}{\rho_L g} \right) = \left(\frac{d_p^3 \, \rho_L^2 \, g}{\mu_L^2} \right) \left(1 + \frac{\Delta P/Z}{\rho_L g} \right) \tag{5.2-26}$$

• the variable $(a_G d_p/\varepsilon)^{0.65}$ derived by Specchia and Baldi [54] for the effect of the type of packed bed.

Based on 220 experiments with helium-water, nitrogen-water, nitrogen-ethanol and nitrogen-aqueous 40 % ethyleneglycol within the ranges:

$$2 < \mathrm{Re_L} < 55$$
$$0.02 < u_G < 0.36 \ \mathrm{m.s}^{-1}$$
$$0.33 < \rho_G < 86 \ \mathrm{kg.m}^{-3}$$

$$0 < \frac{\Delta P/Z}{\rho_L g} < 16$$

Wammes *et al.* [33] proposed the following correlation:

$$\text{ß}_{\mathrm{dyn}} = 3.8 \left(\frac{\rho_L u_L d_p}{\mu_L} \right)^{0.55} \left[\frac{d_p^3 \rho_L^2 g}{\mu_L^2} \left(1 + \frac{\Delta P/Z}{\rho_L g} \right) \right]^{-0.42} \left(\frac{a_G d_p}{\varepsilon} \right)^{0.65} \tag{5.2-27}$$

with a mean relative error of 8 %. A comparison between the predicted and measured values of three different gas-liquid systems is given in Figure 5.2–28.

c - Total external liquid hold-up. Larachi *et al.* [37, 38] obtained total liquid hold-up data by the indirect method (RTD technique). The experimental reported results confirm observations made by Wammes *et al.* [31, 34], but over a wider range of operating conditions: at very low gas velocities, the total hold-up is independent of pressure regardless of the type of gas-liquid system (non-foaming, viscous, and weakly foaming). The practical value of the data pertaining to the very low gas-velocity range (< 10 mm.s^{-1}) is evident since recourse to high pressure measurements is unnecessary. However for larger gas superficial velocities, the influence of pressure has to be taken into account.

Moreover, Larachi [36] found that:

• At given gas-and liquid mass flow-rates, the total liquid hold-up increases with pressure, owing to the lower superficial gas velocity as a consequence of the increase in gas density. As an example, Figure 5.2–29 shows the liquid hold-up as a function of the gas mass flow-rate at different values of the total pressure for the system ethanol-nitrogen. At a given mass flow-rate, the total liquid hold-up increases with pressure. This result is not surprising, considering the fact that the gas occupies less space because of its increased density at constant gas flow-rate. Moreover, since the gases superficial velocity decreases when the pressure is increased (with G being fixed), this brings about less important shear at the gas-liquid interface, leading to a larger residence time of the liquid phase.

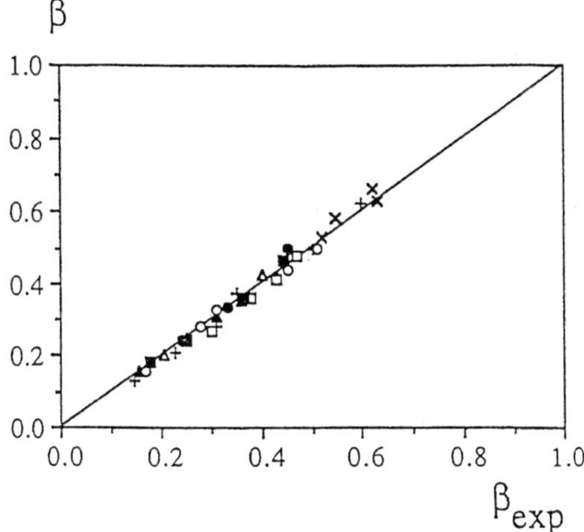

Figure 5.2–28. Comparison between the experimental dynamic liquid hold-up and the predicted values given by Equation (5.2–27) (after Wammes *et al.* [34]).

• The total liquid hold-up, β_t, is reduced when the liquid viscosity decreases. As shown in Figure 5.2–30, at given values of pressure, and of gas-and liquid flow-rates, the total liquid hold-up reduces when the liquid viscosity decreases from ethyleneglycol (19 mPa.s) to water + 40 % sucrose (6 mPa.s), propylene carbonate (2.5 mPa.s), and water (1 mPa.s).

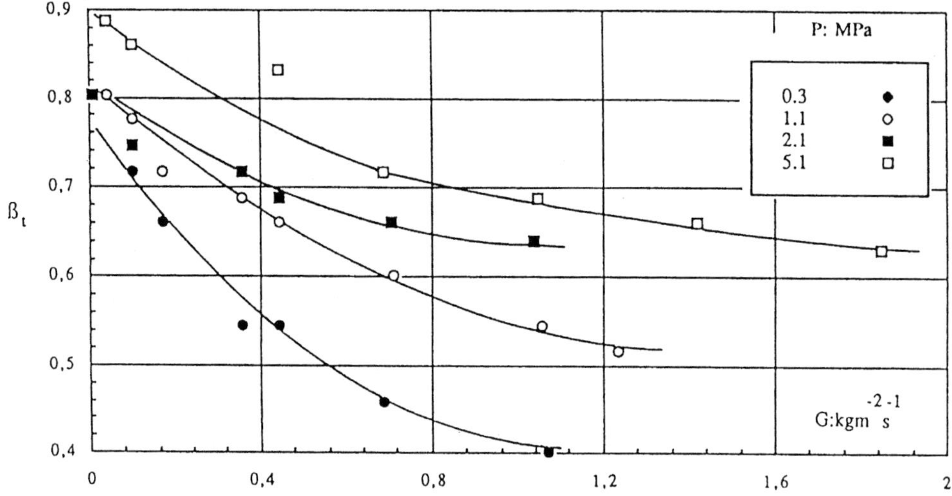

Figure 5.2–29. Influence of the operating pressure on the total liquid saturation as a function of the gas mass flow-rate for the N_2-Ethanol system ($L = 13.2$ kg.m^{-2}.s^{-1}) (after Larachi *et al.* [37])

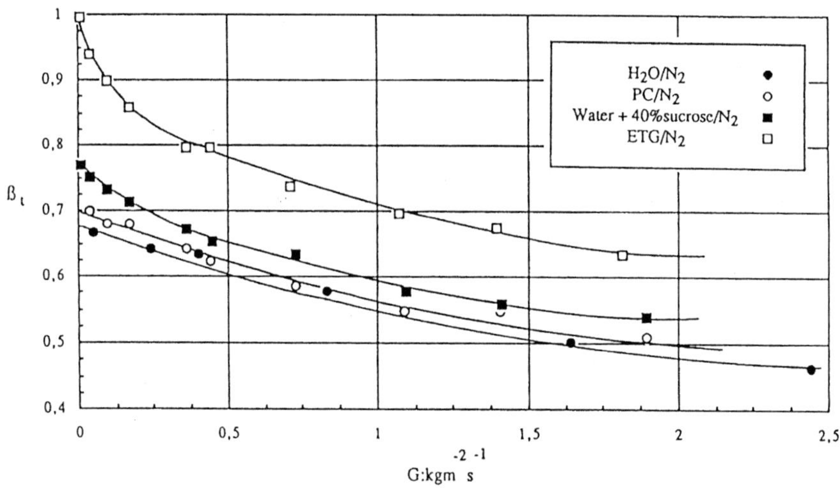

Figure 5.2–30. Effect of the liquid viscosity on the total liquid hold-up at different gas mass flow-rates ($P = 21$ bar; $L = 13.2$ kg.m^{-2}.s^{-1}) (after Larachi *et al.* [37]).

• The total liquid hold-up is much smaller for foaming liquids, regardless of the operating pressure, owing to the high stability of fine gas bubbles adhering to the solid particles. A strong influence of the coalescence behaviour of the liquid is noticed. Figure 5.2–31 shows the liquid holdup as function of the gas flow-rate at a given liquid flow-rate and a pressure of 51 bar for the systems water-nitrogen and water+1 % ethanol-nitrogen as an example.

As for the pressure drop, three types of correlations of β_t appear in the literature [50]:

• Dimensional empirical correlations whose use outside the investigated range of variables is risky.

• Dimensionless correlations based on momentum-or energy-balances and using the Lockhart and Martinelli parameter, χ_G, or some similar parameter. The practical relevance of such correlations is limited since their use requires the knowledge of the single-phase pressure drop of both gas and liquid; furthermore, the influence of the geometry of the bed is not always well described by these single-phase pressure drops alone.

• Dimensionless correlations based on dimensional analysis and on a qualitative analysis of the two-phase flow.

The available correlations in the literature are not able to represent the experimental results derived from 1500 high pressure data on the liquid hold-up. To correlate all the data, the effects of fluid inertia, surface forces, and liquid shear stress have again been accounted for, by using the corresponding dimensionless groups in the following empirical correlation [37]:

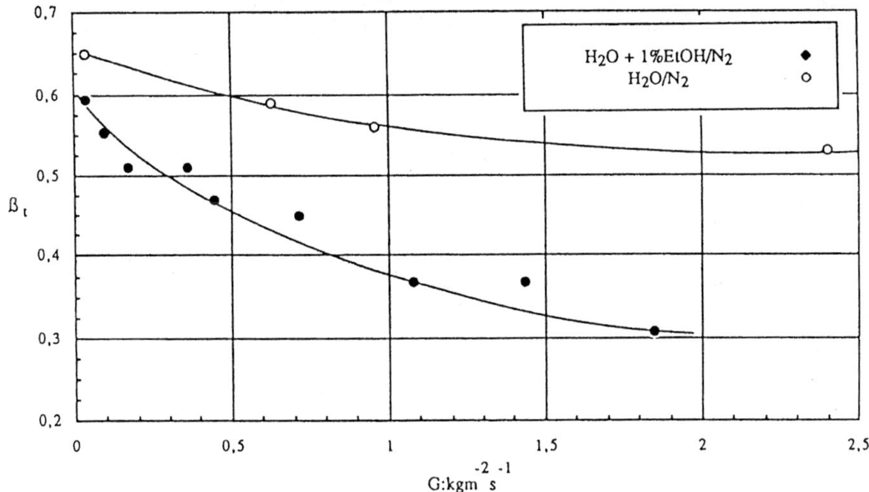

Figure 5.2–31. Influence of foaming on the total liquid hold-up for the N_2-H_2O system (non foaming) and for the N_2-1 % Ethanol-H_2O system (foaming) (P = 52 bar; L = 13.2 kg.m^{-2}.s^{-1}) (after Larachi *et al.* [37]).

$$\beta_T = 1 - 10^{-\Gamma} \qquad (5.2\text{--}28)$$

with

$$\Gamma = 1.22 \frac{We_L^e}{\chi_G^d Re_L^e} \qquad (5.2\text{-}29)$$

where $c = 0.15 \pm 0.016$; $d = 0.15 \pm 0.008$; $e = 0.20 \pm 0.013$.

Figure 5.2–32 compares the correlation proposed here to the experimental data (bias, 0.99; standard deviation, 0.009).

In brief, the total external liquid hold-up is an increasing function of liquid velocity, viscosity and particle diameter. It is a decreasing function of the superficial velocity of the gas, and of the liquid surface tension. Liquid hold-up reduces as the gas density increases, except for very low gas velocities, where it is insensitive to gas density. Non-coalescing liquids exhibit much smaller hold-ups than coalescing liquids. Gas viscosity appears to have a marginal effect on the liquid hold-up.

With the same approach as described for the two-phase pressure drop, Iliuta et al. [47] also proposed a new correlation for the total external liquid hold-up in the high interaction regime, with an average absolute relative error of 13.6 %.

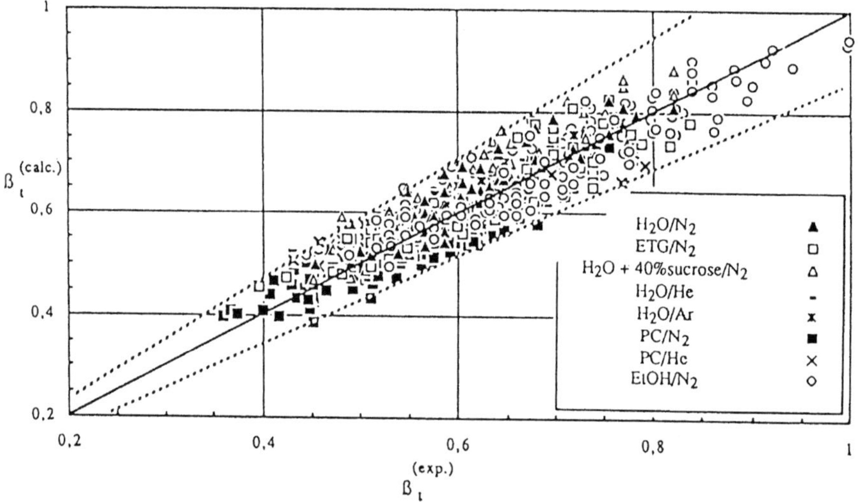

Figure 5.2–32. Parity plot of the correlation (5.2–28) to estimate the total liquid hold-up (after Larachi et al. [37]).

5.2.2.6 Gas-liquid interfacial area

Mass-transfer studies under high pressure are scarce in the open literature for TBR. At present, to our knowledge, only the Twente and Nancy groups have measured the gas-liquid interfacial area concerning TBR operating at elevated pressures.

The gas-liquid interfacial area can be determined by using the chemically enhanced absorption of a gas phase solute, A, into a liquid phase, where it reacts irreversibly with a liquid-phase component, B.

When the reaction between components A and B in the liquid phase is m^{th} order in A, and n^{th} order in B, and the concentration of component B is nearly constant throughout the TBR, the reaction is said to be rapid-pseudo-m^{th} order in A. The condition for this situation is $3 < Ha < E_i$. The rate of absorption is then expressed by:

$$\Phi = \phi a = E\, k_L a\, C_A^* = Ha\, k_L a\, C_A^* = a \left[\frac{2}{m+1} k_{mn}\, D_A\, C_{A*}^{m.1}\, C_B^{n} \right]^{0.5} \qquad (5.2\text{--}30)$$

where Φ is the total mass-transfer rate, ϕ is the average specific mass-transfer flux, E is the enhancement factor, k_L is the liquid-side mass-transfer coefficient, C_A^* is the solute concentration at equilibrium, C_B is the concentration of component B, Ha is the Hatta number, m and n are the reaction orders and D_A is the gas-solute diffusivity.

Thus, the rate of absorption is independent of k_L, that is of the hydrodynamic conditions. Consequently, provided that the average specific rate of absorption, ϕ, is known and C_A^* and C_B are effectively the same in all parts of the system, the specific interfacial area, a, may be determined directly from a measurement of the rate of absorption, Φ.

The following main assumptions are made:

- the gas-phase mass-transfer resistance is negligible compared with the liquid phase resistance,
- the fluids move downward throughout the catalyst bed without any axial dispersion,
- the gas phase in the TBR shows plug flow,
- the average specific rate of absorption, ϕ, is estimated by using the kinetics of the reaction, or is measured experimentally in some laboratory apparatus with a known geometrical interfacial area.

Some aqueous, organic and viscous chemical systems for the determination of the gas-liquid interfacial area in this pseudo-mth order regime are suitable [13].

Table 5.2–7 indicates the chemical systems selected here to investigate the interfacial area of the TBR operating under high pressure.

Table 5.2–7
Gas-liquid chemical systems.

System	Pressure (bar)	Packing	Author
CO_2-N_2-H_2O-2M DEA	< 15	Glass spheres, 3 mm	Wammes
CO_2-N_2-H_2O-ETG 40 %-1.5M DEA	< 60	Al_2O_3 cylinders 3.2 mm × 3.2 mm	*et al.* [32]
CO_2-N_2-H_2O-1.5M DEA	3-31	Glass beads, 0.85 mm	
CO_2-N_2-H_2O-ETG 20 %-1.5M DEA	11	Polypropylene extrudate, 3.37 mm	Larachi *et al.*
CO_2-N_2-H_2O-ETG 40 %-1.5M DEA	3-21	Carbon cylinders 1.6 mm	[55]

Only the trickle-flow regime was investigated, where the mass-transfer resistances are probably the highest compared to other flow regimes. The results are reported mainly qualitatively [32], and only one correlation was given by Larachi et al. [55]. In this, the reduced interfacial area a/a_c grows by increasing both the gas-and liquid superficial velocity.

Figure 5.2–33 shows the dependence of the gas-liquid interfacial area, a, on the gas-and liquid-mass flow-rates for the system CO_2-N_2-H_2O-DEA-20 % ETG at 11 bar through

290

polypropylene extrudates. It appears that a increases faster with gas-mass flow-rate, G, when the liquid holdup is large, i.e. for high liquid-mass flow-rates, L. Note that a behaves as the two-phase pressure drop $\Delta P/Z$.

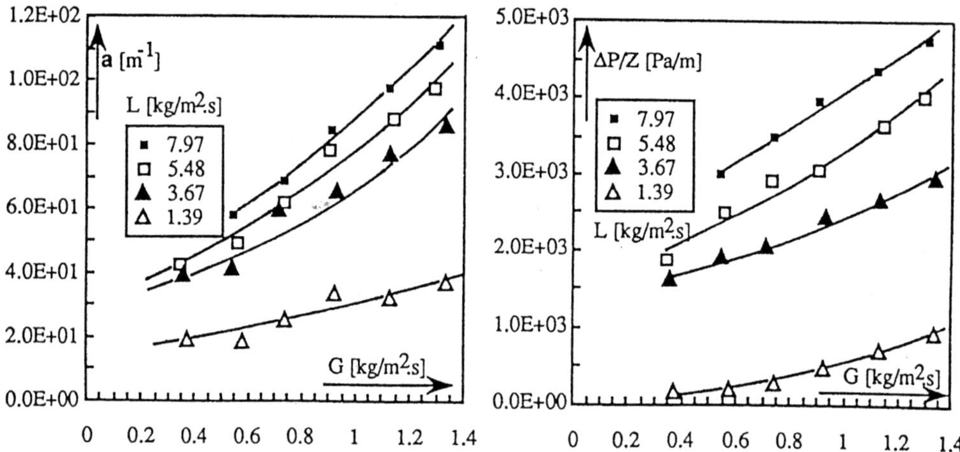

Figure 5.2–33. Gas-liquid interfacial area and two-phase pressure drop versus gas flow-rate for the CO_2-N_2-DEA-H_2O-20 % ETG system (P = 11 bar; hydrophilized polypropylene extrudate) (after Larachi *et al.* [55]).

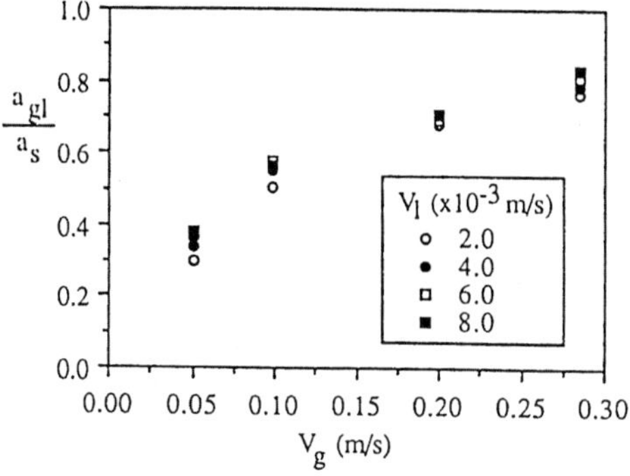

Figure 5.2–34. Effect of the superficial gas velocity at different superficial liquid velocities on the dimensionless interfacial area for the CO_2-N_2-DEA-H_2O-40% ETG system (P = 25 bar, porous Al_2O_3 cylinder) (after Wammes *et al.* [32]).

Figure 5.2–34 indicates with the same chemical system, but more viscous liquids, and with glass spheres, that the increase of the gas-liquid interfacial area with the liquid flow-rate is much less pronounced.

• The results obtained by Wammes *et al.* [32] show that the dimensionless interfacial areas for both liquids of different viscosity are approximately equal under equal operating conditions. The two systems differ by a factor 2.3 in liquid viscosity (Figure 5.2–35).

• Pressure influences the mass-transfer exclusively via the gas density. Therefore it seems to be more general to use the gas density, ρ_G, instead of the total pressure, P, as a natural variable.

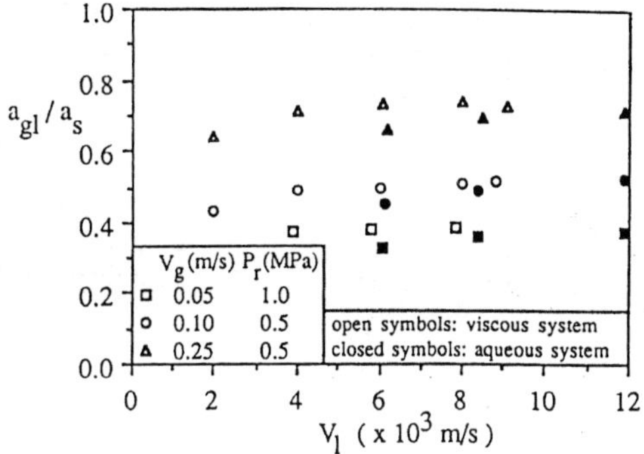

Figure 5.2–35. Comparison of the dimensionless area for aqueous and viscous system (after Wammes *et al.* [32]).

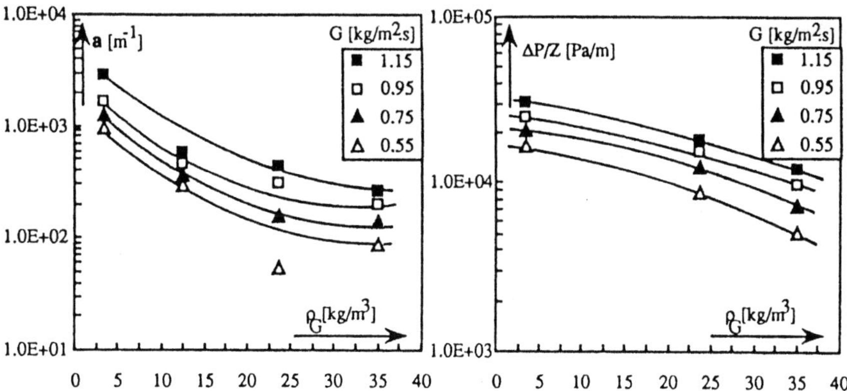

Figure 5.2–36. Gas-liquid interfacial area and two-phase pressure drop versus gas density ($L =$ 3.67 kg.m^{-2}.s^{-1}) (after Larachi *et al.* [55]).

292

In Figure 5.2–36, both the gas-liquid interfacial area and the two-phase pressure drop are plotted versus gas density for different gas mass flow-rates and constant liquid mass flow-rate in the system CO_2-N_2-H_2O-DEA. It can be seen that, for given G and L, both a and $\Delta P/Z$ are decreasing functions of gas-density. The effect of gas-density on a is considerable: for instance a reduction from 800 m^{-1} (for ρ_G = 3.4 kg m^{-3}) to 100 m^{-1} (for $\rho_G \simeq$ 35 kg m^{-3}) at G = 0.55 kg.m^{-2}.s^{-1}, i.e., a factor of 8 for the interfacial area for a factor of 10 on ρ_G.

Expressing a versus the gas superficial velocity, u_G, at different gas densities, as illustrated in Figure 5.2–37, shows that the gas-liquid interfacial area, a, increases when ρ_G is increased above gas-superficial-velocities of 0.02 m.s^{-1}, L and u_G being kept unchanged. A factor of 10 on ρ_G brings about an increase of a from 250 m^{-1} (3 bar) to 630 m^{-1} (31 bar), i.e., a factor of 2.5, rather than 10 as previously for a constant gas-mass flow rate.

To sum up, the gas-liquid interfacial area, a, increases with liquid and gas flow rates and liquid viscosity; a for non-coalescing liquids exceeds that for coalescing ones; a for organic liquids does not differ significantly from that for ionic liquids; a is higher for non-spherical packings and a improves at elevated pressures.

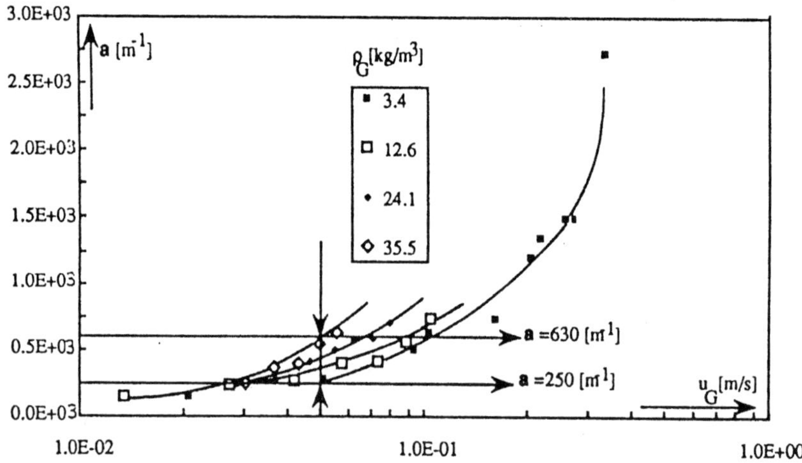

Figure 5.2–37. Gas-liquid interfacial area versus gas superficial velocity at different gas densities (L = 3.67 kg m^{-2} s^{-1}) (after Larachi *et al.* [55]).

The influence of gas density on the gas-liquid interfacial area could be related to the flow patterns and to the interpenetration between gas and liquid. It is probable that the gas-liquid interface results from two distinct mechanisms. The first one is based on the extent of the solid surface where liquid films could develop (wetting of particles), virtually controlled by fluid velocities and liquid properties. The second mechanism depends on the kinetic energy content of the gas phase. The more important the gas inertia, the more important is the contribution of fine gas bubbles penetrating liquid films.

Most often, at atmospheric pressure the interfacial area correlations use pressure drop as a variable because of their similarities. For example, Midoux *et al.* [56] proposed the following correlation for the cocurrent downflow in TBR:

$$\frac{a}{\varepsilon} = 374 \left(\frac{\varepsilon^2}{a_c} \frac{\xi_{LG}}{L/\rho_L + G/\rho_G} \right)^{0.65} \qquad (5.2\text{--}31)$$

where ξ_{LG} is the power dissipated per unit volume of intergranular void.

We tested this correlation with the data obtained for downflow at high pressure. Figure 5.2–38 indicates an acceptable agreement, within the 60% limits of the correlation (Lara-Marquez *et al.* [57]).

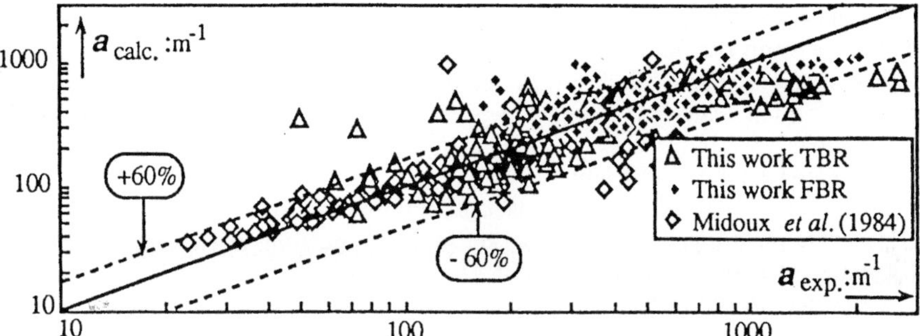

Figure 5.2–38. Comparison between the experimental results under elevated pressures and the correlations (5.2–31) (after Lara-Marquez *et al.* [57]).

Other correlations have been summarized in the literature [41, 51, 58]. The proposition of Iliuta *et al.* [59] is, at present, far better than all the previous published correlations. The average absolute relative error is 28.1 with a standard deviation of 37%.

5.2.2.7 Liquid-side mass-transfer coefficient

There is practically nothing about the high-pressure liquid-side-mass transfer coefficient, k_L, in TBR in the open literature. The only paper published was that of Lara-Marquez *et al.* [57]. The values of $k_L a$ are determined by using the following chemical absorption systems in the slow reaction regime:

• O_2 absorption into aqueous sodium sulfite solutions with $CoSO_4$ as a catalyst,

• CO_2 absorption into DEA in pure ethyleneglycol.

The influence of the gas-mass flow-rate and gas density on $k_L a$ is shown in Figure 5.2–39. As expected, $k_L a$ is an increasing function of G for a given gas density and liquid flow-rate. It can also be seen that when G and L are constant, $k_L a$ decreases when gas density is increased. The effect of gas density on $k_L a$ can be appreciated otherwise by expressing them versus the gas superficial velocity, u_G. A plot of $k_L a$ versus u_G in Figure 5.2–40, for the same data as Figure 5.2–39, shows that increasing gas density leads to an increase of volumetric liquid-side mass-transfer coefficients. No correlations are given by the authors.

In a recent overview, Iliuta *et al.* [59] proposed a new correlation, which estimates the liquid Sherwood number with an average absolute relative error of 22.3% and a standard deviation of 25.3%.

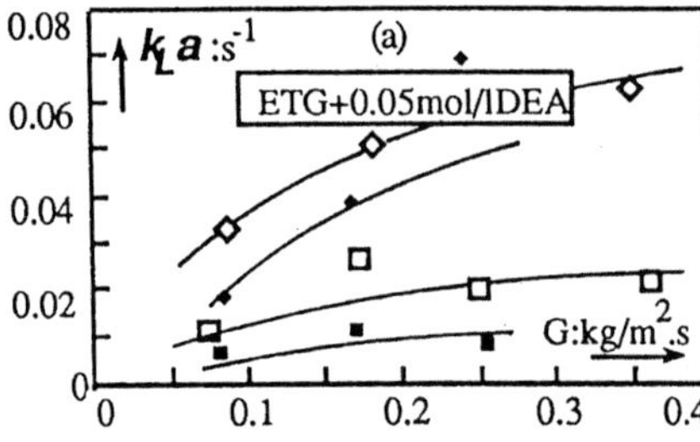
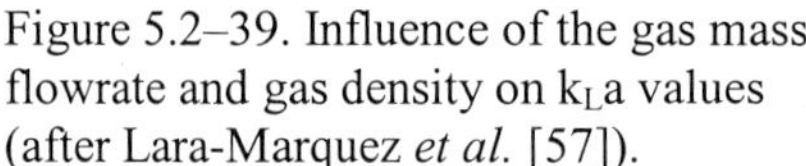

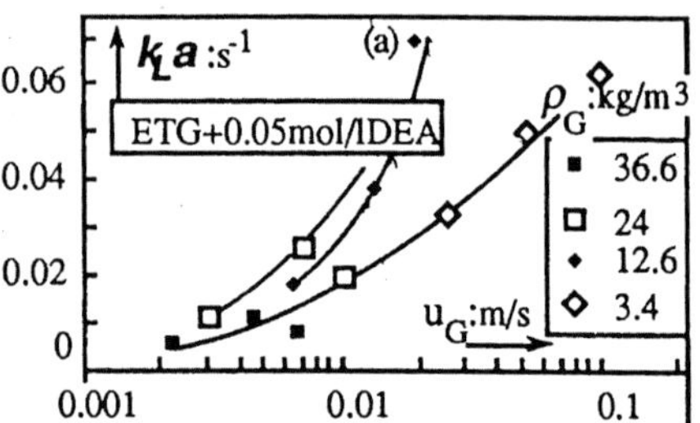

Figure 5.2–39. Influence of the gas mass flowrate and gas density on $k_L a$ values (after Lara-Marquez *et al.* [57]).

Figure 5.2–40. Influence of the gas superficial velocity and gas density on $k_L a$ values (after Lara-Marquez *et al.* [57]).

5.2.2.8 Gas-side mass-transfer coefficient

Regrettably, for our knowledge there is a lack of information in the open literature on values of the gas-side mass-transfer coefficient, k_G, in TBR operating at elevated pressures and especially on the effect of the gas density on it.

Note that at elevated pressures the gas side mass transfer coefficient might reduce because the diffusion coefficient in the gas phase is inversely proportional to the gas density. Then, the penetration theory predicted that k_G was inversely proportional to the square root of the total pressure. Disregarding this fact can lead to overestimated overall gas-liquid mass transfer coefficients at high pressure and high temperature, especially for a mixture of gases involving a highly soluble reactant with small and efficient catalyst packing, when the mass transfer resistances in the two phases become comparable.

5.2.3 Some examples of industrial applications of gas-liquid-solid fixed beds

The refining of petroleum and the production of environmentally acceptable fuels, petrochemical intermediates and manufactured finished products are an area propitious for the industrial penetration of some elevated pressure processes.

Thus the term hydroprocessing describes all the different processes in which hydrocarbons react with hydrogen. Hydrotreating is used to describe those hydroprocesses dealing predominantly with impurety removal from the hydrocarbon feedstock. Hydrocracking is used to describe the processes which accomplish a significant conversion of the hydrocarbon feedstock into lower boiling products.

The hydrotreating process removes objectionable materials from petroleum distillates by selectively reacting these materials with hydrogen in a catalytic fixed bed at elevated temperature and pressure. The concerned materials included sulphur (desulfurization), nitrogen (denitrification), metals (demetalation), olefin, and aromatics. Lighter fractions such as naphta are generally treated for subsequent processing in catalytic reforming units and the heavier distillates, ranging from jet fuel to heavy vacuum gas-oils, are treated to meet strict product-quality specifications or for use as feedstocks elsewhere in the refining.

For hydrodesulfurization operations, the catalyst is typically a cobalt-molybdenum on alumina support. In denitrification operations, the nickel-molybdenum is more common, and is also good for desulfurization. The metals-removal catalyst are designed specifically for the purpose of removing metals from the feed so that they do not affect the hydrotreating capability of the hydroprocessing catalyst. Normally, either of these catalysts provides adequate activity for the saturation of olefins and aromatics. The catalyst pellets are usually small, typically 0.8 to 1.3 mm in diameter, because the reaction kinetics are often diffusion-limited.

The conditions under which a hydroprocessing unit operates is a strong function of feedstock. Typical processing conditions are shown in Table 5.2–8 for a variety of hydroprocesses [60].

Table 5.2–8
Typical hydroprocessing operating conditions [60].

Process	Temperature (°C)	Pressure (bar)	LHSV	Hydrogen consumption Nm^3H_2/m^3 liquid
Naphta hydrotreating	260-343	14-35	2-5	1.7-8.3
Light oil hydrotreating	288-399	17-55	2-5	17-50
Heavy oil hydrotreating	343-427	138-207	1-3	67-177
Residuum hydrotreating	343-427	69-138	0.15-1	100-200
Residuum hydrocracking	399-427	138-207	0.2-1	200-267
Distillate hydrocracking	260-482	35-207	0.5-10	167-400

5.2.3.1 Hydrodesulfurization process

Light oils are invariably hydroprocessed in gas-liquid-solid catalyst trickle-bed reactors (TBR). In these reactors, both the hydrogen and hydrocarbon streams flow down through one or more catalyst beds. A typical schematic diagram is shown in Figure 5.2–41 as an example of hydrodesulfurization process [60, 61].

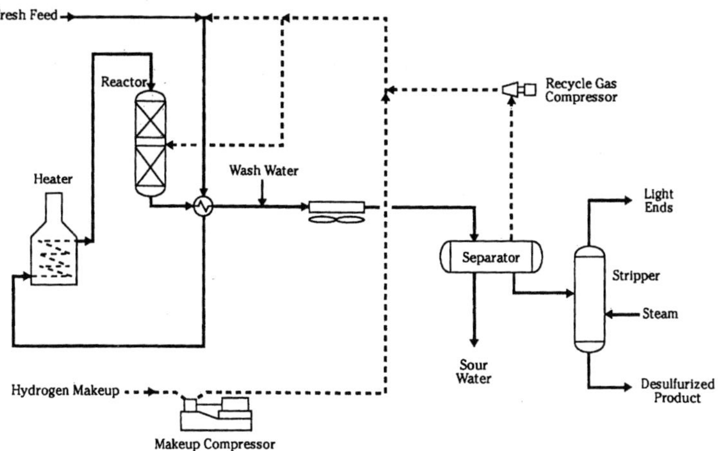

Figure 5.2–41. Typical hydrodesulfurization process flow (after Meyers [60]).

The feedstock mixed with recycle hydrogen is exchanged with the TBR effluent and then, heated to the reaction temperature (about 350°C) in a furnace. The combined feed then flows under a pressure of 26 bar through the TBR, which contains the catalyst that will accelerate the reaction. The reactor effluent is cooled in exchange with the feed, and then, in a series of coolers before being separated in a vapour-liquid separator at 20 bar. The vapour fraction is recompressed combined with fresh hydrogen, and returned to the reactor feed. The liquid part is fed to a fractionator, where it is stripped of light ends, hydrogen sulfide and ammonia on the top, and of desulfurized product at the bottom.

The main processes marketed with this typical generic flow scheme reduce for example the sulfur content of a naphta feed type from 200-1000 ppm to between 0.5-1 ppm or less [61].

5.2.3.2 Hydro-isomerization selective hydrogenation

One of the two main methods currently marketed to primarily extract or convert 1–butene contained in the feedstock C_4 cut is the hydro-isomerization of α-olefins to internal olefins. The isomerization of 1–butene to 2–butenes is recommended for two reasons [61]:

- for petrochemicals, it serves to simplify the schemes for subsequent separations of the components of C_4 cuts.

- in refining, the alkylates produced with isobutane lead to higher octane numbers than those resulting from the direct alkylation of the initial hydrocarbon feedstock.

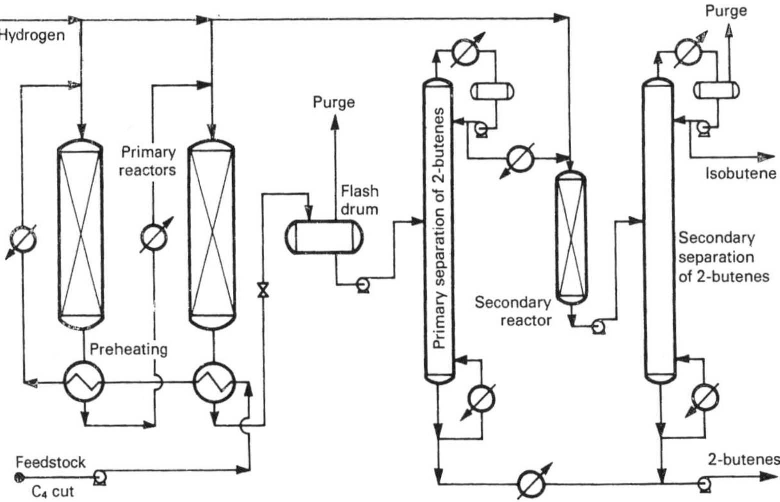

Figure 5.2–42. Hydro-isomerization of 1–butene to 2–butenes (after Chauvel *et al.* [61]).

In principle, the flowsheet of an industrial facility is similar to those of the different hydrotreating processes (Figure 5.2–42) [61]. The feedstock C_4 cut rid of water is pressurized to about 15 to 20 bar by pumping, injected with a hydrogen-rich gas and then, preheated by heat exchange with the reaction effluent and by steam. In a downflow stream, it then enters the reactor, which operates in a gas-liquid mixed phase with one or more catalyst beds (palladium, rhodium on inert alumina). After cooling, the isomerization products are flashed

Figure 5.2–43. Selective hydrogenation unit IFP (after Euzen *et al.* [62]).

298

to remove excess hydrogen gas.

A photography of a selective hydrogenation unit designed by the Institut Français du Pétrole is shown in Figure 5.2–43 [62].

5.2.3.3 Manufacture of cyclohexane

Cyclohexane is an essential intermediate for the synthesis of nylon–6,6. The purity level required for the use of cyclohexane, especially for its oxidation, is higher than 99%. This purity can be obtained by the benzene hydrogenation technique. The conversion is highly exothermic and is favored by low temperature, and high hydrogen partial pressure.

The UOP Hydrar process (currently HB Unibon) is one of the industrial process in which the reaction takes place in the liquid-phase (Figure 5.2–44).

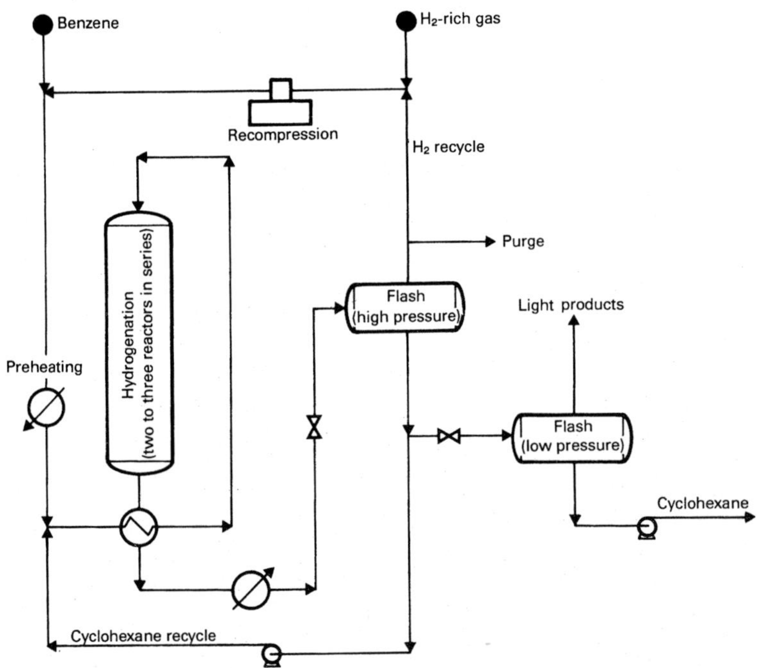

Figure 5.2–44. Cyclohexane production-UOP process (after Chauvel *et al.* [63]).

It uses a fixed bed of platinum-based catalyst promoted by a lithium salt which can tolerate sulfur contents up to 300 ppm in the benzene and whose LSHV in relation to liquid benzene is about 1.5 [63].

In this process, the benzene feed and a cyclohexane recycle, mixed with fresh and recycled hydrogen after compression to the required pressure are preheated and then introduced into a series of two-or three reactors, whose temperature are staged between 200 and 300°C, and which operate under about 30 bar.

The reactor effluent cooled by heat exchange with the feed is first flashed. After purging, to avoid the accumulation of inerts, the gases are recycled. Part of the liquid is also returned as a diluent to facilitate temperature control. The remainder undergoes a renewed flashing to remove hydrocarbon.

5.2.4 Conclusion

This review on fixed beds operating at elevated pressures describes some knowledge concerning hydrodynamics and mass transfer. Further development of the research should be promoted to improve information on all the effects of elevated pressures on fixed beds. The proposal of new research in pilot-scale and commercial units is necessary to verify the existing data and to extend the application range. A much work at high temperatures especially for gas-liquid and liquid-solid mass transfer, heat transfer, effective solid wetting in the presence of high pressure is essential.

This information would allow a better and safer design of fixed bed gas-liquid-solid reactors operating at high pressure.

References
1. Y.T. Shah, Gas-liquid-solid reactor design, McGraw Hill, New-York (1979).
2. C. Mozenski and E. Kucharski, Hydraulics of packed columns under elevated pressure, Inzyniera Chemiczna I Procesowa, 3 (1986) 373-384.
3. H. Krehenwinkel und H. Knapp, Experimentelle Untersuchungen der Fluiddynamik und der Stoffübertragung in berieselten Füllkörperkolonnen bei hohen Drücken, Chem. Ind. Tech., MS 1523/86, 58, 9 (1986).
4. H. Krehenwinkel and H. Knapp, Pressure drop and flooding in packed columns operating at high pressures, Chem. Eng. Technol., 10 (1987) 231-242.
5. B. Benadda, K. Kafoufi and M. Otterbein, Hydrodynamic behaviour of packed column under high pressure, High Pressure Chemical Engineering, Process Technology Proceedings 12, 679-685, by Ph. Rudolf von Rohr and Ch. Trepp, Elsevier, Amsterdam (1996).
6. R. Stockfleth and G. Brunner, Hydrodynamic behaviour of aqueous systems in countercurrent columns at high pressure, 279-283, GVC-Fachausschuss High Pressure Chemical Engineering, Karlsruhe, March 3-5 (1999).
7. P.C. Simoes, M.J. Cebola, R. Ruivo and M. Nunes da Ponte, Hydrodynamics in countercurrent packed columns at high pressure conditions, 259-262, GVC-Fachausschuss High Pressure Chemical Engineering, Karlsruhe, March 3-5 (1999).
8. B. Benadda, M. Otterbein, K. Kafoufi and M. Prost, Influence of pressure on the gas-liquid interfacial area a and the coefficient k_La in a countercurrent packed column, Chem. Engng. Processing, 35 (1996) 247-253.
9. B. Benadda, K. Kafoufi, P. Monkam and M. Otterbein, Hydrodynamics and mass transfer phenomena in countercurrent packed column at elevated pressures, Chem. Engng. Science, in press.
10. A. Gianetto and P.L. Silveston, Multiphase chemical reactors, Hemisphere Publishing Corporation, New-York (1986).
11. G.A. L'Homme, Chemical engineering of gas-liquid-solid catalyst reactions, Cebedoc, Liège (1979).
12. A. Gianetto and V. Specchia, Trickle-bed reactors: state of art and perspectives, Chem.

Engng. Science, 47 (1992) 3197-3213.

13. J.C. Charpentier, Advances in chemical engineering, Volume 11, Drew and Vermeulen, Academic Press, New-York (1981).

14. J.C. Charpentier and M. Favier, Some liquid holdup experimental data in trickle bed reactors for foaming and non-foaming hydrocarbons, AIChE Journal, 21 (1975) 1213-1218.

15. G. Tosun, A study of cocurrent downflow of non-foaming gas-liquid systems in packed beds, Ind. Eng. Chem. Proc. Des. Dev., 23 (1984) 29-35.

16. O. Baker, Simultaneous flow of oil and gas, Oil Gas Journal, 53 (1954) 185-195.

17. W.J.A. Wammes, Hydrodynamics in a cocurrent gas-liquid-trickle-bed reactor at elevated pressure, Ph. D. Thesis, Enschede (1990).

18. W. Hasseni, A. Laurent, N. Midoux and J.C. Charpentier, Régimes d'écoulement dans un réacteur catalytique à lit fixe arrosé fonctionnant sous pression (0.1-10 MPa) à co-courant de gaz et de liquide vers le bas, Entropie, 137 (1987) 127-133.

19. F. Larachi, A. Laurent, G. Wild and N. Midoux, Effet de la pression sur la transition ruisselant-pulsé dans les réacteurs catalytiques à lit fixe arrosé, The Can. J. Chem. Engng., 71 (1993) 319-321.

20. K.M. Ng, A model for flow regime transitions in cocurrent downflow trickle-bed reactors, AIChE Journal, 32 (1986) 115-122.

21. J.B. Wijffels, J. Verloop and F.J. Zuiderweg, Wetting of catalyst particles under trickle flow conditions, ACS Monograph Ser., 133 (1974) 151-163.

22. K.A. Grosser, R.G. Carbonell and S. Sunderasan, The transition to pulsing flow in trickle beds, AIChE Journal, 34 (1988) 1850-1856.

23. D.C. Dankworth, I.G. Kevrekidis and S. Sunderasan, Dynamics of pulsing flow in trickle beds, AIChE Journal, 36 (1990) 605-621.

24. D.C. Dankworth and S. Sunderasan, Time dependent vertical gas-liquid flow in packed beds, Chem. Engng. Science, 47 (1992) 337-346.

25. S. Whitacker, Advances in theory of fluid motion in porous media, Ind. Eng. Chem., 61, 12 (1969) 14-28.

26. A.E. Saez and R.G. Carbonnel, Hydrodynamic parameters for gas-liquid cocurrent flow in packed beds, AIChE Journal, 31 (1985) 52-62.

27. W.J.A. Wammes, S.J. Mechielsen and K.R. Westerterp, The transition between trickle flow and pulse flow in a cocurrent gas-liquid trickle-bed reactor at elevated pressures, Chem. Engng. Science, 45 (1990) 3149-3158.

28. G. Wild, F. Larachi and A. Laurent, The hydrodynamics characteristics of cocurrent downflow and cocurrent upflow gas-liquid-solid catalytic fixed bed reactors: the effect of pressure, Revue de l'Institut Français du Pétrole, 46 (1991) 467-490.

29. R.A. Holub, Hydrodynamics of trickle bed reactors, Ph. D. Thesis, Washington University, Saint-Louis (1990).

30. R.A. Holub, M.P. Dudukovic and P.A. Ramachandran, A phenomenological model for pressure drop, liquid hold-up and flow regimes transition in gas-liquid trickle flow, Chem. Engng. Science, 47 (1992) 2343-2348.

31. W.J.A. Wammes and K.R. Westerterp, The influence of the reactor pressure on the hydrodynamics in a cocurrent gas-liquid trickle-bed reactor, Chem. Engng. Science, 45 (1990) 2247-2254.

32. W.J.A. Wammes, J. Middelkamp, W.J. Huisman, C.M. Debaas and K.R. Westerterp, Hydrodynamics in a cocurrent gas-liquid trickle bed at elevated pressures, Part 1: gas-

liquid interfacial areas, AIChE Journal, 37, 12 (1991) 1849-1854.

33. W.J.A. Wammes, J. Middekamp, W.J. Huisman, C.M. Debaas and K.R. Westerterp, Hydrodynamics in a cocurrent gas-liquid trickle bed at elevated pressures, Part 2: liquid hold-up, pressure drop, flow regimes, AIChE Journal, 37, 12 (1991) 1855-1862.

34. W.J.A. Wammes, S.J. Mechielsen and K.R. Westerterp, The influence of pressure on the liquid hold-up in a cocurrent trickle-bed reactor operating at low gas velocities, Chem. Engng. Science, 46 (1991) 409-417.

35. W.J.A. Wammes and K.R. Westerterp, Hydrodynamics in a pressurized cocurrent gas-liquid trickle-bed reactor, Chem. Engng. Technol., 14 (1991) 406-413.

36. F. Larachi, Les réacteurs triphasiques à lit fixe à écoulement à co-courant vers le bas et vers le haut de gaz et de liquide. Etude de l'influence de la pression sur l'hydrodynamique et le transfert de matière gaz-liquide, Ph. D. Thesis, Institut National Polytechnique de Lorraine, Nancy (1991).

37. F. Larachi, A. Laurent, N. Midoux and G. Wild, Experimental study of a trickle-bed reactor operating at high pressure: two-phase pressure drop and liquid saturation, Chem. Engng. Science, 46 (1991) 1233-1246.

38. F. Larachi, A. Laurent, N. Midoux and G. Wild, Liquid saturation data in trickle beds operating under elevated pressure, AIChE Journal, 37 (1991) 1109-1112.

39. F. Larachi, A. Laurent, G. Wild and N. Midoux, Some experimental liquid saturation results in fixed-bed reactors operated under elevated pressure in cocurrent upflow and downflow of the gas and the liquid, Ind. Engng. Chem. Res., 30 (1991) 2404-2410.

40. M.H. Al-Dahhan and M.P. Dudukovic, Pressure drop and liquid hold-up in high pressure trickle-bed reactors, Chem. Engng. Science, 49 (1994) 5681-5698.

41. M.H. Al-Dahhan, F. Larachi, M.P. Dudukovic and A. Laurent, High-pressure trickle-bed reactors: a review, Ind. Engng. Chem. Res., 36 (1997) 3292-3314.

42. M.H. Al-Dahhan, M.R. Khadilkar, Y. Wu and M.P. Dudukovic, Prediction of pressure drop and liquid hold-up in high pressure trickle-bed reactors, Ind. Engng. Chem. Res., 37 (1998) 793-798.

43. I. Iliuta, F. Larachi and B.P.A. Grandjean, Pressure drop and liquid hold-up in trickle flow reactors: improved Ergun constants and slip correlations for the slit model, Ind. Engng. Chem. Res., 37 (1998) 4542-4550.

44. I. Iliuta and F. Larachi, The generalized slit model: pressure gradient, liquid hold-up and wetting efficiency in gas-liquid trickle flow, Chem. Engng. Science, 54 (1999) 5039-5045.

45. A. Attou, C. Boyer and G. Ferschneider, Modelling of the hydrodynamics of the cocurrent gas-liquid trickle flow through a trickle-bed reactor, Chem. Engng. Science, 54 (1999) 785-802.

46. A. Attou and G. Ferschneider, A two-fluid model for flow regime transition in gas-liquid trickle-bed reactors, Chem. Engng. Science, 54 (1999) 5031-5037.

47. I. Iliuta, A. Ortiz-Arroys, F. Larachi, B.P.A. Grandjean and G. Wild, Hydrodynamics and mass transfer in trickle-bed reactors: an overview, Chem. Engng. Science, 54 (1999) 5329-5337.

48. F. Larachi, I. Iliuta, M.A. Al-Dahhan and M.P. Dudukovic, Discriminating trickle flow hydrodynamic models - some recommendations, Ind. Engng. Chem. Res., 39 (2000) 554-556.

49. M.J. Elmann, N. Midoux, A. Laurent and J.C. Charpentier, A new, improved pressure drop correlation for trickle-bed reactors, Chem. Engng. Science, 43 (1988) 2201-2206.

50. M.J. Elmann, N. Midoux, G. Wild, A. Laurent and J.C. Charpentier, A new, improved liquid hold-up correlation for trickle-bed reactors, Chem. Engng. Science, 45 (1990) 1677-1684.

51. A.K. Saroha and K.D.P. Nigam, Trickle-bed reactors, Reviews in Chemical Engineering, 12 (1996) 207-347.

52. W.P.M. Van Swaaij, Residence time distributions in Raschig ring columns at trickle flow, Ph. D. Thesis, Eindhoven University (1967).

53. J.C. Charpentier, C. Prost, W.P.M. Van Swaaij and P. Le Goff, Etude de la rétention de liquide dans une colonne à garnissage arrosée à co-courant et à contre-courant de gaz et de liquide, Génie Chimique, Paris, 99 (1968) 803-826.

54. V. Specchia and G. Baldi, Pressure drop and liquid hold-up for two-phase cocurrent flow in packed beds, Chem. Engng. Science, 32 (1977) 515-523.

55. F. Larachi, A. Laurent, G. Wild and N. Midoux, Pressure effects on gas-liquid interfacial areas in cocurrent trickle-flow reactors, Chem. Engng. Science, 47 (1992) 2325-2330.

56. N. Midoux, B.I. Morsi, M. Purwasasmita, A. Laurent and J.C. Charpentier, Interfacial area and liquid-side mass transfer coefficient in trickle-bed reactors operating with organic liquids, Chem. Engng. Science, 39 (1984) 781-794.

57. A. Lara-Marquez, F. Larachi, G. Wild and A. Laurent, Mass transfer characteristics of fixed beds with cocurrent upflow and downflow. A special reference to the effect of pressure, Chem. Engng. Science, 47 (1992) 3485-3492.

58. F. Larachi, M. Cassanello and A. Laurent, Gas-liquid interfacial mass transfer in trickle-bed reactors at elevated pressures, Ind. Engng. Chem. Res., 37 (1998) 718-734.

59. I. Iliuta, F. Larachi, B.P.A. Grandjean and G. Wild, Gas-liquid interfacial mass transfer in trickle-bed reactors: state-of-the-art correlations, Chem. Engng. Science, 54 (1999) 5633-5645.

60. R.A. Meyers, Handbook of petroleum refining processes, Second Edition, McGraw Hill, New-York (1997).

61. A. Chauvel and G. Lefebvre, Petrochemical processes, Volume 1, Technip, Paris (1989).

62. J.P. Euzen, P. Trambouze et J.P. Wauquier, Méthodologie pour l'extrapolation des procédés chimiques, Technip, Paris (1993).

63. A. Chauvel and G. Lefebvre, Petrochemical processes, Volume 2, Technip and Gulf Publishing Company, Paris and Houston (1989).

5.3 Slurry catalytic reactors

5.3.1 Introduction

High pressure catalytic processes are developed and carried out in both preformed and powdered catalysts. Preformed catalyst are useful for fixed bed operation. Preformed catalyst pellets, are used as packing in multiphase trickling flow reactors. Trickling flow reactors have been described in detail in another part of this book (see Laurent). In this section we deal with slurry catalytic reactors, where the catalyst is used in powdered form.

The slurry catalytic reactor is particularly relevant during the development of both the catalyst and the process itself. Further, the operating pressure, kinetics, mass transfer, and superficial velocities, etc., are studied and selected most often in a mechanically stirred slurry unit. For example, Inga [1] studied the high pressure Fischer-Tropsch process by scaling-up a 2-L slurry reactor results, to a 2-m high bubble slurry column unit. Generally laboratory slurry catalytic reactors are a must in the development of a high pressure processes.

In high pressure work, slurry reactors are used when a solid catalyst is suspended in a liquid or supercritical fluid (either reactant or inert) and the second reactant is a high pressure gas or also a supercritical fluid. The slurry catalytic reactor will be used in the laboratory to try different catalyst batches or alternatives. Or to measure the reaction rate under high rotational speeds for assessing intrinsic kinetics. Or even it can be used at different catalyst loadings to assess mass transfer resistances. It can also be used in the laboratory to check the deactivating behaviour.

In industrial high pressure reactions, slurry catalytic reactors with powdered catalyst are found in the following processes: medium and high pressure hydrogenation of vegetable and mineral oils (over Ni and Pd catalysts), oxidation of liquid pollutants contained in a slurry, polymerization of olefins (ethylene, propylene) in a catalyst slurry using a lower alkane as a suspending agent (see Phillips process or metallocene process, also see Luft section 9.5). Except for the case of wet air oxidation, the above processes are medium pressure processes.

Novel high pressure with near critical solvents have been developed by Take and co-workers for the hardening of fats and oils in supercritical carbon dioxide [2], Härröd and co-workers for fatty acid ester high pressure hydrogenation (see section 9.3) and also in the pioneering work of Poliakoff and co-workers on chemical organic syntheses at high pressure (section 9.1).

Slurry catalytic reactors are the choice for a high pressure process whenever:
1 A wide range of possible operating pressures (5-150 bar)
2 Absoption of reaction heat is required, such that isothermal conditions can be approached.
3 A high heat transfer flux per unit reactor volume is needed.
4 Low pressure drop across reactor is desired.
5 Wetting of the external catalyst surface is desired as to delay catalyst fouling.

5.3.2 Processes carried out in slurry catalytic reactors

In industrial practice, three-phase catalytic reactors are often used, with gases like such as H_2, H_2O, NH_3 or O_2 as reactants. The process can be classified on the basis of these gases as hydrogenation, hydration, amination, oxidation, etc [3]. Among these processes, hydrogenation is by far the most important multiphase catalytic reaction. Recently, liquid-phase methanol synthesis and the Fischer-Tropsch process were commercialized respectively by Air Products (USA) and Sasol (South Africa) [4], BSCR were used. Some examples of the application of MSSR and BSCR are given in Table 5.3-1 (see also chapter 5.2 by Laurent).

Table 5.3-1
Examples of systems using slurry reactors [3,5]

Feedstock	Product	Catalyst	T (K)	P (MPa)
		MSSR		
Acetone	Methyl isobutyl ketone	Pd on acidic exchange zeolite or Zr phosphate	393 to 433	2 5
Dinitro-toluene	Diaminotoluene	Raney Ni or Pd/C	373	>5
Fatty ester or fatty acids	Fatty alcohols	Copper chromite	473 to 573	9.7 10
Toluene	Methylcyclohexane	Raney Ni	404 to 445	10
Propylene (in liquid hexane or heptane)	Polypropylene	Chromium oxide	343	3.4
		BSCR		
CO/H_2	Fischer-Tropsch synthesis	Cobalt, nickel, iron	531	1.2
CO/H_2	Methanol	$Cu/ZnO/Al_2O_3$	500	7
Benzene	Cyclohexane	Raney Ni Noble metal	473 to 500	5
		Pt/Li Ni	473 to 573	3
Glucose	Sorbitol	Raney Ni	393 to 473	3 to 5

In the following we will describe mechanically agitated slurry reactors and slurry bubble columns.

Bubble slurry column reactors (BSCR) and mechanically stirred slurry reactors (MSSR) are particular types of slurry catalytic reactors (Fig. 5.3-1), where the fine particles of solid catalyst are suspended in the liquid phase by a gas dispersed in the form of bubbles or by the agitator. The mixing of the slurry phase (solid and liquid) is also due to the gas flow. BSCR may be operated in batch or continuous modes. In contrast, MSSR are operated batchwise with gas recirculation.

The main disadvantage of slurry reactors when compared with fixed beds is the considerable amount of backmixing, which reduces the performance of these reactors by reducing the conversion. On the other hand, high liquid hold-up may cause the liquid-phase diffusional resistance to the gaseous reactant to be an important factor affecting the global reaction rate. Some advantages are evident, for example, no moving part is used to mix and suspend the catalyst in BSCR, while in MSSR a mechanically driven stirrer induces the agitation of the slurry phase.

The MSSR presents the same advantages as BSCR, such as high efficiency of heat-and mass-transfer and minimal intraparticle diffusional resistance, and are convenient for use in batch processes. For these reasons, the slurry-agitated reactors are also suitable for kinetic studies in the laboratory. Some of their major drawbacks are:

- large power requirement for mechanical agitation,

- significant back-mixing,
- difficulties of separation of catalyst in continuous operations.

If we compare the sparged (BSCR) and mechanically stirred (MSSR) reactors, we can stress that the slurry bubble column may be operated in a semi-batch mode, does not use any moving parts, and requires minimal maintenance, smaller floor space and has lower power consumption.

The main disadvantages of BSCR compared to MSSR are the rapid reduction in specific interfacial area with height for height/diameter ratios above 10 [6]. Other aspects of the two types of reactors are compared in Table 5.3-2.

Table 5.3-2
Comparison between mechanically stirred reactors and slurry bubble column reactors [7]

MSSR	BSCR
Maintenance	
More	Less
Initial cost	
Slightly lower first cost	Slightly higher first cost
Axial mixing	
Approach well mixing	Gas in plug-flow
Heat transfer	
Higher and better characteristics	Lower characteristics
Sealing problems	
Problem for very high pressure	Advantageous
Bubble size (interfacial area)	
Better control	Bad control
Height of reactor	
Multiple impellers on long shaft are required	Well adapted
Reactor volume	
$<30 \text{ m}^3$	$<100 \text{ m}^3$
Operating pressure	
<40 bar	<100 bar

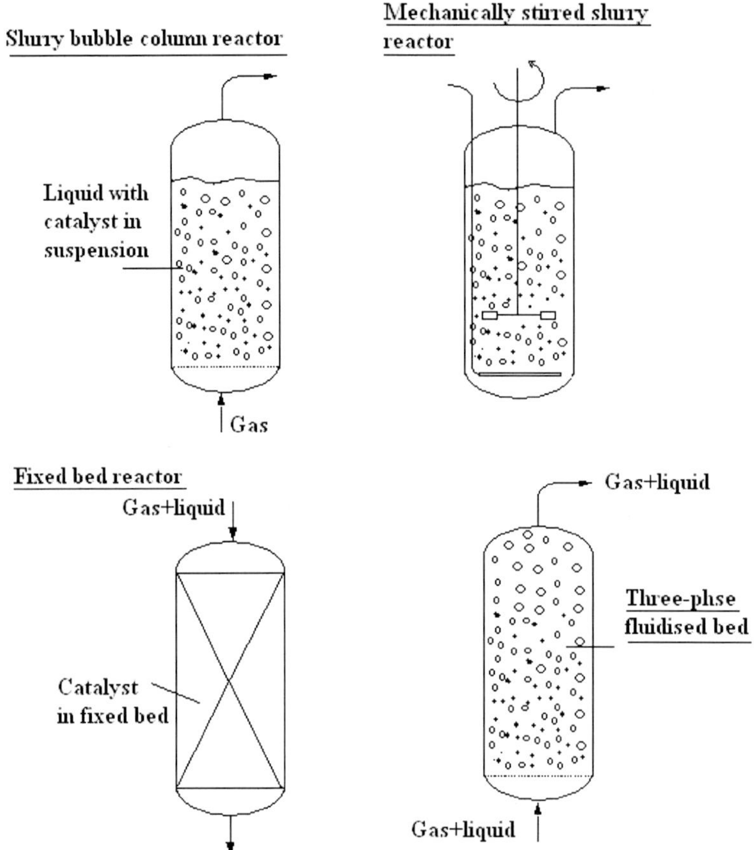

Figure 5.3-1. Schematic representation of some types of three-phase reactors [8].

5.3.3 Process design issues

We review here the decisions to be made in choosing the relevant variables in the design of a slurry catalytic reactor. A fundamental selection refers to the raw materials. These include the catalyst and gas sources and their qualities.

Some decisions regarding operational variables are simply taken from the catalyst manufacturer (for example, catalyst loading) while others, such as pressure-level, usually involve a selectivity study and simple cost and safety considerations. During the experimental stage, attention will be paid to the following variables:

Agitation speed requirements (MSSR) or sparger design (both MSSR and BSCR);
Reaction temperature;
Pressure;
Catalyst loading;
Size of catalyst powder;

Superficial gas velocity.

These factors are introduced in the experimental plan at the bench-scale before the pilot-
-plant stage. On the bench-scale, glass or steel laboratory reactors of about 1 to 2 L will be
used for MSSR and a 30 cm diameter and 2 m height for BSCR.
Once the factors given before have been determined, the following design must be addressed
in order to make a final design for the reactor. These refer to specifying the elements of the
reactor with the auxiliaries for the catalytic reaction. Table 5.3-3 summarises these issues.

Table 5.3-3
Slurry catalytic reactor design features

BSCR	MSSR
Gas sparger design	Gas sparger design/ hollow shaft arrangement
Number of holes in sparger ring	Number of holes in sparger ring.
Minimum speeds of suspension, re--suspension and flooding	Impeller design, number of impellers, rotation speed.
	Selection of agitator drive and speed reducer
	Mechanical assessment of shaft. Vibrations
Mass transfer at bubble/liquid, and liquid/solid interfaces	Mass transfer at bubble/liquid, and liquid/solid interfaces
Heat transfer design	Heat transfer design

In parallel, an experimental procedure has to be run in the pilot plant to produce a set of
experimental conditions to be met by the full-scale reactor. The criteria for the scale-up for
MSSR will be discussed later.

5.3.4 Interface mass transfer and kinetics

A key feature of catalytic slurry reactors is that the particles are small (~0.1 mm), so it is
relatively easy to promote suspension by the mechanical action of the impeller. Moreover,
because of their small size they travel together with the liquid, and therefore a significant
mass transfer resistance develops at the liquid/solid interface that cannot be removed
completely with the standard impellers. Also, because of the liquids large Prandtl number, the
catalyst and the liquid are at the same temperature, so hot spots do not occur in multiphase
slurry reactors.

In general, the transport of dissolved gas from bubbles to liquid is limited by the available
area, so mass-transport resistance in the gas/liquid interface may be large. It should be noted
that for bubble sizes in the range of 1 to 3 mm and of catalyst about 0.1 mm, the mass transfer
area at the liquid/solid interface may be as large as 5 to 20 times that of the bubbles
(depending on catalyst loading). So the gas/liquid mass transfer area may become limiting. In
general, for estimation purposes in stirred slurry reactors, the rising velocity of a bubble is
about 20 to 22 cm/s and its diameter about 1 to 3 mm.

In the reactor (either MSSR or BSCR) it is assumed that the liquid and the solid are well
mixed. In this case the following processes will take place in series.
1. Mass transfer of gas from the bubble to the bubble/liquid interface.
2. Mass transfer from the stagnant liquid film of the bubble to the bulk of the liquid.

3. Mass transfer of dissolved gas from the bulk liquid to the outer surface of the solid catalyst.
4. Reaction on the catalyst, and diffusion of products to the liquid phase.
5. In reversible reactions, or when the concentrations of intermediates affect the product distribution (as in edible-oil hydrogenation [9]), mass transfer of the products from the particle to the liquid phase may strongly affect the selectivity.

Assume that reaction is first-order in the gas. In the steady-state, the above processes all will occur at the same rate, so in the case of bubbling pure gas we can write:

$$R = k_L a_g \left(\frac{C_g}{H} - C_L \right) \tag{5.3-1}$$

$$R = k_C a_C \left(C_L - C_S \right) \tag{5.3-2}$$

$$R = \eta k a_C C_S \tag{5.3-3}$$

$$C_{iL} = \frac{C_g}{H} \tag{5.3-4}$$

In Eqn. 5.3-1, η is the effectiveness factor of the catalyst with respect to the dissolved gaseous reactant and the temperature of the outer surface. The rate of reaction within the catalyst pores is comprised in η. R is the reaction rate expressed in moles of gaseous reactant, A, per unit of bubble-free liquid, per unit of time. Reaction is irreversible. In equation (1) it has not been assumed that the gas is pure gas A, its concentration in the bubbles being C_g. Also, Henry's law for the gas is assumed and written as in Eqn. 5.3-4. Using Henry's law, Eqn. 5.3-4, the intermediate concentrations (C_S, C_L) can be eliminated using the above system of equations. This provides an expression of the global rate in terms of an apparent constant, k_0, that contains the various kinetic and mass transfer steps. Therefore, the observed rate can be written as:

$$R = k_0 a_C C_g \tag{5.3-5}$$

And the overall resistance will be the sum of three resistances in series, the bubbles, the solid, and the reaction as:

$$\frac{1}{k_0} = \frac{a_C H}{a_g k_L} + H \left(\frac{1}{\eta k} + \frac{1}{k_c} \right) \tag{5.3-6}$$

where the specific surface areas can be calculated in terms of the catalyst loading, m, (kg/m^3 bubble-free liquid) and the bubble hold-up, ε_g, as:

$$a_C = \frac{6m}{\rho_s d_P} \qquad (5.3\text{-}7)$$

$$a_S = \frac{6\varepsilon_g}{d_B} \qquad (5.3\text{-}8)$$

Where ρ_p is the catalyst density (kg/m³), d_p is the catalyst particle size (in m) and d_B is the bubble size (in m). To calculate the latter, the bubble hold-up, ε_g, is required. Both are usually estimated from the working regime of the impeller: we will discuss this later. The above equations assume that catalyst particles and bubbles are spherical.

The estimation of the gas hold-up is an important practical variable, since it provides increase in the liquid level of the tank caused by the gas sparger, that is, the minimum foaming height that will form on top of the reactor.

From the above equations the following conclusions can be drawn.

(1) At low catalyst loadings, (1 to 2 %) the solid/liquid interface area will be small, and so it becomes limiting, hence k_o depends only on k_c and on the intrinsic rate, $k\,\eta$. Then the global rate is proportional to the catalyst loading, m, as can be seen from Eqn. 5.3-5
(2) At high loadings (30 to 40%), the observed rate is independent of catalyst loading, since gas transport from the bubbles becomes limiting. Catalyst particles just compete for the little gas that is being transferred from bubbles.
(3) The same occurs with the stirring speed. A gas-starved condition can be achieved by a very low stirring speed. This may be desirable in some hydrogenations where it is necessary not to allow too much hydrogen to be transferred into the liquid (for selectivity reasons). Then, if stirring is increased, the reaction rate increases up to a point where it does not increase further. Since the gas/liquid resistance has already been lowered, the reaction rate becomes the maximum possible.
(4) Another effect is due to the liquid solute being reacted. As has been assumed in the above equations, the liquid substrate does not affect the rate, as long as the dissolved gas is limiting. However, during the last stages of a hydrogenation batch, for example, the concentration of reactant liquid may become limiting, making the chemical rate become second-order with respect to both liquid and gas.

In a liquid phase with mechanical agitation and reactant-gas bubbling, the operation is characterized by the following factors.

1) The reaction rate is proportional to the partial pressure of the gas.
2) Reaction rate is directly proportional to catalyst-loading at low loadings. However, it becomes independent of catalyst concentration at high loadings.
3) The effect holds for the stirring speed. At low rpm, the rate is proportional to stirring speed. At a certain value of the rpm, no further increase in rate is observed.
4) The same holds for the substrate concentration.
5) In most regimes the plot of *1/R vs. 1/m* is a straight line. This can be verified by using Eqn. 5.3-5 with Eqn. 5.3-6 and taking the reciprocal of Eqn. 5.3-5. The plot, as shown in Fig.

310

5.3-1, is useful for determining the gas/liquid mass transfer coefficient as a function of the stirring speed from the ordinate (see Eqn. 5.3-9. If knowledge about k_c is available from stirring correlations, the plot provides also the intrinsic kinetic constant from the slope, as is seen from Eqn. 5.3-9. Figure 5.3-2 depicts the data of Smith [10] for the oxidation of SO_2 in a slurry of carbon particles, for two particle sizes. For very large loading, the slope becomes essentially horizontal.

$$\frac{C_g}{H \times R} = \frac{1}{k_L a_g} + \frac{d_p \rho_p}{6m}\left(\frac{1}{k_c} + \frac{1}{\eta k}\right) \tag{5.3-9}$$

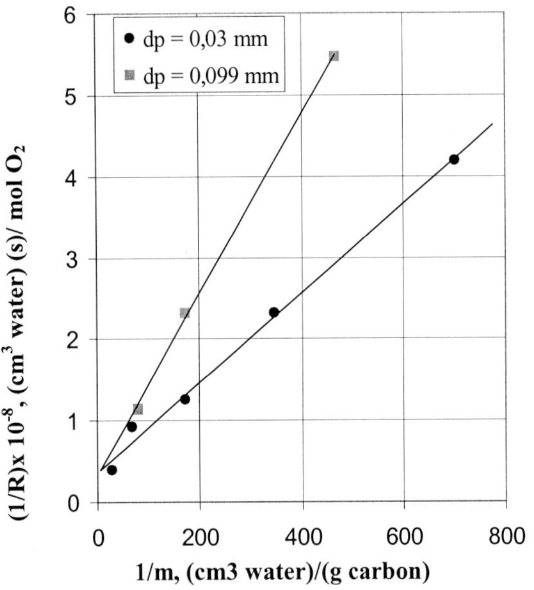

Figure 5.3-2. 1/R *vs.* 1/m, Smith's data [10] for the oxidation of SO_2 in a slurry of carbon particles, for two particle sizes.

5.3.5 Mechanically agitated tanks and three-phase sparged reactors

Design of the agitator and the sparging system for MSSR

The stirring system for slurry reactors consists of one or several turbines held on the same shaft. Usually the gas is fed below the bottom turbine which is of the Rushton-type, to break the large bubbles. The sparging device is usually a perforated ring, a cross with four perforated arms, or a special sparger. For small reactors (less than 2 to 3 m^3) a proprietary type of sparger like that show in Fig. 5.3-3 is used. This consists of a shallow shaft that sucks gas (H_2) from the top of the reactor that is pumped and dispersed at the bottom through the holes. Figure 5.3-4 summarizes the most common methods for feeding the gas. As will be shown, one important function of the sparger is to provide enough energy to give a sufficient gas hold-up and small bubbles (see Eqn. 5.3-8), and thus a high interface surface area.

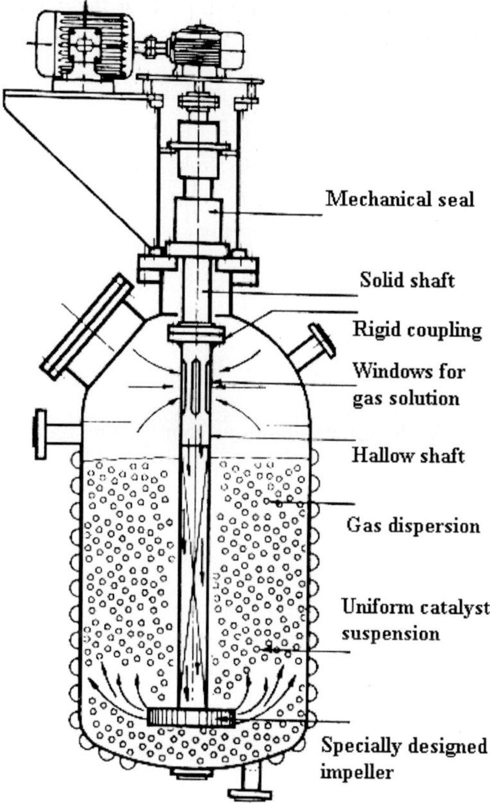

Figure 5.3-3. Gas-induction reactor (with permission of OMEGA-KEMIX).

We give here the design of the sparger ring in some detail to show how the variables are related. The sparger ring has a diameter of about 80% of that of the agitator, and is located 2 to 5 cm underneath. In the design, the following aspects must be considered:
- Design of the sparger ring. Number of holes;
- Number of turbines;
- Usual gas flow-rates;
- Minimum rotational speed of impeller (flooding, solids suspension);
- Mechanical power drawn by agitator;
- Bubble diameter;
- Bubble hold-up, interfacial gas/liquid mass transfer area;
- Mass transfer coefficients (G/L,L/S);
- Heat-transfer coefficients. Slurry-to-walls, slurry-to-tubes coefficients.

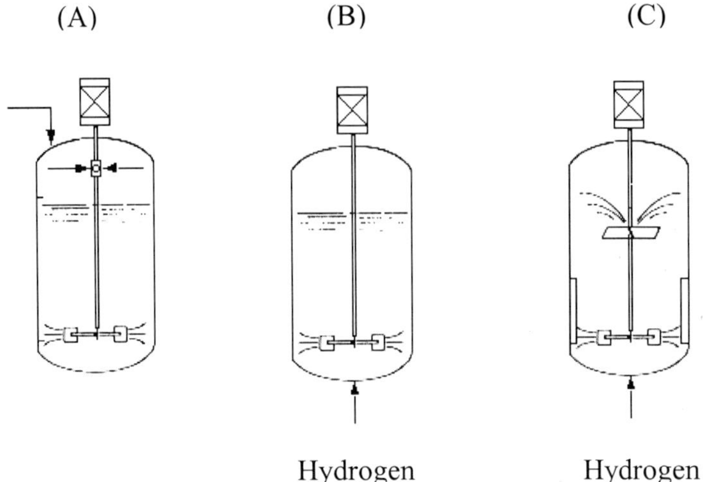

Figure 5.3-4. Hydrogen addition methods: (A), hydrogen addition from the top, recirculation through hollow shaft; (B), hydrogen addition from the bottom; (C), hydrogen addition using a turbine in the upper part of the shaft.

Design of sparger ring. Number of holes
The method given by Treybal is used [11]. Holes of 3 to 6 mm will be considered, depending on the diameter of the ring. It is taken that the Reynolds number across the hole is at least 10000, and a mass flow-rate for each hole, w/n_0, in which w is mass-flow of gas in kg/s. Hence the number of holes, n_0, is found from the following equation:

$$\frac{4w}{\pi d_0 \mu_g n_0} > 10000 \tag{5.3-10}$$

Where:
w = gas mass flow rate in kg/s
n_0 = number of holes
d_0 = hole diameter, in m
μ_g = gas viscosity, kg/(m s)

In these conditions, the bubble that exits a hole has the following diameter (in m):

$$d_B = 0.0071 Re_0^{-0.05} \tag{5.3-11}$$

Therefore, the spacing between two holes on the ring will be made larger than this diameter to avoid early coalescence. The bubble given by Eqn. 5.3-11 is not the final one, because there is the shear stress imposed by the stirrer. The final design diameter will be calculated later.

Number of impellers on the shaft
The following rule will be applied. If the liquid level at rest is larger than 1.25 times the tank's inner diameter, then two impellers are to be used. The upper one is located 1/3 of the tank's diameter from the surface level at rest, and the lower one at a distance from the bottom head equal to 1/6 of the tank's diameter.
For heat-and mass transfer it is very important to have a high superficial velocity for the gas. The maximum attainable value in a stirred tank is 0.08 $m^3/(m^2.s)$. A suitable design value would be 0.06 m/s. Therefore, if the cross-section of the tank must be designed, this value will be chosen, taking into consideration the pressure and temperature conditions of the gas in the tank for correcting gas flows.

Minimum stirrer speed
The speed of the stirrer should chosen with the two requirements of overcoming flooding, and suspending particles.
For the first case the following non-dimensional equation for the minimum speed, given by Westerterp *et al.* [12], is:

$$\frac{N_i D}{\left(\sigma g / \rho_L\right)^{0.25}} = 1.22 + 1.25 \frac{T}{D} \tag{5.3-12}$$

Where:
N_i = minimum speed, rps
D = impeller diameter, m
$g = 9.81$ m/s²
σ = surface tension, N/m (air/water = 0.072 N/m)
ρ_L = liquid density (kg/m³)
T = reactor diameter, m
There is a minimum speed for particle suspension that was originally derived by Calderbank, and is given by the following dimensional equation (see also the present, more detailed data and theory of Nienow [13]).

$$N_{min} = \frac{\beta d_P^{0.2} \mu_L^{0.1} g^{0.45} \left(\rho_P - \rho_L\right)^{0.45} m^{0.13}}{\rho_L^{0.55} D^{0.85}} \tag{5.3-13}$$

where:

$$\beta = 2\left(\frac{T}{D}\right)^{1.33}$$

N_{min} = Minimum speed, rps
d_p = catalyst diameter, cm
μ_L = liquid viscosity, g/ (cm) (s)
$\rho_p - \rho_s$ = density difference solid-liquid, g/cm³
m = catalyst concentration, in % mass
D = impeller diameter, cm

The stirring speed chosen should be higher than these two speeds, N_i and N_{min}. If a special suction impeller is designed, there is also a minimum speed for bubble suction from the upper head space. van Dierendonck and Nelemans [14] have studied in detail this type of aeration. It draws an additional power for sparging as shown in Fig. 5.3-5 for the case of hydrogen.

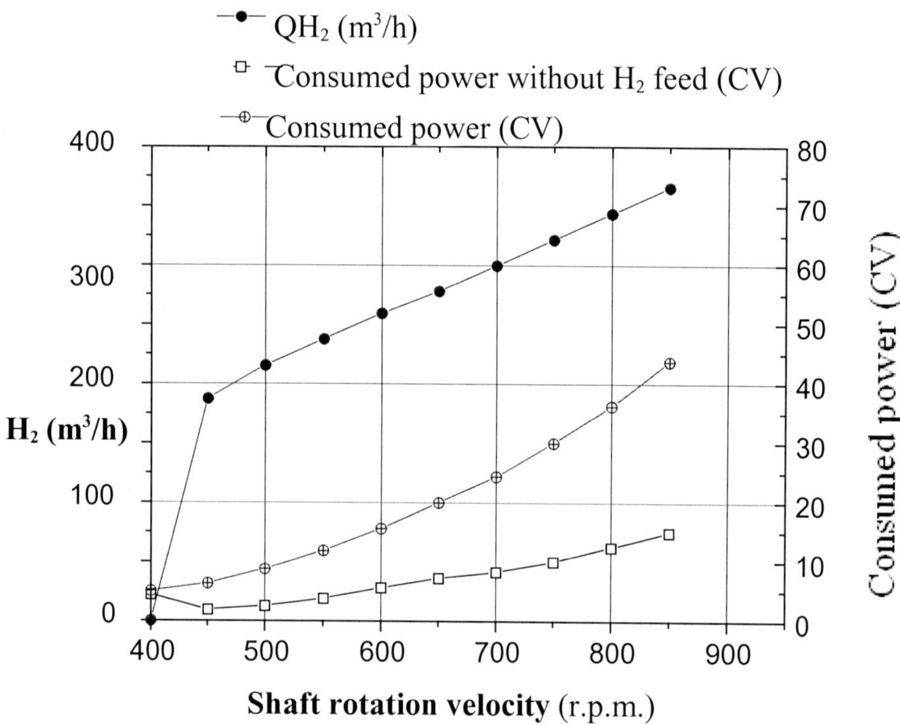

Figure 5.3-5. Agitator rotation velocity *vs.* hydrogen flow and power consumption [15].

Agitator power
When gas is bubbled below the impeller, the power drawn by the agitator decreases relative to the non-gassed liquid. This is because the gas/liquid mixture creates a zone of lower density below the stirrer, that strongly reduces the power consumption. From the point of view of design, however, it should be noted that the drive-motor- should be designed for the most demanding case, that is, the ungassed condition.

The gassed condition is important in mass transfer calculations. In general, the ungassed- and gassed-power are related by the equation $P_G = \psi \, P$. The value of P is calculated from the power correlation for Rushton turbines [16]. The correction factor, ψ, is calculated from the following equations:

$$\psi = 1 - 1.22 \, (Q_G / ND^3) \qquad \text{for } (Q_G / ND^3) < 0.037$$

$$\psi = 0.62 - 1.85 \, (Q_G / ND^3) \qquad \text{for } (Q_G / ND^3) > 0.037$$

Where Q_G is the volumetric gas flow-rate under the reactor conditions. In the case where two co-axial impellers are used, the installed power should be doubled.

Bubble diameter, bubble hold-up and interfacial area

The impeller Reynolds number is used:

$$Re = \frac{ND^2 \rho_L}{\mu_L} \qquad (5.3\text{-}14)$$

Surface aeration is a term that describes the suction action of stirrer from the gas above it. Surface action does not occur in a stirrer if $Re^{0.7} \, (N \, D / u_G)^{0.3} < 30\,000$. In such a case, the interfacial area in the gas slurry mixture depends on the power input per unit volume due to the stirrer and on the gas superficial velocity, as:

$$a_g = 1.44 \left(\frac{P_G}{V_L} \right)^{0.4} \left(\frac{\rho_L}{\sigma^3} \right)^{0.2} \left(\frac{u_G}{u_t} \right)^{0.5} \qquad (5.3\text{-}15)$$

Where u_t is the terminal rising velocity of the bubbles given later.

On the other hand if the conditions above are not met, and $Re^{0.7} \, (N \, D / u_G)^{0.3} > 30\,000$, there is significant surface aeration. In such a case, the area of bubbles increases significantly:

$$\frac{a}{a_g} = 8.33 \times 10^{-5} \left(\frac{ND}{u_G} \right)^{0.3} - 1.5 \qquad (5.3\text{-}16)$$

Where a is the increased area, and a_g is obtained from Eqn. 5.3-15.

The bubble-diameter is related to the gas hold-up, ε_g, by the following equation:

$$d_B = K \left(\frac{V_L}{P_G} \right)^{0.4} \left(\frac{\sigma^3}{\rho_L} \right)^{0.2} \varepsilon_g^{\,n} \left(\frac{u_G}{u_L} \right)^{0.25} \qquad (5.3\text{-}17)$$

which is written in SI units (with $K = 1.90$ and $n = 0.25$ for organic liquids). Bubble hold-up is calculated from Eqn. 5.3-17, with u_t given by Eqn. 5.3-19 below:

$$\varepsilon_g = \left[0.24 K \left(\frac{u_G}{u_L} \right)^{0.25} \left(\frac{u_G}{u_t} \right)^{0.5} \right]^{\frac{1}{1-n}} \qquad (5.3\text{-}18)$$

This is valid for bubble sizes > 1mm, and

$$u_t = \sqrt{\frac{2\sigma}{d_B \rho_L} + \frac{g d_B}{2}}$$

(5.3-19)

The equations given before require iterative solution. Because, in order to calculate the interfacial area, the gas hold-up and the bubble diameter are required, and this depends on the terminal velocity of the bubbles, which in turn depends again on their diameter. The procedure for doing this has been given [11]. The iteration is started with $u_t = 21$ cm/s, which is close to its final value, and then one should proceed using the following algorithm:

1) Assume $u_t = 0.21$ m/s.
2) Calculate ε_g from Eqn. 5.3-18.
3) Calculate d_B with Eqn. 5.3-17.
4) Calculate u_t with Eqn. 5.3-19.
5) Check if this is equal to that assumed. If not, iterate.
6) Write u_t, d_B, ε_g and a_g.

In order to see the magnitude of the variables, consider the aeration of a cylindrical tank 1m deep and 1 m diameter. Nitrogen is sparged to desorb oxygen from the liquid. Water flows continuously across the tank at a rate of 1 L/s. The nitrogen flow-rate is 0.061 kg/s, and the temperature is 20°C. The results of the outlined calculation of a suitable sparger ring, are as follows:

Ring diameter = 240 mm
$Re_0 = 35\ 000$ (in the holes)
$d_0 = 1/8\ " = 3$ mm
$n_0 = 39$ holes
Spacing between holes = 20 mm > 3 mm
$N_{min} = 5$ rps = 300 rpm (for no flooding of a Rushton turbine)
Ungassed power = 1.60 kW (turning at 5 rps)
Correction factor $\psi = 0.45$
Gassed power = 0.72 kW
$u_G = 0.063$ m/s
$u_t = 0.245$ m/s
$\varepsilon_g = 0.0225$
$d_B = 0.63$ mm
$a_g = 213$ m^2/m^3

The foaming owing to bubbling produced a 2.25% increase in volume. If the liquid is self--foaming the volume increase can be larger (maybe 20%), and make the operation more difficult, or impossible.

Mass transfer coefficients

There are two mass transfer coefficients that are significant, the gas/liquid mass transfer coefficient and liquid to solid coefficient. If the gas is a mixture, the gas/liquid coefficient is more difficult to estimate.

Bubble-to-liquid mass transfer

There are several correlations. One is based on the Rayleigh number, valid for bubbles of less than 2.5 mm (spherical bubbles)

$$Sh = 2 + 0.31\, Ra^{1/3} \qquad\qquad (5.3\text{-}20)$$

where Sh is the Sherwood number, defined as $Sh = k_L\, d_B/ D$, where D is the diffusivity of the dissolved gas in the liquid and which can be estimated. The expression for Ra is :

$$Ra = \frac{d_B^3\,\Delta\rho g}{D\mu_L} \qquad\qquad (5.3\text{-}21a)$$

Other correlations are given in terms of the power dissipated by the stirrer per unit mass of liquid, which is readily calculated. The following equation provides k_L directly:

$$k_L = 0.592 D^{1/2} \left(\frac{\dfrac{P_G}{\rho_L V_L}}{\nu} \right)^{1/4} \qquad\qquad (5.3\text{-}21b)$$

where the power is expressed in the cgs system qs erg/s/g, the diffusivity in cm²/s, and the mass transfer coefficient in cm/s. The kinematic viscosity, ν, is expressed in cm²/s.

Mass transfer coefficient at the liquid/solid interface

The basis of the correlation for k_c is the theory of isotropic turbulence. According to this theory, the Reynolds number is defined in terms of the energy dissipation rate.
For an eddy-size greater than the particle diameter $\zeta > d_p$:

$$Re = \left(\frac{\dfrac{P_G}{\rho_L V_L}\, d_P^4}{\nu^3} \right)^{1/2} \qquad\qquad (5.3\text{-}22)$$

and for $\zeta < d_p$:

$$Re = \left(\frac{\dfrac{P_G}{\rho_L V_L}\, d_P^4}{\nu^3} \right)^{1/3} \qquad\qquad (5.3\text{-}23)$$

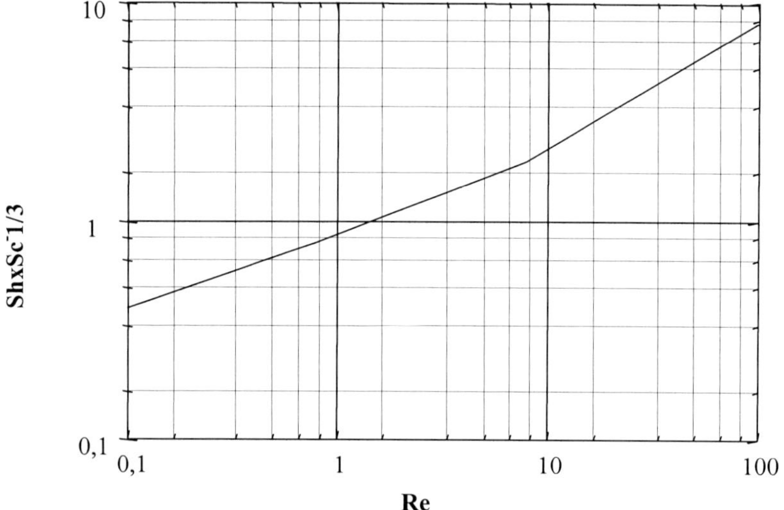

Re

Figure 5.3-6. Correlation of fluid/particle mass transfer coefficients in MSSR (after [10]).

Data from some authors (see [10]) suggest that the fraction $Sh/Sc^{1/3}$ vs. Reynolds number, as defined above, correlate well (see Fig. 5.3-6). The eddy size from the isotropic turbulence theory is calculated from the following expression:

$$\zeta = \left(\frac{\nu^3}{\dfrac{P_G}{V_L \rho_L}} \right)^{1/4}$$
(5.3-24)

where the units are in the cgs system. In all the equations, $\nu = \mu_L/\rho_L$.

Heat transfer in mechanically slurry catalytic reactors

There is not information about multiphase slurry reactors at high solid loading. Murthy [16] suggests to use the following equations for hydrogenation reactors. Heat transfer to jacket (process side coefficient) is given by:

$$\frac{hD_t}{\lambda} = 1.15\, Re^{0.65}\, Pr^{0.33} \left(\frac{\mu}{\mu_w} \right)^{0.24} \left(\frac{H_A}{D_t} \right)^{0.4} \left(\frac{H_L}{D_t} \right)^{-0.56}$$
(5.3-25)

Where, D_t is the tank diameter, Re is for the impeller, H_A is the height of impeller over bottom and H_L the height of liquid over bottom. Heat transfer to a single coil is correlated with the equation:

$$\frac{hd_c}{\lambda} = 0.17\, Re^{0.67}\, Pr^{0.37} \left(\frac{\mu}{\mu_w}\right)^a \left(\frac{D_i}{D_t}\right)^{0.1} \left(\frac{D_c}{D_t}\right)^{0.5} \tag{5.3-26}$$

Where d_c is the tube diameter, D_c the diameter of the coil, and a constant that depends on viscosity, $a = 0.97$ (for a viscosity of 0.3 cP) and $a = 0.18$ (for a viscosity of 1000 cP). For multiple coils, the correlation is:

$$\frac{hd_c}{\lambda} = 0.026\, Re^{0.65}\, Pr^{0.28} \left(\frac{D_t}{D_c}\right) \tag{5.3-27}$$

5.3.6 Design of bubble slurry column reactors (BSCR)

5.3.6.1 Hydrodynamic characteristics of BSCR

Effective properties of BSCR

We can consider two cases, the first one is when a steep solid concentration exist due to the high density of the particles and/or the wide range of the particle sizes. In such cases the slurry is treated as a heterogeneous solid/liquid two-phase system.

The second case is considered when a low density of solid phase and/or small sizes and concentration of solid is used. This allows considering the solid concentration uniform along the whole column and the slurry considered as a homogeneous phase with modified fluid properties. According to Deckwer [17] with $d_p < 50\mu m$ and $C_{solid} < 16\%$ (in mass), we can use such simplification. The equivalent density can be calculated using the following expression:

$$\rho_{LS} = \varepsilon_L \rho_L + \varepsilon_s \rho_s \tag{5.3-28}$$

The effective viscosity depends on the solid hold-up, on particle size and distribution, on the surface properties, on the particle shape and density, on the properties of the liquid (ρ, μ, σ), on temperature, and the shear stress in the column. Depending on the solid concentration encountered in BSCR, we can classify the suspensions into "dilute" and "concentrated" groups.

Dilute suspension

Generally, these behave as Newtonian fluids and, for the case of an extremely dilute suspension of spherical non-interacting particles having a density equal to that of the continuous medium, we can apply the Einstein formula for a suspension of spheres:

$$\frac{\mu_{LS}}{\mu_L} = 1 + C\alpha_s \tag{5.3-28 a}$$

Where α_S is solid volume fraction in suspension $\left(\dfrac{\varepsilon_S}{\varepsilon_S + \varepsilon_L}\right)$

320

For $\alpha_S \leq 0.1$ according to Fortier [18], $C = 2.5$. However, values of $C = 5.5$ have been reported, depending on the nature of the particles and the interaction between them. Jinescu [19] suggested $C = 2.5$ for $0.003 \leq \alpha_S \leq 0.05$, and $C = 4.5$ to 4.75 for $0.15 \leq \alpha_S \leq 0.18$.

Concentrated suspension

These generally behave as non-Newtonian fluids. The Einstein theory has been extended to this case by considering the hydrodynamic interaction between particles. Formulae of the following type are proposed:

$$\mu_{LS} = \mu_L\left(1 + C_1\alpha_S + C_2\alpha_S^2 + C_3\alpha_S^3 + ...\right). \tag{5.3-29}$$

We can recommend the expression proposed by Thomas [20], valid for solid concentration up to the maximum volume fraction attainable (≈ 60 vol. %):

$$\mu_{LS} = \mu_L\left(1 + 2.5\alpha_S + 10.05\alpha_S^2 + 2.73.10^{-3} exp(16.6\alpha_S)\right). \tag{5.3-30}$$

The range investigated was: 0.099 μm $\leq d_S \leq 435$ μm and $\alpha_S \leq \alpha_{S,\ max} = 0.6$
The slurry conductivity can be calculated using the expression proposed by Tareef [21] and recommended by Deckwer *et al.* [17]:

$$\lambda_{sl} = \lambda_l \frac{2\lambda_l + \lambda_s - 2\psi_s\left(\lambda_l - \lambda_s\right)}{2\lambda_l + \lambda_s + 2\psi_s\left(\lambda_l - \lambda_s\right)}, \tag{5.3-31}$$

where , ψ_s is the solid volume fraction in the slurry.
The heat capacity of the suspension is given by [22]:

$$C_{p,sl} = \phi_s C_{p,s} + \phi_l C_{p,l} \tag{5.3-32}$$

where ϕ_l, ϕ_s are, respectively the weight fractions of the liquid and solid in the slurry.

Flow regimes

Flow regimes in BSCRs is studied through the size, shape and velocity of the bubbles as a function of the operating conditions such as the gas-and/or liquid-flow-rates, solid concentration, particle diameter, liquid velocity, etc. Owing to the slurry's physical and hydrodynamic characteristics, the flow regimes in slurry bubble column reactors can be classified into: homogeneous and heterogeneous bubble flows, churn-turbulent or slug-flow regimes. The homogeneous bubble-flow occurs at low gas velocities (<5 cm/s), and a narrow bubble-size distribution is observed. At higher gas velocities (>7 cm/s) either heterogeneous or churn-turbulent states may occur, depending on the column diameter (Fig. 5.3-7). The slug-flow regime is generally limited to small column diameters (<0.15m).

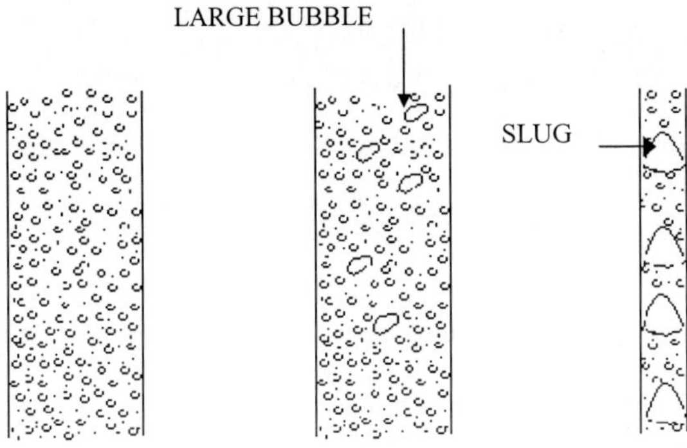

b) HOMOGENEOUS BUBBLE FLOW, $u_G = < 5$ cm/s

a) HETEROGENEOUS FLOW, $u_G = > 7$ cm/s

c) CHURN-TURBULENT OR SLUG- FLOW, Column diameter < 0.15

Figure 5.3-7. Scheme of the flow regimes in bubble slurry column [6].

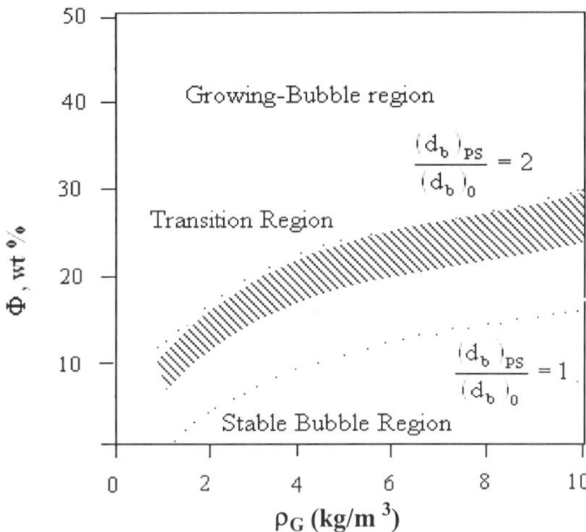

Figure 5.3-8. Details of the churn-turbulent flow regime of BSCR according to Inga [1], $(d_b)_0$-gas bubble size at atmospheric conditions in the absence of solid particles, $(d_b)_{PS}$- bubble size at the operating pressure and catalyst concentration. Column height = 2.8 m, internal diameter = 0.316 m. P ≤ 8 bars. Organic media, catalyst diameter < 100 μm.

The works of Fukuma *et al.* [23], Morooka *et al.* [24] and Clark [25] has indicated that the presence of solid particles in the reactor favours the transition to the heterogeneous flow regime. A critical solid hold-up exists, beyond which the coalescence of bubbles is more frequent. This critical value is higher with smaller particles. Clark [25], studied operation at pressure above 100 bar, and inferred that
- high pressure stabilizes bubbles, and therefore homogeneous flow may exist at larger superficial gas velocities than under atmospheric pressure,
- as at atmospheric pressure the presence of solid particles favours coalesced bubble flow,
- the flow-regime changes from homogeneous to heterogeneous between the bottom and the top of the column.

Fig. 5.3-8 [22] represents the influence of two important parameters for the bubble behaviour-the presence of solid catalyst, and the gas density (or pressure), in the case of an industrial application of the BSCR (large column-diameter, not pure liquid phase). The churn-turbulent flow regime is separated into two regions, based in the bubble growth. There stable bubble region and the growing-bubble region. The stable bubble region, characterized by a large degree of mixing and a very wide bubble-size-distribution. In the growing-bubble region, a shift of the bubble size towards large bubbles is due to coalescence.

Distributor design

Wilkinson *et al.* [26] showed that for sparger hole diameters larger than 1 to 2 mm, the sparger design does not influence the gas hold-up. Even though spargers with small hole diameters (<1 mm) lead to formation of smaller bubbles, and thus to higher gas hold-up, they are not used in industry, owing to the risk of clogging. In the literature [6] for static distributors (including sintered plates, perforated tubes, crosses and plates), an expression is proposed for calculating the diameter of the holes:

$$We_{G0} = \frac{u_0^2 d_0 \rho_G}{\sigma} \geq We_{G0,crit} = 2 \tag{5.3-33}$$

Where u_0 is the velocity of gas in the hole.

In the practice of the sparger design, it is very important to maintain equal flows through each orifice. For a perforated-pipe sparger the following three factors have to be calculated [27]:
- the kinetic energy of the gas at the inlet of the pipe;
- the total pressure-loss along the pipe, and
- the pressure loss across each orifice.
A detailed calculation method is given in [27].

Critical gas velocity for complete particles suspension

To calculate the critical gas velocity needed to suspend the solid catalyst in the reactor, we recommend the correlation published by Koide *et al.* [28], which refers either to flat-bottom or conical-bottom columns for a bubble column without liquid motion:

$$\frac{u_{gc}}{u_s} = 0.8\left(\frac{\rho_S - \rho_L}{\rho_L}\right)^{0.6}\left(\frac{C_S}{\rho_S}\right)^{0.146}\left(\frac{\sqrt{gd_R}}{u_t}\right)^{0.24}\left[1 + 807\left(\frac{\rho\mu^4}{\rho_L\sigma_L^3}\right)^{0.578}\right] \tag{5.3-34}$$

where,
u_s, settling velocity of a single solid particle in the liquid (m/s);
d_R, column diameter (m);
ρ_L, liquid density (kg/m³);
ρ_S, solid density (kg/m³);
C_S, solid concentration (kg/m³ of slurry);
μ_L, liquid viscosity (kg/m³);
σ_L, liquid surface tension (kg/s⁻²).

Gas hold-up

The gas hold-up depends on the following parameters: the gas velocity, particle size, concentration, liquid properties, and the aspect ratio, L/d_T, of the column and the gas distributor. As shown in Fig. 5.3-9, the presence of particles can alter the gas hold-up, depending on the particle size. On the other hand, in continuous operations the gas hold-up is usually lower in the presence than in absence of particles [29,30] owing to the increase of the apparent viscosity of the slurry.

High-pressure conditions favour a smaller bubble size and narrower bubble-size distribution, and therefore lead to higher gas hold-up in BSCR, except in systems operated with porous plate distributors and at low gas velocities. For design purposes in BSCR at high pressure, where the liquids operate in the batch mode, Luo *et al.* [31] proposed the following formula for the calculation of the gas hold-up, based on their proper experimental data and those of many other authors [1,26,31-34] for various systems of gas, liquid and solids:

$$\frac{\varepsilon_g}{1-\varepsilon_g} = \frac{2.9\left(\frac{u_g^4}{\sigma_g}\right)^\alpha \left(\frac{\rho_g}{\rho_{sl}}\right)^\beta}{\left[\cosh\left(Mo_{sl}^{0.054}\right)\right]^{4.1}} \tag{5.3-35}$$

where Mo_{sl} is the modified Morton number for the slurry phase,

$$Mo_{sl} = \frac{(\xi\mu_l)^4 g}{\rho_{sl}\sigma^3}, \text{ and}$$

$$\alpha = 0.21\, Mo_{sl}^{0.0079} \text{ and } \beta = 0.096\, Mo_{sl}^{-0.011}.$$

The quantities ρ_g and σ are, respectively, the gas density and gas/liquid surface tension at operating pressure and temperature. The effective slurry density can be calculated by the formulae given above. K is a correlation factor, which accounts for the effect of particles on slurry viscosity:

$$ln\,\xi = 4.6\varepsilon_s\left\{5.7\varepsilon_s^{0.58} sinh\left(-0.71 exp\left(-5.8\varepsilon_s\right)ln\,Mo_l^{0.22}\right)+1\right\} \tag{5.3-36}$$

324

where Mo_l is the Morton number of the liquid, $Mo_l = \dfrac{\left(\xi\mu_l\right)^4 g}{\rho_l \sigma^3}$.

The error of the prediction by the Eqn. 5.3-35 is 40% at ambient pressure and only 10 at high pressure. Table 5.3-4 resumes the applicable ranges of this correlation.

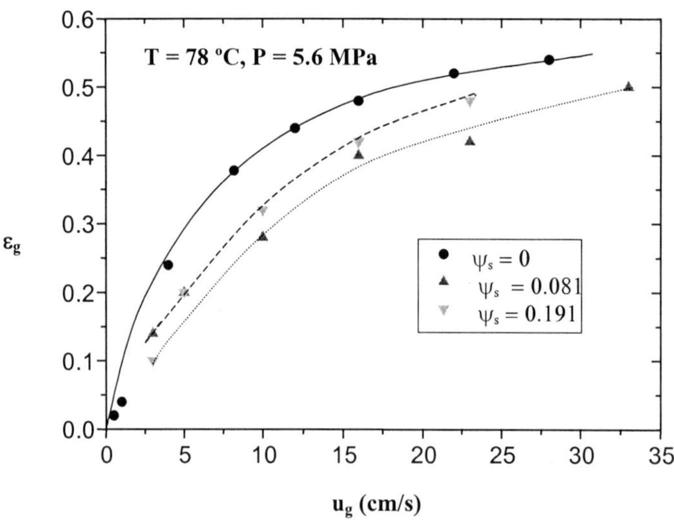

Figure 5.3-9. Hold-up ε_G vs. superficial gas superficial velocity u_g, with ψ_s the volume fraction of solid in the BSCR [31].

Table 5.3-4
Applicable range of the gas hold-up correlation

Parameter	Range
$\rho_l\,(kg/m^3)$	668 to 2965
μ_l (mPa.s)	0.29 to 30
σ_l (N/m	0.019 to 0.073
ρ_g (kg/m^3)	0.2 to 90
ε_S	0 to 0.4
d_p (µm)	20 to 143
ρ_S (kg/m^3)	2200 to 5730
u_g (m/s)	0.05 to 0.69
u_l (m/s)	0 (batch liquid)
D_c (m)	0.1 to 0.61
H/D_c	>5
Distributor types	perforated plate, sparger, and bubble cap

Deckwer *et al.* [17] proposed a simple expression for the calculation of the gas hold-up in a Fischer-Tropsch synthesis:

$$\varepsilon_g = 0.053u_g^{1.1} + 0.015 \tag{5.3-37}$$

Where u_g is the gas velocity in cm/s.

Gas/liquid mass transfer

As the expressions for the calculation of k_La cited in the literature are deduced for low pressure conditions [see, for example, ref. 6] and knowing that the gas hold-up is of the first importance in the design of BSCR, we recommend calculation of the gas hold-up using the formula proposed by Luo *et al.* [31]. The interfacial area, supposing the bubbles to be spherical, is then calculated using the expression given before (Eqn. 5.3-8).

Inga [1] suggested a modified expression of the Wilkinson equation (cited in ref. 1) to calculate the mean bubble diameter in the presence of catalyst particles and at different pressures and gas velocity:

$$\overline{d}_B = 3g^{-0.44} \sigma^{0.34} \mu_{SL}^{0.22} u_g^{-0.22} \rho_{SL}^{-0.45} \rho_G^{-0.47} exp(6.38C_S) \tag{5.3-38}$$

where C_S is the solid concentration, in kg/m^3.

The liquid-phase mass transfer coefficient is then calculated using the Calderbank and Moo-Young correlation [35]:

for $\overline{d}_B < 2.5$ mm,

$$k_L = 0.31\left(\frac{g\mu_L}{\rho_L}\right)^{1/3}\left(\frac{D_L\rho_L}{\mu_L}\right)^{2/3} \tag{5.3-39}$$

and for $\overline{d}_B > 2.5$ mm,

$$k_L = 0.42\left(\frac{g\mu_L}{\rho_L}\right)^{1/3}\left(\frac{D_L\rho_L}{\mu_L}\right)^{1/2}. \tag{5.3-40}$$

Inga [1] compared the k_La in BSCR and MSSR (Fig. 5.3-8) and observed that, at similar values of agitation-energy input per unit slurry volume, BSCR gave higher values of k_La than in a stirred slurry reactor. He claimed that this is because of the difference in interfacial areas in these two reactors. Indeed, he obtained interfacial areas in BSCR that where considerably higher than in a stirred slurry reactor.

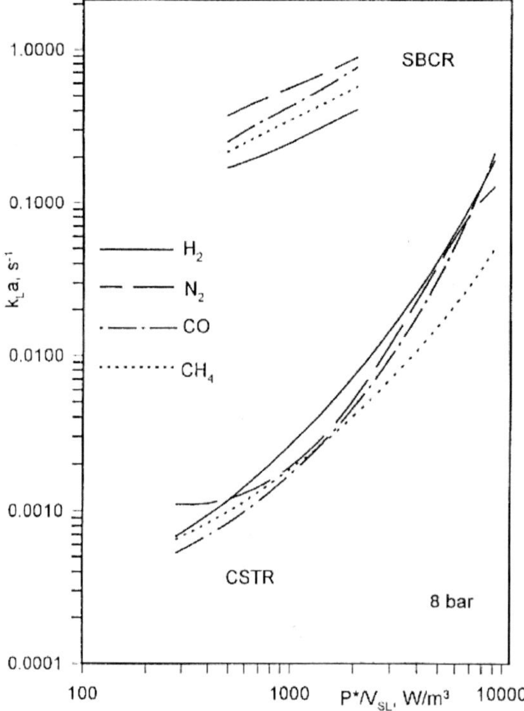

Figure 5.3-10. Effect of power input per unit volume on $k_L a$ [1], $\Phi_s = 25$ wt.%, 8 bar, 300 K: $V_{BSCR} = 4$ L; liquid, a hydrocarbon mixture, $P_G/V_{SL} = \rho_{SL} u_G g$.

Liquid/solid mass transfer

We can mention the equation proposed by Sano *et al.* [36]:

$$\frac{k_S d_p}{D_L F_C} = 2 + 0.4 \left(\frac{e d_p^4 \rho_L^3}{\mu_L^3} \right)^{1/4} \left(\frac{\mu_L}{\rho_L D_L} \right)^{1/3} \tag{5.3-41}$$

Where e is the power supplied to the liquid per unit mass of the liquid, and may be calculated as $e = P_G/(V_L \rho_L)$, and F_C the shape factor of the solid.

Mixing or axial dispersion in gas, liquid and solid in BSCR

The liquid and solid phases are often supposed to be well-mixed, and a perfect mixing model can be applied to model the species-concentration profile. However, many expressions for the description of axial dispersion in the liquid phase in a BSCR are given in the literature [37,38].

The gas phase can be considered in a plug flow regime, and the axial dispersion can be calculated by the expression suggested by Mangartz and Pilhofer [39]:

$$D_{EG} = 5 \times 10^{-4} d_T^{1.5} \left(\frac{u_G}{\varepsilon_g} \right)^3 ,$$
(5.3-42)

with
D_{EG} in cm^2 s^{-1},
u_G, the supercritical gas velocity in cm s^{-1}
d_T, the reactor diameter in cm.

Heat transfer

In BSCR heat can be removed through the walls of immersed heat exchanger's tubes. The heat transfer coefficient in BSCR can be correlated, supposing that the hydrodynamic conditions may be satisfactorily represented by u_g, μ_{sl}, ρ_{sl}, $C_{p,sl}$, and λ_{sl} as:

$$h = f\left(u_g, \mu_{sl}, \rho_{sl}, C_{p,sl}, \lambda_{sl} \right)$$

Several workers (Kolbel *et al.* [40, 41], Deckwer *et al.* [17], Michael and Reicheit [42]) have investigated the heat transfer in BSCR *versus* solid concentration and particle diameters. Deckwer *et al.* [17] applied Kolmogoroff's theory of isotropic turbulence in combination with the surface renewal theory of Higbie [43] and suggested the following expression for the heat transfer coefficient in the Fischer-Tropsch synthesis in BSCR:

$$St_{sl} = 0.1 \times Pr_{sl}^{-0.5} \times \left(\frac{u_g^3 \rho_{sl}}{\mu_{sl} g} \right)^{-0.25} ,$$
(5.3-43)

Where, $u_g \leq 0.1$ m/s and $Pr_{sl} = \dfrac{C_{p,sl} \times \mu_{sl}}{\lambda_{sl}}$

5.3.6.2 Design models for slurry bubble reactors

Axial dispersion model

The most popular model for describing the liquid- and gas mixing in BSCR is the axial dispersion model. This formulation allows one to introduce large degrees of back-mixing, as those occurrs in slurry bubble columns. Indeed, the axial dispersion model (ADM) is a method for describing flow patterns of a gas, liquid, and a solid catalyst that exist between the ideal limits of plug-flow and complete back-mixing. However the knowledge of the BSCR's hydrodynamics brought into question some aspects of applicability of ADM, such as [5]:
- the use of a unique dispersion coefficient for the description of the macroscopic circulation and axial and radial flow of continuous phase, especially under churn-turbulent flow (frequent in the industrial applications),
- the ability to account for different classes of bubbles that may exist in a BSCR.
However, in the absence of a clear and fully developed model, this approach may be useful to describe the performance of bench- and pilot plant-reactors having large length-to-

328

diameter ratios (L/d_T). Many correlations are given in the literature for calculating the transport parameters. Extensive studies have been performed by Deckwer and his co-workers. [37,44 to 46].

Other models, more adequate for BSCR flow patterns, are proposed in the literature. The most attractive seems to be the two-compartment convective-diffusion model [47]. An extensive study utilizing computer-aided radioactive particle tracking (CARPT) have revealed that, at sufficiently high superficial gas velocities and in columns having large aspect ratios, a large-scale liquid circulation cell occupies most of the column height, with liquid ascending along the central core region and descending along the annular region between the core and the walls. Liquid recirculation is driven by non-uniform radial gas hold-up profiles (there is more gas in the centre then at the walls). Superimposed on this recirculation are turbulent fluctuations in the axial-, radial-, and azimuthal directions, owing to eddies induced by the wake of the rising gas bubbles. This model is schematically presented in Fig. 5.3-11.

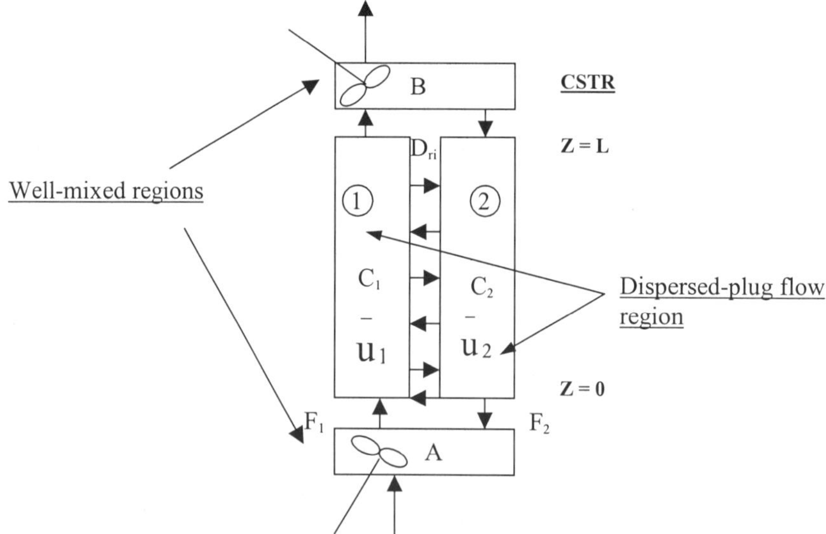

Figure 5.3-11. Scheme of the two-compartment convective-diffusion model and equations for the end zones in BSCR (tested with a liquid tracer) [47].

Where, C- tracer concentration (mol/cm^3), D_{ri}-radial eddy diffusivity at inversion point (cm^2/s), F-liquid volumetric flow rate (cm^3/s), L-length of the recirculation region (m), $\bar{u}$-liquid interstitial mean velocity (m/s), and Z-axial coordinate, cm

5.3.7 Scale-up of slurry catalytic reactors

5.3.7.1 Scale-up of mechanically stirred reactors (MSSR)

Usually, the scale-up problem of a MSSR is posed in the following terms. Suppose that you want to scale-up a catalytic slurry reactor from pilot plant stage, for example, 50 L, to an industrial unit of 2 m^3. We want to know which factors are to be kept constant, and which must be changed. These are common situations found in chemical process scale-   -up. The first thing to hold constant is the reactor geometry and internals. If many reaction runs are planned in the pilot stage, then geometrical similitude is necessary, because scale- -up rules are often valid only for geometrically similar vessels [16].

The second factor to consider is the catalyst conditions and the catalyst-particles, environment. From the point of view of the effectiveness factor, the size, shape, porosity, metal-loading, and other internal characteristics, must be the same. Otherwise, the intrinsic reaction rate and selectivity will be compromised. In practice there will be a trend to reduce costs by using less metal. This will change the process markedly, and other adjustments will be required.

The third factors to control and keep constant are the gas pressure and superficial gas velocity. This probably will involve gas recirculation with either a small compressor, or through a hollow shaft or some other pumping device. As seen before, the bubble diameter, the mass transfer area, the gas hold-up, and the terminal bubble-rise velocity, all depend on the superficial velocity of the gas and the power input per unit volume. When these are kept constant, the various mass transfer resistances in the pilot plant and in the large unit will be the same, hence the global rate will be conserved. The last factor is the input power to the agitator. As required for mass transfer, the scale-up must be made on the basis of constant power input per unit volume. If turbulent conditions and geometrical similarity prevail, this rule imposes the following relationship :

$$\frac{N_1}{N_2} = \left(\frac{D_2}{D_1}\right)^{2/3} \qquad\qquad (5.3\text{-}44)$$

where N_1 and N_2 are the stirring speeds (in rps) in the pilot-and in the industrial unit, respectively and D are the impeller diameters (1 = small, 2 = large). If the reactor operates in the laminar regime and/or the two units are not geometrically similar, the exponents of Eqn. 5.3-25 may differ.

Using the above criteria, the other variables that determine reactor performance are affected. For example, using the scale-up rule of Eqn. 5.3-25, the process-side heat-transfer coefficient in the large reactor may be lower by 25% or so. Since the reaction may be strongly exothermic, the new heat transfer area should be scaled up, with more area per unit volume installed in the large reactor than in the pilot plant unit.

In summary, in the scale-up of a stirred slurry catalytic reactor, a certain number of adjustments may be necessary, as follows,
a) In principle, the catalyst powder and catalyst loading will be the same.
b) Also, the operating pressure and the superficial velocity will be maintained.
c) In order to attain the same diffusional regime in the industrial unit and in the pilot plant, the mass transfer coefficients, the mass transfer areas, and the superficial velocity of the gaseous reactant, should be kept constant.
In increasing the scale of operation, other problems will intrude. These are,
a) resuspension of deposited catalyst, after some time of resting in wet condition;

b) accumulation of spent catalyst from previous batches.

5.3.7.2 Scale-up of BSCR

The technical goals in scaling-up are to attain the desired process results [48]. Scale-up generally involves one of two approaches: principles of similarity, or mathematical modelling. The principles of similarity are based on the premise that, if the rules of similarity are met, a large-scale reactor will perform the same, in terms of technical goals, as a small-scale one. This approach does not predict the reactor performance. On the other hand, mathematical modelling is capable of predicting the reactor performance, but requires more fundamental knowledge about the reaction system. For these reasons, in industrial practice, all or parts of the similarity method are widely used, but mathematical modelling is less common because it is more demanding.

To illustrate the similarity approach, we take an example of scaling-up of a BSCR by a factor of 'SF'. Consequently, the following similarity rules should be respected:

For a Batch process

$SF = V_{R,large} / V_{R,small}$

For a continuous process

$SF = Q_{L,large} / Q_{L,small}$

Table 5.3-5

Summary of some similarity constraints in BSCR scale-up.

	Similarity constraints			
Process	Geometric	Kinematic	Dynamic	Stoichiometry
Both		$d_{C,large} > 0.15\ m$		$u_{G,small} = \dfrac{4 Q_{G,large}}{\pi d_{C,large}^2}$
Semi-batch Continuous	$\dfrac{H_{large}}{d_{C,large}} = \dfrac{H_{small}}{d_{C,small}}$ $\dfrac{\pi}{4} d_{C,large}^2 . H_{large} = SF.V_{R,small}$		$\theta_L = \dfrac{\dfrac{\pi}{4} d_{C,large}^2 . H_{large}}{Q_{L,large}}$	

Finally, we can add some important points to keep in mind:
- Back-mixing (axial dispersion is not considered in dynamic similarity), for both the liquid and the gas, increases with increasing column diameter and gas superficial velocity. Consequently, if PFR exist in a small reactor it will be lost, and therefore the reactor performance will be reduced. One remedy is to extend the batch cycle-time.

- The gas space-time $\theta_G = \dfrac{\varepsilon_G V_R}{Q_G}$, should remain constant, and V_R and Q_G be scaled to the same factor.

- The sparged gas power per unit volume, and volumetric gas flow per sparger opening, must be maintained constant.

It is also important to maintain the same gas and liquid used in the bench scale BSCR, the same physical and transport properties, pressure, temperature, and the concentration of reagents in the gas and liquid phases.

Recently, Inga [1] recommended the following rules (Table 5.3-6) for the scale-up and scale-down of the BSCR

Table 5.3-6
Adjustable and non-adjustable parameters in the scale-up and scale-down of BSCR

Parameter	BSCR	Remarks
Diameter	Minimum 0.3 m	Most of the calculations point to 0.15 m
Length	$H/D \approx 4$ to 6	Economic consideration point toward large columns
Internals	Cooling coils	Most applications will require the input of heat or its removal
P	Dependent	Should be kept the same
T	Dependent	Should be kept the same
Catalyst suspension	Very important	No commercial application will allow catalyst-settling
Flow rate	Its importance depends on the hydrodynamic regime	Depend strongly on the gas velocity below 0.1 m/s
Catalyst activity	Dependent	
Reaction kinetics	Dependent	
Reactor model	Gas; PFR Liquid; CSTRs	Degree of mixing in BSCR depends on u_G and geometry of the reactor

5.3.8 Examples

Example 1

It is desired to calculate the needed volume of a cylindrical BSCR of 1 m inner diameter, where the hydrogenation of acetylene to ethane is performed. Hydrogen and acetylene are bubbled in stoichiometric quantities in an octane suspension of metal catalyst. The sparger is a plate of 50 % porosity with a mean diameter of 250 μm. The total gas flow is 30.10^{-3} m³/s. The BSCR operates at $T = 77$ °C and $P = 10$ atm. At these conditions the overall rate of the reaction is controlled by the mass transfer in the gas/liquid interface. The gas hold-up in the BSCR is 12 %, the slurry phase is considered well mixed, and the gas phase flow towards the top of the column supposed in plug flow regime. It is also supposed that the column is high enough to reach a conversion of acetylene of 80 %.

Data for acetylene in octane at 77 °C:
$D_m = 7.5.10^{-9}$ m²/s, Henry's low constant for acetylene in octane $H = 0.26$ atm/mol/m³. The mean bubble diameter can be taken as 0.13 mm. The density and viscosity of octane at 77 °C are respectively 684 kg/m³ and 3.2×10^{-4} kg/m s (example adapted from [49]).

Solution

For the plug-flow of bubbles, the conversion can be written as:

$$\frac{V_r}{F_{A0}} = \int_0^{x_A} \frac{dx_A}{R} \qquad (5.3\text{-A1})$$

Where R is the volumetric rate-constant, mol/(m³ s)

In this expression the acetylene is taken as the key component, A, and the reaction rate is calculated knowing that the limiting step in the mass transfer is the gas/liquid interface. The expression of the reaction rate may be expressed as:

$$R = k_g a_g \left(p_{Ag} - p_{Ai} \right)$$
$$R = k_L a_g \left(C_{Ai} - C_{AL} \right) \tag{5.3-B1}$$
$$R = k_s a_s \left(C_{AL} - C_{AS} \right)$$

If we make the assumption of a first-order irreversible catalytic reaction, the rate of reaction at the catalyst surface may be written as:

$$r_A = k a_s C_{AS} \tag{5.3-C1}$$

The resolution of the system of equations (5.3-B1) and (5.3-C1) to eliminate the unknown concentration and express the reaction rate only as a function of p_{Ag} is:

$$R = \left(\dfrac{1}{\dfrac{1}{k_g a_g} + \dfrac{H}{k_L a_g} + \dfrac{H}{k_s a_s} + \dfrac{H}{k\, a_s}} \right) p_{Ag} \tag{5.3-D1}$$

Knowing that the reaction is mass transfer limited at the gas/liquid interface, the expression is simplified to:

$$R = \dfrac{k_L a_g}{H} p_{Ag} \tag{5.3-E1}$$

The catalytic reactions occur at the solid catalyst surface and can be written as:
$$C_2H_2 + 2H_2 \rightarrow C_2H_6$$
The number of moles of each component at a conversion x_A is:

$$H_2 = 2F\,(1-x_A)$$
$$C_2H_2 = F\,(1-x_A)$$
$$C_2H_6 = F\,x_A$$
$$\text{Total moles} = F\,(3-2x_A)$$

Where F is the total molal feed rate of hydrogen or acetylene (moles of gas per second). The partial pressure of acetylene in the gas mixture can be expressed as:

$$p_{Ag} = P_T y_A = \dfrac{\left(1-x_A\right)}{\left(3-2x_A\right)} P_T \tag{5.3-F1}$$

Substituting (5.3-F1) in (5.3-E1) and then (5.3-E1) in (5.3-A1) we obtain the expression:

$$\frac{V}{F_{A0}} = \frac{H_A}{k_L a_g P_T} \int_0^{x_A} \frac{3 - 2x_A}{1 - x_A} dx_A = \frac{H_A}{k_L a_g P_T} \left[2x_A - ln(1 - x_A)\right]$$ (5.3-G1)

To calculate the bubble-free volume needed to reach the conversion of 80 % of acetylene, we have to calculate the gas-liquid acetylene mass transfer coefficient, k_L, and the specific area, a_g.

The gas/liquid mass transfer coefficient can be calculated using the Calderbank's formula:

$k_L = 3.275\ 10^{-4}\ m/s$

The gas-liquid interfacial area is calculated as:

$$a_g = \frac{6\varepsilon_g}{d_B} = 553\ m^2/m^3$$

where d_B is the mean bubble diameter ($d_B = 0.13 .10^{-2}$ m), ε_g is the gas hold-up, $\varepsilon_g = 0.12$.

The free-flow rate of the acetylene is $F_{A0} = Q_0 y_{A0} \dfrac{P_T}{RT} = 3.484$ moles/s. Finally, the volume of reactor needed to reach a conversion of acetylene of 80 % is 1.6 m^3, and therefore the reactor height should be about 2 m^3.

Example 2

This example refers to the hydrogenation capacity of a high-pressure slurry reactor and is based on the illustration given by Satterfield [50]. The oil to be hydrogenated is a hydrocarbon that will be reacted with gaseous hydrogen at 589 K and 5.52 MPa using a suspended catalyst (loading m = 8 kg catalyst/m^3 liquid oil). Assuming that the liquid is saturated with the hydrogen, we want to calculate the maximum possible space-velocity (expressed as liquid hourly space velocity or LHSV expressed in m^3 oil/h / m^3 reactor), provided that the hydrogen uptake of the oil is 89 m^3 of H_2 (measured at 288 K, 0.1 MPa) per cubic meter of liquid feed.

Other data are as follows (589 K, 5.52 MPa):

Molecular weight of hydrocarbon	= 170 kg/kmol
Density of oil	= 510 kg/m^3
Viscosity of oil	= 0.07 mPa s
Solubility of H_2 in oil (sat'd conc)	= 20 mol%
Diffusivity of H_2 in oil	= 5x10^{-8} m^2/s
Size of catalyst, d_p	= 10 microns
Particle density, ρ_p	= 3000 kg/m^3

334

Since the liquid is saturated with hydrogen, only the liquid-to-particle mass transfer coefficient and the intrinsic rate constant will be significant. In the case that the reaction is fast, the reaction rate will depend only on the liquid-sold mass transfer resistance. Since the particle are very small (10 micrometers), and the loading is moderate (0.8% mass), the Sherwood number will be that of lonely spheres, so $Sh = 2$. For this case we can take $Sh = =4$ [50], rather safely.

$$Sh = 4 = \frac{k_s d_p}{D_{12}} \qquad (5.3\text{-}A2)$$

$$k_s = \frac{4D_{12}}{d_p} = \frac{4 \times 5 \times 10^{-8}}{10 \times 10^{-6}} = 0.02 \; m/s \qquad (5.3\text{-}B2)$$

The available mass transfer area for the catalyst loading is given by equation (5.3-7), that is

$$a_s = \frac{6m}{\rho_s d_p} = \frac{6 \times 8}{3000 \times 10^{-5}} = 1600m^2/m^3 \qquad (5.3\text{-}C2)$$

We next calculate the saturation H_2 concentration. Then we will be able to calculate the maximum mass transfer rate.

$$C_{sat} = 0.2mol/mol = \frac{0.2}{136} \times 510 = 0.7kmol/m^3$$

The volumetric mass transfer rate will be:

$$r = k_s a_s (C_{sat} - C_s) = 0.02 \times 1600 \times (0.7 - 0) = 21.2kmol/m^3 s \qquad (5.3\text{-}D2)$$

So the space velocity in terms of LHSV, will be:

$$LHSV = \frac{v_0}{V} = \frac{r}{XC_0}, \qquad (5.3\text{-}E2)$$

where r is the reaction rate (kmol H_2/m^3/h) and C_0 is the maximum hydrogen uptake per cubic meter of oil total conversion (C_0 = 89 m³ at STP conditions expressed in mol/m³). The numerical value for the LHSV is:

$$LHSV = \frac{21,1 \times 22,4 \times 288 \times 3600}{89 \times 273} = 20264h^{-1}$$

This is a very high rate indeed (liquid space velocities are usually in the range of 1-50 h⁻¹ in the petroleum industry), but according to Satterfield the value found is within the right order of magnitude for catalytic hydrogenation. In practice, the following factors will limit the hydrogenation rate: 1) Hydrogen saturation of the liquid phase is not achieved because gas-liquid mass transfer resistance may be limiting. 2) Hydrogen concentration on the particle surface may not be zero as has been supposed on equation 5.3-D2, 3) The intrinsic rate is not so high due to a little lower metal loading, and 4) Catalyst loading somewhat less than 0.8%.

References

1. J.R. Inga, Scaleup and Scaledown of Slurry Reactors: A New Methodology. Ph.D. Thesis, University of Pittsburgh, 1997.
2. T. Take, S.Wieland and P. Panser,Procedings of the 3rd International Symposium of High Pressure Chemical Engineering, Eds. R. Von Rohr and Ch. Trepp, Elsevier, Zürich, 1996, p.17.
3. Y.T. Shah Gas-Liquid-Solid Reactor Design, McGraw-Hill, New York, 1979.
4. J.R. Inga and B.I. Morsi, Ind. Eng. Chem. Res., 38 (1999) 928.
5. K.D.P. Nigam and Schumpe (eds.), Three-Phase Sparged Reactors, Amsterdam, Gordon and Breach Sci. Publ., 1996.
6. P.A. Ramachandran, and R.V. Chaudari, Three-Phase Catalytic Reactors, New York, Gordon and Breach Sci. Publ., 1983.
7. H.F. Rase, Chemical Reactor Design of Process Plants, Vol. 1, John Wiley and Sons, New York, 1977.
8. P. Trambouse, H. van Landeghem and J-P Wauquier, Chemical Reactors, Design/Engineering/Operation, Technip, Paris, 1983.
9. C.N. Satterfield, Mass Transfer in Heterogeneous Catalysis, Krieger, 1981
10. J.M. Smith, Chemical Engineering Kinetics, 3rd Edtn., McGraw-Hill, New York, 1981.
11. R.E. Treybal, Mass Transfer Operations, 3rd Edtn, McGraw-Hill, 1980
12. K.R. Westerterp, L.L. Van Dierendonck, and J.A. de Kraa, Chem.Eng.Sci., 18 (1963) 157.
13. A.W. Nienow in Harnby, Edwards, Nienow (Eds), Mixing in the Process Industries, Chapter 16, Butterworth-Heinemann, 1990.
14. L.L. Van Dierendonck and J. Nelemans, in the Proc. 2nd ISCRE (5th European) Elsevier, Amsterdam, 1972, p.B6-45.
15. AFORA, Plantas de Hidrogenaciónb Para Quimica Fina, LGAI, Bellaterra-Barcelona, SPAIN
16. A.K.S. Murthy, Notes on the Design of Agitated Tank Reactors, The BOC Group, Murray Hill, NJ, USA.
17. W.-D. Deckwer,Y. Louisi, A. Zaidi and M. Ralek, Ind. Eng. Chem. Proc. Des. Dev., 19 (1980) 699.
18. A. Fortier, Mécanique des Suspensions. Masson et Cie, Paris, 1967 (cited in 6).
19. V.V. Jinescu, Int. Chem. Eng., 14 (1974) 397.
20. D.G. Thomas, J. Colloid Sci., 20 (1965) 267.
21. Tareef, Colloid J USSR, 6 (1945) 545, cited in [20].
22. L-S. Fan, Gas/Liquid/Solid Fluidization Engineering, Butterworth Publishers, Stoneham, 1989.
23. M. Fukuma, K. Muroyama and A. Yasunishi, J. Chem. Eng. Jap., 20 (1987) 321.
24. S. Morooka, K. Uchida and Y. Kato, J. Chem. Eng. Jap., 15 (1982) 29.
25. K.N. Clark, Chem. Eng. Sci., 45 (1990), 2301.
26. P.M. Wilkinson, A.P. Spek, L.L. van Dierendonck, AIChE J., 38 (1992) 544.
27. Perry's Chemical Engineers Handbook, 7th Edn., McGraw-Hill, New York, 1997.
28. K. Koide,T. Yasuda, S. Iwamoto and E. Fukuda, J. Chem. Eng. Jap., 16 (1983) 7.
29. Y. Kato, A. Nishiwaki, T. Fukuda and S. Tanaka, J, Chem. Eng. Jap., 5 (1972) 112.
30. S. Kara, B.G. Kelkar and Y.T. Shah, Ind. Eng. Chem. Proc. Des. Dev., 21 (1982) 584.
31. X. Luo, D.J. Lee, R. Lau, G. Yang and L-S Fan, AIChE J., 45 (1999) 665.

32. K. Akita and F. Yoshida, Ind. Eng. Chem. Proc. Des. Dev., 12 (1973) 76.
33. H.F. Bach and T. Philofer, Ger. Chem. Eng., 1 (1978) 270.
34. K. Koide, A. Takazawa, M. Komura and H. Matsunaga, , J. Chem. Eng. Jap., 17 (1984) 459.
35. P.H. Calderbank and M.B. Moo-Young, Chem. Eng. Sci., 16 (1961) 39.
36. Y. Sano, N. Yamaguchi and T. Adachi, , J. Chem. Eng. Jap., 7 (1974) 255.
37. W.-D. Deckwer, Y. Serpemen, M. Ralek and B. Schmidt, Ind. Eng. Chem. Proc. Des. Dev., 21 (1982) 222.
38. T. Matsumoto, N. Hidaka and S. Morooka, AIChE J., 35 (1989) 1701.
39. R.H. Mangartz, and T. Pilhofer, Chem. Eng. Sci., 34 (1979) 1425.
40. H. Kolbel, E. Borchers, and J. Martins, Chem. Ing. Tech, 30 (1958) 400.
41. H. Kolbel, E. Borchers, and J. Martins, Chem. Ing. Tech, 32 (1960) 84.
42. R. Michael, and K.H. Reichert, Can. J. Chem. Eng., 59 (1981) 602.
43. Highbie, R., Trans. Inst. Chem. Engrs., 31, (1935) 365.
44. W.-D. Deckwer, Y. Serpemen, M. Ralek and B. Schmidt, Ind. Eng. Chem. Proc. Des. Dev., 21 (1982) 231.
45. W.-D. Deckwer and A. Schumpe, Chem. Eng. Sci., 48 (1993) 889.
46. Y.T. Shah, B.G. Kelkar, S.P. Godbole and W.-D. Deckwer, AIChE J., 28 (1982) 353.
47. S. Degaleesan and M.P. Dudukovoc, Ind. Eng. Chem. Res., 36 (1997) 4670.
48. S-Y. Lee and Y. P. Tsui, Chem. Eng. Progress, July (1999) 23.
49. J. Santamaría, J. Herguido, M.A. Menéndez and A. Monzón, Ingeniería de Reactores, Madrid, Editorial Sintesis, 1999.
50. Satterfield, C.N., Mass transfer in heterogeneous catalysis, MIT Press, 1970.

5.4 Catalytic reactors for olefin polymerizations

5.4.1   History, catalysts, polymers and process elements

In June 1953, the first small amount of linear high-density polyethylene HDPE was made by Ziegler's group in Mulheim, Germany, using a chromium catalyst under moderate conditions (100°C, 70 bar). A few months later, a titanium catalyst was found to polymerize ethylene so rapidly that experiments even at low pressure and temperature provided a surprisingly high yield of HDPE. The proclamation of one of Ziegler's researchers who found this high activity first "Es geht in Glas!" stands for a most spectacular breakthrough in the history of polymerization technology. Julio Natta's group used this type of catalyst for propylene polymerization in 1954, and isotactic polypropylene, iPP, was characterized by Natta and started to conquer a significant part of the world plastics market.

In the mid 1970s another breakthrough happened with the invention of highly active Ziegler catalysts which use internal and external electron donors and allow, for example, control of the isotacticity of PP. New catalysts [1, 2, 8], improved polymerization technology [3 - 5], and better understanding of structure-properties relationships [4, 5] have accelerated the trend towards "tailor-made polymers" [3, 8]. At the end of the 1980s , the development of single-site catalysts (see fig. 5.4-1), with their improved control over the microstructure of polymers, offered a new potential for polymer producers, and this still holds true, despite the fact that the annual production of mPOs (metallocene-based polyolefins) is of minor importance currently.

The mobility of the ligands, the type of the bridge (here silyl dimethyl), and the nature of the metal, are not the only factors determining the activity of the activated metal site and the structure of the polymer formed. There is also a remarkable influence of the co-catalyst and the nature of the support used; see, for example, [8].

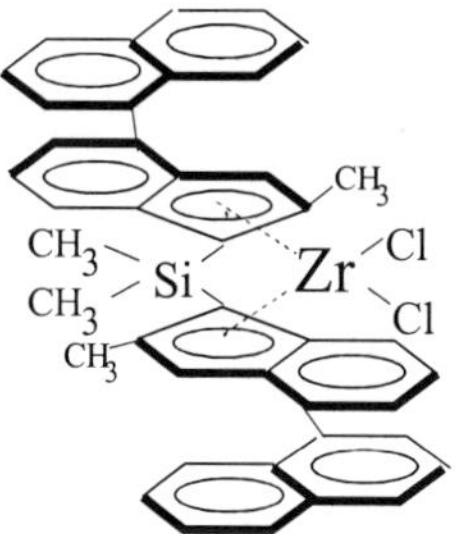

Fig. 5.4-1 A highly active homogeneous single-site zirconocene catalyst for isotactic PP

Nowadays, the leading two polyolefin companies are producing more than 20 million metric tons of polyethylene (PE) and polypropylene (PP) annually. This is about 35% of the world production of all POs. The demand for high impact polypropylene HIPP, and linear low density polyethylene, LLDPE, especially (see Fig. 5.4-2) is permanently growing at about 7% per year. "Distance holders", such as the methyl groups and the butene branch shown in Fig. 5.4-2, help to control the degree of crystallinity, and consequently the processability.

338

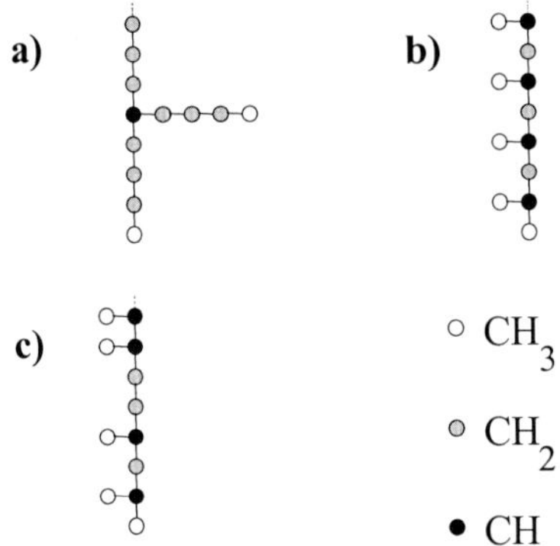

Fig. 5.4-2: Segments of basic polymer structures a), LLDPE with a but-1-ene branch;   b) , isotactic PP;   c), isotactic PP with one mis-insertion

The use of homogeneous catalysts in processes where the (insoluble) polymer precipitates in the early stage of polymerization can cause significant technological and economic problems, especially

-    polymer deposits on the reactor wall ("wall sheeting");
-    uncontrolled particle agglomeration;
-    the particles formed are much too small ("fines");
-    the need of a high amount of expensive co-catalysts, for example methyl aluminoxane, MAO.

Therefore, in such heterogeneous polymerizations, almost all industrial catalysts are supported, for example on silica, whereas the typical Ziegler's titanium catalysts are by definition supported on magnesium chloride. These catalysts are adsorbed at the surface or incorporated into the crystal structure of the support. Other catalysts, such as Phillips chromium catalysts, can be coupled at the support surface by a chemical bond.

These heterogeneous catalysts lead to a typical fragmentation process during the early stage of polymerization. The first polymer produced fills the pores, creates a hydraulic pressure, and depending on the mechanical stability of the support breaks the pores stochastically. At the beginning, the polymer is distributed within the support material, but after a (very small) critical conversion, when the amount of polymer exceeds the volume fraction of the support, the fragmented support is more-or-less homogeneously distributed within the polymer phase that keeps the fragments together. So called "pre-polymerization" procedures slowing down the polymerization rates can help to control the primary particle structure. Depending on the monomers used, and on the process conditions, different polymer morphology results from the fragmentation step. As one can see from Fig. 5.4-3, after fragmentation, the active sites are separated from each other, covered by some polymer, and form the intra-particle morphology. Often, this "early-stage" structure is replicated more-or-less perfectly during the following polymerization process leading to the final internal structure, (see [7]).

There is no doubt that this changing morphology influences the intra-particle heat-and mass-transfer [9]. How strong this effect can be, for a given particle, is primarily a question of the local reaction conditions within a given reactor. A small amount of the particle volume, for example, the micropores below 10 nm diameter, might be filled with condensed gases, owing to the capillary condensation, in accord with Kelvin's equation.

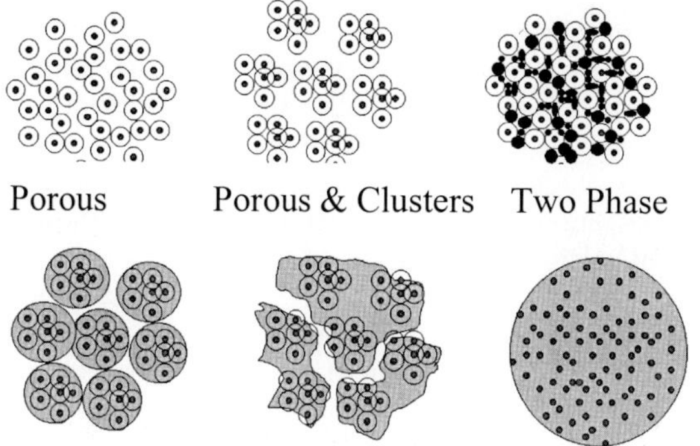

Fig. 5.4- 3 Morphology of polymer particles resulting from supported catalysts, (see [7]).

One should note, that in a real industrial process there are de-pressurization processes accompanied by fast intra-particle pressure drop, and the morphology of particles might be changed by stress created by fastly evaporating liquid gases stored in the pores and sorbed by the polymer.

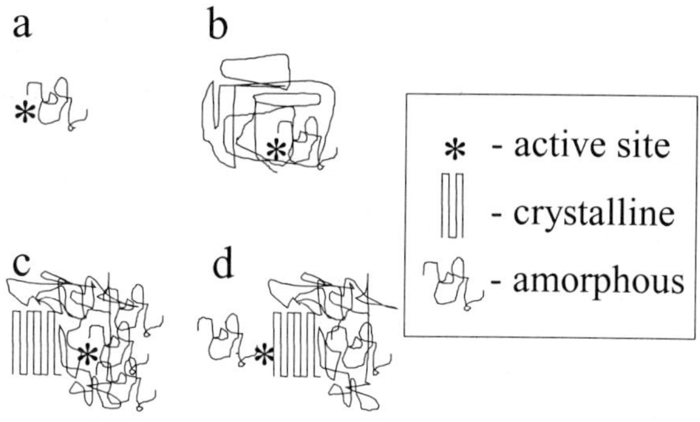

Fig. 5.4-4 Micro-conditions around active sites during catalytic polymerization;
a, single active site producing a polymer chain in the early stage of polymerization;
b, active site surrounded by amorphous, swollen polymer;
c, active site covered by crystalline and amorphous polymer;
d, active site separated from the polymer phase.

From the chemical engineering point of view, there are three basic PO polymerization processes:

340

- (homogeneous) solution;
- (heterogeneous) slurry;
- gas phase,

with a clear trend to simpler and cheaper gas-phase processes. In ethylene polymerizations, the modern trend is set by the so-called "world-scale plants", which are currently designed for an annual capacity of about 500 kt. PP plants are smaller, typically 250 kt/a.

Because of the (chain-length dependent) restricted solubility of polymers in hydrocarbons, solution polymerization requires higher temperatures and / or higher pressures. Technologically, solution polymerization processes are similar to the high-pressure LDPE process, see chapter 3, and will not be discussed here.

In comparison to solution-polymerization processes, slurry- and gas-phase processes do not show a dramatic viscosity rise with increasing monomer conversion. All three types of polymerization processes can differ regarding the micro-conditions around the catalytically active sites; see Fig. 5.4-4. Figure 5.4-4(a) shows an active site that produces a polymer chain by incorporating monomer units between the activated metal and the growing polymer chain. The typical time scale for forming one polymer chain consisting of a few thousand monomer units is below one second. Consequently, within parts of a second, sufficient polymers are available to form a whole ("instantaneous") molecular-weight distribution even within a single particle.

The initial polymerization rate of a freely accessible site, for example at the surface of a macro pore of the silica used, is determined by the monomer concentration a d s o r b e d at the surface. However, after a few polymers are formed, this active site is covered by its own polymer, which is more or less swollen with the monomer and / or swollen with any other type of hydrocarbons used - see Fig. 5.4-4(b). Consequently, the monomer concentration near the active site is changing during a very small initial time period from a d s o r p t i o n - equilibrium to a b s o r p t i o n - equilibrium. This is , of course, only true if the sorption processes are much faster than the consumption of the monomer by reaction. Under common polymerization conditions, say 20 bar pressure and 80°C, the polymer starts crystallizing immediately after it is formed, a process which should reduce the diffusion rates of all components around the active sites. However, because crystallization needs more time, the assumption that the swelling equilibrium of the system "monomer + a m o r p h o u s polymer" determines the polymerization rate seems to be quite reasonable, at least when the polymerization rates are not too high. Furthermore, it is of special interest that the crystal structures formed in-situ during the polymerization process can differ significantly from those formed after melting and recrystallization, for example in an extrusion process. Figure 5.4-5 shows a DSC diagram of an isotactic PP that demonstrates a much lower degree of crystallization of the in-situ polymer compared to the re-crystallized one.

This example shows that, for modelling purposes, the in-situ structure of the polymer formed is to be taken into account and, for example, sorption measurements using recrystallized polymers can be misleading if one wants to model heat- and mass- transfer of the polymerizing particle.

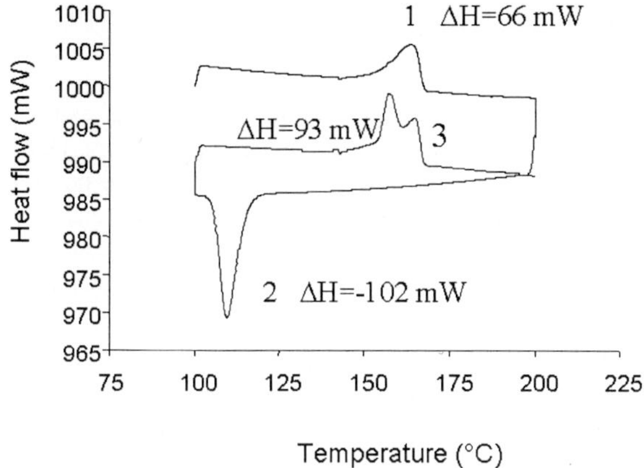

Fig. 5.4-5 DSC diagram of a polypropylene sample; 1, melt peak of the in-situ polymer; 2, crystallization of the polymer, 3, melt peak of the recrystallized polymer.

The discussion above explains why basic information on sorption and diffusion under the reaction conditions, especially at elevated pressures, is required for kinetic and mass- and heat- transfer modelling of catalytic polymerization reactors. If such information is sufficiently available, one should be able, for example, to compare the kinetics of gas-phase and slurry-processes directly by taking into account both gas solubilities in swollen polymers and the hydrocarbons used in slurry processes.

In another case, depending on the reaction conditions, thermodynamic phase separation of the active-site-containing phase might occur during the polymerization process. In this case, active sites would be separated from the polymer, would not be covered by the polymer produced, and would be directly accessible on the surface of polymer particles, see Fig. 5.4-4(d). In this case, the surface concentration of the monomer, instead of the monomer concentration in the swollen polymer, is to be considered as the driving force of the polymerization process. If such a separation process is combined with capillary condensation then a direct contact of active sites with, for example, liquid monomer is enabled yielding high polymerization rates.

The real situation is even more complex if one takes the corresponding thermal effects into account, see [9,18], because olefin polymerizations are highly exothermic. Theoretically, their adiabatic temperature rise is between 1000 and 1900 °C. Using pure monomer gas leads correspondingly to an adiabatic temperature rise of 10–19 K per % monomer conversion. Poor powder mixing accompanied by limited heat-transfer in large-volume gas-phase reactors therefore easily causes non-isothermal effects. On the other hand, the polymer quality is controlled by the molecular weight of the polymer, which is a strong function of temperature. Therefore, controlled heat transfer from the reactor is needed, but usually heat transfer limits the productivity of industrial reactors. This problem can be partially compensated by the so-called "condensed mode" operation, with the injection of liquid propylen, propane or similar components, see below.

342

5.4.2 Fundamentals of modelling

5.4.2.1 Modelling of polymerization kinetics

Despite the complex interaction between the components of a catalyst recipe, for example consisting of catalyst, co-catalyst, electron donors (internal and external), monomers, chain-transfer agents such as hydrogen, and inert gases and the catalyst support, the local polymer production rate rate (polymerization rate) in a given volume, R_P (kg polymer hr^{-1}), can often be described by a first-order kinetic equation with respect to the local monomer concentration near the active site, c_M (kgm^{-3}), and is first order to the mass of active sites ("catalyst") in that volume, m^* (kg):

$$R_P = k_p c_M m^*$$
(5.4-1)

Depending on the goal of the simulation, the "volume" mentioned here can be a given volume of the

- polymer;
- particle, or
- reactor.

For so-called multi-site catalysts, Ziegler catalysts for example, kinetic constants might vary with the type of active sites, see [10]. In such a case, one should summarize Eqn.(5.4-1) over all contributions of different active sites, or k_p is interpreted as an average value of all contributing sites.

If catalyst decay over the reaction time can be seen as a decreasing number of active sites, then often a first-order rule holds true

$$\frac{dm^*}{dt} = -k_d m^*$$
(5.4-2)

If the deactivation constant, k_d, remains constant over the reaction time, equation (5.4-2) can be integrated and Eqn. 5.4-3 results.

$$m^* = m_0^* \exp(-k_d t)$$
(5.4-3)

By combining equations (5.4-3) and (5.4-1) one obtains the description of the polymerization rate as a function of time when pre-activated catalysts are used:

$$R_P = k_p c_M m_0^* \exp(-k_d t)$$
(5.4-4)

The initial polymerization rate can be derived as

$$R_{P,0} = k_{p,0} c_{M,0} m_0^*$$
(5.4-5)

At constant monomer pressure, and without any kind of mass-transfer limitation, c_M equals $c_{M,0}$. Furthermore, assuming that k_p does not change with time, the polymerization rate can be expressed as

$$R_P = R_{P,0}\exp(-k_d t) \tag{5.4-6}$$

It becomes obvious that a logarithmic plot of the measured R_P as a function of time provides a linear curve if the model assumptions are reasonable, with $R_{P,0}$ as the intercept of the ordinate, and k_d the negative slope of this curve.

The monomer concentration near the active sites, c_M, depends on the reaction conditions. As a working hypothesis, one could assume that the differences in c_M is the major kinetic difference between gas- versus slurry polymerizations. Working out this hypothesis requires some more modeling effort. For example, we can consider the following two cases:

- Case A, Gas phase polymerization;
- Case B, Slurry polymerization;

with the sub-cases

- Case 1, There is no polymer layer between the active sites and the reaction mass;
- Case 2, Active sites are covered by the polymer formed.

For example, "A1" characterizes a gas-phase polymerization with active sites outside the polymer phase, see above.

CASE A1 (see Fig. 5.4-4(a,d))
The driving force of the polymerization reaction is the monomer concentration at the surface of the polymer. In this case Langmuir's isotherm (equation 5.4-7) could be used to describe the monomer concentration quantitatively or, more simply, Henry's sorption rule (Eqn 5.4-8) holds true

$$c_M = \frac{c_{M,\max}k_L P_M}{1+k_L P_M} \tag{5.4-7}$$

$$c_M = H_M P_M \tag{5.4-8}$$

with monomer pressure, P_M; maximum monomer concentration sorbed, $C_{M,\max}$; Langmuir constant, k_L, and Henry's constant, H_M.

The most simple model uses the gas-phase concentration of the monomer as driving force. Alternatively, if capillary condensation of monomers is to be taken into account, the monomer concentration can be calculated directly from the liquid density. A sorption balance, for example, should be used to measure the constants of Equations (5.4-7) and (5.4-8), because of their dependence on the changing polymer structure formed in the polymerization process.

Model discrimination, for example by experimental investigation of the dependence of polymerization rate versus pressure, is required if one wants to use the model over a wide pressure range.

CASE A2 (see Fig. 5.4-4(b,c))
Because of the extra mass-transfer resistance through the polymer layer, the monomer concentration may depend on the intra-particle morphology. Micro-porosity, crystallinity, and the thickness of the polymer layer around active sites should influence the transport processes

344

to the active sites. This is generally the case. However, usually the diffusion (or better the permeation, because of the partial pressure drop in the transport direction; see [11]) through micro-porous layers is much faster than the consumption rate caused by polymerization. Even non-porous layers do not diminish the polymerization rate too much, because the diameters of the primary particles do not greatly exceed 2 microns usually. Therefore, for highly permeable polymer layers and catalysts, which are not too active, a thermodynamic ("swelling") equilibrium is reached all over the primary particles' volume. The Flory-Huggins equation can be used to model the equilibrium monomer concentration for monomer-polymer systems in this case:

$$\ln \frac{P}{P^0} = \ln \phi + (1 - \phi) + \chi(1 - \phi)^2 \qquad (5.4\text{-}9)$$

At the pressure P, knowing the interaction parameter, χ, and the vapour pressure of the monomer, P_0, the volume-fraction, ϕ, of the monomer sorbed can be estimated from Equation 5.4-9. Figure 5.4-6 shows the pressure- and temperature dependence of the volume fraction of absorbed propylene in a given amorphous polypropylene [12].

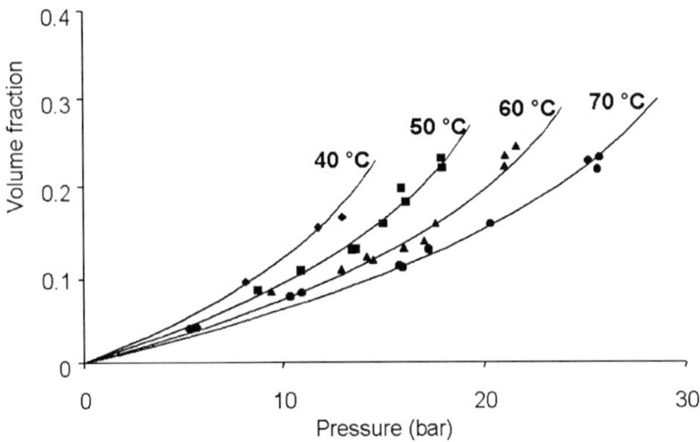

Fig. 5.4-6 Sorption of propylene in amorphous PP

CASE B1

Compared to case A1, the only difference is that the polymer is more or less swollen with the suspending agent, and the reacting gases, monomer, hydrogen, etc., are solved in the suspending agent. Therefore, two thermodynamic equilibria exist between gas phase and suspending agent, and suspending agent and surface of the polymer phase respectively.

A multi-component gas solubility model and a multi-component surface adsorption model are generally required to estimate the monomer concentration at active sites. If the latter equilibrium can be neglected then the gas-solubility in the suspending agent determines the monomer concentration near the active site, which changes significantly with temperature and pressure.

CASE B2

Compared to B1, the swelling of the polymer phase by the suspending agant is more important during the polymerization process. The diffusion coefficient of low molecular weight components may increase by several orders of magnitude because of the reduced viscosity of the polymer phase. Both a multi-component gas solubility model and a multi-component sorption model are required, and should include their temperature- and pressure-dependence. In presence of inert components, these components flow together with the monomer into the monomer-consuming particles, but are not consumed. This leads to the enrichment of these components, which might change the local reaction conditions.

With reliable data on solubility and sorption, one should be able to compare directly the kinetics of slurry- and gas-phase-polymerizations. Unfortunately, there is a great lack of data in this research field. One representative example is given in ref. [12].

By knowing $R_{P,0}$ and $c_{M,0}$, one can easily estimate the factor $k_{p,0}$ m^*_0 from equation (5.4-5). Note that the molar-number of active sites is proportional of the catalyst mass used and that only portion of the metallic catalyst species are activated; sometimes a few percent of the catalyst or less are activated.

5.4.2.2 Modelling of the molecular weight distribution

With a constant temperature and constant concentration of reacting components near the active site the molecular weight distribution of the polymer formed by a single-site catalyst can be described by a Flory-Schulz equation, that can be derived easily from the polymerization mechanism:

$$y_j = j^2 q e^{-jq} \tag{5.4-11}$$

The mass fraction of polymer with a chain length j within a certain interval dj is given by $y_j dj$. The chain-termination probability, q, can be estimated from Mayo's equation:

$$q = k_m + k_h \frac{c_{H_2}}{c_M} + \sum_k k_k \frac{c_k}{c_M} \tag{5.4-12}$$

The transfer constants to monomer, hydrogen, and other components, k_m, k_h, k_k, are depending on reaction temperature. Because of the magnitude of k_h, the MWD can be controlled via the hydrogen concentration, c_{H_2}.

Under same reaction conditions, multi-site catalysts such as Ziegler catalysts provide a MWD that can be considered as a superposition of a number of Flory-Schulz distributions (see ref. 10). Current results show that even metallocene catalysts can be interpreted as "two-site" catalysts [13-15]. Often, it is not clear whether a certain shift of the molecular weight over the reaction time (at constant reaction conditions) can be attributed to small changes of the hydrogen concentration.

A single polymer particle contains a statistically large number of polymers, of the order of 10^{13}. The corresponding MWD can be described with the Flory-Schulz equation only if the single-site type is guaranteed, and under constant reaction conditions, i.e., constant temperature, pressure, and concentrations near the active sites. However, in an industrial polymerization process, a particle often encounters different conditions over its lifetime. In such a case, the final MWD of the particle is a composition of all the MWDs produced, weighted with the corresponding polymer production rates. The overall MWD of all particles is again a composition of all single-particle contributions. Consequently, one has generally to integrate over the stochastic life of the particles, respecting their residence-time distribution in the different reactors or reactor segments. Monte-Carlo methods are very useful for such calculations.

5.4.2.3 Single particle modelling

Catalyst fragmentation depends mainly on two factors: the (time dependent) hydraulic forces created by the polymer produced and the mechanical resistance of the pores filled with the polymer. Both factors may vary with the radial position within a particle. Fragmentation results in a dispersion of catalyst fragments within the continuous polymer matrix, see Fig 5.4-7. For modelling purposes the catalyst is often assumed to be prepolymerized and fully fragmented, because currently there is no detailed mathematical model available that would allow to describe the fragmentation as structure-forming process quantitatively in the very early stage of polymerization.

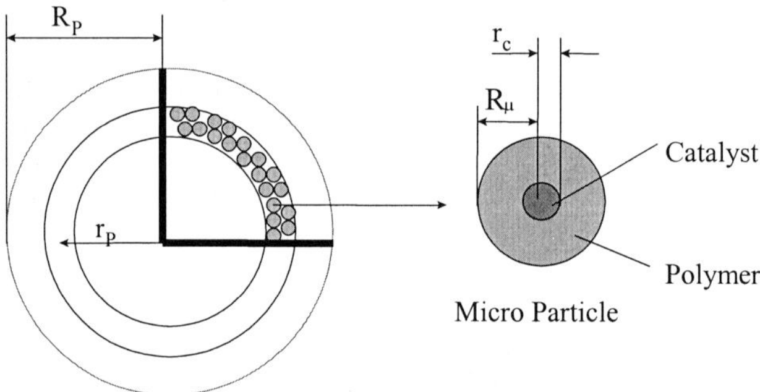

Fig. 5.4-7: Physical background of the multi-grain model MGM

Even after fragmentation, only strongly simplified mathematical models are available to describe the growth process, for example the polymeric flow model, PFM, and the multi-grain model, MGM. Another group of models, core-shell models, CSM, describe the particle growth of an unfragmented catalyst core, which is surrounded by a polymer layer. Note that the CSM is part of the MGM, which makes use of the CSM for micro-particles growth. The PFM can be seen as a simplified MGM that assumes a quasi-homogeneous porous polymer system with only one efficient diffusion coefficient for a given component. In both models, uniformely distributed active sites are assumed, see Fig.5.4-7.

Most authors agreed upon the fact that mass transfer limitation is linked to the early stage of polymerization, and significant particle overheating occurs only if high local polymerization rates exist, for details see [16, 18] and [7]. As can be seen from Fig. 5.4-7, in the MGM the macroparticle is assumed to consist of microparticles having a size of $1 - 2$ microns finally, each of them containing an uniform catalyst fragment in the center. This refers to "case A2", see above. Two different sets of diffusion coefficients are used to quantify the mass transfer through the porous surroundings of the microparticles and through the polymer layer respectively. The size of the centralized active fragment is often set to zero. Using these models, a large number of simulations have been carried out so far and – despite the strict simplifications - useful information for better understanding of single particle behaviour resulted from these studies. For some engineering aspects and a large number of catalysts and recipies currently used in industries, this model is able to adapt the real heat and mass transfer processes within restricted fields of operating conditions using fitted sets of model parameters.

However, some critical remarks should be made to realize that currently used quantitative modelling is quite far from a precise description of real mass and heat transfer processes in such porous structure-creating and structure-replicating processes, which are characterized by moving boundary conditions at different scales of geometry. For example, we consider the homo-polymerization of pure ethylene at 10 bar pressure and 70 °C, in the absence of inert gases, and without special effects such as capillary condensation, and without too much swelling of the polymer by monomer sorption. At high catalyst activities, above 10 kg/(gCat hr), using diffusion coefficients of ethylene in POs measured without reaction, one would calculate an enormous partial pressure drop of the monomer through the particle. Sometimes, for pure monomer and high catalyst activities, such simulations of the early stage of the polymerization indicate vacuum in the center of the particle. Such a large pressure drop is unrealistic, because convective monomer flow through pores of 1 micron diameter , see [17], requires a much lower pressure drop to provide the amount of monomer consumed by reaction under the same conditions. However, if inert gases are present and catalyst activities are below 5 kg/gCat hr, diffusion models based on MGM or PFM are often able to adapt convective effects. It can be expected that, in the future, models like the dusty-gas model, DGM, will be used more frequently, see [19].

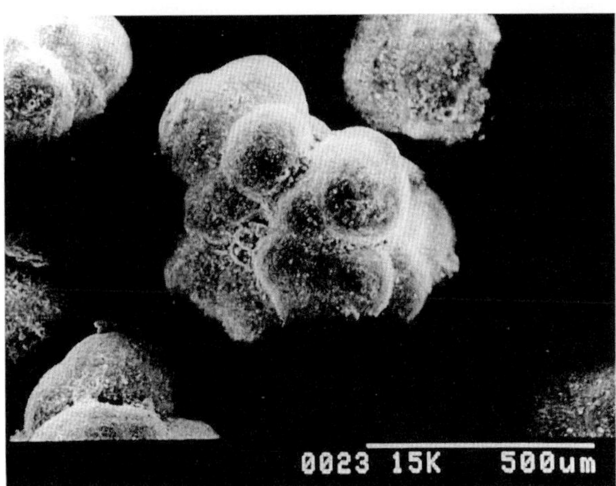

Fig. 5.4-8: Non-spherical polymer particles from an industrial catalyst

Another critical remark concerns the spherical shape assumed for both micro- and macro-particles in the MGM. Often this assumption should be qualified as a rough simplification, for examples for agglomerates, see Fig. 5.4-8. However "real-morphology" models are hardly available, and non-symmetric growth is difficult to compute.

Compared to the MGM, the easier to calculate PFM is less flexible, but provides often practically the same information compared to the PFM, because of the fit of model parameters and due to the fact that concentration and temperature gradients within the microparticles are often neglectible.

5.4.3   The SPHERIPOL process

This process is an illustrative example for a large number of PO processes used today. It is characterized by combining
- loop reactors and fluid-bed reactor technology
- liquid pool process steps and gas phase polymerization
- pre-polymerization, homopolymerization, and co-polymerization.

SPHERIPOL demonstrates an important technology for reactive compounding by filling porous particles with a sticky rubbery polymer that improves impact properties. Figure 5.4-8 shows the process scheme. Not too much public information is available concerning the first "baby"-loop reactor, and specific know-how is required to reach optimum primary morphology of the isotactic PP homopolymer. The second loop reactor is operated in "liquid pool" mode too - the reactors are filled with liquid propylene. High flow rates and high surface-volume ratios ensure well controlled heat transfer from the reaction mass and fast mixing of the feed components with the circulating reaction mass. The large (length-diameter) aspect ratio helps to keep the reactor wall relatively thin and investment low. Furthermore, the residence time distribution is easy to calculate.

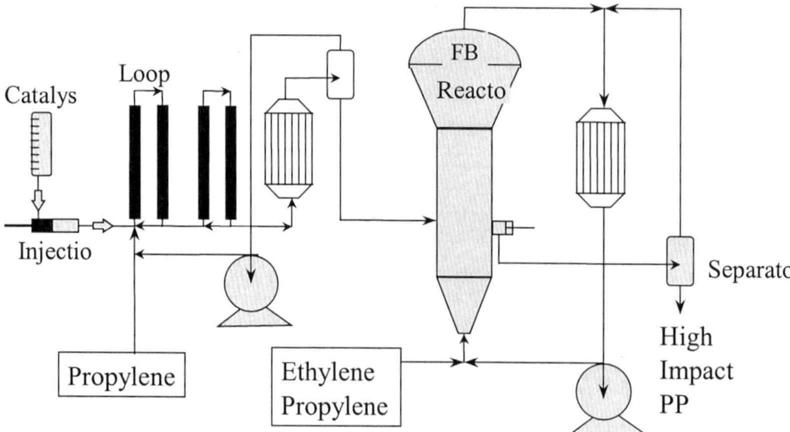

Fig. 5.4-9: The currently most used PP process: SPHERIPOL

The liquid pool operation together with controlled catalyst preparation result in highly porous isotactic polypropylene, iPP. The molecular weight of the polymer matrix can be controlled by adding some hydrogen. The system usually operates under vapor pressure of the liquid propylene, between 20 and 40 bar.

In a next process step after the liquid pool operation, the monomer is evaporated, separated from the polymer powder, and recycled after compressing and cooling. The remaining porous iPP powder still contains some propylene according to the swelling equilibrium, see above. Afterwards, the catalyst surface can be deactivated by adding a very small amount of an alcohol. This procedure can help to avoid sticky material polymerizing at the outside particle surface, which could result in uncontrolled particle agglomeration in the following gas phase reactor.

Afterwards, the PP powder is fluidized by a mixture of ethylene, propylene and inert gas (propane) in the fluid bed reactor, FBR, which is cooled by convective gas flow. The monomer conversion per gas circulation is limited to a few percent, because of the adiabatic temperature rise of more than 10 K per % conversion. Neglecting the heat loss through the reactor wall and through the product withdrawal, and assuming a well mixed reactor, the heat removal capacity of a steady-state FBR limits the productivity and can be expressed by

$$\dot{Q}_{rem} = \phi_g c_{pg} (T - T_{g0}) + \phi_l (\Delta h_v + c_{pl}[T - T_{l0}]) \qquad (5.4\text{-}13)$$

with gas mass flow, ϕ_g; specific heat capacity of the circulating gas, c_{pg}; gas temperature at the reactor exit, T; gas temperature at the reactor entry, T_0; liquid gas mass flow, ϕ_l; specific heat of evaporation, Δh_v; specific heat capacity of the liquid gas, c_{pl}; liquid gas temperature at the reactor entry, T_{l0}. The first term corresponds to the convective heat transfer by the circulating feed gas, whereas the second term concerns the heat removal by so called "condensed mode" operation, which is characterized by injection of liquid (evaporizing) gases like propane. It becomes clear that fluidization, heat production and heat removal capacity are combined by the operating pressure. Increasing the pressure at constant fluidization velocity results in higher mass flow rates of circulating gases. This improves the convective heat removal. At the same time, minimum fluidization velocities that are required for sufficient powder mixing do not decrease too much at higher pressures. Consequently, increasing the pressure is one possibility to improve the heat removal, however, because of condensation phenomena and due to economic reasons, the operating range of pressures is restricted. Therefore, common FBRs are operated between 15 and 30 bar.

After an iPP particle reached the FBR, co-polymerization of ethylene-propylene starts preferrably inside the porous PP matrix. Depending on the individual residence time, the particle will be filled with a certain amount of ethylene-propylene rubber, EPR, that improves the impact properties of the HIPP. It is important to keep the sticky EPR inside the pre-formed iPP matrix to avoid particle agglomeration that could lead to wall sheeting and termination of the reactor operation. Ideally a "two phase" structure, see Fig.5.4-3, is produced. Finally, a "super-high impact" PP results that contains up to 70% EPR. How much EPR is formed per particle depends on three factors: catalyst activity in the FBR, individual particle porosity, and individual particle residence time in the FBR. All particle properties are therefore influenced by the residence time distribution, and finally, a mix of particles with different relative amounts of EPR is produced – a so called "chemical distribution"; see, for example, [6].

References

1. E.Albizatti, U.Giannini, et al., Catalysts and Polymerizations, in E.P. Moore (ed), Polypropylene Handbook, Hanser Publishers, Munich, 1996, ISBN 3-446-18176-8.
2. K.B. Sinclair and R.B. Wilson, Metallocene Catalysts: A revolution in Olefin Polymerisation, Chemistry & Industry, Nov. 1994.
3. K.W. Swogger, Enhanced Polyethylene using INSITE^{TM} Process Technology, MetCon'96, Houston TX, 1996.
4. M Rätzsch, and W. Neißl, Bimodal Polymers Based on PP and PE (German), in Aufbereiten von Polymeren mit Neuartigen Eigenschaften, VDI Verlag, Düsseldorf 1995, ISBN 318 234 191 X.
5. J. Scheirs, L.L. Boehm et al., PE100 Resins for Pipe Applications: Continuing the Development into the 21st Century, Elsevier 1996, TRIP, Vol. 4, No. 12.
6. Chemical distribution, Glen Ko.
7. J.A. Debling "Modelling Particle Growth and Morphology of Impact Polypropylene Produced in Gas Phase" PhD Thesis, Univ. of Wisconsin, Madison, 1997.
8. K. Soga, M. Terano (Editors), Studies in Surface Science and Catalysis, "Catalyst Design for Tailor-made Polyolefins", Elsevier 1994.
9. T. F. McKenna, R. Spitz, D. Cokljat, AIChE Journal, 45 (11), 1999, 2392.
10. Kissin, PE with Ziegler multi-site cat.
11. G.Weickert, G.B. Meier, J.T.M. Pater and K.R. Westerterp, Chem. Engng. Sci. 54 (1999) 3221
12. G.B. Meier, G.Weickert, W.P.M. van Swaaij (1st +2nd publication).
13. G.B. Meier, PhD thesis
14. Montell, personal communication on two-stage metallocenes.
15. Busico, NMR measurements on two-site metallocenes.
16. S. Floyd, K.Y. Choi, T.W. Taylor, W.H. Ray, J.Appl.Polym.Sci.,32,2231 and 2935 (1986).
17. W.H. Hedley, J. Rheol. 22(2) (1978) 91T.F. McKenna, J. DuPuy, R. Spitz, J. Appl. Polym. Sci., 57, (1995) 371.
19. N. Benes, Mass Transport in Thin Supported Silica Membranes" PhD Thesis, Univ. of Twente, Enschede, The Netherlands, 2000

CHAPTER 6

SEPARATION OPERATIONS AND EQUIPMENT

A. Laurent[a]: sections 6.1, 6.2, 6.3, 6.4, 6.5

E. Lack[b], **T. Gamse**[c], **R. Marr**[c]: section 6.6

T. Gamse[c], **R. Marr**[c]: section 6.7

[a] Ecole Nationale Supèrieure des Industries Chimiques (ENSIC)
B. P. No. 451, 1. Rue Granville F-54001 Nancy Cedex, France

[b] NATEX GmbH Prozesstechnologie
Hauptstrasse, 2 A-2630 Ternitz, Austria

[c] Institut fur Thermische Verfahrenstechnik und Umwelttechnik Erzherzog Johann Universität
Infeldgasse, 25 A-8010 Graz, Austria

A survey of distillation and extraction processes at high pressure, together with contactors suitable for different applications, is provided in this chapter. Well-known methods used at lower pressures are reviewed and peculiarities arising at high pressure are discussed. Correlations are critically examined and suggestions are reported for specific cases.

The rest of the chapter is devoted to the most common use of supercritical fluids for separation, i.e. the extraction of solid and liquid materials by means of dense gases. For this application the good physical properties of supercritical fluids (low viscosity, high density, high diffusion coefficient) and the simple separation of the extraction gas from the produced extract give great advantages if compared to extraction with liquid, in most cases organic, solvents. A number of possible applications, mainly related to natural products, and currently either used in production units or tested at laboratory scale, are considered.

352

6.1 Pressure distillation[*)]

6.1.1 Introduction

Mass-transfer processes involving two fluid streams are frequently carried out continuously using cross-, counter-current-, or cocurrent flow in a column apparatus. Three common examples of processes of mass transfer are as follows.

• In a plate or a packed *distillation* column.

Here a vapour stream is rising against the downward flow of a liquid reflux, and a state of dynamic equilibrium is set up in a steady state process.

The more volatile constituent is transferred under the action of a concentration gradient from the liquid to the interface, where it evaporates and then diffuses into the vapour stream. The less volatile component is transferred in the opposite direction and, if the molar latent heats of the components are the same, equimolecular counter-diffusion takes place.

• In a packed *absorption* column.

The flow pattern is similar to that in a packed distillation column, but the vapour stream is replaced by a carrier gas together with one or more solute gas(es). The solute diffuses through the gas phase to the liquid surface, where it dissolves and is then transferred to the bulk of the liquid. In this case, there is no mass transfer in the opposite direction and the transfer rate is supplemented by bulk flow.

• In a liquid-liquid *extraction* column.

Here the process is similar to that occuring in an absorption column, with the exception that both streams are liquid and the lighter liquid rises through the denser one.

The design of commercial equipment for carrying out these unit operations continuously requires numerous experimental data.

Three main steps are involved in the design of these mass-transfer processes:

• Data on the vapour-liquid equilibrium relationships for the system.

• Equilibrium data and material-and heat-balances to determine the number of equilibrium stages (theoretical plates or transfer units) required for the desired separation. The required height of the column can be calculated if data are available for the specific rate of transfer of material between the two phases, expressed in terms of the plate-efficiency or the height of one transfer unit.

• Data on the liquid- and vapour-handling capacity of equipment to determine the required cross-sectional area and diameter of the equipment.

What is the influence of the total system pressure on the design of these unit operations?

Many fluid systems perform differently at higher pressures than at lower ones (Figure 6.1–1). *This make some high-pressure mass-transfer processes different, also.*

For the liquid-liquid extraction, the effects of total system pressure can be ignored for all classical practical purposes except for the SCF extraction.

Since distillation is the method used most widely for separating mixtures of liquids in the chemical-and hydrocarbon processing industries, we will examine this unit operation particularly and make some critical observations for the gas-liquid absorption.

[*)] References at the end of section 6.5

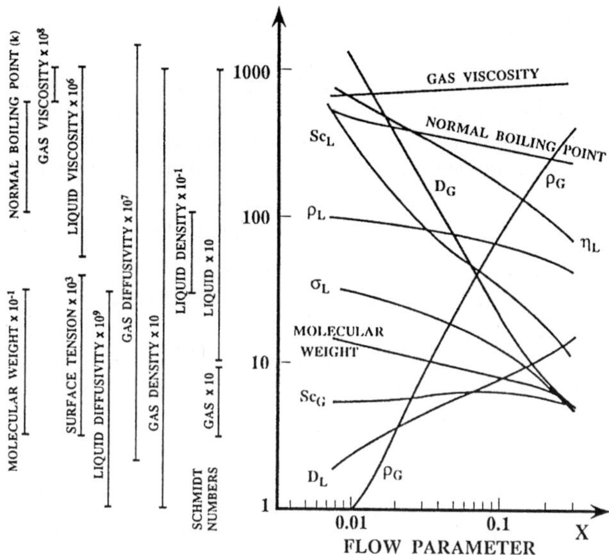

Figure 6.1-1. Physical properties versus flow parameters [1].

6.1.2 Examples of pressure distillation

Distillation operations will be classified on the basis of the column's operating pressure [2].

High pressure distillations normally use a top column pressure of 6 bar or greater. These high pressures produce a greater vapour density that may represent a significant percentage of the liquid density. The accompanying increase in boiling temperature not only lowers the liquid density, but also reduces the surface tension and the viscosity of the liquid phase. The relative volatility, in general, will reduce as the pressure is increased. Since separation becomes more difficult with reduced relative volatility, the number of theoretical stages required is greater for the same operation at higher pressure. Even through the capacity of a column increases at higher pressure, owing to the greater vapour density, the number of theoretical stages or the reflux ratio required will also be greater.

All these factors complicate distillation column design, and require somewhat-modified design techniques. Usually a distillation column should be designed for the lowest operating pressure, which is economically feasible.

Distillation is usually carried out under pressure to permit the condensation of low-boiling materials at ambient temperatures.

For example, propylene and ethylene of polymer grade have been distilled at pressures as low as 4 bar and 6 bar absolute, respectively. However, the condenser must be operated at temperatures of −12.2°C for propylene, and −68°C for ethylene. This latter very low temperature will be produced by a costly refrigeration system.

Propylene is usually distilled at a pressure of at least 16 bar, so that the condenser can be cooled with water. Ethylene is normally distilled at a pressure of about 20 bar, which permits

354

operation of the condenser at a temperature of $-29°C$ and allows the use of less expensive propane-or ammonia refrigeration.

Very low-boiling hydrocarbons, such as methane, are normally distilled at pressures of about 70% of their critical pressure. Of course, all distillations must be carried out below the critical temperature in order to provide liquid reflux. Ethylene and ethane are usually distilled at 40%-55% of critical pressure, while propylene and propane are distilled at 35%-50% of critical pressure.

Near the industrial high-pressure distillation of the light hydrocarbons, C_1 to C_4, another family that includes some low-molecular-weight compounds is the halogenated hydrocarbons, especially the fluorinated ones. These, however, have an exceptionally high liquid density. It is questionable whether they will show the same performance characteristics as the low hydrocarbons. In the absence of evidence, it would be wise to be cautious.

Another example of high-pressure distillation is the technique of breaking azeotropes with Pressure Sensitive Distillation (PSD) [3-4]. The presence of a liquid azeotrope severely limits the use of standard distillation. However a simple change in pressure can also alter relative volatilities; in some cases this results in a significant change in the azeotropic composition and allows the recovery of feed components without adding a separative agent. In the basic concept, PSD exploits this change in composition of a binary azeotrope with pressure [4]. Well known commercially viable applications of PSD include the separation of tetrahydrofuran (THF)-water, acetonitrile-water, methanol-methyl ethyl ketone, methyl acetate-methanol, benzene-n-propanol, and methyl ethyl ketone-cyclohexane. For example, a dual-column PSD process operates respectively at 1 bar and 6 bar for the two systems THF–water and methanol-dichloromethane [4]. At present, PSD is under-utilized because of its unfamiliarity in chemical engineering, but PSD may be an attractive alternative.

6.1.3 Interphase mass transfer and two-film theory

6.1.3.1 Two-film theory for distillation and dilute systems

In each of the above mass transfer processes, material is transferred across a phase boundary. Because no material accumulates there, the rate of transfer on each side of the interface must be the same, and therefore the concentration gradients automatically adjust themselves so that they are proportional to the resistance to transfer in the particular phase. Furthermore, if there is no resistance to transfer at the interface, the concentrations on each side will be related to each other by the phase-equilibrium relationship.

The mass-transfer rate between two fluid phases will depend on the physical properties of the two phases, the concentration difference, the interfacial area, and the degree of turbulence. Mass-transfer equipment is therefore designed to give a large area of contact between the phases and to promote turbulence in each of the two fluids.

The two-film theory is one of the mechanisms suggested to represent the conditions in the region of the phase boundary. This model suggested that the resistance to transfer in each phase could be regarded as lying in a thin laminar film close to the interface (Figure 6.1–2).

The transfer across these films is regarded as a steady-state process of molecular diffusion. The turbulence in the bulk fluid is considered to die out at the interface of the films. Although it does not closely reproduce the conditions in most practical equipment, it gives expressions which can be applied to the experimental data which are generally available, and for this reason it is still extensively used.

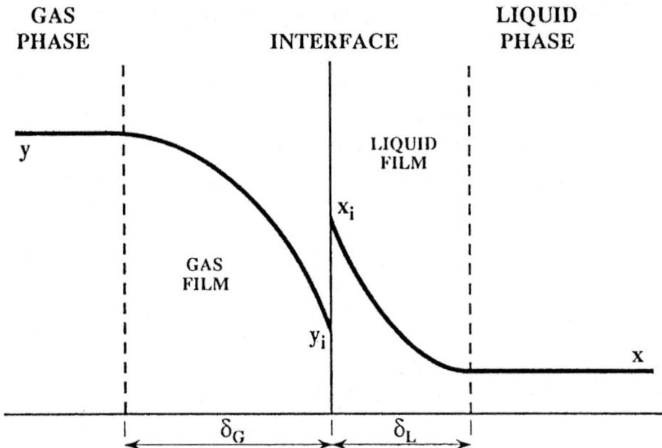

Figure 6.1-2. Two-film theory.

The rate of transfer per unit area, N_A, is given as:

$$N_A = k_{G,p}\,(p - p_i) = k_{L,c}\,(c_i - c) \tag{6.1–1}$$

where $k_{G,p}$ is the gas-phase mass-transfer coefficient (driving forces: pressure), p and p_i are the solute partial pressures respectively in the bulk gas-phase and at interface, $k_{L,c}$ is the liquid–phase mass-transfer coefficient (driving forces: concentration), c and c_i are the solute concentrations respectively in the bulk liquid phase and at interface.

An alternative expression for the rate of transfer is given by:

$$N_A = k_{G,y}\,(y - y_i) = k_{L,x}\,(x_i - x) \tag{6.1–2}$$

where $k_{G,y}$ is the gas-phase mass-transfer coefficient (driving forces: mole fraction), y and y_i are the solute mole fractions respectively in the bulk gas-phase and at interface, $k_{L,x}$ is the liquid–phase mass-transfer coefficient (driving forces: mole fraction), x and x_i are the solute mole fractions respectively in the bulk liquid-phase and at interface.

The mass-transfer coefficients, by definition, are equal to the ratios of the molar mass flux to the concentration driving forces. The mass-transfer coefficients are related to each other as follows:

$$k_{G,y} = k_{G,p}\,P_T \tag{6.1–3}$$

$$k_{L,x} = k_{L,c}\,\bar{\rho}_L \tag{6.1–4}$$

where P_T is the *total system pressure* employed during the experimental determinations and $\bar{\rho}_L$ is the average molar density of the liquid phase.

The coefficient $k_{G,y}$ is relatively independent of the total system pressure, and is therefore more convenient to use than $k_{G,p}$, which is inversely proportional to the total system pressure.

356

The above equations may be used for finding the interfacial concentrations corresponding to any set of values of x and y, provided the ratio of the individual mass-transfer coefficients is known. Thus:

$$\frac{y - y_i}{x - x_i} = -\frac{k_{L,x}}{k_{G,y}} = -\frac{k_{L,c}\,\overline{\rho_L}}{k_{G,p}P_T} \tag{6.1–5}$$

Equation (6.1–5) may be solved graphically if a plot is made of the equilibrium vapour-and liquid-compositions, and a point representing the bulk concentration x and y is located on this diagram (Figure 6.1–3).

In the design of equipment, it is necessary to estimate the rate of mass transfer from known or predicted values of the mass-transfer coefficients and the bulk concentrations. This may be done by solving Equation (6.1–5) simultaneously with the equilibrium curve to obtain x_i and y_i. The rate of transfer may then be calculated from Equation (6.1–2).

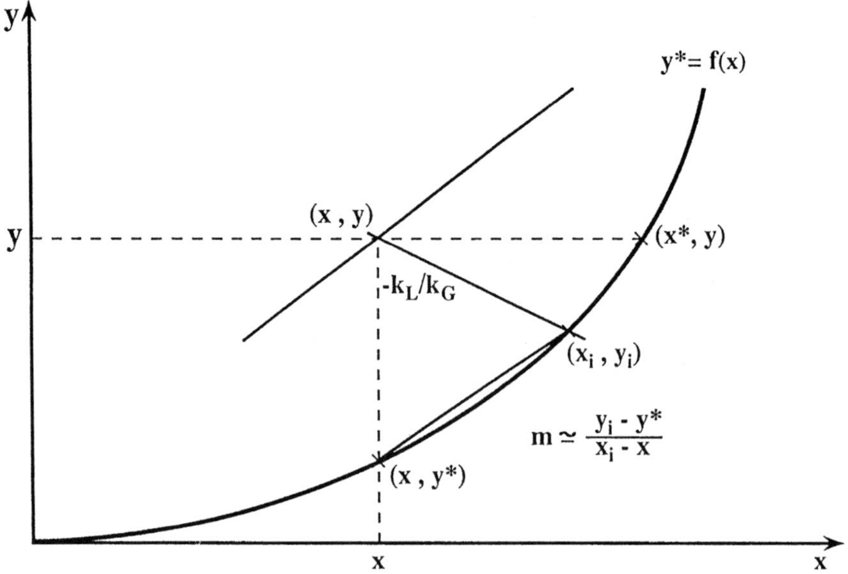

Figure 6.1-3. Graphical identification of the different concentrations and driving forces.

If the equilibrium relationship $y^* = f(x_i)$, is a straight line, not necessarily through the origin, the rate of mass transfer is proportional to the difference between the bulk concentration in one phase and the concentration (in that same phase) which would be in equilibrium with the bulk concentration in the second phase. One such difference is $y-y^*$ and another is x^*-x. In this case, the rate of mass transfer may be defined by the equation:

$$N_A = K_{G,y}\,(y - y^*) = k_{G,y}\,(y - y_i) = K_{L,x}\,(x^* - x) = k_{L,x}\,(x_i - x) \tag{6.1–6}$$

where $K_{G,y}$ and $K_{L,x}$ are respectively the overall gas-phase and liquid-phase mass-transfer coefficients.

This Equation (6.1–6) can be rearranged to the formula:

$$\frac{1}{K_{G,y}} = \frac{1}{k_{G,y}}\left(\frac{y - y^*}{y - y_i}\right) = \frac{1}{k_{G,y}} + \frac{1}{k_{G,y}}\left(\frac{y_i - y^*}{y - y_i}\right) \tag{6.1–7}$$

$$\frac{1}{K_{G,y}} = \frac{1}{k_{G,y}} + \frac{1}{k_{L,x}}\left(\frac{y_i - y^*}{x_i - x}\right) \tag{6.1–8}$$

Comparison of the last term in the parentheses with the diagram of Figure 6.1–3 shows that it is equal to the slope of the chord connecting the points (x, y^*) and (x_i, y_i). If the equilibrium curve is a straight line, then it is the slope m. Thus:

$$\frac{1}{K_{G,y}} = \frac{1}{k_{G,y}} + \frac{m}{k_{L,x}} \tag{6.1–9}$$

Considering the gas-absorption unit operation, for dilute concentrations of many gases and over a fairly wide range for some gases, the equilibrium is given by Henry's law with the following two equations:

$$p = He\, x \tag{6.1–10}$$

$$\text{or } p = He'\, c \tag{6.1–11}$$

where He is expressed in SI units in kPa per mole fraction solute
He' is expressed in SI units in kPa per kmole per cubic metre.
When Henry's law is valid, the slope, m, can be computed according the relationship:

$$m = \frac{He}{P_T} = \frac{He'\, \bar{\rho}_L}{P_T} \tag{6.1–12}$$

If it is desired to calculate the rate of mass transfer from the overall concentration-difference based on bulk-liquid compositions $(x^* - x)$, the appropriate overall coefficient, $K_{L,x}$, is related to the individual mass-transfer coefficients by the equation:

$$\frac{1}{K_{L,x}} = \frac{1}{k_{L,x}} + \frac{1}{mk_{G,y}} \tag{6.1–13}$$

Conversion of these equations to a $k_{G,p}$ and $k_{L,c}$ basis can be accomplished readily by direct substitution of Equations (6.1–3) and (6.1–4):

$$\frac{1}{K_{G,p}} = \frac{1}{k_{G,p}} + \frac{He'}{k_{L,c}} \tag{6.1-14}$$

$$\frac{1}{K_{L,c}} = \frac{1}{k_{L,c}} + \frac{1}{He'k_{G,p}} \tag{6.1-15}$$

where $K_{G,p}$ and $K_{L,c}$ are respectively the overall gas–phase and liquid–phase mass–transfer coefficients.

The $k_{L,c}$ and $K_{L,c}$ values are reported in SI units of m s^{-1} or kmole m^{-2} s^{-1} (kmole m^{-3})$^{-1}$.

When $k_{G,p}$ and $K_{G,p}$ values are indicated in SI units of kmole m^{-2} s^{-1} (kPa)$^{-1}$, one must be careful in converting them to a mole–fraction basis to multiply by the total pressure actually employed in the original experiments and NOT by the total pressure of the system to be designed. This conversion is valid for systems in which Dalton's law of partial pressures is valid [5].

When the equilibrium curve is not straight, there is no logical basis for the use of an overall transfer coefficient, since the value of m will be a function of position in the equipment. In such cases, the rate of mass transfer must be calculated by solving for the interfacial compositions as described above.

Nevertheless, experimentally observed rates of mass transfer in industrial equipment are often expressed in terms of overall mass-transfer coefficients even when the equilibrium lines are curved. This procedure is purely empirical, since the theory indicates that, in such cases, the rate of transfer may not vary in direct proportion to the overall bulk-concentration differences $(y–y^*)$ and $(x^*–x)$ at all concentration levels, even though the rates may be proportional to the concentration difference in each phase taken separately, $(x_i–x)$ and $(y–y_i)$.

In most types of mass-transfer equipment, the interfacial area, a, that is effective for mass transfer cannot be determined accurately. For this reason, it is customary to report experimentally observed rates of transfer in terms of mass-transfer coefficients based on a unit volume of the apparatus, rather than on a unit of interfacial area. Calculation of the overall coefficients from the individual volumetric coefficients is made practically, for example, by means of the equations:

$$\frac{1}{K_{G,y}a} = \frac{1}{k_{G,y}a} + \frac{m}{k_{L,x}a} \tag{6.1-16}$$

$$\frac{1}{K_{L,x}a} = \frac{1}{k_{L,x}a} + \frac{1}{mk_{G,y}a} \tag{6.1-17}$$

$$\frac{1}{K_{G,p}a} = \frac{1}{k_{G,p}a} + \frac{He'}{k_{L,c}a} \tag{6.1-18}$$

$$\frac{1}{K_{L,c}a} = \frac{1}{k_{L,c}a} + \frac{1}{He'k_{G,p}a} \tag{6.1-19}$$

6.1.3.2 Two film-theory for concentrate systems

The mass-transfer principles presented above are derived for equimolecular counter-diffusion in distillation, and for dilute systems in gas absorption. It will be remembered that the mass-transfer rate in a steady-state absorption process, in which mass-transfer is taking place through a second gas which is not undergoing transfer, is greater than that for equimolecular counter-diffusion because of the effects of bulk flow. This problem is important in a process of gas absorption where the concentration of the soluble gas is high.

When solute concentrations in the gas and/or liquid phases are large, the equations obtained for dilute systems are no longer applicable.

The correct equations to use for concentrated systems are as follows:

$$N_A = \overline{k}_{G,y} \frac{y - y_i}{y_{BM}} = \overline{k}_{L,x} \frac{x_{i-x}}{x_{BM}} \tag{6.1–20}$$

$$N_A = \overline{K}_{G,y} \frac{y-y^*}{y^*_{BM}} = \overline{K}_{L,x} \frac{x^*-x}{x^*_{BM}} \tag{6.1–21}$$

$$\text{where } y_{BM} = \frac{(1-y) - (1-y_i)}{\ln\,[(1-y)/(1-y_i)]} \tag{6.1–22}$$

$$y^*_{BM} = \frac{(1-y) - (1-y^*)}{\ln\,[(1-y)/(1-y^*)]} \tag{6.1–23}$$

$$x_{BM} = \frac{(1-x) - (1-x_i)}{\ln\,[(1-x)/(1-x_i)]} \tag{6.1–24}$$

$$x^*_{BM} = \frac{(1-x) - (1-x^*)}{\ln\,[(1-x)/(1-x^*)]} \tag{6.1–25}$$

and where $\overline{k}_{G;y}$; $\overline{k}_{L;x}$; $\overline{K}_{G;y}$ and $\overline{K}_{L;x}$ are the individual-and overall-gas-phase and liquid-phase mass-transfer coefficients for concentrated systems.

The factors y_{BM} and x_{BM} arise from the fact that in the diffusion of a solute through a second stationary layer of insoluble fluid, the resistance to diffusion varies in proportion to the concentration of the insoluble stationary fluid approaching zero as the concentration of the insoluble fluid approaches zero.

The factors y^*_{BM} and x^*_{BM} cannot be justified on the basis of the kinetic theory of fluids since they are based on overall resistances.

Substitution of Equations (6.1–22) through (6.1–25) into Equations (6.1–20) and (6.1–21) results in the following simplified formulae:

$$N_A = \overline{k}_{G;y} \ln\,[(1-y_i)/(1-y)] \tag{6.1–26}$$

$$N_A = \overline{K}_{G,y} \ln\,[(1-y^*)/(1-y)] \tag{6.1–27}$$

$$N_A = \overline{k}_{L;x} \ln\,[(1-x)/(1-x_i)] \tag{6.1–28}$$

$$N_A = \overline{K}_{L;x} \ln\,[(1-x)/(1-x^*)] \tag{6.1–29}$$

The equation for computing the interfacial-gas and liquid compositions in concentrated systems is:

$$\frac{y - y_i}{x - x_i} = -\frac{\overline{k}_{L,x} y_{BM}}{\overline{k}_{G,y} x_{BM}}$$
(6.1–30)

Note that when $\overline{k}_{L,x}$ and $\overline{k}_{G,y}$ are given, the equation must be solved by trial and error, since x_{BM} contains x_i and y_{BM} contains y_i.

The overall gas-phase and liquid-phase mass-transfer coefficients for concentrated systems are computed according to the following equations:

$$\frac{1}{\overline{K}_{G,y}} = \frac{1}{\overline{k}_{G,y}} \cdot \frac{y_{BM}}{y_{BM}^*} + \frac{1}{\overline{k}_{L,x}} \cdot \frac{x_{BM}}{y_{BM}^*} \left(\frac{y_i - y^*}{x_i - x} \right)$$
(6.1–31)

$$\frac{1}{\overline{K}_{L,x}} = \frac{1}{\overline{k}_{L,x}} \cdot \frac{x_{BM}}{x_{BKm}^*} + \frac{1}{\overline{k}_{G,y}} \cdot \left(\frac{x^* - x_i}{y - y_i} \right)$$
(6.1–32)

When the equilibrium curve is a straight line, the terms in parentheses can be replaced by the slope m as before.

All these equations reduce to their dilute-system equivalents as the inert concentrations approach unity in terms of mole-fractions of inert concentrations in the fluids. In dilute systems, the logarithmic-mean insoluble-gas and non-volatile-liquid concentrations approach unity.

In distillation, in which both components diffuse simultaneously, these log-mean factors should be omitted.

6.1.3.3 Additivity of resistances
In the Equation (6.1–16) for example:

$$\frac{1}{K_{G,y}a} = \frac{1}{k_{G,y}a} + \frac{m}{k_{L,x}a}$$
(6.1–16)

The two terms on the right represent, respectively, the separate resistances in the vapour- and liquid phases.

The relative importance of each resistance is determined directly by three factors: each of the individual transfer coefficients, $k_{G,y}a$, and $k_{L,x}a$, and the distribution-coefficient m.

The size of each individual transfer coefficient is determined by eddy and molecular diffusivities of its phase. The molecular diffusivity can be changed only by altering the system or changing the temperature and the *pressure*. The eddy diffusivity is controlled by the fluid dynamics, which are functions of the flow of each phase, the viscosity, and the density.

Typical gas-phase controlling systems are:
• distillation of close boiling mixtures,
• high-pressure distillation,

• absorption or stripping of highly soluble gases.

Typical liquid-phase controlling systems are:

• absorption or stripping of nearly insoluble gases in liquids,

• distillation of systems with very high liquid viscosities,

• at the bottom of the stripping section of a distillation column, when the value of m is high.

Most of the distillation systems fall into the class where either the gas phase is controlling or the gas-phase resistance is the more important one.

6.1.4 Transfer unit concept

6.1.4.1 HTU = Height equivalent to one Transfer Unit

Frequently, the values of the individual coefficients of mass transfer are so strongly dependent on flow-rates, that the quantity obtained by dividing each coefficient by the flow-rate of the phase to which it applies is more nearly constant than the coefficient itself.

The quantity obtained by this procedure is called the *height-equivalent to one transfer unit*, since it is expresses, in terms of a single length dimension, the height of apparatus required to accomplish a separation of standard difficulty. The height of a transfer unit is that depth of packing which produces a change in composition equal to the mass-transfer driving force causing that change.

The following relationships between the transfer coefficients and the values of HTU apply:

$$H_G = \frac{G_M}{k_{G,y}\, a y_{BM}} = \frac{G_M}{k_{G,y}\, a} \tag{6.1--33}$$

$$H_L = \frac{L_M}{k_{L,x}\, a x_{BM}} = \frac{L_M}{k_{L,x}\, a} \tag{6.1--34}$$

$$H_{OG} = \frac{G_M}{K_{G,y}\, a y^*_{BM}} = \frac{G_M}{K_{G,y}\, a} \tag{6.1--35}$$

$$H_{OL} = \frac{L_M}{K_{L,x}\, a x^*_{BM}} = \frac{m G_M}{L_M} \frac{x_{BM}}{y^*_{BM}} H_L \tag{6.1--36}$$

where H_G and H_L are the heights of respectively one gas–phase transfer unit and one liquid–phase transfer unit, H_{OG} and H_{OL} are the heights of respectively one *overall* gas–phase transfer unit and one *overall* liquid–phase transfer unit, G_M and L_M are respectively the molar gas–mass–velocity and the molar liquid–mass–velocity.

This equilibrium-stage concept based on a transfer unit has been proposed for such continuous-contacting devices as, for example, packed towers.

The equations that express the addition of individual resistances in terms of HTUs, applicable to either distillation and dilute systems or concentrated systems are:

$$H_{OG} = \frac{y_{BM}}{y^*_{BM}} H_G + \frac{m\, G_M}{L_M} \frac{x_{BM}}{y^*_{BM}} H_L \tag{6.1--37}$$

$$H_{OL} = \frac{x_{BM}}{x_{BM}^*} H_L + \frac{L_M}{m \, G_M} \frac{y_{BM}}{x_{BM}^*} H_G \tag{6.1-38}$$

These equations are strictly valid only when m, the slope of the equilibrium curve is constant.

In Equation (6.1–37), the two terms represent respectively the separate resistances in the vapour- and liquid-phases. If the relative magnitudes are such that the value of one term is negligible with respect to the other, the system is considered to be single-phase resistant.

6.1.4.2 HETP = Height Equivalent to one Theoretical Plate

HETP is another quantity that is used to express the efficiency of a device for carrying out a separation, particularly in which mass is transferred by a stage-wise action rather than a differential contact. For example, in a tray column, the HETP value is the tray spacing divided by the fractional overall tray efficiency.

The HETP of a column, valid for either distillation or dilute-gas absorption and stripping systems in which constant molar overflow can be assumed, and in which no chemical reactions occur, is related to the height of one overall gas-phase mass-transfer unit, H_{OG}, by the equation:

$$HETP = H_{OG} \frac{\ln\,(m \, G_M/L_M)}{(m \, G_M/L) - 1} \tag{6.1-39}$$

For gas-absorption systems in which the inlet gas is concentrated, the correct equation is:

$$HETP = \left(\frac{y_{BM}^*}{1-y}\right)_{AV} H_{OG} \frac{\ln\,(m \, G_M/L_M)}{(m \, G_M/L_M) - 1} \tag{6.1-40}$$

where the correction term, $y_{BM}^*/(1-y)$, is averaged over each individual theoretical plate. The compositions are used in conjunction with the local values of the gas-and liquid-flow-rates and the equilibrium slope, m, to obtain the values for H_G, H_L and H_{OG} corresponding to the conditions on each theoretical plate, and to compute the local values of the HETP.

6.1.4.3 NTU = Number of Transfer Units

The NTU required for a given separation is closely related to the number of theoretical stages or plates required to carry out the same separation in a stage-wise or plate-type apparatus.

Figure 6.1–4 shows a differential element in a column with the fluid streams.

In this differential section, dz, we can equate the rate at which solute A is lost from the gas-phase to the rate at which it is transferred through the gas-phase to the interface, as follows:

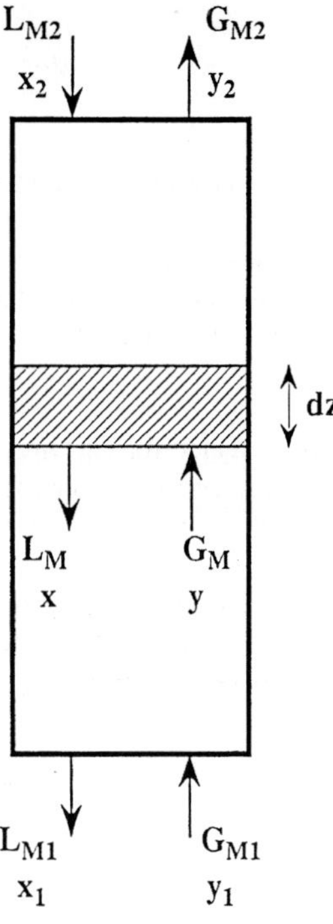

Figure 6.1-4. Schematic representation showing differential element over which a mass balance is made.

$$-d(G_M y) = -G_M dy - y\, d\, G_M = N_A\, a\, dz \qquad (6.1\text{–}41)$$

When only one component is transferred:

$$d\, G_M = -N_A\, a\, dz \qquad (6.1\text{–}42)$$

Substitution and rearranging yields:

$$dz = -\frac{G_M\, dy}{N_A\, a\, (1-y)} \qquad (6.1\text{–}43)$$

For this derivation, we use, for example, the gas-phase rate relationship:

$$N_A = k_{G,y}\, (y - y_i) \qquad (6.1\text{–}2)$$

We integrate over the tower to obtain the height, Z, of the column:

$$Z = \int_{y_2}^{y_1} \frac{G_M \, dy}{k_{G,y} \, a(1-y)(y-y_i)}$$

(6.1–44)

Multiplying and dividing by y_{BM}, we obtain:

$$Z = \int_{y_2}^{y_1} \left[\frac{G_M \, k_G}{y \, a \, y_{BM}} \right] \left[\frac{y_{BM} \, dy}{(1-y)(y-y_i)} \right]$$

(6.1–45)

The first term under the integral can be recognized as the HTU term (here H_G); the second term under the integral defines the number of individual gas-phase transfer units, N_G, required for changing the composition of the vapour stream from y_1 to y_2:

$$N_G = \int \frac{y_{BM} \, dy}{(1-y)(y-y_i)}$$

(6.1–46)

The general expression given by Equation (6.1–45) is more complex than normally required, but it must be used when the mass-transfer coefficient varies from point to point, as may be the case when the gas is not dilute, or when the gas velocity varies as the gas dissolves and changes the total flow-rate of gas at different sections.

Similar equations of definition are obtained with the other transfer units:

$$N_L = \int \frac{x_{BM} \, dx}{(1-x)(x_i-x)}$$

(6.1–47)

$$N_{OG} = \int \frac{y_{BM}^* \, dy}{(1-y)(y-y^*)}$$

(6.1–48)

$$N_{OL} = \int \frac{y_{BM}^* \, dx}{(1-x)(x^*-x)}$$

(6.1–49)

where N_L, N_{OG} and N_{OL} are respectively the number of liquid–phase transfer unit, the number of *overall* gas–phase transfer unit and the number of *overall* liquid–phase transfer unit.

Equation (6.1–48) is the more useful one in practice: it requires either actual experimental H_{OG} data, or values estimated by combining individual measurements of H_G and H_L.

It is easy to define the number of transfer units for equimolar counter–diffusion (distillation) and for dilute systems, when the mole fraction of the solute is so small that the inclusion of terms, such as *(1−y)* and y_{BM}^* for example, can be neglected.

As for the values of HTU (see Equation 6.1–37), it is possible to write the equations that express the addition of individual resistances in terms of the number of transfer units. For example, when m, the slope of the equilibrium curve, is constant, the equations applicable to either dilute or concentrated systems are:

$$\frac{1}{N_{OG}} = \frac{1}{N_G} + \frac{1}{N_L}\frac{m\,G_M\,x_{BM}}{L_M\,y_{BM}^*} \tag{6.1–50}$$

$$\frac{1}{N_{OL}} = \frac{1}{N_L} + \frac{1}{N_G}\frac{L_M\,y_{BM}^*}{m\,G_M\,x_{BM}} \tag{6.1–51}$$

Equation (6.1–50) is another equation for the additivity of resistances. It is subject to the same considerations as Equation (6.1–16) plus some others.

It may be helpful to remember the genesis of the concept of the number of transfer units. For a gas or a vapour, it is essentially the change in concentration of the gas stream, dy, divided by the driving force *(y−y*)* or *(y−y$_i$)*. Similar considerations hold for the liquid.

The flow-rates G_M and L_M entered because a material balance was made, giving the amount transferred for constant molal flows as $G_M\,dy$ and $L_M\,dx$.

The change in concentration of the phase would depend on the flow-rate, even if the transfer coefficients and rate of transfer did not. This equation is not truly one for the additivity of resistances, for each resistance has a scale factor, L_M or G_M. The ratio of G_M to L_M is usually dictated by the plant production; the individual values G_M and L_M are normally controlled by the physical capacity of the equipment rather than by mass transfer considerations, which control the contacting length rather than the flow-area of the equipment.

6.1.4.4 Efficiency

A tray model is shown in Figure 6.1–5. This tray model has the inlet and outlet vapours well mixed, with compositions of y_{N+1} and y_N respectively. It is further assumed that the inlet–and outlet liquids are well mixed and of constant and uniform composition, but not necessarily the liquid on the tray between [8-12].

The *overall* tray, or Murphree efficiency, E_{MV}, is defined as:

$$E_{MV} = \frac{y_N - y_{N+1}}{y_N^* - y_{N+1}} \tag{6.1–52}$$

where y_N^* is the composition of the vapour that would be in equilibrium with the composition, x_N, of the liquid leaving the tray.

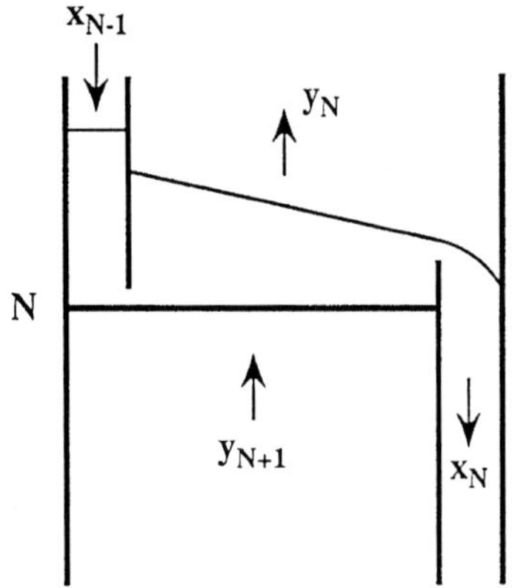

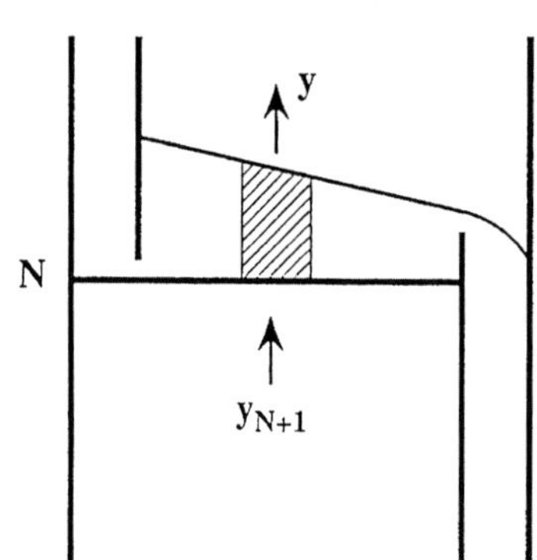

Figure 6.1-5. *Overall* tray model.   *Point* tray model.

Overall Murphree vapour efficiencies may exceed 100%. Thus y_N^* can be less than y_N, because the latter is an average of the $y's$ all across the stage, most of which are in contact with liquid richer than the exit liquid. Thus, the numerator in Equation (6.1–52) would be larger than the denominator, and E_{MV} would exceed 100%.

The *point* efficiency can be defined similarly. Here, the outlet vapour is not mixed although the inlet vapour is assumed to be well mixed and of uniform composition, y_{N+1}. The local-or point-efficiency, E_{OG}, is defined as:

$$E_{OG} = \frac{y - y_{N+1}}{y^* - y_{N+1}} \tag{6.1–53}$$

Here, point efficiencies do not exceed 100%, because y^* is the maximum value y can achieve at the point under consideration.

Similar equations may be written for the liquid-phase efficiencies.

By definition of the number of transfer unit, for example in distillation, the rearrangement of Equations (6.1–44) and (6.1–48) gives:

$$- \ln \frac{y^* - y}{y^* - y_{N+1}} = \frac{K_{G,y} aZ}{G_M} = N_{OG} \tag{6.1–54}$$

The left-hand side of Equation (6.1–54) is the vapour *point* efficiency, and E_{OG} is then related to the overall transfer unit, N_{OG}, by the equation:

$$- \ln (1 - E_{OG}) = N_{OG} \tag{6.1–55}$$

N_{OG} can be related to the individual-phase number of transfer units from Equation (6.1–50).

6.1.5 Effects of the total pressure

The application of Fick's law for the cases of an equimolecular counter-diffusion and a steady-state unidirectional diffusion of a component through an inert-gas film in an ideal gas system gives the two equations for the rate of mass-transfer:

$$N_A = \frac{D_{AB}\,P_T}{RT\,\delta_G}\,(y - y_i) \tag{6.1--56}$$

$$N_A = \frac{D_{AB}\,P_T}{RT\,\delta_G}\,\frac{(y-y_i)}{y_{BM}} \tag{6.1--57}$$

where D_{AB} is the diffusion coefficient of solute A, R is the gas constant, T is the absolute temperature, and δ_G is the gas-film thickness.

The influence of the total system pressure on the rate of mass transfer has been shown to be the same as would be predicted from the stagnant-film theory, where:

$$k_{G,y} = \frac{D_{AB}P_T}{RT\delta_G} \tag{6.1--58}$$

$$\overline{k}_{G,y} = \frac{D_{AB}P_T}{RT\delta_G} \tag{6.1--59}$$

Since the quantity $D_{AB}P_T$ is known to be relatively independent of the pressure, it follows that the rate-coefficients $\overline{k}_{G,y}$; $k_{G,y}\,y_{BM}$; $k_{G,p}\,P_T\,y_{BM}$; $k_{G,p}\,p_{BM}$ and $k_{G,y}$ do not depend on the total pressure.

To determine the mass-transfer rate, one needs the interfacial area in addition to the mass-transfer coefficients. The literature on tower packings, for example, normally reports $k_{G,p}a$ values measured at very low inlet-gas concentrations, so that $y_{BM} \simeq 1$, and at a total pressure close to 1 atmosphere. Thus, the correct rate coefficient for use in packed-tower design involving the use of the driving force $(y-y_i)/y_{BM}$ is obtained by multiplying the reported $k_{G,p}a$ values by the value of P_T employed in the actual test unit (here 1 atm = 101.3 kPa) and *not* the total pressure of the system to be designed [5].

From another point of view one can correct the reported values of $k_{G,p}a$ in units of kmole s^{-1} m^{-3} kPa^{-1} valid for a pressure of 101.3 kPa to some other pressure by dividing the quoted values of $k_{G,p}a$ by the design pressure and multiplying by 101.3 kPa [5]:

$$(k_{G,p}a \text{ at design pressure } P_T) = (k_{G,p}a \text{ at 1 atm}) \cdot \frac{101.3}{P_T}$$

One way to avoid a lot of confusion on this point is to convert the experimentally measured $k_{G,p}a$ *values to values of* $\overline{k}_{G,y}a$ *straightaway, before beginning the design calculations.*

It should be noted that these corrections are proposed with the assumption of a constant value of the interfacial area, a, as a function of the total pressure. It is well known that a can change significantly with the hydrodynamics, and particularly with the pressure conditions.

A design based on the rate-coefficient $\overline{k}_{G;y}a$ and the driving force $(y-y_i)/y_{BM}$ will be independent of the total system pressure, with the following limitations: caution should be employed in assuming that $\overline{k}_{G;y}a$ is independent of total pressure for systems having significant vapour-phase non-idealities, for systems that operate in the vicinity of the critical point where the reduced density ρ_L/ρ_c becomes larger than unity, or for total pressures higher than about 30 to 40 bar [5].

For the liquid-phase mass-transfer coefficient, $\overline{k}_{L;x}$, the effects of the total system pressure can be ignored for all practical purposes [13-14].

Thus, when using $\overline{k}_{G;y}$ and $\overline{k}_{L;x}$ for the design of mass-transfer processes the primary pressure effects to consider will be those which affect the equilibrium curves and the values of m.

6.2 Packed towers: random and structured packings[*)]
Traditionally, packings have not been used in distillation duties, except for small columns of less than 1 m diameter, and even less for high pressure service. The situation has changed over the last 10-15 years, partly through a better understanding of how packed towers work, particularly with respect to liquid distribution, but also through the development of new proprietary packings.

6.2.1 Maximum column capacity
The literature gives few methods that deal with the capacity of random packings in high-pressure duties and there appear to be significant differences among them [15, 16].
Even an approach based on the traditional Generalized Pressure Drop-Correlation (Figure 6.2–1) seems doomed, because high-pressure duties show pressure-drops *two to three times higher* than predicted by such a correlation [17, 32].
The abscissa of this correlation is known as the flow parameter, X:

$$X = \frac{L}{G}\left(\frac{\rho_G}{\rho_L}\right)^{0.5}$$

$$(6.2-1)$$

where G and L are respectively the gas–mass velocity, and the liquid–mass velocity, ρ_G and ρ_L are respectively the gas density, and the liquid density.
This flow parameter is the square root of the ratio of liquid-kinetic-energy to gas-kinetic-energy.
The ordinate includes the flooding gas-flow-rate, G_F, the gas-and liquid densities, ρ_G and ρ_L, the packing factor, F, and a liquid viscosity term, μ_L/μ_E.

---

*) References at the end of section 6.5

The flow parameter, X, primarily depends on vapour-density, and hence on pressure. Thus, vacuum distillations have low values ($X < 0.02$), while high-pressure distillations have high values. The high-pressure region starts at about 10 bar, equivalent to an X of about 0.2. Values of X greater than 1.0 are produced either by operations involving very high liquid–to gas–mass–flow ratios or by high gas–to liquid–density ratios. The stripping of light hydrocarbons under high pressure in an example of the latter case.

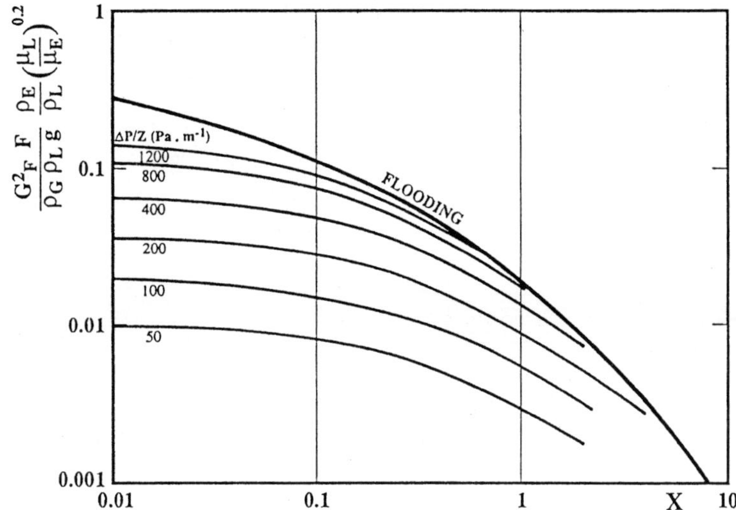

Figure 6.2–1. Generalized flooding-and pressure-drop-correlations for packings versus flow parameter.

An alternative approach is to plot the capacity factor, C_s, (Figure 6.2–2) [18]:

$$C_S = \frac{G_F}{[\rho_G\,(\rho_L - \rho_G)]^{0.5}} \qquad (6.2\text{–}2)$$

as a function of the flow parameter X. The C_S factor is based on the overall column-cross-sectional area. This maximum C_S value must be adjusted for the effect of liquid viscosity, μ_L, and surface tension, σ_L, to give a reduced capacity factor:

$$\frac{C_S}{\left[\left(\frac{\sigma_L}{20}\right)^{0.20}\left(\frac{\mu_L}{0.2}\right)^{-0.12}\right]} \qquad (6.2\text{–}3)$$

The recommended design procedure for an approximate evaluation utilizes a final design vapour-rate, C_S, which is a percentage of the reduced maximum operational capacity factor (usually between 80% and 87%). Note that in high-pressure operations, the usable hydraulic capacity of the tower packing may be reached because of excessive liquid hold-up before the maximum operational C_S is attained.

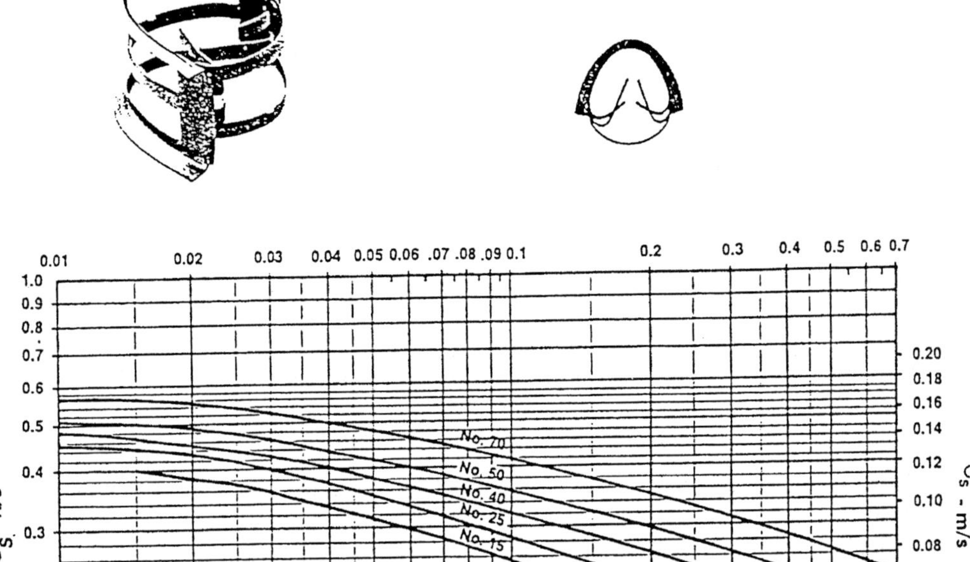

Figure 6.2–2. Maximum operational capacity for IMTP *Norton* packings versus flow parameter [18].

6.2.2 Efficiency

The efficiency of tower packings in distillation is usually considered in terms of HETP values. No method yet exists for the prediction of HETP with a high degree of confidence at elevated pressures. The designer should use caution in determining the design-HETP value. At present, the best of the published rigorous models for predicting the efficiency of a structured packing is the model of Bravo, Rocha and Fair [19-21]. However, this is quite complex to use, and is only validated for distillation systems at operating pressures ranging from 0.02 to 4.14 bar. The model fails to represent the data at higher pressures (for example at 20.4 bar for the butane systems).

M.J. Lockett [22] recommends a simple shortcut method for HETP, based on the preceding model.

Information in the public domain is scarce, and it is usually necessary for operating companies to rely upon the specialist knowledge of the packing manufacturers. The reason is that it is only in recent years that packings have been used in high-pressure distillation duties. And all packings worth considering are proprietary products about which little or no independent performance information is available for this region. The problem is the same for the structured packings which have also been installed in high pressure distillation. Experience at the lower end of the range (10-20 bar) has been good, but there have been a number of spectacular failures at higher pressures [33].

It may be that the random or structured packing efficiency is degraded by the need to limit the bed-height space required for liquid collection redistribution, and by maldistribution and channelling in the packing [33].

6.3 Tray columns[*]

Tray columns are the best choice for high-pressure distillation, especially above 20 bar. An important factor in trays' favour is that they have been used for many years in high-pressure distillation, while it is only in the last 10-15 years that packings have been used significantly in such duties, and especially for structured packings above about 15 bar.

6.3.1 Flow regimes

Three flow regimes are commonly encountered in industrial columns: spray, froth, and emulsion.

In high pressure distillation, tray operation is usually in the *emulsion* regime. In small diameter (less than 1.5 m) columns, or at low liquid loads, or the low end of the pressure range (towards 10 bar), however, the froth-and spray regimes can be found.

In the spray regime, flooding (usually called jet flooding) is caused by excessive entrainment of liquid from an active area to the tray above. It increases the tray pressure-drop, and the entrained liquid recirculates to the tray below. The larger liquid load in the downcomer and the increased tray-pressure-drop together cause the downcomer to overfill; so the tray floods.

In the emulsion regime, there is little entrainment of liquid by the vapour. Instead, the high liquid load causes the downcomer to overfill and the tray to flood.

In the froth regime, which is between the spray and emulsion ones, flooding may be by either mechanism, depending on the tray spacing and the particular combination of vapour and liquid loads.

6.3.2 Downcomer flooding and flooding

As the operating distillation pressure is increased, the mass flow of vapour will be much greater than at atmospheric pressure, because of the high vapour density. While at the same reflux ratio, the mass flow of liquid will also be greater than for an atmospheric distillation. Therefore, liquid flow-rates per unit of column cross-sectional area will be higher as the operating pressure increases. The capacity of the fractionating device at high pressure may be dependent on its ability to handle the high flow-rates.

In a tray column, the tower cross-sectional area is the sum of the active area, plus the areas

---

[*] References at the end of section 6.5

of the downcomer transferring liquid to the tray below, and the downcomer receiving liquid from the tray above. The active area required for trays (in which the vapour bubbles through the liquid phase) is usually determined by the vapour flow rate. The downcomer handles a mixture of froth, aerated liquid, and clear liquid.

The downcomer area required for trays not only increases with the liquid-flow-rate, but also with the difficulty in achieving separation between the vapour and the liquid phases. The volume required for the downcomer (downcomer residence time) increases at a lower surface tension and a smaller density difference between vapour and liquid. Because of the large downcomer area required to handle the high liquid flow rates typical of high-pressure distillations, a trayed column cross-sectional area may be 40% to 80% greater than the active tray area calculated from the vapour flow rates for such distillations. Thus, the downcomer area becomes a significant factor in the determination of the diameter of a tray column.

The mechanisms causing the downcomer to overfill are not easy to define.

Downcomer operation is often described in terms of a non-flooded downcomer, where complications arise from the cascade of froth flowing over the weir onto a froth layer in the downcomer, causing entrainment into-, and gross circulation of the froth layer, in a manner analogous to a waterfall.

In fact, the operation of a downcomer at the point of flooding can be simply illustrated: froth from the active area flows over the weir onto the downcomer froth. The bulk of the vapour disengagement occurs probably in a very short time, and only the small bubbles in the froth are entrained downwards. Within the downcomer, the bubbles coalesce until they are large enough to rise out from the froth. If coalescence is not fast enough, some of the vapour is carried down to the tray below. Even with this simpler picture, the process is not easily modelled.

In current design practice for downcomers, three parameters are commonly considered: liquid residence time, liquid velocity, and downcomer backup.

6.3.3 Liquid residence time

The residence-time concept is commonly misunderstood. The residence time is defined here as the downcomer volume divided by the liquid flow-rate. Typically, it is said that a liquid-residence time is required to allow adequate disengagement of vapour. Generally, two mechanisms are at work in a downcomer to provide the separation of vapour from liquid. The more obvious one is the relative velocity of the phases. If the downward velocity of the liquid exceeds the bubble-rise velocity, it does not matter how much residence time is provided, separation will not occur. This is true unless there is coalescence (the second mechanism), which there always is, to some extent. Coalescence is time dependent and therefore a residence-time criterion has some relevance.

The literature recommends residence times of 4 s for hydrocarbon systems at medium pressure (> 7 bar) and 5 s at high pressure (> 20 bar). In principle, however the two mechanisms cannot be accounted for by a single number, and the use of residence-time criteria for high-pressure systems is not recommended [1].

6.3.4 Liquid velocity

Liquid-velocity correlations are generally based on one of two concepts, and it is not always clear which one is the basis of a given correlation. One concept relies on the principle that vapour should not be entrained to the tray below. The other is grounded on the phenomenon of downcomer-inlet-choking caused by the inability of the low-density froth leaving the active area of the tray to flow into the downcomer.

Various empirical correlations are available for the limiting liquid velocity, but there is poor agreement among them. One of these correlations, compared with published FRI data, appears to be reliable:

$$Q_{DMAX} = 7.5 \sqrt{s(\rho_L - \rho_V)SF} \tag{6.3-1}$$

where Q_{DMAX} is the design velocity (in gallons min^{-1} ft^{-2})

    s is the tray spacing (in inches)

    ρ_L and ρ_V are the liquid-and vapour densities (in lb ft^{-3})

    SF is a system derating factor (usually ignored !).

6.3.5. Downcomer backup

In high-pressure systems (> 10 bar), the trays often are limited by downcomer froth-backup [1].

Liquid flows through the downcomer from a lower to a higher pressure, and consequently backs up in the downcomer. The primary requirement of a downcomer is that its height must be sufficient to accommodate this backup. It is generally acknowledged that the froth in the downcomer should not extend beyond the weir on the tray above. If it does, it increases the clear liquid height on the tray above, which boosts the tray pressure-drop, which extends the froth-height in the downcomer, and so on, to a potentially unstable situation that may eventually lead to flooding.

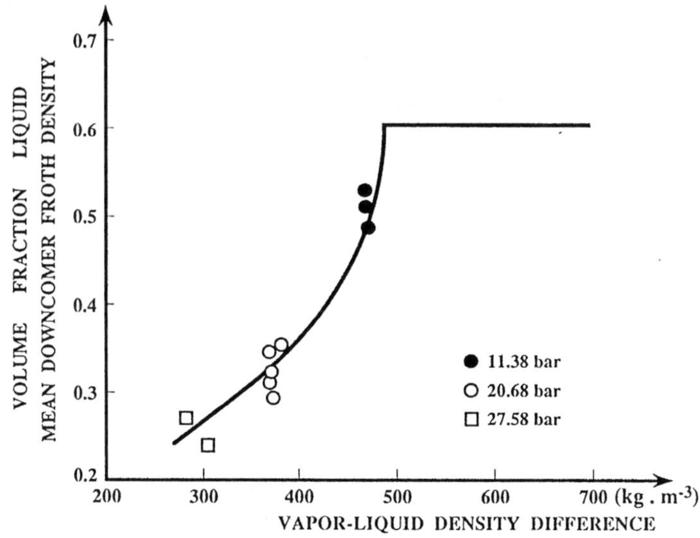

Figure 6.3-1. Empirical correlation for downcomer froth-density [1].

Calculation of the froth height in the downcomer requires consideration of two aspects, the clear liquid back-up and, the so called froth-density, which is actually the volume fraction of liquid in the froth.

The clear liquid back-up is obtained from a tray-pressure balance and is normally taken to be the sum of the tray-pressure-drop, the clear liquid height on the active area of the tray, and the pressure-drop of liquid flowing under the downcomer apron onto the active area.

In the absence of a fundamental understanding, the prediction of froth density relies on empirical correlation. Figure 6.3–1 shows a simple graphical correlation of mean downcomer froth density as a function of the density difference between liquid and vapour. Brierley [1] indicates that the correlation was produced from relatively few data. Therefore, it is preferable to obtain a value from operating data close to flooding on the same system.

The clear liquid backup is divided by the froth-density to give the froth height; if this exceeds the tray spacing plus the outlet weir height, the tray is deemed to be flooded.

6.3.6 Tray efficiency

As might be expected, the vapour phase may offer the controlling resistance to mass transfer in high pressure distillations. Values for tray efficiencies at elevated pressure are scarce [23, 24]. The prediction of tray efficiency may be approached in several ways. One way is to utilize field performance data taken for the same system in very similar equipment. Unfortunately such data are seldom available. When they are available, and can be judged as accurate and representative, they should be used as a basis for efficiency specification [25]. Another way is to utilize laboratory-or pilot-plant efficiency data. For example a small laboratory-Oldershaw tray-column can be used with the same system. Of course, the results must be corrected for vapour-and liquid mixing effects to obtain overall tray efficiencies for large-scale design [26]. Another approach is the use of empirical or fundamental mass-transfer models [27-30].

A notable feature of high-pressure distillation is the high efficiency that is usually obtained on trays. Figures close to 100% are not uncommon. However, the efficiency of trayed columns has been shown to increase only from atmospheric pressure up to a pressure of 11.5 bar. At higher operating pressures, the efficiency of the trays decreases with increasing pressure. There is an entrainment of vapour in the liquid phase which is carried back down the column. For example, for a C_4-hydrocarbon separation the tray efficiency will be reduced by 16% as the pressure is raised from 11.5 bar to 27.6 bar.

At high pressures, especially approaching the critical pressure, the accuracy of the vapour-liquid equilibrium data are questionable. Direct measurement is not easy; prediction with equations of state is risky [1].

A question to be resolved in predicting efficiency concerns the liquid-flow pattern. It is usual practice to assume that the vapour is fully mixed, but there is a diversity of treatments of the liquid phase. The two limiting cases are completely-mixed-liquid and plug-flow-liquid. Achieved efficiencies on well designed trays usually fall between these cases. The assumption of a well-mixed tray liquid is only valid for the smallest trays (pilot scale).

6.4 Trays or packings? [*]

Practical guidance on how to get the best choice between trays and packings in high-pressure distillations is very difficult [33].

---

*) References at the end of section 6.5

In a number of cases, it has been possible to uprate columns substantially by redesign of the trays and by replacing of trays with packing. Nevertheless, the choice between trays and packings clearly is usually close. R. Brierley advises that one should be very prudent until detailed schemes have been evaluated, unless practical experience can be called upon particularly with the various manufacturers [1]. R.F. Strigle [2] on the other hand indicates trustfully that tower packings can be used instead of trays or to replace trays in high pressure distillations, with some examples, to accomplish the following objectives:

- To increase the capacity of downcomer-limited tray columns.
- To improve product-purity or-yield by providing a greater number of theoretical stages than developed by the trays.
- To conserve energy by providing a greater number of theoretical stages, in order to reduce the reflux ratio.
- To permit the use of vapour recompression due to the reduced pressure-drop produced by tower packings, as compared to that developed by trays in the separation of close-boiling materials.
- To reduce energy-use by operating columns in parallel, which, because of the low pressure-drop of tower packings, permits the overhead vapour from the first column to supply reboiler heat to the second column.
- To lower the residence time for materials that degrade or polymerize.
- To reduce the inventory in the column of flammable or hazardous materials.
- To increase product recovery in batch distillations, as a result to the lower liquid hold-up in a packed column.

6.5 Conclusions for pressure distillation

Optimizing the design of high-pressure distillation columns is not easy and often requires several solutions to solve a specific distillation challenge.

Pressure and temperature are the first operating conditions to be set. They are related: the higher the pressure, the higher are the boiling temperatures of the feed and the products. In the case of high-pressure distillation it is economical to select the lowest operating pressure, which will permit satisfactory condensation of the distillate, and reflux at normal cooling-water temperatures [31].

In another way, the effects of pressure on the cost of a distillation column are complex:

- As pressure increases, the temperatures increase. Relative volatilities are reduced as temperatures rise, and this change increases the required reflux and/or the equilibrium stages. On the other hand, the contact-efficiency between vapour and liquid improves with increasing temperature.
- The vapour-handling capacity of a given column (constant diameter) increases roughly with the square-root of the pressure.
- The thickness required for the column-shell increases directly with the pressure and with the column diameter. For a given vapour-handling capacity, increasing the total pressure, P_T, reduces the required diameter (the diameter varies inversely with $P_T^{0.25}$), but increases the thickness of the column shell (the thickness varies directly with $P_T^{0.75}$) [31].

Because some of these effects work in opposite directions, the capacity of a given column to make a specific separation may either increase or decrease as the pressure is raised. The optimum design pressure is usually lower than the pressure for maximum capacity.

Because increasing the pressure raises boiling temperature, it also increases the differences between overhead condensing temperatures and the temperature of the cooling medium, but at the same time reduces differences between the bottom vaporization temperatures and the temperature of the heating medium. Consequently, the size required for the condenser is reduced and the size for the reboiler is increased as the pressure is raised [31].

References of sections 6.1, 6.2, 6.3, 6.4 and 6.5

1. R.J.P. Brierley, High pressure distillation is different, Chem. Engng. Progress, July, (1994) 68-77.
2. R.F.Jr. Strigle, Random packings and packed towers. Design and applications, Gulf Publishing Company, Houston (1987).
3. B.A. Horwitz, Optimize pressure sensitive distillation, Chem. Engng. Progress, April, (1997) 47-51.
4. T.C. Frank, Break azeotropes with pressure sensitive distillation, Chem. Engng. Progress, April, (1997) 52-63.
5. R.H. Perry and D.W. Green, Perry's Chemical Engineers' Handbook, Seventh Edition, McGraw Hill, New-York (1997).
6. J.M. Coulson and J.F. Richardson, Chemical Engineering, Third Edition, Pergamon Press, New-York (1983).
7. A.F. Foust, L.A. Wenzel, C.W. Clump, L. Maus and L.B. Andersen, Principles of unit operations, John Wiley and Sons, New-York (1980).
8. R. Billet, Distillation engineering, Ed. Chemical Publishing Co, New-York (1979).
9. C.D. Holland, Fundamentals of multicomponent distillation, McGraw Hill, New-York (1981).
10. Institution of Chemical Engineers, Distillation, EFCE Publications Series 3, 3rd International Symposium, London (April 1979).
11. Institution of Chemical Engineers, Distillation and absorption '92, EFCE Publications Series 94, Symposium Birmingham (September 1992).
12. M. Van Winkle, Distillation, Ed. McGraw Hill, New-York (1967).
13. K.L. Philips, Proposed explanations for apparent dependence of liquid phase mass transfer coefficients on pressure, Canadian J. Chem. Engng., 51 (1973) 371-374.
14. F. Yoshida and S.I. Arakawa, Pressure dependence of liquid phase mass transfer coefficients, AIChE Journal, 14 (1968) 962-963.
15. C. Mozewski and E. Kucharski, Hydraulics of packed columns under elevated pressure, Inzymiera Chemiczna i Procesowa, 3 (1986) 373-384.
16. H. Krehenwinkel and H. Knapp, Pressure drop and flooding in packed columns operating at high pressures, Chem. Engng. Technol., 10 (1987) 231-242.
17. P. Copigneaux, Distillation et absorption : colonnes garnies, Techniques de l'Ingénieur, J 2626-1 à 21, Paris (1980).
18. Norton's design packing tower booklet.
19. J.L. Bravo, J.A. Rocha and J.R. Fair, Mass transfer in gauze packings, Hydroc. Proc., Janvier, (1985) 91.
20. J.L. Bravo, J.A. Rocha and J.R. Fair, A comprehensive model for the performance of

columns containing structured packings, I. Chem. E. Symp. Ser. 128, A 439, Rugby (1992).

21. J.L. Bravo, J.A. Rocha and J.R. Fair, Distillation columns containing structured packings- A comprehensive model for their performance: mass transfer model, Ind. Chem. Res., 35 (1996) 1660-1667.

22. M.J. Lockett, Easily predict structured packing HETP, Chem. Engng. Progress, January, (1998) 60-66.

23. M.J. Lockett, Distillation tray fundamentals, Cambridge University Press, Cambridge (1986).

24. W. Witt and H. Knapp, Tray efficiencies in high pressure absorption columns: experiments and correlation, German Chem. Engng., 8 (1985) 158-164.

25. M.J. Lockett and I.S. Ahmed, Tray and point efficiencies from a 0.6 meter diameter distillation column, Chem. Engng. Res. Des., 61 (1983) 110-118.

26. J.R. Fair, H.R. Null and W.L. Bolles, Scale-up of plate efficiency from laboratory Oldershaw data, Ind. Engng. Chem. Process Des. Dev., 22 (1983) 53-58.

27. AIChE, Bubble tray design manual, Prediction of fractionation efficiency, AIChE, New–York (1958).

28. G.X. Chen and S.T. Chuang, Prediction of point efficiency for sieve trays in distillation, Ind. Engng. Chem. Res., 32 (1993) 701-708.

29. M.M. Dribika and M.V. Biddulph, Scalingup efficiencies, AIChE Journal, 32 (1986) 1864-1875.

30. H. Chan and J.R. Fair, Prediction of point efficiencies on sieve trays. Binary systems, Ind. Engng. Chem. Process Des. Dev., 23 (1984) 814-819.

31. R.J. Hengstebeck, Distillation: principles and design procedures, Reinhold Publishing Co., New-York (1961).

32. H.Z. Kister and D.R. Gill, Predict flood point and pressure drop for modern random packings, Chem. Engng. Progress, 87(2) (1991) 32-42.

33. D.P. Kurtz, K.J. McNulty and R.D. Morgan, Stretch the capacity of high pressure distillation columns, Chem. Engng. Progress, 87(2) (1991) 43-49.

378

6.6 Extraction from solids

The most widespread application of supercritical fluid technology is for extraction purposes.

6.6.1 Fundamentals

The basic principles for using a supercritical fluid as an extraction medium are the solubility and phase equilibrium of substances in the compressed gas. First, the compounds which have to be extracted must be soluble in the supercritical fluid at moderate pressure and temperature. The solubility and phase-equilibria can be varied over a wide range by changing the pressure and temperature. However, beside the solubility behaviour mass-transfer plays an essential role for extraction processes. Fig. 6.6-1 shows a typical trend of extraction-yield over extraction-time. At the beginning, the extraction efficiency is limited by the solubility in the available amount of fluid. Higher solubilities, and therefore shorter extraction times, can be achieved by increasing the extraction pressure, because with a higher fluid density, the solvent power also increases. The same is normally found also by increasing the extraction temperature, because at higher pressure levels the increase in vapour pressure of the substances to be dissolved is more effective than the reduction in the fluid density at higher temperatures. The second phase of extraction is controlled by diffusion, and this, especially, leads to long extraction times. Therefore, the aim is to reach the proposed extraction yield within the solubility phase, because otherwise the process will not be economical.

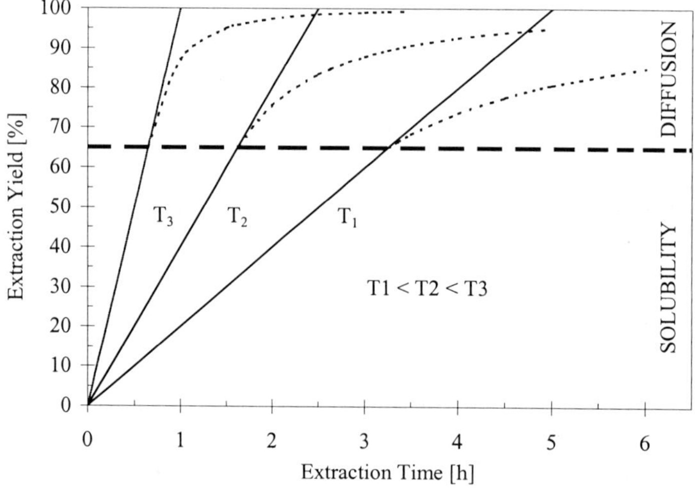

Figure 6.6-1. Typical trend of extraction lines

For calculation of the extraction yield in the solubility phase, Brunner [1] gives the following correlation for solid raw materials:

$$E = k_S * a_S * V_t * \Delta c_m \qquad (6.6\text{-}1)$$

where k_S is the mass-transfer coefficient (m/s); a_S, the specific interfacial area (m²/m³); V_t, the bed volume (m³); and Δc_m, the mean concentration gradient. The mass-transfer coefficient, k_S, can be calculated by a correlation with the Sherwood number

$$Sh = k_S * d \, / \, D_{12} = 2 + 1{,}1 * Sc^{1/3} * Re^{0,6} \qquad \text{for } 3 < Re < 3.000 \qquad (6.6\text{-}2)$$

It is obvious that the mass-transfer coefficient, k_S, can be influenced by the diffusion coefficient, D_{12}, which is also included in the Sherwood number, Sh, as in the Schmidt number, Sc. Diffusion can be increased by shortening the diffusion length. For solid materials this is achieved by smaller particle sizes, which further leads to a higher specific interfacial area, a_S. However, there is a limit for reducing the particle size because if the particles are too fine, the problem of channelling arises, so an optimum has to be found.

A further possibility for increasing extraction yields is to use higher flow-rates. This causes an increase of the Reynolds number, Re, which is defined as:

$$Re = \frac{v * d}{v}, \qquad (6.6\text{-}3)$$

because at same extractor diameter, d, the velocity, v, is increasing. Higher flow-rates also give a larger mean concentration gradient, Δc_m, because the equilibrium concentration is constant at a given pressure and temperature, and the loading of the extraction fluid is lower. Increasing the velocity, v, also gives a better mixing effect in the extractor, which results, under certain conditions, in a fluidized bed.

If pure supercritical solvents have too low a solubility for substances of interest, the addition of small amount of entrainers to the extraction gas is another possibility for increasing the extraction efficiency by reducing the extraction time. The effect of the entrainer has to be considered. On the one hand, the entrainer may influence the polarity of the fluid and therefore dramatically increase the solubility of the more polar substances. Supercritical CO_2 is a particularly non-polar solvent. Adding small amounts of alcohol or even water, for example, changes the polarity much more than increasing the pressure to high levels. On the other hand, so-called matrix modifiers are used which influence the availability of compounds in the solid matrix so that they can be extracted. An example is caffeine, which is soluble in pure CO_2, but for the decaffeination of coffee and tea, water-saturated CO_2 has to be used. Otherwise, the caffeine is incorporated too strong in the matrix of the raw material and cannot be extracted. The negative effect of using a modifier is that no solvent-free products are achieved, because both the extract and the entrainers are collected in the separator. Either the product is useful in this form, or a subsequent separation step has to be included to provide solvent-free extracts and to recover the entrainer and recycle it in the extraction process.

After sufficient extraction the dissolved substances have to be separated from the fluid in a following step. *Decreasing the pressure* at constant temperature reduces the fluid-density, and therefore the solvent-power of the fluid. The extracted substances are collected at the bottom of the separator, as shown in Fig. 6.6-2. To close the solvent cycle, the fluid has to be recompressed to extraction pressure.

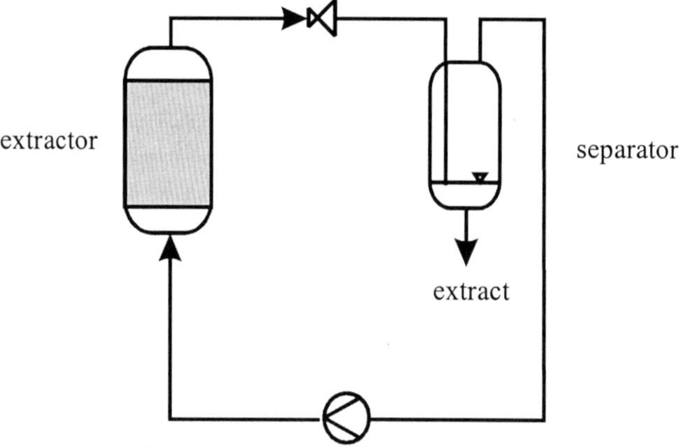

Figure 6.6-2. Separation by pressure reduction

A density reduction of the extraction gas can also be achieved by *increasing the temperature* (see Fig. 6.6-3). However, for this kind of separation the solubility behaviour of the dissolved substance has to be taken into account. A temperature increase also results in a higher vapour pressure, which may result in higher solubility so that no separation can occur. No further complete separation is achieved, which results in a residual loading of the extraction gas and in longer extraction times and higher residual concentrations because of the smaller concentration difference between the entering fluid and the extractive material. The advantage of this separation method is that the process can be operated under isobaric conditions so that no compression step is necessary after separation and the pump only cycles the fluid at a constant pressure level.

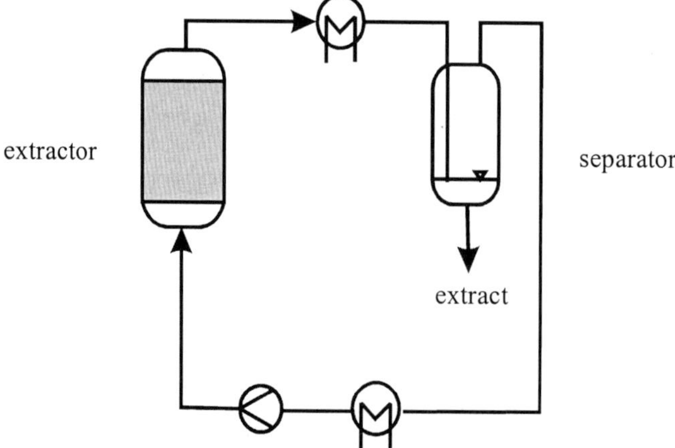

Figure 6.6-3. Separation by temperature change

Most applications use a *combination of pressure- and temperature change*. After pressure-reduction the extraction gas is heated so that it reaches the gaseous state. In this phase, no solvent-power of the extraction gas is present for any substances and therefore complete separation of extracted substances takes place. In cycling the extraction gas after separation, it has to be condensed, under-cooled to prevent cavitation of the pump, and recompressed and heated up to extraction temperature.

A further method separates the extracted substances by *absorption*. Basic for this method is that there should be a high solubility of extracted substances in the absorption material, and that the solubility of absorption substance in the circulation solvent should be as low as possible. Further, the absorption material must not influence the extract in a negative way and a simple separation of extract and absorption material has to be available. An ideal absorption material is therefore a substance which is present in the raw material. Most plant-materials contain water, which can act as a very successful absorption material. An ideal example is the separation of caffeine for the decaffeination of coffee and tea. On the one hand, water has a low solubility in CO_2, and on the other, water-saturated CO_2 is necessary for the process. The extracted caffeine is dissolved into water in the separator and caffeine can be produced from this water-caffeine mixture by crystallization. One advantage of this separation method is that the whole process runs under nearly isobaric conditions.

Analogous to above mentioned method, the *adsorption* of extract on a suitable solid material, is possible (see Fig. 6.6-4). The disadvantage of adsorption is that regeneration of the adsorbent is difficult or impossible. Therefore, no extract is produced and adsorption can only be used if the residual material is the product and not the extract.

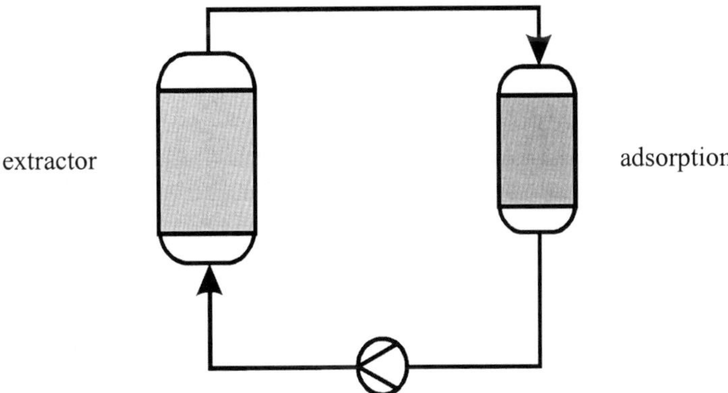

Figure 6.6-4. Separation by adsorption

An industrial-scale application is the decaffeination of coffee and tea where a direct separation of the extracted caffeine in the extractor is realized. A layer of activated carbon follows a layer of raw material, and so on. In this way, the loaded extraction fluid is directly regenerated in the adsorption layer and enters as pure solvent into the next stage of raw material. The great advantage of this method is that no further high-pressure vessel is necessary for separation, which reduces investment costs dramatically.

Adding a substance with low solvent power can also reduce the solvent power of a supercritical gas. Adding nitrogen, for example, reduces the solubility of caffeine

382

dramatically, and lasts to a separation. In a subsequent step this substance has to be removed from the extraction gas before entering the extractor again, and this can be achieved by membrane techniques.

6.6.2 Design criteria

In this Chapter, fundamentals of design criteria in relation to processes and equipment are reviewed for dense-gas-extraction from solid matrices. Although, as mentioned in previous chapters, numerous dense gases can be used as solvents. In the following discussion we concentrate on the most extensively used gas-carbon dioxide. The reason for this is its non-toxic, non-flammable and inert nature, the possibility of gentle treatment of thermally sensitive materials, and the fact that it is inexpensive and an environmentally acceptable material.

Successful engineering design presupposes knowledge of reliable data on mass transfer, including thermodynamic properties, such as phase-equilibrium data and solubility, within a technically and economically possible range of pressure and temperature.

For the development of a new process or application, or the design of a plant, the following main parameters must be known or obtained from the engineer concerned:

- Specific basic data;
- Thermodynamic conditions for extraction and separation;
- Mass transfer;
- Energy consideration/consumption by means of the TS-diagram;
- Specific separation problems;
- Selection and design of proper plant components.

As will be explained later, the first two factors influence mainly the size of extractors and separators, while mass-transfer determines the CO_2 circulation system, and consequently the energy consumption and the size of heat exchangers and the piping system.

6.6.2.1 Specific basic data

The design engineer must have a clear picture of the service requirements of the plant, which presuppose knowledge of:

- Raw material specifications;
- Desired plant size;
- Final product specification;
- Plant location, with local prevailing conditions.

The extraction of vegetable plant-material requires careful selection of raw materials, in connection with chemical analysis, because the content of the vegetables to be extracted can vary substantially. Also the possibility of high concentrations of undesired substances - either contained within them, or contaminations such as pesticides - can influence the product quality enormously. Exhaustive extraction is not always possible with acceptable effort, owing to the vegetable structure, and the extraction efficiency using low-quality raw materials may be too poor to allow economic processing.

A more detailed review in Chapter 8 shows that the plant size can have a strong effect on its feasibility.

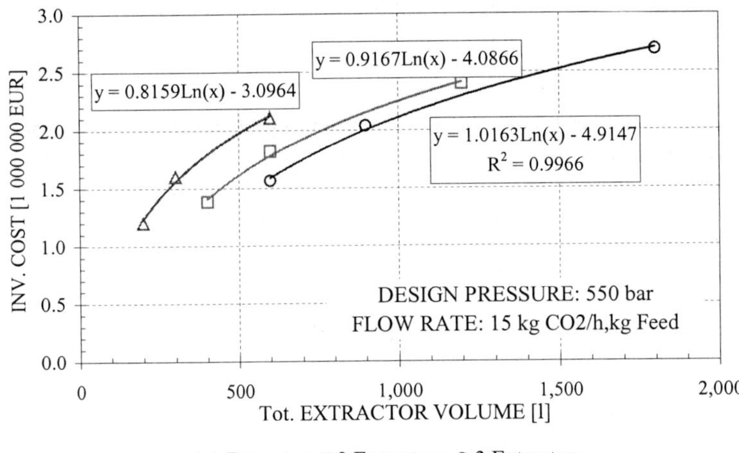

Figure 6.6-5. Investment cost for multi-purpose plants designed for 550 bar

It is obvious from Fig. 6.6-5 that the investment costs for a specific type of plant can be determined accurately enough from the following equation:

$$\text{Investment Cost} = C \cdot \ln V_T - A \tag{6.6-4}$$

With: A and C = Plant Specific Constants
        V_T = Total Extraction Volume [L]

The final product specification can influence the process with regard to processing conditions (pressure, temperature, pretreatment of feed material), and may require fractionated separation or further treatment such as concentration, purification or demoisturization.

The plant location can determine the mechanical construction - for example, in case earthquake factors are to be considered, and climatic conditions can have an impact on machinery sizing and dimensions particular for cooling machines and electrical drives.

Extractable substances

Natural material to be extracted can be divided into two categories.

A. Raw materials, whose geometry must be maintained during the process, and/or only undesired substances are removed.

This group represents the largest one in terms of tonnage and comprises the following applications:
- Decaffeination of green coffee, black- and green tea;
- Defatting of cocoa press cake and nuts;

- Removal of plant protective materials from cereals, particularly rice and pharmaceuticals (for example, ginseng);
- Reducing the alcohol content in beverages.

Such processes require a high selectivity for the substances to be removed, in order to maintain the flavour, appearance, smell, and shape of the treated feed material which represents the main product. Because CO_2 has selective solubility properties, which can be altered to some extent, its use may often be feasible. By-products, such as caffeine can - if recovered - improve the economics.

B. Materials which allow pretreatment and where the extracted substances represent the main product.

For such processes the extract's revenues must cover the raw-material's treatment cost, and therefore only highly valuable extracts with a corresponding yield justify the application of the CO_2 technology. Furthermore, it must be remembered that extracts produced using different solvents, can vary in their chemical composition and appearance, owing to the different selectivities.

Table 6.6-1
List of bulk densities for various feed materials

Pepper, ground:	470 kg/m³	Camomile:	175 kg/m³
Paprika, ground:	450 kg/m³	Laurel, ground:	130 kg/m³
Caraway, ground:	400 kg/m³	Vanilla, ground:	350 kg/m³
Coriander, ground:	400 kg/m³	Mace, ground:	450 kg/m³
Clove, ground:	450 kg/m³	Nutmeg, ground:	470 kg/m³
Juniper, ground:	400 kg/m³	Ginger, ground:	450 kg/m³
Cardamom, ground:	370 kg/m³	Rice:	820 kg/m³
Rosemary, ground:	360 kg/m³		

Pretreatment of raw materials

Higher extraction volumes in high-pressure processes represent higher costs and attention must be paid to limiting them as far as possible. The main impact comes from the bulk density of the feed material (see Table 6.6-1).

The ground feed-material should have average particle sizes of 0.4 to 0.8 mm. Smaller particles, although better for mass transfer, reduce the fluidized bed velocity, can cause clogging of filters, and tend to channelling. Ground raw materials with bulk densities below 200 to 250 kg/m³ should, if possible, be pelletized. Such pretreatment can be advantageous in the destruction of cell matrices, thereby possibly lowering mass-transfer resistances. For example, the pelletizing of hops increases the bulk density from about 150 to 500 kg/m³ and improves the yield.

The extraction result is further influenced by the moisture content. For spices and herbs this concentration should be in general between 8 and 15 %. At higher values the water is extracted preferentially and/or the co-extraction of polar substances is at a low level. Low moisture contents cause shrinking of the cell structure and consequent hindrance to diffusion, which reduces the yield. The optimum moisture levels should be determined by means of extraction tests. The best moisture content for black tea is between 20 and 27 %, for green coffee between 35 and 45 %, and for the removal of pesticides from rice, between 15 and 20 %. For paprika extraction the moisture should be as low as possible, to avoid water enrichment at the opposite side of the CO_2 entrance, which results in agglomeration and clogging of filters at the CO_2 exit.

For rosemary, a high moisture content yields a solid, crystalline extract, while the raw material with a high essential-oil content yields a viscous, brownish and homogenous product. Essential oils act as entrainers for other valuable substances in other biomaterials, also.

6.6.2.2 Thermodynamic conditions

No other solvent extraction process other than the CO_2 technology allows such a strong influence on loading, phase equilibrium, and selectivity. Unfortunately, the solubility of extracted substances in CO_2 is relatively low, compared with the usual solvents which give absolute miscibility with the extracted valuable materials in most cases. The determination of solubility and solvent ratios is therefore important for the economy of the process.

The target for the design engineer is to obtain the highest possible loading, with consideration of the selectivity and product quality, within the extraction step, and the lowest possible one for the separation to ensure optimum precipitation and to avoid carry-over of extract with the solvent.

The solubility is mainly dependent on the density of the dense gas, whereas the extraction pressure is limited for technical and economic reasons. Industrial-scale plants use pressures between 45 and 60 bar for the separation step. Lower pressures would increase the energy consumption substantially, especially in the pump-driven process, because condensation is to be performed at lower temperature levels.

The separation temperature should be chosen to be at least 5 to 10°C above saturation temperature to prevent floating with the liquid CO_2 and to ensure that the precipitated extract is under liquid conditions.

For the modelling of thermodynamic conditions, many useful models have been published in the literature [4] to describe the experimental equilibrium data and to enable fast optimization of the parameters. The well-known models are based on the three equation systems: empirical equations of state; semi-empirical equations of state; and derivations from the association laws and/or from the entropies of the components.

Semi-empirical equations of state are used to fit binary systems, and thermodynamic data for the solute and the solvent have to be known. For systems with more than two components, the mathematical solution is very difficult. A useful equation of state for the description of simple binary systems is the Peng-Robinson equation [5], applicable for large pressure and temperature ranges.

For unknown solute properties, a derivation from the association laws is used to adjust experimental results; Chrastil has published the most popular one [6]. An example is given in Fig. 6.6-6, comparing measured and calculated equilibrium data for rapeseed oil in CO_2, showing the dominating influence of the solvent density. This equation is very useful for describing the equilibrium distribution of natural substances in dense gases.

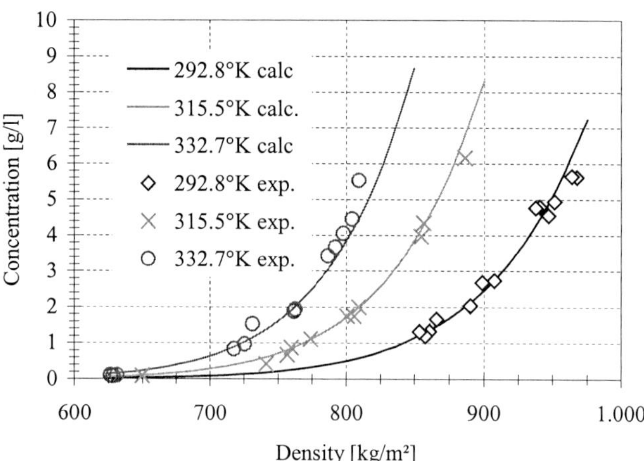

Figure 6.6-6. Comparison of measured and calculated equilibrium data (Chrastil) for rapeseed oil

6.6.2.3 Mass transfer

The target of this investigation is to find the optimized specific solvent mass-flow-rate, in order to obtain the highest extraction efficiency, that means, the highest possible yield in the shortest time, with the lowest energy input. The investigation should yield the amount of CO_2 per kg of raw material and hour, the extraction time, and the length-to-diameter ratio of the extractor.

As in conventional processes the mass transfer at HPE is determined from diffusion and hydrodynamic data.

Diffusion

The diffusion depends on the size of particles and the initial distribution of the substances to be extracted within a solid matrix. The valuable substance can be located within the plant cell, on the surface, or absorbed in the solid matrix. First, the location of the substances must be determined. Further, it is necessary to know if the desired substance is bound in any chemical reaction to the matrix or bound to a complex. Literature searches are necessary to find out more information about how the substance is embedded in the matrix.

For non-chemically bonded substances the diffusion can be influenced by the following processes:
- Changing the particle size to shorten the diffusion paths and destruction of the plant cells.
- Breaking of the cell walls through pelletizing.
- Swelling of the cell structure through moisturizing of the raw material.
- Breaking of complex bonding through moisturizing or changing the pH-value.

Hydrodynamics

The hydrodynamics are mainly influenced by the size and form of the particles and by the particle size distribution, which in most cases should be in a range between 0.4 and 0.8 mm. Too small particle sizes can cause channelling which, for example, can occur during the extraction of paprika powder, egg-yolk powder, cocoa powder, and algae powder. Fibrous feed materials such as ginger, black- and green-tea, and paprika have a tendency to swell and block the filters at the outlet of the extractor. The mechanism of transport in the solid phase proceeds in the following parallel and consecutive steps.

- Absorbing the solvent in the plant matrix, swelling the cell structure, and dissolving the soluble substances.
- Transport of dissolved substances to the outer surface by means of diffusion and passing through.
- Transport from the surface layer into the solvent.

The particle-size distribution and the shape of the particles also influence the ratio of length-to-diameter of the extractor. As the costs for extractors depend not only on volume, but mainly on the diameter, the proper selection of the extractor height/diameter ratio must be made, with a preference for slim vessels. So far, only for the decaffeination of coffee beans with a particle size of about 7 mm can large ratios of 9:1 (length to diameter) be applied. For large particle sizes it is of advantage to feed the solvent from top to bottom, in order to avoid back-mixing. For the usual particle-size-distribution, ratios of 6:1 should be used. For raw material which tends to swell, like black tea or paprika, the ratio should be only 3:1, or if the extractor is equipped with baskets, the baskets should be equipped with multiple distribution.

Increasing the mass-transfer through agitation is, in most technical applications, not feasible.

6.6.2.4 Process optimization by means of the TS-Diagram

The fourth main parameter deals with the estimation of the optimized process cycle. There are various possibilities for running the cycle in the TS-diagram.

For the energy- and cost-calculation of the HPE-process a program has been developed, allowing the determination of the various cycle processes under selected conditions for extraction and separation. In principle, the cycle is either in the pump- or in the compressor-mode.

Pump process

The pump process commences (see Fig. 6.6-7) in the storage tank, with dense gas under liquid conditions (1). To avoid cavitation, the solvent is precooled (2) ahead of the circulation pump, where compression to the required extraction pressure (3) takes place. The temperature for extraction is adjusted by means of a heat exchanger (4). The dissolved substances in the solvent leaving the extractor are to precipitate in the separator, in our case by changing the pressure and temperature by means of a control valve, according to an isenthalpic expansion. The liquid part of the solvent (5 to 7), which is now in the two-phase region (5), is evaporated and the complete gaseous phase superheated (8). The solvent power is substantially reduced and the dissolved substances can be separated. The gaseous solvent is condensed afterwards, thereby closing the cycle process.

388

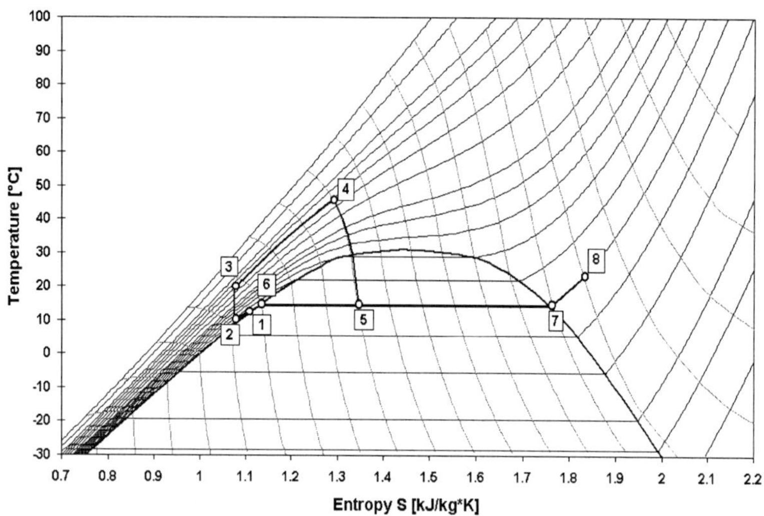

Figure 6.6-7. Pump process in the temperature-entropy diagram

The main energy consumers in the pumping mode are the cooling unit, with the highest demand for electrical energy, the circulation pump, the CO_2 pre-heater, and the evaporator.

The electrical power consumption of the refrigerator depends only on the separation condition. The pumping mode is optimal for high extraction pressures (more than 150 bar), low extraction temperatures (40 to 80°C), and separation pressure between 45 and 60 bar.

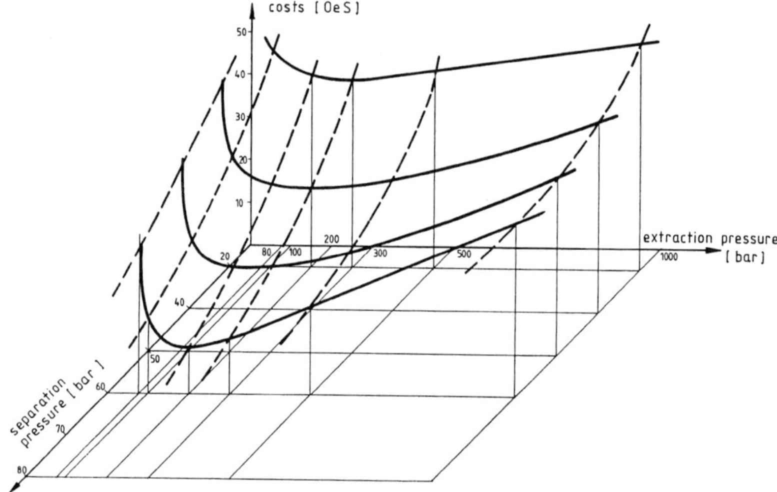

Figure 6.6-8. Energy consumption of pump process

Compressor process

The starting point for the compressor process (see Fig. 6.6-9) is the condition prevailing within the separator, that means, the gaseous solvent enters directly into the compressor (1) and is pressurized to the extraction pressure (2). The use of intermediate cooled compressors is necessary, otherwise low efficiency and extremely high outlet temperatures would be reached. After compression, the extraction fluid has to be cooled to the extraction temperature (3), and under these conditions extraction takes place. Analogous to the pump process, the loaded fluid is expanded in the two-phase region (4) and heated up to gaseous conditions (1) where separation takes place.

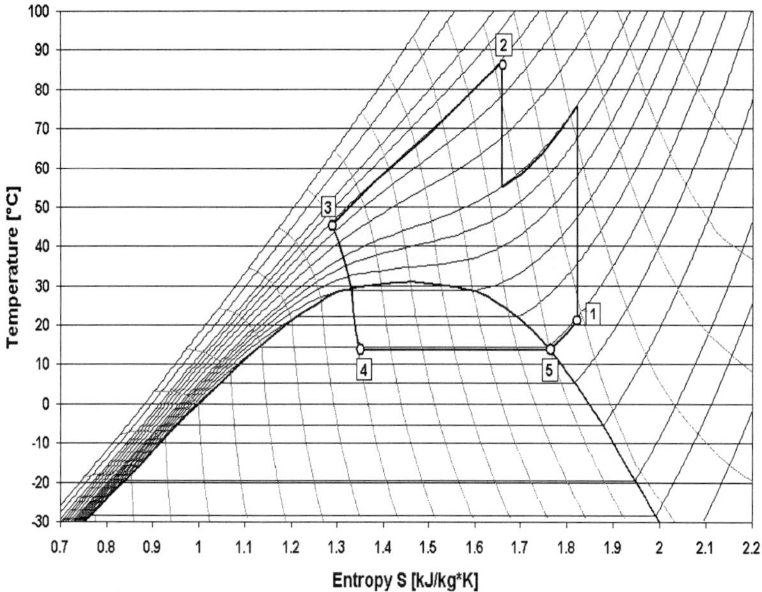

Figure 6.6-9. Compressor process in the temperature-entropy diagram

The advantages of using compressors are that for the solvent circulation no condenser and undercooler are necessary, which reduces investment and running costs.

For this process, almost all of the energy is provided from the compressor. Compression ratios greater than 3.5 require multistage compressors, where intermediate cooling steps must be installed. Compressor processes have low energy consumption for low extraction pressures (80 to 200 bar) and low separation pressures (30 to 40 bar). Owing to the high temperature after the compressor (80 to 120°C) it is also recommended for high extraction temperatures. In the food industry bone-dry compressors are normally used, in which the discharge pressure is limited to 200 bar. Also high separation pressures - above 60 bar - can be maintained with the compressor process. Under these conditions CO_2 cannot be compressed with a pump, owing to the high compressibility. In the case where a single-stage compressor is sufficient, the investment costs are lower as well.

Isobaric process

Processes with high mass-flow rates, for example more than 40 t/h, have energy demands of a very high level. Especially for the decaffeination processes, in which several hundred - up to a thousand tons of CO_2 are in circulation, isobaric processes were developed. In these processes, the extraction step and the separation step have nearly the same pressure and temperature. The separation of the dissolved substance from the CO_2 in circulation is maintained by adsorption on activated charcoal, with an ion-exchanger, or by absorption in a washing column.

6.6.2.5 Separation of dissolved substances

Most of the industrially applied separation processes use precipitation by means of reduced solvent power by changing the pressure/temperature in one or more steps. In most cases, as for the production of oleoresins from spices, or the extraction of hops, a single-step separation is sufficient. Double- or triple-step separations are applied for spice extraction, in case enrichment of pungency, colour or essential oils are desired - as, for example, with pepper, with precipitation of piperin, the pungent substance of the pepper, in the first step, and essential oil in the second step.

For highly volatile aromas of fruits or wine, the single- or double step separation based on pressure is not sufficient, and needs expensive precipitation at very low temperatures. About minus 50°C are necessary for sufficient recovery of aroma components. For this application an aroma rectification, with multiple withdrawal at different temperatures, is appropriate.

The separation in the isobaric decaffeination processes is executed with absorption of caffeine, that means, the caffeine dissolved in CO_2 is carried over into water by means of a packed washing column, or by adsorption with activated charcoal, but without recovery therefrom. Other separation methods under investigation are the use of membranes, since the difference in molecular weight between extract and solvent is high enough, or by the addition of substances of low solvent power. It is questionable whether the advantage of the possible isobaric process can compensate for the investment for the additional process steps required.

6.6.3 Cascade operation and multi-step separation

Up to now, no appropriate methods are available for continuous feeding and emptying devices for solid materials which are applicable for high pressure plants. In different research studies the possibility of screw presses, or moving cells through a high pressure vessel are described, but the technical realization for industrial plants is too difficult. Therefore, solid materials are extracted by cascade operation which means that more than one extractor is installed (see Fig. 6.6-10). In this way, a quasi-continuous flow of the solid material can be achieved because the extractors are connected in series. The compressed extraction gas passes through the extractors in the sequence 1-2-3-4 until extraction in extractor 1 is finished. This extractor is switched off from the cycle, decompressed, emptied, refilled with fresh material, and added as the last one to the sequence which is now 2-3-4-1. In the next step, extraction in extractor 2 is finished and this one will be added again as last one in the sequence, and so on. This sequence has the advantage that extremely high extraction yields are achieved because the pure solvent-flow is contacted with pre-extracted material so that a concentration difference is present, and on the other side the pre-loaded solvent flow enters the extractor with fresh material, so the maximum solvent capacity is used.

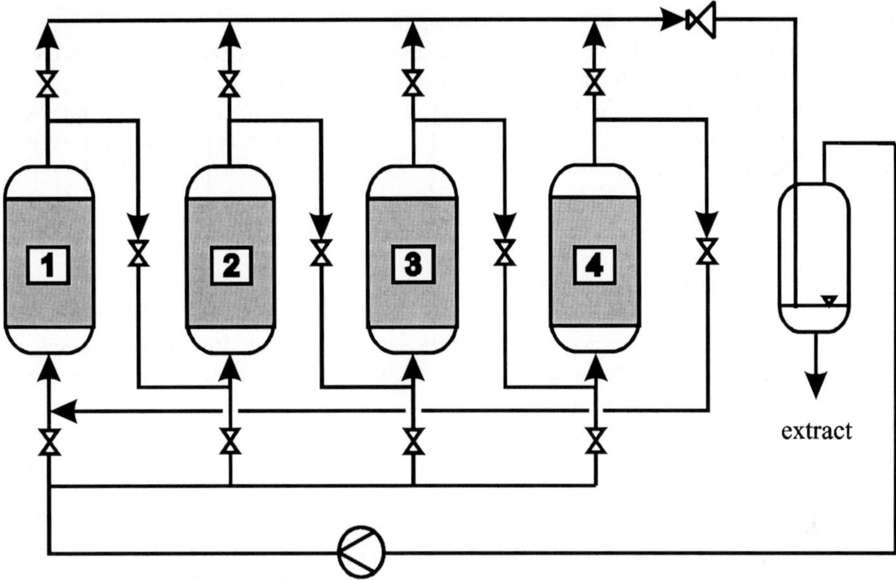

Figure 6.6-10. Cascade operation for extraction of solid materials

The number as well as the volume of extractors has to be optimized and fitted to the extraction problem. Larger extractor volumes result in lower separation times from the sequence, but the investment costs increase with increasing extractor volume. Extractors which are too small cause longer dead-times because the number of separations from the sequence increases.

The basis of a multi-step separation (see Fig. 6.6-11) is, that the solvent power of the compressed fluid can be influenced by changing pressure and temperature. This is one of the advantages of the high-pressure technology, that in the down-stream phase a simple fractionation of the product can be achieved. Normally, little pressure reduction is performed in the first separator, together with temperature change, to remove this fraction with the lowest solubility in the compressed fluid. All other substances have to remain in the compressed gas and enter the next separator where, again, change in pressure and/or temperature takes place. The last separator must guarantee that all light volatile substances are separated, and leave at the outlet a pure extraction fluid which is recycled. Further loading of the fluid will cause a reduction in extraction efficiency. In this way different fractions of the total extract are produced so that different products can be obtained from one raw material in a single process. This becomes more and more common because in most cases certain fractions are of interest but not the total extract.

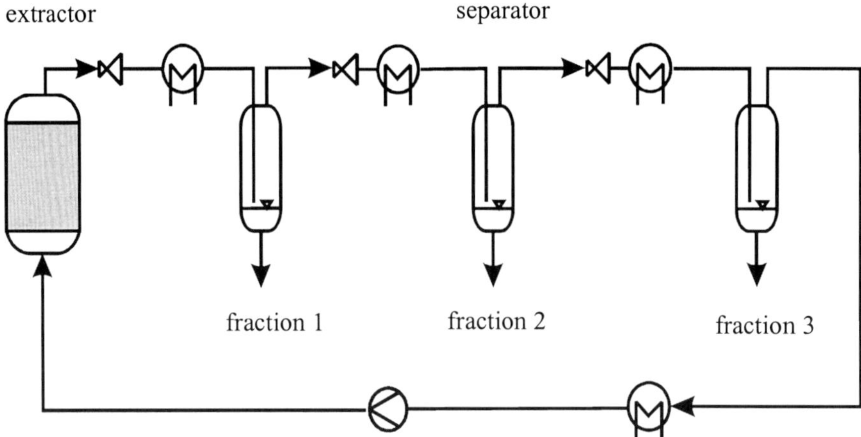

Figure 6.6-11. Multi-step separation

6.6.4 Main applications

With the permission of the Author: Schütz Consulting, D-83308 Trostberg, 2000.

As one can see from Table 6.6-2 the decaffeination of coffee and tea is the largest application for supercritical fluid extraction, in terms of annual capacities and investment costs. Since the beginning of the 1970s, to the early 1990s, nearly 50% of the whole production capacity for decaffeination of coffee and tea changed to the supercritical extraction process. As the market for decaffeinated coffee is stable, no further plants have been installed within the past eight years.

The second largest application is in the extraction of hops. In the last twenty years nearly all producers of hop extracts have changed to the supercritical extraction process. Even in the Eastern European countries the methylene chloride process was stopped several years ago.

The extraction of spice oleoresins is relatively new, and industrial plants have been in operation for about the last ten years. Because the CO_2 extracts are different to the conventional processed oleoresins, the acceptance in the food industry is growing slowly. The spice plants are much smaller than the decaffeination and hop plants, and use extractors of between 200 and 800 l. The same is true for medical herbs and high value fats and oils, which are more or less at the beginning of development.

One of the latest industrial applications is the reduction of plant protectives from cereals, which started commercial production at the end of 1999 in Taiwan.

Table 6.6-2
Commercial Plants for Supercritical Fluid Extraction

Product Group	Total Number of plants	America	Asia/ Australia	Europe	Total Capacity [1000 t Input/ Year]
Coffee and tea decaffeination	5	1	0	4	100
Hops (including coca defatting)	7	4	1	2	60
Nicotine from Tobacco	3	1	1	1	n.a.
Chemistry	5	3	1	1	n.a.
Environmental purposes including cleaning of foodstuff and medicinal plants	5	2	3	0	n.a.
Spices	12	1	5	6	9
Fats and oils including Lecithine	8	0	3	5	3
Medicinal plants	7	0	1	6	3
Flavours	7	0	3	4	3
Total:	59	12	18	29	>178

Explanation: of the caffeine-extraction plants, most of the commercial plants produce at least two different products. Plants were consolidated to what is thought to be the main product. Capacities are estimated on a twelve months' operation basis, which is - at least for hops - higher than reality. The order of product groups is in terms of decreasing total extraction volume. The product groups "Chemistry" and "Environmental Purposes" contain both products from natural and artificial origins. Several plants with the same product at one site are counted as one.

6.6.5 Specific application processes

Beside the typical extraction purposes, different applications using supercritical fluids - mainly carbon dioxide - are tested in the laboratory as well as on an industrial scale.

Decontamination of soils using supercritical fluids is an attractive process compared to extraction with liquid solvents because no toxic residue is left in the remediated soil and, in contrast to thermal desorption, the soils are not burned. In particular, typical industrial wastes such as PAHs, PCBs, and fuels can be removed easily [7 to 21]. The main applications are in preparation for analytical purposes, where supercritical fluid extraction acts as a concentration step which is much faster and cheaper than solvent-extraction. The main parameters for successful extraction are the water content of the soil, the type of soil, and the contaminating substances, the available particle-size distribution, and the content of plant material, which can act as adsorbent material and therefore prolong the extraction time. For industrial regeneration, further the amount of soil to be treated has to taken into account, because there exists, so far, no possibility of continuous input and output of solid material for high pressure extraction plants, so that the process has to be run discontinuously.

The neutralization and impregnation of paper is of great interest for all libraries, because acid degradation results in a reduction in the pH and weakening of the mechanical properties of the paper. For this reason, in the first step the degradation products are extracted with supercritical carbon dioxide, which increases the pH. In the following step, neutralizing substances are dissolved in the CO_2 stream, and impregnation of the paper takes place [22].

Supercritical CO_2 is used in a process for bone-tissue treatment to obtain a novel bone-substitute for human surgery. The supercritical extraction step results in de-lipidation of bones, and therefore reduces infection effects. Then efficient enzymatic deproteination can be performed [23,24].

As halogenated solvents are widely banned, either for health- or environmental reasons, the degreasing and cleaning of mechanical and electronic parts becomes a worldwide issue. Carbon dioxide, mainly in the liquid state, is becoming one of the most attractive substitutes for chlorinated solvents [25,26].

In the field of polymer recycling and/or disposal, new techniques are required. Most of the electronic waste contains flame retardants, mainly halogenated organic substances. In many recycling processes, plastics are incinerated and the formation of halogenated dibenzodioxins and dibenzofurans cannot be avoided. One promising way to separate halogenated flame retardants from polymer matrices uses the extraction with supercritical carbon dioxide. The advantage of this process is that the polymer as well as the flame retardant can be recycled, especially because flame retardants are relatively high-priced products [27].

References of section 6.6

1.   G. Brunner, Ber. Bunsenges. Phys. Chem. 88 (1984) 887.
2.   R. Eggers, R. Tschiersch, Chem.-Ing.-Techn. 50 (1978) Nr. 11, 842.
3.   R. Eggers, Chem.-Ing.-Techn. 53 (1981) Nr. 7, 551.
4.   E. Lack, Theses, TU Graz, (1985).
5.   D.V. Peng u. D.B. Robinson, Ind. Eng. Chem. Fundam. 15 (1976) 59.
6.   J. Chrastil, J. Phys. Chem. 86 (1982) 3016.
7.   B.O. Brady, C.P.C. Kao, K.M. Dooley, F.C. Knopf, R.P. Gambrell, Ind. Eng. Chem. Res. 26(2), (1987) 261.
8.   I.J. Barnabas, J.R. Dean, W.R. Tomlinson, S.P. Owen, Anal. Chem. 67(13), (1995) 2064.
9.   J. Puiggene, M.A. Larrayoz, E. Velo, F. Recasens, Proc. 3rd Int. Symp. Supercrit. Fluids, Strasbourg, (ed. G. Brunner, M. Perrut), (1994) Vol.2, 137.
10. G.K.C. Low, G.J. Duffy, S.D. Sharma, M.D. Chensee, S.W. Weir, A.R. Tibbett, Proc. 3rd Int. Symp. Supercrit. Fluids, Strasbourg, (ed. G. Brunner, M. Perrut), (1994) Vol.2, 275.
11. G. Markowz, G. Subklew, Proc. 3rd Int. Symp. Supercrit. Fluids, Strasbourg, (ed. G. Brunner, M. Perrut), (1994) Vol.2, 505.
12. L. Barna, J.M. Blanchard, E. Rauzy, C. Berro, Proc. 4th Meeting Supercrit. Fluids, I.N.S.A.-Villeurbanne (ed. D. Barth, J.M. Blanchard, F. Cansell), (1997) 63.
13. R. Zaragoza, J.M. Blanchard, L. Barna, Proc. 4th Meeting Supercrit. Fluids, I.N.S.A.-Villeurbanne (ed. D. Barth, J.M. Blanchard, F. Cansell), (1997) 145.
14. J. Thauront, J.C. Fontan, R. Zaragoza, S. Ruohonen, L. Parvinen, J.M. Blanchard, L. Barna, M. Carles, C. Perre, Proc. 4th Meeting Supercrit. Fluids, I.N.S.A.-Villeurbanne (ed. D. Barth, J.M. Blanchard, F. Cansell), (1997) 151.
15. L. Barna, J.M. Blanchard, E. Rauzy, C. Berro, Proc. 4th Meeting Supercrit. Fluids, I.N.S.A.-Villeurbanne (ed. D. Barth, J.M. Blanchard, F. Cansell), (1997) 157.
16. J. Hawari, A. Halasz, L. Dusseault, J. Kumita, E. Zhou, L. Paquet, G. Ampleman, S. Thiboutot, Proc. 5th Meeting on Supercrit. Fluids: Materials and Natural Products Processing, Nice, (ed. M. Perrut, P. Subra), (1998) 161.

17. A. Izquierdo, M.T. Tena, M.D.L. de Castro, M. Valcarcel, Chromatographia 42 (3/4), (1996) 206.
18. W. Sielschott, C. Frischkorn, M. Schwuger, Chem.-Ing. Tech. 65 (4), (1993) 434.
19. A. Schleussinger, I. Reiss, S. Schulz, Chem.-Ing. Tech. 68 (12), (1996) 1602.
20. Y. Yang, S.B. Hawthorne, D.J. Miller, J. Chromatogr. A, 699, (1995) 265.
21. D.R. Gere, C.R. Knipe, W. Pipkin, L.G. Randall, Int. Environ. Techn. 5 (4), (1995) 20.
22. C. Perre, A. Beziat, M. Carles, A.C. Brandt, Proc. 3rd Int. Symp. Supercrit. Fluids, Strasbourg, (ed. G. Brunner, M. Perrut), (1994) Vol.2, 19.
23. J. Fages, A. Marty, J.S. Condoret, Proc. 3rd Int. Symp. Supercrit. Fluids, Strasbourg, (ed. G. Brunner, M. Perrut), (1994) Vol.2, 253.
24. J. Fages, A. Marty, D. Combes, J.S. Condoret, French Patent 2 699 408, European Patent EP 0603920 A 1.
25. N. Dahmen, J. Schön, H. Schmieder, E. Dinjus, Proc. Int. Meeting GVC-Fachausschuß "Hochdruckverfahrenstechnik", Karlsruhe, (1999) 293.
26. V. Perrut, M. Perrut, Proc. Int. Meeting GVC-Fachausschuß "Hochdruckverfahrenstechnik", Karlsruhe, (1999) 289.
27. F. Steinkellner, T. Gamse, R. Marr, P. Alessi, I. Kikic, Proc. 5th Int. Symp. Supercrit. Fluids, Atlanta, cd-rom 5340939.PDF.

6.7 Extraction from liquid mixtures

6.7.1 Introduction

The extraction of liquid mixtures with supercritical fluids is comparable to liquid-liquid extraction, with the compressed gas acting instead of an organic solvent. While in liquid-liquid extraction the pressure influence is negligible, it plays an important role in high pressure extraction, changing as well the physical properties of the fluid like density, viscosity, surface tension, etc. as the phase equilibria. One advantage is that depending on the feed material the density difference between the two countercurrent flowing phases can be adjusted, which can be done by liquid-liquid extraction only by changing the temperature and/or the solvent.

Another advantage of high pressure extraction is the simple solvent regeneration. For liquid-liquid extraction solvent regeneration includes, in most cases, a re-extraction or distillation step, which is energy consuming and therefore cost intensive, and causes problems if heat sensitive materials have to be treated. For a high pressure extraction plant the solvent regeneration is achieved by changing pressure and/or temperature after the extraction step, thus changing the density and with it the solvent power of the extraction gas, which can be easily recycled after separation closing the solvent cycle.

Extraction of liquid feed materials has the great advantage, compared to solid materials, that liquids can be continuously introduced in and withdrawn from the high pressure separation unit. This gives the benefit of a continuous operating process which is easier to handle and to design than a discontinuous one; in addition, higher throughputs can be achieved because of no dead times, which are required by the discontinuous process (vessel emptying and refilling).

The basic for developing a high pressure liquid extraction unit is the phase equilibrium for the (at least) ternary system, made up of compound A and compound B, which have to be separated by the supercritical fluid C. Changing pressure and temperature influences on one hand the area of the two phase region, where extraction takes place, and on the other hand the connodes, representing the equilibrium between extract and raffinate phase.

6.7.2 Operation methods

Compared to liquid-liquid extraction different operation methods for a continuous supercritical extraction are available. Among others, note that the possibility of adsorption-desorption of compounds of the liquid feed mixture on a supporting material will not be treated in this chapter.

6.7.2.1 Single stage extraction

We refer to Fig. 6.7-1. Reaching once equilibrium between the supercritical fluid *SCF1* and the feed in the extractor *E1* is enough for separation. By changing pressure and temperature the produced extract *EX1* and raffinate *R1* concentrations can be varied following the ternary phase equilibrium. The supercritical solvent-to-feed flow rate ratio affects the amounts of products obtained from a given feed. The apparatus required to apply this method are a normal stirred reactor, where contact of the two phases takes place, followed by a separator eliminating the extract from the extraction gas, which is recycled back to the extractor.

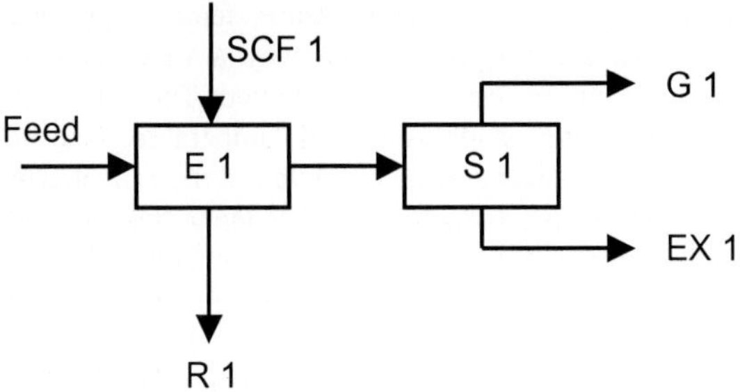

Figure 6.7-1. Single stage extraction

6.7.2.2 Multistage cross-flow extraction

For this kind of extraction (fig. 6.7-2) the number of extraction steps can be varied. In the first extraction step *E1* the feed is contacted with the supercritical fluid *SCF1*. The supercritical fluid - extract mixture enters the separator *S 1* and the extract of the first extraction step *EX1* is separated from the gas *G1*. The raffinate *R1* enters the next extraction step *E2* where it is contacted with new supercritical fluid *SCF2* and so on. The raffinate *Rn* leaves the last extraction step *En,* and the extract *Exn* the separator *Sn*, as well

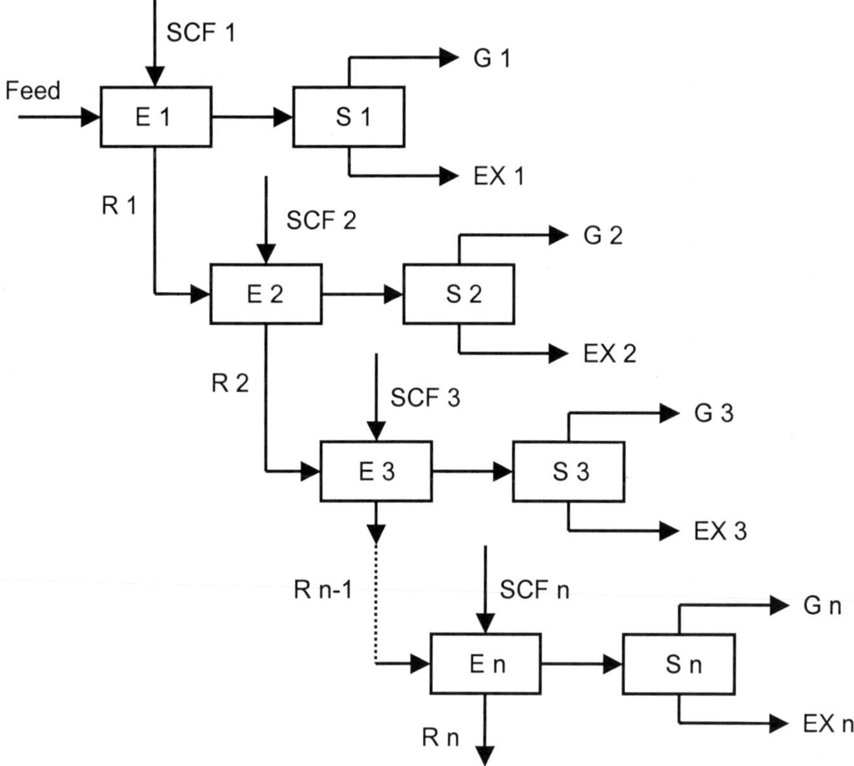

Figure 6.7-2. Multistage cross-flow extraction

The advantage of this extraction method is that the parameters pressure, temperature and solvent to feed ratio can be varied in each extraction step. By this way a very accurate fractionation of the different compounds included in the feed can be achieved. The solubility of the compounds in the supercritical fluid, depending on pressure and temperature, can be changed in each extraction step. The highly soluble substances are extracted in the first step at low fluid density. Increasing the density in the following extraction steps leads to the removal of the less soluble substances. Further, the flow rate of the supercritical fluid can be adjusted in each extraction step, either constant flow for each step or different flow rates, depending on the separation to be achieved.

At the industrial scale this extraction method is not installed yet for supercritical fluids. The reason is that the investment costs are relatively high because of the needed number of high pressure vessels; in addition, for the case of different pressure levels and flow rates, in each extraction step one high pressure pump or compressor is necessary.

6.7.2.3 Multistage countercurrent extraction

The countercurrent extraction method is the most common one for separation of liquid mixtures with supercritical fluids as solvents. The extraction is performed in pressure columns where the liquid feed and the supercritical fluid flow in countercurrent way.

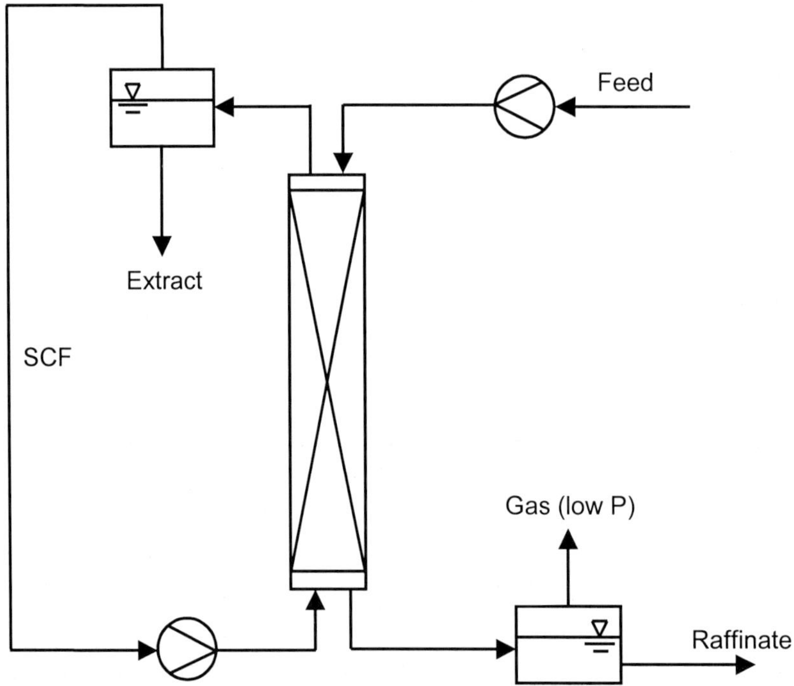

Figure 6.7-3. Continuous operating high pressure extraction column with feed inlet at the top of the column

Normally the supercritical fluid phase is the light one and enters therefore at the bottom of the column. The liquid phase, the heavy one, can be inserted either at the top or somewhere in the middle of the column. For the case of liquid feeding at the top (fig. 6.7-3) the extraction column operates as a stripping column at given pressure and temperature. The raffinate is withdrawn at the bottom and separated from the solved gas by pressure reduction to atmospheric pressure in a separator. The supercritical fluid with the extracted substances leaves at the top of the column and enters a separator, where pressure and temperature are changed to separate the extract from the supercritical fluid.

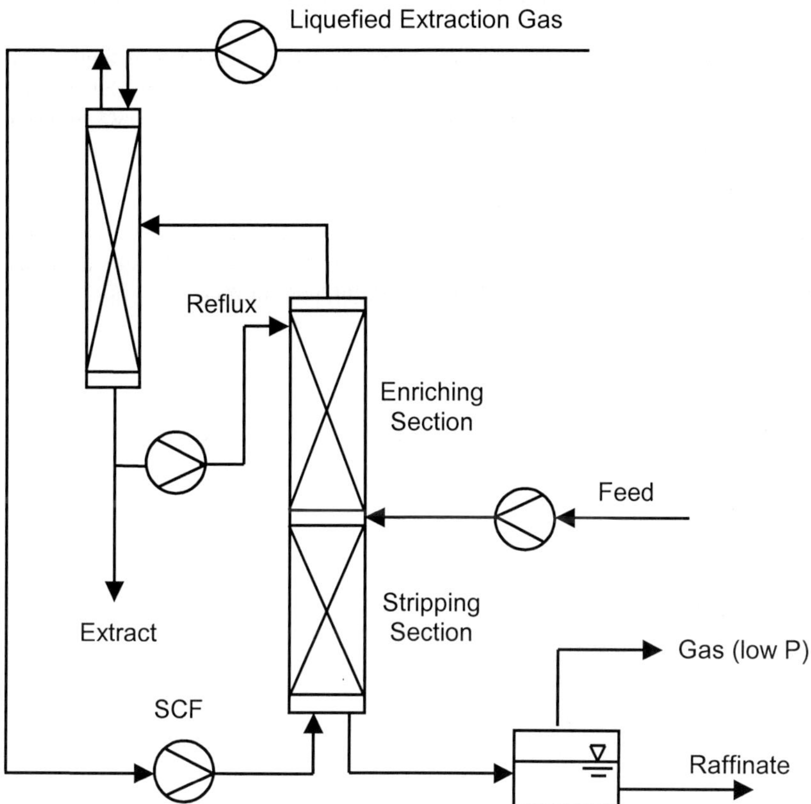

Figure 6.7-4. Extraction column with reflux and separation of extract in regeneration column

In laboratory scale units an enrichment can be achieved by changing the temperature over the column height and producing by this way an internal reflux. This method is not adopted for large scale columns, because a heating or cooling from the wall side is not sufficient enough for temperature change of the whole content. For columns of industrial size an external reflux has to be used, as given in figure 6.7-4. For this case the column is divided into an enrichment and a stripping section, separated by the entrance of the feed material. At the top of the column the supercritical fluid together with the solved substances exit and enter a separator. The separator can be designed either like the apparatus for separation in high pressure extraction plants for solid materials or a second countercurrent column, a regeneration column, can be used. Extraction gas in liquefied form can be introduced at the top of the regeneration column to increase separation efficiency and to produce a absolutely pure fluid at the outlet, which is recycled.

A specified part of the separated extract, the reflux, is pumped back to the top of the extraction column and the rest of the extract is withdrawn as top product of the extraction column.

6.7.3 Modelling of countercurrent high pressure extraction

The gas extraction column can be calculated according to the basic equations on phase equilibria, mass and energy balance and the kinetic equations for mass transfer. Based on these equations the number of theoretical steps or the number of transfer units can be determined for a given separation problem. Then, having the height of a theoretical plate or of a transfer unit allows to calculate the height of the column needed for separation. Calculation of compound concentration along the extraction columns is also possible. For more details on the modelling of high pressure extraction columns from liquids, we refer to the extensive book by Brunner [1].

6.7.4 Types of extraction columns

For a good separation efficiency an intensive contact of both phases along the column height is necessary to ensure good mass and heat transfer. In general the columns can be divided in four groups:

- columns without internals
- plate columns
- packed columns
- columns with energy input

6.7.4.1 Extraction columns without internals

In this case the column operates as a bubble column. Either the heavy phase forms droplets (dispersed phase) moving countercurrent to the continuous supercritical phase from the top to the bottom or the supercritical phase is dispersed in form of drops or bubbles moving going up in the continuous liquid phase. For both cases the drop sizes and the drop size distribution is essential for separation efficiency. The smaller the drop sizes the larger is the mass transfer based on the higher specific surface area.

From the viewpoint of construction, devices for producing fine, uniform drop sizes are necessary.

6.7.4.2 Plate columns

In plate columns the two phases are intensively mixed on each plate and separated between each plate (Fig. 6.7-5). For the distribution of the light phase through the liquid a lot of devices were developed. The simplest one is a perforated sieve tray, where the supercritical phase can pass through. To avoid weeping of the liquid through the holes different devices like bubble caps or valves (Fig. 6.7-6) were developed.

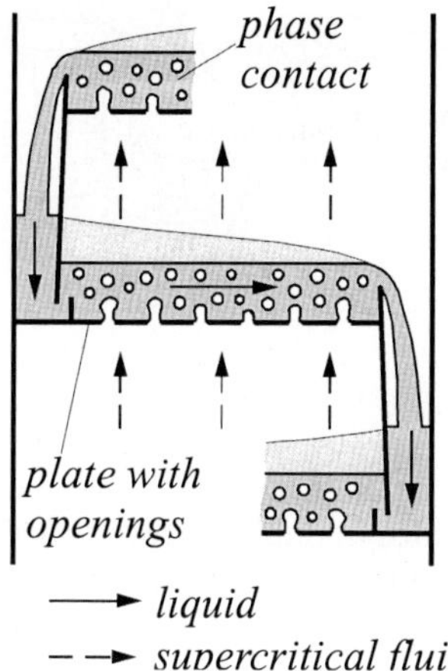

Figure 6.7-5. High-pressure extraction plate column

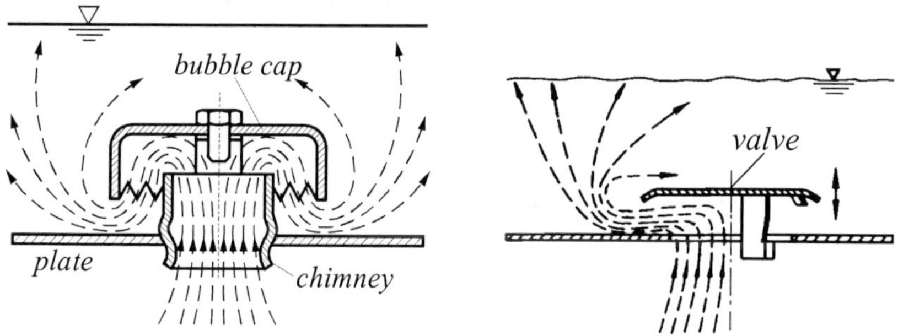

Figure 6.7-6. Bubble-cap tray (a) and valve tray (b)

6.7.4.3 Packed columns

A large number of packings of different forms are used for packed columns either as random or structured packings (Fig. 6.7-7). The liquid feed wets the surface of the packings and the supercritical gas passes over this thin liquid film. By this way a very large phase boundary area is available where mass transfer takes place. The material of the packings (plastic, metal, ceramic, steal or glass) influences the degree of wetting.

The packing cannot be higher than a maximum value because the liquid has the tendency to pass to the wall of the column. Therefore at given sections the liquid has to be collected and re-distributed to the next layer of packings. A maldistribution of the liquid results in low separation efficiency.

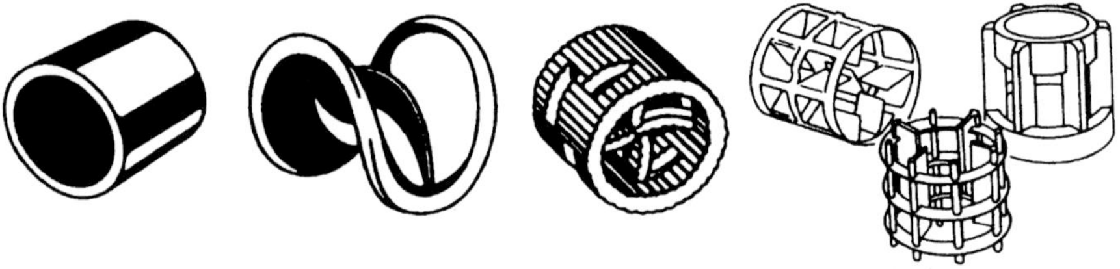

Figure 6.7-7. Examples of different random packings

6.7.4.4 Columns with energy input
The energy input can be performed either by rotation or by pulsation.
Extraction columns with rotating internal parts are divided into mixing and separation zones. On a vertical shaft stirrers are mounted in the mixing zones giving an intensive contact of the countercurrent flowing phases. In the separation zones the phases coagulate before entering the next mixing zone.
Pulsed columns are either sieve tray or packing columns. The content of the column is brought into vibration so that the moving phases are finely distributed on the sieve trays or on the packings. This pulsation causes a higher separation effect compared to the same columns without pulsation. For producing the vibration either extra devices are used or the pulsation of the high pressure pumps is great enough to produce the vibration. For sieve tray columns another possibility is to move the trays up and down.

6.7.5 Applications
Most of the application are up to now in laboratory scale, but the interest in this separation method shows that this process is a very promising way for the future.
Brunner et al [1, 2] investigated separations of fatty acids according to chain length, using methyl esters of different carbon chain length from C14 to C18, separation of tocopherols from a by-product of the edible oil production and separation of fish oil esters [3]. Stahl et al [4] proposed the supercritical fractionation of orange peel oil and Reverchon et al [5,6] of an orange flower concrete. Different authors treated citrus peel oil [7,8] and citrus oil [9-12].
The separation of alcohol from different raw materials like cider [13], wine fermentation broth [14], alcohol-water mixtures [15-18] and wine [19-22] is an interesting alternative to distillation processes especially if heat sensitive substances are present. Separex company separated the aroma substances out of whisky and cognac, which gives a very interesting product for food industry.
Further investigations were done for cleaning of wastewater. For an industrial scale the high pressure extraction will only be sufficient if a high contamination in a relatively low waste water stream is available. Otherwise this separation method is too expensive, as for flow rates in the range of some m^3/h the investment and operation cost for such a plant become uneconomic.

References of section 6.7

1.   G.Brunner, Gas Extraction, Steinkopf Darmstadt Springer New York, 1994.
2.   G.Brunner, T.Malchow, K.Stürken, T.Gottschau, J.Supercrit.Fluids 4 (1991) 72.

3.   V.Riha, G.Brunner, J.Supercrit.Fluids 17 (200), 55.

4.   E.Stahl, K.W.Quirin, D.Gerard, Verdichtet Gase zur Extraktion und Raffination, Springer, Berlin, 1987.

5.   E.Reverchon, G.Della Porta, G.Lamberti, R.Taddeo, Proceedings 4th Italian Conference on Supercritical Fluids and their Applications Capri, (1997), 47.

6.   E.Reverchon, G.Della Porta, G.Lamberti, J.Supercrit.Fluids 14 (199), 115

7.   C.Perre, G.Delestre, L.Schrive, M.Carles, in: M.Perrut, G.Brunner (Eds.), Proceedings 3rd International Symposium on Supercritical Fluids, Strasbourg, vol.2, (1994), 465.

8.   M.Sato, M.Goto, A.Kodama, T.Hirose, Proceedings 4th Italian Conference on Supercritical Fluids and their Applications, Capri, (1997), 39.

9.   M.Sato, M.Goto, T.Hirose, in: M.Perrut, G.Brunner (Eds.), Proceedings 3rd International Symposium on Supercritical Fluids, Strasbourg, (1994), vol.2, 83.

10.   M.Sato, M.Goto, T.Hirose, Ind.Eng.Chem.Res. 34, (1995), 3941.

11.   G.Brunner, M.Budich, T.Wesse, V.Leibküchler Proceedings 4th Italian Conference on Supercritical Fluids and their Applications, Capri (1997), 27.

12.   M.Sato, M.Kondo, M.Goto, A.Kodama, T.Hirose, J.Supercrit.Fluids 13, (1998), 311.

13.   I,Medina, J.L.Martinez, in: M.Perrut, G.Brunner (Eds.), Proceedings 3rd International Symposium on Supercritical Fluids, Strasbourg, (1994), vol.2, 401.

14.   S.Fernandes, J.A.Lopes, A.S.Reis Machado, M.Nunes da Ponte, in: M.Perrut, G.Brunner (Eds.), Proceedings 3rd International Symposium on Supercritical Fluids, Strasbourg, (1994), vol.2, 491.

15.   M.Randosz, Fluid Phase Equilibria 29 (1986), 515.

16.   M.Randosz, J.Chem.Eng.Data 31 (1986), 43.

17.   G.Brunner, K.Kreim, Ger.Chem.Eng. 9 (1986), 246.

18.   G.Brunner, K.Kreim, Chem.-Ing. Tech. 57(6) (1985), 550.

19.   T.Gamse, I.Rogler, R.Marr, Proceedings 4th Italian Conference on Supercritical Fluids and their Applications, Capri (1997), 191.

20.   T.Gamse, I.Rogler, R.Marr, J.Supercrit.Fluids 14 (1999), 123.

21.   T.Gamse, F.Steinkellner, R.Marr, Proceedings International Meeting of the GVC-Fachausschuß "Hochdruckverfahrenstechnik" on High Pressure Chemical Engineering, Karlsruhe, (1999), 255.

22.   T.Gamse, R.Marr, in: M.Perrut and E.Reverchon (Eds.), Proceedings 7th Meeting of Supercritical Fluids, Antibes / Juan-Les-Pins, vol. 2, 579.

High Pressure Process Technology: Fundamentals and Applications
A. Bertucco and G. Vetter (Editors)

405

CHAPTER 7

SAFETY AND CONTROL
IN HIGH PRESSURE PLANT DESIGN AND OPERATION

G. Luft[a], **S. Maier**[b]: section 7.1

G. Luft[b]: section 7.2

E. Lack[c], **H. Seidlitz**[c]: section 7.3

[a] Department of Chemistry, Darmstadt University of Technology
Petersenstr. 20, D-64287 Darmstadt, Germany

[b] Römerstr. 3, D-67136 Fußgönheim, Germany

[c] NATEX GmbH Prozesstechnologie
Hauptstrasse, 2 A-2630 Ternitz, Austria

The risk in manufacturing plants (especially chemical ones) results mainly from storage, handling, and reaction of compounds which are often flammable and toxic. High pressure increases the risk by increasing concentration of compounds, changing safety-influencing properties, and by the high energy density which requires additional safety measures.

The general safety aspects of high-pressure plants are presented in Chapter 7.1. The protective design of the various parts of a plant including the buildings is considered in detail together with testing procedures, safe plant operation, and inspection.

The safety requirements are then explained regarding high-pressure polymerization and extraction.

Loss prevention of polyethylene plants is outlined in Chapter 7.2. The major hazard that can occur is the runaway of the high-pressure reactor and decomposition of ethylene besides fire and disintegration of high-pressure separators, pipes, and compressors. The critical conditions for runaway and ethylene decomposition during homo- and copolymerization are revealed together with the influence of decomposition sensitizers. Relief devices and venting systems are described.

The protection of extraction plants run at pressures of up to 48 MPa by safety valves and rupture discs is shown in Chapter 7.3. In this context the use of interlocking and computerized control systems is demonstrated.

7.1 General safety aspects in high-pressure facilities

At the beginning of the 20th. century chemical processes were developed producing chemicals, such as ammonia and 20 years later polyethylene, using high pressure as a key-process parameter. Already since the first experiences with high pressure, engineers and chemists are forced to realise that high-pressure units contain higher potential energy than comparable low-pressure units. This fact becomes dangerous only if the stored energy is uncontrolled released. Severe disasters could happen if high-pressure facilities disintegrate. The higher risk consists first in the energy content stored in the compressed fluid, in a further part the spring energy stored in the pressure containment walls and last but not least in the stored energy of the process chemicals which are often in a most important way flammable, toxic and explosives. Pressure and temperature are chosen so that the reacting system is in a condensed state. If pressure is suddenly released the reactor content is immediately evaporated and forms dangerous explosive clouds with sometimes dramatically high energy content.

These additional risks to keep the impounded energy under control must be taken into account dealing with questions as construction, construction materials, proofed handling of the materials, design of the needed buildings, erection of the different parts of a plant, plant operation, maintenance and initial as well as operational inspections. The optimal training of involved people is a further aspect for a safe plant design and operation. Today high-pressure engineering has been developed very largely by the chemical industry, which in a high level acquired a huge stock of knowledge and experience of these features and the involved danger. National bodies and safety organisations have developed a lot of common rules and proposals to work with high pressure. All persons involved in such processes have to observe the so-called state of technique. Actually the chemical industry has not remained to be only user of high-pressure technology. Also geologists, biochemists, deep sea scientists, new material scientists, mechanical engineers and others take benefits of this knowledge. Metallurgical and ceramic industries use high-pressure technique for hydrostatic extrusion of materials and for the compacting of metallic-or ceramic powders in isostatic presses. Special materials, for example leather or cardboard, were cut with high-pressure jet streams. All these processes must obey the additional safety aspects of high pressure. All the newcomers in the high-pressure field should study the design and operation details of such facilities before the start-up. This chronological order is an urgent requirement for a safe plant operation.

7.1.1 Safety aspects in design and operation

The most urgent objectives working with high-pressure facilities, are the design of a chemical high-pressure plant, the operation of those units, the plant maintenance and the testing procedures for all components of the plant. In Chapter 4 of this book information and references are given referring to the main problems when designing components for high-pressure applications. The most important methods to calculate the dimensions of such parts are given. The main parameters for details of the shape of such parts are also mentioned as well as the importance of selecting a qualified construction material and its safe handling during fabrication and operation. For a safe plant-and component design the details of the chemical reaction and their possible deviations from the so-called normal operating standard of the process should be known. The knowledge of the maximum pressure and temperature during the life time of a process is very important and has a strong influence on the design parameters and the selection of the materials. For a good process design or if new chemicals are used, the designer must obey the process parameters and look for the worst conditions which could occur during the operational period. All parameters of a chemical reaction must be tested during the preliminary experiments. Best results will be achieved in so-called integrated pilot-plants or in a lower scale with mini-plants [1]. Such experimental units are necessary to evaluate the relevant safety parameters. Attempts to shorten this process, in order to safe money and time, proceeding directly from laboratory scale into a technical unit maybe lead to uncontrolled and unsafe situations and sometimes even to a disaster. During the development phase of a process in an experimental unit it is important to install sufficient process instrumentation for measuring all the relevant parameters and their dynamic behaviour. This knowledge is a part of the guarantee to find the proper parameters indicating very quickly the deviation of a reaction and helps to develop a well tested strategy for safe process operation. Most accidents in the past could only happen because such important parameters were not available before starting the process.

Another important point for the plant design is the appropriate arrangement of the individual components. This point is of considerable influence on a safe plant operation. The pressure relevant parts should be arranged in isolated areas which are seldom contacted by operators. The components with the highest amount of stored energy should be placed in their own compartments made of thick reinforced concrete walls. In case of a detonation after the damage of a pressure vessel such barricades do not suppress an induced shock wave. Special barricades which are resistant against shock waves are extremely expensive. The best solution for problems with shock waves is to move to remote areas with no neighbours and no other facilities [2]. The observation of the parts under high pressure is realised by a suitable instrumentation or with electronic cameras. These instruments allow the observation from a safe distance. In simple cases also a mirror could be used for observation of instruments from behind a safe barricade.

Valves can be safely handled by means of long extension shafts which are led through the wall of a barricade. The stems must be secured against the possible blow-out during an uncontrolled pressure release for example in the case of a broken stem at the inner side of the valve. Valve extension handles should be arranged in sleeves welded to the valve stem inside the barricade [3].

Compression and pumping equipment is more difficult to shield because maintenance and control inspection must be done from time to time. Another aspect is the vibration of oscillating machines and the connected pipes and instrumentation. Suitable dampers should be installed to reduce the danger caused by vibrations. Modern instrumentation allows a continuos observation of deviations during the operation of machines and plant components. Those installations represent a further contribution toward a safe plant. Large machines should be operated only with correctly designed instrumentation to detect the influence of vibration. Another aspect is the chemical reaction or decomposition of components like ethylene or acetylene in an insufficiently cooled compressor stage. A correct monitoring of the temperatures inside the individual stages is necessary to avoid a damage of the machine or perhaps a greater disaster. The deficiency of an important lubricant exhibits a severe fault and could cause the damage of the machine. Large units of chemical plants are more and more equipped with sophisticated monitoring systems. These systems are installed to guarantee a safe operation and to avoid an uncontrolled release of dangerous chemicals.

Designers and operators of pumps should know that it could be dangerous to operate the machine for example when an outlet valve at the discharge site of the pump is closed. Such a situation must be avoided by a suitable instrumentation and monitoring of that deviation from the safe condition.

All components of a high-pressure plant must be further protected against the overload by suitable pressure relief valves. Relief valves and the connected pipes must be mounted to carry also the reacting forces during venting of the valve. This requires a proper design and construction of the mounting parts. To avoid air pollution the venting stream is sometimes fed to a catch tank to collect the vented chemicals. A lot of information for those systems is given in literature [4]. An other point is the calculation of the mass flow through a relief valve and its correct design. If the cross-section of a valve is too large a quick opening and closing of the valve could result in so-called pumping of the entire system. That effect often causes heavy vibrations which are so powerful that the valve with the connected pipe could be pulled out of the fixing. This could lead to a breakout of chemicals and the risk of explosion or hazardous fire.

The leakage of high-pressure systems is dangerous. For example, a expanding stream of compressed hydrogen is heated up and mixed with the surrounding air could be self-ignited and forming a colourless flame which is invisible for operating personal. To avoid this danger the flanges of the pipe connections are wrapped with metallic shields. This kind of protection is also suitable for all cases of leakage which are dangerous for operating persons. Sometimes permanently burning small flames are installed in the neighbourhood of those connections to ignite the vented components. With this technique an explosion of a larger cloud will be prevented.

7.1.2 Safety aspects due to changing properties with high pressure

High pressure can change a variety of safety relevant physical and chemical properties. In the following some selected examples are considered.

In many processes combustible gases are used together with oxidizing agents. Mixtures of those compounds can also be generated by malfunction of the equipment or faults of operation. The range in which the mixtures are flammable are characterized by the lower (LFT) and the upper (UFL) flammability limit. It is a general experience but not a strict rule that the upper flammability limit increases with increasing pressure. This is shown in Fig. 7.1-1, left, for mixtures of methane and air. At ambient pressure and temperature a UFL-value of 15 vol.% of methane was measured by Vanderstraeten et al. [5] in agreement with literature data [6]. It increases to 43 vol.% at an initial pressure ratio of $p_1/p_0 = 55$. The increase of UFL with pressure was found higher at higher temperatures.

Spontaneous ignition can occur when mixtures of combustible gases and oxygen or air are heated under high pressure. The ignition temperature decreases steeply with increasing pressure as shown in Fig. 7.1-1, right, for methane-air mixtures of different compositions and for stoichiometric methane-oxygen mixtures [7].

By the explosion of flammable fuel- or gas-air mixtures a considerable energy is released. This becomes obvious when the TNT equivalence is considered (Tab. 7.1-1). Most of the paraffins, mono-olefins, and aromatics exhibit a TNT equivalence of 10 - 12 kg TNT/kg flammables. Lower values are reported for ammonia, ethylene oxide, etc. Hydrogen shows an extreme high TNT equivalence.

Pure compressed gases like acetylene or ethylene can decompose explosively when an ignition source like a flash, a hot wire or other hot-spots like hot parts in a compressor or high-pressure reactor are involved. Fig. 7.1-2 shows the critical conditions for ethylene in a range of 10 - 60 MPa and 25 - 300°C [9]. The critical temperature decreases first steeply and then less steeply with increasing pressure, demonstrating that ignition and decomposition become more probable at high pressures.

Table 7.1-1
TNT equivalence of flammable fuel- and gas-air mixtures [8]

Material	Formula	TNT equivalence [kg TNT/kg material]
Paraffins	C_nH_{2n+2}	10.48 – 11.95
Methane	CH_4	11.95
Ethane	C_2H_6	11.34
Propane	C_3H_8	11.07
n-Butane	C_4H_{10}	10.93
Iso-Butane	C_4H_{10}	10.90
Alkylbenzenes		9.59 - 9.99
Benzene	C_6H_6	9.69
Alkylcyclhexanes	C_nH_{2n}	10.36 – 10.47
Cyclohexane	C_6H_{12}	10.47
Mono-Olefins	C_nH_{2n}	10.67 – 11.26
Ethylene	C_2H_4	11.26
Propylene	C_3H_6	10.94
Iso-Butylene	C_4H_8	10.76
Miscellaneous		
Ammonia	NH_3	4.45
Ethyl chloride	C_2H_3Cl	4.58
Vinyl chloride	C_2H_3Cl	4.58
Ethylene oxide	C_2H_4O	6.38
Chlorobenzene	C_6H_5Cl	6.53
Acrolein	C_3H_4O	6.57
Butadiene	C_4H_6	11.22
Hydrogen	H_2	28.65

Spontaneous flames have been observed when oxygen or air was injected into homogeneous dense mixtures of methane or hydrogen with supercritical water at 500°C and 5 - 100 MPa [10]. Fig. 7.1-3 shows a flat diffusion flame observed through a sapphire window when oxygen was injected into a supercritical water/methane mixture at 450°C and 100 MPa. As a rule, polymerization initiators need to be stored at relatively low temperatures of –20 to max. 30°C in order to prevent premature decomposition. For metering into the polymerization reactor by means of pumps or other injection equipment, they are usually dissolved in hydrocarbons, if they are solids, or diluted with hydrocarbons like *n*- or *iso*-paraffins. When metering is carried out at the high pressure which is applied e.g. in polyethylene synthesis, there is the risk that the initiator precipitates from the solution. The crystals deposit onto the valves of the delivery pumps and interfere with its operation, or they deposit in other parts of the injection system. Thus they can cause blockages and can suddenly increase initiator concentration and rate of polymerization when they are entrained into the reactor with the risk of runaway of the reactor.

Fig. 7.1-4, left, indicates the pressure and temperature values recorded at the solubility limits for *t*-butyl perbenzoate in *iso*-dodecane, which is a liquid at ambient pressure and

410

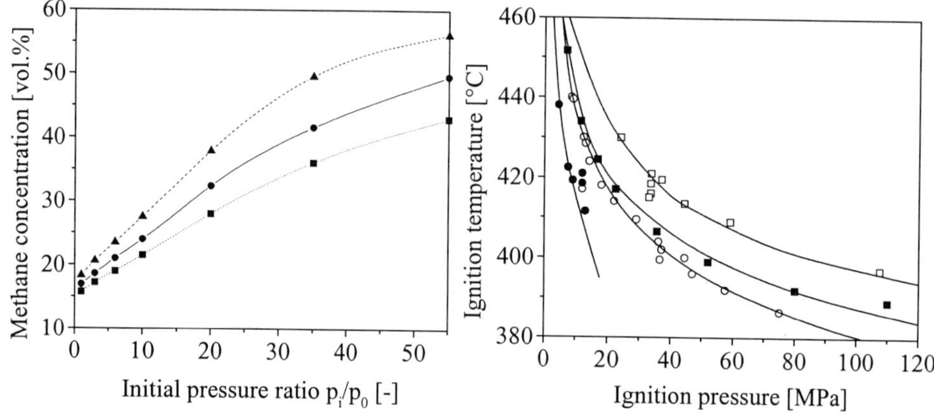

Figure 7.1-1. Compressed methane/air mixtures. Left, pressure dependence of the upper flammability limit; ν, 20°C; λ, 100°C; σ, 200°C; right, ignition temperatures. methane/oxygen (mol/mol): λ, stoichiometric; methane/air (mol/mol): O, 1/1; ν, 1/2; □, 1/3.

temperature. At conditions above the solid line crystals appear which redissolve after pressure reduction below the dashed curve. The critical pressure can be increased at a given temperature by diluting the peroxide [11].

The influence of the pressure on the solubility is important also for the selection of appropriate lubricants for high-pressure compressors besides the change of the viscosity. The

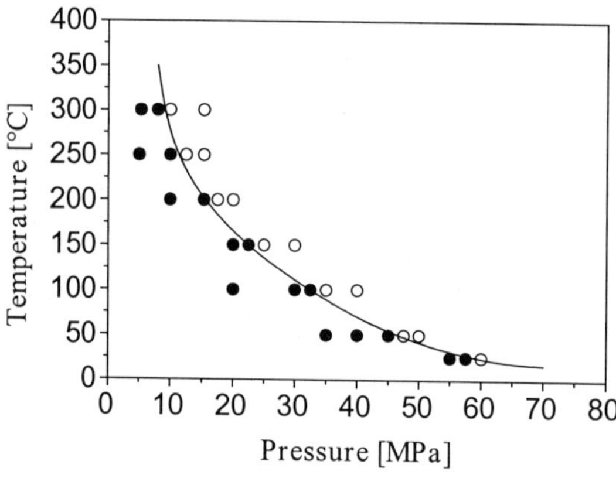

Figure 7.1-2. Critical conditions of decomposition of ethylene starting from hot spot. ●, no decomposition; O, decomposition; monomer: pure ethylene; method: ignition by hot wire at relative low energies.

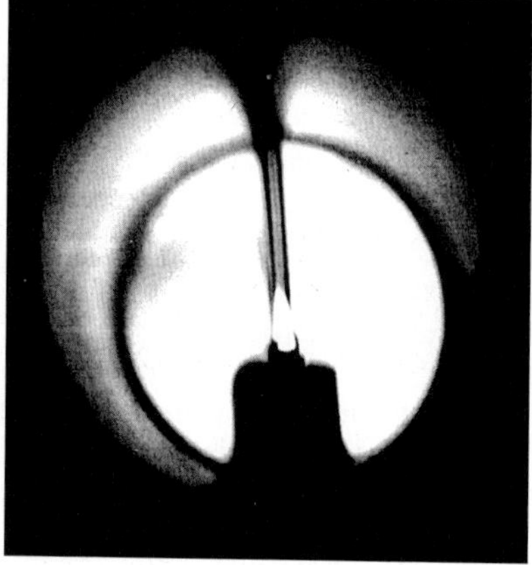

Figure 7.1-3. Steady flame when oxygen is injected into supercritical 70/30 water/methane mixture at 100 MPa and 450°C.

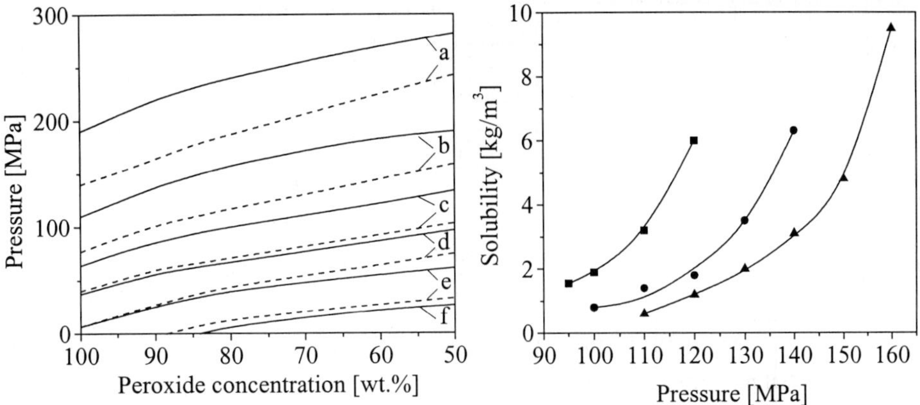

Figure 7.1-4. Change of solubility with pressure. Left: *t*-butyl perbenzoate in *iso*-dodecane. solid line, crystallisation curve; dashed line, solubility curve; a, 40°C; b, 30°C; c, 20°C; d, 10°C; e, 0°C; f, -10°C; right: synthetic lubricant oil in ethylene; σ, 60°C; ●, 80°C; ■, 110°C

solubility of a lubricant in the medium which is compressed should be as low as possible to avoid the entrainment of the lubricant into the chemical process.The solubility of a number of lubricants used for the high-pressure compressors in the synthesis of low-density polyethylene was investigated by Gebauer et al. [12]. In Fig. 7.1-4, right, the solubility of a synthetic oil in compressed ethylene is presented for pressures up to 160 MPa and temperatures of 60 - 110°C. It increases steeply with increasing pressure and temperature. On the other hand the solubility of compressed ethylene in the lubricant increases too with increasing pressure demonstrating that the suitability of a lubricant decreases with increasing pressure.

Blockages of valves and pipes can also occur by gas hydrates. Such adducts can be formed by a number of gases with water. In Fig. 7.1-5 the pressure-temperature diagram of the system propane/water with an excess of propane is presented. The line, (g), shows the vapour-pressure curve of propane. Propane hydrate can be formed at temperatures below 5.3°C. At pressures below the vapor pressure of propane a phase of propane hydrate exists in equilibrium with propane gas (Fig. 7.1-5, area b). At higher pressures above the vapor pressure of propane and low temperatures a propane hydrate- and a liquid propane phase were found (area d). In order to exclude formation of gas hydrates these areas should be avoided handling wet propane and other compounds like ethylene, carbon dioxide [14], etc.

7.1.3 Protective design and construction

The following points should be considered for a protective design and construction of safe high-pressure components:
- The relevant processing components like the types of reactors (tubular reactor, stirred reactor etc.), and their mode of operation.
- Accurate data for working pressure and -temperature.
- Experimental results for the main reaction as well as side reactions and the chemicals involved.
- Transient operating conditions and their dependency of time.

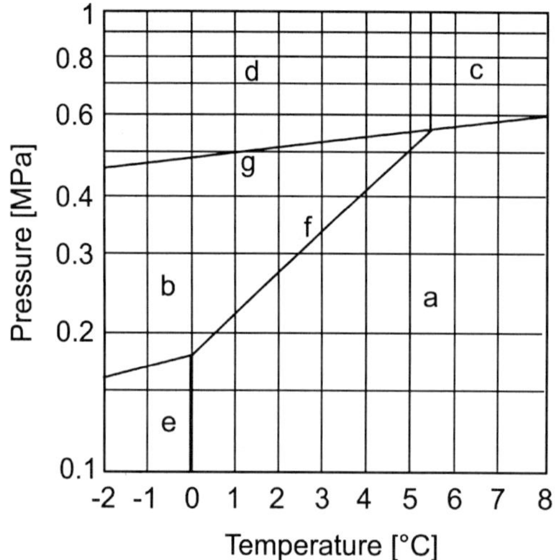

Figure 7.1-5. Pressure–temperature diagram of the system propane/water [13]. a, Propane gas/water; b, propane gas/propane hydrate; c, propane liquid/water; d, propane liquid/propane hydrate; e, propane gas/ice; f, hydrate curve; g, vapor pressure curve of propane.

- Load cycles of the individual parts of the construction.
- Practically proofed details about the individual steps of the manufacturing process and the handling of the materials.

A further point is the design strategy by which all components and especially the reactors are in all parts testable for example by ultrasonic- or X-ray-methods. Fig. 7.1-6 gives an example for a not testable (left) and a testable construction (right) of a high-pressure heat exchanger. The picture at left side shows the older design where the area around the connection between the shell and the plate of the apparatus is not reliably testable. The right side gives the improved version which exhibits best testing conditions.

A further important requirement is the proper selection of the materials for high-pressure components in case of permanent load cycles or an pulsating operation of the unit. The number of cycles can vary in a wide range. Fig. 7.1-7 compares the numbers of cycles of an ammonia reactor, a small pressure vessel for research purposes, an isostatic press, and the housing of a compressor. The base of the figure is an operation time of ten years of the mentioned parts. For the design and manufacturing of those parts an excellent knowledge about the important material parameters is a premise for a safe and long-endurance operation of components in a chemical process unit.

The start-up and runaway scenarios of high-pressure facilities lead to considerable pressure and/or temperature differences in the wall of the individual components causing additional stresses which could damage the structure of the material. The stressed area is normally the place at which small cracks start, and could be the beginning of a rupture which may cause the damage of a component. Rapid changes of pressure and temperatures produce additional mechanical stresses. The scattering of pressure and temperature must be limited to guarantee a safe operation. A special instrumentation can keep the stresses within allowable limits.

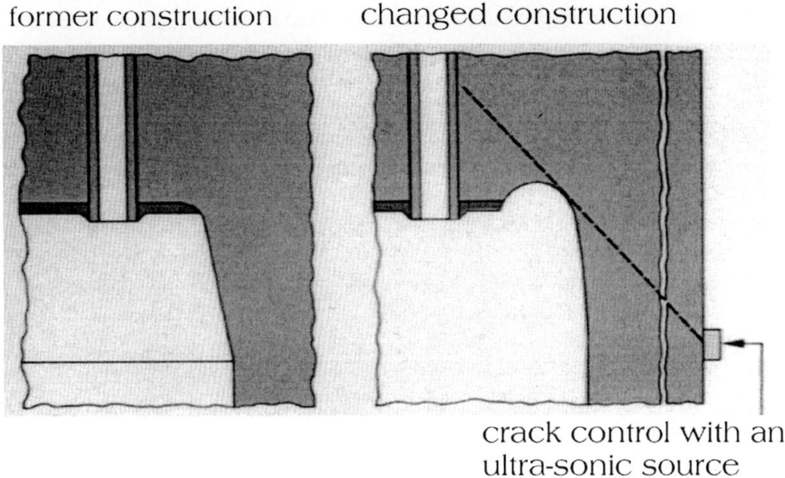

Figure 7.1-6. Construction details of a high-pressure heat exchanger. Left: The connection between shell and the covered plate is not testable; right: The corrected well testable arrangement [15].

Tough materials with sufficient high elongation at rupture - in the minimum 10 % - and with a good ratio between allowed yield strength and the tensile strength-ratio of approximately 0.85 and better should be chosen. The cooperation with a experienced steel manufacturer is an important link in the design process.

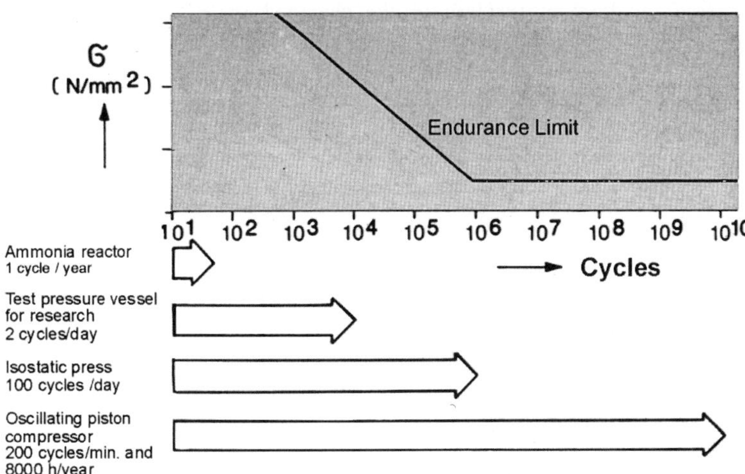

Figure 7.1-7. Load cycles for high-pressure components within 10 years.

All pressurised volumes in high-pressure components must be protected against overload in pressure and temperature. The protection against excess-pressure is reached with several types of relief valves. If it is impossible to install a safety valve, for example because the chemicals are toxic or in case of polymerisation reactions where the relief system could be blocked, the safe operation must be achieved by other methods. The design must regard the maximum obtainable pressure and temperature in the system and regulate and control the process with a suitable instrumentation. Or high-pressure units are designed in a manner to withstand the highest possible temperature and pressure. This design strategy is called "the design of intrinsically-safe pressure systems".

For handling acetylene in chemical processes under pressure a special strategy has been developed to solve the specific problems of this chemical - the decomposition of the substance even under mild conditions. It is known that this reaction is generating a pressure increase of approximately factor 11 above the normal working pressure. Therefore the design pressure of all components of acetylene plants should be selected eleven times higher compared to the operating pressure.

7.1.4 Design criteria of buildings

The degree of protection for people inside and outside the production facilities and the investment which is required for the operation of high-pressure units depends on some important factors:
- The amount of stored energy in the plant or the plant components must be known.
- The size of the individual parts of the plant influences the dimensions of a building.
- The neighbourhood to other plants, facilities, as well as public installations must be taken into account.
- The presence of weakening effects such as fatigue, corrosion of seals, as well as of pumps or compressors must be known.
- An idea about possible dimensions of parts out of the plant which may be catapulted in case of any accident.
- Information about the type of reaction of the chemicals with air (fire, explosion, detonation) when they are released.

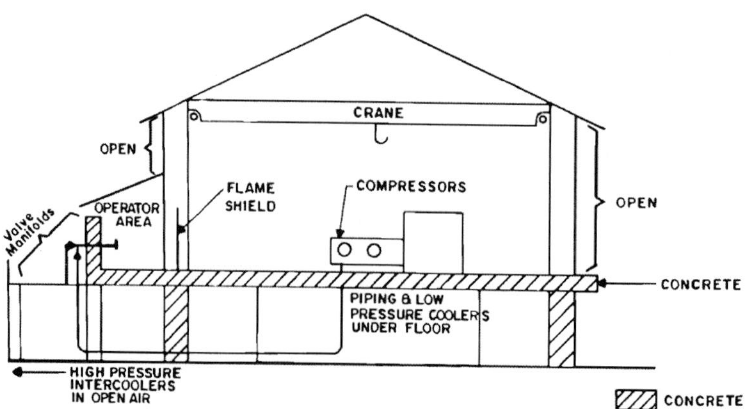

Figure 7.1-8. Cross section of a polyethylene compressor building [16].

All this points should be regarded during the design discussion for a safe building. Preventative actions should be suggested as already mentioned before to minimise the extent of such dangerous situations. The task to find an optimal arrangement of equipment and building in conjunction with sufficient ventilation of the critical plant areas must be solved. Each situation in a plant will require an engineering evaluation of the potentials involved. In average, venting ratios of 5 to 15 are maintained. In [17] a good definition of venting such buildings is given. In Fig. 7.1-9 some typical cases are demonstrated. The criteria of fully vented rooms:

$$\frac{A}{V^{2/3}} \geq 0.6 \qquad\qquad (7.1\text{-}1)$$

with A = open area and V = volume of the containment.

All locations with a ratio smaller than 0.6 are partially vented. In Fig. 7.1-9 some typical cases are demonstrated.
In spite of these recommendations a good operability of the plant is necessary. The adequate arrangement of corridors between the buildings and/or the equipment, combined with additional installations for example so-called vapour barriers are possibilities to improve the safety of plants. The intelligent separation of flammables from potential sources of ignition should be practised where ever possible. Working areas in a building should be located in a safe distance to high-pressure equipment. Manually operated valves should be located in the safest areas of a building of a plant.

The same criteria are recommended for control rooms as well as for the backside instrumentation rooms. Compressors are usually installed in halls with a roof for weather protection with in minimum two sides open to the atmosphere [17].

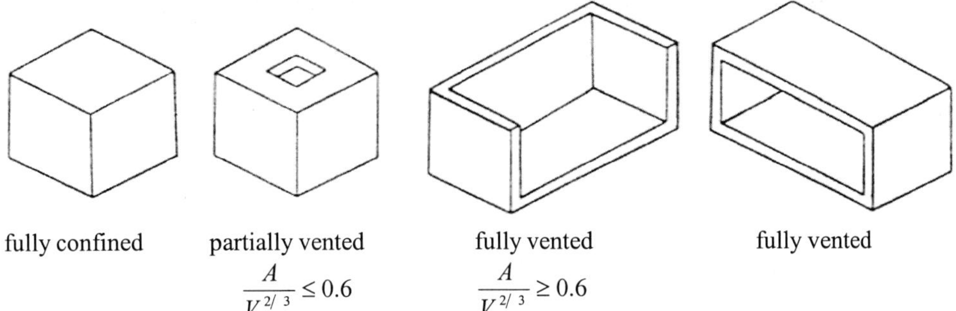

Figure 7.1-9. Different types of vented barricades, definition for fully or not fully vented barricades [17].

For research purposes and a smaller risk special types of barricades are used [18]. In Fig. 7.1-10 and 7.1-11 two typical constructions are presented. Fig. 7.1-10 shows an example with reinforced concrete walls and earth. Fig. 7.1-11 shows a high-pressure chamber which is typical for research laboratories. The high-pressure bunker is often located in the last floor of a building integrated with a with frangible material closed opening in the roof with further safety catch installations to capture projectiles of the beneath chamber in the case of an accident. The use of different kind of barriers should protect against the spread of flammable gases, the radiation or flash burns and the effects of an explosion or shrapnel formed of parts of the plant. The decision for any kind of barricade is based on an assessment of the total energy containment within the system. Barriers have been designed using plastic sheeting, canvas curtains, concrete, masonry, rope mats, wire mats, sheet metal, and any conceivable building material. Most of the mentioned examples provide some degree of protection, but many of them create additional dangers if misused [2].

It is common engineering practice to construct the walls of safety containment of reinforced concrete of sufficient thickness to withstand explosive forces. The designer must know the energy in case of rupture of a plant component additional to the explosion energy. Fig. 7.1-12 gives an impression of the magnitude of stored energy in pressurised high-pressure components. This example considers only the stored physical energy for a gas with an adiabatic coefficient $\kappa = 1.66$ [19].

If, additionally, chemical energy is stored the values are much higher [20]. An example of a total energy consideration is given in Tab. 7.1-2 for a polyethylene reactor with a volume of 0.5 m³ loaded with 238 kg fluid ethylene. The adiabatic coefficient of ethylene is $\kappa = 1.4$. In case of an accident the energy evolution will reach an amount of approximately 20×10^7 Joule. It is assumed that only 10 % of the ethylene decompose. The result demonstrates that only approximately 45 % is stored physical energy. It is common practice to calculate energy contents of systems as an equivalent mass of TNT. In the example given before the energy content is equivalent to approximately 43 kg TNT.

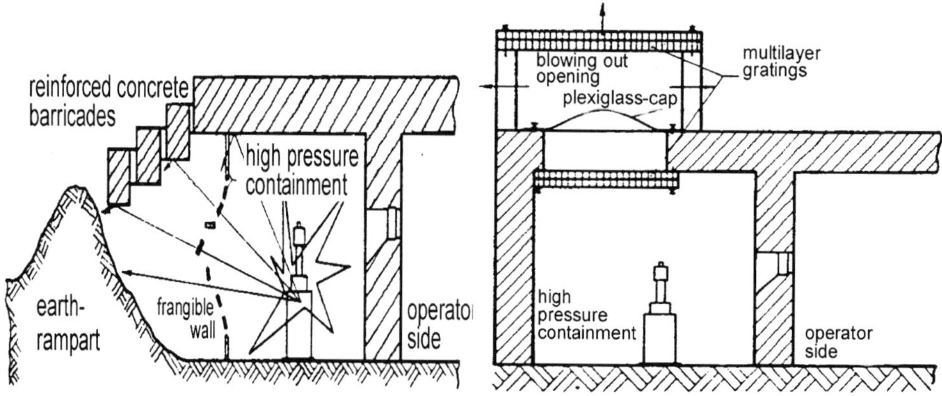

Figure 7.1-10. High-pressure bunker with reinforced concrete walls and an earth wall.

Figure 7.1-11. Typical high-pressure bunker in research laboratories.

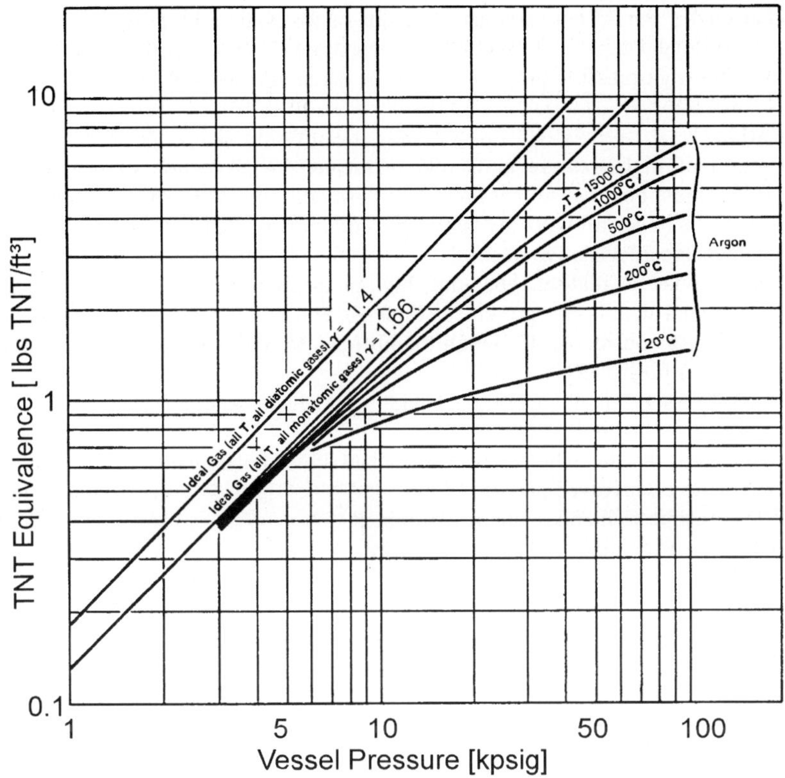

Figure 7.1-12. Energy stored in a gas pressurised vessel as function of pressure and the energy equivalent in mass of TNT according to a volume of 1 ft³ and a gas with an adiabatic coefficient $\kappa = 1.66$ (Argon). Argon data are based on the Redlich-Kwong - Equation of State. These curves should only be used as a guide. Variation of temperature within a vessel must be considered [19].

Table 7.1-2
Energy balance for a bursting polyethylene reactor during ethylene decomposition. Pressure: 200 MPa; volume: 0.5 m³ [20]

	Energy evolved	
	Joules x 10^7	ft lbf x 10^7
Adiabatic expansion of gas	8.65	6.38
Decomposition of 10 % of gas	10.65	7.86
Elastic strain energy of container	0,019	0.014
Total Energy	19.32	14.25

The installed barricades should not serve as an incubator for disaster by providing a small area for escaping volatile and explosive vapours to concentrate. This is a further aspect when designing a building for high-pressure processing must be taken into account.

7.1.5 Plant operation

Safety in operation of high-pressure units mainly means the protection of people and investment. Operation must be done so that nobody is harmed. Equipment specifications and installation plans must be in accord with any applicable codes or regulations. These may be in the form of State and/or local legislative rules.

Some typical facts must be realised operating high-pressure facilities. In the following some important additional details are presented. First general aspects for people operating high-pressure facilities are considered. Common thumb rules for the application of technical installations are presented further [21].

Employees in high-pressure plants must obey some important recommendations to do safe work.

- The responsibilities for the whole work in a plant must be defined exactly and be communicated with all involved people.
- All employees must be well and permanently trained.
- Employees must be experienced in operating chemical plants.
- Each person must observe carefully all deviations in a plant and discuss the changes with the responsible person.
- The employees must be attentive during the job.
- All employees must make use of all the special safety rules for working with pressurised facilities.
- First step before opening any part of an equipment is to convince himself that pressure is released.
- If a worker has caused a safety relevant fault he must announce it to his supervisor.
- Each failure should be thoroughly investigated and reported. The causes for this failure must be exactly located so that corrective actions can be started to prevent recurrence.
- No failure should be dismissed as insignificant. Repeated similar failures are the indicators for severe safety problems.
- Don't enter areas with high pressure without a special licence for inspection, maintenance, etc.
- Difficult procedures should be intensively trained before doing it in the plant.

Main thumb rules when handling pressurised parts:

- Parts of a high-pressure system could be clogged by product and pressurised gas or liquid could be stored in small sectors of the facility.
- All connected pressurised systems must be closed and disconnected to avoid the pressurising of components which are alleged pressure-free at the beginning of the work.
- When flanges or the cover of vessels should be opened it is recommended to remove the nut of bolts only turn by turn and to prove that there is no pressurised content behind the flange or the cover.
- Short time after pressure release in the plant an elastic polymer seal is often internally pressurised by dissolved gas and contains stored energy which leads to an explosion-like expansion and damage of the seal.

- If defect parts must be replaced it must be assured that the new part has the same specifications as the old one.
- All spare parts must be controlled before their storage and use in the plant.
- Particular attention should be paid to fatigue failures as they may reveal design deficiencies rather than material defects.

The reported facts are basic and general, but the meaning of the single points should be realised at nearly each situation by working with high pressure. A good philosophy and the best intention is an important factor starting the work in such plants and is also the source for a responsible and safe doing.

7.1.6 Testing procedures and inspections

The last step in the manufacturing process are the test procedures in the workshop as well as in the plant after its erection. After a period of operation the regular inspections have to be completed by additional tests in case of repair or after larger maintenance work. The several national bodies give the manufacturer and the operating companies detailed targets how high-pressure vessels and high-pressure components have to be handled.

After manufacturing the facility must be pressure tested. First step of testing is usually the water-test wherein the pressure vessel must withstand the design pressure and a pressure which is multiplied with an defined safety factor. Common safety factors are in a range of approximately 1.5. During this part of the testing procedures the maximum stresses caused by the construction or the manufacturing process are diminished because the material can be stressed up to the yield strength at those locations. The material then is plastically deformed and the stresses are brought down to an allowable level. The water-tests must be accurately prepared. First it is important to load the apparatus totally with water. It is necessary to remove all gas from the system due to its high stored energy. One percent dissolved gas in the test liquid represents the same stored energy as in the volume of the fluid itself.

When pressure vessels will be tested and operated later on in very cold climates the lower limit of the transition temperature zone of the impact strength of the material applied must be considered because the full load pressure during the test or the start up may cause an unsafe situation if the vessel has not been heated up sufficiently.

The pressure vessel must be placed at a remote area that in case of a rupture nobody will be harmed. In former times a number of vessels failed during this procedure. It must be noted here that already by a small leakage a water jet stream as generated by high pressure is strong enough to penetrate the body of a person. For this reason people involved in the test must be protected by safety shields. After a successful water-test the gas-test will follow. The tests will be repeated with the design pressure for detecting leaks. First at the end of the test procedures the plant is ready for service. The test results should be reported in a certificate. All the procedures will guarantee a safe plant operation. Normally after a so-called water campaign the start-up with the original chemicals follows. To maintain safe operation it is very important to install an inspection- and maintenance schedule. The example given in Fig. 7.1-6 is one of the important features, that the individual apparatus and components can be tested readily during the service in the plant. The periodical test campaigns are an excellent step to guarantee the safe operation over the life span of a high-pressure plant.

References

1. S. Maier, G. Kaibel, Chem.-Ing.-Tech. 62 (1990) Nr. 3, 169-174.
2. A.W. Guill, Safety in High Pressure Polyethylene Plants, American Inst. of Chem. Engng. 1973, 49-52.
3. W.L. Frank, E.H. Perez, J.Y.Yeung, CEP, 94,1998, No.10, 71-81.
4. V. Pilz, W. van Herck, Proc. 3rd Int. Symp. "Large chemical plants", Antwerpen, Oct. 1976.
5. B. Vanderstraeten, D. Tuerlinckx, J. Berghmans, S. Vliegen, E. Van't Oost and B. Smit, Journal of Hazardous Materials 56 (1997) 237.
6. K. Nabert and G. Schön, Sicherheitstechnische Kennzahlen brennbarer Gase und Dämpfe, Deutscher Eichverlag, Braunschweig 1963.
7. J.U. Steinle and E.U. Franck, Ber. Bunsenges. Phys. Chem. 99 (1995) 66.
8. H.A. Potho, M.J. Googin, UCC ND for the United States Department of Eneergy: The Ad Hoc Commitee for High Pressure Safety, Oak Ridge, Tennessee, April 1981.
9. Th. Zimmermann and G. Luft, Chem.-Ing.-Tech. 66 (1994) 1386.
10. G.M. Pohsner and E.U. Franck, Ber. Bunsenges. Phys. Chem. 98 (1994) 1082.
11. G. Luft and H. Seidl, Angew. Makromol. Chem. 86, No. 1326 (1980) 93.
12. M. Gebauer, W. Clausnitzer and R. Linden, Chem. Technik 45, No. 4 (1993) 208.
13. C.F. Winans, Dechema-Monogr. 47 (1962) 805 and 840.
14. A. Fredenhagen and R. Eggers, Chem.-Ing.-Tech. 72 (2000) 1221.
15. H. Hauser, K. Nagel, H. Spähn, VGB Kraftwerkstechnik, 54, 1974, 123-134.
16. S.J. Tunkel, CEP September 1983, 50-55.
17. P.E. Hatfield, Safety in High Pressure Polyethylene Plants, American Inst. of Chem. Engng. 1973, 57-69.
18. I. Meyn, GIT Fachz. Lab. 25 (1981) 1, 25-31.
19. H.A. Potho, Jour. of Pressure Vessel Technology, May 1979, Vol. 101,165-170.
20. W.R.D. Manning, S. Labrow, High Pressure Engineering, Leonard Hill, London, 1971.
21. K.L. O'Hara, M.B. Poole, Safety in High Pressure Polyethylene Plants, American Inst. of Chem. Engng. 1973, 53-56.

7.2 Runaway of polyethylene reactors

7.2.1 General remarks

The major hazard that can occur in the high-pressure polyethylene process is a runaway of the reactor and decomposition of ethylene as well as fires, explosion, and disintegration of high-pressure parts. Although the last incidents are well understood, the reasons for runaway and ethylene decomposition have been evaluated only recently. Experience over twenty years has shown that decomposition mostly takes place in the reactor and in the high-pressure separator, but decompositions have also been reported from ethylene-feed and product lines.

In stirred autoclaves, decompositions start frequently from the bottom zone where the temperature reaches a maximum and the concentration of the polymer is high. When the motor of the agitator is placed inside the autoclave, as is done in some processes, decompositions were also observed starting from the motor. In tubular reactors, decompositions are located in the area of steeply rising temperature, down-stream of the initiator injection. They can also occur close to an initiator injection point when the temperature profile is occasionally shifted down-stream so that the temperature at the injection point is unexpectedly high. Furthermore, decompositions can take place in areas with stagnant flow which exist near the tube flange, if the thermocouples arranged in the lens-rings disturb the flow of the ethylene–polyethylene mixture.

Decompositions in the high-pressure separator can start in the tails-line of the reactor, after the letdown valve, when a temperature rise occurs by the reverse Joule–Thomson effect and proceeds into the separator. Also, decompositions which begin in the bottom zone of stirred autoclaves or the cooling zone of a tubular reactor can propagate through the tails-line into the high-pressure separator when the throughput is high.

The frequency of decompositions depends on the process, whether the plant is equipped with a tubular reactor, a multi-chamber autoclave, or single autoclaves. Important factors are the pressure, temperature, and residence time in the reactor and in the high-pressure separator. The frequency of decompositions is also influenced by the raw materials and the resins which are manufactured. Last, but not least, the frequency of decompositions depends on the shape of the plant, on the training of the operators, and on the know-how of the company or the assistance of the licensor.

In a carefully operated and well maintained plant equipped with a tubular reactor or a multi-chamber autoclave, on average one decomposition takes place per annum. The number of decompositions is lower when the reactor is well mixed and exhibits only small temperature gradients as in single stirred autoclaves with a lower length-to-inside-diameter, or when two well stirred single autoclaves are arranged in series. Experience has shown that the frequency of decomposition is also lower when the reactor is run at lower pressures and temperatures and when only homopolymers having a higher melt index are manufactured.

7.2.2 Decomposition reaction

The raw material, ethylene, belongs to the group of endothermic compounds which are inherently stable and can decompose explosively even when oxygen is not present. Under high temperatures and pressures a decomposition results in a steep increase of pressure and temperature. In a closed bin, extremely high pressures and temperatures were reached.

The course of the pressure during a decomposition occurring in a vessel of 10 m^3 volume is presented in Fig. 7.2-1, left [1]. Within 1 s the pressure increases from 18.5 to 30.8 MPa. The peak pressure exceeds the range of the pressure-recording device. Extrapolation of the data

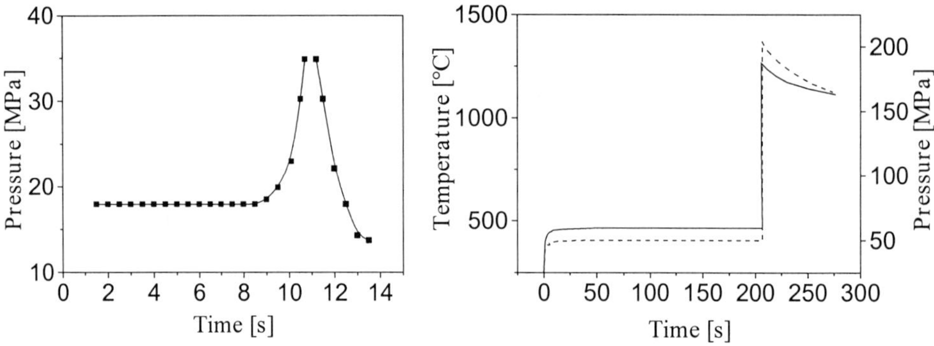

Figure 7.2-1. Course of pressure (solid line) and temperature (dashed line) during decomposition.

indicates a peak pressure of around 50 MPa. When the initial pressure is higher, the pressure increases more steeply, as shown in Fig. 7.2-1, right.

During decomposition, ethylene decomposes mainly into carbon, methane, and hydrogen:

$$C_2H_4 \rightarrow (1+z)\ C + (1-z)\ CH_4 + 2z\ H_2$$

Under the conditions of high-pressure polymerization of ethylene, the parameter z is between 0.05 and 0.16 [2]. The composition of the gaseous decomposition products is shown in Fig. 7.2-2 as a function of pressure for two initial temperatures of 200 and 300°C. The concentration of methane was found to be in the range of 60 - 80% by volume, increasing with the initial pressure and decreasing with the initial temperature, whereas the fraction of hydrogen was found to be higher with decreasing initial pressure.

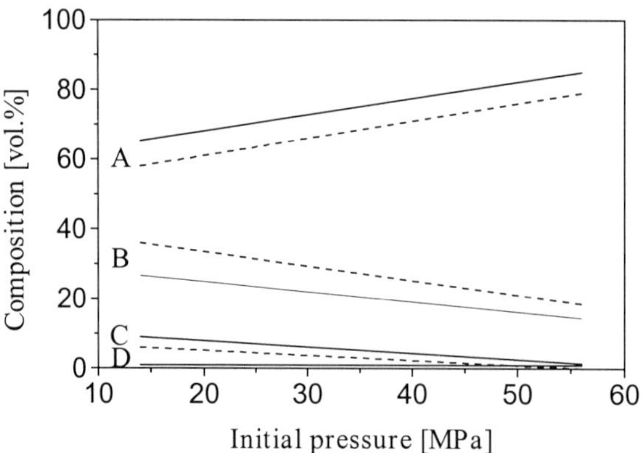

Figure 7.2-2. Composition of gas from decomposition during homopolymerization of ethylene [3]. A, CH_4; B, H_2; C, C_2H_4; D, C_2H_6; solid line, 200°C, dashed line, 300°C.

When ethylene–vinyl acetate copolymers are manufactured, the decomposition products also contain carbon monoxide and carbon dioxide. When decomposition takes place in a tubular reactor or in a multi-chamber or cascade of autoclaves, up to 50% of the decomposition gases can consist of undecomposed ethylene.

7.2.3 Critical conditions

The critical conditions for runaway and decomposition during polymerization were evaluated from tests at 5 - 230 MPa and 275 - 450°C. The tests were performed in an autoclave of 200 ml capacity. Besides the homopolymerization of ethylene its copolymerization with different co-monomers was studied.

7.2.3.1 Homopolymerization

The critical conditions for runaway during polymerization, which were evaluated from the course of temperature and pressure, are shown in Fig. 7.2-3. Four different areas can be distinguished. At pressures and temperatures below the curve at the bottom, no reaction takes place. In the area above, a stable polymerization occurs. At conditions above the dashed line, decomposition can start during polymerization after an induction period. At higher temperatures and pressures a spontaneous decomposition can be observed.

At a given pressure the critical temperature for runaway during homopolymerization decreases, first steeply and then less steeply, with increasing pressure (Fig. 7.2-3, dashed line). At 5 MPa the critical temperature is around 450°C. It reduces to 350°C when the pressure is 50 MPa, and to 290 - 300°C at the operation pressure of LDPE-reactors of 150 - 300 MPa. In this range of pressure, spontaneous decomposition can take place when the temperature is above 330°C.

7.2.3.2 Influence of co-monomers

Co-monomers can reduce or increase the critical temperature for runaway and decomposition. As an example, the influence of vinyl acetate, which is often used as a co-monomer in the high-pressure polymerization of ethylene, is shown in Fig. 7.2-4. For this purpose, the critical temperature for decomposition during runaway from polymerization is

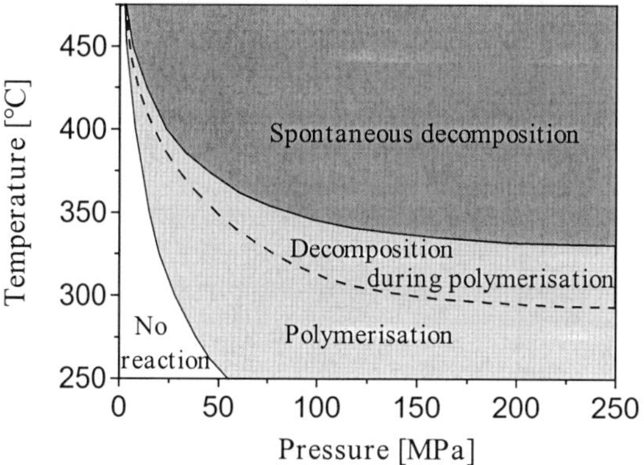

Figure 7.2-3. Critical conditions for runaway during homopolymerization of ethylene [4].

424

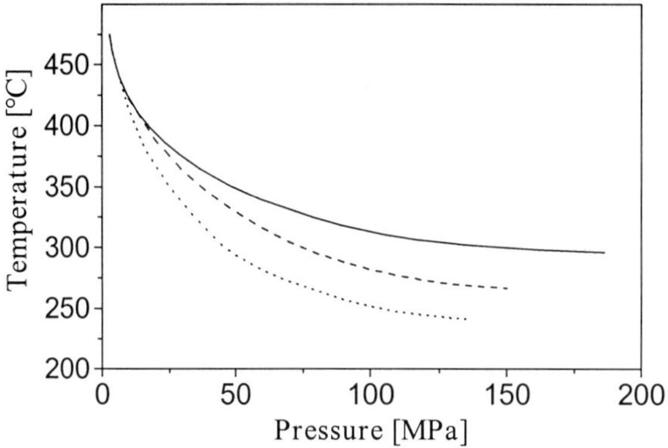

Figure 7.2-4. Influence of vinyl acetate [5]. Solid line, 0 wt.%; dashed line, 10 wt.%; dotted line, 30 wt.%.

plotted versus the pressure for 10 wt.% (dashed line) and 30 wt.% vinyl acetate (dotted line) in the mixture with ethylene. For comparison, the critical temperatures and pressures of the homopolymerization of ethylene are incorporated (solid line). Also, the critical temperatures of copolymerization decrease, first steeply and then less steeply, with increasing pressure. The influence of vinyl acetate is low when the pressure is below 10 MPa. At the operating pressures of industrial LDPE reactors the critical temperature of the copolymerisation with 30 wt.% vinyl acetate is lower by around 60°C.

7.2.3.3 Influence of oxygen

It is well known that runaway of LDPE reactors occurs when the concentration of peroxide or oxygen initiator is too high, because of malfunction of the initiator-metering device. It can

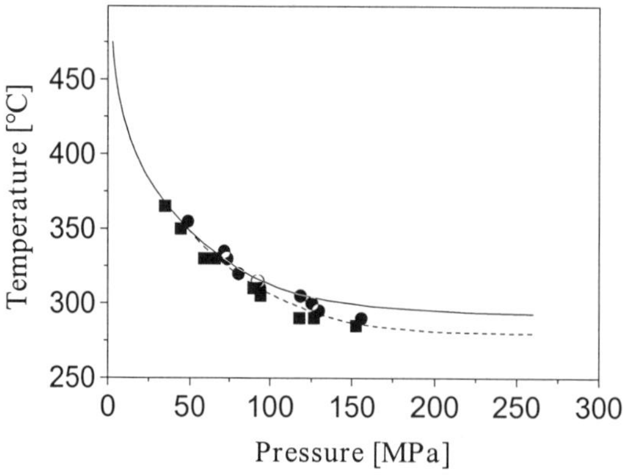

Figure 7.2-5. Influence of oxygen. Dashed line, excess of oxygen 20 wt. ppm; solid line, without excess of oxygen; ■, polymerization; ●, decomposition during polymerization.

also happen that the ethylene feed contains oxygen. For this reason the influence of oxygen was investigated. The results with excess of 20 wt. ppm oxygen are presented in Fig. 7.2-5. Again, the critical temperature is plotted versus the pressure. The tests in which runaway occurred are marked by circles. Squares show the tests with stable polymerization. For comparison, the critical data of the homopolymerization are again shown (solid line).

By the addition of an excess of oxygen the critical temperature for runaway decreased. At pressures above 150 MPa the critical temperature was found lower by 10 - 15°C.

7.2.3.4 Influence of decomposition sensitizers

Sensitizers are compounds which are able to initiate or to trigger a decomposition. Decomposition can also take place in the "safe" area of stable polymerization considered in Fig. 7.2-3. The potential of a variety of compounds which can be entrained into the reactor by monomers and additives, or can be formed from impurities and construction materials, to induce decomposition in the reactor or high-pressure separator was investigated. It was found that the risk of decomposition induced by a sensitizer depends not only on the pressure and temperature, but also on its concentration.

As an example, in Fig. 7.2-6 the influence of copper(II) acetate under a pressure of 195 MPa and 220°C is presented. When Cu(II) acetate was injected into an autoclave reactor

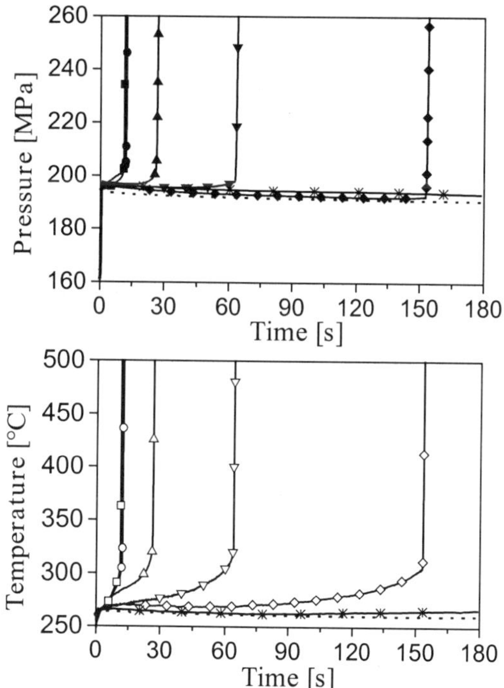

Figure 7.2-6. Decomposition induced by sensitizer. Concentration of CuAc$_2$ (mol ppm): □, 650; O, 325; Δ, 160; ∇, 32; ◊, 20; *, 15; dashed line, 0.

426

during polymerization in a concentration of 650 mol ppm, a spontaneous decomposition was observed. At concentrations below 20 mol ppm decomposition did not occur.

7.2.4 Increase of pressure and temperature during decomposition

Experience from decomposition in industrial LDPE-plants as well as from laboratory experiments shows high maximum pressures and temperatures, together with extremely high rates of pressure-rise during decomposition. As can be seen from Fig. 7.2-7, top, the maximum pressure in closed bins increases steeply with increasing density of the mixture from which a decomposition starts. The maximum pressure is higher when the decomposition starts from a higher pressure and lower temperature.

Simultaneously, the temperature inside a polymerization reactor can rise to 1400 - 1500°C for some seconds (Fig. 7.2-7, bottom). The increase of the temperature is higher when the initial pressure and temperature are higher. The time during which the high temperature is maintained is only short. Black rust is not formed and the strength of the construction materials is not affected.

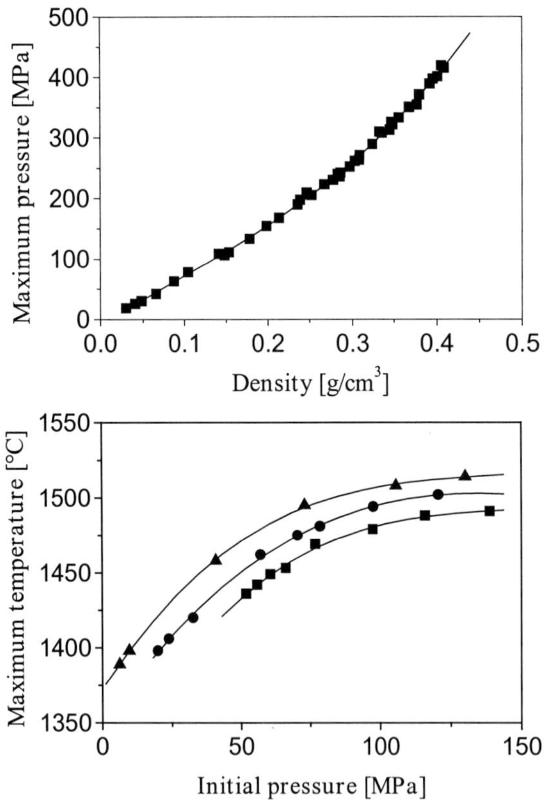

Figure 7.2-7. Maximum pressure (top) and temperature (bottom: ■, 350°C; ●, 400°C; σ, 450°C) during decomposition.

The rate of pressure-rise increases with increasing initial pressure and decreasing initial temperature (Fig. 7.2-8). Below the operation-pressure of relief devices, rates of pressure-rise up to 1000 MPa/s can be measured during fast decompositions.

The rate of pressure-rise in closed vessels is steeply influenced by the volume of the vessels. The relationship between the maximum rate of pressure-rise, dp/dt, and the volume, V, of the vessel can be described by a cubic law [6]:

$$(dp/dt)_{max} \cdot V^{1/3} = \text{const.} = K_G$$

Following the above expression, the maximum rate of pressure-rise reduces with increasing volume of the vessel. The parameter K_G depends on the nature of the decomposing compounds and on the initial temperature and pressure from which the decomposition starts. Values of K_G are given in [7]. For example, K_G is 820 for decomposition during runaway from polymerization at 150 MPa and 300°C. With a volume of 2 m^3 of an autoclave used for the production of 100,000 t/a, a maximum pressure-rise of 65 MPa/s results.

7.2.5 Loss prevention

Decomposition should be averted for both safety and environmental reasons. In case of decomposition, multiple levels of protection are applied in appropriate sequences of procedural controls, instrument controls, interlocks, and relief devices to minimize damage to the plant and impairment of environment. Because the discharge of ethylene and its decomposition products into the air involves considerable risk, precautions for safe venting must also be considered.

7.2.5.1 Relief devices

For relief-protection dump valves, relief valves, and finally rupture discs are used. The sizing of the relief devices is based on the rate of pressure-rise. The flow area should be large enough to stop the pressure rise as soon as the set pressure is reached.

It is known from Fig. 7.2-3 that the pressure should be reduced to 100 or 50 MPa to avoid decomposition. When a relief valve is activated the rate of pressure-decay should be fast, so

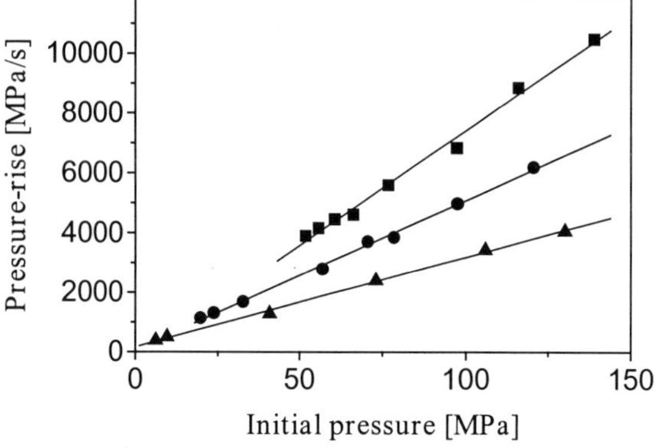

Figure 7.2-8. Rate of pressure-rise during decomposition. ■, 350°C; ●, 400°C; ▲, 450°C.

that the time to reach the above-mentioned low pressure levels is fairly well below the induction period of decomposition. For this reason, the flow-diameter of relief valves is only a little lower then the inside diameter of a tubular reactor. These dump valves, arranged at the reactor entrance and the outlet, can discharge 80% of the gas within 50 - 100 s.

Rupture discs do not re-close, and are used therefore as the last defense. The illustration in Fig. 7.2-9 shows a rupture-disc port assembly. The holder is provided with a liner made of Inconel and internally mounted. The effective area is typically in the range of 100 – 200 · 10^{-4} m^2.

7.2.5.2 Venting systems

When gases from the reactor or the high-pressure separator are discharged through the chimney into the air, explosive mixtures can be formed. The explosion limits of mixtures of ethylene with air, as well as methane, hydrogen, and vinyl acetate with air are listed in Table 7.2-1 together with the ignition temperature.

From Table 7.2-1 it can be seen that hydrogen is the most critical compound regarding the wide range of explosion-limits. Also, ethylene is critical because of its low ignition temperature.

In order to avoid explosion, the gases should be cooled and diluted by injection of water or steam before release into the chimney. A water-injection device is shown schematically in Fig. 7.2-10. It consists of a water tank (1), a Venturi device (2) in the discharge line (3), nozzles (4) to inject the water into the gas stream, a dip-tube (5), and a line (6) which connects the water tank and the inlet of the Venturi device. The water tank is equipped with a

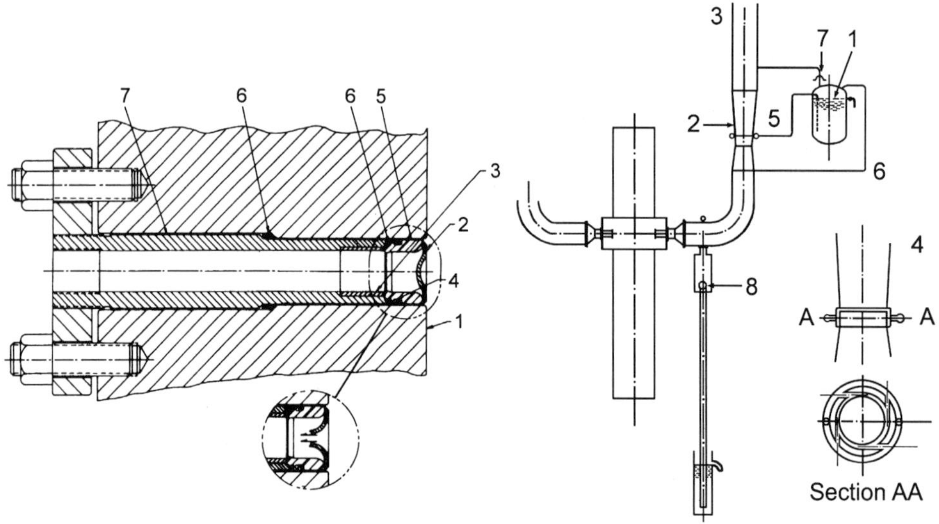

Figure 7.2-9. Rupture disc port assembly [8].
1, vessel wall; 2, disc; 3, Inconel liner;
4, retaining cab; 5, rupture nose; 6, seal
assembly; 7, rupture sleeve.

Figure 7.2-10. Water injection device [9].
1, water tank; 2, Venturi device;
3, discharge line; 4, nozzle; 5, dip tube;
6, line; 7, rupture disc; 8, drain line.

Table 7.2-1
Conditions for explosive mixtures [10]

Compound	Lower explosion limit [% vol.]	Upper explosion limit [% vol.]	Ignition temp. [°C]
Ethylene	2.7	28.6	435
Methane	5.0	16.0	595
Hydrogen	4.0	75.6	560
Vinyl acetate	2.6	13.4	385

heater and a rupture disc (7). The excess water is discharged by a drain line (8) at the bottom.

References

1. J.F. Sullivan and D.I. Shannon, PVP Vol. 238, Codes and Standards and Applications for High Pressure Equipment, ASME (1992) 191.
2. Th. Zimmermann and G. Luft, Chem.-Ing.-Tech. 66 (1994) 1386.
3. J. Albert and G. Luft, AIChE Journal 45 (1999) 2214.
4. H. Bönsel and G. Luft, Chem.-Ing.-Tech. 67 (1995) 862.
5. J. Albert and G. Luft, Chemical Engineering and Processing 37 (1998) 55.
6. W. Bartknecht, Explosionen, Ablauf und Schutzmaßnahmen, 1. Auflage, Springer Verlag, Berlin, 1978.
7. Decomposition in LDPE plants, Brochure Maack Business Services, Au/Zürich 1999.
8. R.G. Ziefle, Designing Safe High Pressure Polyethylene Plants, Chem. Eng. Progress Technical Manual, American Institute of Chemical Engineers, New York, 1973.
9. J.J.L. Marzais, A Water Injection Device for High Pressure Gases, Chem. Eng. Progress, Technical Manual, American Institute of Chemical Engineers, New York, 1973.
10. G Hommel, Handbuch der Gefährlichen Güter, Springer Verlag, Berlin, 1993.

7.3 Safety in high-pressure extraction plants

The main service requirements in safety are, first of all, the protection of persons inside and outside the plant, precautions to prevent injurious influences on the environment, and to avoid damage to property and the plant itself. The plant has to be designed, erected, and operated to fulfil these. Numerous laws and regulations try to create common standards for plant operators, employees, the authorities, and the public.

The fulfilment of all laws and regulations is, in most cases, not sufficient for safe design and operation. On the one hand, the corresponding legal framework always leaves a certain freedom of action and, on the other hand, the legislative authority cannot cover all the diverse technical risks. To reduce the risks to a minimum the design must be executed by experienced professionals with a sense of responsibility and with careful systematic planning.

Safety of plant design [1]

For the design of new plants and processes it is important to incorporate as much as possible of known and approved systems or process steps. The operational experiences of such known systems or process steps are to be considered for the design, as well.

For the design of plants which are to operate under elevated pressures the following items are relevant:
- Within each single process step it must be ensured that the corresponding operational pressure cannot be exceeded.
- Deviations from the set points, and possible malfunctions, must be guarded against by proper precautions such as controls, shut-off valves, shut-down devices, safety valves, rupture discs, etc.
- There should be a safe removal of all materials leaving a safety device.
- In case safety devices carry still a too high restend risk, additional measures in construction like bunkers or dividing walls are necessary.
- In case of malfunctions, the supply of cooling agents must be ensured.
- The gap between the operation- and design temperatures must be sufficient.
- Peak-values in temperature, owing to compression energy at the time of gaseous pressurizing in autoclaves, are to be considered.
- The material selected for construction must be resistant against corrosion and erosion, or has to be subject to control on a regular basis, or has to be changed.
- One should investigate whether if a higher quality steel is used the plant safety can be improved.
- One should investigate whether certain design measures, such as cladding, reinforcement, or higher corrosion-allowance will improve the safety.

For dense gas extraction plants the building design shall consider, that:
- The engine-hall is as far as possible soundproof.
- The engine-hall is adequately vented, and CO_2 leakages from pumps and compressors are removed.
- Exhaust of CO_2 in the areas of quick acting closures is removed.
- Exhaust installations are available at the lowest floor levels where there may be CO_2 concentrations.

- The distance between the control room and regular operating areas is to be as short as possible.
- The running of cables and the compressed-air system, is as far as possible, to be protected against fire.

Safety in plant operation

Safety must always have top priority at commissioning and in operation. The responsible plant operators have to take measures so that at no point or time are the life and health of humans inside and outside the plant, endangered. No economic or other constraints should have influence in this regard.

The most important points are:
- The proper qualification of the personnel.
- Regular training of operators in regard to safety and operation.
- The start-up and operation of the plant must occur only when it is in a technically trouble-proof condition.
- The plant operators are to be properly familiar with all plant components.
- That failures and malfunctions are corrected immediately.
- That maintenance, service, and repair are carried out under all safety conditions, as most accidents happen from such activities.

7.3.1 Protection of individual pressure ranges

Within pressure plants, especially high-pressure extraction plants, different pressure levels are always given. The highest pressure in SFE plants is within the extraction section, followed by the separation step, and the CO_2 storage.

The pressure ranges themselves are determined from DIN regulations, IG-norms or other regulations.

The most important pressure ranges in SFE plants are:
according to DIN: PN25, 40, 64, 100, 160, 250 and
according to IG: 325, 500, 700, 1600 bar.

The most important target for the design engineer is to minimize the applied pressure ranges, and to obtain as little as possible connections between the various pressure ranges. Each connection between different pressure ranges embeds a high potential for danger. The safety precautions between connected pressure ranges require exacting requirements for safety devices, interlocking systems, and the safety analysis itself.

Design and selection of pressure ranges

The following examples explain the link between thermodynamic conditions within the plant and the necessary pressure ranges to be used.

Table 7.3-1
For a multipurpose plant for spices and herbs the following operating conditions and pressure ranges are applicable

Section	Operation condition bar/°C	Pressure range, piping system bar
Extraction	Up to 480/80	IG 700
1. Separator	Up to 220/50	PN 250
2. Separator	Up to 60/25	PN 100
Condenser		PN 100
CO_2 working-vessel		PN 100
CO_2 recycling		PN 100
CO_2 make-up	22/minus 20	PN 40

Table 7.3-2
For an isobaric decaffeination process the following data apply

Section	Operation condition bar/°C	Pressure range, piping system bar
Extraction	260/85	IG 325
Washing column or absorber	260/85	IG 325
Condenser		PN 100
CO_2 working-vessel	60/22	PN 100
CO_2 make-up	22/minus 20	PN 40

While the CO_2 make-up system and the CO_2 working-vessel are connected, in most cases, by a single piping line, and therefore simple protection is adequate, the numerous connection lines between the extraction section and the peripheral equipment have to undergo an accurate safety evaluation.

The simplest way would be to design the whole plant for the highest pressure range, but this is excluded by economic reasons.

Critical aspects to consider are:
- By-pass lines for pumps and compressors, with a connecting line from the pressure-side to the vessel on the suction-side. This safety risk is eliminated in most cases by using a check-valve in the discharge line after the branch to the by-pass valve.

- Depressurizing valves, reducing the extraction-pressure down to the separation-pressure. Safety valves located in front of -, or directly on the separator, must be large enough to remove the complete possible mass-flow.
- All pipes and valves necessary for discontinuous operations within a batch process, and the overall necessity for pressurizing and expansion of the extractors.

For normal extraction times, two manifolds with a corresponding set of valves located on the top and bottom of each extractor are adequate. For short extraction times three manifolds may be necessary. Small and medium-sized plants are equipped with control valves for discontinuous operation steps. From the technical- and economic points of view, high-pressure control valves are limited with K_{VS} values between 6 and 10, especially for pneumatically driven valves. If such valve sizes are too small, high-pressure ball valves must be used, thereby substantially increasing the costs for the interlocking system and the safety requirements.

7.3.2 Use of safety valves and rupture discs

The calculation of safety valves is regulated by the "AD-Merkblatt" A2, Edition September 1998, and must be used by plant suppliers. Included in this regulation is the set-up and the piping system for the safety valves, also. The support of the safety valves must consider the potentially enormous reaction forces.

The most difficult task for the safety-valve sizing is the achievement of the maximum flow-rate to be discharged. In case the safety valve is used to save the operation of a pump or compressor, their maximum mass-flow-rate is the basis for the sizing. Much more effort is necessary in systems where vessels, complex piping systems, pumps, and thermal expansion can activate the safety valve. In most cases, some discussions with the relevant authorities are necessary to clarify all assumptions for risks relevant to safety.

The specification of safety valves on vessels must consider:
- The determination of all normal feed-streams into the vessel.
- Calculation of the maximum possible mass-flow; for example, the mass-flow through a ball valve resulting from a malfunction or mal-operation which is located between two systems with different pressure ranges. This mass-flow depends on the pressure-drop within the connecting piping system.
- Mass-flow resulting from wrong opening of a control valve, for which full opening of the valve, at the highest possible pressure difference is to be assumed.
- Mass-flow owing to pumps and/or compressors.
- Mass-flow owing to thermal expansion, that means, heating sources within the corresponding system, such as heat exchangers, heating coils or jackets within vessels, trace-heating, or from ambient heat. Such influences increase the system-pressure gradually and cause - compared to other reasons - relatively low mass-flows, and therefore have only low contributions to the safety-valve sizing.

Overall, wrong operation of ball valves can create enormous mass-flows, which cannot be exhausted economically any more by safety valves. Safety valves with more than a 500 mm² opening section are hard to incorporate, because of their size, weight, position, and tightness. For such cases, rupture-discs are used. The calculation is given in the "AD Markblatt" A1,

Edition of January 1995. The evaluation of relevant mass-flows is according to safety valves. The use of rupture-discs requires some experience from the design engineer concerning positioning and reaction forces. The exhaust of all safety valves and rupture discs should be connected to a common blow-off tube, at a level above the building.

7.3.3 Interlocking systems

For larger plants, especially where ball-valves are used, interlocking systems are very important.

Special attention must be paid to:

- Interlocking for quick-acting closure systems of extractors. With discontinuous-charge processes the extractors must be opened for charging several times a day, depending on the extraction times, and therefore a reliable interlocking system for the closures is very important. It must be ensured that, before opening, the extractor is without any pressure, all connected inlet- and outlet valves remain securely closed, all available quick-shut-off valves are in their closed position, and all valves located between extractor and atmosphere are open, and remain in this position. The control system must periodically control the limit switches and their correct position is needed to release the opening mechanism.
- The extractor must be protected against overheating during pressurizing, and against forced chill during expansion.
- Ball valves require a certain low pressure-difference before being opened, in order to avoid shock pressures in the piping system.
- Valves connecting two different pressure systems may be activated only when the pressure within the elevated system is below the set point of the corresponding safety valves or rupture discs.
- The pressure drop within the extractors may not exceed a certain defined level to secure the filter cartridges or plates.
- Switching of valves in the continuous CO_2 circulation system is allowed only when all valves for the discontinuous processes are closed.
- In case the allowed operating pressures are exceeded, the by-pass of machinery must be opened and the relevant machines shut-off.

With increasing levels of automation, the interlocking system must be guaranteed to a higher degree of safety, as well.

7.3.4 Safety analysis

From the numerous HAZOP analyses that are known none is specifically adjusted for high-pressure extraction. NATEX in cooperation with the relevant authority TÜV-Vienna adjusted the "PAAG-Methode" for the design and execution of their high-pressure extraction plants. The plant is thereby split into diverse basic operations such as, for example CO_2 storage, CO_2 circulation, extraction, separation, CO_2 recovery, cooling system, heating system, and so on. In each section the foreseen duties are described and their sector limits established. Afterwards, for each individual section, all imaginable deviations, existing measures to avoid the same, and possible additional measures are defined.

As an example of this procedure the CO_2 make-up system is reviewed, with regard to the following deviations:

- Pressure too low or too high
- Temperature too low or too high
- Level too high or too low
- Pressure elevated or reduced
- Temperature elevated or reduced
- No flow
- Elevated or reduced flow
- Decreasing flow

After such deviations are defined, for each individual point the possible reasons and effects are to be considered. Finally, it can be ascertained whether the available measures are sufficient, or additional ones are required. Important for the efficiency of such an analysis is the experience of the team in charge. The members of this team should consist of experts in production, process engineering, control systems, and relevant authorities such as the TÜV. Carefully executed safety analyses are decisive for the elaboration of control systems, interlocking systems, malfunction lists and, in general, for the operation handbook.

7.3.5 Controls and computerized systems

The duties of the control engineer include the programming of all necessary process operations, interlocking- and safety precautions into a control program, and avoiding the possibility of incorrect operations by the operators. The major part of the malfunctions which still remain should be detected during the commissioning of the plant.

Finally it should be mentioned that carefully executed safety analyses can also simulate rare and unlikely malfunctions, and that adequate measures can then prevent the corresponding incidents. Reducing the remaining risk must be the aim, and should always be present in the mind of the plant-operating executives.

References

1.   Ullmann, Encyklopädie der Technischen Chemie, 4. Auflage, Verlag Chemie GmbH, D-6940 Weinheim (1981) Band 6, 721.

437

CHAPTER 8

ECONOMICS OF HIGH PRESSURE PROCESSES

E. Lack[a], H. Seidlitz[a]: section 8.1

G. Luft[b]: section 8.2

A. Bertucco[c], A. Striolo[c], F. Zanette[c]: Section 8.3

[a] NATEX GmbH Prozesstechnologie
Hauptstrasse, 2 A-2630 Ternitz, Austria

[b] Department of Chemistry, Darmstadt University of Technology
Petersenstr. 20, D-64287 Darmstadt, Germany

[c] Dipartimento di Principi e Impianti di Ingegneria Chimica (DIPIC) Università di Padova
Via Marzolo, 9 I-35131 Padova Italy

Economical considerations are investigated in this section, with reference to high pressure extraction plants, to the production of polyethylene, and to the precipitation by supercritical anti-solvent process.

In the first case, comparison with conventional low pressure solvent plants are not included, because of differences in the legal situation within diverse countries, which can result in higher or lower investment and processing costs due to regulations in regard to environmental protection, allowed remaining solvent residue levels, usage of diverse solvents and safety precautions.

About the production of polyethylene it is shown that the total costs of the low-pressure process are similar to those of the high pressure one.

Finally, the industrial application of a technique based on the precipitation by a supercritical anti-solvent is discussed: it can be economically attractive only with high-value products, such as pharmaceutical ones.

Anyway, before setting up a high pressure plant, pre-feasibility studies have always to be carried out in order to ascertain which parameters have greater influence on investment and processing cost.

8.1 High-pressure extraction plants

Within Chapter 6.6.2 the main parameters to be considered for the design of a high-pressure extraction plant have been discussed in detail. In brief mass-transfer and thermodynamic conditions with their related parameters play the most important roles for an optimized process. In this Chapter the influence of such selected parameters, design features, and the kinds of project financing are investigated with a view to production costs and return on investment (R.O.I.).

8.1.1 Description of standardized units

The simplest way of cutting costs is to standardize extraction plants. In batch processes with solid feed materials it is advisable to standardize the plants by using the payload volume of feed-material and calculating the capacity corresponding to the given bulk density and the elaborated cycle time. To justify these efforts a certain market size for the materials in question must be given, and the number of plants of the selected size that can be sold. The individual sizes depend on the availability of individual plant components, such as pumps, compressors, piping and armatures. For standardized pressures see Chapter 7.3.1.

According the size the following classification is possible:

8.1.1.1 Laboratory units

Laboratory units can be divided into bench-scale, research-, and pilot-scale. Bench-scale units, available in sizes of 20 to 500 ml (up to 400 bar) are used for screening tests, because one can obtain a relatively fast overview of the influence of the diverse major parameters, and only small quantities of raw material are necessary. Optimization duties require research units in sizes of 2 to 10 l, and are available in the pressure range between 325 and 700 (1000) bar. Such units enable quantity- and quality analyses. An illustration of one commercially available unit is shown in Fig. 8.1-1. For scale-up purposes, pilot plants ranging from 20 to 100 l, with design pressures up to 550 (700) bar, are recommendable for use with new products.

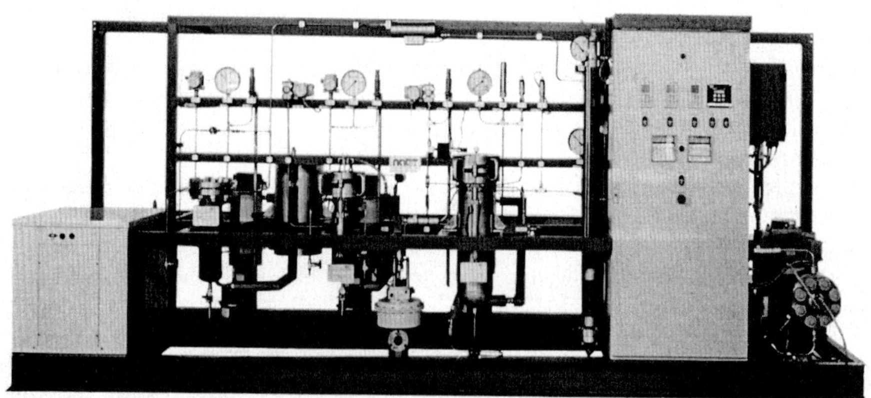

Figure 8.1-1. Picture of a 5 litre/1000 bar research unit (archives NATEX)

8.1.1.2 Medium scale units

These plants, with the greatest number of industrial application so far, are used mainly for the extraction of plant materials such as spices and herbs. The extraction volumes are in the range of 200 to 800 l in which, in most cases, cascade-mode operation with 2 to 4 extractors is applied. The design pressure-range is up to 550 (800) bar. Fig. 8.1-2 illustrates a multipurpose extraction unit for spices and herbs.

Figure 8.1-2. Multipurpose extraction unit for spices and herbs (archives NATEX)

8.1.1.3 Large-scale units

Such plants, with vessel volumes above 5 m^3, are used for decaffeination of coffee, for tobacco, and recently for reduction of plant protective materials. Fig. 8.1-3 shows a large-scale unit for rice, with a daily capacity of over 90 t. Such plants are operated at pressures up to 325 bar.

Figure 8.1-3. Plant protective reduction plant (archives NATEX)

Larger individually designed plants, used for hops, decaffeination of tea, and de-fatting or de-oiling are in a volume range between 1 and 5 m³ giving a throughput capacity between 1500 and 5000 tons/annum and operated at pressures up to 450 bar.

8.1.2 Feasibility studies

From investigations reported in Chapter 6.6, it is obvious that the production-costs will mainly be dominated by the overall energy consumption, which will depend on the solvent flow-rate, extraction-, and separation conditions, and the process design features such as the isobaric- or non-isobaric, single- or cascade-mode working, CO_2 recovery, the individual execution of particular plant components, and the degree of automation.

The influence of diverse parameters on production costs will now be reviewed, based on the following assumptions given in Table 8.1-1.

Table 8.1-1
Common basis for feasibility investigations

Category	Dimension	Value	Remarks
Production hours	hrs/annum	8,000	47.6 weeks : 7x24 hrs
Labour Cost	EURO	45,000	Average per man-year
Electrical Power	EUR/KWh	0.10	
Cooling Water	EUR/m³	0.50	
Drinking Water	EUR/m³	0.80	
Steam	EUR/t	22.0	
CO_2	EUR/kg	0.25	
Other Auxiliaries	EUR/annum	80,000	Spare parts, lubricants
Administration	EUR/annum	50,000	
Capital Reserve	%	3	From credit
Operational Capital Reserve	%	20	
Interest Rate	%	6.0	
Depreciation	-	Linear	Building 20 years Equipment 10 years

8.1.3 Influence of design

One of the most important duties of the design engineer is to optimize the process in regard to investment- and production costs, based on elaborated parameters. Particularly at larger plant-capacities, attention must be paid to energy consumption.

As a case-study, the decaffeination process for tea will be considered, where in every case the extraction parameters (pressure, temperature and extraction cycle) are maintained. In the following, a production-cost comparison between a non-isobaric process and an isobaric one is discussed. The separation of caffeine in the isobaric process can be achieved by means of:

A) Use of a countercurrent washing column;

B) Using activated charcoal as adsorbent.

In case "B" two alternatives are looked at, in which the adsorbent is either within the extractor and the CO_2 flows through alternating layers of tea (extracting caffeine) and adsorbents (purifying the CO_2), or is situated externally in an adsorber. The first option has the advantage of a multiple, reduced CO_2 mass-flow, depending on the number of tea/adsorbent layers, but owing to the adsorbent has a lower payload volume.

A comparison for the different cases of the production costs and dependence on the annual capacity is given in Fig. 8.1-4. The results are based on the cascade-operation mode, with three extractors, extraction at 280 bar and 65°C, cycle-times of 7.5 hours, and a separation pressure of 60 bar for the non-isobaric process.

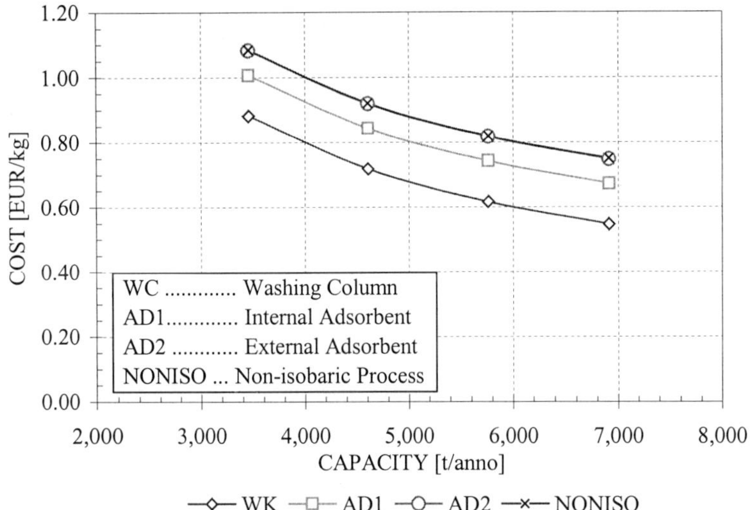

Figure 8.1-4. Cost comparison between different isobaric processes, and the non-isobaric process.

At first view, it may be surprising that the non-isobaric process and the isobaric one, using an external adsorbent, have the same production costs (pre-supposing benefits from the caffeine recovery for the non-isobaric design). Comparing both processes using adsorbents shows that placing the adsorbent within the extractor gives a slightly higher economy, in our case, in the range of 7 to 10%, with higher values for higher capacities. The lowest production costs can be obtained with an incorporated washing column, from which the savings are up to about 28%. A break-down by diverse categories is given in Fig. 8.1-5 for a plant of size 3x4 m³. To achieve the same plant capacities the payload volume for tea is used in each case. As expected, the total production costs for the non-isobaric process are highest, but owing to

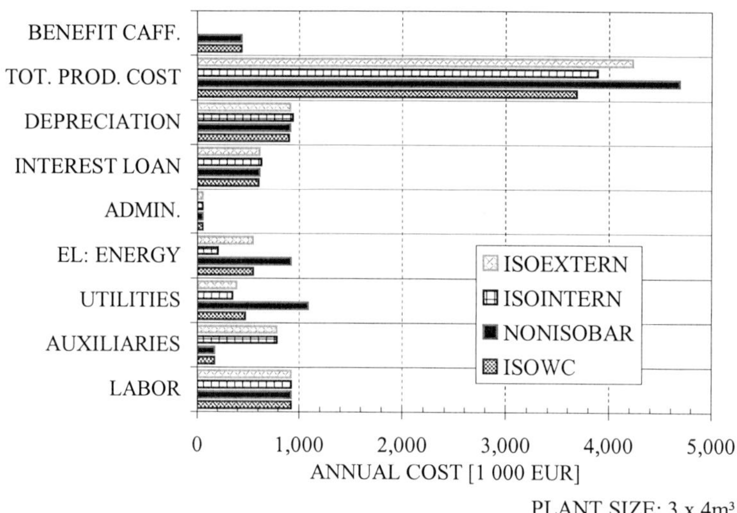

Figure 8.1-5. Annual production cost comparison between isobaric processes with the non-isobaric one, for a plant size of 3x4 m³.

the possible caffeine recovery, a reduction of about 10% could be obtained. The washing column design also enables caffeine recovery, which reduces the production costs shown by about 12% (from the caffeine).

The fixed costs are, in every case, more or less the same, but for the internal adsorbent process ("isointern") they are slightly higher, owing to the larger gross extractor volume that is necessary. Electrical energy and utilities (CO_2, steam, water, lubricants) are highest for the non-isobaric design and, owing to the much lower specific mass-flow rate, considerably less for the "isointern" process. The auxiliary consumption (activated charcoal, spare parts) is about 4.7 times higher for the adsorbent process.

Another decision for the design engineer is the selection of the operation mode, whereby he can choose between the single- and cascade-mode of operation. In principle this is valid for multipurpose plants of medium scale, such as plants to extract spices and/or herbs. Fig. 8.1-6 gives data on the different production costs for a plant with a total extraction volume of 600 l, operated in different modes, but with the same capacity. The investigation is based on equal batch times (in our case 4 hours), equal mass-flow per kg of raw material and, of course, equal extraction- and separation conditions.

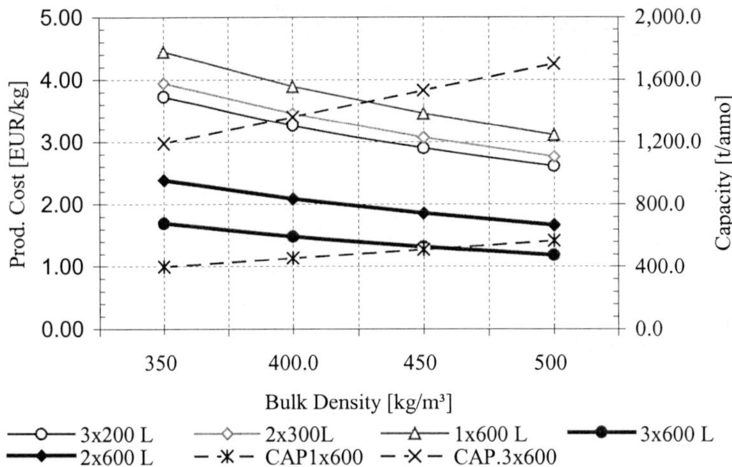

Figure 8.1-6. Comparison of production costs for various modes and enlarged capacity for multipurpose plants designed for 550 bar.

From Fig. 8.1-6, the influence of the bulk density is also obvious, and one can see that the production costs at the highest values (500 kg/m³) decrease by about 30% compared to the lowest one considered (350 kg/m³). Comparing the single- with the cascade-mode omits the preference for incorporating the highest possible number of extractors. Using two extractors, which is, in our case, 2 x 300 l (payload volume), increases the production costs by only about 6%, while the single mode (1 x 600 l) creates over 20%. The engineer must give attention in some cases to the possibility of future extension of capacity.

In our case, the 2 x 300 l cascade could be enlarged with one additional extractor, thereby increasing the capacity by 50%, or the single mode (1 x 600 l) capacity could be doubled or tripled by adding one or two extractor units. In Fig. 8.1-6 the effect on production costs of adding two extractors is projected, while Table 8.1-2 informs about the other possibilities.

Table 8.1-2
Reducing production costs by means of capacity enlargement.

Plant Size [l]	Bulk Density [kg/m³]			
	350	400	450	500
2x300	100.0%	87.5%	77.8%	70.0%
3x300	70.1%	70.1%	70.1%	70.1%
1x600	100.0%	87.5%	77.8%	70.0%
2x600	53.6%	53.6%	53.6%	53.6%
3x600	38.1%	38.1%	38.1%	38.1%

444

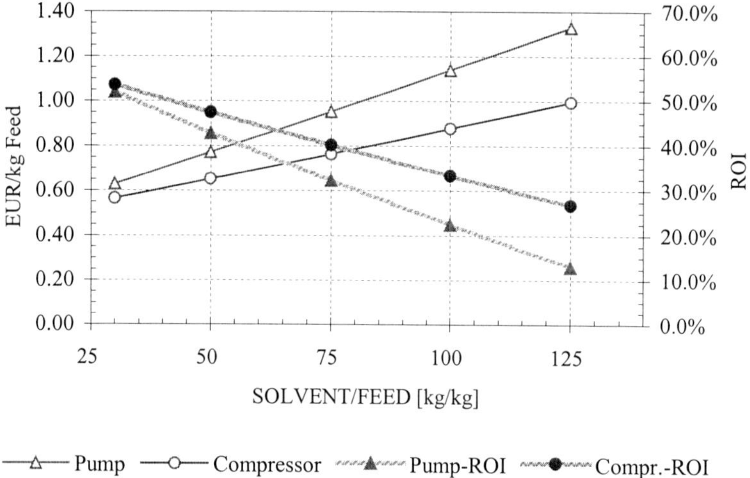

Figure 8.1-7. Comparison between compressor- and pump-mode operation for fish oil extraction with different solvent/feed ratios.

As a result of this investigation it can be mentioned that changing the single-mode to a cascade of two extractors reduces the production costs remarkably, down to about 54%. Further enlargement to three extractors reduces them to about 38%, always compared with the single-mode costs. Enlarging a cascade from two- to three extractors reduces production costs by about 30%.

Another possibility for the design engineer is the selection of pump-, or compressor-mode operation, but with the limitation as described in Section 6.6.2.4. As a case-study the fractionation of fish-oil esters by a continuous countercurrent process is considered, in which EPA and DHA should be separated from other components (C-14 to C-18). Assuming a column with 40 theoretical stages will give acceptable results, the influence of the solvent/feed (S/F) ratio for both alternatives is investigated. Fig. 8.1-7 shows that preference should be paid to the compressor process, especially at higher S/F rates. Further, the strong influence of the S/F ratio is demonstrated, whereby the pump-mode has a stronger increase in production costs. A four-fold increase doubles the production costs when a pump is used, while for the compressor mode the increase is about 70%. In the same Figure, ROI (Return on Investment) values under certain conditions (raw material cost = 1.2 DEM/kg, revenues = 10 DEM/kg, feed = 400 kg/h) are plotted as well. To optimize the plant, the benefit is to set in relation to the S/F ratio (as measured for production costs) and to product revenues (parameter for quality), as illustrated in Fig. 8.1-8. Such charts allow one to estimate whether a higher quality and/or yield, obtainable at the higher S/F ratios, justify the related higher investment- and production costs.

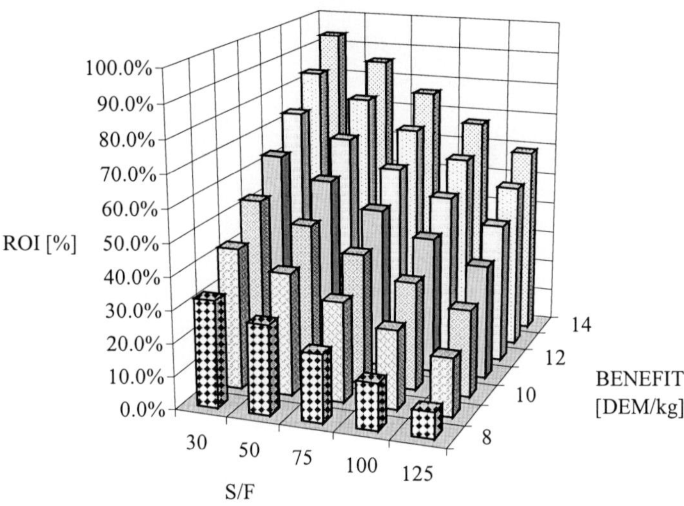

Figure 8.1-8. Return on Investment (ROI) for the compressor process of fish-oil fractionation at different solvent/feed ratios, depending on variable benefits per kg of product.

8.1.4 Influence of process parameters

The influence of process parameters such as pressure, temperature, and possibly the use of modifiers, are to investigated by the design engineer, so that their effects can be evaluated by means of statistical tests. A result of such a method is given in Fig. 8.1-9, for a test series with the target gaining knowledge about the possibility of extracting organo-pesticides from ginseng. As one can recognize for ground ginseng the influence of temperature is highly significant, independently of the type of pesticide, the density, and whether a modifier is used or not. The influence of density is different, that means, probably significant for HCB and quintocen, with a stronger effect at lower temperatures. If water is used as modifier the effect is negative. The same method was used for whole- or larger pieces of ginseng, and shows a different result from the use of water as matrix-modifier, which improves the diffusivity of the fluid to active centers of pesticide adsorption. Elevated temperatures to accelerate diffusion and desorption, a much higher CO_2 input, and swelling of the cell structure is absolutely necessary.

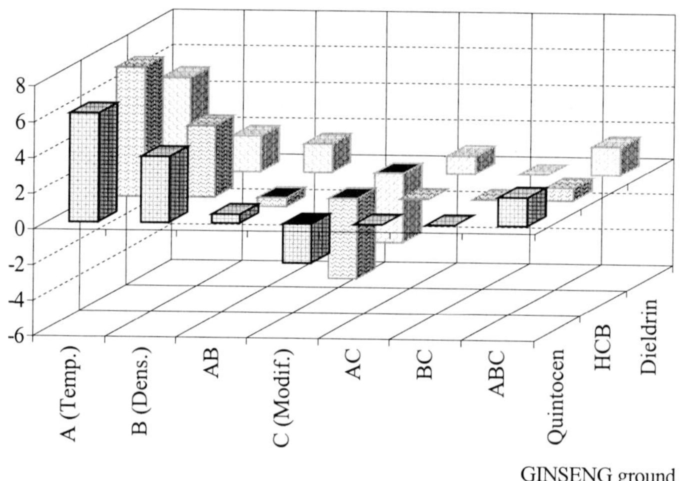

Figure 8.1-9. Influence of various parameters for the removal of organo-pesticides from ginseng.

Figure 8.1-10 illustrates the test results for ground ginseng (initial pesticide concentration of about 130 ppm) with two different CO_2 densities of 450 kg/m³ (105 bar, 50°C) and 750 kg/m³ (271 bar, 70°C) in the dependence on total CO_2 consumption, and with the latter

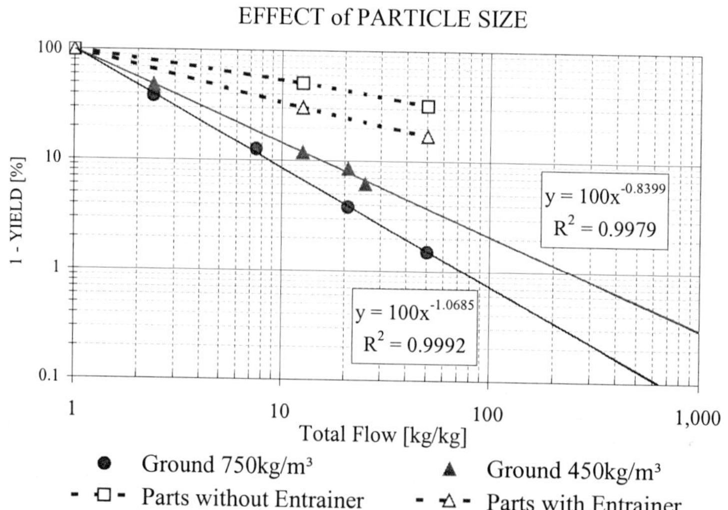

Figure 8.1-10. Test results for the extraction of organopesticides from ginseng at two different CO_2 densities, for ground and larger particle sizes, with and without entrainer.

conditions used on larger ginseng parts, with- and without-matrix modification (increased moisture content). From the CO_2 consumption-rate obtained, and the assumption that the maximum flow velocity inside the extractor is 5 mm (avoiding fluidization) Fig. 8.1-11 is obtained, predicting the necessary extraction times which depend on the required extraction efficiency. With an elevated CO_2 density of 850 kg/m³ (420 bar, 70°C) within about 12 hours 99.8 % of pesticides can be removed, whereas extraction conditions of 271 bar and 70°C (750 kg/m³) require 20 hours to give the same yield. Operating with 450 kg/m³ would theoretically require about 160 hours, which is not acceptable at all.

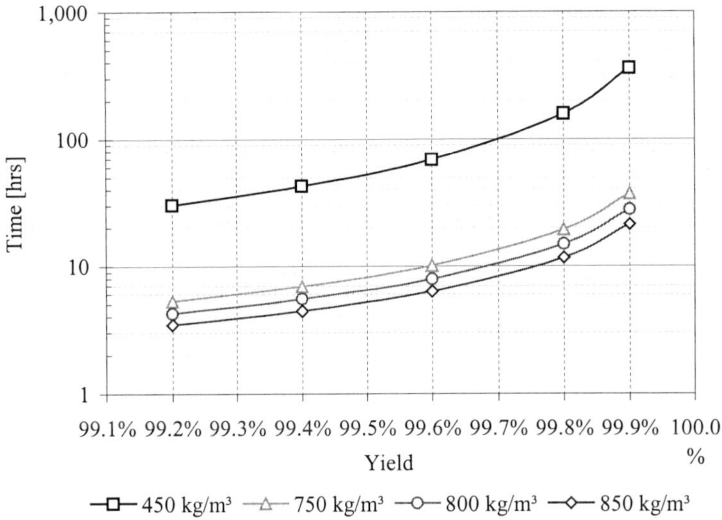

Figure 8.1-11. Necessary extraction times as a function of the degree of pesticide removal at various CO_2 densities.

The dependence of the production costs on sizes of the plant and the extraction conditions are given in Fig. 8.1-12, showing that when the higher CO_2 density of 850 kg/m³ is applied the production costs are 65% of those given with a density of CO_2 of 750 kg/m³.

The influence of pressure at a constant temperature is considered in the next example, the extraction of hops. The temperature is maintained the same, to give the same product quality. The separation conditions are varied, say, precipitation is performed either at 60 bar to obtain a total extract, or in a double-stage separation to provide an α-acid rich fraction (up to 50%) at 60 bar and an aroma rich product (α-acid content below 20%) at 45 bar. The extraction pressures investigated are 280 and 350 bar, and the plant consists in every case of three extractors with a payload volume of 5 m³ each. The production costs for the above-mentioned cases are given in Fig. 8.1-13. Owing to the higher solubility at 350 bar, which is more than double that at 280 bar (both at temperatures of 70 to 80°C) [1] the extraction time for the latter case is longer (5 hours), despite the higher mass-flow-rate, resulting in a higher capacity for the same plant size for the first conditions. The production cost, in case the higher

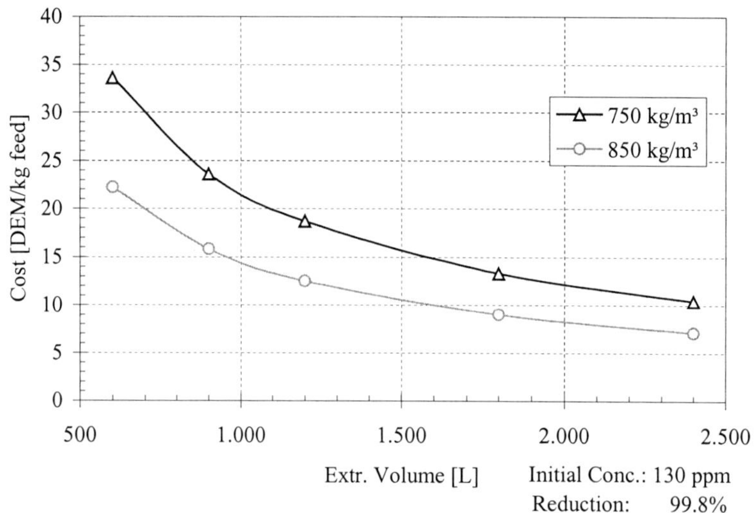

Figure 8.1-12. Production costs for removal of pesticides from ground ginseng at two different CO_2 densities, for various extraction volumes.

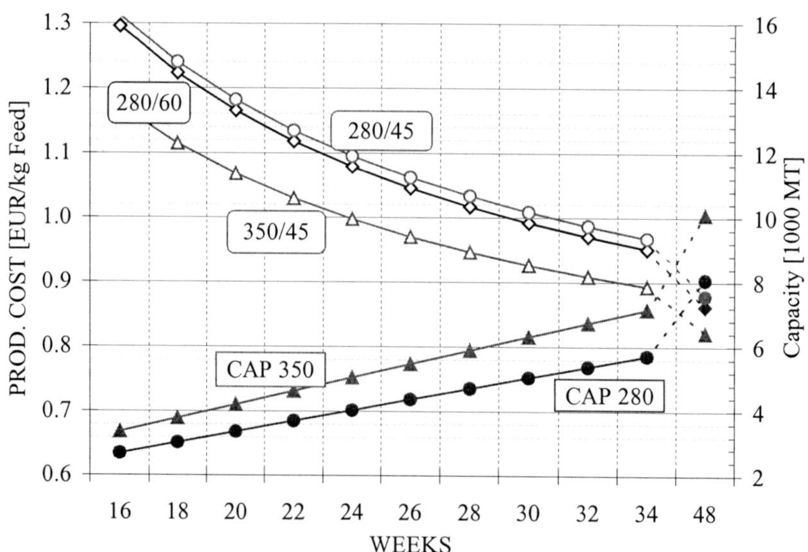

Figure 8.1-13. Production cost for hop-extraction under different processing conditions: dependence on processing-weeks.

extraction pressure of 350 bar is used, amount only about 80% of the production cost for the same plant size, but operated at the lower pressure of 280 bar. The influence of separation conditions, in our case 60 or 45 bar, is not significant and is less than 2% (in favour of the higher pressure) and independent from the applied extraction pressure.

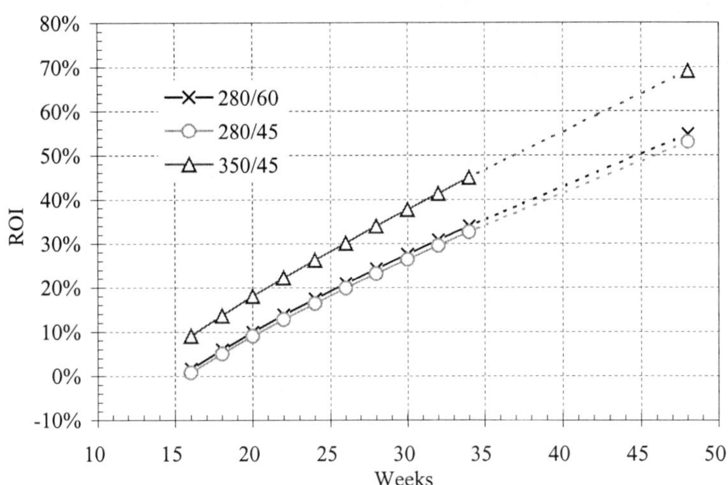

Figure 8.1-14. Return on Investment (ROI) for hops production under different processing conditions, and dependence on processing-weeks.

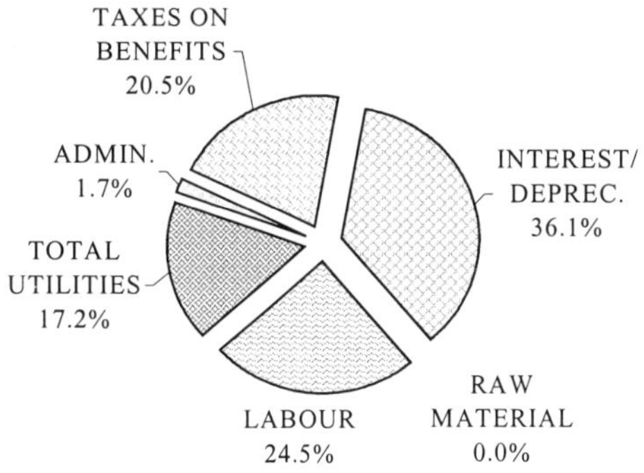

Figure 8.1-15. Cost breakdown for the case 350/45 and 20 processing weeks

Extraction of hops can also be achieved under liquid conditions (preferably 72 bar, 20°C) with production costs about 10% lower compared to the process with the supercritical condition (extraction pressure 350 bar, separation pressure 45 bar). However, because hops-extraction plants are operated only four to six months/year, extractions under supercritical conditions are preferred, as such a design provides more flexibility, by allowing the processing of other materials such as tea or cocoa.

The return on investment (ROI) is illustrated in Fig. 8.1-14, with the assumption of toll-processing at DEM 2.5 per kg of feed material.

Fig. 8.1-15 gives information about the cost breakdown. One can see that interest and depreciation play the major role, amounting to about 36% of production cost (or 50% before taxation), followed by labour costs, with about 25%, and about 20% for taxes on benefits. Thus, it is clear that the design engineer must carefully select the plant components and degree of automation, in order to save investment and labour costs.

8.1.5 Influence of financing
In this Chapter the influence of financing is discussed.

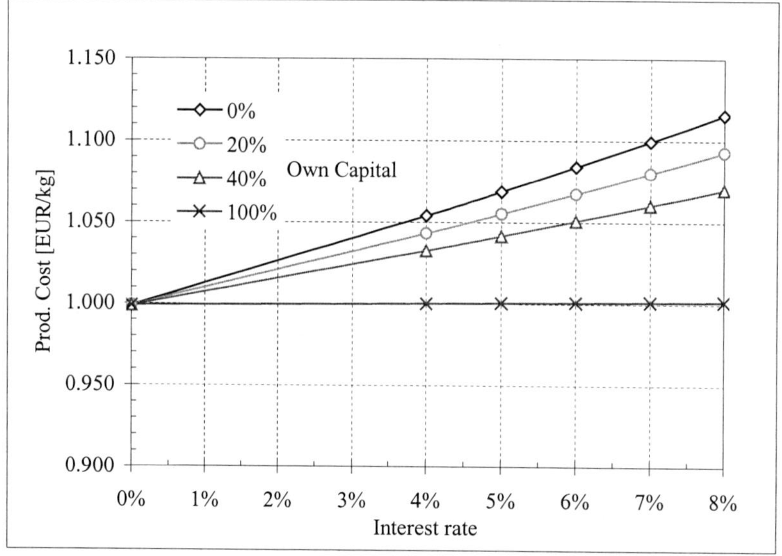

Figure 8.1-16. Influence of interest rate and own-capital input on production cost, based on 20 processing weeks.

The following assumptions are considered:
- No interest rate on the owner's capital.
- Toll production, without considering costs for raw-material.

As an example, the hops plant described before (8.1.4.) is used again, and the 350 bar, double-stage separation process is chosen. From Fig. 8.1-15 it is obvious that interest and depreciation contribute the largest part of the production costs. For this reason, both values are considered separately. In Fig. 8.1-16 the influence of diverse interest rates and the own-capital input is shown. The interest-rate influence on the running cost is, as one might expect, not significant, as no raw material pre-financing is required. With increasing own-capital input (0, 20, 40%) the production-cost increase per percent of elevated interest-rate is smaller, amounting to about 2.75, 2.3 or 1.8%.

The influence of investment costs in dependence of interest rates on production costs is illustrated in Fig. 8.1-17. The savings in production costs increase in a non-linear fashion with falling investment costs for the SFE section, and are higher at higher interest rates.

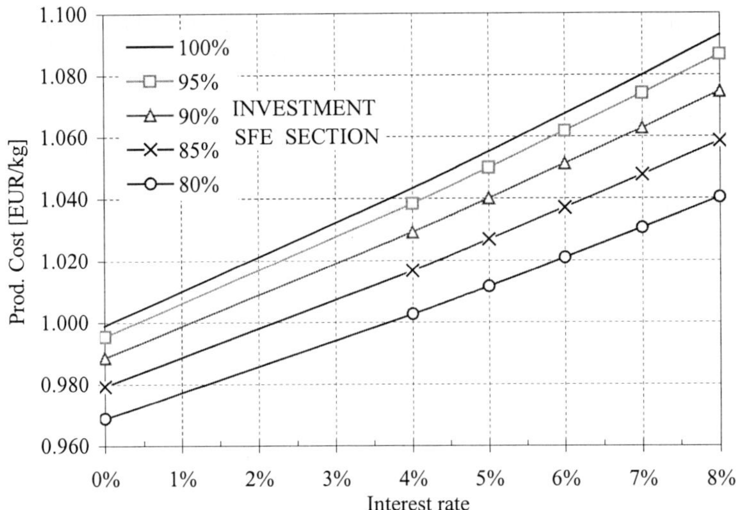

Figure 8.1-17. Influence of interest rate, and reduced investment cost (of the SFE section only), on production cost with a processing period of 20 weeks and 20% own-capital input.

The graphical representation of such percentage savings, compared to the estimated initial investment cost (100%) is given in Fig. 8.1-18. The relatively low values are due to the depreciation over 10 and 20 years, respectively, for equipment and building, but mainly because of the low plant exploitation of only 20 weeks/year.

A graph based on the same considerations as above, but with reduced labour costs, is given in Fig. 8.1-19. The savings in production costs increase in a non-linear fashion with reduced labour costs, but fall with higher interest rates.

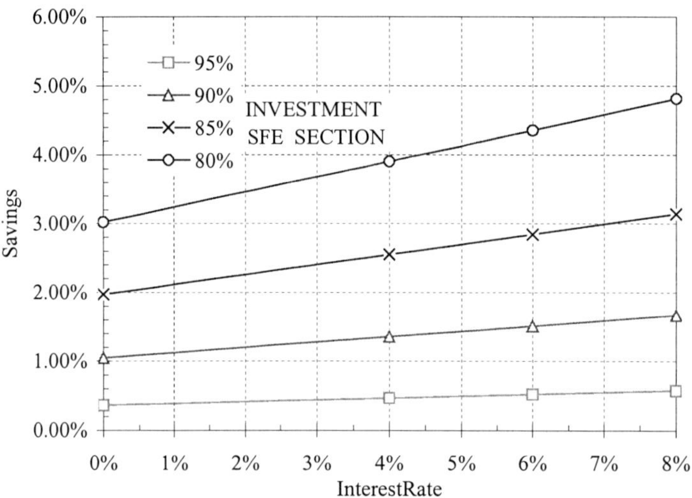

Figure 8.1-18. Savings on production costs in dependence on different interest rates for the case where the investment in the SFE section can be reduced (processing period 20 weeks, 20% own capital input).

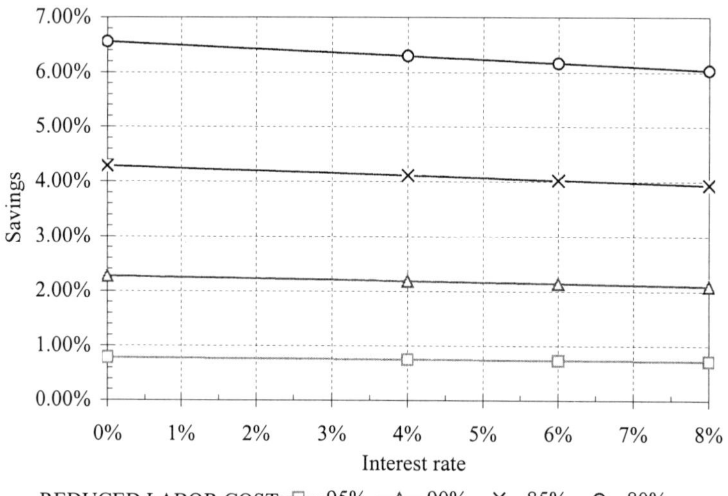

Figure 8.1-19. Savings in production costs depending on different interest rates, where labour costs can be reduced (processing period, 20 weeks; own capital input, 20%).

References

1. E. Stahl, K.W. Quirin, D. Gerard, Verdichtete Gase zur Extraktion und Raffination, Springer-Verlag, (1987) 184.

8.2 High-pressure polymerization of ethylene

As in any other process, so also in the high-pressure polymerization of ethylene, do capital costs, utilities, maintenance, manpower, and costs of raw materials contribute to the production costs of low-density polyethylene (LDPE). The cost structure is typical for the production of bulk chemicals but is strongly influenced by the requirements of a high-pressure process.

Economic data for a LDPE plant equipped with a tubular reactor are given as an example. The industrial unit is described in detail in Chapter 5.1. In order to show the influence of plant capacity the capital and production costs of two units for the production of 100,000 and 150,000 t/a are presented. Then the costs of the high-pressure process are compared to those of a low-pressure unit.

8.2.1 Consumption of polyethylene in Western Europe

Polyethylene is produced by various processes whose synthesis conditions and technology differ widely and which give products with varying technological properties. The polymerization at high pressures gives polymers of lower density (LDPE), which are suitable for transparent films because of their flexibility. At low pressures and using catalysts, high-density polyethylene (HDPE) is obtained, having higher strength and rigidity, and is mainly processed by injection moulding. By co-polymerization of ethylene and α-olefins such as 1-butene, a linear low-density polyethylene (LLDPE) without long-chain branches could be obtained whose density covers the range of density of LDPE and HDPE. A recent development is the application of single-site catalysts to produce metallocene-based polyethylene (mPE) which promises very interesting characteristics.

The consumption of polyethylene has developed rapidly and amounts today to approx. 11 million tons in Western Europe (Fig. 8.2-1). In this year LDPE will have a share of around 48% in WE.

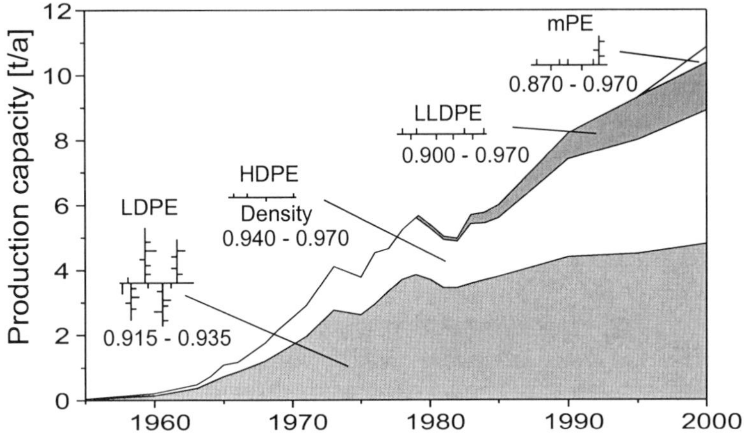

Figure 8.2-1. Consumption of polyethylene in Western Europe.

454

8.2.2 General remarks

The costs assume the production of pellets on 1997 German prices. Polymerization-grade ethylene is available at 5 MPa. The on-stream time is 8,000 h/a. The tubular reactors equipped with multiple feeds of ethylene and peroxide initiators are operated at 200 MPa. The initiators are a mix of dicyclohexyl peroxy dicarbonate, *t*-butylperoxy pivalate, *t*-butylperoxy 2-ethylhexanoate, and di(*t*-butyl)peroxide, which is fed in after the heating zone and at two further locations downstream.

The pressure of the high-pressure separator is 25 MPa. A twin-screw extruder with a side extruder to process LDPE of 0.918 - 0.939 g/ml density and 0.3 - 2.0 g/10 min melt-flow index is used.

8.2.3 Capital costs

The costs of the main apparatus and machines, provided by suppliers, are collected in Table 8.2-1. More than 80% of the total costs for apparatus and machines come from the primary- and secondary compressor, extruder, and silos.

The total costs of the plant are higher by a factor of four than the costs of the main apparatus and machines (Table 8.2-2). They amount to 109.6 million DM for the plant of 100,000 t/a and 144.8 million DM for the 150,000 t/a plant. This means that 1,096 DM per ton are required for the smaller-, and 965 DM per ton per annum for the larger plant.

This reduction in capital costs per ton of LDPE was made possible by the increase in reactor capacity. Fig. 8.2-2 shows the development of name-plate capacity of LDPE reactors. A steep increase occurred since 1980. Whereas in this year the maximum stream-size was in the range of 80,000 - 120,000 t/a, the maximum capacity of today's reactors is 300,000 t/a.

The effect of plant capacity on investment can be seen from Fig. 8.2-3. Costs of around 245 million DM, or only 817 DM/t on the 1997 basis, are required for plants equipped with maximum-stream-size reactors of 300,000 t/a. The data can be extrapolated to actual (1998) costs by means of the price index of 1.01 published in [1]. These plants with large design capacities have significantly reduced specific investment costs.

Table 8.2-1

Costs of main apparatus and machines in million DM, on 1997 prices

	Capacity [t/a]	
	100,000	150,000
Reactor	1.6	2.0
Polymer separation	2.2	2.8
(high- and low-pressure separator, wax separator, recycle gas coolers)		
Gas compression	8.7	11.5
(primary- and secondary compressor, booster compressor)		
Extruder	7.0	9.3
Silo	7.3	10.1

Table 8.2-2
Battery limits investment in million DM, on 1997 prices

	Capacity [t/a]	
	100,000	150,000
Main apparatus and machines	26.8	35.7
Piping	11.4	15.1
Control and instrumentation	7.2	9.5
Electrical equipment	6.7	8.9
Installation	28.5	37.9
Coating and insulation	7.2	9.5
Buildings	11.5	15.2
Traffic zone	5.3	7.1
Remaining costs	5.0	5.9
Battery limits investment	109.6	144.8

8.2.4 Production costs and total costs

The total costs can be divided into fixed-capital- and variable costs. Fixed-capital costs have to be taken into account also when a plant is out of operation, whereas variable costs depend on the production rate. Fixed-capital costs are depreciation, maintenance, manpower, overheads, etc. Variable costs are utilities, monomers, and other chemicals such as initiators, modifiers, or stabilizers. The costs are presented in Table 8.2-3. The total costs are 1,491 DM/t LDPE for the lower-, and 1,431 DM/t LDPE for the higher-plant capacity.

On the one hand, smaller plants need the same number of men to run the unit as bigger plants, so they tend to have higher fixed costs per ton of product. On the other hand, the quantity of off-grades during the switch from one resin to the other is smaller and also off-stream time.

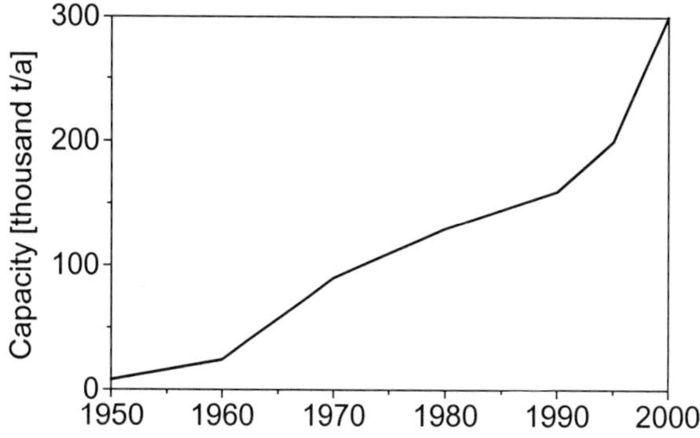

Figure 8.2-2. Name-plate capacity of LDPE reactors.

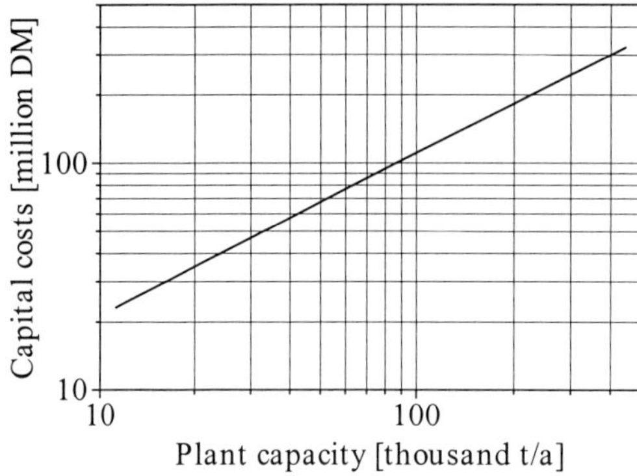

Figure 8.2-3. Effect of plant capacity on capital costs, on 1997 prices.

The production costs are dominated by the costs of ethylene, whose ratio is 69.4% for a 100,000 t/a plant (Fig. 8.2-4). Also, depreciation (7.4%), utilities (6.2%), overheads (5.3%), and maintenance (4.5%) contribute much to the costs. Labour costs (2.8%) play a minor role. If a plant is already depreciated, the costs reduce by around 110 DM/t, if the maintenance costs have not increased.

Table 8.2-3
Production costs in DM/t LDPE, on 1997 prices

	Capacity [t/a]	
	100,000	150,000
Depreciation (10% of investment)	109.6	96.5
Interest (4% of investment)	43.8	38.6
Insurance (1% of investment)	11.0	9.6
Maintenance (6% of investment)	67.0	58.0
Labour costs (6 labourers and 1 foreman/shift, 1 supervisor)	42.0	28.0
Plant overhead (50% of labour and maintenance)	54.5	43.0
Administration (45% of labour)	16.0	11.2
Quality control and laboratory (20% of labour)	8.4	5.6
Electricity (0.10 DM/kWh)	88.0	90.0
Other utilities (steam, water)	4.0	4.0
Ethylene (1,020 DM/t)	1,035.0	1,035.0
Initiators (10 to 20 DM/kg)	6.0	6.0
Other chemicals	6.0	6.0
Total costs	1,491.3	1,431.5

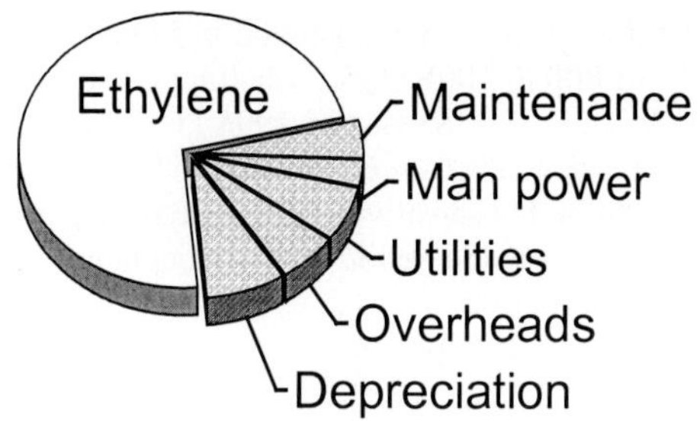

Figure 8.2-4. Composition of total costs. Plant capacity: 100,000 t/a.

8.2.5 Sensitivity analysis

Between 1976 and 1996 the average price of polymer-grade ethylene was about 500 US$ per ton. As shown in Figure 8.2-5, top, the global price of ethylene varies greatly. During this period, a first maximum of about 750 US$/t was observed in Western Europe in 1980. The ethylene feedstock prices increased to 900 US$/t in 1991. A minimum of 300 US$/t in 1976, followed by two further minima of 350 and 400 US$/t in 1986 and 1994, were reported [2].

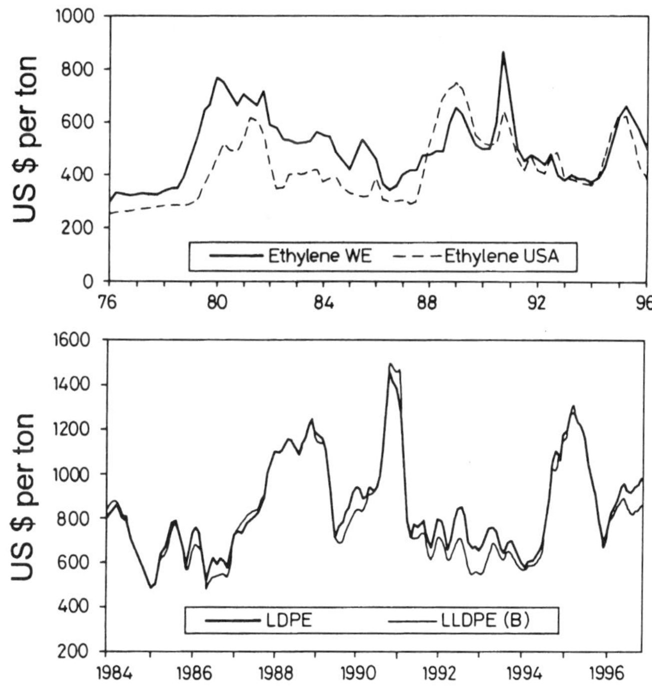

Figure 8.2-5. Global price of ethylene (top) and polyethylene (bottom).

458

The variation of the sales-price of polyethylene since 1984 is demonstrated in Figure 8.2-5, bottom. The average global price was in the range of 800 to 1000 US$/t. Maxima can be seen in the period of 1988 to 1989 (1,250 US$/t), in 1991 (1,500 US$/t), and 1995 (1,300 US$/t). Low prices in the range of 500 to 800 US$/t have been noted between 1985 and 1996.

In view of these variations it is of interest to determine the sensitivity of the payback period to changes in ethylene- and product prices. For this purpose, the period t_R was calculated from the capital investment, I, and the annual cash flow, R, as below,

$$t_R = f \cdot I / R \qquad \text{with } f = 1.1 \qquad (8.2\text{-}1)$$

which can be obtained from the net margin plus depreciation. In Fig. 8.2-6, t_R is plotted versus the changes in ethylene- (solid line) and LDPE- (broken line) prices for the plant of 100,000 t/a capacity. The reference is a payback period of 7 years calculated on 1997 prices of 1,020 DM/t ethylene and 1,700 DM/t LDPE. It increases, first less steeply and then more steeply when the price of ethylene feedstock increases, for a fixed sales-price of the product (solid line). It reduces moderately when the price of ethylene reduces, e.g., with 20% lower costs of ethylene the payback period reduces to 3.5 years.

At a constant ethylene price, t_R increases steeply when the sales price of the polymer reduces, and vice versa, in that t_R reduces when the market for LDPE improves. An increase of 20% in the sales-price of LDPE gives a payback period of only 3 years.

8.2.6 Comparison of economics of high- and low-pressure process

In many cases, the linear low-density polyethylene (LLDPE) produced in low-pressure processes competes for the same market as LDPE. For this reason, in Figure 8.2-7 capital- and operation costs of the high-pressure polymerization are compared with those of a low-pressure solution process having the same capacity. Also, the production costs of the low-pressure process are dominated by the costs of the monomer, but some differences can be noted which are typical for the economics of low- and high-pressure processes.

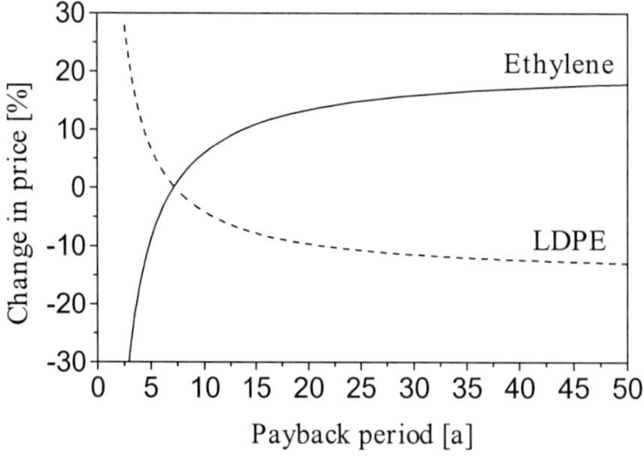

Figure 8.2-6. Influence of ethylene- (solid line) and LDPE- (broken line) price on the payback period. Plant capacity: 100,000 t/a, on 1997 prices.

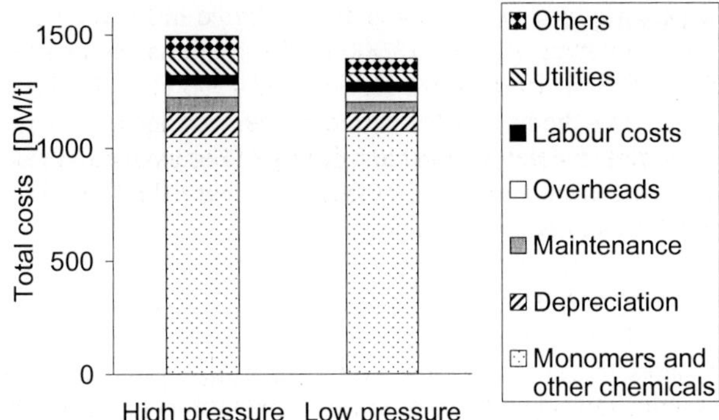

Figure 8.2-7. Comparison of costs of high- and low-pressure processes. Capacity: 100,000 t/a, on 1997 prices, production of pellets.

The percentage of monomer costs is higher in the low-pressure plant, owing to the extra costs of butene co-monomer whose price in Western Europe exceeds that of ethylene. The capital investment is lower, mainly because of the lower costs of the compressor and reactor, resulting in a lower depreciation. The low-pressure process requires less electrical energy which explains its lower costs for utilities.

Above all, the total costs of the low-pressure solution process are 1,393 DM/t compared to 1,491 DM/t of the high-pressure polymerization, which is a difference of 7%. However, it must be mentioned that the average European sales prices of standard film-grades LDPE in recent years have been higher by 100 to 200 DM/t than those of LLDPE [3].

References

1. H. Kölbel and J. Schulze, Europa Chemie 21 (1999).
2. D. Oxley, Chem. Systems, Lecture at the Symposium "LDPE Outlook", Aztec Peroxides Inc., Houston 1997
3. K. Hahn, Elenac, The Renaissance of LDPE, Lecture at Chem. Systems Annual European Seminar, London 1999.

8.3 Precipitation by supercritical anti-solvent

8.3.1 Rationale
As reported later in Chapter 9.9 of this book, active–ingredient–containing polymeric micro-particles are widely used in technological and medical applications. For example, these particles are suitable as drug–delivery devices and can control the pharmaceutical release–rate over time. The particle size is absolutely important when dealing with drug–delivery devices. Very small particles can be inhaled, while larger ones can be injected into the blood stream. Therefore, it is important to control the microparticle size in the production.

Preliminary studies [1,2] suggest that the use of supercritical CO_2 allows size– and shape–control of polymeric particles, in particular in the production of protein–loaded micro-particles [3]. However, to realise an industrial application of such a production, a reasonable economic income has to be expected.

The goal of this brief discussion is to understand the scenarios in which a supercritical anti–solvent process (SAS) could be economically feasible.

8.3.2 Process description
The precipitation of polymeric microparticles by continuous SAS techniques is described later in Chapter 9.8. The typical semi-continuous plant to which we refer in this Section is illustrated in Fig. 8.3–1.

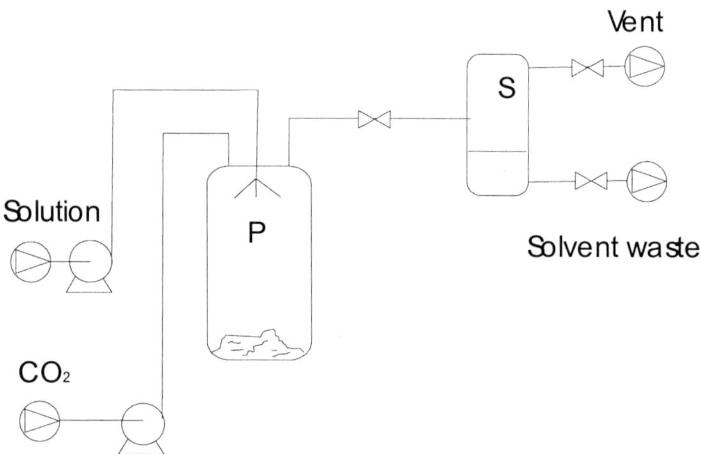

Fig. 8.3–1. Scheme of operating plants commonly used for SAS processes

The polymer solution ("Solution") is sprayed through a micrometric nozzle into the precipitator (P) together with pressurised CO_2 from a cryogenic tank. Vessel P is operated at constant temperature and at a pressure that ensures precipitation of the desired product. The mixture CO_2–solvent coming from vessel P enters the separator (S) where the solvent is recovered. After precipitation, more CO_2 can be flowed through the vessel, in order to reduce the amount of residual organic solvent in the product down to the desired level. In this discussion, we refer to a particular solvent–CO_2–drug system used for academic purposes by

Zanette [4]. The organic solvent considered was tetrahydrofuran (THF), the polymer was Eudragit®, and the drug was a compound supplied by a pharmaceutical company. However, the procedure discussed here can be applied easily to any SAS process of this type.

In the plant used for academic purposes [4], both the solvent and exhausted CO_2 are wasted. In an industrial plant both streams should be recycled after purification, for obvious economic reasons. The precipitator size and plant–flow–rates are obtained by increasing 80–fold the relative quantities used in the pilot plant [4]. This scale factor was suggested by the company that supplied the drug. Two vessels, P, in parallel are needed: while the former is running, the latter can be cleaned and the solid product can be recovered. Cleaning and product–recovery expenses are not directly evaluated in this example. In the pilot plant, the flow of THF–polymer–drug solution was 0.072 kg/h, and the CO_2 flowed in the quantity of 1.08 kg/h (the ratio CO_2 to solution equals 15). The precipitator was a 0.4–liter vessel. The actual precipitator scale–up is not considered here. The main factor to consider in scaling–up the precipitator is the nozzle scale–up. The nozzle–size, nozzle–shape, and number of nozzles per reactor volume, determine the precipitate size in a complex and still incompletely understood way [5–8]. It is assumed that issues related to the injectors are already solved.

Note that at this level of description, a detailed plant design is not needed. It will eventually follow from the specific application for which the plant may be needed.

Micro–encapsulation, as obtained by continuous SAS techniques, is a physical process, guided both from thermodynamics and kinetics. The entire process involved is not clear. Mass–transfer kinetics and thermodynamic equilibria related to polymer–particle precipitation from a solution expanded by supercritical CO_2 are currently being investigated [9,10]. Many empirical observations are now available, suggesting that for a given polymeric solution, both pressure and temperature play an important role in determining the precipitated particles' morphology.

The operating conditions chosen in this example (120 bar, 60°C) looked promising for the production of homogeneous particulate from the given solvent–polymer–drug system [4]. Our attention is then focused on the operating conditions of pieces of equipment other than the precipitator, which may be changed for economic reasons.

8.3.3 Process simulation

To study different operating conditions in the pilot plant, a steady–state process simulator was used. Process simulators solve material– and energy–balance, but they do not generally integrate the equations of motion. The commercially–available program, Aspen Plus T_M, was used in this example. Other steady–state process simulators could be used as well. To describe the CO_2–solvent system, the predictive PSRK model [11,12], which was found suitable to treat this mixture, was applied. To obtain more reliable information, a model with parameters regressed from experimental data is required.

Figure 8.3–2 shows the simplified flow–sheet used in the simulation program. For plant–modelling, the solution was considered to be constituted of only CO_2 and THF. The dissolved polymer and drug were neglected. Because of the low concentration of these compounds in the feed, this approximation should not have significant consequences.

462

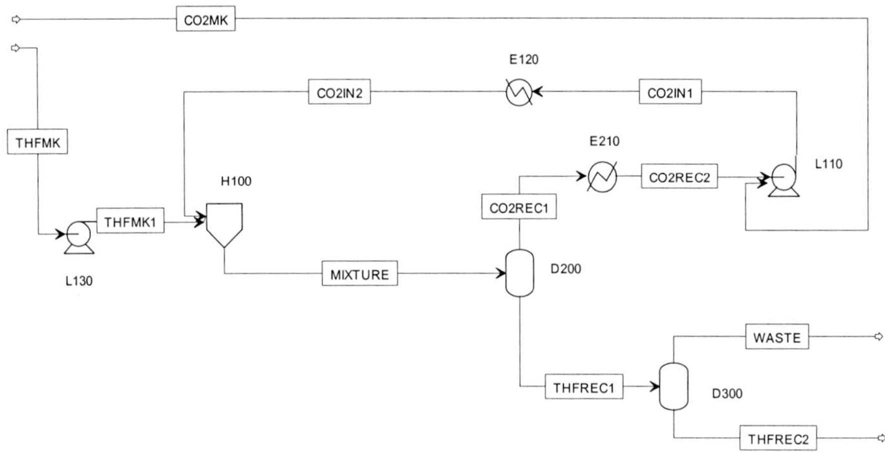

Fig. 8.3–2. Aspen-Plus flow sheet used for simulation comparisons

The equipment list is reported in Table 8.3–1.

Table 8.3–1
Equipment list for flow-sheet represented in Figure 8.3–2

Symbol	Description
H100	Precipitator
L110	Pump
E120	Heater
L130	Pump
D200	Flash-type separator
E210	Condenser
D300	Flash-type separator

The precipitator, H100, was simulated as a mixer, because solute precipitation was not reproduced. The operating conditions in the precipitator were set to 120 bar and 60°C according to the investigations carried out in the pilot plant. Operating conditions in the flash D300 were initially set to 1 bar and 20°C (room conditions) so the stream THFREC2 could be recycled to the solution–preparation mixer, which is not included in Figures 8.3–1 or 8.3–2. The cooler E210 condensed the recycled CO_2 so that it can be integrated with CO_2 make-up and pumped into the precipitator. The outlet conditions from cooler E210 were set equal to the conditions characterising the CO_2 make–up stream. By changing the pressure and temperature in D200, different quantities of solvents can be recycled, and different amounts of energy are required in the various pieces of equipment. To obtain the heating and cooling power required in the separators, the temperature and pressure in each drum were fixed as an input to the program. The process simulator gives as a result the heat duties required to maintain those operating conditions at steady states. The amount of THF needed to run the plant is the one in stream THFMK1 minus the one in THFREC2, which is stored in a vessel (not shown, for clarity) and recycled.

Table 8.3–2 shows heat duties in the different pieces of equipment, and the solvent make–ups needed to run the plant, as a function of pressure and temperature in the flash drum, D200. The data reported are computed by assuming efficiencies for heaters and coolers equal to 0.7, and 0.6, respectively, to account for thermal inefficiencies [4]. From analysis in the pilot plant, which is probably more accurate than the results from the process simulator, an accurate estimate was that 20 kW is needed to pump all the fluids in the final plant.

Table 8.3–2
Results obtained by simulating different operating conditions in the D200 separator

Case number	1	2	3	4	5	6	7	8	9	10
P (bar)	60	50	50	60	50	60	70	70	70	70
T (°C)	50	50	60	60	40	40	40	50	60	70
CO_2 make–up(kg/h)	2.97	2.00	0.99	1.46	3.58	6.12	13.72	4.48	2.00	0.62
THF make–up(kg/h)	0.99	0.65	0.35	0.48	1.18	2.09	4.31	1.56	0.71	0.18
Heating power (kW)	11.11	11.50	12.08	11.68	10.93	10.62	10.50	10.62	11.28	12.03
Cooling power (kW)	12.68	13.35	14.32	13.74	12.39	11.42	10.06	11.80	13.06	14.22

Figure 8.3–3 shows the CO_2– and THF–make–ups, together with the heating and cooling powers, required to run the plant in each case studied. The power needed to pump fluids is not represented, for clarity.

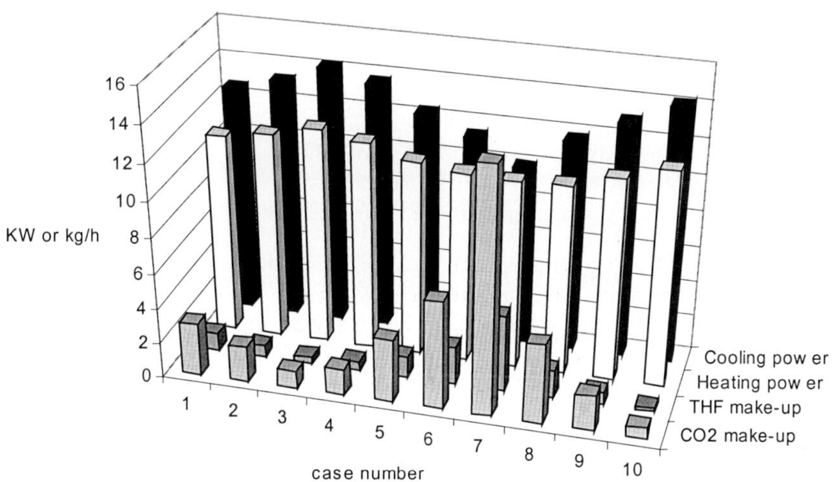

Fig. 8.3–3. Cooling power, heating power, THF make–up, and CO_2 make–up needed to run the plant for each operating set of conditions tested for flash drum D200.

Figure 8.3–3 suggests that Cases 6 and 7 are equivalent on a price base. However, the operating conditions of 60 bars and 40°C in the flash drum D200 minimise solvent wastes.

464

More convenient operating conditions would probably be obtained by lowering the temperature in the D300 flash separator. In this way, the THF lost would be drastically reduced and a more efficient recycle could be obtained. By testing different sets of conditions in the D300 drum, we found that running conditions corresponding to 2 bar and –15°C allow one to recover 97% of THF in stream THFREC1 for recycling. We do not show the different conditions tested because the procedure followed is exactly the same as described to optimise the D200 flash drum.

Table 8.3–3 presents the energy– and mass–balances relative to the optimised running conditions. The D200 separator is set at 60 bar and 40°C, while the D300 is running at 2 bar and –15°C. Running– and fixed–costs analysis in the following paragraphs have been carried out with reference to this particular case.

Table 8.3–3

Heat– and mass–balance table relative to the flow diagram reported in Figure 8.3–2, with the flash drums D200 and D300 operating at 60 bar and 40°C, and 2 bar and –15°C respectively.

Stream ID	CO2IN1	CO2IN2	CO2MK	CO2REC1	CO2REC2	MIXTURE
T (°K)	258.8	333.2	255.2	313.2	255.2	336.7
P (atm)	123.365	118.431	59.215	59.215	59.215	118.431
Vap Frac	0.000	0.000	0.000	1.000	0.000	1.000
Mol Flow (kmol/h)	1.985	1.985	0.139	1.846	1.846	2.043
Mass Flow (kg/h)	87.979	87.979	6.117	81.863	81.863	92.160
Volume Flow (l/min)	1.520	3.639	0.108	8.932	1.441	3.494
Enthalpy (MMkcal/h)	-0.193	-0.188	-0.014	-0.174	-0.179	-0.191
CO_2 (kmol/h)	1.963	1.963	0.139	1.824	1.824	1.963
THF (kmol/h)	0.022	0.022	–	0.022	0.022	0.080

Stream ID	THFMK	THFMK1	THFREC1	THFREC2	WASTE
T (°K)	293.2	315.6	313.2	258.2	258.2
P (atm)	0.987	118.431	59.215	1.974	1.974
Vap Frac	0.000	0.000	0.000	0.000	1.000
Mol Flow (kmol/h)	0.058	0.058	0.197	0.064	0.133
Mass Flow (kg/h)	4.181	4.181	10.297	4.383	5.914
Volume Flow (l/min)	0.087	0.087	0.219	0.086	23.456
Enthalpy (MMkcal/h)	-0.003	-0.003	-0.016	-0.004	-0.012
CO_2 (kmol/h)	–	–	0.139	0.007	0.132
THF (kmol/h)	0.058	0.058	0.058	0.056	0.002

From Table 8.3–3 it is evident that the recycled CO_2 (stream CO2IN1) is not pure. To assure a THF flow rate of 5.76 kg/h in the precipitator, the THF make–up has to be less than this amount. As a consequence, the polymer solution injected in the precipitator is more concentrated than was tested in the pilot plant [4]. In this particular case, the different inlet concentration should not make any difference in the precipitate morphology. However, in general, changing the polymer concentration can lead to the formation of either microparticles of small size or to polymer fibres [5].

Finally, if the organic–solvent concentration in the recycled CO_2 stream is too high for the particular system considered, a small distillation tower may be needed. Preliminary studies show that a tower with only six ideal trays, operating at 50 bars, can reduce the THF percentage in the CO_2 recycled stream to 0.1 % [4]. This option should also be applied if CO_2 is to be used to remove the residual THF content from the solid product.

8.3.4 Capital–cost evaluation

From the flow sheet of Figure 8.3–2 it is possible to determine roughly the chemical–plant price. For a rather precise estimate, a detailed design of each piece of equipment would be needed [13]. However, at a first stage of economical evaluation, we do not need to go into these details.

The *fixed–capital* cost is the price that would be paid to buy an operating processing plant [14]. This cost is several times greater than the sum of bare–equipment prices. Actually, it is necessary to add different costs to the purchase prices. Purchase prices depend mainly upon the size, material, and special operating conditions (such as high pressures, high or low temperatures, and corrosive materials). Installation costs, auxiliary–services costs, piping costs, site–development costs (generally, one buys the land, and has to prepare it to host a chemical plant), offices and dressing–rooms construction costs, etc., need to be added to the bare–purchase prices. There are also some indirect expenses to be evaluated. For example freight and insurance, which will contribute to what is known as 'overhead cost'. All these sorts of expenses are generally evaluated from the bare–equipment purchase costs through indexes [13–15]. The sum of all these costs constitutes the chemical plant fixed–capital cost.

It is also possible to understand the fixed capital cost as the sum of *direct* and *indirect* costs. Direct costs are the installed–equipment costs plus the price of plants built separately from the production line, where utilities, such as steam, are produced. Indirect costs contain engineering costs, fees, supervision costs, etc. [15] that the investor needs to sustain during plant construction.

Once the plant is finished, there is the need to begin the production. *Working capital* is the money needed to get the plant into production. This includes, for example, raw materials and supplies carried in stock, finished products in stock, and available cash for monthly payments. This sum is usually equal to 10–20% of the fixed–capital costs. However, this money is not part of the fixed–capital costs, and cannot be accounted for in depreciation [13,14].

In our example, we obtained a rather precise fixed–capital cost estimate from a private industry with high–pressure–technology experience. This estimate, updated to year 2000, was of 600,000 EUR for the plant sketched in Fig. 8.3–2, including all costs listed in the above paragraph. In case a distillation tower was needed to obtain pure CO_2 for recycling, the chemical plant would be more expensive because a flash–type separator would be substituted with a distillation tower, with boiler and condenser.

We have neglected costs related to land, plant–housing, construction of facilities–providing plant, etc., because the plant is thought as being part of an existing industrial site.

A reasonable guess was attempted for working–capital costs.

8.3.5 Manufacturing costs evaluation

To operate the chemical plant, some direct expenses must be sustained. Manufacturing costs are the expenses connected with the manufacturing operation of the process plant. These costs can be divided into *direct production costs*, *fixed charges*, *plant overhead costs* and *general expenses*. Direct manufacturing costs are expenses strongly associated with the manufacturing operations.

Among direct production costs, we can list raw materials, utilities, direct operation labour, maintenance, catalysts, royalties, etc. In our case, solvents (THF and pressurised CO_2), polymer, and active ingredient are needed on a daily basis. The corresponding quantities were determined using the process simulator, as described in paragraph 8.3.3. In this discussion, additional solvent (THF) losses are supposed to equal 5% of the total flow rate entering the precipitation vessel. This value takes into account the residual solvent in the solid precipitate and the losses during system–depressurisation. Raw–material prices are known only for compressed CO_2 and THF, while they are reasonably guessed for the polymer and drug. Additionally, compressed air in the amount of 20 Nmc/hr is needed for automatic–control and safety systems, and electrical power for heating, cooling and pumping is needed. The plant is simple to run, and only one full time employee is going to be needed. Over a 24–hour period, three shifts are required.

All the basic assumptions to compute operating costs are listed in Table 8.3–4.

Table 8.3–4
Basic assumptions to compute manufacturing costs.

CATEGORY	DIMENSION	VALUE	REFERENCE
Production hours	Hrs / year	8,000	330 working days
Labour cost	EUR / man year	45,000	Average per–man year
Electrical power	EUR / kWh	0.10	
Cooling power	EUR / kWh	0.04	
Heating power	EUR / kWh	0.03	
CO_2	EUR / kg	0.25	
THF	EUR / kg	2.5	
Polymer	EUR / kg	17	
Pharmaceutical	EUR / kg	32	
Compressed air	EUR / Nmc	0.015	Safety and automatic control
Inflation rate	%	3.0	

Fixed charges are costs that do not change significantly with production rate. Among these costs we can list depreciation, property taxes, insurance, etc. The original investment must be repaid during the plant life. The original investment is repaid over an extended period of time through depreciation. Depreciation can be seen in various ways. For example, it could be understood as a tax allowance, an additional operating cost, a mean of building up a fund to finance plant replacement, or a measure of falling value [14]. We refer to specialised texts for a more detailed discussion [13]. Only the fixed capital investment can be claimed as depreciation. In this example we are adopting a 'straight–line' depreciation [14]. With this method, depreciation can be determined from the equation:

$$D = \frac{FCI - S}{t} \quad . \tag{8.3–1}$$

In Equation 8.3–1, D is depreciation, FCI is fixed-capital investment, t is the number of years over which the depreciation is accounted for, and S is the salvage value. Thus S is the value the plant could be sold for after the t years of operation. In this discussion we assume that the plant lasts for ten years, and that its salvage value is zero. With the straight–line method, annual depreciation costs are constant. Other methods, such as the sum–of–the–years–digits method, determine depreciation costs to be greater in the early years of the property than in the later years. Local regulatory laws generally define which method can be used to determine depreciation costs.

Plant overhead costs do not vary widely with changes in the production, either. Generally, these costs take into account medical services, general plant maintenance, quality–control laboratories, storage facilities, etc. In this discussion we do not consider them, because we think the plant as part of an existing industry.

Table 8.3–5
Manufacturing–cost summary for the chemical plant sketched in Figure 8.3–2.

		units	quantity	price (EUR/year)
Fixed capital investment:	600000	EUR		
Working capital	90000	EUR		
Total capital investment:	690000	EUR		
Manufacturing expenses:				
Direct costs				
	Raw materials			
	CO_2	kg/year	48447	12112
	THF	kg/year	1010	2525
	Polymer	kg/year	1370	23293
	Drug	kg/year	1370	43845
	Operating labour	employee/year	3	135000
	Utilities			
	Heating power	kWh/year	79200	2376
	Cooling power	kWh/year	90288	3612
	Pumping power	kWh/year	158400	15840
	Total power	EUR/year	21828	
	Compressed air	Nmc/year	158400	2376
	Maintenance and repair	EUR/year	20700	20700
	Operating and supplies	EUR/year	2000	2000
Total direct costs		EUR/year		263678
Fixed charges				
	Overheads	EUR/year	6000	6000
	Local taxes	EUR/year	400	400
	Depreciation	EUR/year	60000	60000
Total fixed charges		EUR/year		66400
General expenses				
	Administrative costs	EUR/year	12000	12000
	Research and development	EUR/year	10000	10000
	Distribution and advertising	EUR/year	250	250
Total general expenses		EUR/year		22250
Revenue from sales				
	Product	kg/year	2740	602870
Net annual profit		EUR		250543
Income taxes		EUR		125271
Net annual profit after taxes		EUR		125271
After–tax rate of return		%		18.16

468

Finally, general expenses are connected to a company's industrial operation. These expenses take care of administration, distribution and marketing of the final product, research and development of new products, and financing expenses. In the following analysis no influence from financing is considered. This means we are assuming that the investment is made from own capital, and therefore that no interest is paid. In case the original capital came from an external financing society, the money spent as interest on a yearly basis should be added to the general expenses.

Table 8.3–5 presents the costs we assumed to be associated with the chemical plant sketched in Fig. 8.3–2, with operating conditions of the flash drums D200 and D300 optimised at 60 bar and 40°C, and 2 bar and –15°C, respectively.

The total manufacturing expenses are the sum of direct costs, indirect costs, general expenses, and depreciation.

Figure 8.3–4a shows a pie–chart giving all the costs connected with the plant's production. The costs are grouped into four significant categories: direct–manufacturing costs, fixed charges, general expenses, and income taxes (see Table 8.3–5). The incidence of fixed charges, in particular of depreciation, over the total manufacturing expenses, is a function of the number of working days per year and of the number of operating hours per day. If the shifts–per–day are reduced from three to one, the incidence of depreciation would be roughly three times as much as it is with our assumptions.

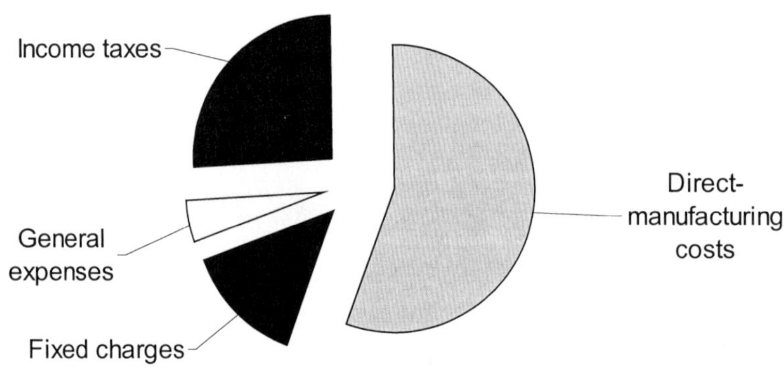

Fig. 8.3–4a. Pie–chart relating to total manufacturing expenses

Figure 8.3–4b shows a pie–chart of the direct–manufacturing costs. Heating power, cooling power, and power required to pump fluids are grouped under the term 'power' for clarity. It can be concluded that the cost for THF and CO_2 make–ups are not significant. Not even the costs for electricity, drug, and polymer contribute significantly to the total manufacturing expenses. More than 50% of the total costs are due to the operating labour. Because an increase in the production rate would not require an additional worker, this analysis suggests that a larger production would lower the impact of direct labour on the manufacturing costs.

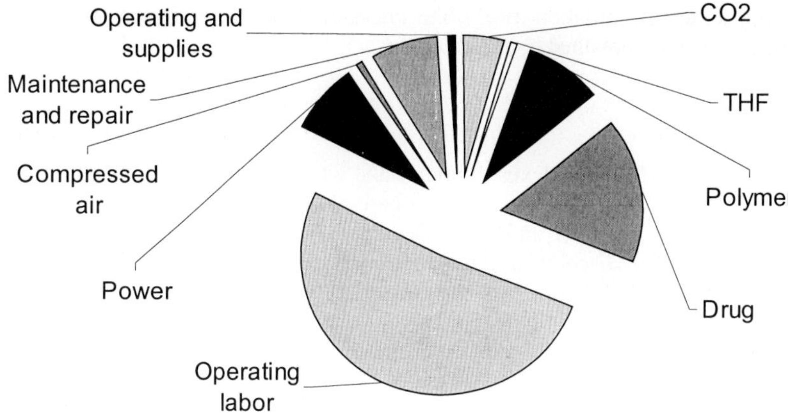

Fig. 8.3–4b. Pie–chart relating to direct–manufacturing costs.

The chemical plant considered here is capable of producing 0.346 kg/hr of pharmaceutical-containing microparticles. To complete the discussion about economic profitability we need a price for the final product. In Table 8.3–5 we assumed a value of 220 EUR/kg.

The net annual profit is the difference between the revenue from sales and the total manufacturing expenses. Out of this profit, the company has to pay income taxes, which are supposed to equal 50% of the net annual income. The after–tax net annual income is the difference between the net annual income and the income taxes.

It is now possible to compute the after–tax rate of return, which is the net annual profit after tax, divided by the total capital investment (fixed capital costs plus working capital), multiplied by 100. In the case presented here, the after–tax rate of return is 18.16%.

When dealing with ordinary industrial operations, profits cannot be predicted with extreme accuracy. In any type of investment a certain amount of risk is always involved, and a chemical venture is an investment. At least moderate risks are involved in most industrial ventures. In general, a 20% return before income taxes would be a minimum acceptable return for such an investment proposition. Therefore the plant proposed here looks promising, and further economic analysis is strongly suggested.

8.3.6 Cash–flow analysis

To define the economic performance of a manufacturing venture, an analyst must predict various sources and sinks of money throughout the lifetime of a project. In fact the investors invest a huge amount of money when they begin to build the chemical plant, but they will earn money from sales only when the plant is finished and operating. Owing to inflation and devaluation, "future" money is different from "present" money [13,15]. Therefore, a future income needs to be discounted to its present value in order to evaluate the expected profitability of a chemical plant. Of course, when dealing with future events, nobody can be absolutely positive about prices, inflation rates, etc. Therefore many assumptions need to be made. Considering an annual inflation rate, i, the present value, PV, of future money, FM, earned after y years is readily obtained by the equation:

$$FM = PV \cdot (1 + i)^y. \tag{8.3–2}$$

It is possible to define different indexes to understand whether a venture will be profitable [13]. Here we consider the 'discounted break–even period' [14]. This index is defined as the time that must elapse after start–up until the discounted cumulative cash–flow repays the fixed capital investment. The discounted cumulative cash–flow is the sum of expenses and gains sustained or earned in the lifetime of a chemical plant. All these sums of money need to be discounted to their present value.

The construction of a chemical plant is an expense and therefore corresponds to a negative cash–flow at the beginning of the financial venture. In this example, we suppose that the plant is completed and ready to run in one year. Once the plant is finished and operating, the final product can be sold and a net profit can be generated on a yearly basis (see Table 8.3–5). In the traditional balance–sheet presented in the previous paragraph, depreciation is treated as a manufacturing expense. In reality, it returns to investors as net profit after taxes. From the investor's point of view, depreciation is a profit over which no taxes are paid. Therefore the sum of depreciation plus net profit–after–tax for each future year needs to be discounted to its present value to determine the discounted cumulative cash–flow. After the ten years over which the depreciation is accounted for, it can no longer be listed as an expense in the manufacturing costs summary, and therefore the investors need to pay income taxes on this sum of money also.

Figure 8.3–5 shows the discounted cumulative cash–flow as a function of elapsed years. With the assumptions contained in this qualitative discussion, we found that the discounted break–even period is 4 years when the product can be sold at 220 EUR/kg, it is 9 years when the price is 150 EUR/kg, but it is more than 20 years if the price falls to 120 EUR/kg. These results are obtained assuming an inflation rate of 3% per year.

8.3.7 Conclusions

The cost of supercritical–fluid units increases roughly as the square root of the capacity [16], and therefore capital amortisation, even when plants are built according to GMP requirements, decreases sharply when the plant–capacity increases. However, in our example depreciation constitutes only 15–20% of the total manufacturing expenses (see Table 8.3–3), and therefore the operating costs per kilogram of product would not be reduced significantly by using a larger processing plant. However, the operating labour constitutes more than 50% of the direct manufacturing expenses, and therefore, it would be possible to manufacture a cheaper product by increasing the production rate to as much as one worker per shift is able to handle.

Considering the plant sketched in Figure 8.3–2, which could manufacture 0.346 kg/h of product, with the D200 separator operating at 50 bar and 40°C, and the D300 separator operating at 2 bar and –15°C, the technological process would be industrially interesting when the product price is at least 150–220 EUR/kg. This provides an added–value of a least 100 EUR/kg over the raw–material price (drug and polymer), which in this discussion is assumed equal to 49 EUR/Kg. Therefore, at present, it appears that only high–value products can be manufactured through supercritical–CO_2–anti–solvent processes.

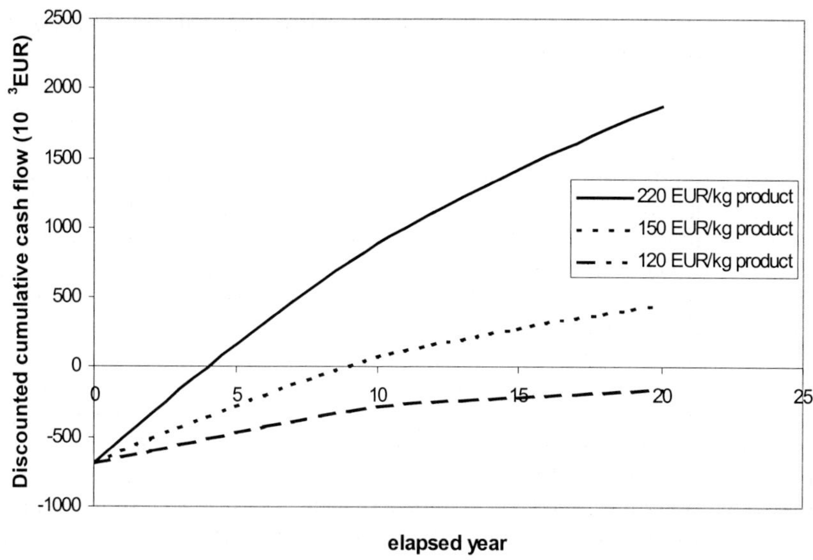

Fig. 8.3–5. Discounted cumulative cash-flow as a function of the elapsed years. Three different prices for the product were considered.

References

1. D.J. Dixon; K.P. Johnston; R.A. Bodmeier, AIChE J. 39 (1993) 127.
2. A.K. Lele; A.D. Shine, AIChE J. 38 (1992) 742.
3. N. Elvassore, A. Bertucco, P. Caliceti, submitted to Ind. Eng. Chem. Res. (1999).
4. F. Zanette, Undergraduate Thesis, University of Padua 1999.
5. M. Baggio, Undergraduate Thesis, University of Padua, 1998.
6. L. Benedetti, A. Bertucco, P.Pallado, Biotech. Bioeng. 53 (1996) 232.
7. D.J. Dixon, K.P. Johnston, J. App. Pol. Sc. 50 (1993) 1929.
8. E. Reverchon, G. Della Porta, A. Di Trolio, Proceedings of the 4th Italian Conference on Supercritical Fluids and Their Applications, Italy, 1997.
9. M. Lora, A. Bertucco, I. Kikic, Ind. Eng. Chem. Res. 39 (2000) 1487.
10. E. Reverchon, G. Della Porta, M.G. Falivene, J. Supercritical Fluids 17 (2000) 239.
11. O. Redlich, J.N.S. Kwong, Chem. Rev. 44 (1949) 233.
12. G. Soave, Chem. Eng. Sci., 27 (1972) 1197.
13. M.S. Peters; K.D. Timmerhaus, Plant Design and Economics for Chemical Engineers, McGraw Hill, International Ed., New York, 1991.
14. G.D. Ulrich, A guide to Chemical Engineering Process Design and Economics, John Wiley and Sons, New York, 1984.
15. J.M. Douglas, Conceptual Design of Chemical Processes, McGraw Hill, Inc., New York, 1988.
16. M. Perrut, Supercritical Fluid Applications: Industrial Developments and Economic Issues, Proceedings of the 5th International Symposium on Supercritical Fluids, Atlanta, Georgia, USA, 2000.

High Pressure Process Technology: Fundamentals and Applications
A. Bertucco and G. Vetter (Editors)

CHAPTER 9

APPLICATIONS

In this charter a number of interesting applications of high pressure technology in the process and chemical industries are presented in detail. By no way they are exhaustive of all possibilities and potentials currently investigated to exploit the use of high pressure in general and supercritical fluids in particular. Only, they reflect the expertise of the authors and are aimed to stimulate more research and technological interest in this field by academia and industry.

The chapter is arranged in a different way from the previous ones, as the name and affiliation of authors are reported under the title of each section. Eleven sections are included overall. Sections 9.1 through 9.5 are related to reactions at high pressure (both chemical and biochemical, hydrogenations, polymerisations and oxidation with supercritical water). Section 9.6 reports about extraction with dense CO_2, section 9.7 deals with processing polymers under pressure, section 9.8 about precipitation and micronization techniques by supercritical fluids. Last three sections refer to more specific topics, i.e. the potential of dense gases in the production of pharmaceuticals (9.9), the action of high pressure on living organisms (9.10) and the dry-cleaning by supercritical CO_2 (9.11).

With the examples reported it should be evident that the use of high pressure in the process and chemical industry can be profitably extended not only from the economical but also from the environmental point of view.

9.1 Chemical Reactions in Supercritical Solvents (SCFs)

M. Poliakoff, N. Meehan

School of Chemistry, University of Nottingham
University Park, Nottingham   NG7 2RD England

9.1.1 Introduction

"There is no point in doing something in a supercritical fluid just because it's neat. Using the fluids must have some real advantage"

This statement, made in 1988, by the "supercritical pioneer" Val Krukonis is still true, but the interest in supercritical fluids, SCFs, is now much greater for a wide variety of chemical applications. These include solvents for chemical synthesis, [1] inorganic chemistry, [2] homogeneous [3] and heterogeneous catalysis, [4] polymerization, [5,6] and materials processing [7,8]. In addition, near-critical fluids, particularly nc-H_2O, are being identified as promising and "greener" solvents for organic chemistry [9]. Although SCFs are not new, it is only recently that a whole range of drivers have begun simultaneously to favour SCFs. These drivers include the urgent need to replace halogenated solvents, the need to reduce use of volatile organic compounds (VOCs), the increasing cost of waste disposal, an increasing interest in process intensification and a greater awareness of the environmental consequences of chemical processes. These drivers have been further reinforced by funding aimed specifically at cleaner synthesis and processing, both nationally and internationally (EU, NATO, etc). Most important of all has been the realization by academic researchers that small-scale high-pressure SCF experiments can be carried out safely in normal chemical laboratories with commercially available equipment.

This Chapter is not a review nor is it comprehensive. Rather, it aims to show how SCFs can give real advantages to chemistry and to indicate some of the applications. Questions of equipment are largely left to other chapters but it should be stressed that miniaturization of apparatus is very important for laboratory-scale experiments. Paradoxically, this small-scale equipment can sometimes pose greater engineering challenges than larger equipment because miniaturization complicates the handling of solids and slurries and increases the chances of blockages. The chapter is divided into six sections: (i), SCFs as reactants; (ii), SCFs as catalysts; (iii), SCFs as solvents; (iv), SCFs as an aid to separation; (v), Reactions involving gases and; (vi), Continuous organic reactions in SCFs.

9.1.2 SCFs as Reactants

One of the earliest chemical applications of SCFs was the polymerization of ethylene, C_2H_4. As pointed out by McHugh and Krukonis, [10] this process remains an important landmark in supercritical chemistry because the operating temperatures and pressures are much higher than those proposed in most new applications. Therefore, the C_2H_4 polymerization demonstrates that engineering problems are unlikely to be the major obstacle to the introduction of new SCF processes.

The main reasons for using an SCF as a reactant in a reaction is either to avoid using an additional solvent or to maximize the concentration of reactant. Examples include the formation of C_2H_4 complexes, [11,15] C-H activation, [16,17] transient formation of noble-gas complexes [18,19] and the activation of CO_2 itself [3,20,21]. Apart from CO_2, these reactions have all been carried out on a very small scale, often without isolation of solid compounds from the SCF solution, as for example in the activation [17] of scCH$_4$ by (C$_5$Me$_5$)Ir(CO)$_2$, see Figure 9.1-1.

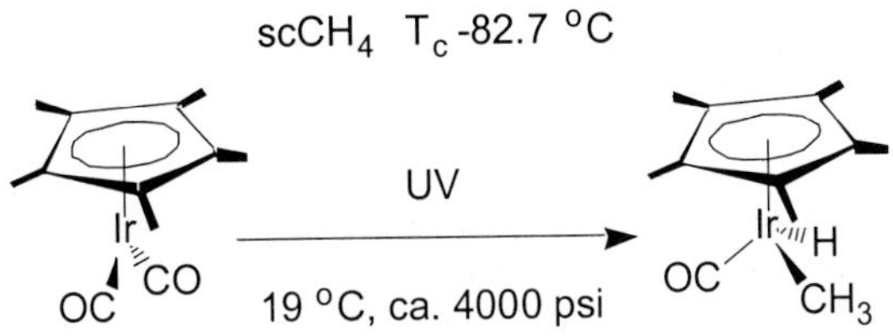

scCH$_4$ T$_c$ -82.7 °C

UV

19 °C, ca. 4000 psi

OC CO   Ir   OC Ir H CH$_3$

Figure 9.1-1. Using SCFs as reactants: activation of scCH$_4$ by (C$_5$Me$_5$)Ir(CO)$_2$.

Many of the compounds were previously unknown at room temperature and it is only the high concentration of free ligand in the SCF, which has stabilized the compounds for sufficiently long to be characterized. Although such small-scale experiments are not of industrial relevance they remain important because the reactions are the first introduction for many chemists to SCFs.

The activation of CO$_2$ is somewhat different since the conversion of CO$_2$ into more useful chemicals is a major environmental goal. The first major breakthrough was the work of Jessop, Ikariya and Noyori [20] who used a ruthenium catalyst to react H$_2$ with scCO$_2$ to form HCOOH. The interest of this work was the fact that the rate of reaction was greatly enhanced in scCO$_2$ compared to the conventional reaction in THF. This enhancement was attributed, at least in part, to the higher concentration of H$_2$ in scCO$_2$ (see Section 9.1.6 below). The reaction is, in fact, an equilibrium and for high yields the equilibrium needs to be shifted towards the product by sequestering the HCOOH with base (*e.g.*, Et$_3$N). The work has since been substantially extended and reviewed [3,20]. The reaction has also been carried out using immobilized catalysts which reduce the problems of catalyst separation [23,24]. Shifting of an equilibrium is also a key feature in the industrial synthesis of butan-2-ol in sc-Butene: see Section 9.1.4.

9.1.3 SCFs as Catalysts

High temperature and supercritical H$_2$O have greatly enhanced acidity (and basicity) compared to room-temperature water because the dissociation constant increases as water is heated towards the critical point [25,27]. At the same time, there is a marked drop in the dielectric constant which reduces the solubility of polar compounds but increases the solubility of non-polar compounds, and of organic compounds in particular.

Thus, the combination of increased acidity and increased solvent power can be exploited for cleaner acid-catalysed reactions of organic compounds, for example Friedel-Crafts alkylation, [31] or hydrolysis of esters [9,28]. The acidity of high-temperature D$_2$O can be used to deuterate organic compounds but the process is more efficient with an additional heterogeneous acid catalyst [32]. A related application, but not a catalytic one, exploits the acidity of sc-H$_2$O for the *in situ* generation of H$_2$ from metallic Zn: the H$_2$ can then be used for the selective reduction of organic nitro-compounds [29,30], see Figure 9.1-2.

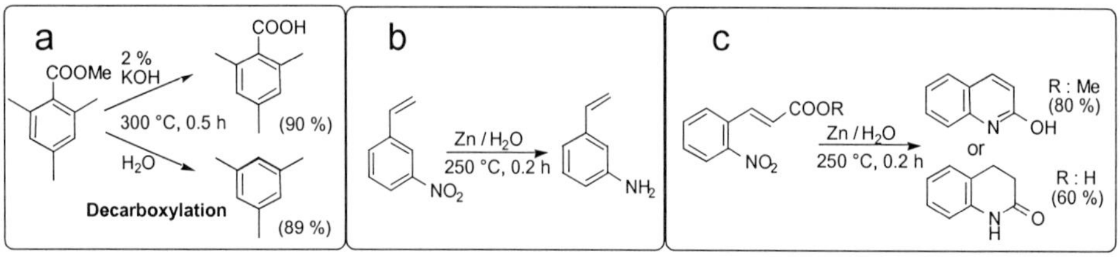

Figure 9.1-2. Reactions of organic compounds in high temperature H_2O: (a), saponification of esters [28]; (b), selective reduction of nitro-aromatics [29]; (c), the synthesis of quinolones [30].

The high acidity/basicity of high-temperature H_2O can also be used to accelerate the hydrolysis of transition-metal salts [33,36]. Hydrolysis leads to the transient precipitation of metal hydroxides which are rapidly *dehydrated* to form the corresponding metal oxides, usually as nano-particulate crystals. Recently, it has been shown [37] that mixtures of metal salts could be hydrolysed in high temperature H_2O to form nanoparticles of mixed metal oxides, *e.g.*, $Ce_{1-x}Zr_xO_2$. These reactions have yielded *highly crystalline* oxide nanoparticles at temperatures 200 - 700 °C lower than would be needed for more conventional ceramic routes. The significance of such work is that the oxide materials have potential as catalysts for selective oxidation. The characteristics responsible for the catalytic activity (*e.g.*, the concentration and distribution of the active centres, surface structure, phase-composition, microstructure, etc.) are very sensitive to the technique used to prepare the catalyst. Therefore, even well-known catalysts may exhibit unique properties when they are prepared using new techniques or novel treatment procedures. Indeed, tests on La_2CuO_4 have shown that, for the oxidation of CO, "supercritical" La_2CuO_4 is *ca.* 8 times more active than La_2CuO_4 prepared conventionally [38].

9.1.4 SCFs as Solvents

Supercritical CO_2 has been by far the most popular SCF solvent, both on grounds of cost and acceptability. However, $scCO_2$ is a comparatively poor solvent [10]. Broadly, it has a solvent power similar to a light alkane such as hexane. However, there is a number of compounds, particularly fluorinated organic compounds which are unexpectedly soluble in $scCO_2$.

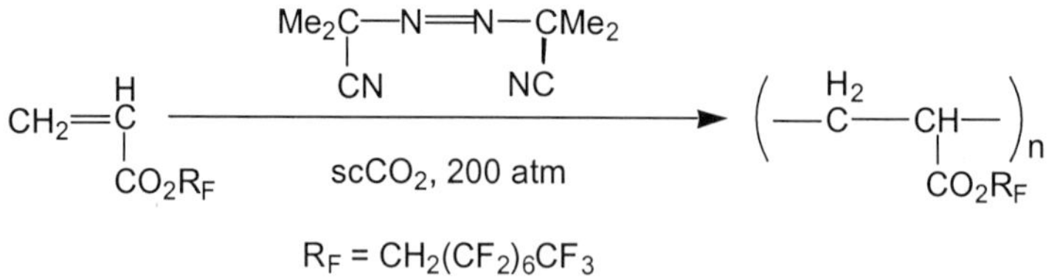

Figure 9.1-3. Reaction investigated by DeSimone and coworkers [41]

There is continuing discussion [39] over the precise origin of this enhanced solubility which derives, at least in part, from favourable interaction between F and CO_2 [40]. Whatever the origin of this so-called "CO_2-philicity", it has been widely exploited in chemical reactions where $scCO_2$ acts as a solvent. One of the first examples was an elegant experiment by DeSimone and coworkers [41], in which $scCO_2$ replaced CFC solvents in the polymerization of fluoroacrylates (see Figure 9.1-3). The work was then expanded enormously to include the development of a whole range of stabilizers for emulsion polymerization [5]. DeSimone's original work is now being commercialized by DuPont [42], which if successful will be one of the first examples of industrial scale-reactions in $scCO_2$.

$scCO_2$ has considerable attractions as a solvent for many other reactions, especially where there are problems of product separation (see Section 9.1.5), since merely a reduction in pressure is often sufficient to separate products from CO_2.

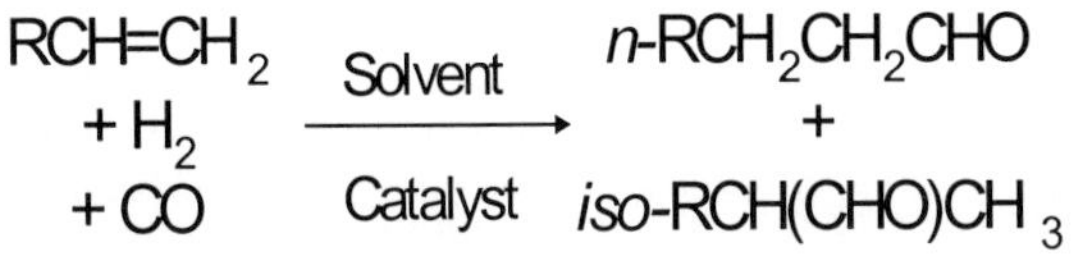

$$RCH{=}CH_2 + H_2 + CO \xrightarrow[\text{Catalyst}]{\text{Solvent}} n\text{-}RCH_2CH_2CHO + iso\text{-}RCH(CHO)CH_3$$

Figure 9.1-4. Alkenes hydroformylation reaction.

An example of particular interest is the hydroformylation of alkenes where, in the case of alkenes higher the C_7, it is difficult to remove conventional solvents without destroying the homogeneous catalyst [3] (see Figure 9.1-4). Such difficulties could be removed by use of $scCO_2$. Unfortunately, most of the conventional catalysts are almost insoluble in $scCO_2$. Although there is one report that the reaction will still proceed in the presence of undissolved catalyst [43], many workers in this field have tried to modify the catalysts to improve their solubility in $scCO_2$. The most popular strategy has been to fluorinate the alkyl groups on the phosphine ligands of the catalyst [44-51] An alternative approach is to immobilize the catalyst on a heterogeneous support [52] (see Section 9.1.7).

9.1.5 SCFs as a Tool for Product Separation

One of the main attractions of SCF solvents is the ease of separating products at the end of the reaction. For products which are liquids, phase separation can be achieved merely by reducing the pressure. However, it should be remembered that some of the SCF will still remain dissolved in the liquid phase. This may not be a problem in reactions carried out in $scCO_2$ but it could present difficulties for products isolated on a large scale from flammable SCFs, where outgassing of flammable vapour from the liquid product could occur.

Solid products require somewhat different treatment. In general, rapid expansion of SCFs [35] is the most effective approach, at least for small-scale reactions. RESS precipitation separates the product as fine particles, free for traces of the SCF and other volatile components of the reaction mixture. RESS is particularly useful for the precipitation of organometallic compounds with relatively weakly bound ligands (*e.g.*, C_2H_4, $\eta^2\text{-}H_2$) [12,13]: see Figure 9.1-5.

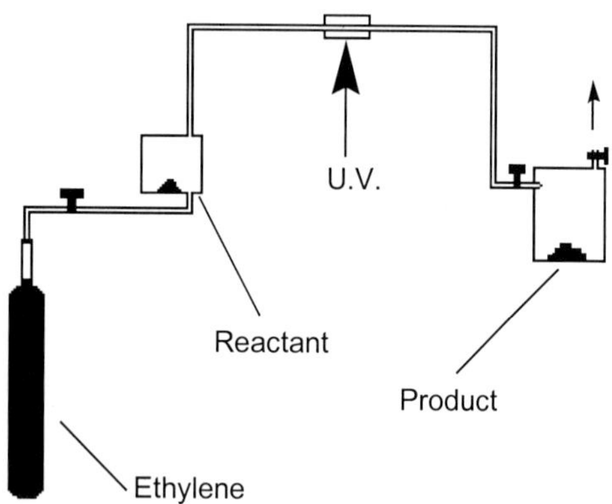

Figure 9.1-5. Schematic view of a continuos reactor for generation of C_2H_4 complexes with precipitation by RESS.

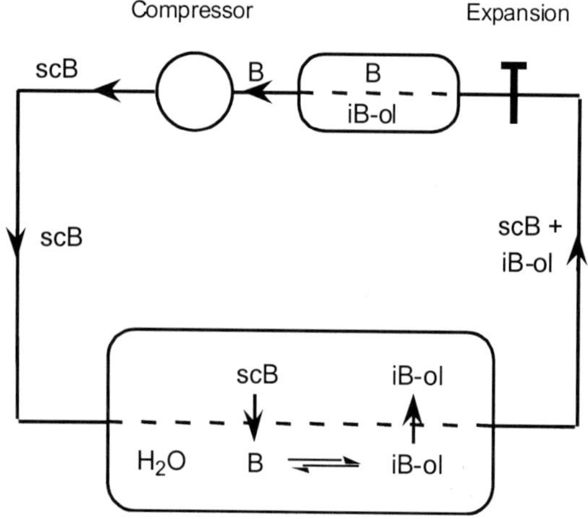

Figure 9.1-6. Simplified outline of the Idemitsu Petoleum Co. Process. scB, sc-butene; iB-ol, butan-2-ol.

The product is separated without the use of either heat or vacuum, conditions which might otherwise lead to loss of the weakly-bound ligand. In addition, the product is cooled somewhat by expansion of the SCF, thereby reducing the rate of decomposition. In reactions where the SCF is also a reactant (*i.e.* scC_2H_4 in the preparation of C_2H_4 complexes) the product, once precipitated is held under a flow of the reactant gas, again minimizing the chances of decomposition [12].

A somewhat different application of SCFs in separation is the industrial process for the production of butan-2-ol from sc-Butene. This process, operated by the Idemitsu Petroleum Company has been described in some detail elsewhere [1]. Briefly, the reaction between butene and H_2O takes place in an aqueous phase. sc-Butene is only partially miscible with H_2O and an equilibrium is set up between butene and butan-2-ol in the aqueous phase but, under these conditions, butan2-ol is more soluble in sc-Butene than in H_2O.

$$CH_3CH_2CH=CH_2 + H_2O = CH_3CH_2CH(OH)CH_3$$

The butan-2-ol is extracted into the supercritical phase, thereby shifting the equilibrium in the aqueous phase further towards butan-2-ol. The butan-2-ol is separated from the sc-Butene, which is then recycled (see figure 6.1-6).

9.1.6 Reactions Involving Gases

Since SCFs are dense gases, they are in general, fully miscible with permanent gases such as H_2, N_2 and O_2. Since these gases, particularly H_2, are only sparingly soluble in conventional solvents, SCFs offer considerable potential advantages as solvents for such reactions. One of the first examples was the oxidation of CH_4 in scH_2O [25] and subsequently McHugh and his co-workers attempted the oxidation of cumene in scXe [10]. At Nottingham, a series of experiments was undertaken for the synthesis of organometallic dihydrogen and dinitrogen compounds in a range of SCFs [2,54-56]. In this case, the chemical problem is not so much any difficulty in forming the compounds, but rather the stabilization of the compounds once they are formed. The most striking example was the photochemical synthesis [54] of $(C_5H_5)Re(N_2)_3$ from $(C_5H_5)Re(CO)_3$ and N_2 in scXe or scC_2H_6. (see Figure 9.1-7).

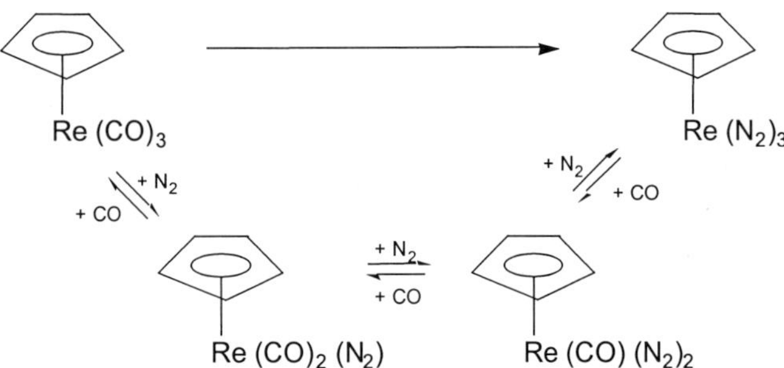

Figure 9.1-7. Photochemical synthesis of $(C_5H_5)Re(N_2)_3$.

480

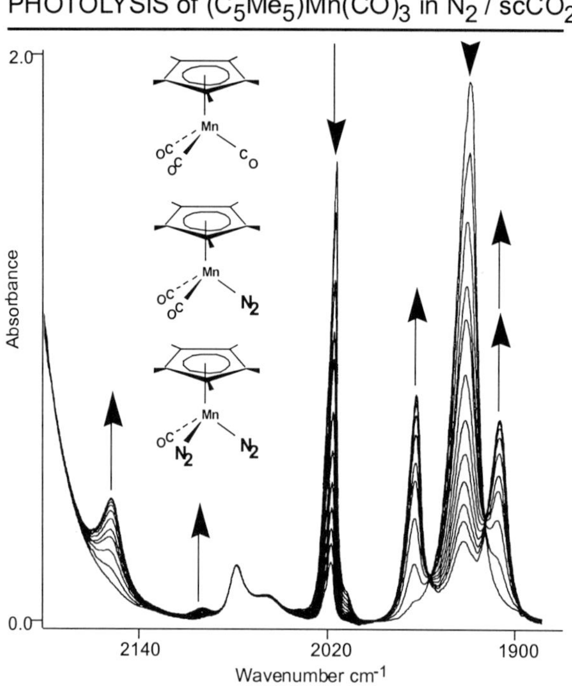

Figure 9.1-8 IR spectra recorded during the photochemical generation of dinitrogen complexes in scCO$_2$.

Dinitrogen compounds are usually highly unstable, easily decomposing by loss of the N$_2$ ligand. The high concentration of N$_2$ in the SCF solution stabilizes the (C$_5$H$_5$)Re(CO)$_{3-x}$(N$_2$)$_x$ (x = 1 or 2) intermediates sufficiently long for the reaction to proceed to (C$_5$H$_5$)Re(N$_2$)$_3$. This reaction is particularly interesting because it occurs quite efficiently in scXe or scC$_2$H$_6$, but not in sCO$_2$ where it stops, largely at the formation of (C$_5$H$_5$)Re(CO)$_2$(N$_2$). Further photolysis in scCO$_2$ merely leads to decomposition, possibly via O-transfer from CO$_2$ to the Re centre.[57] Figure 9.1-8 shows a similar effect in the photolysis of (C$_5$Me$_5$)-Mn(CO)$_3$ in scCO$_2$, where only a small quantity of the *bis*-dinitrogen complex is formed.

The synthesis of dihydrogen complexes only proceeds as far as the mono-substituted complex. The reason for this is almost certainly that η^2-H$_2$ is bonded less strongly than N$_2$ to the metal centre. Indeed, complexes such as (C$_5$H$_5$)Mn(CO)$_2$(H$_2$) were expected to be thermally extremely labile. Therefore, it was quite unexpected that such complexes could be isolated from a continuous reactor [12] similar to that in Figure 9.1-5. One of the more important SCF reactions involving gases is hydrogenation, which is the subject of a separate (section 9.3.). The high solubility of H$_2$ in SCFs reduces the mass-transport limitations which are frequently encountered in heterogeneously catalysed hydrogenation (see Figure 9.1-9).

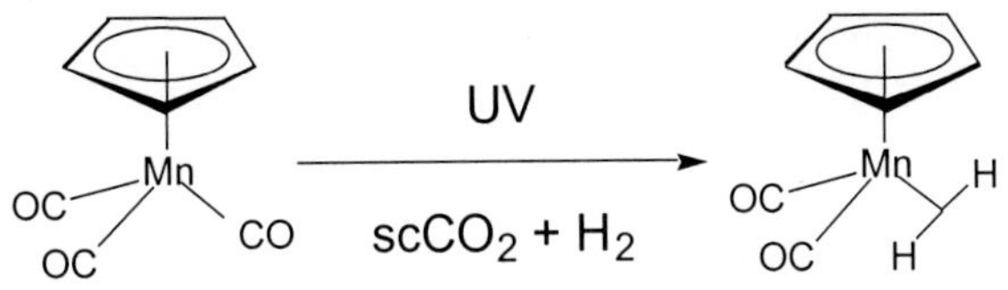

Figure 9.1-9. Example of hydrogenation in SCFs.

The field has recently been reviewed [4]. However, there is still some argument over whether SCF hydrogenation does necessarily involve a single-phase system. Härröd and co-workers suggest that single-phase behaviour is crucial [58-60], and a detailed analysis [61] of the supercritical hydrogenation of cyclohexene suggests that this reaction almost certainly does involve a single phase.   A quite different explanation for the effectiveness of supercritical hydrogenation has been put forward by Bertucco and co-workers [62], who suggested that, for higher molecular weight compounds, the $scCO_2$ dissolves in the organic substrate to form an "expanded liquid" which may contain up to 80% CO_2: the high rates of hydrogenation arise from the solubility of H_2 being much higher in this expanded liquid than in the normal liquid substrate. They have termed this process *three-phase* hydrogenation (the three phases are gas, liquid, and the catalyst) [62]. This view is supported by very recent work of Jessop and coworkers, who have found that, above T_c of CO_2, even subcritical pressures of CO_2 can accelerate both homogeneously- and heterogeneously-catalysed hydrogenations, including the hydrogenation of a model unsaturated fatty acid, oleic acid [63].

9.1.7   Continuous Organic Reactions

Supercritical hydrogenation is just one example of continuous reactions which can be carried out in $scCO_2$ solution. Other reactions which have been carried out successfully include Friedel-Crafts alkylation of aromatics by alcohols [64], the dehydration of alcohols to form ethers [65] (using acid catalysts), and the hydroformylation of alkenes [52] (using rhodium catalysts immobilized on SiO_2). In each of these reactions, it is possible to obtain a selectivity which is at least as good, and often better, than with conventional solvents. However, the precise role of the $scCO_2$ in these reactions is not as obvious as in supercritical hydrogenation.

Figure 9.1-10 shows a greatly simplified continuous fixed-bed catalytic reactor for organic reactions. The attraction of such a reactor for SCF chemistry is that it is safe, because only a very small volume of material is under supercritical conditions at any one time, yet the overall productivity can be very high [61]. The reactor is highly versatile because it can be used for a wide range of reactions merely by changing the catalyst. Furthermore, the reactions are often highly selective, not because the catalysts are inherently better under SCF conditions but

482

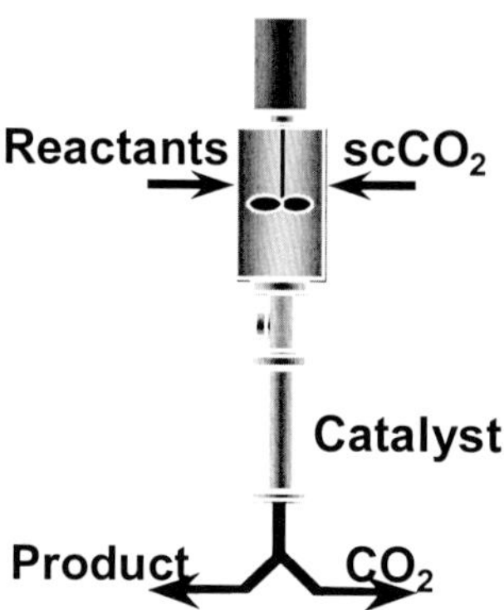

Figure 9.1-10. A schematic continuous reactor for organic reactions.

because these conditions provide improved control of the reaction parameters and, hence, greater optimization of the process.

Finally, such SCF processes can be environmentally cleaner through simpler product separation and reduced by-products, which may remove the need for downstream purification of the products. Currently, a plant is being built to commercialize these reactions. It will be important to establish whether the advantages already demonstrated in the laboratory will survive when scaled up to a full-size plant.

9.1.8 Future Developments

The pace of development is accelerating in supercritical chemistry. Increasing numbers of chemists are being attracted to the field and new reactions and applications are being reported all the time. Two particularly exciting developments have been in the control of reactions and in the development of new polymer stabilizers for CO_2.

Clifford, Rayner and co-workers at the University of Leeds [66,67] have used the density of scCO$_2$ to control the stereoselectivity of reactions in a novel way. Previous workers have observed an influence of density on selectivity in reactions very close to the critical point. The novelty of the Leeds work is that the effects are observed at densities considerably above ρ_c and temperatures higher than T_c. This selectivity has no obvious counterpart in reaction chemistry in conventional solvents. Effects have been observed in a whole range of reactions, and this approach may well have widespread applicability.

The importance of stabilizers for SCF polymerization was briefly outlined in Section 9.1.4. The drawback with existing stabilizers, however, is that most of them are based on fluorocarbons or siloxanes, which are high-cost chemicals. Cheaper polymeric stabilizers are usually only soluble in scCO$_2$ at pressures too high to make viable their widespread use. Very recently, Beckman and co-workers reported [68] a totally new approach to the problem polymers were prepared by co-polymerization of propene oxide and scCO$_2$. These polymers are not only much cheaper than fluorinated polymers but are *more soluble* than these materials in scCO$_2$. The polyether polymers are likely to have widespread applicability, not only as building blocks for stabilizers for SCF polymerization, but also as the basis of inexpensive surfactants for generating inverse aqueous micelles and microemulsions in scCO$_2$. Such micelles are themselves key to whole areas of SCF processing and chemistry [69-71].

References

1. *Chemical Synthesis Using Supercritical Fluids*; Jessop, P. G.; Leitner, W., Eds.; Wiley-VCH: Weinheim, 1999.
2. Darr, J. A.; Poliakoff, M. *Chem. Rev.* 1999, *99*, 495-541.
3. Jessop, P. G.; Ikariya, T.; Noyori, R. *Chem. Rev.* 1999, *99*, 475-493.
4. Baiker, A. *Chem. Rev.* 1999, *99*, 453.
5. Kendall, J. L.; Canelas, D. A.; Young, J. L.; DeSimone, J. M. *Chem. Rev.* 1999, *99*, 543-563.
6. Cooper, A. I. *J. Mater. Chem.* 2000, *10*, 207.
7. Cooper, A. I.; Howdle, S. M. *Materials World* 2000, *In Press*.
8. Eckert, C. A.; Knutson, B. L.; Debendetti, P. G. *Nature* 1996, *383*, 313.
9. Eckert, C. A.; Liotta, C. L.; Brown, J. S. *Chemistry & Industry* 2000, 94.
10. McHugh, M. A.; Krukonis, V. J. ; Butterworth-Heinmann, Boston, MA.:, 1994; 2nd Ed.
11. Banister, J. A.; Howdle, S. M.; Poliakoff, M. *J. Chem Soc., Chem. Commun.* 1993, 1814.
12. Banister, J. A.; Lee, P. D.; Poliakoff, M. *Organometallics* 1995, *14*, 3876.
13. Lee, P. D.; King, J. L.; Seebald, S.; Poliakoff, M. *Organometallics* 1998, *20*, 524.
14. Linehan, J. C.; Yonker, C. R.; Bays, J. T.; Autry, S. T.; Bitterwolf, T. E.; Gallagher, S. *J. Am. Chem. Soc.* 1998, *120*, 5826.
15. Yonker, C. R.; Wallen, S. L.; Linehan, J. C. *J. Microcolumn Separations* 1998, *10*, 153.
16. Jobling, M.; Howdle, S. M.; Healy, M. A.; Poliakoff, M. *J. Chem Soc., Chem. Commun.* 1990, 1287.
17. Banister, J. A.; Cooper, A. I.; Howdle, S. M.; Jobling, M.; Poliakoff, M. *Organometallics* 1996, *15*, 1804.
18. Sun, X. Z.; Grills, D. C.; Nikiforov, S. M.; George, M. W.; Poliakoff, M. *J. Am. Chem. Soc.* 1997, *119*, 7521.
19. Sun, X. Z.; George, M. W.; Kazarian, S. G.; Nikiforov, S. M.; Poliakoff, M. *J. Am. Chem. Soc.* 1996, *118*, 10525.
20. Jessop, P. G.; Ikariya, T.; Noyori, R. *Nature* 1994, *368*, 231.
21. Jessop, P. G.; Hsiao, Y.; Ikariya, T.; Noyori, R. *J. Am. Chem. Soc.* 1994, *116*, 8851.
22. Jessop, P. G.; Ikariya, T.; Noyori, R. *Science* 1995, *269*, 1065.
23. Kröcher, O.; Köppel, R. A.; Baiker, A. *Chem. Commun.* 1996, 1497.
24. Krocher, O.; Koppel, R. A.; Froba, M.; Baiker, A. *Journal of Catalysis* 1998, *178*, 284-298.
25. Shaw, R. W.; Brill, T. B.; Clifford, A. A.; Eckert, C. A.; Franck, E. U. *Chem. Eng. News* 1991, *69*, 26.
26. Savage, P. E. *Chem. Rev.* 1999, *99*, 603.
27. Savage, P. E.; Gopalan, S.; Mizan, T. I.; Martino, C. J.; Brock, E. E. *A. Inst. Chem. Eng. J.* 1995, *41*, 1723.
28. Aleman, P.; Boix, C.; Poliakoff, M. *Green Chemistry* 1999, *1*, 65.
29. Boix, C.; Poliakoff, M. *J. Chem. Soc. Perkin. 1* 1999, 1487.
30. Boix, C.; Martinez de la Fuente, J.; Poliakoff, M. *New J. Chem.* 1999, *23*, 641.

484

31. Chandler, K.; Deng, F.; Dillow, A. K.; Liotta, C. L.; Eckert, C. A. *Ind. Eng. Chem. Res.* 1997, *36*, 5175.
32. Boix, C.; Poliakoff, M. *Tetrahedron Lett.* 1999, *40*, 4433.
33. Adschiri, T.; Kanazawa, K.; Arai, K. *J. Am. Ceram. Soc.* 1992, *75*, 1019.
34. Adschiri, T.; Yamane, S.; Onai, S.; Arai, K. Proc. 3rd Int. Symp. on Supercritical Fluids, Strasbourg, France, 1994; p 241.
35. Hakuta, Y.; Terayama, H.; Onai, S.; Adschiri, T.; Arai, K. Proc. 4th Int. Symp. on Supercritical Fluids, Sendai, Japan, 1997; p 255.
36. Smith, R. L.; Atmaji, P.; Hakuta, Y.; Kawaguchi, M.; Adschiri, T.; Arai, K. *J. Supercritical Fluids* 1997, *11*, 103.
37. Cabanas, A.; Darr, J. A.; Lester, E.; Poliakoff, M. *Chem. Commun.* 2000, 901.
38. Galkin, A. A.; Kostyuk, B. G.; Lunin, V. V.; Poliakoff, M. *Angew. Chem.* 2000, in press.
39. Kirby, C. F.; McHugh, M. A. *Chemical Reviews* 1999, *99*, 565-602.
40. Dardin, A.; DeSimone, J. M.; Samulski, E. T. *J. Phys. Chem.* 1998, *102*, 1775.
41. DeSimone, J. M.; Guan, Z.; Carlsbad, C. S. *Science* **1992**, *257*, 945.
42. McCoy, M. *Chem. Eng. News* 1999, *June 14th*, 11.
43. ColeHamilton, D. J.; Sellin, M. F. *Abstracts of Papers of the American Chemical Society* 1999, *218*, 344-INOR.
44. Leitner, W.; Buhl, M.; Fornika, R.; Six, C.; Baumann, W.; Dinjus, E.; Kessler, M.; Kruger, C.; Rufinska, A. *Organometallics* 1999, *18*, 1196-1206.
45. Morita, D. K.; Pesiri, D. R.; David, S. A.; Glaze, W. H.; Tumas, W. *Chem. Commun.* 1998, 1397-1398.
46. Palo, D. R.; Erkey, C. *Industrial & Engineering Chemistry Research* 1998, *37*, 4203-4206.
47. Palo, D. R.; Erkey, C. *Industrial & Engineering Chemistry Research* 1999, *38*, 3786-3792.
48. Kreher, U.; Schebesta, S.; Walther, D. *Zeitschrift Fur Anorganische Und Allgemeine Chemie* 1998, *624*, 602-612.
49. Kayaki, Y.; Noguchi, Y.; Iwasa, S.; Ikariya, T.; Noyori, R. *Chem. Commun.* 1999, 1235-1236.
50. Koch, D.; Leitner, W. *J. Am. Chem. Soc.* 1998, *120*, 13398-13404.
51. Francio, G.; Leitner, W. *Chem. Commun.* 1999, 1663-1664.
52. Meehan, N. J.; Sandee, A. J.; Reek, J. N. H.; Kamer, P. J. C.; van Leeuwen, P. W. M. N.; Poliakoff, M. *Chem. Commun.* 2000, 1497-1498.
53. Debenedetti, P. G.; Tom, J. W.; Kwauk, X.; Yeo, S.-D. *Fluid Phase Equilibria* 1993, *82*, 311.
54. Howdle, S. M.; Grebenik, P.; Perutz, R. N.; Poliakoff, M. *J. Chem. Soc., Chem. Commun.* 1989, 1517.
55. Howdle, S. M.; Poliakoff, M. *J. Chem Soc., Chem. Commun.* 1989, 1099.
56. Howdle, S. M.; Healy, M. A.; Poliakoff, M. *J. Am. Chem. Soc.* 1990, *112*, 4804.
57. Webster, J. M. Ph.D. Thesis, University of Nottingham, UK, 2000.
58. Härröd, M.; Møller, P. In *High Pressure Chemical Engineering*; Rudolph von Rohr, P., Trepp, C., Eds.; Elsevier: Netherlands, 1996, p. 43.
59. Härröd, M.; Macher, M.-B.; Högberg, J.; Møller, P. Proc. 4th Conf. on Supercritical Fluids and Their Applications, Capri, Italy, 1997; p 319.
60. van den Hark, S.; Härröd, M.; Møller, P. *J. Am. Oil Chemists Soc.,* 1999, *76*, 1363.

61. Hitzler, M. G.; Smail, F. R.; Ross, S. K.; Poliakoff, M. *Org. Proc. Res. Dev.* 1998, *2*, 137.
62. Devetta, L.; Giovanzana, A.; Canu, P.; Bertucco, A.; Minder, B. J. *Catalysis Today* 1999, *48*, 337-345.
63. Jessop, P. G.; Wynne, D. C.; DeHaai, S.; Nakwatase, D. *Chem. Commun.* 2000, 693.
64. Hitzler, M. G.; Smail, F. R.; Ross, S. K.; Poliakoff, M. *Chem. Commun.* 1998, 359.
65. Smail, F. R.; Gray, W. K.; Hitzler, M. G.; Ross, S. K.; Poliakoff, M. *J. Am. Chem. Soc.* 1999, *121*, 10711.
66. Oakes, R. S.; Heppenstall, T. J.; Shezad, N.; Clifford, A. A.; Rayner, C. M. *Chem. Commun.* 1999, 1459-1460.
67. Oakes, R. S.; Clifford, A. A.; Bartle, K. D.; Petti, M. T.; Rayner, C. M. *Chem. Commun.* 1999, 247-248.
68. Sarbu, T.; Styranec, T.; Beckman, E. J. *Nature* 2000, *405*, 165.
69. Johnston, K. P.; Bright, F. V.; Randolph, T. W.; Howdle, S. M. *Science* 1996, *272*, 1726.
70. Johnston, K. P.; Harrison, K. L.; Clarke, M. J.; Howdle, S. M.; Heitz, M. P.; Bright, F. V.; Carlier, C.; Randolph, T. W. *Science* 1996, *271*, 624-626.
71. Clarke, M. J.; Harrison, K. L.; Johnston, K. P.; Howdle, S. M. *J. Am. Chem. Soc.* 1997, *119*, 6399.

9.2.   Enzymatic reactions

Ž. Knez[a], T. Gamse[b], R. Marr[b]

[a] Department of Chemical Engineering University of Maribor
P.O. Box 222, Smetanova 17, SI-2000 MARIBOR, Slovenia

[b] Institut für Thermische Verfahrenstechnik und Umwelttechnik Erzherzog Johann Universität
Infeldgasse 25, A-8010 GRAZ, Austria

9.2.1   Introduction

The production of fine chemicals results in the generation of considerable volumes of waste, as the syntheses generally include a number of steps. The yield of each of these steps is usually 60%-90%, but 10% is not unusual. Based on these data we can conclude, that typically 1 kg of end-product leads to the generation of 15 kg of wastes. Most of them are solvents and by-products from solvents and intermediates. Therefore, ideally several reactions should be performed in water or in dense gases.

The basic research in the future should be oriented towards novel materials for use in carbon dioxide as a solvent. These materials may have important applications in the synthesis of polymers, pharmaceuticals and other commodity chemicals, in the formation of thin films and foams, in coatings and extracts, and in the manufacture of microelectronic circuits. The full deployment of these applications would result in significant reductions in both volatile emissions and aqueous- and organic liquid wastes in manufacturing operations.

Manufacturing processes in liquid and supercritical CO_2 also tend to have advantages in terms of energy reductions, ease of product recovery, and reduction in side reactions.

Catalysis is becoming important and, in the future, biocatalysis (using cells, cell debris, enzymes) may be the most efficient way of producing fine chemicals.

Since the first reports on the use of supercritical fluids (SCFs), as reaction media several studies on enzymatic oxidation, hydrolysis, trans-esterification, esterification, inter-esterification, and enantio-selective synthesis have proved the feasibility of enzymatic reactions in supercritical fluids. The advantages of using supercritical carbon dioxide as a medium for enzyme-catalyzed reactions, have been well documented. Frequently, the temperature ranges used for employing supercritical carbon dioxide in processing are compatible with the use of enzymes as catalysts. An additional benefit of using supercritical fluids along with enzymatic catalysis is that it provides a medium for the recovery of products or reactants. However, a limitation of the process may arise from the non-polarity of carbon dioxide, which preferentially dissolves hydrophobic compounds. However, recent advances in the understanding of the chemical properties of materials that are soluble in carbon dioxide have permitted the development of novel surfactants that allow dissolution of both hydrophilic and hydrophobic materials in CO_2. This has made it possible to consider the use of CO_2 as a solvent in a wide variety of manufacturing processes. For example, CO_2-soluble surfactants can be used to carry out dispersion polymerization reactions in supercritical CO_2, and in the dry-cleaning of garments in liquid carbon dioxide. On the other hand, several other gases (Freon R23, propane, butane, dimethyl ether, SF_6) have been used recently.

The use of supercritical fluids reduces mass-transfer limitations because of the high diffusivity of reactants in the supercritical medium, the low surface tension, and because of the relatively low viscosity of the mixture. The Schmidt number (Sc= η /ρ·D, where η is dynamic viscosity, D is diffusivity, and ρ is density) for CO_2 at 200 bar is 45 times lower than for water at 1 bar and 20°C. The high diffusivity of SCFs and their low surface tension lead to reduced internal mass-transfer limitations for heterogeneous chemical or biochemical catalysis.

9.2.2 Enzymes

Enzymes are classified according to the types of reactions which they catalyse. The major classes are [1]:
1. Oxidoreductases – catalyse oxidation-reduction reactions;
2. Transferases – catalyse group-transfer reactions;
3. Hydrolases – catalyse hydrolytic reactions;
4. Lyases – catalyse reactions involving a double bond;
5. Isomerases – catalyse reactions involving isomerisation;
6. Ligases or synthetases – catalyse reactions involving joining two molecules coupled with the breakdown of a phosphate bond.

9.2.2.1 Enzyme stability in supercritical fluids

The use of enzymes in supercritical fluids presents the problem that the number of parameters influencing the stability of the enzymes increases dramatically. This is the reason why, up to now, no prediction can be made of whether an enzyme is stable under supercritical conditions and if the equivalent of even higher activity and selectivity is available compared with reactions in organic solvents. In the following chapters the influence of parameters on the enzyme stability will be given. Although much research has been done [2-5] it is very difficult to compare these results because non-standard methods have been applied.

9.2.2.2 Effect of water activity

Most of the enzymes include water, sometimes in bonded form. Treating these enzymes with pure CO_2 leads to removal of the water, which is soluble in CO_2 depending on the pressure and temperature.
The solubility of water in supercritical CO_2 can be calculated by Chrastil equation [6]

$$c = \rho^{k} \cdot \exp(\frac{a}{T} + b) \tag{9.2-1}$$

where c is the solubility (g/L), ρ is the CO_2 density (g/L) and T the temperature (K). The calculated water parameters are $k = 1.549$, $a = -2826.4$ and $b=-0.807$.

If the bonded water is extracted by dry CO_2 the enzyme is denatured and loses its activity. Therefore a certain amount of water is necessary in the supercritical fluid because acting with water-saturated CO_2 again causes an inhibition of the enzyme and consequent loss of activity. The optimal water concentration has to be determined for each enzyme separately. Table 9.2-1 shows the residual activity of lipase from *Candida cylindracea*, esterase from *Mucor mihei*, and esterase from *Porcine liver* after a incubation time of 22 hours in supercritical CO_2 at 40°C. It is obvious that higher water concentrations cause a strong reduction in the residual activity compared to the activity of the untreated enzyme, which was set as 100 %. Further, the system-pressure has an influence because at higher pressures the activity-loss is lower with a larger amount of water in the system [7,8].

9.2.2.3 Effect of pressure

The influence of the system-pressure on the stability of enzymes is not so significant for pressure ranges up to 300 bar. This is of great advantage because this means that the solvent-power of the supercritical fluid can be adjusted for running reactions. On the one hand, the solubility of substances increases with higher pressures because of a higher fluid-density and this is essential to bring the initial products in the reactor and remove the end-products from the reactor. On the other hand higher pressure results normally also in higher reaction rates. Therefore a pressure increase is in most cases positive for enzymatic reactions.

Table 9.2-1

Residual activity at different CO_2 humidity at 40°C, 22 hours incubation time and 170 mL total volume of the system (I, lipase from *Candida cylindracea*; II, esterase from *Mucor mihei*; III, esterase from *Porcine liver*)

		Residual activity		
	µl H_2O	I	II	III
150 bar	0	90.4 ± 3.2 %	77.7 ± 4.3 %	92.4 ± 2.3 %
	50	89.8 ± 1,9 %		
	100	95.1 ± 3.5 %	87.0 ± 5.6 %	81.6 ± 9.7 %
	200	85,2 ± 0.8 %	76.2 ± 2.8 %	86.9 ± 5.3 %
	300	53,0 ± 2.5%	41.6 ± 1.5 %	36.8 ± 6.8 %
300 bar	0	104.3 ± 5.1%	101.8 ± 5,1 %	89.2 ± 2.6 %
	50		100.5 ± 4.7 %	
	100	91.5 ± 3.8 %	82.0 ± 5.6 %	93.1 ± 6.3 %
	200	91.6 ± 2.1 %	74.4 ± 1.8 %	94.5 ± 3.4 %
	300	78.9 ± 2.8 %		71.7 ± 3.5 %

9.2.2.4 Effect of temperature

The temperature effect is much more significant than the pressure effect. For the enzyme stability, a temperature increase above certain levels, depending on the enzyme, results in deactivation of the enzyme. In Table 9.2-2, the residual activities of various enzymes after one hour incubation time, in supercritical CO_2 at 150 bar, are given. It is obvious that temperatures over ca. 75°C reduce enzyme activity dramatically. However, no correlation for the stability with the temperature for different types of enzymes is yet available [8-10].

This limitation causes the temperature-limits for the solubility of the initial and end products of the reaction. Normally, at higher pressure levels an increase in temperature also results in higher solubilities of substances in supercritical fluids because the increase in the vapour-pressure of the compounds to be dissolved overcomes the reduction in density.

9.2.2.5 Number of pressurization-depressurization steps

When using batch reactors the influence of pressurization-depressurization steps on the enzyme activity is of importance [11]. Bringing the enzyme under pressure does not play an important role, but the depressurization steps especially have an influence on the residual

Table 9.2-2
Residual activity of different enzymes after 1 hour treatment in supercritical CO_2 at 150 bar (I, lipase from *Candida cylindracea*; II, lipase *Amano AY*; III, lipase from *Pseudomonas sp.*; IV, esterase EP10 from *Burkholderia gladioli*)

Temperature	Residual activity [%]			
[°C]	I	II	III	IV
25	100.0	100.0	100.0	100.0
35	112.5	102.6	95.1	101.8
45	112.9	95.6	119.1	83.3
55	111.0	92.7	115.3	83.3
65	99.6	90.1	97.1	96.3
75	111.4	76.9	102.3	87.1
85	94.3	78.0	101.0	29.6
95	57.9	48.7	89.1	1.9
105	3.0	35.5	25.5	1.9

enzyme activity. The supercritical fluid enters the enzyme by diffusion, which is normally relatively slow. After a certain time the enzyme is saturated with the fluid. When there is too fast an expansion the fluid cannot leave the enzyme fast enough, which causes yet a higher fluid pressure in the enzyme than in the system. Overcoming a certain pressure difference results in an effect similar to cell-cracking, where the cell membranes are broken by the resulting overpressure inside the cells. This causes an unfolding of the enzymes and therefore destroys the structure which is of importance for the activity and selectivity. Unpublished experiments (of the Institut fuer Thermische Verfahrens- und Umwelttechnik, TU Graz) have shown that a depressurization from supercritical-CO_2 conditions is much smoother than entering the two-phase region and expanding the liquid part of the fluid. This can be explained in combination with the change in density which is continuous at supercritical conditions. Expanding liquid CO_2 causes evaporation of the fluid accompanied by a large change in density, and this volume-expansion causes unfolding of the enzyme.

9.2.2.6 Inhibition of enzymes

Inhibition of the enzymes can be caused by water, as already shown in Chapter 9.2.2.2. Therefore esterification reactions are very difficult to handle, because the produced water has to be removed from the reactor to avoid enzyme inhibition. Increasing the CO_2 flow-rate would be one possibility for removing the produced water from the reactor, but this results in higher investment and operation costs.

Initial products as well as end products can cause enzyme-inhibition if these compounds block the active centre of the enzyme so that no activity and selectivity is available. Therefore, tests have to be performed to determine the enzyme activity depending on the and amount of initial- and end-products to optimize reaction conditions.

490

9.2.3   Enzyme reactors

9.2.3.1 Process schemes and downstream processing schemes
   Batch-, stirred-tank-, extractive semibatch-, recirculating batch-, semicontinuous flow-, continuous packed-bed-, and continuous-membrane reactors have been used as enzyme reactors, with dense gases used as solvents.

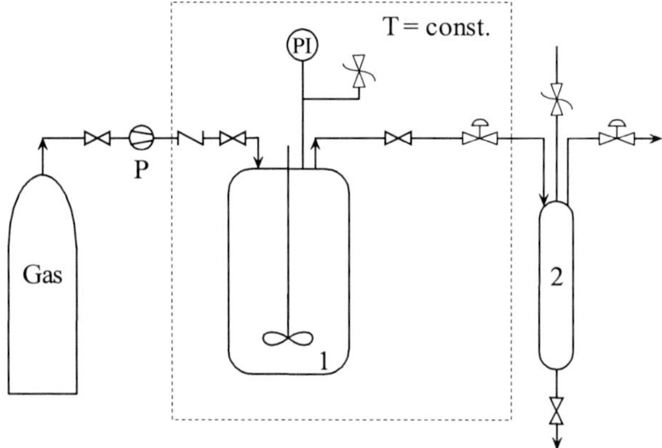

Figure 9.2-1. Design of experimental batch-stirred-tank apparatus for synthesis under high pressure: 1, reactor; 2, separator; P, high pressure pump; PI, pressure indicator [17].

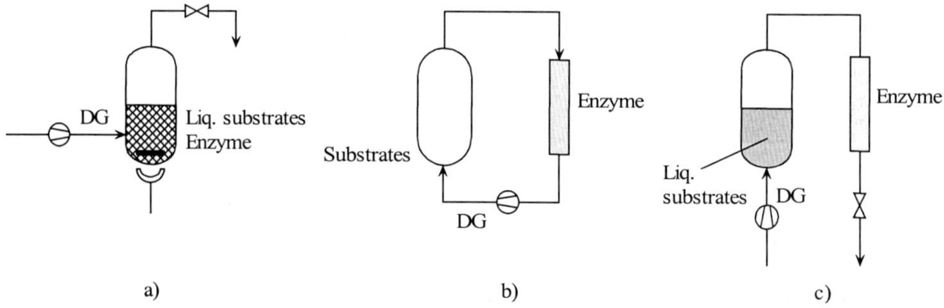

Figure 9.2-2. Operating principle of dense-gases enzymatic reactor types: a), extractive semibatch; b), recirculating batch; c), semicontinuous flow.

   Batch-stirred-tank reactors [12-21] are usually used for screening enzymatic reactions in dense gases. The design of the system is shown in Figure 9.2-1. Initially, the reaction mixture was pumped into the reactor and then the enzyme-preparation was added. Finally, dry gas was pumped into the reactor, up to the desired pressure. The initial concentration of the reactant never exceeded its solubility-limit in the gas.

Extractive batch reactors (Figure 9.2-2) [22] are used for continuous extraction of products from reaction mixtures (containing liquid substrates and an enzyme preparation), which shifts the reaction equilibrium towards formation of the product.

Recirculating batch reactors (Figure 9.2-2) [23,24] are mainly used for kinetic studies of enzymatic reactions performed in dense gases.

In semicontinuous-flow reactors (Figure 9.2-2) [25] the dense gas is saturated with substrates in an autoclave and then fed continuously through the enzyme bed. The advantage of this type of reactor is that the pure gas should be compressed. The disadvantage of semicontinuous flow reactors is that the concentration of substrates in dense gases cannot be varied, and that with changes of pressure and temperature, precipitation of substrates or products in the reactor can occur.

A continuous packed-bed reactor [15, 16, 26-29] is presented in Figure 9.2-3. The system has a high pressure pump for the delivery of gas into the system. The gaseous fluid was dried by passing through the columns packed with molecular sieves. The flow-rate of gas during these runs was varied. The substrates were pumped into the system, using a HPLC. The gas and the substrates were equilibrated in the saturation column. The initial concentration of the reactant never exceeded its solubility-limit in the gas.

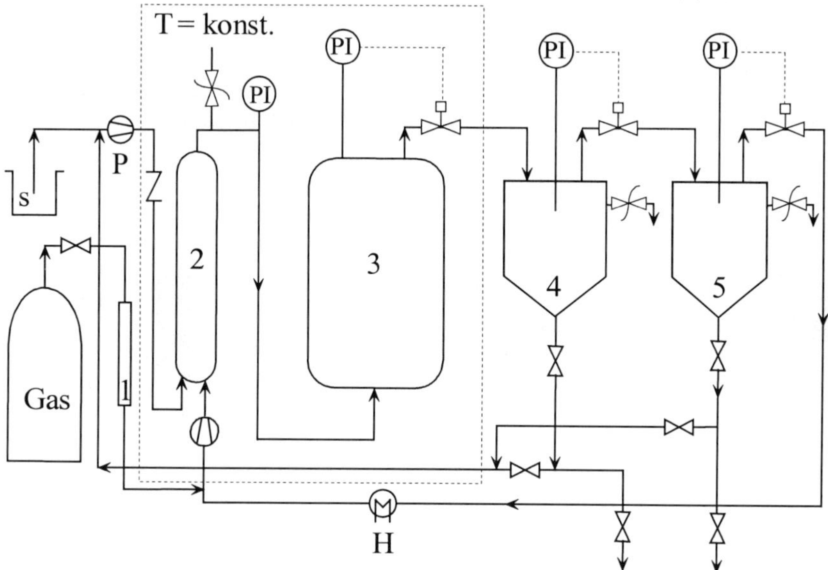

Figure 9.2-3. Design of continuous experimental apparatus for synthesis under supercritical conditions: S, substrates; 1, molecular sieves; 2, saturation column; 3, packed-bed enzymatic reactor; 4,5, separators; P, high-pressure pump; PI, pressure indicator; TI, temperature indicator; H, heat exchanger [17].

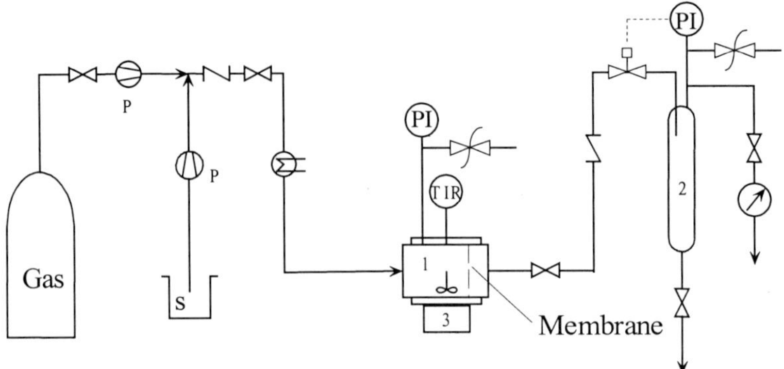

Figure 9.2-4. High-pressure continuous stirred tank membrane reactor; S, substrates; 1, reactor; 2, separator; 3, magnetic stirrer; P, high pressure pump; TIR, temperature regulator and indicator; PI, pressure indicator.

Supercritical carbon dioxide is depressurized through the expansion valves into separator columns 4 and 5, where the product and the unreacted substrates are recovered. The substrates are collected in column 5 and recycled (added to the feed through the pipe connecting column 5 with the feed vessel). The gas phase is finally vented into the atmosphere after flow-rate measurement through a rotameter. The gas can be condensed and recycled on a pilot- or industrial scale.

The continuous high-pressure enzyme membrane reactor [30] is shown in Figure 9.2-4. The membrane with 35 mm diameter is placed between two sintered plates and fitted in the reactor. A certain amount of the catalyst (hydrated enzyme preparation) is put in the reactor which is electrically heated, with a heating jacket, to constant temperature. The substrates and the gas are pumped into the membrane reactor with the high-pressure pump. The products and unreacted reactants are collected in the separator. The catalyst remains in the reactor (behind the membrane).

One of the main advantages of the use of dense gases as a solvent for enzyme-catalysed reactions is the simple downstream processing. The physico-chemical properties of dense gases are determined by their pressure and temperature, and are especially sensitive near the critical point. Usually the single-phase reaction mixture is contacted with enzyme preparation, and later changes in P and/or T cause the less soluble substance to precipitate.

By reducing the solvent-power of a dense gas in several stages, fractionation of the product and unreacted reactants is possible. Fractionation is also possible by extracting the mixture, usually with the same dense gas as used in reaction, but under different process conditions. In all downstream processing schemes, various particle-formation techniques [31] or chromatographic techniques can be integrated.

9.2.3.2 Processing costs

Table 9.2-3 shows the specification of commercially available jojoba oil and its bio-synthetically analogous product-oleyl oleate.

Economic evaluations were made for the enzyme-catalyzed production of oleyl oleate in a high-pressure batch-stirred-tank reactor (HP BSTR) and in a high-pressure continuously operated reactor (HP COR). Table 9.2-3 shows details. It was assumed that the reaction mixture is completely precipitated from CO_2. The production costs are strongly dependent on the solubility of the substrates in dense gas, the enzyme-lifetime, and the productivity of the enzyme catalyst.

The annual production would be 550 t of oleyl oleate with market price 5 EURO/kg. As biocatalyst, an immobilized *Rhizomucor miehei* lipase - Lipozyme IM - product from NOVO Nordisk was used.

Table 9.2-3
Quality of jojoba oil and bio-synthetically produced analogue

	Commercial jojoba oil	Oleyl oleate produced by biosynthesis in SCF
Acidity value	Below 1	Below 1
Iodine value	90-97	93
Saponification value	100-110	109
Solidifying point	Below 100°C	Below 100°C
Density (g/mL)	0.86-0.88	0.88
Refractive index	1.4640-1.4650	1.4650

Assumptions	HP BSTR	HP COR
Production rate	550 000 kg/year	550 000 kg/year
Enzyme productivity	1.7 kg product/ g enzyme	1.6 kg product/ g enzyme
Running time	4800 hours/year	4800 hours/year
Conversion (per pass)	90%	88%
Reaction time	5 hours	

Resulting process	HP BSTR	HP COR
Required amount of enzyme in reactor	8 kg	8.5 kg
Volume of the reactor	900 L	40 L

Cost	HP BSTR	HP COR
Total equipment cost	280 500 EURO	336 600 EURO
Notional capital estimate summary:	442 170 EURO	438 243 EURO
Annual operating cost	1 123 020 EURO	904 740 EURO
Return investment	0.84 year	0.89 year

9.2.4 Conclusions

The application of SCF as reaction media for enzymatic synthesis has several advantages, such as the higher initial reaction rates, higher conversion, possible separation of products from unreacted substrates, over solvent-free, or solvent systems (where either water or organic solvents are used). Owing to the lower mass-transfer limitations and mild (temperature) reaction conditions, at first the reactions which were performed in non-aqueous systems will be transposed to supercritical media. An additional benefit of using SCFs as reaction media is that they give simple and ecologically safe (no heat and solvent pollution) recovery of products.

However, for some specific reactions, solvent-free systems are preferred because of their higher yields. The main area of development should be related to the hydrolysis of glycerides, transesterification, esterification and inter-esterification reactions. As lipases have high and stable activity in SC CO_2 (even at the high temperature) an intensive development is expected.

Only a little research on the separation and synthesis of chiral compounds has been published so far. Because enzymes have extremely high selectivity, and owing to the great importance of enantioselective synthesis or enantiomeric resolution in the pharmaceutical industry, the most intense research in this area can be expected, along with minimizing the use of substances and maximizing their effect.

References

1. T.W. Randolph, D.S. Clark, H.W. Blanch and J.M. Prausnitz., Proc. Natl. Acad. Sci. USA, 85 (1988) 2979.
2. K. Nakamura, M.Y. Chi, Y. Yamada and T. Yano., Chem. Eng. Commun. 45 (1986) 207.
3. T.W. Randolph, H.W. Blanch, J.M. Prausnitz and C.R. Wilke, Biotechnol. Lett. 7(5) (1985) 325.
4. D.A. Hammond, M. Karel and A.M. Klibanov, Appl. Biochem. Biotech. 11 (1985) 393.
5. S.V. Kammat, E.J. Beckman and A.J. Russell, Crit.Rev.Biotechnol. 15 (1995) 41.
6. J. Chrastil, J.Phys.Chem. 86 (1982) 3016.
7. D.J. Steinberger, T. Gamse and R. Marr, 5th Conf.Supercritical Fluids and their Applications, Garda (1999) 339.
8. T. Gamse and R. Marr, 5th Int.Symp.Supercritical Fluids, cd-rom (2000) 7265469.
9. J.K.P. Weder, Z.Lebensm.Unters.Forsch.Forsch, 171 (1980) 95.
10. J.K.P. Weder, Food Chem., 15 (1984) 175.
11. D.J. Steinberger, A. Gießauf, T. Gamse and R. Marr, Proc. GVC-Fachausschuß "High Pressure Chemical Engineering", Karlsruhe (1999) 209.
12. J.C. Erickson, P. Schyns and C.L. Cooney, AIChEJ 36(2) (1990) 299.
13. T. Dumont, D. Barth, C. Corbier, G. Branlant and M. Perrut, Biotechnol. Bioeng. 39 (1992) 329.
14. Ž. Knez and M. Habulin, Biocatalysis 9 (1994) 115.
15. Ž. Knez, M. Habulin and V. Krmelj, J. Supecrit. Fluids, 14 (1998) 17.
16. Ž. Knez, M. Habulin and V. Krmelj, J Am. Oil Chem. Soc., 67 (1995) 775.
17. Ž. Knez, V. Rižner, M. Habulin and D. Bauman, J. Am. Oil Chem. Soc., 72 (1995) 1345.
18. M. Rantakylä and O. Aaltonen, Biotechnol. Lett., 16 (1994) 825.
19. H. Noritomi, M. Miyata, S. Kato and K. Nagahama, Biotechnol. Lett., 17 (1995) 1323.
20. M. Habulin, V. Krmelj and Ž. Knez, J. Agric. Food. Chem., 44 (1) (1996) 338.
21. M. Habulin, V. Krmelj and Ž. Knez, Proc. High Pressure Chemical Engineering, Ph. Rudolf von Rohr, Ch. Trepp (eds.), Elsevier, Amsterdam (1996) 85.
22. H. Gunnlaugsdottir and B. Sivik, J. Am. Oil Chem. Soc., 72 (1995) 399.
23. Z. R. Yu, S. Rizvi and J.A. Zollweg, Biotechnol. Prog., 8 (1992) 508.
24. D. C. Steytler, P. S. Moulson and J. Reynolds, Enzyme Microb. Technol., 13 (1991) 221.
25. D. A. Miller, H. W. Blanch and J. M. Prausnitz, Ind. Eng. Chem. Res., 30 (1991) 939.
26. T. Dumont, D. Barth and M. Perrut, J. Supercrit. Fluids 6(2) (1993) 85.

27. K. Nakamura, in Supercritical Fluid Technology in Oil and Lipid Chemistry, J. W. King, G. R. List (eds.), AOCS Press, 1996.
28. M. A. Jacson and J. W. King, J. Am. Oil Chem. Soc., 73 (1996) 353.
29. A. Marty, D. Combes and J.S. Condoret, Biotech. Bioeng., 43 (1994) 497.
30. Ž. Knez, V. Krmelj and M. Habulin, High Pressure Membrane Bioreactors, GVC – Fachausschusses "Hochdruck – Verfahrenstechnik", Berlin (2000) 14.
31. Ž. Knez, Supercritical fluids as a solvent for industrial scale enzymatic reactions – yes or no? : technical lecture invited by DECHEMA for the 2nd COST (Cooperation in Science and Technology – Commission of the European Communities) Workshop on chemistry under extreme or non-classic conditions, Lahnstein, 1995.

9.3 Hydrogenation under supercritical single-phase conditions

M. Härröd[a], M.-B. Macher[a], S. van den Hark[a], P. Møller[b]

[a]Department of Food Science, Chalmers University of Technology
P.O. Box 5401 SE-40229 Göteborg, Sweden

[b]Augustenborggade 21B, DK-8000 Aarhus C, Denmark

Supercritical fluids may combine gas- and liquid properties in a very favourable way. Using the new supercritical single-phase hydrogenation processes, extremely fast reactions can be achieved, and the time-scale for the reactions is seconds compared to hours in the traditional processes. This can be utilized to reduce investment- and production-costs and to improve product quality.

In this chapter we put the new supercritical single-phase hydrogenation process into a general context. To create the basis for understanding the new technology, a short overview of the most important aspects of the traditional processes and the new technology are given. Finally, the impact of using the new technology will be described. To do this, concentration profiles and phase diagrams will be used.

9.3.1 Introduction

Hydrogenation is a major type of chemical process. In such processes, the hydrogen and the substrate are mixed and brought into contact with a homogeneous or heterogeneous catalyst. In this chapter we focus on heterogeneous hydrogenation processes. They can be divided into two groups: gas-phase hydrogenation and gas - liquid hydrogenation.

In gas-phase hydrogenation the substrate is gaseous. Therefore it is possible to mix enough hydrogen for the reaction with the substrate and maintain the gas-phase condition. The transport properties are very good in gases, and in combination with the single-phase condition it is possible to have enough hydrogen at the catalyst surface. This leads to very high reaction rates. However, only small molecules can be transferred to a gas phase at reasonable pressures and temperatures. Therefore, the excellent reaction properties of the gas-phase process can only be utilized for small, volatile molecules.

Larger molecules will form a liquid under the reaction conditions. In these cases, hydrogen is mixed with the liquid and brought in contact with a suitable catalyst. The main problem is the low solubility of hydrogen in the liquid, i.e., it is not possible to dissolve enough hydrogen for the reaction in the liquid, and a two-phase mixture of a gas-phase and a liquid-phase has to be used. In combination with low mass-transport this leads to a lack of hydrogen at the catalyst surface. Thus, in gas- liquid-phase hydrogenation, the hydrogen concentration at the catalyst becomes the limiting factor for both reaction rate and selectivity.

In the new supercritical single-phase hydrogenation, a solvent is used to dissolve the substrate, and the hydrogen into one single phase. This can only be achieved if the whole mixture of substrate, solvent, and hydrogen is brought to a near-critical- or supercritical state. Just as in the gas-phase process, it is possible to mix enough hydrogen for the reaction with the substrate and maintain the single-phase condition. The transport properties are very good under supercritical conditions, and in combination with the single-phase condition it is possible to have enough hydrogen at the catalyst surface. This leads to very high reaction rates.

Thus, when the substrate can be dissolved in a suitable solvent at reasonable concentrations, pressures, and temperatures, the excellent reaction properties of the supercritical single-phase process can be utilized also for large molecules.

9.3.2 Traditional hydrogenation processes

In hydrogenation processes, hydrogen is added to the substrate molecule. The bond to be hydrogenated is, in most cases, a double bond, but it can be a single bond or a triple bond. Using different catalysts, different bonds can be hydrogenated and many different atoms can be involved in the reactions, e.g., C, N, S, O. These hydrogenation processes are performed in many different areas, e.g., in petrochemical-, fine-chemical-, food-, and the pharmaceutical industry [1] to achieve a desired product quality.

The wide range of substrates being hydrogenated leads to a variety of processes (ranging from, e.g., slurry-batch to counter-current fixed-bed reactors), each having its typical reaction conditions and advantages. Common to all catalytic processes is that an intimate contact between the reactants and the catalyst is important for the reaction rate. If the consumption of reactants at the catalyst cannot be compensated by the physical mass-transport of fresh reactants to the catalyst, the latter reduces the reactant concentration at the catalyst and limits the reaction rate. Good reviews on the state-of-the-art in heterogeneous catalysis are available, e.g., Ertl et al., [2]. Below, we will describe the most important features of gas-phase and gas - liquid hydrogenation processes and their most important parameters.

9.3.2.1 Gas-phase hydrogenation

Figure 9.3-1 shows a general concentration profile for a gas-phase hydrogenation. In the gas-phase process the concentrations of substrate and hydrogen can be chosen freely. This means that a surplus of hydrogen at the catalyst surface, compared to the stoichiometric need, can be achieved.

All the reactants, including the hydrogen, are in a gas phase with a very high diffusivity and low viscosity. Depending on the balance between the catalytic activity and the diffusion of the gases into the catalyst pores, an internal concentration gradient (see Fig. 9.3-1) may be present. Thus, high concentrations of both reactants are possible at the catalyst surface. Because of these high concentrations, the reaction rate can be extremely high. The chemical kinetics, which in this case determine the reaction rate, can be fully utilized because the excess of hydrogen or substrate is arbitrary.

Examples of these favourable reaction conditions are seen in ammonia production (N_2/H_2) [3] and ethylene hydrogenation (C_2H_2/H_2) [4]. The reaction rates in these processes can be very high, 1 - 10 mmol/g_{cat}min [3,4] (i. e., about 2 000 - 20 000 $kg_{product}/m^3_{reactor}h$).

There are two major drawbacks of the gas-phase processes: first, the limited heat-transport properties of gases make adequate temperature control impossible for the exothermic hydrogenation reactions [4]; and, secondly, the use of gas-phase processes is limited to small volatile molecules.

Many interesting organic molecules are large (10 - 60 carbon atoms) and they have a very low vapour pressure at the reaction conditions (e.g., the vapour pressure of C18:0-fatty alcohol is 0.27 bar at 300 °C [5]). Thus, large molecules can only be hydrogenated at very low

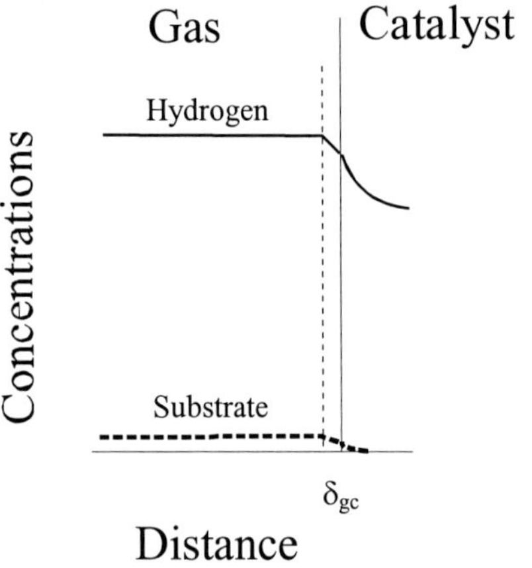

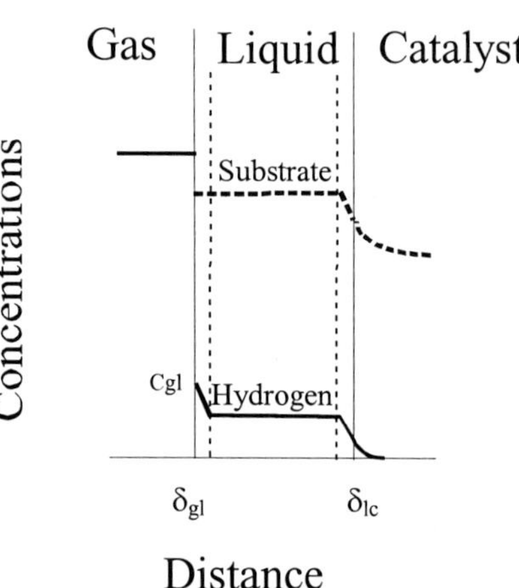

Figure 9.3-1. General concentration profiles in a *gas-phase* hydrogenation.

--- Substrate

— Hydrogen

δ_{gc} Stagnant film layer at the gas - solid catalyst interface.

Figure 9.3-2. General concentration profiles in a *gas - liquid phase* hydrogenation.

--- Substrate

— Hydrogen

δ_{gl} Stagnant film layer at the gas - liquid interface.

δ_{lc} Stagnant film layer at the liquid - solid catalyst interface.

c_{gl} Equilibrium conc. of H_2 in the liquid.

concentrations, making the gas-phase process uneconomic. At higher concentrations these substrates create a liquid phase, which will co-exist with the gas phase. This results in a completely different system, which will be discussed next.

9.3.2.2 Gas-liquid hydrogenation

When the substrate is liquid, or a solid dissolved in a liquid, the reaction condition becomes more complex. The amount of hydrogen that can be dissolved in a liquid is generally much lower than the stoichiometric need for the reaction. The hydrogenation is completely based on mass-transport from the gas phase through the liquid and to the catalyst. This transport is described with the concentration profiles in Fig. 9.3-2.

At the surface of the liquid, the hydrogen gas is in equilibrium with the liquid phase (see c_{gl}, Fig. 9.3-2). Most hydrogenation catalysts are very active. In combination with the poor mass-transport properties in the liquid, large concentration-gradients occur at the gas - liquid and liquid-catalyst interfaces (see δ_{gl}, δ_{lc}, Fig. 9.3-2). Thus, the mass-transport of hydrogen is the limiting factor for the actual hydrogen concentration at the catalyst surface, and there is always a lack of hydrogen at the catalyst. Hence, the reaction rates are always limited by the low hydrogen concentration at the catalyst surface.

Note the difference between the gas-phase and the gas - liquid-phase systems. The concentrations of hydrogen and substrate at the catalyst are inverted in the gas - liquid system relative to the gas-phase system (cf. Fig. 9.3-1 and Fig. 9.3-2).

There is also another concentration gradient, from the catalyst surface into the pores, which can be more severe in gas - liquid-phase systems than in gas-phase systems owing to the much lower diffusivity in liquids than in gases.

Detailed overviews of all aspects of multiphase mass transport coupled with reactions are given elsewhere [6].

A typical example of the gas-liquid hydrogenation processes is the hardening process of vegetable oils for the production of margarine and shortening. The standard process takes place in large batch reactors (5 - 20 m^3), where oil, hydrogen and catalyst are mixed for long times, at high temperatures and low pressures (2 h, 140 - 200 °C, 1 - 3 bar) [7]. This gives a productivity of about 400 $kg_{oil}/m^3_{reactor}$h. The low solubility of hydrogen in oil under these conditions, together with the transport resistances, lead to low concentrations of hydrogen at the catalyst surface and very low reaction rates [8]. Commercial oils, which are partially hydrogenated, normally contain 30 - 40% *trans*-fatty acids. According to the half-hydrogenation theory, the *trans*-fatty acid formation increases when the hydrogen concentration at the catalyst surface decreases [9]. For oil hydrogenation this has been verified in several studies (e.g., Hsu et al.,1989 [10]). The *trans*-fatty acid content has become the subject of many health discussions related to ingestion of fats and oils [11] and today the *trans*-fatty acid content in table spreads should be below 5% in several countries [12].

Thus, if the hydrogen supply to the catalyst surface could be increased sufficiently, partially-hydrogenated oils with acceptable levels of *trans*-fatty acids could be produced. However, this is not possible using the traditional technology.

9.3.2.3 Important process parameters

The process conditions and design of catalyst and reactor can be used to improve the reactor productivity. Most important is to increase the hydrogen concentration at the catalyst. This mass-transport can be increased by increasing the concentrations of the reactants (= driving force) or by improving the contact between the phases (= reduced transport resistance). A wide range of catalysts is available. The effect of these parameters will be discussed below.

Pressure

The system pressure affects the hydrogenation in different ways. Most prominent is the increase of the hydrogen solubility in liquids with increasing pressure (e.g., in vegetable oils the hydrogen solubility is 2 mol% at 3 bar and 20 mol% at 100 bar [13]). It is important to note that the stoichiometric need for full hydrogenation of this oil is about 85 mol% of hydrogen. Increased solubility of hydrogen leads to an increased mass transport, because of an increased driving force, and hence the entire concentration profile of hydrogen in Fig. 2 will be at a higher level. This results in increased reaction rates and it may also influence the selectivity. For example, in partial hydrogenation of oils the concentrations of *trans*-fatty acid isomers are reduced in favour of *cis*- with increasing hydrogen pressure [10].

Besides the effect on the solubility in multi-phase systems, the pressure can also directly increase the reactant concentrations in gas-phase reactions through the compressibility of gases. The reaction-rate increases because of the increased concentrations. The compression might also directly influence the rate constants. In this way, pressure can change the reaction equilibrium and selectivity, and thus the product quality.

The benefits of enhanced reaction rate and improved selectivity have to be seen in relation to the costs of working at higher pressures. The technical/economical limit for standard materials is somewhere around 300 bar.

Temperature

The temperature range used is determined mainly by the catalyst used, and whether formation of side-products will occur. Each catalyst has a specific ignition temperature at which it becomes active for the desired reaction. This temperature has to be exceeded, otherwise no catalytic reaction will occur. Above this temperature, the reaction rate increases only slowly at increasing temperature (cf. the Arrhenius function). In general, the reaction rate is much more temperature-sensitive than is the mass-transfer rate. Thus, in reactions where the mass-transfer determines the reaction rate, as in gas - liquid reactions, a temperature rise above the ignition temperature has only a minor effect on the reaction rate.

The temperature may influence the selectivity. For example, in the hydrogenation of oils, a temperature rise causes the hydrogen concentration at the catalyst surface to decrease. In turn, this leads to increased formation of *trans*-fatty acids [10]. The mechanism can be summarized by stating that a temperature rise causes:
- the reaction rate to increase and the consumption of hydrogen to increase;
- the solubility of hydrogen in oils to increase slightly [8,13] (however, this effect on the hydrogen transport is too small to balance the increased consumption);
- the adsorption of reactants on the catalyst to decrease [14].

Thus, the hydrogen concentration at the catalyst surface decreases when the temperature increases.

Another effect is that increasing temperature might shift the reaction equilibrium towards the reactants. This applies especially to hydrogenation processes, since most of them are exothermic.

In several processes it is important also to study the formation of side-products. Increasing temperatures frequently give more side-products (e.g., over-hydrogenation of the product, or hydrogenation of other functional groups within the substrate molecule, or polymerization reactions or coke formation). The formation of these side-products usually restricts the maximum temperature at which the hydrogenation can be performed successfully.

However, for some small molecules the temperature might be increased so far that the liquid substrate evaporates under the reaction conditions used. This results in a change from a gas - liquid process into a gas-phase process. This will increase the reaction rate greatly (see, e.g., the hydrogenation of cyclohexene to cyclohexane [15]). If the coke formation and the formation of side-products can be neglected, very good results can be achieved with strongly increased temperature.

Reactor and catalyst design

Much research is done in the field of catalyst-design and reactor engineering in order to improve mass- and heat-transport properties [7]. Another purpose is to improve the selectivity. This is, in some cases, strongly related to the transport properties, but in other cases the choice of catalytic material and design is essential.

It is always difficult to compare different processes. The reaction-rate and the product-quality have to be compared to the overall economy of the whole process. In this chapter we will use the reactor productivity as a simplified way to find the major differences between different reactors. We will express the productivity as $kg_{product}/m^3_{reactor}h$.

In a slurry-batch reactor the hydrogen and the liquid are mixed intensively together with the solid catalyst. The catalyst concentration is normally a few wt%. The reaction time is typically a few hours. This means that the productivity becomes in the range of 100 - 1 000 $kg_{product}/m^3_{reactor}h$. The transport resistance between the liquid and the catalyst ($*_{lc}$ in Fig. 9.3-2) is normally the restricting factor.

A fixed-bed reactor is filled with a porous catalyst. Thus, the catalyst concentration becomes much higher, and the contact between the catalyst and the fluids is much better. This normally means that the resistance between the hydrogen and the liquid ($*_{gl}$ in Fig. 9.3-2) becomes the restricting factor, and that the productivity increases strongly. Typical values for the reaction-rate are in the range of 1 000 - 10 000 $kg_{product}/m^3_{reactor}h$.

The design of the catalysts is very important for the hydrogenation processes. During the years, great improvements have been achieved. However, it is beyond the scope of this chapter to discuss these issues. A good introduction can be found in the brochures from the various catalyst manufacturers.

Solvent

The objectives of using solvents are diverse, e.g., to dissolve a solid substrate, to limit catalyst deactivation, to improve selectivity, or to enhance mass-transport. The solvents are selected depending on the substrate and the desired effect. Hence, they range from water, alcohols, ethers, or low alkanes, to CO_2 . The effects of the solvent on phase-behaviour, viscosity, and density at different concentrations, temperatures and pressures can explain much about the effect of the solvent on the reaction.

To our knowledge, the only commercial application of solvents in hydrogenation processes is that where a solid substrate should be brought into contact with a solid catalyst. One example is in the hydrogenation of sugars. Alcohol or water is used to dissolve the sugar crystals [16]. Another example is that crystalline sterols are dissolved in alkanes, alcohols, or ethers prior to the hydrogenation process [17]. Without the solvents, these reactions cannot be performed. After the addition of solvents these processes are typical gas - liquid-phase hydrogenation processes.

Carbon dioxide has been investigated as a solvent for hydrogenation processes in several research papers. The first investigator was Zosel [18]. His substrate consisted of triglycerides, whose solubility in carbon dioxide is below 1 wt.% (0.05 mol%) at 80°C and 300 bar [19]. Supercritical CO_2 is miscible with H_2 [20]. Solubility, viscosity, and density data for some triglyceride/hydrogen/CO_2-systems have recently been measured in Erlangen [21,22]. The authors found that slightly increased hydrogen concentrations can be achieved using CO_2 and that at relevant temperatures, pressures, and concentrations for hydrogenation processes, a two-phase system always occurs [22]. The observed reaction rates for triglycerides in a slurry system [18], and for fatty-acid methyl esters in a fixed-bed system [23] are the same as the corresponding traditional gas - liquid processes without CO_2. Thus, CO_2 has only a marginal effect on the hydrogenation rate on FAME or TG under technical / economic conditions.

In commercial processes for edible oils, solvents are not used because the improvement of productivity is not enough to compensate for the extra costs of adding a solvent [8].

Carbon dioxide has also been used for hydrogenation of other substrates [15, 24 - 26]. For a recent review, see Baiker 1999 [27]. Improved reaction rates are frequently reported. However, the reaction rates are always below 10 000 $kg_{substrate}/m^3_{reactor}h$.

One group of researchers has presented the phase-behaviour of their system. They conclude that their system was operated under two-phase conditions because, from an industrial point of view, the pressure required to form a single-phase system becomes too expensive [24].

Unfortunately, many authors in the area do not mention the compositions of their reaction mixtures [15,23,25]. Hence the number of phases in their systems is not clear. Even if the composition is known, the number of phases can be difficult to determine. The low reaction rate reported indicates that they have performed their reactions in gas - liquid-phase systems.

9.3.3 The supercritical single-phase hydrogenation

The most striking feature of supercritical single-phase hydrogenation processes is the tremendous reaction rate. The whole reaction is completed in a few seconds. Below, we will describe the supercritical single-phase hydrogenation process in detail, compare it with the traditional processes, and look at the impact of using the new technology.

9.3.3.1 Single-Phase Conditions

By choosing a suitable solvent and suitable conditions (concentrations, temperatures, and pressures), the solvent, the substrate, the product, and the hydrogen can form a single phase. When such a single phase is fed to a fixed-bed reactor, extremely high reaction rates have been achieved for very large molecules [28 - 30]. These reaction rates, expressed on a molar basis, are the same as those achieved in gas-phase reactors for small molecules [30].

The single-phase condition eliminates the transport resistance at the gas - liquid interface (cf., Fig. 9.3-2 and Fig. 9.3-3). This resistance is the restricting factor in the traditional fixed-bed reactors. Compared to the gas - liquid process, the resistances at ($*_{lc}$ in Fig. 9.3-2) and in the catalyst are strongly reduced. The reasons are the reduced viscosity and the increased diffusivity in the supercritical fluid. The single-phase condition makes it possible to feed the catalyst with hydrogen in excess. Compared to the gas - liquid reaction, much higher hydrogen concentrations at the catalyst surface are possible at the single-phase reaction   (cf., Fig. 9.3-2 and Fig. 9.3-3).

The concentration profiles for the supercritical single-phase reaction are similar to those in gas-phase reactions (cf., Fig. 9.3-3 and Fig. 9.3-1). The difference is that, with the solvent, these conditions can be achieved for much larger molecules. These supercritical single-phase conditions can be achieved for suitable solvents and at certain conditions. The catalyst used determines the temperature-range for the reaction, as previously discussed.

Suitable solvents can be found in the group of compounds which has its critical point close to this temperature range. With regard to the catalyst, one has to ensure that the solvent is inert and does not influence the hydrogenation process. It is desirable that the solvent has a very good solvent-power for the substrate and the product. If the solvent is supercritical, it is always miscible with hydrogen.

As an example, we will illustrate the technology with oils/propane/hydrogen. The oils are triglycerides and fatty-acid methyl esters. The phase behaviour of soybean oil/propane/H_2 has been measured under suitable reaction conditions [22] (see Fig. 9.3-4). The phase behaviour always depends on the composition, temperature and pressure [31].

The solubility of hydrogen in oil without propane can be seen at the baseline between the oil and the hydrogen in Fig. 9.3-4. The same point is illustrated by "c_{gl}" in Fig. 9.3-2. The stoichiometric need of hydrogen depends on the reaction, but generally it is above 50 mol%. Full hydrogenation of a soybean oil requires more than 85 mol% hydrogen. To maintain the chemistry in the reaction, this ratio of oil to H_2 has to be maintained also when a solvent is added (see the dotted line in Fig. 9.3-4). For technical reasons, the stoichiometric need has to be exceeded to some extent. Thus, the composition of the feed to the reactor must be to the right of the dotted line, i.e., "need" in Fig. 9.3-4.

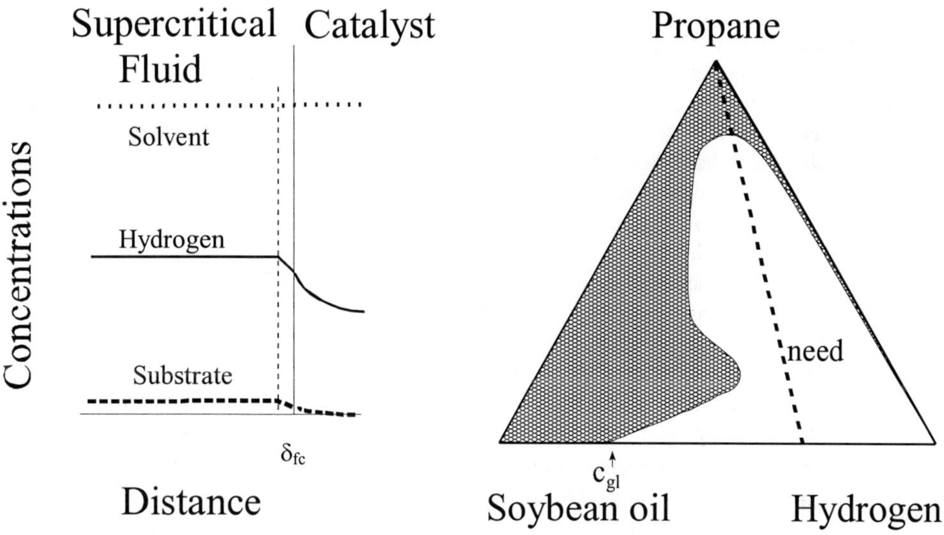

Figure 9.3-3. General concentration profiles in a *supercritical single-phase* hydrogenation.
- - - Substrate
— Hydrogen
.... Solvent
δ_{fc} Stagnant film layer at the fluid-solid-catalyst interface.

Figure 9.3-4. Phase-diagram for the system soybean oil/propane/hydrogen at 180 bar and 130 °C.
Shaded area measured single-phase [22].
c_{gl} Equilibrium concentration of H_2 in the liquid phase at these conditions, without using a solvent [13].
- - - Limit for the hydrogen needed for full conversion of the substrate.

504

To apply the supercritical single-phase concept successfully, two criteria have to be fulfilled. First of all, a homogeneous phase is essential (i.e., the shaded area in Fig. 9.3-4). Secondly, the hydrogen concentration in this phase should be at least as high as required by the reaction stoichiometry (i.e., to the right of the dotted line in Fig. 9.3-4). This means that supercritical single-phase hydrogenation processes can only be performed when the concentration of the reaction mixture is in the area that lies to the right of this dotted line and is, at the same time, in the single-phase area (i.e., in the shaded area to the right of the dotted line in Fig. 9.3-4).

The phase behaviour described in Fig. 9.3-4 is valid for a specific oil, pressure, and temperature. Data are available for some other temperatures and pressures [22]. From these data we can see that it is possible to increase the loading (i.e., the oil concentration) and still maintain the single-phase conditions if we reduce the temperature and/or increase the pressure.

This concept has been applied, and extremely high reaction rates were achieved. For example, a residence time of less than one second was enough for partial hydrogenation of fatty-acid methyl esters (FAME). This resulted in a reactor productivity of 240 000 $kg_{FAME}/m^3_{reactor}h$ [29]. Compared to batch processes, this reaction-rate is about 500 times higher (cf., the example in the gas - liquid section). In the hydrogenation of FAME (C_{18}) to fatty alcohols (FOH), we have achieved reaction rates in the same range as obtained in gas-phase reactions of similar but much smaller molecules (methyl acetate, C_2), when the reaction rate is expressed in $mol/g_{cat}h$. When the reaction rate is expressed in $kg_{FOH}/m^3_{reactor}h$ the reaction rate with the supercritical single-phase process becomes 10 times higher than with the gas-phase process [30].

9.3.3.2 Measurement of phase behaviour in complex reaction mixtures

The phase-behaviour depends on the composition, temperature and pressure. The composition of the reaction mixture can, in some cases, become rather complex and therefore difficult to measure with traditional methods. One simpler method of determining single-phase conditions is to observe the reaction rate. The difference in reaction rate between two-phase and single-phase conditions is dramatic. The reaction times go from hours to seconds [28 - 30]. For a given system, the reaction rate is proof of the presence or absence of a liquid phase. When comparing different systems, the reaction rate is a strong indication of the presence or absence of a liquid phase.

Another way to obtain information on the phase-behaviour is to study the pressure-drop over the reactor. A detailed study on this subject has been made by van den Hark et al. [32].

9.3.3.3 Connecting the different reaction systems

Starting from a gas - liquid system at high pressure, single-phase conditions can be achieved in two ways: either by increasing the temperature, or by adding a suitable solvent. The concentration profiles are very similar for single-phase systems, i.e., for the gas-phase system and for the supercritical single-phase system (cf., Fig. 9.3-1 and Fig. 9.3-3). However, the concentration profiles are very different for the gas - liquid system compared to the single-phase systems (cf., Fig. 9.3-2 versus Fig. 9.3-1 and Fig. 9.3-3).

An example of increasing the temperature is given in the hydrogenation of cyclohexene. Here single-phase conditions were achieved by increasing the temperature from about 100^oC to 300^oC. The productivity became as high as 240 000 $kg_{product}/m^3_{reactor}h$ [15]. From a traditional point of view, this reaction is identical to commercial gas-phase reactions such as ammonia production and hydrogenation of ethylene. All components are in a gaseous phase. From a supercritical point of view, they are supercritical because the temperature and the pressure of the systems are above the critical point for the mixtures. To transfer a gas - liquid process to a single-phase process by increasing temperature is only practicable if both the substrate and the product have a critical point in the range of possible reaction conditions, and if no side- reactions occur at the high temperatures.

The other alternative for achieving single-phase conditions is to add a suitable solvent. The advantage of the supercritical single-phase technology is that the reaction temperature can remain unchanged. However, the pressure, the solvent and the composition of the reaction mixture have to be selected carefully to ensure single-phase conditions. If this is not the case, a multiphase system remains [28 - 30,32].

9.3.3.4 Impact of using supercritical single-phase hydrogenation technology

Performing reactions in a supercritical single phase using solvents opens a number of new possibilities.

First, extremely high reaction rates have been achieved, even for very large molecules. The reaction time is in the range of seconds; therefore only continuous-flow reactors are suitable for this type of reactions.

Secondly, with the supercritical single-phase hydrogenation we have gained a new tool to control the reaction selectivity (i.e., the product quality) because:

- The concentrations at the catalyst surface (both hydrogen and substrate) can be controlled independently of other process conditions. The unique feature is that very high concentrations of hydrogen can be achieved; this leads, for example, to the suppression of *trans*-fatty acids in partial hydrogenation of methylated rapeseed oil [29].
- The high concentration of hydrogen at the catalyst surface ensures a high reaction rate and makes it possible to adjust other process settings (e.g., to reduce the temperature) to suppress unwanted side-reactions [29,30].
- Extremely high degrees of conversion can be achieved by increasing the reaction time greatly. However, the reactor volume will still be very small because of the extremely high reaction rate.
- The short residence-times in the reactor give less time-thermal-dependent degradation of heat-sensitive products and/or substrates.
- The addition of solvent makes it possible to control the temperature in the reactor despite the exothermic reactions and high reaction rates. The reactor operates nearly adiabatically, but the temperature rise in the reactor can be controlled, because the solvent acts as an internal cooling medium. The concentration of substrate determines the maximal temperature rise and therefore, by controlling the concentration, the maximal temperature rise is controlled. In this way the amount of unwanted side-products can be reduced.
- The catalyst life might be improved. Several studies on isomerization and polymerization processes indicate that supercritical solvents can dissolve coke-precursors on the catalyst surface, and remove them before they can form actual coke, and this improves the catalyst life [33,34]. Since coke formation also occurs in hydrogenation processes, it is reasonable to believe that catalyst life can be improved also for supercritical single-phase hydrogenation.

Thirdly, the economy of the whole process seems to be very favourable. The tremendous reaction rate makes the reactor very small. The concentration of the substrate in the solvent (i.e., the loading) is crucial, as in any other solvent-based process. Single-phase conditions have been achieved at a loading of 15 - 20 wt.% for different lipids at a total pressure of 150 bar [29,35]. This would allow a moderate solvent recirculation. Based on these results, a company has decided to build a pilot plant, which will be put into operation in Göteborg, Sweden, during the spring of 2001.

Regarding safety, there is no *additional* risk in using high pressures and possibly flammable solvents. The risk of using high pressure is compensated by the smaller volume of the plant (risk = pressure * volume), and the explosion - risk is already present in all plants using hydrogen. The technology for handling these solvents and risks is well known in the petrochemical industry.

The solvents used have to be safe, from the toxicological and physiological points of view. In this context, we can again give propane as an example. It is allowed for unlimited use in the production of foodstuffs [36] and is already being used commercially for de-oiling of lecithins at SKW Trostberg in Germany [37].

9.3.3.5 Outlook

In our studies, the hydrogenation under supercritical single-phase conditions has shown very promising results. High reaction rates and reduced formation of by-products are some of the big advantages of this technique. However, there is still much basic research to be done in order to fully understand the process. For example, more solubility data in the region of interest for different substrate/solvent/hydrogen mixtures are required to optimize the process for each reaction. Product-separation and solvent-recovery need to be developed for scale-up. Further investigations on the kinetics of the different reactions are necessary to optimize the product quality for each case. This is also a question of catalyst and reactor design. Some of these aspects are being studied in our laboratory apparatus, and we will also run the pilot plant which is under construction. However, close co-operation of specialists in the different areas is necessary to bring this promising technology into industrial use.

References

1. Rylander, P. N. *Catalytic Hydrogenation in Organic Syntheses*, Academic Press: NewYork, NY, 1985.
2. Ertl, G., Knözinger, H., Weitkamp, J., (eds.), *Handbook of Heterogeneous Catalysis*, VCH: Weinheim, Germany, 1997, vol. 3.
3. Appl, M. *Ammonia, Principles and Industrial Practice*, Wiley - VCH: Weinheim, Germany, 1999.
4. Weiss, A. H., Tsaih, J. -S., Antoshin, G. V., Kobayashi, M., Nakagawa, A., Tien, N.D., Tahn, T. T. Heat Transfer Control of Selectivity in Acetylene Hydrogenation. *Catalysis Today,* 1992, 13, 697 - 698.
5. Lide, D. R., (ed.), *CRC Handbook of Chemistry and Physics*, 77th ed., CRC Press: Boca Raton, 1996, 6 - 113.
6. Mills, P. L., Ramachandran, P. A., Chaudhari, R. V. Multiphase Reaction Engineering for Fine Chemicals and Pharmaceuticals. *Reviews in Chem. Eng.* 1992, 8, ch. 4.1, 59 - 159.

7. Hastert, R. C. Hydrogenation. In *Bailey's Industrial Oil and Fat Products*, Hui, Y. H., (ed.), Wiley: New York, NY, 1996, 4, 213 - 300.

8. Veldsink, J. W., Bouma, M. J., Schöön, N. H., Beenackers, A. A. C. M. Heterogeneous Hydrogenation of Vegetable Oils: A Literature Review. *Catal. Rev.-Sci. Eng.* 1997, 39, 253 - 318.

9. Horiuti, I., Polanyi, M. Exchange Reactions of Hydrogen on Metallic Catalysts. *Transactions of the Faraday Society.* 1934, 30, 1164 - 1172.

10. Hsu, N., Diosady, L. L., Rubin, L. J. Catalytic Behaviour of Pd in the Hydrogenation of Edible Oils II. Geometrical and Positional Isomerization Characteristics. *J. Am. Oil Chem. Soc.* 1989, 66, 232-236.

11. Wahle, K. W. J., James, W. P. T. Isomeric Fatty Acids and Human Health. *Eur. J. Clin. Nutr.* 1993, 47, 828 - 839.

12. Fitch Haumann, B. Widening array of spreads awaits shoppers. *INFORM* 1998, 9, 6 - 13.

13. Wisniak, J., Albright, L. F. Hydrogenating Cottonseed Oil at Relatively High Pressure. *Ind. Eng. Chem.* 1961, 53, 375 - 380.

14. Nieuwenhuys, B. E., Ponec, V., van Koten, G., van Leeuwen, P. W. N. M., van Santen, R. A. Bonding and Elementary Steps in Catalysis. In *Catalysis, An Integrated Approach to Homogeneous, Heterogeneous and Industrial Catalysis*, Moulijn, J. A., Van Leeuwen, P. W. N. M., van Santen, R. A., (eds.), Elsevier: Amsterdam, The Netherlands, 1993, 79, ch. 4, 89 - 158.

15. Hitzler, M. G., Smail, F. R., Ross, S. K., Poliakoff, M. Selective Catalytic Hydrogenation of Organic Compounds in Supercritical Fluids as a Continuous Process. *Organic Process Research & Development.* 1998, 2, 137 - 146.

16. Tukas, V. Glucose Hydrogenation in a Trickle-Bed Reactor. *Collect. Czech Chem. Commun.* 1997, 62, 1423 - 1428.

17. Ekblom, J. Process for the Preparation of Stanol Esters. *Int. Patent Appl.* PCT/FI98/00166. 1998.

18. Zosel, K. Process for the Simultaneous Hydrogenation and Deodorization of Fats and Oils. *US Patent 3,969,382.* 1976.

19. McHugh, M., Krukonis, V. *Supercritical Fluid Extraction, 2nd ed.*, Butterworth - Heinemann: Stoneham, MA, 1994, 300.

20. Tsang, C. Y., Streett, W. B. Phase Equilibria in the H_2/CO_2 System at Temperatures from 220 to 290 K and Pressures to 172 MPa. *Chem. Eng. Sci.* 1981, 36, 993 - 1000.

21. Weidner, E., Richter, D. Influence of Supercritical Carbon Dioxide and Propane on the Solubility and Viscosity of systems containing Hydrogen and Triglycerides (Soybean Oil). In *Proceedings 6th Meeting on Supercritical Fluids, Nottingham, April 1999*, I.S.A.S.F.: Nancy, France, 1999, 657 - 662.

22. Weidner, E. Thermo- und Fluiddynamische Aspekte der Hydrierung in Gegenwart nahekritischer Lösungsmittel. *Presented at: 30 Sitzung des Arbeitsausschusses: "Technische Reaktionen", Frankfurt, Germany*, 18th January 2000.

23. Tacke, T. Fetthärtung mit Festbettkatalysatoren. *Die Ernährungsindustrie.* 1995, 12, 50 - 53.

24. Bertucco, A., Canu, P., Devetta, L. Catalytic Hydrogenation in Supercritical CO2: Kinetic Measurements in a Gradientless Internal-Recycle Reactor. *Ind. Eng. Chem. Res.* 1997, 36, 2626 - 2633.

25. Pickel, K.H., Steiner, K. Supercritical Fluids Solvents for Reactions. In *Proceedings 3rd Int. Symposium on Supercritical Fluids, Strasbourg*, I.S.A.S.F.: Nancy, France, 1994, 3, 25-29.

26. Minder, B., Mallat, T., Pickel, K.H., Steiner, K., Baiker, A. Enantioselective Hydrogenation of Pyruvate in Supercritical Fluids. *Catalysis Letters.* 1995, 34, 1-9.

27. Baiker, A. Supercritical Fluids in Heterogeneous Catalysis. *Chemical Reviews.* 1999, 99, 453 - 473.

28. Härröd, M., Møller, P. Hydrogenation of Substrates and Products Manufactured According to the Process. *US Patent 5,962,711.* 1999.

29. Macher, M.-B., Högberg, J., Møller, P., Härröd, M. Partial Hydrogenation of Fatty Acid Methyl Esters at Supercritical Conditions. *Fett/Lipid.* 1999, 101, 301 - 305.

30. van den Hark, S., Härröd, M., Møller, P. Hydrogenation of Fatty Acid Methyl Esters to Fatty Alcohols at Supercritical Conditions. *J. Am. Oil Chem. Soc.* 1999, 76, 1363 - 1370.

31. McHugh, M., Krukonis, V. *Supercritical Fluid Extraction, 2nd ed.*, Butterworth - Heinemann: Stoneham, MA, 1994, 27 - 84.

32. van den Hark, S., Härröd, M., Møller, P. Hydrogenation of Fatty Acid Methyl Esters to Fatty Alcohols at Supercritical Single-Phase Conditions: Reaction Rate and Solubility Aspects. In *Proceedings of 5th Int. Symposium on Supercritical Fluids, Atlanta, April 2000*, Teja, A., Eckert, C. (eds.), Atlanta, GA, 2000.

33. Bochniak, D. J., Subramaniam, B. Fischer-Tropsch Synthesis in Near-Critical n-Hexane: Pressure-Tuning Effects. *AIChE J.* 1998, 44, 1889 - 1896.

34. Tiltscher, H., Wolf, H., Schelchshorn, J., Dialer, K. Process for Restoring or Maintaining the Activity of Heterogeneous Catalysts for Reactions at Normal and Low Pressures. *US Patent 4,605,811.* 1986.

35. van den Hark, S., Härröd, M., Camorali, G., Møller, P. Production of Fatty Alcohols at Supercritical Conditions. In *Proceedings, 6th meeting on Supercritical Fluids, Nottingham, April 1999*, I.S.A.S.F.: Nancy, France, 1999, 273 - 278.

36. Amaducci, S. Processing aids. In *Eurofood Monitor*, Agra Europe, London, 1993, pp.F1 - F16.

37. Heidlas, J. De-oiling of Lecithins by Near-Critical Fluid Extraction. *Agro-Food-Industry Hi-Tech.* 1997, 1, 9 - 11.

9.4 Supercritical Water Oxidation (SCWO). Application to industrial wastewater treatment

M. J. Cocero

Departamento de Ingenierìa Quìmica, Universidad de Valladolid
Prado de la Madalena SP-47005 Valladolid, Spain

9.4.1 Introduction

Irrespectively of their origin, industrial or urban, the most used way for organic materials elimination from wastewater is complete oxidation to carbon dioxide and water. Biological oxidation processes using microorganisms, although widely used, are limited to biodegradable compounds from feed-streams that do not contain toxic or inhibitory products for the biological process. Otherwise, chemical oxidation processes are preferred.

For aqueous mixtures and/or solutions, conventional oxidation processes operate at slightly elevated pressure and temperature in order to keep the oxidant in the liquid phase, so that oxidation takes place. This process is known as "wet air oxidation" or " wet oxidation". It is a gas liquid reaction process where the limiting factor is the oxygen concentration within the reaction phase, that is, the liquid phase. Wet oxidation efficacies are generally satisfactory for upgrading the process to the industrial scale, although they are not sufficient to allow effluent emission to the ambient conditions. Generally, a subsequent treatment process, such as a biological process, is needed. As usual, to avoid these complexities, other processes have been proposed and claimed to be adequate for their substitution. In our case, this is supercritical water oxidation.

As is well known, if one takes a mixture of liquid and gas in equilibrium conditions, and the pressure and temperature increase, thermal expansion causes the liquid to become less dense. At the same time, the gas becomes denser as the pressure rises. At the critical point, the densities of the two phases become identical and the distinction between them disappears. Thus, above the critical temperature, T_c, the system is simply described as a fluid, called a supercritical fluid. In supercritical conditions any fluid is miscible with any non-polar organic compound in all proportions, and also shows complete miscibility with "permanent" gases, such as *oxygen*. So supercritical water is an ideal medium for oxidation of organics, since the oxygen and waste can be brought into contact in the same homogenous phase, without interface transport constraints and with reasonably fast kinetics, the oxidation proceeds rapidly to completion [1]. The phase diagram of water is shown in Fig. 9.4-1.

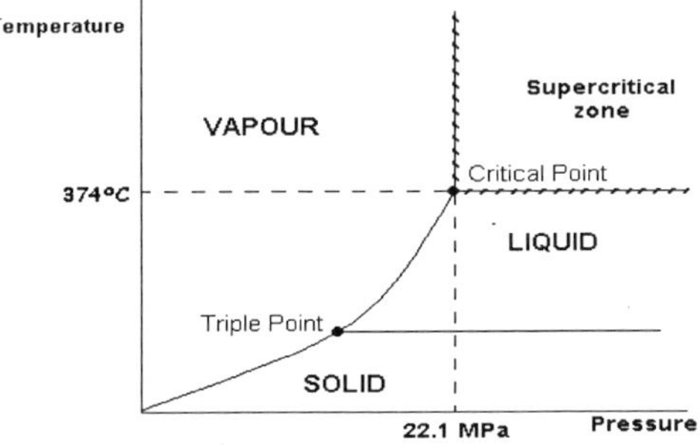

Figure 9.4-1. Phase-diagram of water

510

9.4.2 Supercritical Water as Reaction Media

9.4.2.1 Physical Properties of Supercritical Water

In the region near to the critical point, the properties of water are quite different from those of normal liquid water, or steam. For example, at the critical point the density of water, 300 Kg/m^3, is intermediate between liquid water, (1000 kg/m^3) and water vapour, (1 kg/m^3). The water density as a function of temperature at a supercritical pressure of 25.0 MPa is shown in Fig. 9.4-2. The ion product of water pKw ranges from 14 at 25°C, to 22 at 450°C [3]. The solvation properties of water also change dramatically correlating directly with density changes. The static dielectric constant of water drops from a room-temperature value of around 80 to about 5-10 in the near-critical region and, finally, to around 2 at 450 °C [2] (similar to gaseous hydrocarbons). Raman spectra of deuterated water in this region indicate little, if any, residual hydrogen bonding [3]. As a result, supercritical water acts as a non-polar dense gas, and its solvatation properties resemble those of a low-polarity organic compound. This is the reason why typically non-polar compounds are completely soluble in water under supercritical conditions, while salts are insoluble under supercritical conditions. For example, the solubility of NaCl at 25 MPa drops from about 37 % at 300°C to about 120 ppm at 550 °C [4]. The fact that inorganics are practically insoluble is consistent with a low dielectric constant of about gases, such as nitrogen, oxygen, and carbon dioxide 2. Supercritical water also shows complete miscibility with "permanent" gases, such as N_2, O_2, CO_2 [5].

The combination of the solvation and physical properties makes supercritical water an ideal medium for the oxidation of organics. When organic compounds and oxygen are dissolved in water above the critical point they are immediately brought into intimate molecular contact in a single homogenous phase at high temperature, with no interface-transport limitations and, for sufficiently high temperatures, the kinetics are fast, and the oxidation reaction proceeds rapidly to completion.

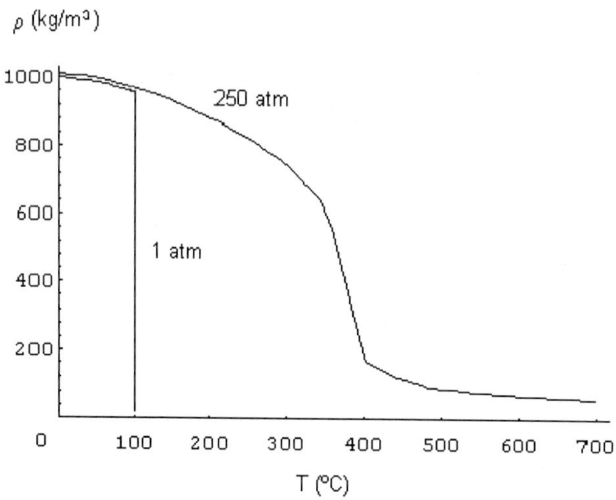

Figure 9.4-2. Density of supercritical water

9.4.2.2 Oxidation reactions in SCWO

The products of hydrocarbon oxidation in SCWO are carbon dioxide and water. Heteroatoms are converted to inorganic compounds, usually acids, salts, or oxides in high oxidation states. Phosphorus is converted into phosphate, and sulphur to sulfate; nitrogen-containing compounds are oxidized to nitrogen gas. Working at temperatures below 700 °C, no NO_x is formed.

$$(C_XO_YH_Z) + O_2 \longrightarrow CO_2 + H_2O + \Delta Q$$

$$(C_VO_XH_YN_X) + O_2 \longrightarrow CO_2 + H_2O + N_2 + \Delta Q$$

$$(C_VO_XH_YCl_X) + O_2 \longrightarrow CO_2 + H_2O + HCl + \Delta Q$$

$$(C_VO_XH_YS_X) + O_2 \longrightarrow CO_2 + H_2O + H_2SO_4 + \Delta Q$$

SCWO kinetics have been studied for various organic compounds [6]. These SCWO kinetic data show three general trends:

First, in most cases, the oxidation rate is independent of, or weakly dependent on, the oxidant concentration.

Second, pseudo-first-order kinetics with respect to the concentration of starting compounds, appears to be a reasonable assumption for SCWO.

Third, the activation energy ranges from about 30 KJ/mol to 480 KJ/mol.

Studies on reaction mechanisms and by-product analysis have indicated that short-chain carboxylic acids, ketones, aldehydes, and alcohols are the major oxidation intermediates under near-critical conditions, but at supercritical conditions, with T above 650 °C, no intermediate compounds have been found [7].

9.4.2.3 Catalysis

As in every process where chemical reactions are involved, SCWO can be favoured by using catalysts. When the catalysts are selected, high-temperature and high-pressure operating conditions should be borne in mind, as well as the oxidant atmosphere. The most used catalysts are Ce, Co, Fe, Mn, Ti, and Zn oxides and, as supports, Al, Hf, Zr, or Ti oxides [8].

These catalysts have been used in air wet-oxidation processes, where catalysts are essential to obtain high efficiencies [9]. In SCWO processes these efficiencies can be obtained by increasing temperature, and therefore they are not considered necessary. Catalysts studies have been performed on synthetic mixtures, and poisoning studies should be made when using wastewater.

The main advantage of using catalysts in SCWO processes is to reduce the operating temperature. A complete review of catalytic oxidation in supercritical water can be found in reference [8].

9.4.3. SCWO Process description

9.4.3.1. Feed preparation and pressurization

Organic waste materials in an aqueous medium are pumped and compressed from atmospheric pressure to the reactor's working pressure. Air is always compressed to the

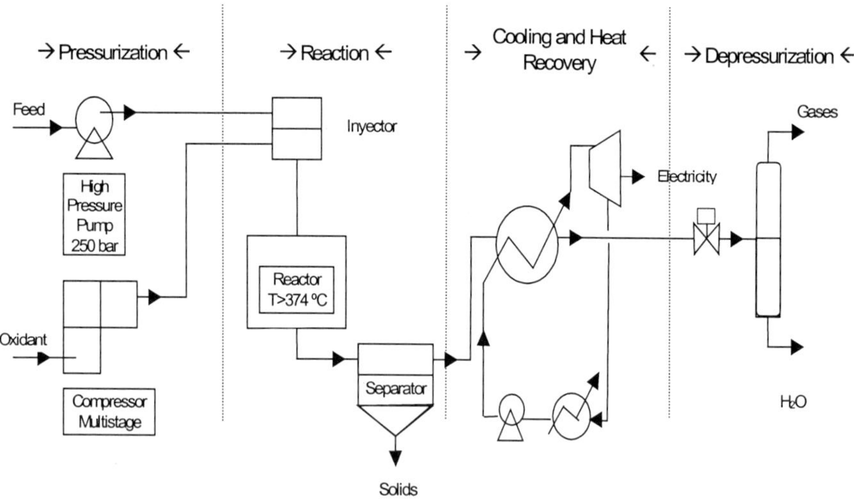

Figure 9.4-3. SCWO Flow Diagram.

operating pressure before entering the reactor. A flow-diagram of the process is in Figure 9.4-3. Alternatively, oxygen, stored as a liquid, can be pumped to the pressure of the reactor and then vaporized. The choice of oxidant (air or oxygen) is dictated by economics. Oxygen compression is considerably less expensive than air (on an equivalent-oxygen basis), but represents an additional raw-material cost. In bench-scale facilities, it may be advantageous to use hydrogen peroxide, but the commercial utility of this oxidant is limited, owing to its high cost.

The feed to the process is controlled to an appropriate upper limit of heating value of 4600 KJ/kg by adding dilution water or blending higher-heating-value waste material with lower-heating-value waste material prior to feeding it to the reactor. When the aqueous waste has a heating value too low, the feed could be preheated by exchange with effluent leaving the reactor.

9.4.3.2 Reaction

Mixing of the oxidant and organic waste streams with the hot reactor content initiates the exothermic oxidation reaction. Although for water, T_c is 374 °C and P_c is 22 MPa, as the reaction temperature increases, the efficiency also increases and the residence time decreases. According to some experimental studies [9], supercritical oxidation of any kind of high-molecular-weight material produces an almost instantaneous transformation to small molecules, which are responsible for the global kinetic process, since their stability is higher than the precursors. From these small molecules, acetic acid and ammonia are the most refractory intermediates.

A reaction temperature around 650°C has been found to be necessary for ammonia and acetic acid oxidation with higher than 99% efficiency and residence time of 50 seconds [7,10]. No significant effect was found on pressure variation, and therefore a reaction pressure slightly higher than the critical pressure is usually selected to ensure that the process takes place in only one phase [11].

9.4.3.3 Cooling and heat recovery

The gaseous products of reaction, along with the supercritical water, leave the reactor. In some processes, effluent is used to preheat the feed to operating temperature, either by a heat exchanger, or by recirculating part of it towards the reactor. The remaining effluent is cooled and taken as emission at atmospheric pressure.

Excess thermal energy contained in the effluent can be used to generate steam for external consumption, to produce electricity with high efficiency, or for the heating needs of industrial processes at high temperature. On a larger scale, system energy recovery may potentially take the form of power generation by direct expansion of the reactor products through a supercritical steam turbine. Such a system would be capable of generating significant power in excess of that required for air compression, or oxygen pumping, and feed pumping.

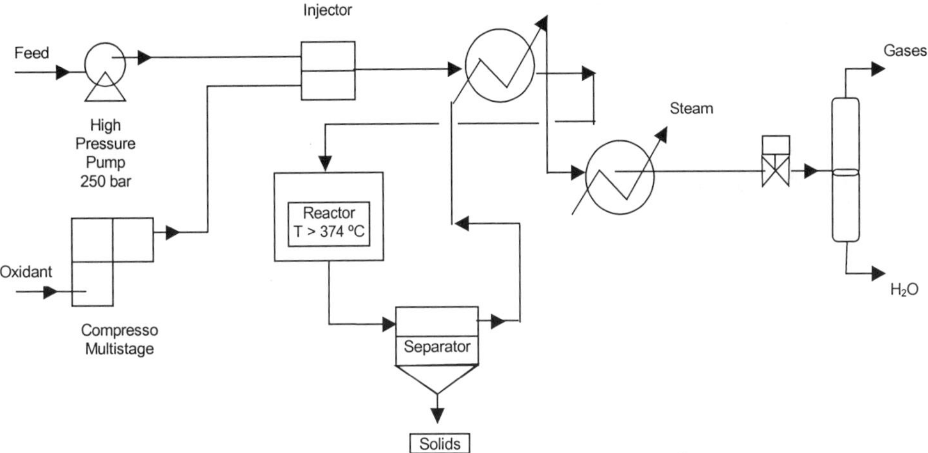

Figure 9.4-4. Flow diagram for heat recovery.

For dilute aqueous wastes, with low heating value, it is possible to use a heat exchanger to preheat the waste until operation conditions (see Figure 9.4-4).

Once cooled, the effluent from the process separates into a liquid-water phase and a gaseous phase, the latter containing mainly carbon dioxide along with oxygen which was in excess of the stoichiometric requirements, and nitrogen. The separation is carried out in multiple stages in order to minimize erosion of valves, as well as to maximise the separation due to phase-equilibrium constraints.

9.4.4 Design Considerations

9.4.4.1 Reactor Configuration

As well as the operating conditions inside the reactor, the design features have a powerful influence on reactor performance. The type of reactor selected has an influence on efficiencies, on corrosion endurance, solids-operation feasibility, or even reactor reliability. The most important SCWO reactor configurations are listed in the following.

514

Reactor vessel

MODAR Inc. (Massachusetts. USA) developed the first reactor vessel [13]. It comprised an elongated, hollow cylindrical pressure-vessel, capped at both ends so as to define an interior reaction chamber. Defined within the reaction chamber are a supercritical temperature zone, in the upper region of the reactor vessel, and a subcritical temperature zone in the lower region of the reactor vessel. Oxidation of organics and oxidizable inorganics takes place in the supercritical temperature zone. Dense matter, such as inorganic material initially present and formed by reactions, if insoluble in the supercritical-temperature fluid, falls into the liquid phase provided in the lower-temperature, subcritical zone of the vessel. A perimeter curtain of downward-flowing subcritical-temperature fluid is established about a portion of the interior of the cylindrical wall of the vessel to avoid salt-deposits on the walls of the reactor vessel.

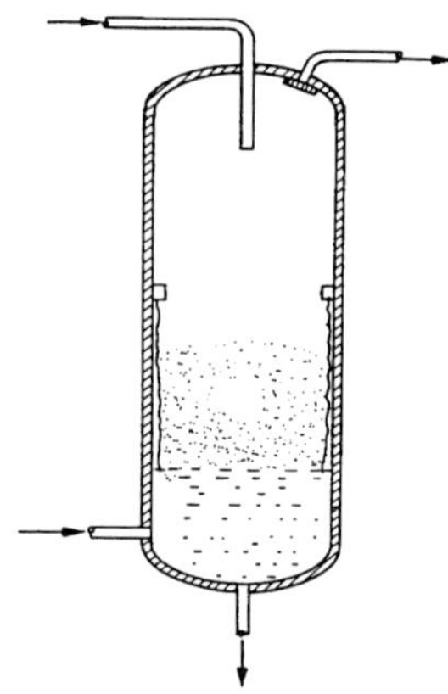

Figure 9.4-5. MODAR Tank Reactor.

Tubular reactor

Tubular reactors with many oxygen feedpoints (Fig. 9.4-6), are supplied for solution-treatment by Eco Waste Technology (EWT), Texas, USA [12]. Tubular reactors are designed with small diameters, so fluid circulation is high and salts-deposition is avoided. Through the design, the deposition of solids already present in the feed can be avoided, but precipitated salts formed inside the reactor have a tendency to adhere themselves to the reactor walls.

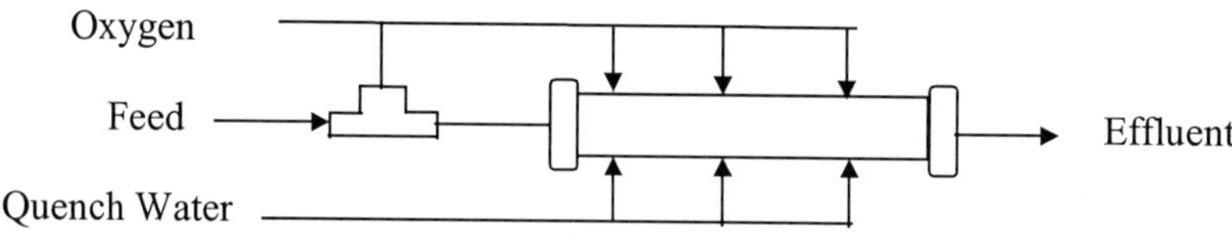

Figure 9.4- 6. Tubular Reactor

Transpiring wall reactor

Developed by Summit Research Corp. (Santa Fe, Nuevo Méjico, USA), the reactor is made with a porous Inconel interior wall. Through it, a wall "cleaning" fluid circulates, to avoid salts encrusting onto it. Part of the reactor effluent can be used as cleaning fluid at the end of the salts elimination. In this way salts deposition on the wall is avoided and there is a gain from feed preheating [14]. This reactor is sketched in Fig. 9.4-7.

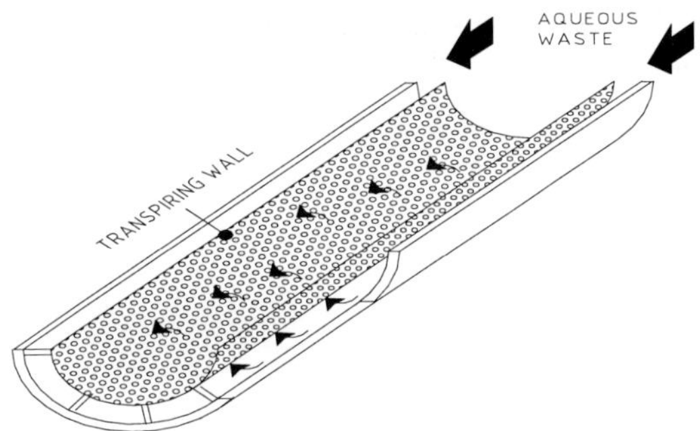

Figure 9.4-7. Transpiring Wall Reactor.

Cooled wall reactor

This was developed at the Chemical Engineering Department of Valladolid University, Spain (see Fig. 9.4-8). In this type of reactor the temperature and pressure effects are isolated. This is achieved by using a cooled wall vessel, which is maintained near 400 C, and a reaction chamber where the reactants are mixed and reaction takes place. This reaction chamber is made of a special material to withstand the oxidizing effect of the reactants at a maximum temperature of 800 °C and a pressure of 25 MPa. It is enclosed in the main vessel, which is pressurized with the feed-stream before entering the reaction chamber, so that it works at about 400 °C and does not suffer from the oxidizing atmosphere. It is made of relatively thin stainless steel [15].

9.4.4.2 Materials of construction

Corrosion

Since the most influential parameter for the choice of material is corrosion under operating conditions, we shall discuss this first.

The SCWO environment is favourable to corrosion. Extremes in pH, high concentrations of dissolved oxygen, ionic inorganic species, and high temperature and pressure variations may increase corrosion [16].

516

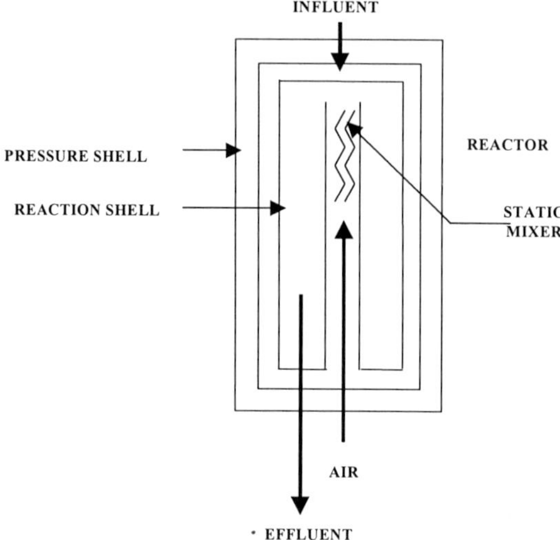

Figure 9.4-8. Cooled Wall Reactor.

In general, corrosion may be electrochemical or chemical in nature. Electrochemical corrosion takes place via the electrode reactions:

Anode   Me $\longleftrightarrow$ Me^{n+} + ne$^-$
Cathode   nH$^+$ + ne$^-$ $\longleftrightarrow$ n/2 H$_2$

The combination of these two separate reactions yields:

Me + nH$^+$ $\longleftrightarrow$ Me^{n+} + n/2 H$_2$

Taking the rate of electrochemical corrosion as:

$$\log(R/R^0) = \log(m_{H+}/m_{H+}{}^0) + 2 \cdot \log(\rho/\rho^0) + (-Ea/RT + Ea^0/RT) \cdot 1/2.303 \qquad (9.4\text{-}1)$$

The corrosion rate increases nearly up to the critical point, going downwards at higher temperatures. The reason is that water's dielectric constant decreases to around 2, so that dissociated ionic species decrease, the H$^+$ concentration being very low. So electrochemical corrosion will be important in the zone about 300-390 °C, that is in heat exchangers but not in the reactor [17].

Chemical corrosion takes place in the presence of dry gases, such as HCl and Cl$_2$ or water-free organic liquids via radical reactions. The presence of these species in water is increased enormously under supercritical conditions as the temperature increases, owing to a reduction in the dielectric constant. As the concentration increases, the corrosion rate increases, so chemical corrosion will be important in the reactor.

Taking these corrosion processes into account, it would be interesting to operate under supercritical conditions, in order to avoid species dissociation, but at a temperature near to the critical point, to avoid chemical corrosion.

Typical ways in which corrosion can be presented in SCWO conditions [18] are:

• Uniform or general attack.

Attack proceeds at about the same rate over the entire surface of the metal because there are no clearly defined cathode and anode surfaces.

• Pitting

Pitting is a form of localized corrosion that may perforate the metal. There is much debate about the initiation of pits but, once started, the anode is situated in the pit while the cathode is usually on the surrounding surface.

• Crevice corrosion

This is a form of localized corrosion that takes place in or around crevices or microscopic pockets.

• Intergranular corrosion

This is an attack of the metal at, or adjacent to, grain-boundaries, with relatively little corrosion of the grains: it causes the metal to lose strength or even disintegrate. Stainless steels are susceptible to this form of corrosion. Chromium carbide precipitates at grain-boundaries in the temperature range of 600 to 750° C during heat-treatment or welding, sensitizing the metal. Intergranular corrosion may be minimized by selecting low-carbon metals, or metals with titanium or niobium additions that bind the carbon.

• Stress corrosion cracking (SCC)

The combination of residual or applied tensile stress above a threshold value in the metal, the solution's composition, temperature, metal composition, and metal structure all affect this brittle-fracture-type of cracking. Using steels with a high content of nickel can reduce its appearance.

These are most frequent kinds of corrosion, although erosion corrosion and galvanic corrosion can be encountered.

Alloys exposed to SCWO

Most widely used materials for SCWO equipment construction are:

-Stainless steel, such as 316

-Alloys having a high content of Ni, such as Hastelloy C-276, Inconel 625 or Monel.

-Titanium (grade 2, 9, 12, etc.)

-Ceramic materials (TiO_2, ZrO_2, etc.)

The most influential parameters affecting the corrosion of materials are considered to be the pH of the solution, salts, chloride, and the operating temperature.

• Stainless steel

Stainless steels were the first materials to be tested for possible use as reactor-material for SCWO facilities, because of their corrosion resistance at "normal" temperatures and their relatively low costs. Stainless steel can be used at T<400 °C provided that the feed be free of chloride, or that it is present at low concentration [18]. Acidic pH is known to accelerate corrosion, and at basic pH it suffers "stress-corrosion cracking".

518

- Nickel-based alloys: Hastelloy and Inconel
Nickel-based alloys withstand chlorides in the feed better than does stainless steel but under extreme conditions, pitting, SCC and crevice corrosion is present [19, 20, 21]. The corrosion takes place in non-annealed structures in the as welded condition. Performance of these alloys may be enhanced by post-weld solution annealing [22].
- Monel
Studies performed so far using Monel (nickel-copper alloys) have shown its low corrosion-resistance due to problems associated to selective leaching in the SCWO processes. Normally it is not to be considered in the materials' selection.
- Titanium
Titanium is specially interesting for SCWO equipment construction because of its ability to become "passive", that is, non-reactive, after developing a superficial film of titanium dioxide. Titanium dioxide is refractory and strongly resistant to acids and attack by strong oxidants: it is only destroyed in the presence of anhydrous reducing agents (the presence of water is enough to regenerate the oxide film) and under strongly basic conditions [23, 24]. Since titanium withstands acidic media better than basic ones, it is not advisable to neutralize them when titanium is used in the equipment's construction. When hydrogen peroxide is present, the corrosion rate increases owing to titanium peroxides' formation.
Titanium's drawback is that exhibits poor mechanical resistance, but it is adequate for use as a coating or as a liner, or in reactors provided with pressurized shells.
- Ceramic materials
Silicon-based materials (silicon carbide, silicon nitride) usually form a superficial film of silicon dioxide that protects the material against corrosion in acidic or neutral media. At pH equal to or higher than eleven it suffers corrosion heavily because the silicon dioxide is dissolved [25].
Alumina will suffer intergranular corrosion at pH near to 8, losing some material, whereas at pH above 11 it is dissolved [26].
Zirconium dioxide suffers a phase-transformation at 250°C that produces stress in the material and eventually cracking [26].

9.4.4.3 Solids Separation

Owing to reduced salt solubility, the formation of metal oxides and, eventually, the presence of stable solid-matter particles, these are all present in the SCWO processes. These particles can cause equipment-fouling and erosion. However the reduced solubility of salts under supercritical conditions introduces the possibility of a solid fluid separation.

When dealing with precipitated salts it is important to know the physical state in which the salt is present inside the reactor, either solid or liquid. For example, sodium nitrate has melting point 306.8 °C, lower than the operating temperature. Therefore, once saturating conditions are surpassed, sodium nitrate does not adhere to the reactor walls as a solid but drains itself downwards. Since it is denser than supercritical water [27], care should be taken to prevent its accumulation at the bottom of the reactor.

It is also good practice to have each salt's phase-diagram, in order to know how they would precipitate under supercritical conditions. For example, sodium chloride yields crystals that are larger than sodium sulphate [28], which facilitates its separation by filtration or using cyclones. When a sodium chloride solution is heated, it reaches an L/V equilibrium zone, where water-evaporation takes place, and the salt's concentration in the liquid drops, thus producing formation of larger crystals (10 to100 μm). In contrast, sodium sulphate reaches saturation before entering the L/V equilibrium zone. Above the solubility-limit a rapid nucleation is produced and solution crystallisation produces small crystals (1 to 3μm). The small-size crystals will create additional difficulties in separation processes.

Several other non salt-specific factors, such as pressure and temperature, influence crystallization. In SCWO, the solubility reduces as the temperature increases, owing to the reduction in dielectric constant. For example, sodium chloride's solubility in supercritical water is 824 mg/L at 400°C while at 500 °C it is only 299 mg/L [28]. Pressure increase produces an increase in dielectric constant, and it is not infrequent for the precipitated salt to redissolve itself. For example, Foy *et al.* [24] oxidized chloro-compounds at temperatures around 550°C and 600 bar: in these conditions, the dielectric constant is sufficiently high to avoid chloride deposition.

The "common ion effect" must also be considered, since it affects each one of the crystallization equilibrium, both in supercritical and subcritical water. Adding sodium acetate, for instance, to a solution containing sodium salts would favour sodium bicarbonate precipitation (formed from oxidation of acetate), thus avoiding the precipitation of more corrosive salts, such as the chloride or sulfate [28].

Solids-separation devices
- Hydrocyclone
  This is one of the simplest types of solids separators. It is a high-efficiency separation device and can be used to effectively remove solids at high temperatures and pressures. It is economical because it has no moving parts and requires little maintenance.
  The separation efficiency for solids is a strong function of the particle-size and temperature. Gross separation efficiencies near 80% are achievable for silica and temperatures above 300°C, while in the same temperature range, gross separation efficiencies for denser zircon particles are greater than 99% [29].
  The main handicap of hydrocyclone operation is the tendency of some salts to adhere to the cyclone walls.
- Cross micro-filtration
  Cross-flow filters behave in a way similar to that normally observed in crossflow filtration under ambient conditions: increased shear-rates and reduced fluid-viscosity result in an increased filtrate number. Cross-microfiltration has been applied to the separation of precipitated salts as solids, giving particle-separation efficiencies typically exceeding 99.9%. Goemans *et al.* [30] studied sodium nitrate separation from supercritical water. Under the conditions of the study, sodium nitrate was present as the molten salt and was capable of crossing the filter. Separation efficiencies were obtained that varied with temperature, since the solubility decreases as the temperature increases, ranging between 40% and 85%, for 400 °C and 470°C, respectively. These workers explained the separation mechanism as a consequence of a distinct permeability of the filtering medium towards the supercritical solution, as opposed to the molten salt, based on their clearly distinct viscosities.
  Therefore, it would be possible not only to filter precipitated salts merely as solids but also to filter those low-melting-point salts that are in a molten state.
  The operating troubles were mainly due to filter-corrosion by the salts.

9.4.4.4 Heat exchanger

Heat exchange from reactor effluent to the pressurised feed is the major feature that increases the overall thermal efficiency of SCWO over other oxidation processes for

520

treatment of diluted aqueous waste. The driving force for the heat exchange is the temperature differential created by oxidation of waste in the reactor. When dealing with a feed of low heating-value, either additional fuel could be added, or the heat-exchange area increased, or both. Minimization of the heat-transfer area by operating with the maximum temperature rise is usually preferred.

Shell- and tube-heat exchangers are the most widely used for high-pressure processes, bearing in mind the fact that the shells should be designed as pressure vessels. An alternative to shell- and tube-exchangers than is worth considered is the spiral heat exchanger placed inside a pressure vessel, which can be pressurised by the fluid circulating through the outer spiral of the exchanger. This type of exchanger would be specially indicated for solid-containing feeds.

Heat-exchangers are exposed to thermal and mechanical pressure load cycling and are commonly the site of the most aggressive localised corrosion in the system. So materials of construction should be chosen for high corrosion conditions.

Heat-exchanger fouling, or scale build-up, can be a major factor limiting the kind and concentration of solids in feed that can be processed. The plugging problem can be dealt with by periodically flushing the exchanger with warm water or diluted nitric acid, as in conventional processes.

9.4.5 Other Oxidation Processes of Waste Water at High Pressure

9.4.5.1 Wet air oxidation.

This name is given to an aqueous-phase oxidation process brought about when an organic and/or oxidizable inorganic-containing liquid is mixed thoroughly with a gaseous source of oxygen (usually air) at temperatures of 150 to 325 °C. Pressures between 1.5 and 14 MPa are maintained to promote the reaction and to control evaporation. A basic flow diagram is shown in Figure 9.4- 9.

Despite the fact that the destruction of toxic organic chemicals as fuels can be as high as 99.9%, with adequate residence time, many materials are more resistant (e.g., cholorobenzenes). Total COD reduction is usually only 75-95% or lower, indicating that while the toxic compounds may satisfactorily undergo destruction, certain intermediate products remain unoxidized. Because the wet-air oxidation is not complete, the effluent from the process can contain appreciable concentrations of volatile organics and may require additional treatment such as biological oxidation or adsorption on activated carbon [31].

The volatile organic components that are emitted in the gaseous effluent can be controlled by a variety of technologies including scrubbing techniques, granular-carbon adsorption and fume incineration. The specific technology is selected on a case-by-case basis.

Wet-oxidation has proved to be effective in treating a variety of hazardous wastes, and actually is an operating technology for a variety of industries in the USA and Europe [32].

9.4.5.2 Deep-shaft wet–air oxidation

A new process configuration, that for deep-shaft wet-air oxidation, is available as a more satisfactory and economical engineering approach to the application of the wet-oxidation concept (see Figure 9.4-10). The deep-shaft wet-air oxidation reactor is unique to its purpose. The tube-diameter and length are designed so that sufficient reaction time and the desired pressure can be attained during fluid-waste oxidation. The pressure is developed naturally by

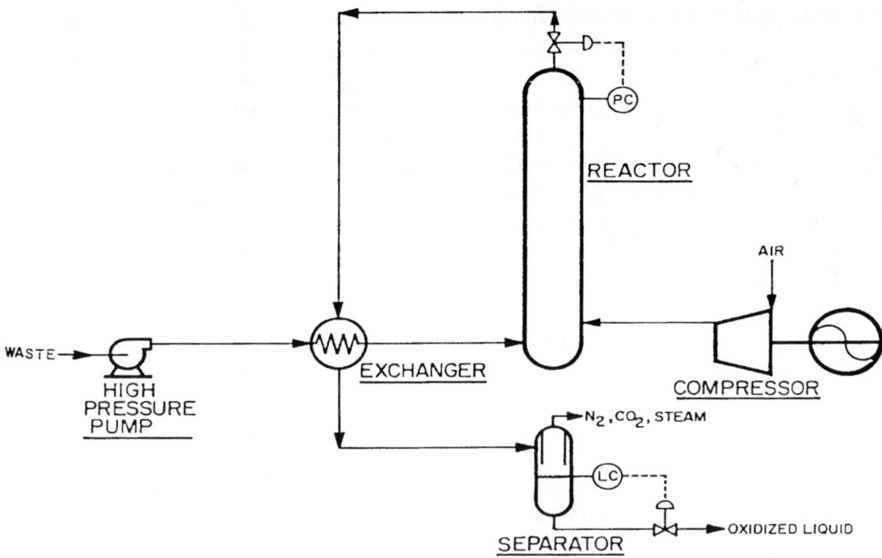

Figure 9.4-9. Wet-air oxidation.

the hydrostatic liquid-head above the waste flowing down the tube. Heat resulting from the exothermic combustion reaction maintains the required down-hole temperature, sustaining the reaction [34]. Deep-shaft wet-air oxidation can be classified as subcritical or supercritical, depending on the maximum operating reactor-temperatures, being below or above the critical point of water.

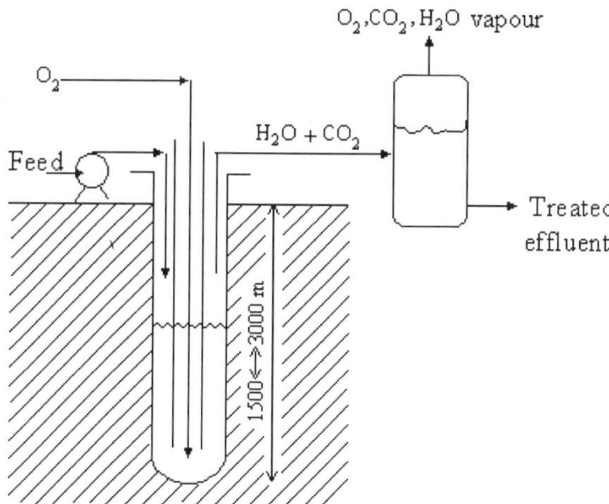

Figure 9.4-10. Deep-shaft wet-air oxidation.

522

9.4.6 SCWO Application to Wastewater Treatment

9.4.6.1 Pilot plant operations

When studying the technical and economic viability of this technology it is necessary to first study waste degradation in a pilot plant, in order to define the operating conditions valid on an industrial scale. Typical features of a pilot plant are presented in Figure 9.4-10. This scheme corresponds to the pilot plant that the Department of Chemical Engineering of Valladolid's University designed and manned.

The pilot plant has a throughput of about 25 kg water/h, and a basic flow-diagram is shown in Figure 9.4-11.

The main characteristics of the plant are:

- Volume of reactor, 15 L (filled with 4 mm diameter alumina spheres, and therefore the effective volume is 9 L);
- Oxidant, Air;
- Reactor adiabatic.

The feed-stream is pressurized to the operation pressure by a metering pump, and the air which is used as oxidant in the oxidation reaction, is compressed to the operational pressure in a four-stage compressor. Both streams are mixed in a static mixer inside the reaction chamber as it is shown in Fig. 9.4-10. The reactor has been described in the section 9.4.4.1. A more detailed description of the pilot plant can be found in reference [7].

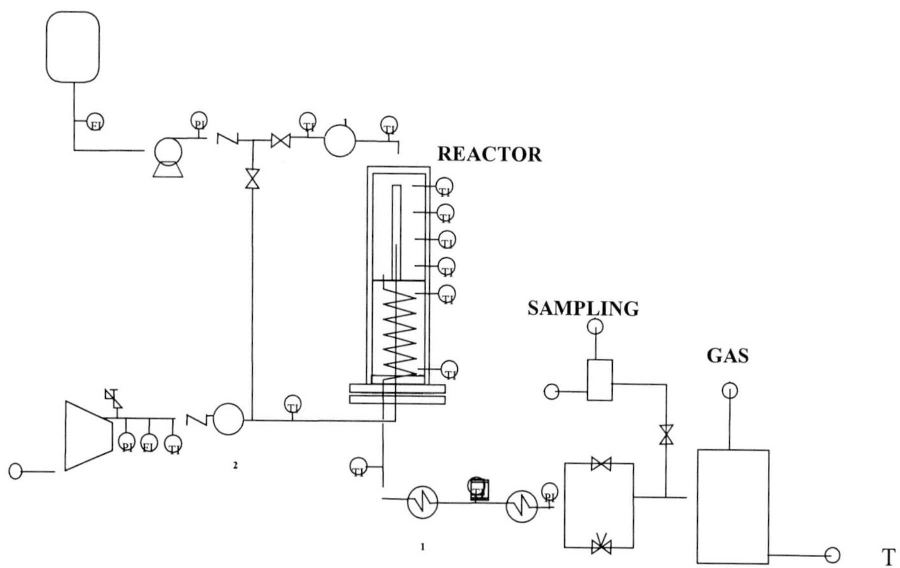

Figure 9.4-11. Basic Flow Diagram of the SCWO pilot plant.

9.4.6.2 Example of operation with an industrial waste: cutting-oil waste.
Cutting-oil is a hazardous and poorly biodegradable waste, composed of two immiscible phases, an aqueous phase that represents ¾ of the total volume, and an oily phase. The waste was diluted with a solution of iso-propanol in water, in order to give a more homogenous feed, and to fix the enthalpy of the waste in the range $4 \cdot 10^3$ KJ/Kg.

Operating conditions:
Waste flow-rate: 15 L/h   Reaction temperature: 625 °C   Residence time: 60 s
Air flow-rate: 20 kg/h   Reaction pressure: 25 MPa
Average results of analysis of feed samples and liquid effluent are shown in Table 9.4-1.

Table 9.4-1.
Analysis of feed samples and liquid effluent.

Sample	TOC(mg/L)	IC(mg/L)	T.S. (mg/L)	Oil/grease (mg/L)	Cl⁻ (mg/L)
Feed	193,125	-	887	20,296	50
Effluent	15	41	132	32	55

In terms of total organic carbon, a removal efficacy of 99.9% was found. From the above results we conclude that SCWO is a suitable process for the treatment of cutting-oils.

9.4.6.3 Commercial process
This process was developed into industrial pilot plants by the pioneering work of MODAR and MODEC, which are integrated into General Atomic. Several plants are in operation in the USA for the treatment of military wastes, supported by the Department of Defence, processing between 30-900 kg/h of waste [14]. A commercial supercritical oxidation plant developed by Ecowaste Technologies at Huntsman's Chemical Plant in Austin (Texas, U.S.A.) has been in operation since the summer of 1994, processing 1,500 kg /h of waste (mainly alcohols and amines). Chematur Engineering (Sweden) has developed the process Aqua Critox, whose diagram is presented in Figure 9.4-12, which corresponds to the plant working at Karlskoga (Sweden), for 250 kg/h-feeds.
The plant is built to be very flexible and to have the possibility to handle different wastes, including sludges and dispersed solids. Chematur has a licence on Japan, Shinko-Pantec Kobe. This company has built a large plant, 1.3 m³/h, to treat of sewage sludge.

9.4.6.4 Economic Features
Variable costs for SCWO are influenced by the throughput and heating value of the waste. For wastes with a heating value of 3,800 KJ/Kg the feed can be introduced, without preheating, directly into the reactor. This approach would provide an effluent at 650°C, which can be used to produce electricity. For diluted aqueous wastes a heat exchanger is used in order to preheat the feed and cool the reactor effluent (Figure 9.4-4). The heating value of the feed would provide for heat losses as well as for differences across the heat exchanger. The lower the heating value of the feed, the lower is the temperature differences across the heat exchanger and the larger the heat exchanger. For example, for a reactor temperature of 550 °C it is possible to work in a thermally self-sufficient reactor with a heating value of 530 KJ/kg (which is equivalent to a 1.7% wt/wt phenol concentration) heating the feed up to 400°C, leaving an effluent at 150°C [34].

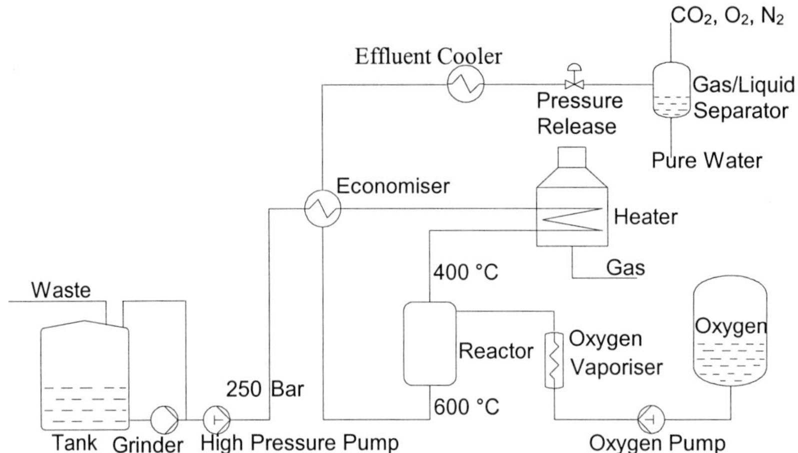

Figure 9.4-12. Basic flow diagram of the SCWO pilot plant.

The treatment cost is also heavily dependent on the capital cost. Chematur has evaluated the treatment-cost for the Aqua Critox process as between 36 $/Ton, for fine chemicals wastewater (3.6 MT/h), and 96 $/Ton, for industrial sludges (1 MT/h). For de- inking sludge the treatment cost could be as low as 9 $/Ton after the income from recovered paper filler is deducted.

References

1. R.W. Shaw, A.A. Brill, C.A. Clifford, E.U. Eckert, Supercritical Water. A Medium for Chemistry, Chemical Engineering News. Dec 23, (1991),26.
2. M. Vematsu and E.U. Franck, Static Dielectric Constant of Water and Steam, J. Phys. Chem., Ref. Data 9, (1980), 1291.
3. W. L. Marshall and E. U. Franck, Ion Product of Water Substance, 0°C to 1000°C, 1 bar to 10000 bar, New International Formulation and its Background, J. Phys. Chem. Ref. Data., 10, (1981), 295.
4. K.S. Pitzer and R.T. Pabalan, Thermodynamics of NaCl in Steam, Geochem. Cosmochem., Acta 50, (1986), 1445.
5. M.L. Japas and E.U. Franck, High Pressure Phase Equilibria and PVT-Data of the Water-oxygen System including Water-air to 673 K and 25.0 MPa, Ber. Bunsenges. Phys. Chem., 89, (1985), 1268.
6. P.E. Savage, S. Gopalan, T.J. Mizan, J. Martino, E.E. Brock, Reactions at Supercritical Conditions: Applications and Fundamentals, AIChE Journal, 41, (1995), 1723.
7. M. J. Cocero, E. Alonso, D. Vallelado, R. Torío, F. Fdz-Polanco, Supercritical Water Oxidation in Pilot Plant of Nitrogenous Compounds-Isopropanol Mixtures in the Temperature Range 500-750°C, Ind. Eng. Chem. Res., 39, (2000), 3707.
8. Z. Ding, M.A. Frisch, L. Li, E.F. Gloyna, Catalytic Oxidation in Supercritical Water, Ind. Eng. Chem. Res., 35, (1996), 3257.

9. V. Mishra, V. Mahajani, J.B. Joshi, Wet Air Oxidation, Ind. Eng. Chem. Res., 34, (1995), 2.

10. M. J. Cocero , E. Alonso, D. Vallelado, R. Torío, F. Fdz-Polanco, Optimization of Operational Variables of a Supercritical Water Oxidation (SCWO) Process, Water Sci. Tech. 42, (2000), 107.

11. M. J. Cocero, E. Alonso, D. Vallelado, R. Torío, F. Fdz-Polanco, Supercritical Water Oxidation in Pilot Plant with Energetically Self-sufficient Reactor, ECCE 2, Montpellier, (1999).

12. J.F. Brennecke, New Applications of Supercritical Fluids, Chemistry & Industry, Nov 4, (1996), 831.

13. Ch.Y. Huang, H. Barner, J. Albano, W. Killilea, G. Hang, *Patent WO92/21621 EEUU.* (1992).

14. R.W. Shaw and N. Dahmen, Destruction of Toxic Organic Materials using Supercritical Water, Supercritical Fluid Fundamentals and Applications, E. Kiran Ed., 425 NATO ASI, Blackie Academic (2000).

15. M.J. Cocero, J.L. Soria, F. Fdez-Polanco, Behaviour of a Cooled Wall Reactor for Supercritical Water, High Pressure Chemical Engineering, Ph. Rudolf Ch Trepp Eds. 121, Elsevier (1996).

16. L.B. Kriksunov and A.A. Mc Donald, Corrosion in Supercritical Water Oxidation Systems: a Phenomenological Analysis, J. Electrochemical Society, 142, (1995), 4069.

17. P. Kritzer, N. Bonkis, E. Dinjus, The Corrosion of Alloy 625 in High-Temperature, High-Pressure Sulfate Solutions. 54, (1998), 689.

18. T .A. Danielson, Corrosion of Selected Alloys in Sub-and Supercritical Water Oxidation Environments, Thesis, University of Texas USA, (1995).

19. D. K. Russell, Pick the Right Materials for Wet Oxidation, Chem. Eng. Progress, March (1999), 51.

20. P. Kritzer, N. Bonkis, E. Dinjus, Corrosion of Alloy 625 in Aqueous Solutions Containing Chloride and Oxygen. Corrosion 54, (1998), 824.

21. P. Kritzer, N. Bonkis, E. Dinjus, The Corrosion of Alloy 625 (NiCr22Mo9Nb;2.4856) in High-Temperature, High-Pressure Aqueous Solutions of Phosphonic Acid and Oxygen, Materials and Corrosion, 49, (1998), 1.

22. J. L. Soria, Oxidación en agua supercrítica diseño y operación a escala piloto, PhD, Thesis, Universidad de Valladolid, Spain, (1998).

23. G.L. D'Arcy, Corrosion Behaviour of Ti Alloys Exposed to Supercritical Water Oxidation Conditions, Thesis, University of Texas USA, (1995).

24. B.R. Foy, K. Waldthausen, M.A. Sedillo, S. Buellow, Hydrothermal Processing of Chlorinated Hydrocarbons in a Titanium Reactor, Env. Sci., Technol. 30, (1996), 2790.

25. R.J. Morin, Ceramics for Corrosion Resistance in Supercritical Water Environments, Thesis, University of Texas USA, (1993).

26. N. Bonkis, N. Clanssen, K. Ebert, R. Janssen, M. Schacht, Corrosion Screening Tests of High-Performance Ceramics in Supercritical Water Containing Oxygen and Hydrochloric Acid. J. of European Ceramic Society 17, (1997), 71.

27. P.C. Dell'Orco, E.F. Gloyna, S. Buelow, Oxidation Processes in the Separation of Solids from Supercritical Water, Supercritical Fluids Engineering Science, Fundamentals and applications, ACS Symposium Series, 514, (1993), 314.

28. F.J. Armellini and J.W. Tester, Precipitation of Sodium Chloride and Sodium Sulfate in Water from Sub-to Supercritical Conditions: 150 to 550ºC, 100 to 300 bar, J. Supercrit. Fluids, 7, (1994), 147.
29. P.C. Dell'Orco, L. Li, E.F. Gloyna., The Separation of Particles from Supercritical Water Oxidation Processes, Separation Science and Technology, 28, (1993), 625.
30. M.G.E. Goemans, F. Tiller, L. Li, E.F. Gloyna, The Separation of Inorganic Salts from Supercritical Water by Cross-flow Microfiltration, J. Membrane Science, 124, (1997) 129.
31. W.M. Copa and W.B. Gitchel, Wet Oxidation, Standard Handbook of Hazardous Waste Treatment and Disposal, H.M. Freeman Ed., 8.77-8.90, Mc Graw Hill (1988).
32. A.R. Willhelmi and P.V. Knopp, Wet Air Oxidation. An Alternative to Incineration, AIChE Journal (1979), 46.
33. J.M Smith, Deep-Shaft Wet-air Oxidation, Standard Handbook of Hazardous Waste Treatment and Disposal, H.M. Freeman Ed., 8.137, Mc Graw Hill (1988).
34. M. J. Cocero, T. Sanz., F. Fdez-Polanco, Study of Alternatives for the Design of a Mobile Unit for Wastewater Treatment by Supercritical Water Oxidation, J. Tech. Biotech. (in press).

9.5 High-pressure polymerization with metallocene catalysts

G. Luft

Department of Chemistry, Darmstadt University of Technology
Petersenstr. 20, D-64287 Darmstadt, Germany

In recent years metallocene catalysts have been introduced into low-pressure gas-phase-, solution-, and slurry-processes to manufacture polyethylene and polypropylene. The new technology extends not only the range of conventional materials but generates new speciality polymers. Some companies have also retro-fitted high-pressure reactors to make use of the advantages of metallocene catalysts.

9.5.1 Advantages of high-pressure polymerization with metallocenes

The advantages of the new process result from the excellent properties of metallocene-based polyethylene (mPE). One of the main features of mPE is the uniform molecular-weight distribution (Fig. 9.5-1, left, curve b). It is generated by the *single-site* metallocene catalyst. In contrast, polyethylenes from the Ziegler-Natta processes show a broad molecular-weight distribution (Fig. 9.5-1, left, curve a). Ziegler-Natta catalysts exhibit sites of different activity which contribute to the broad range of molecular weights. These catalysts are known as *multi-site* catalysts.

Fig. 9.5-1, right, compares the co-monomer incorporated into the polymer, on a molecular basis. With the Ziegler-Natta, multi-site catalysts less monomer is incorporated in the high-molecular-weight fraction. The low-molecular-weight fraction is rich in co-monomer content as shown by the negative slope of curve a. In a metallocene-based polyethylene the co-monomer is uniformly distributed (curve b).

The narrow molecular-weight distribution and the uniform incorporation of co-monomers lead to improved product properties such as high impact strength, transparency, and heat seal, together with less stickiness and blocking of films. Furthermore, mPE exhibits a smaller fraction of extractables. On the other hand, mPE suffers from a more difficult processibility, low melt strength, and higher melt fracture. A main deficiency is the inability of metallocene

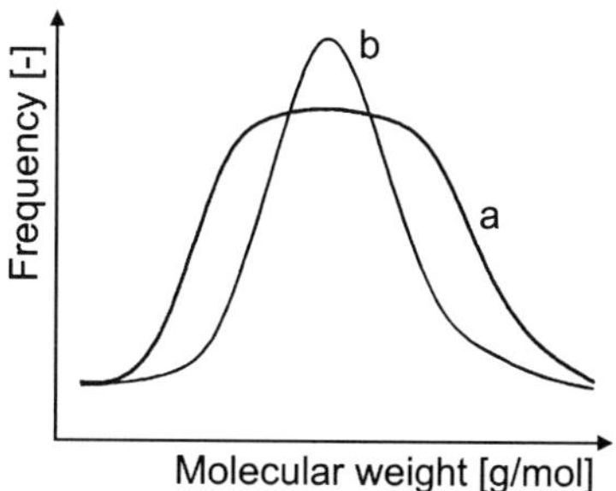

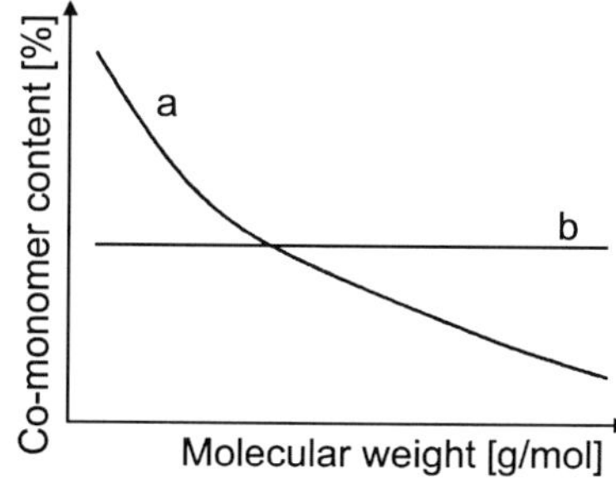

Figure 9.5-1. Molecular-weight distribution and incorporation of co-monomers. a, Ziegler-Natta catalyst; b, metallocene catalyst.

528

catalysts to incorporate polar co-monomers and co-monomers containing heteroatoms such as vinyl acetate and methyl acrylate.

The advantages known from the production of low-density polyethylene (LDPE) become obvious also when metallocene catalysts are used under high-pressure conditions. The compressed monomer can dissolve the polymer which is formed during polymerization, which means that no additional solvent is required for the polymer. The high-pressure polymerization proceeds with a high rate, which requires a short residence time and small reactor volume. Established technology, with stirred autoclaves as well as tubular reactors, can be applied.

9.5.2 Catalyst and co-catalyst

The structure of metallocene catalysts developed for use in polymerization under low- and also high-pressure is shown in Fig. 9.5-2 together with the co-catalyst. Here, M is a transition metal of group IVb. The cyclopentadienyl rings can be bridged by linear or cyclic alkyl groups or a silyl group (R_2). The substituents R_1 are often n-, iso-, or cyclo-alkyls. The groups X are halogen atoms or alkyl, aryl, or benzyl. The cyclopentadienyl ring can also be part of an indenyl structure. Both rings must not be identical. Monocyclopentadienyl structures also exist in which one ring has been replaced by a heteroatom such as nitrogen, attached to a bridging atom. Examples of commercially available metallocene catalysts and laboratory products are shown in Fig. 9.5-3. These are an unbridged zirconocene (a) and bridged, so-called *ansa*-metallocenes, differing in stereochemistry (b, c). These catalysts were prepared for the first time by Brintzinger [1]. Whereas *meso*-Me$_2$Si(indenyl)$_2$ZrCl$_2$ (b) produces atactic polypropylene (PP), the chiral catalyst (c) with C_2-symmetry gives isotactic PP, and by use of achiral catalyst (d) having C_s-symmetry, syndiotactic PP is obtained in low-pressure polymerization. Syndiotactic PP could not be produced before the discovery of metallocenes. A patent application was recently filed by BASF claiming catalysts (e, f) to make mPE with a bimodal molecular-weight distribution in low- and high-pressure polymerization [2]. Complexes in which metals of group IVb, such as catalyst (g), or of the lanthanide series are bound to linked cyclopentadienyl-amido-ligands have beeen developed by Dow and Exxon [3].

The *ansa*-sandwich complexes, sometimes called constrained-geometry catalysts, exhibit a good ability to incorporate co-monomers. The catalysts (a) to (g) are soluble in solvents such as toluene. In the catalyst (h) the metallocene is supported on silica. Such solid-supported catalysts can be suspended in hydrocarbons for metering into the polymerization reactor [4].

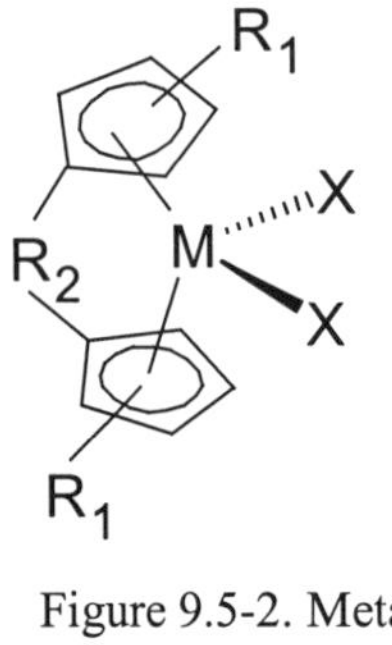

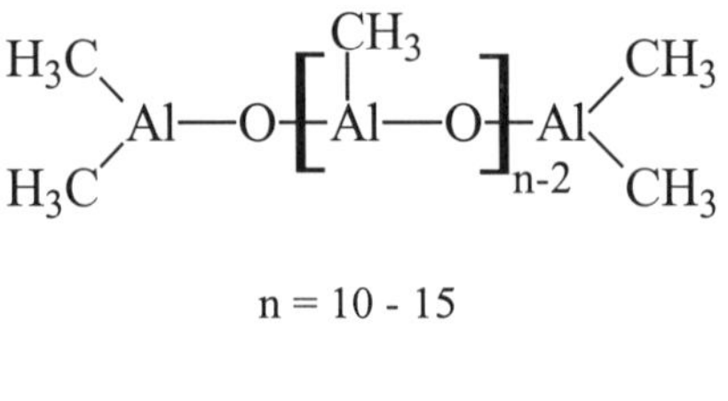

$$n = 10 - 15$$

Figure 9.5-2. Metallocene catalyst (left) and alumoxane co-catalyst (right).

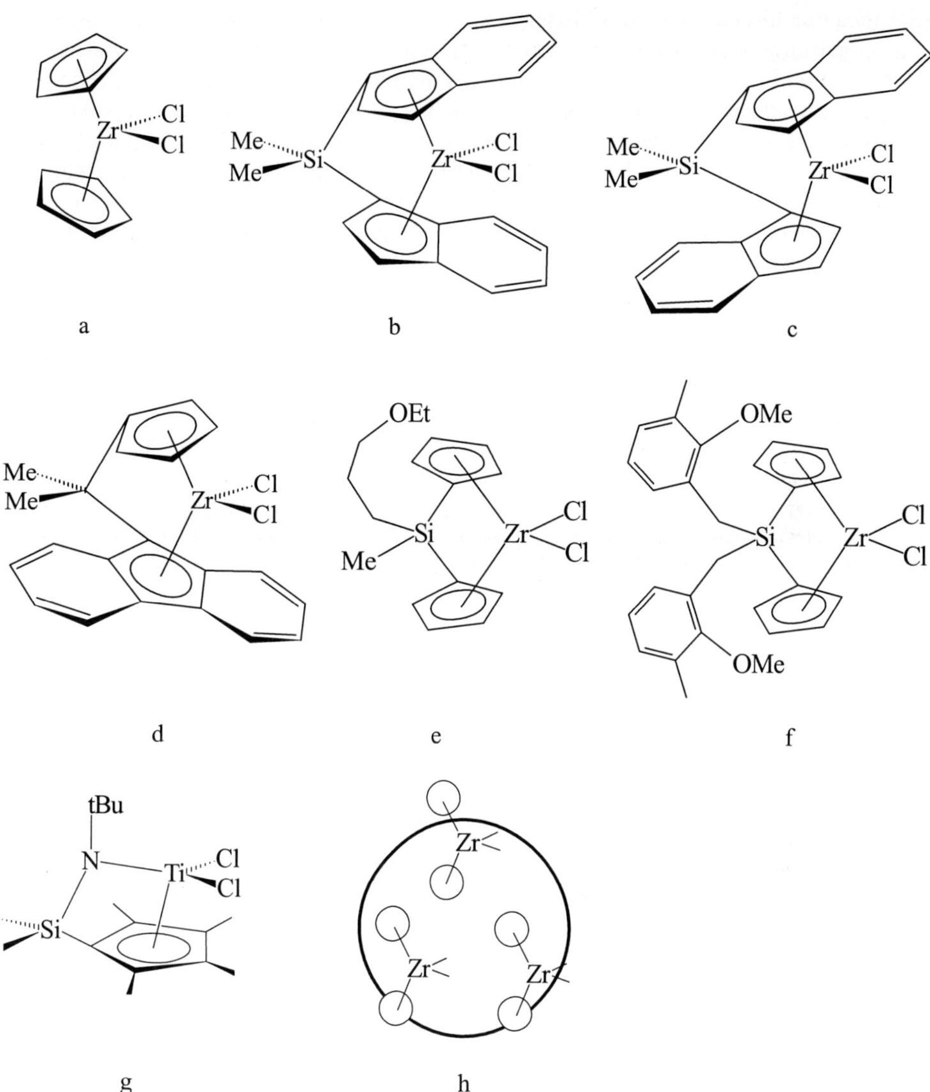

Figure 9.5-3. Different types of metallocene catalysts.

The metallocene catalysts must be first activated by an aluminoxane co-catalyst, e.g., tetramethylaluminoxane (MAO) which is an oligomer, n being 10 - 15 (Fig. 9.5-2, right). Because a high excess of MAO is required for activation, a binary co-catalyst system was developed consisting of an organoborate and tri-isobutylaluminium (TiBA). Organoborate can be used in a stoichiometric ratio which reduces the costs and the residue of activator products in the polymer [5].

530

9.5.3 Reaction mechanism and kinetics

It is generally adopted that the catalytically active species in the metallocene-catalysed polymerization is a 14-electron cation. As an example, the mechanism of activation of an unbridged zirconocene catalyst is presented in Fig. 9.5-4, top. In the first two steps the activation by MAO, resulting in the 14-electron cation, is shown. The same cation can be generated by N,N'-dimethylanilinium-tetrakis(pentafluorophenyl)borate and methylated metallocenes. As side-products methane and an amine are formed. TiBA can also be involved in the activation, which is not shown in Fig. 9.5-4, bottom. On the other hand, TiBA acts as a scavenger in the polymerization. The above-mentioned reactions take place in the absence of the monomer and are performed before the catalyst is used in the polymerization process.

Activation by MAO

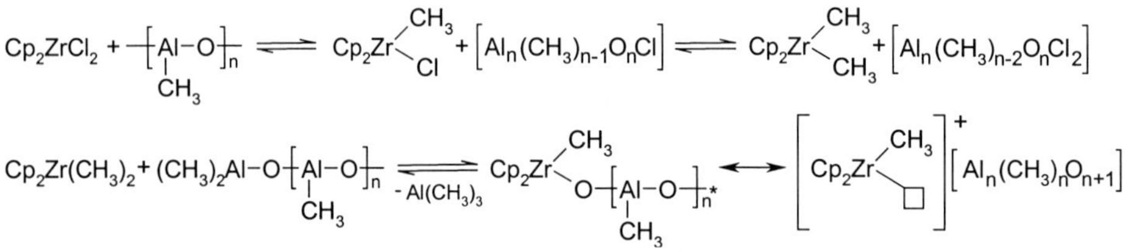

$$Cp_2ZrCl_2 + \left[Al\text{-}O\right]_n \rightleftharpoons Cp_2Zr\overset{CH_3}{\underset{Cl}{<}} + \left[Al_n(CH_3)_{n\text{-}1}O_nCl\right] \rightleftharpoons Cp_2Zr\overset{CH_3}{\underset{CH_3}{<}} + \left[Al_n(CH_3)_{n\text{-}2}O_nCl_2\right]$$

$$Cp_2Zr(CH_3)_2 + (CH_3)_2Al\text{-}O\left[Al\text{-}O\right]_n \underset{-Al(CH_3)_3}{\rightleftharpoons} Cp_2Zr\overset{CH_3}{\underset{O\left[Al\text{-}O\right]_{n^*}}{<}} \longleftrightarrow \left[Cp_2Zr\overset{CH_3}{\underset{\square}{<}}\right]^{+} \left[Al_n(CH_3)_nO_{n+1}\right]^{-}$$

Activation by organoborate/TiBA

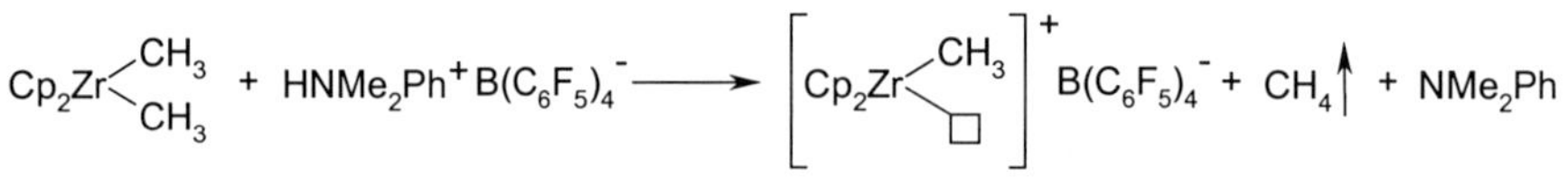

$$Cp_2Zr\overset{CH_3}{\underset{CH_3}{<}} + HNMe_2Ph^+B(C_6F_5)_4^- \longrightarrow \left[Cp_2Zr\overset{CH_3}{\underset{\square}{<}}\right]^{+} B(C_6F_5)_4^- + CH_4\uparrow + NMe_2Ph$$

Figure 9.5-4. Mechanism of activation of metallocene catalyst.

Chain propagation

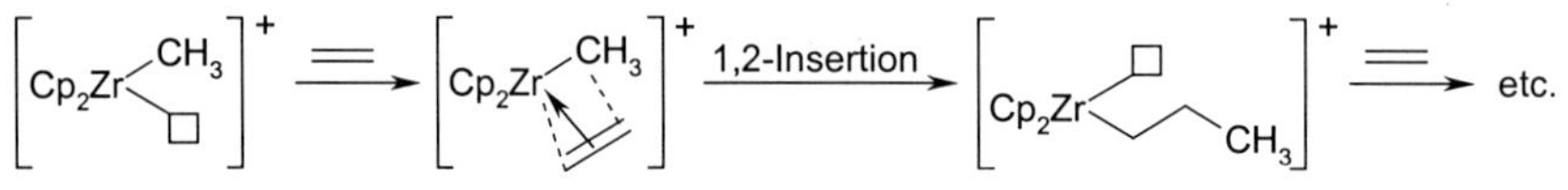

$$\left[Cp_2Zr\overset{CH_3}{\underset{\square}{<}}\right]^{+} \rightleftharpoons \left[Cp_2Zr\overset{CH_3}{<}\right]^{+} \xrightarrow{1,2\text{-Insertion}} \left[Cp_2Zr\overset{\square}{\underset{CH_3}{<}}\right]^{+} \rightleftharpoons etc.$$

Chain termination

$$\left[Cp_2Zr\overset{\square\quad H}{\underset{CH\text{-}R}{<}}\right]^{+} \longrightarrow \left[Cp_2Zr\overset{H}{\underset{\square}{<}}\right]^{+} + H_2C=CH\text{-}R$$

$$\left[Cp_2Zr\overset{\quad H}{\underset{CH\text{-}R}{<}}\right]^{+} \longrightarrow \left[Cp_2Zr\overset{CH_3}{\underset{\square}{<}}\right]^{+} + H_2C=CH\text{-}R$$

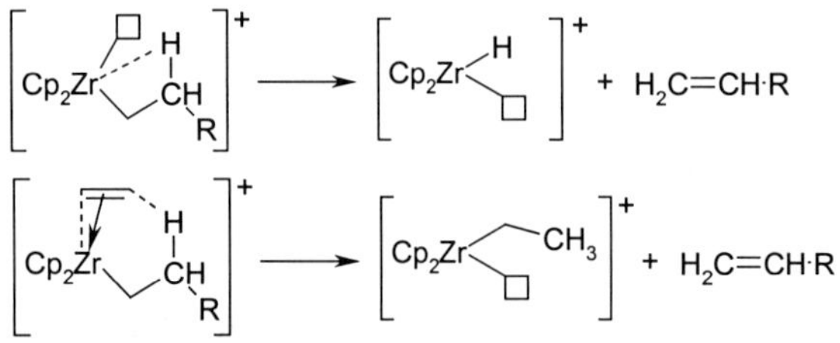

Figure 9.5-5. Mechanism of ethylene polymerization with metallocene catalyst.

In the polymerization reaction the monomer is coordinated at a vacant site of the metal, followed by insertion into the metal-carbon bond (Fig. 9.5-5). Finally, the chain-propagation is terminated by transfer of a hydrogen atom in a β-position to the metal or to the coordinated monomer.

From the mechanism presented in Fig. 9.5-5 and from results of polymerisation tests at high pressures the following rate equation was evaluated [6]:

$$r_{br} = k_{br} \cdot [\text{Ethylene}]^a \cdot [\text{Cat}^\#]^b \quad \text{with} \quad k_{br} = k_{br}^0 \exp(-[E_{br} + \Delta v_{br}^\# \cdot (p - p_0)]/R \cdot T) \qquad (9.5\text{-}1)$$

In all tests, rac-Me$_2$Si[Ind]$_2$ZrCl$_2$ activated by [Me$_2$PhNH]$^+$[B(C$_6$F$_5$)$_4$]$^-$/TiBA with molar ratios of [Al]/[Zr] = 200 and [B]/[Zr] = 1.3 was used. The overall rate, r_{br}, of high-pressure polymerization of ethylene depends on the concentration, [Ethylene], of the monomer and on that of the active species which is assumed to be proportional to the concentration, [Cat$^\#$], of the catalyst. The overall rate constant, k_{br}, consists of the rate constants of chain-propagation and termination. Its dependence on the temperature, T, is described by the activation energy, E_{br}. The influence of the pressure, p, is taken into account by the activation volume, $\Delta v_{br}^\#$, see Chapter 3.2.

From polymerization tests under the high pressure of 150 MPa, and varying concentrations of ethylene, Rau, Schmitz, and Luft [6] found that chain-propagation is the rate-limiting step and depends linearly on the concentration of ethylene (a = 1). This is different from polymerization at low pressure where an order of 1.2 - 2 was observed [7]. In agreement with results from low-pressure polymerization of ethylene, the overall rate at high pressure also depends linearly on the concentration of metallocene catalyst (b = 1).

At the high temperatures of the high-pressure process, T is 140 - 210°C, the rate of polymerization reduces with increasing temperature because of the deactivation of the

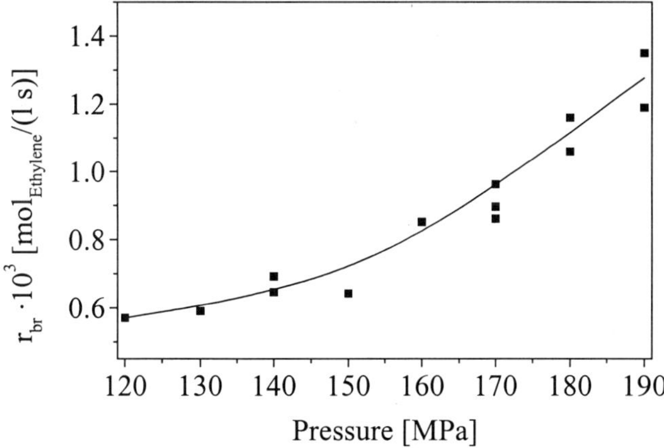

Figure 9.5-6. Influence of pressure on the rate of polymerization. Temperature, 160°C; Zr in feed, 0.012 mol ppm; Al/Zr, 5,000 mol/mol; residence-time, 240 s; n-hexane in feed, 10 mol.%.

532

catalyst. An activation energy of E_{br} = -73 kJ/mol was determined, which means that the influence of the temperature is governed by the deactivation of the catalyst.

In order to investigate the influence of the pressure, polymerization tests were run at pressures of 120 - 190 MPa. As can be seen from Fig. 9.5-6, the rate of polymerization increases from (0.58 to 1.3) $\cdot 10^{-3}$ mol ethylene/(l s). When r_{br} is plotted on a logarithmic scale versus the pressure, a value of activation volume of -32.5 ml/mol can be evaluated from the slope of the resulting straight line. The negative value is characteristic for polymerization reactions because the volume reduces in the transition state (see Chapter 3.2).

9.5.4 Productivity

In order to estimate the quantity of catalyst which is required to produce a given amount of polyethylene, the productivity can be considered. The productivity of the high-pressure polymerization of ethylene with a zirconocene catalysts is in the range of 400 - 6,000 kg PE/g Zr. It depends steeply on the ratio of co-catalyst to catalyst, as well as on pressure and temperature.

Independently of the co-catalyst, the productivity is only low when the ratio of Al to Zr is small. The productivity increases steeply with an increase in this ratio, to reach a constant level. A further increase of Al/Zr does not further increase the productivity. When organoborate is the co-catalyst, less aluminium compound is required to reach a high productivity, as with MAO.

When the pressure is moderately high, e.g., 120 MPa the productivity is 2,350 kg PE/g Zr at a polymerization temperature of 160°C and a ratio of 5,000 mol Al/mol Zr in the reactor. It is twice this value when the pressure is increased to 190 MPa.

In Fig. 9.5-7 the productivity is plotted versus the temperature for a pressure of 150 MPa. Again, the molar ratio of Al to Zr is 5,000. The decrease in the number of active sites at high temperatures steeply reduces the productivity. At a temperature of 140°C a high productivity of nearly 6,000 kg PE/g Zr can be observed. It decreases to around 400 kg PE/g Zr when the

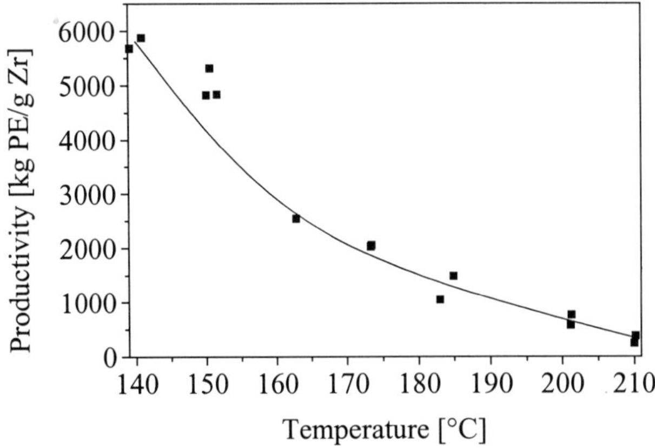

Figure 9.5-7. Influence of the temperature on the productivity. Pressure, 150 MPa; for other data, see Fig. 9.5-6.

polymerization is run at 210°C.

In the co-polymerization, the productivity also reduces with increasing concentration of the co-monomer in the feed. With some co-monomers a rate-enhancement effect can be observed upon the addition of small quantities of co-monomer. The effects of the co-monomer are shown in Fig. 9.5-8, in which the productivity of the same rac-Me$_2$Si[Ind]$_2$ZrCl$_2$ catalyst is plotted versus the concentration of 1-hexene-co-monomer in the feed.

9.5.5 Properties of metallocene-based polyethylene

The basic properties of mPE, such as the average molecular weight, molecular-weight distribution, or density depend mainly on the structure of the metallocene catalyst and its concentration in the polymerization reactor. They also depend strongly on the polymerization temperature and can be varied by incorporation of co-monomers. The influence of the pressure is only small.

The number-average molecular weight of mPE samples prepared at 160°C and pressures of 120 - 190 MPa was found to be in the range of 28,000 - 25,000 g/mol, decreasing slightly with increasing pressure. The density measured on the same samples was 0.966 g/cm^3.

At a pressure of 150 MPa, the number-average molecular weight, M$_n$, decreased from 50,000 to 28,000 g/mol when the temperature was increased from 140 to 210°C (Fig. 9.5-9). The polydispersity determined from the ratio of weight- to number-average molecular weight was in the range of 4 - 6 at polymerization temperatures of 140 - 160°C. At higher temperatures it increased steeply. The samples prepared at temperatures above 190°C showed a bimodal molecular-weight distribution resulting in polydispersities above 20.

As can be seen from Fig. 9.5-10, the melting point reduces first less-, and then more steeply with increasing polymerization temperature. In the range of 140 - 160°C the melting point is around 133°C. At the high temperature of 205°C the melting point is only 127°C.

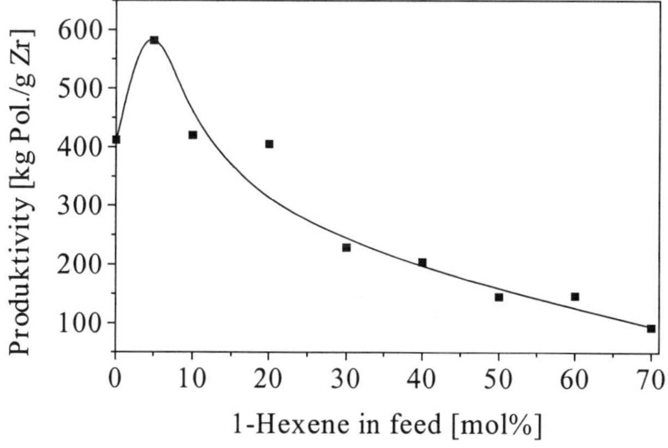

Figure 9.5-8. Effect of co-monomer 1-hexene on the productivity. Pressure, 150 MPa; temperature, 210°C; Zr in feed, 0.2 - 1.1 mol ppm; Al/Zr, 200 mol/mol; toluene in feed, 10 mol.%.

534

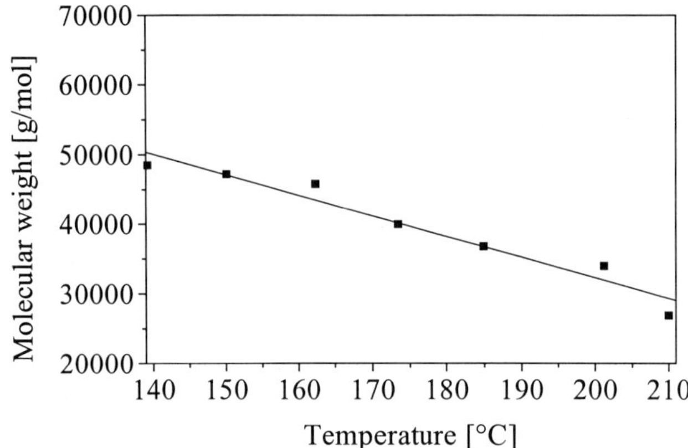

Figure 9.5-9. Influence of temperature on the number-average molecular weight. Pressure, 150 MPa; Zr in feed, 0.012 mol ppm; Al/Zr, 5,000 mol/mol; n-hexane in feed, 10 mol.%.

The properties of mPE can be changed over a wide range by incorporation of co-monomers. The effect of a co-monomer was proved in the co-polymerisation with 1-hexene. By the addition of 60 mol.% 1-hexene to the feed, M_n could be reduced to nearly half the value for homopolyethylene. Also, the density and melting point are reduced steeply by the incorporation of 1-hexene, whereas the polydispersity is not influenced.

9.5.6 Technology of the process

Metallocenes are so-called *drop-in* catalysts. Only minor changes are required to retrofit commercial-scale reactors when metallocene catalysts are to be used. By the high-pressure

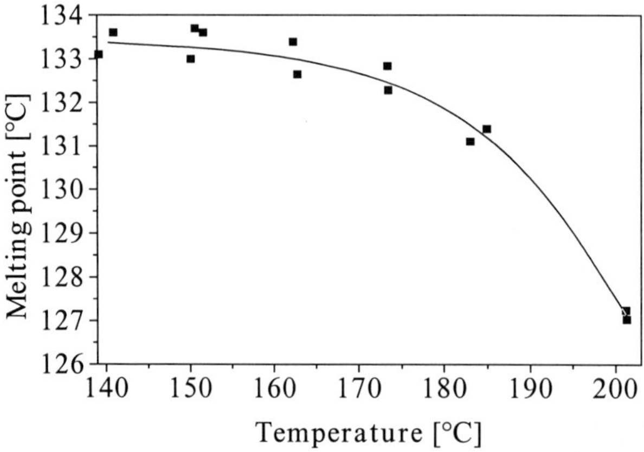

Figure 9.5-10. Influence of polymerization temperature on the melting point. Pressure, 150 MPa; Zr in feed, 0.012 mol ppm; Al/Zr, 5,000 mol/mol; n-hexane in feed, 10 mol.%.

process metallocene-based polymers can be manufactured either in autoclave- or tubular-reactors of existing high-pressure LDPE plants. Before the metallocene is fed into the reactor it must be activated by MAO or organoborate/TiBA.

For activation, the metallocene is first dissolved in toluene to obtain a solution of 1.8 g/l. Afterwards, 23.5 l of a 30% solution of MAO in toluene is added and stirred during 30 minutes. When organoborate/TiBA is the activator system, the metallocene (e.g., 1 g rac-Me$_2$Si[Ind]$_2$ZrCl$_2$) is dissolved in 5 l toluene and stirred for 20 minutes at room temperature under the addition of 0.1 l TiBA. Then the activator (2.5 g [Me$_2$PhNH]$^+$[B(C$_6$F$_5$)$_4$]$^-$) dissolved in 5 l toluene is added.

The activated catalyst can be metered by membrane pumps. The reactor must additionally be equipped with tanks for the preparation and storage of the catalyst. Furthermore, devices for injecting catalyst deactivators, such as methanol, may be required after the reactor to terminate the polymerization.

1-Alkenes or dienes can be used as co-monomers in the metallocene-catalysed process. The reactors are run at pressures of 120 - 200 MPa and temperatures of 120 - 200°C. The residence times are preferably between 200 and 400 sec. On the industrial scale, the quantity of aluminium compound in the reactor should be selected to give a ratio of 200 to 1,000 mol Al per mol of Zr.

9.5.7 Further development

It is expected that around 500,000 tons mPE will be manufactured in 2000 in Western Europe. Resins from high-pressure- as well as from low-pressure processes suffer more difficult processability. Their narrow molecular-weight distribution results in an unfavourable viscosity behaviour [8]. As with linear low-density polyethylene (LLDPE), the viscosity of mPE is independent of the shear-rate over a wide range and reduces only at high shear-rates (Fig. 9.5-11). Therefore, energy consumption is high when mPE is processed with extruders which are designed for LDPE.

Processability can be improved by blending mPE with LDPE, which exhibits an excellent processability. Efforts are being made to increase the polydispersity of mPE by use of two or

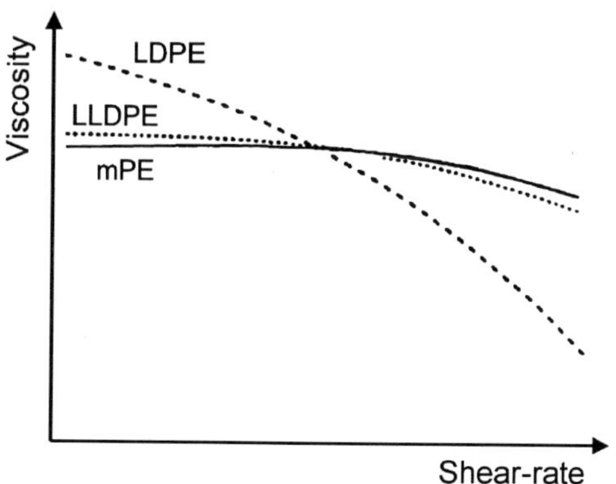

Figure 9.5-11. Viscosity behaviour of various polyethylene resins. Solid line, mPE; dashed line, LDPE; dotted line, LLDPE.

536

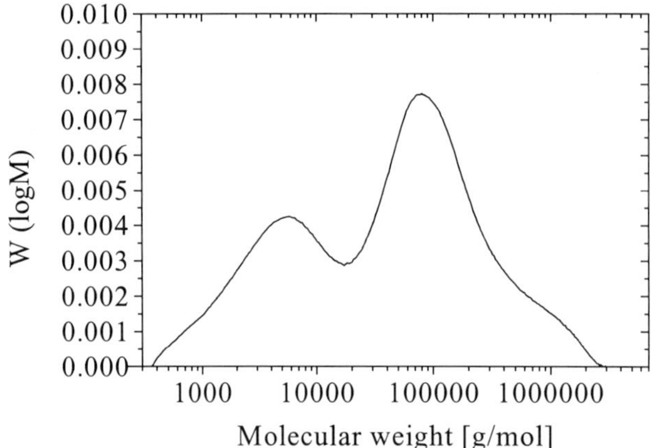

Figure 9.5-12. Bimodal molecular-weight distribution. Pressure, 150 MPa; temperature, 210°C; Al/Zr, 200 mol/mol; toluene in feed, 10 mol.%.

more catalysts in the same- or in different reactors. It is also claimed that resins from solution-processes with monocyclopentadienyl catalysts should have better processability because of a long-chain branching. Furthermore, catalysts with a special architecture are under development [2], which generate bimodal molecular-weight distributions shown in Fig. 9.5-12.

Last, but not least, the development of ternary catalyst systems consisting of a metallocene, an organoborate, and an aluminium compound will reduce the manufacturing costs of mPE and will improve its ability to compete with other polymers.

References

1. H.H. Brintzinger, F.R.W.P. Wild, L. Zsolnai and G. Huttner, J. Organomet. Chem., 232 (1982) 233.
2. A. Rau, G. Luft, T. Wieczorek, S. Schmitz, A. Dyroff, A. Gonioukh and R. Klimesch, Patent filed 1999, German Patent Office, 19903783.3.
3. (a) J.C. Stevens, F.J. Timmers, D.R. Wilson, G.F. Schmidt, P.N. Nickias, R.K. Knight and G.W. Lai (Dow Chemical Co.), Eur. Pat. Appl. 416815, 1991.
   (b) J.A.M. Canich (Exxon Chemical Co.), Eur. Pat. Appl. 420436, 1991.
   (c) J.A.M. Canich, G.G. Hlatky and H.W. Turner (Exxon Chemical Co.), PCT Appl. 5026798, 1991.
4. Ch. Bergemann and G. Luft, Chem.-Ing.-Techn. 70 (1998) 174.
5. A. Rau, G. Luft, S. Schmitz and C. Götz, Chem.-Eng.-Techn. 12 (1998) 954–958.
6. A. Rau, S. Schmitz and G. Luft, Chem.-Ing.-Techn., 73 (2001) 846-851
7. H.H. Brintzinger, M.H. Prosenc and F. Faber, in W. Kaminsky (Ed.) Metalorganic Catalysts for Synthesis and Polymerization, Springer, Berlin, 1999 p.223.
8. M. Lux and W.F. Müller, Kunststoffe 88 no.8 (1998) 1130-1134.

9.6 Supercritical Fluid Extraction and Fractionation from Solid Materials

E. Lack[a], B. Simándi[b]

[a] NATEX Prozesstechnologie GmbH, Hauptstrasse 2, A-2630 Ternitz

[b] Budapest University of Technology and Economics, Department of Chemical Engineering
Budapest, XI., Müegyetem rkp. 3. K. ép.mfsz. 56. H-1521, Budapest, Hungary

9.6.1 Decaffeination of coffee and tea and extraction of hops

The decaffeination of green coffee beans was one of the first industrial processes in the
CO_2-extraction technology [7]. In principle, water is a very good solvent for caffeine, but
water has a very low selectivity. The water-soluble substances are about 20% of the green
coffee beans, and if tea is extracted with water, the final beverage is produced. Therefore it
needs special processes to keep the process-losses at a low level. With the development of
high-pressure extraction, new solutions for such extraction problems were found. From the
beginning of this technology, carbon dioxide was found to be the optimal gas for this process
and it has also the best selectivity.

9.6.1.1 Decaffeination of green coffee beans

As one of the first processes, the decaffeination of green coffee beans was applied on the
industrial scale [6]. Zosel proposed three possibilities for the decaffeination. In the first, the
moistened green coffee beans are extracted in a pressure vessel at 160 to 220 bar. The caffeine
diffuses from the beans into the CO_2, which is fed into the washing-tower [8].

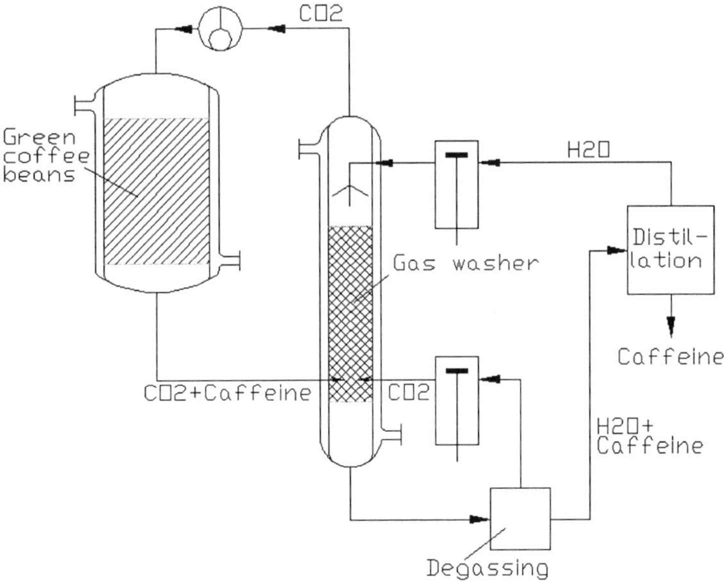

Figure 9.6-1. Zosel's suggestion for the decaffeination of green coffee beans [8]

538

In the washing-tower the caffeine is washed out from the CO_2 with demineralized water at 70 to 90°C. After 8 to 12 hours, depending on the coffee raw material, all of the caffeine is in the washing water, which is degassed and the caffeine recovered by distillation of the washing water. Thereby, the caffeine concentration in the beans is reduced from the initial 0.9 to 1.8% down to 0.08%. For the decaffeination of one kg raw coffee about 3 to 5 liter washing water is necessary.

In the second process, the caffeine-loaded CO_2 flows through an activated-carbon bed and the caffeine is adsorbed. Usually the activated carbon is recycled by heat treatment up to 600°C and the caffeine is destroyed.

In the third variation a mixture of moisturized green coffee beans and activated carbon is filled into the extractor, and the activated carbon pellets used are just big enough to fit between the beans. For 3 kg of coffee beans, 1 kg of activated carbon is needed. At 220 bar and 90°C the caffeine in the supercritical CO_2 diffuses directly out of the beans into the activated carbon. A CO_2 circulation is not necessary. The required degree of decaffeination is reached after 6 to 8 hours. After extraction, the beans and activated carbon are separated by a vibrating sieve.

Schoeller-Bleckmann started about 1979 with CO_2 extraction and focused his research work in 1983 on decaffeination processes [9,104]. From the beginning of this development work great attention was paid to the idea that caffeine can be recovered according to the first process proposed by Mr. Zosel.

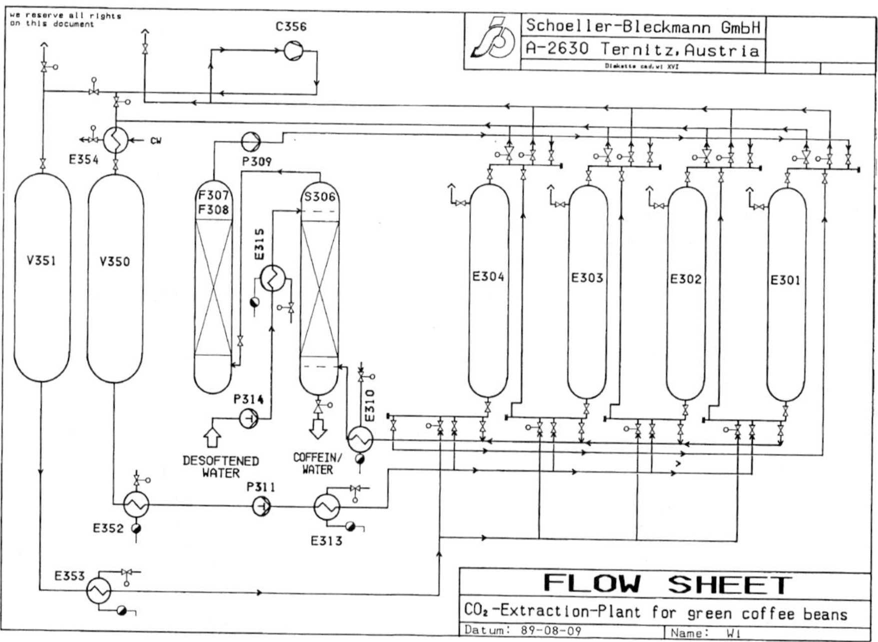

Figure 9.6-2. Flow sheet for the decaffeination process by Schoeller-Bleckmann

The pre- and after-treatment of the coffee beans is analogous to the usual decaffeination processes. The pre-treatment mainly consists of storing, cleaning and moistening. The moistening is effected by vapour, water, or a combination of both. A water content of 40 to 45% is optimal. After the extraction, the coffee beans have to be dried back to the initial moisture content of 10 to 12%.

Fig. 9.6-2 shows a flow-sheet for an extraction process for green coffee beans. The plant is designed with four extractors (E301-E304): three of them are always in operation, and one is switched off the CO_2 cycle to be discharged and charged with fresh beans. The process works isobarically and isothermally, which means that in the extraction step the same thermodynamic conditions are in the washing column S306 and in the adsorber F307 and F308. With the circulation pump, P309, the regenerated CO_2, which is nearly free of caffeine, is pumped through the extractors, which are connected in series. The loaded CO_2 flows through the heat-exchanger E310, into the washing-column S306, and is regenerated by being washed in a countercurrent with fresh demineralized water. For better CO_2 regeneration, and therefore a lower final caffeine concentration in the beans, (which is necessary, for example, for the US market) a final purification of CO_2 with activated carbon is necessary.

One of the four extractors is disconnected from the CO_2 cycle for CO_2 pressure-release, discharging, and refilling with coffee beans, and then the extractor is again pressurized with CO_2 by means of pump P311. Then this extractor is connected to the CO_2 cycle and another extractor is disconnected. This cascade operation has the advantage that the caffeine-concentration in the washing water is relatively constant, between 1 and 3.5 g/l. This concentration is rather low, and it needs much energy to recover the caffeine by distillation. With membrane processes such as reverse osmosis, the recovery can be done with a lower energy-input.

Continuous process

For a long time nobody dared to develop a continuous high-pressure extraction process. It was already known from coal gasifying that continuous high-pressure extraction processes, in which solid materials have to be fed into and out of the pressure reactor, are very difficult to effect. Moreover, extremely high development costs and protracted test series were necessary. The General Foods Corporation decided to take this step, and describe in the European Patent Application Nr. 0331852 A2 [10] a process of extracting caffeine from green coffee beans whereby a supercritical fluid essentially free of caffeine is continuously fed to one end of an extraction vessel, and the green coffee beans and caffeine-laden supercritical fluid is continuously withdrawn from the opposite end.

Fig. 9.6-3 is a schematic illustration of the extraction vessel. The periodic charging and discharging of coffee into the extraction vessel is maintained through the use of intermediate pressure vessels, known as blow cases. The blow cases are merely smaller pressure vessels, of about the same volume as the portions of coffee that are periodically charged and discharged, and which are isolated at both ends by valves, typically ball valves. One blow case is situated on the top of the extraction vessel for charging and one at the bottom for discharging. The supercritical CO_2 flows from the bottom to the top of the extraction vessel and then the caffeine-laden supercritical fluid is fed to a countercurrent water-absorber. The preferred weight ratio of supercritical fluid to coffee is between 30 and 100 kg CO_2/kg coffee.

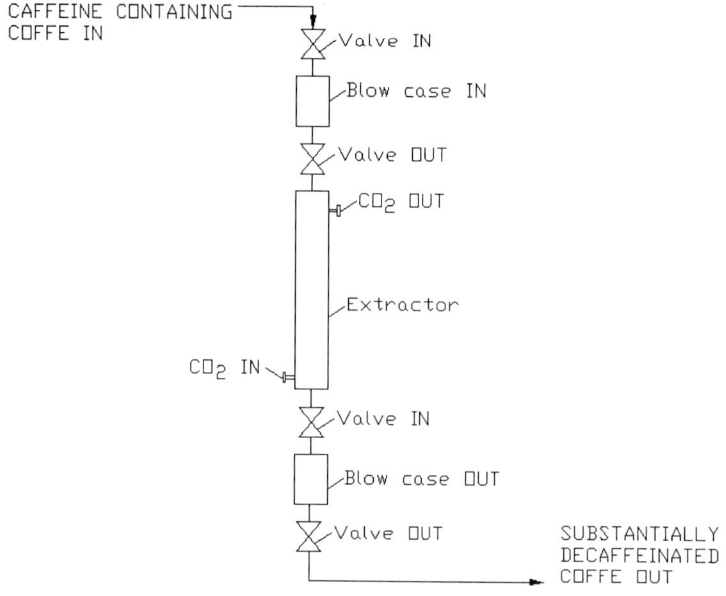

Figure 9.6-3. Nearly continuous supercritical fluid extraction of green coffee beans according to EPA Nr. 0331852A2 [10]

9.6.1.2 Decaffeination of tea

In contrast to the decaffeination of coffee, which is primarily executed with green coffee, black tea has to be extracted from the fermented aromatic material. Vitzthum and Hubert have described a procedure for the production of caffeine-free tea in the German patent application, 2127642 [11]. The decaffeination runs in multi-stages. First, the tea will be clarified of aroma by dried supercritical carbon dioxide at 250 bar and 50°C. After decaffeination with wet CO_2 the moist leaf-material will be dried in vacuum at 50°C and finally re-aromatized with the aroma extract, removed in the first step. Therefore, the aroma-loaded supercritical CO_2 of 300 bar and 40°C will be expanded into the extractor filled with decaffeinated tea. The procedure also suits the production of caffeine-free instant tea, in which the freeze-dried watery extract of decaffeinated tea will be impregnated with the aromas extracted before.

A considerable reduction of the extraction time, and of the CO_2-mass-flow, is reached if the CO_2 flowing through the extractor is already regenerated in the extractor. Such a procedure is described in European Patent Specification 0200150 of Klima, Schütz and Vollbrecht [12]. The tea is moistened up to 15 to 50%wt. with water, and extracted with wet CO_2 at 255 to 350 bar and 50 to 80°C. Subsequently the caffeine is separated from CO_2 by means of an adsorbent; for example, activated carbon, and the caffeine-free CO_2 introduced to the next layer of tea. Consequently, the extractor is filled with tea and adsorbents in layers. Before the CO_2 stream leaves the extractor it will be conducted through a pure adsorbent-bed. The adsorbent material can also be filled with gas-permeable materials, for example, textile bags or tubes made of textiles. This facilitates the discharging of tea and adsorbents after extraction. By the quick successive charging and regeneration of CO_2, the extraction time may be reduced to 1 to 4 hours. It has already been mentioned that in Europe and North America only fermented black tea can be extracted, as fresh green-tea leaves usually will not be transported. Nevertheless, in Patent Application DE 3414767 a procedure is described [13] where green leaves in a moistened condition are treated at 20 - 80°C and a pressure up to 300 bar, with CO_2. The procedure also suits for decaffeination of the so-called green tea.

9.6.1.3 Preparation of hop extracts with CO_2

Hop is a natural product, which is used worldwide by the brewing industries to bitter the beer. After harvesting, the hops must be dried immediately to give better stability. Even then, the storage-life of hops is relatively short. To maintain a continuous supply for the brewing industries, more than 25% of the world production of hops is extracted.

Nearly all extraction plants in US, Europe and Australia operate with CO_2. No other product has changed completely to the CO_2 process in this way.

Until 1980 most of the extraction plants operated with organic solvents such as methylene chloride, although some operated with ethanol. The resulting extracts contained residual solvents, which were partly lost during the traditional brewing process, but the other parts remained in the beer.

Now nearly all hop-extraction plants operate with CO_2. In principle, it would be possible to reach the desired limits and to meet public health requirements with respect to level of residual solvents, but some solvents such as methylene chloride are always subject to doubt and public discussion.

At the beginning of the 1980s the European Community published a clear statement regarding the use of CO_2 as solvent:

Forum der Brauerei: 1/1983 [14].

CO_2 is a natural occurring substance, and everyone is in contact with CO_2 through the atmosphere, food and beverages. It is not necessary to fix any residual limits.

The advantages of hop extracts are:
- Easy and cheap storage for a long time (2 years). Hops and hop pellets have to be stored under cool conditions, and pellets in an oxygen-free atmosphere.
- Easy and cheap transport in cans. Raw hops have a very large volume.
- They are more independent of price fluctuation.
- The demand for α-acids is a little lower than for pellets.
- Extracts are easy to dosify, and produce a stable quality of beer.

Content of hops [15]

Dried raw hops contain about 13 to 23% bitter substances, as
- 4 to 12 % α-acid
- 4 to 6 % β-acid
- 3 to 4 % soft resins
- 1.5 to 2 % hard resins
- as well as cellulose, lignin, sugar, pectin, protein, organic acids, enzymes, tannins, fats, waxes and hop oil.

CO$_2$ Process

There are already several large industrial plants in operation which process hops with carbon dioxide. These plants are located in Germany, Great Britain, Australia, and in the US.

Great Britain and Australia use mainly the extraction with liquid CO$_2$. In Germany, supercritical CO$_2$ is mainly used, which allows a better and faster extraction of all the important substances.

In former days in the US the subcritical process was used, but in the new plants, which are located in the hops areas, the supercritical process is applied.

Liquid carbon dioxide extraction

Liquid CO$_2$ has a limited solvent power for hop resins, which for α-acids is about 0.8 wt.% at 7°C and 50 bar, and for β-acids, about 0.13 wt.% at 20°C [16].

Industrial-hops-extraction's are maintained at 60 bar and 8°C.

To reach an extraction efficiency of 95% for α-acids, a solvent ratio of at least 50 kg CO$_2$ per kg of raw hops is necessary.

The volatile hop oil is nearly completely extracted without danger of decomposition owing to the low extraction temperature.

Because of their low extraction-pressure, subcritical extraction plants have lower investment costs than supercritical plants.

Supercritical carbon dioxide extraction

The solubility of hops components is better under supercritical conditions, which means that a lower flow rate of CO$_2$ is necessary, and a shorter extraction time is required.

Industrial plants are working at 300 to 350 bar and 40 to 65°C. Under these conditions all the hop bitter-principles are extracted quantitatively. The content of total resins in the extract is about 90% and a part of the hard resins is extracted, which is not the case with subcritical conditions.

Above 400 bar, most of the hard resins are extracted. At 300 bar and 80°C the solubility of hops-extract in CO$_2$ is around 3.2%wt. The supercritical extraction has many advantages compared to the subcritical extraction, so that even though the investment costs are higher, all new plants have been built for supercritical extraction.

Preparation of hops for extraction

As the bulk density of raw hops is very low (around 125 kg/m³), the hops are ground or powdered, and pelletized. With this procedure a bulk density of about 450 to 500 kg/m³ can be reached, which is very important for the economy of a high-pressure plant.

Larger plants have their own pelletizing equipment, but smaller plants buy hops already pelletized on the market.

The quality of beer produced with CO$_2$-extract compared to pellets and to extracts produced with other solvents [17,18,19]

The yields of bitter substances in the CO$_2$ extracts are similar to those obtained with pellets and dichloromethane extracts. The long term stability of the beer is similar to the others.

The foam values are better because CO$_2$ does not extract any tannins. Therefore, the precipitation of protein, which creates complexes with the tannins, is reduced.

The taste-tests yielded positive results, although in beers produced using polyphenol-free CO$_2$-extracts a somewhat less full-bodied note, with equivalent bitterness, was found. This lack of quality can be equalized by adding of a small quantity of pellets during wort-boiling, which increases the polyphenol content of the beer. The fresh beers showed a slight improvement in flavour compared with the control samples and, after maturation, the beers produced with CO_2 extracts were frequently more stable.

Pesticide content in CO_2 extracts [20]

One of the pesticides most used in hop cultivation is "Dithiocarbamate", which is not soluble in CO_2 and therefore, in most cases, is not found in the extract.

High contents of pesticides in the raw hops, of 527 ppm or 487 ppm, show strongly reduced levels in the extracts, at only 6.7 ppm or 7.0 ppm, respectively.

Dithiocarbamate concentrations lower than 20 ppm in the raw hops yield a CO_2 extract with no detectable pesticide content.

The same is valid for heavy metals. Heavy-metal ions are not soluble in CO_2, and are therefore not normally found in CO_2 extracts.

Because of these advantages the CO_2 extraction is a very successful technology for hops, with the result that nearly all hop-extraction plants have changed to the CO_2 process in the last twenty years [21].

9.6.2 Extraction of spices and herbs

Introduction

Spices and herbs come from the aromatic parts of plants. Those can be roots, barks, leaves, flowers, parts of flowers, fruits or seeds. People started using spices to improve the taste of their food since several thousand years. For centuries spices have been harvested in the countries of origin, transported and traded in the developed countries. They have increasingly become commodities and the falling prices made them affordable for a wider range of people. The European Union is the second biggest spice market in the world behind the USA, with an estimated potential of 170,000 tons/year (1995). Spices can be used as fresh products, but in order to allow their transport over longer distances, and to store them for a longer period of time, they are dried for preservation. Commercially used spices and herbs are available in whole, broken, rubbed or ground form [26,30,31].

The whole- and ground spices do not fulfil the requirements of uniform, consistent, sterile and shelf-stable food products. The food and flavour industry has therefore used plant extracts for many years. The oleoresins are the solvent extracts of spices, containing both the essential oils and other non-volatile components. The production of oleoresins involves the selection and grinding of raw materials, solvent extraction, removal of the solvent, and finishing the oleoresin by standardization. The advantages of oleoresins are: consistency in flavour and aroma, sterility, stability in storage, reduced storage space, and economy. A variety of solvents can be used effectively, but regulation requires the removal of organic solvents to a prescribed residual level. The thermal instability of the valuable components may limit the use of elevated temperatures during the extraction and evaporation processes. In addition, the higher-boiling impurities are retained in the final products [32,33].

The current trend of consumer preference towards natural products requires new proccessing methods for spice-oils and extracts, without the addition of external material. In recent years there has been an increased interest in supercritical and subcritical extraction [26,27], which use carbon dioxide as a solvent [34,35,36]. Carbon dioxide (CO_2) is an ideal solvent for the extraction of natural products because it is non-toxic, non-explosive, readily available, and easily removed from extracted products. The use of supercritical fluid extraction (SFE) instead of steam distillation or extraction with classical solvents (hexane, ethanol), which are the traditional spice extraction methods, gives several advantages: the problem of toxic residual solvent in the products is eliminated, the lower temperatures involved lead to less deterioration of the thermally labile components in the extract, SFE is more selective than extraction with commonly used solvents which also extract unwanted components (like tannins, chlorophylls, minerals), and SFE extracts processed with CO_2 own nearly the same organoleptic characteristics of the spice itself [7,37].

From a technical and practical point of view, the components of plant extracts can be classified as: essential oils, fatty oils, pungent constituents, colours (pigments) and natural antioxidants.

One of the first articles on this subject was published in 1974 [22], describing a list of spices which were extracted with liquid CO_2 in Russia in the years 1966 to 1972. In the year 1963, at the Max Planck Institut, systematic investigations were made for various natural products, followed by the granting of many patents. Further investigations were made by Peter Hubert and Otto G. Vitzthum at HAG AG [23], Mrs. Hedi Brogle at Nova Werke AG [24], Prof. Siegfried Peter at TU-Erlangen and Prof. Stahl at the University of Saarbrücken. Although many spices and herbs were tested since 1966 the first industrial plant was installed in the early beginning of the eighties [25].

Information about experimental solubility and equilibrium data are important, even when complex mixtures are extracted. Reviews of high-pressure phase-equilibrium data have been published by several authors, for example, by Knapp *et al.* [38] covering the period 1900 to 1980, by Fornari *et al.* [39] covering 1978 to 1987, and by Dohrn and Brunner [40], between 1988 to 1993.

The spices and herbs most commonly used in Europe are listed in Table 9.6-1, indicating the characteristic components of each as well. There are several commercial plants using carbon dioxide to extract flavour, aroma and other valuable components from plants, such commercial production is also indicated in Table 9.6-1.

The classical method for the production of spice oleoresin consists of two steps, as for example for ginger.

- In the first step, the essential oil is separated by steam distillation.
- In the second step the involatile valuable substances are extracted with organic solvents.

The extraction solvent must be selected carefully to provide the right types of extracts which are desired for flavouring purposes. After extraction, the solvent must be completely removed under vacuum. Generally, the oleoresin consists of essential oil, organically soluble resins, and non-volatile fatty oils and waxes. In some cases, where little or no volatile essential oil is present in the raw material, as with Paprika and Capsicum, only the second step is necessary. Spice oleoresins have found increasing usage in the food industry, especially in automatic meat-processing, or generally in the large food industry sector. To ease dosifying, the oleoresin can be mixed with salt or dextrose, which gives a powdered product.

The advantages of oleoresins can be summarized as follows, in that they possess:

- Uniformity of flavour.
- Higher stability compared to natural spices and herbs, which lose the volatile, essential oil by evaporation, polymerization and oxidation.
- Constant spice-equivalents.

545

Table 9.6-1
Most common spices used in Europe

Spice/ herb	Botanical name	E	F	P	C	A	O	CP	References
Allspice	*Pimenta dioca* L.	X				X			41
Anise	*Pimpinella anisum* L.	X	X					X	22,42
Basil	*Ocimum basilicum* L.	X						X	43
Bay leaves	*Laurus nobilis* L.	X					X		22
Caraway seed	*Carum carvi* L.	X	X					X	42,44 to 46
Cardamon	*Elettaria cardamomum* White et Mason	X						X	47
Celery seed	*Apium graveolens* L.	X	X					X	46,48
Chervil	*Anthriscus cerefolium* Hoffm.	X							49
Cinnamon	*Cinnamomum zeylanicum* Blume	X				X		X	50
Cloves	*Syzygium aromaticum* L.	X				X		X	47,50,51
Coriander	*Coriandrum sativum* L.	X	X					X	46,48,52
Cumin seed	*Cuminum cyminum* L.	X	X						47
Dill seed	*Anethum graveolens* L.	X	X						22,
Fennel seed	*Foeniculum vulgare* Mill.	X	X						47,53
Fenugreek	*Trigonella foenum graecum* L.		X				X		
Ginger	*Zingiber officinale* Rosc.	X		X				X	47,54,55
Horseradish	*Armoracia rusticana* G. M. Sch.			X					
Mace	*Myristica fragrans* Houtt.	X						X	22,47
Marjoram	*Origanum majorana* L.	X						X	46,56
Mustard flour	*Sinapis alba* L.		X	X					57
Nutmeg	*Myristica fragrans* Houtt.	X	X					X	22,23
Oregano	*Origanum vulgare* L.	X				X			58,59
Paprika	*Capsicum annuum* L.			X	X			X	47,60 to 64
Parsley	*Petroselinum hortense* Hoffm.	X	X						22,46,47
Pepper, black and white	*Piper nigrum* L. *Piper album* L.	X		X				X	22,23,46,50 65 to 67
Rosemary	*Rosmarinus officinalis* L.	X				X		X	68 to 70
Saffron	*Crocus sativus* L.	X			X				71
Sage	*Salvia officinalis* L.	X				X		X	70,72,73
Savory	*Satureja hortensis* L.	X							74
Star anise	*Illicum verum* Hooker fil.	X						X	
Tarragon	*Artemisia dracunculus* L.	X							
Thyme	*Thymus vulgaris* L.	X				X		X	75 to 77
Turmeric	*Curcuma longa* L.	X			X				47

E, Essential oil; F, Fatty oil; P, Pungent constituent; C, Colour; A, Antioxidant; O, Other; CP, Commercial production.
By preparing emulsions, water-dispersible mixtures can be produced.

- Very low microbiological contaminations.
- Ease of further processing in the food industry.

One of the disadvantages of conventional oleoresins is their contamination with small amounts of organic solvent. Further, by removing this solvent under vacuum, a part of the initial original flavour is lost.

By using the CO_2-extraction technology the two operations, production of essential oil and of the non-volatile constituents, can be achieved in one step. Moreover, the process works at low temperatures, of 40 to 60°C, which avoid the decomposition of heat-sensitive aroma compounds.

As no organic solvents are brought into the process no pollution of air or waste-water streams occurs.

- By the use of a low extraction-pressure (around 100 to 120 bar) and a low temperature (40°C) the CO_2-processed extracts are similar to the classical steam distillates.
- By the use of higher extraction-pressures (around 350 to 450 bar) and 40 to 70°C, the CO_2-processed extracts contain all CO_2-soluble components, including resins, colorants, fatty oils and waxes, and such extracts are different to solvent-extracts.

9.6.2.1 Description of a spice plant

For the extraction of spices, herbs and aromas medium-sized plants are used. These plants are equipped with two or three extractors with a payload volume ranging from 200 to 800 l

SFE MULTIPURPOSE EXTRACTION UNIT

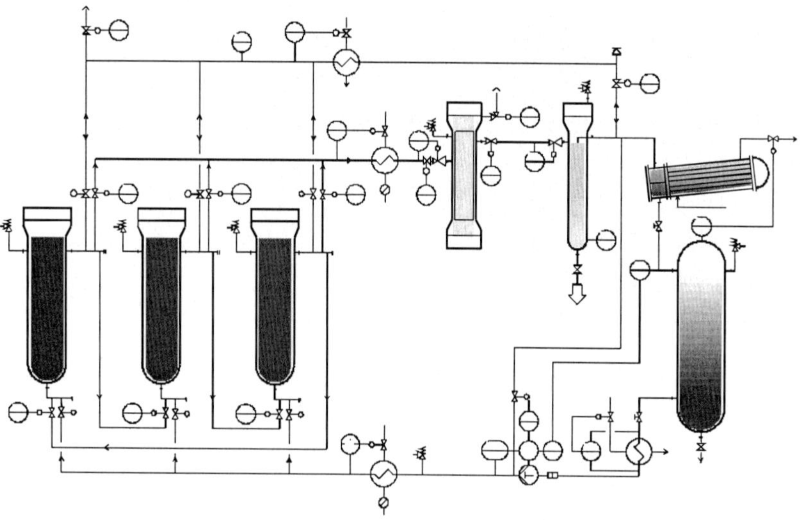

Figure 9.6-4. shows a flow-sheet of a multipurpose spice and herb plant

per extractor. The applied pressures vary from 300 up to 500 bar, depending on the product-mix.

Multipurpose plants are equipped with a two-step separation system, to enrich the resin and the essential oil fraction. Plants which process highly viscous, sticky extracts (such as pepper), are provided with two separators switched in parallel for the first separation step.

The plant consists of a CO_2 working-tank, a precooler, a circulation pump, a preheater, three extractors, a first and second separator, and a condenser.

The smallest commercial units for the processing of herbs and aromas are equipped with two extractors, each of 200 l payload and a total yearly capacity of 150 metric tons. For the extraction of spices used in the food industry, industrial sizes of 2 x 300 l, 3 x 300 l, up to 3 x 500 l, are the most economical ones. For the extraction of the most used spices, such as paprika, chili, coriander and caraway, larger sizes up to 3 x 800 l have an advantage with regard to processing costs. Because of the low solubility of piperin, colorants of paprika, and fatty oils at 300 bar, spice plants should be designed for pressures of 500 or 550 bar. The extractors are designed for operation with baskets in order to allow quick and easy raw material handling.

Especially for multipurpose application, proper attention must be paid to simple and efficient cleaning.

Figure 9.6-5. shows the spice extraction plant at NOVO AGRITECH near Hyderabad, India (archives NATEX)

9.6.2.2 Extraction of essential oils

Essential oils are complex mixtures of volatile compounds, conventionally obtained by steam distillation. These consist of hydrocarbons, including terpenes ($C_{10}H_{16}$), sesquiterpenes

($C_{15}H_{24}$) and diterpenes ($C_{20}H_{32}$), and oxygenated hydrocarbons such as alcohols, esters, aldehydes, ketones, ethers, and phenols.

Table 9.6-2
Comparative chemical compositions of the volatile oils of different plants, obtained by hydrodistillation and SFE (% of peak area)

Plant	Compound	Hydrodistillation	SFE products
Thyme	p-Cymene	12.9	13.5
	γ-Terpinene	11.2	6.8
	Thymol	57.0	64.3
	Carvacrol	6.4	4.1
	β-Caryophyllene	1.9	6.7
Oregano	p-Cymene	3.9	1.5
	γ-Terpinene	1.6	0.1
	Linalool	1.8	1.7
	Borneol	1.3	1.2
	Thymol	1.4	1.3
	Carvacrol	83.8	90.5
Rosemary [68]	α-Pinene	25.2	8.3
	1,8-Cineole	20.6	20.0
	Camphor	10.3	15.3
	Borneol	13.7	15.6
	Verbenone	4.8	8.4
	β-Caryophyllene	1.0	2.2
Clary sage	Linalyl acetate	10.3	8.2
	Linalool	14.9	0.9
	α-Sclareol	-	50.0
Dill	Limonene	55.7	42.6
	Dihydrocarvone	4.7	6.1
	d-Carvone	36.5	50.2
Fennel	α-Pinene	2.5	0.5
	Limonene	3.2	0.9
	Fenchone	17.3	13.9
	Methyl chavicol	3.5	3.1
	trans-Anethole	71.4	77.9
Plant	Compound	Hydrodistillation	SFE products
Coriander	Pinene	15.3	7.7
	Linalool	68.5	75.5
Celery	Limonene	50.5	33.4
	3-Butylphthalide	23.6	40.6
Parsley	α-Pinene	24.0	1.5
	β-Pinene	21.9	3.0
	Myristicin	7.4	4.0
	Apiole	38.5	84.9

Other compounds, containing sulphur or nitrogen, may be present in some oils (like onion, garlic).Investigations on the solubility of pure essential-oil components in carbon dioxide showed that the separation of terpene hydrocarbons and oxygenated derivatives by fractionated extraction is difficult, since their solubility behaviours and vapour pressures are almost the same. Saturating the CO_2 with water as modifier can increase the differences in solubility [78].

There are a number of published applications in which spices are extracted with liquid carbon dioxide to isolate a flavour- or aroma concentrate [22,48,54]. Liquid CO_2 dissolves the essential oils and lighter fractions of the oleoresins. Supercritical CO_2 is generally a better extracting-solvent than liquid CO_2, because higher densities, equivalent to higher solubility, can be achieved by raising the pressure.

In general, volatile components can be extracted from the dried raw materials under conditions close to the critical point of carbon dioxide. Temperatures should be within the range of 32 to 60°C. However, some heat-sensitive components may decompose, even below this range. Extraction pressures should be between 74 and 120 bar, since at higher pressures the increased solvent power of CO_2 also increases the solubility of unwanted components. The yields obtained by SFE are very similar to those found by steam distillation. However, even under mild extraction conditions, some small amount of cuticular waxes is co-extracted with the volatiles. The major constituents of the waxes are n-paraffins, ranging from C_{25} to C_{33}.

The extraction, in combination with fractional separation of extracts in a multi-stage separation system, results in high-quality volatile oils [42,44]. Recently, this fractional separation technique has been refined by Reverchon and co-workers for selective precipitation of cuticular waxes and volatile oils [43,56].

Major components and compositions of distilled oils and SFE extracts of selected plants are shown in Table 9.6-2.

Alteration of essential-oil components during distillation can be recognized by comparing the oils obtained by steam distillation and by SFE. The hydrolysis of esters (like linalyl acetate) to the corresponding alcohols was observed in clary sage oils. The hydrolysis of thymol bound in glycosides resulted in different thymol concentrations in distilled thyme oils, which was proved by appropriate treatments (acidic and enzymatic) of the previously CO_2-extracted plant material.

Comparison of the volatile constituents of the oils resulted in the conclusion that the concentrations of terpene hydrocarbons (like pinene, limonene) were higher in the distilled oils, while those of the oxygenated derivatives (like thymol, borneol, carvone, apiole) were higher in SFE products.

Mace and nutmeg

Mace and nutmeg are grown on large evergreen trees native to the Moluccos Islands and the East Indian Archipelago. The fruit of the tree has quite a large seed, about 4 cm long and 2 cm wide, brown, and an oval shape. This is the nutmeg seed. The mace covers the shell like a net and has a yellow-to-red color.

The extraction yield during CO_2 extraction is mainly dependent on the origin of the raw material. Raw material from China or Vietnam results in the highest yields.

CO_2 extraction yields at 420 bar and 60°C.
Nutmeg (Vietnam and China):   30 to 34 %wt.

550

Mace (Vietnam and China):   35 to 37 %wt.
Nutmeg (India):   27 to 32 %wt.
Mace (India):   22 to 25 %wt.

The oleoresin of mace is reddish-orange and liquid-to-semi-solid at room temperature. The oleoresin of nutmeg is yellow to dark yellow, and solid at room temperature.

Coriander

Coriander is an annual herb of the parsley family, native to Europe and the Mediterranean region. For CO_2 extraction, only ground coriander seed, the dried fruit of the plant, is used. Coriander seed contains 0.8 to 1.5% volatile oil. The principal component is d-linalool, present at about 60 to 70% in the volatile oil. The CO_2 extracts of Austrian coriander (450 bar, 60°C) contain about 8 to 10% volatile oil.

The corresponding results at various pressures are given in Fig. 9.6-6, where Austrian coriander with a content of about 10% fatty oil and 1% volatile was extracted.

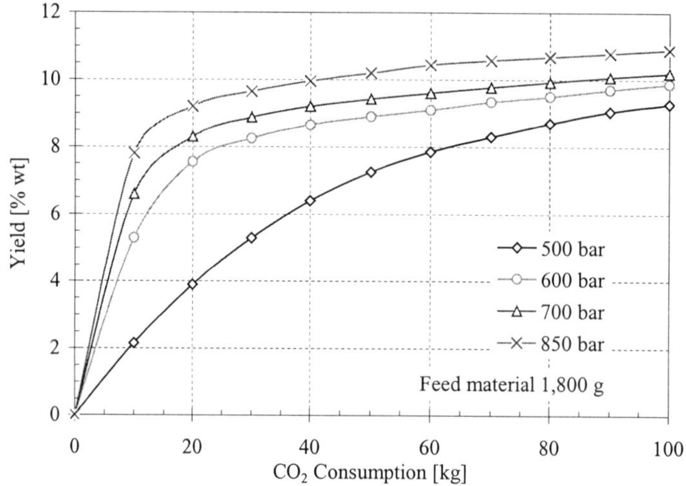

Figure 9.6-6. Extraction yield of coriander extract in dependence of CO_2-throughput

As the triglycerides of the fat consist mainly of long-chain fatty acids, most of them are saturated, higher extraction pressures are necessary to obtain reasonable loadings in CO_2.

The advantages of using higher pressures are remarkably high comparing to the necessary CO_2 throughput for obtaining a 10% yield. For 500 bar one needs, therefore, about 75 kg CO_2/kg feed, 75% of that for 600 bar, about 50% for 700 bar, and only 35% for 850 bar, that means, at same specific mass-flow rate the extraction time is 1/3 of that necessary for 500 bar.

The extract of Austrian-, Polish-, Czech- and Hungarian coriander is pale yellow and solid at room temperature.

The extraction yield of Indian coriander is much higher, about 15%, but the extract is dark greenish brown and has a lower content of volatile oil (about 6%).

Caraway

Caraway seed is the fruit of a herb within the parsley family. Each seed is a half of the fruit and is about ½ cm long, tan to brown, and curved with five lighter coloured ridges along the length of the seed. Caraway is mostly cultivated in Europe and Western Asia.

Caraway seeds can contain about 1.5% to 3.5% of volatile oil and up to 15% of fixed oil. The main component of the volatile oil is carvone, representing about 50 to 60% of the CO_2 extract. If extracted at 350 bar and 60°C, the extraction yield is 8 to 10% for raw materials produced in Austria, Poland, the Czech Republic and Hungary, and 16% for raw materials from Egypt.

The CO_2 extracts from Egyptian raw material contain 15% of essential oil with a *d*-carvone content of 55 to 60%. The extract is dark greenish and liquid at room temperature.

Parsley

The plant is mainly used as a vegetable. As the yield of volatile oil from the freshly harvested plant is not more than 0.06% it is better to produce extractives out of the seed.

The volatile oil in the seed is about 3.5 to 4.5%.

The oleoresin of parsley provides a flavour characteristic of the entire plant, and is used in chicken- and potato-soup bases, salad dressings, meatball seasonings, etc. Extraction by CO_2 of Austrian parsley seed yields about 15% with extraction conditions of 450 bar and 65°C.

By two-step separation the volatile oil can be enriched up to 80% in the second step.

I. Separator	II. Separator
p = 120 bar	p = 45 bar
t = 48°C	t = 20°C
Yield = 10.5%	Yield = 4.5%

Raw material from Italy or Spain can have an extraction yield up to 17%.

Cardamom

Cardamom seed is the dried fruit of a plant from the ginger family. One seed-pod contains 6 to 8 brown-to-black highly flavoured grains, with a diameter of about 2.5 mm.

Whole cardamom is available in three forms:
1. Whole green cardamom;
2. Whole tan-coloured cardamom;
3. Whole bleached cardamom.

Cardamom is native to Southern India and Ceylon, but is also grown in Middle America. Cardamom seeds contain about 3 to 8% volatile oil, with about 25 to 40% of cineol, 30 to 40% alpha-terpinyl acetate and 1 to 2% limonene.

CO_2-Extraction tests of ground cardamom showed that an extraction pressure of 350 bar is sufficient. An extraction pressure of 500 bar gives only a marginally higher extraction yield, but the quality is lower than from the 350-bar extract.

The extraction yield at 350 bar and 42°C for raw materials from Guatemala was 5.5%, and the extract was a very aromatic and mobile oil.

The content of volatile oil in the extract is between 60 and 65%.

Cloves

Cloves are the dried unopen flower bud of an evergreen tree belonging to the Myrtle family. Cloves must be picked just before the blossoms open themselves to have the bud present; then the spice is dried in the sun.

Cloves contain about 20% volatile oil.

CO_2-Extraction of cloves gives a yield of 18 to 25% with extraction conditions of 350 bar and 65°C. The extraction pressure should not be too high, otherwise waxes and bitter substances enter the extract. The content of volatile oil is more than 80%, of which the content of Eugenol is around 70%.

Vanilla

Vanilla planifolia is the only orchid, which is used as spice plant. The plant has its origin in tropical Mexico, but today the main source of vanilla is the Malagasy Republic.

To obtain the characteristic flavourful and aromatic vanilla bean, a long process of controlled fermentation and drying is necessary. The characteristic aroma of vanilla is of vanillin, which is present at a concentration of 1 to 2.5%. Further, vanilla beans contain volatile oils, resins, vanillic acid, *para*-hydroxybenzoic acid, tannins, fixed oils, sugar, gum, waxes and water.

Vanilla is a very expensive and an exclusive flavoured foodstuff and very well investigated by the CO_2 extraction technology. For 30 years, vanilla extraction with CO_2 has been an important part of the CO_2-extraction technology. Before extraction the vanilla beans have to be ground to a particle size of 1 to 3 mm. Owing to the high moisture of the vanilla pods, which is between 26 and 30%, the pods must be ground under cryogenic conditions. Further, vanilla tends to swell during CO_2-extraction.

The favourable extraction conditions are at 350 to 420 bar and 65°C, giving an extraction yield of 6 to 8%.

The extract is of high quality with the pleasant flavour of vanilla beans.

The vanillin content in the extract is between 12 and 22%, dependent on the raw material.

Table 9.6-3
Analytical data of oleoresin vanilla

Source	Vanillin [wt.%]	Vanillic acid [wt.%]	*para*-Hydroxy benzaldehyde [wt.%]	*para*-Hydroxy benzoic acid [wt.%]
INDONESIA				
Grade 1 BEAN	1.13	0.14	0.098	0.041
Grade 1 OLEORESIN	14.3	0.55	0.61	0.019
Split BEAN	1.33	0.12	0.12	0.03
Split OLEORESIN	13.8	0.55	0.8	0.016
MADAGASCAR				
Bourbon BEAN	2.02	0.122	0.15	0.023
Bourbon OLEORESIN	18.9	0.43	0.52	-

Sometimes there are problems in using CO_2 extracts in the usual formulation because of the high concentration of vanillin. The concentration is up to 100 times higher than in alcohol extracts, where it is only 0.2%. Under certain conditions, crystalline vanillin is separated from the water phase of the extract. After drying this crystalline phase, a product with more than 90% vanillin can be produced. By mixing these vanillin crystals back with the oil-phase any desired vanillin concentration can be obtained, which makes the product very useful for the food industry.

9.6.2.3 Extraction of pungent constituents

Pungent components produce a hot sensation on the tongue or at the back of the throat. The most important species of this group are: capsicum, ginger, pepper (black and white), mustard and horseradish.

Ginger

Ginger is the rhizosome of the plant *Zingiber officinale* Roscoe, grown in southern Asia. It was one of the first oriental spices to reach south-eastern Europe in the ancient spice trade.

There are many different varieties on the market, including:

- Jamaica ginger is light in colour. The odour is aromatic, agreeable, pungent and spicy. The taste is aromatic, pungent, and with a volatile oil content of 1.0% vol/wt.
- African ginger, primarily from Sierra Leone, is mainly used in meat-seasonings and has an average volatile oil content of 1.6%.
- Nigerian ginger is light in colour, it is strongly aromatic and camphoraceous, and has a volatile oil content of about 2.5% vol/wt.
- Indian ginger (Cochin) is light brown in colour, has a strong lemon odour, and contains about 2.2% volatile oil.
- Indian ginger (Bombay) is darker and has an earthy odour.
- Japanese ginger is partially peeled, and usually limed with calcium carbonate.
- Chinese and Australian ginger are light brown, and contain about 2.5% volatile oil (vol/wt).

For the CO_2 extraction, usually Nigerian and Indian ginger (Cochin) are used as raw material.

To obtain the highest yield of the pungent substances, the extraction pressure should be 420 to 450 bar and the temperature 60°C. With the CO_2 extraction, it can be ensured that the content of 6-shogoal is at a low level of 1.5 to 2.0% and does not reach the levels of 5% and more as in conventional solvent-extracts.

With the two-step separation, the content of volatile oil and gingerol can be adjusted in the extracts.

Raw material: Indian Ginger Cochin

1.) Extraction parameters: 450 bar, 60°C

1. Separator: 80 bar, 40°C	2. Separator: 40 bar, 15°C
Yield: 3.2%	Yield: 2.3%

554

Sum of gingerol: 22.0%	Sum of gingerol: 11%
6-Shogoal: 1.6%	6-Shogoal: 1.55%
Medium, viscous oleoresin	Thin, liquid extract, mostly volatile oil

2.) Extraction parameters: 450 bar, 60°C

1. Separator: 120 bar, 50°C	2. Separator: 40 bar, 15°C
Yield: 1.7%	Yield: 4.3%
Sum of gingerol: 26.5%	Sum of gingerol: 17.0%
6-Shogoal: 1.2%	6-Shogoal: 1.5%
Hard, viscous, dark resin	Low viscous, oily, reddish brown extract

The total extract of Cochin Ginger has about the following specification:

Yield: 4.5 to 5.5%wt.
Sum of gingerol in the extract: 20 to 24%
Volatile oil: 30 to 33%
Density: 0.9450 to 0.9750

Pepper-white and black

White and black pepper are both from the same plant, *Piper nigrum*. Black pepper is the unripened dried fruit, while white pepper is the natural berry, from which the red hull is removed. Black pepper berries have a round shape and a diameter of 0.25 to 0.5 cm, black to dark brown in colour and shrivelled in appearance. The round berries of white pepper are smooth, and white to grey in appearance.

Black and white pepper have two important main components, the volatile oil and the pungent component, which is known as piperine. Black pepper contains about 0.6 to 2.6% of volatile oil, depending on the source, but it can also contain 2.0 to 5% in good quality pepper berries. White pepper contains 1.0 to 3.0% of volatile oil.

The extraction of pepper with liquid- and supercritical carbon dioxide is one of the most difficult jobs, even though it is often described in the literature [29].

On the one side, the pungent component, piperine, needs a high extraction pressure and temperature (450 to 500 bar, 80°C) to reach sufficient extraction efficiency. On the other side, the volatile oil is very sensitive and should be extracted under gentle conditions.

Further, the volatile oil needs mild separation conditions to give a sufficient separation from the gaseous CO_2 stream.

The piperine crystallizes in the separator, and is very difficult to drain out of it. Further, crystalline piperine is abrasive and blocks pipes and valves.

Under supercritical conditions at 500 bar, 80°C, a crystalline extract with 40 to 65% piperine can be obtained with an extraction efficiency of about 90%. The commercial extracts are standardized to 40% piperine and with 20- or 10% essential oil.

9.6.2.4 Production of natural colorants

The colouring matters of spices and herbs belong to one of the following categories:

1. Tetrapyrrole derivatives: chlorophylls, pheophytins;
2. Isoprenoid derivatives: carotenoids, xanthophylls;

3. Benzopyran derivatives: anthocyanins, flavonoids and related compounds;
4. Other natural pigments including tannins, betalains, leucoanthocyanins, quinones and xanthones.

Pigments of groups 3 and 4 are usually water-soluble, and can be extracted with polar solvents only.

The chlorophylls are responsible for the green colour of nearly all plants. The isomers, chlorophyll a and chlorophyll b exist in a 3:1 ratio in plants. Chlorophylls are magnesium-chelated tetrapyrroles with an esterified 20-carbon alcohol, phytol. Acidic conditions can cause the replacement of magnesium for hydrogen and change the chlorophyll into pheophytins. The pheophytins are brown in colour, and are normally undesirable in most foods.

There have been only a few citations in the literature concerning the extraction and recovery of natural chlorophyll pigments. Chlorophylls, despite their high molecular weights, are soluble in pure CO_2 (like pheophytin a at 500 bar and 55°C), or in the presence of an entrainer (like copper chlorophyllin at 60 bar and 20°C with 5% ethanol entrainer). The extraction of dried grass resulted in 1.56 wt.% yield of green components [79].

Chlorophylls are also present in many oleoresins. By fractionation of fennel oil in two separators in series, the pigments precipitate in the first separator [80]. The effects of the extraction parameters on chlorophyll concentration are shown in Table 9.6-4. Increasing extraction pressure and/or temperature favour the dissolution of the pigments.

Table 9.6-4
Comparison of chlorophyll-contents in fennel extracts obtained with different extraction parameters

Extraction conditions		Pigment content (μ/g)		
Pressure (bar)	Temperature (°C)	Chlorophyll (phytyl-free)	Chlorophyll (a and á)	Pheophytins
238	38	0	0	12.06
238	62	0.72	0	17.17
270	50	2.50	0.98	12.19
302	38	2.64	0	19.72
302	62	3.12	1.42	24.63

The carotenoids are a group of lipid-soluble pigments responsible for the yellow, orange, and red colour of many plant materials. Most foods contain a variety of carotenoids that differ mainly in their content of double bonds and oxygen atoms: over 400 different compounds are known. The most important among these is β-carotene. In addition to its excellent colouring properties, it is a precursor to vitamin A, and has antioxidant activity (which is believed to assist the prevention of certain forms of cancer).

The solubility of β-carotene in supercritical fluids has been studied extensively [81 to 85]. The extraction of β-carotene from a wide varieties of natural sources has also been described: like alfalfa-leaf protein concentrates [86], carrots [34,87], sweet potatoes [88], and algae [89].

556

Capsicum spices *(Capsicum annuum)*

The colour of the mature paprika is caused by carotene and carotenoid compounds, which can be found in pericarpal tissue. The pod when it is finished growing is still green in colour, then during the ripening the green colour slowly turns into red and the mature pod will be purple.

According to literature data, the total pigment content of ripe paprika consists of about 50 to 60 organic compounds, which are stable but different in their structures. Seven of them comprise 90 to 95% of the total pigment content. These are capsanthin, capsorubin, β-carotene, cryptoxanthin, lutein, violaxanthin and zeaxanthin. They mostly consist of 40 carbons and are linear compounds with many conjugated double bonds, and with rings at the ends of the chain. (See Figure 9.6-7)

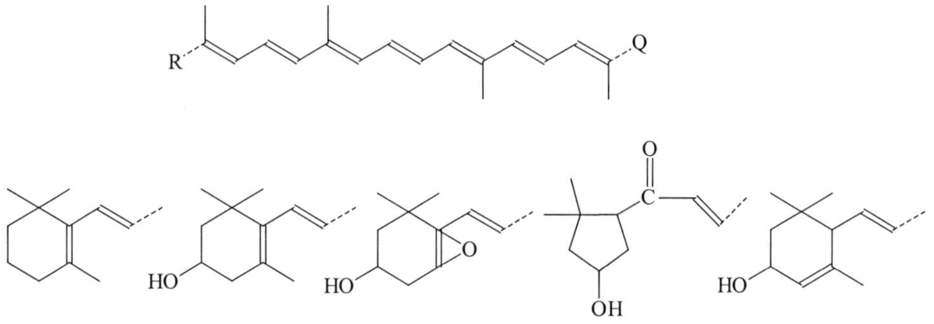

	R	Q
Capsanthin	b	d
Capsorubin	d	d
β-Carotene	a	a
β-Criptoxanthin	a	b
Lutein	b	e
Violaxanthin	c	c
Zeaxanthin	b	b

Figure 9.6-7. Chemical structure of capsicum colour substances

The content of colouring matter reduces during storage. Under the most careful storage, the ground paprika loses from its colour value 0.1 g/kg per month. This process can be delayed if the product is stored at or below a temperature of +5°C.

In the traditional process, the dried and ground paprika is extracted by organic solvents (such as hexane, acetone, benzene, methylene chloride, or dichloroethane) which have low boiling points. These traditional processes have several disadvantages. Most organic solvents like benzene and the chlorinated solvents are toxic, and government food regulations dictate that their residues must be reduced in the oleoresins to very small concentrations, generally in the range of 25 to 30 ppm or less.

If the solvent is not removed totally from the porous extraction residue, it leads to a considerable loss of solvent; on the other hand it leads to a restriction in employing the extraction residue as fodder. A further disadvantage of the solvent extraction is that the colorants, which are susceptible to oxidation, can be damaged during solvent-removal.

The traditional extraction processes do not provide a method for selective fractionation of the oleoresin components. This fractionation must be carried out in additional processing steps, involving further use of organic solvents or costly processes such as molecular distillation.

These disadvantages (mentioned above) can be avoided by using supercritical fluid extraction.

Coenen *et al.* [60,61] proposed a two-step extraction for the separation of pungent compounds and carotenoid fractions. Aroma- and pungent components were recovered at 120 bar and 40°C, and the paprika residue was re-extracted at 320 bar and 40°C to recover carotenoids. The solubility of capsaicin in carbon dioxide was relatively low at a pressure of 120 bar, so a great amount of solvent (for example 130 kg of CO_2 per kg of paprika) was needed to recover the aroma components totally. The extraction time was 6.5 hour. In the separator the pressure was 56 bar and the temperature was 45°C. The orange, paste-like extract recovered in the first step was extremely pungent in taste. It contained water, and the yield was about 15%. In the second step, a relatively great amount of CO_2 (approximately 50 kg/kg) was needed to recover the carotenoids in quantitative yield. The extraction time was 4 hours. The dark red, liquid colour-concentrate is without capsaicinoids. The yield was 2.5%.

Noble sweet, red paprika (not containing capsaicin) was extracted at a pressure of 300 bar and a temperature of 40°C to recover carotenoids and other lipophilic compounds (such as fatty oils). The extraction time was 6 hours, and the consumption of CO_2 was 193 kg/kg. A water-containing, thick, dye-concentrated oil was obtained in 8% yield. The extraction residue was pale yellow in colour, tasteless and odourless. It was an excellent fodder because of its great protein and carbohydrate content.

Paprika can be extracted to recover carotenoids, not only with CO_2 but also with other gases. For example, by using ethane or ethylene, better results were obtained for the yield, extraction time, and quality of product. The solubilities of carotenoids are better in these gases, which is why the consumption of solvent and the extraction time were reduced. Practically water-free dye-concentrate was recovered by supercritical fluid ethane (under the conditions: extraction 250 bar, 45°C; separation 46 bar, 45°C). The separation of pungent substances (capsaicinoids, free fatty acids) from the pigments can be carried out effectively in a continuous, counter-current extraction column with a large number of theoretical plates.

In the patent of the Krupp company [60], ground paprika and oleoresin may be used as the raw material in the SFE process. The solvent can be supercritical fluid carbon dioxide, ethane, ethylene, or a mixture of the last two. As modifiers ethanol, acetone, water, and mixtures of these solvents were proposed.

Capsicum spices *(Capsicum frutescens)*

According to present knowledge, the pungency of paprika is caused by various amides of alkaloid character with the common name, "capsaicinoids". These are produced by glands associated with the placenta in the centre of the pod where the seeds are produced. Seeds are not the sources of pungency, although they may absorb some of these capsaicinoids. In pungent paprika, the content of capsaicinoids is 0.3 to 1.5% [63]. Five naturally occurring capsaicins have been reported: they are capsaicin (C), dihydrocapsaicin (DC), nordihydrocapsaicin (NDC), homocapsaicin (HC), and homodihydrocapsaicin (HDC) (see Fig. 9.6-8).

Structure	Common name	Mol. wt
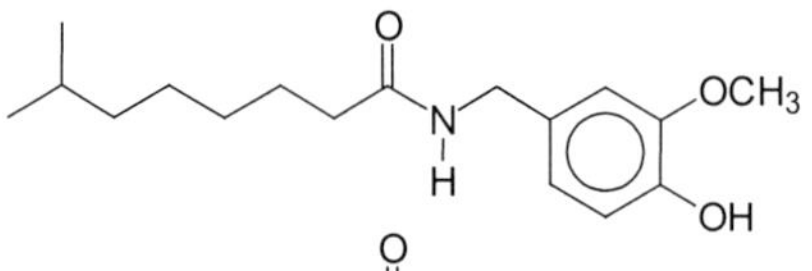	Nordihydrocapsaicin (NDC)	293
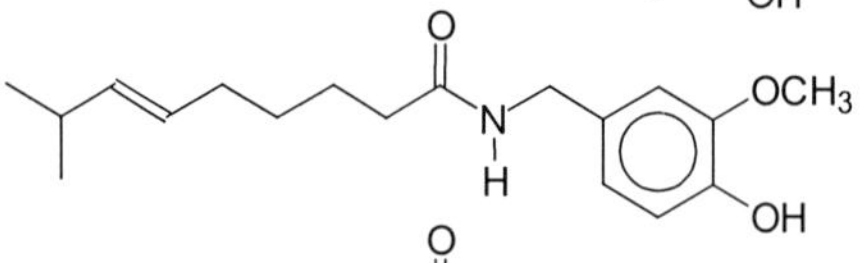	Capsaicin (C)	305
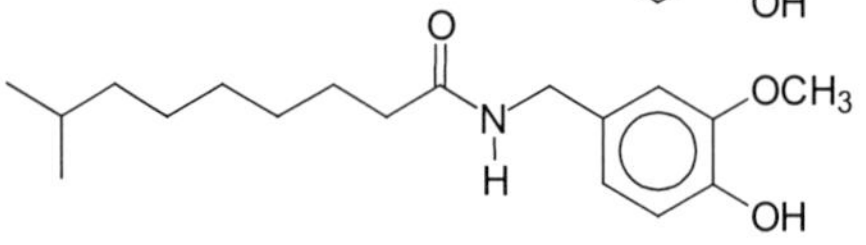	Dihydrocapsaicin (DC)	307
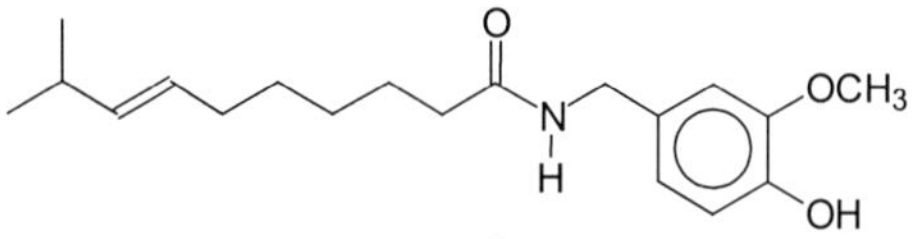	Homocapsaicin (HC)	319
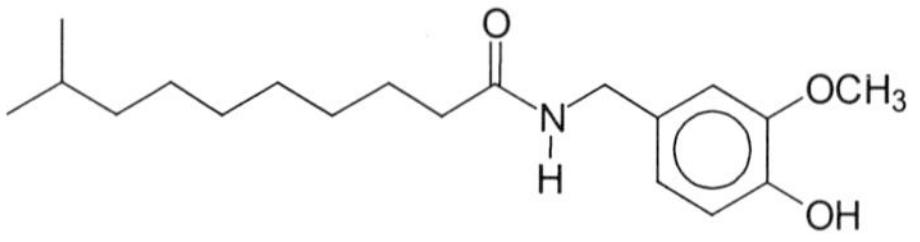	Homodihydrocapsaicin (HDC)	321

Figure 9.6-8. Chemical structures of pungent substances of capsicum.

Capsaicin is the most important of the capsaicinoids. Pure capsaicin, or 8-methyl-*N*-vanillyl-6-nonenamide, is colourless crystals, soluble in alcohol, but insoluble in cold water, which is why drinking water helps to alleviate the burning sensation when it is eaten.

Capsaicin is a powerful local stimulant and when swallowed it produces a sense of heat in the stomach. Capsaicin is applied widely as a condiment and can be employed as a counter-irritant in the treatment of lumbago, neuralgia and rheumatism.

The equilibrium solubilities of capsaicin in carbon dioxide were investigated in the pressure range of 70 to 400 bar at temperatures of 25, 40 and 60°C. The maximum solubility ($3.2 \cdot 10^{-4}$ mol/ml) was obtained at 400 bar and 60°C [63].

Ground chilli was extracted by Hubert and Vitzthum [29] with supercritical fluid CO_2. A red oil was obtained with 4.9% extraction yield, containing 97% of the capsaicin present in the raw material.

The supercritical CO_2 extraction of *Capsicum annuum* var. Scotch Bonnet gave 16.4% of extract, and it contained 3.2 and 0.5% capsaicin and dihydrocapsaicin, respectively, per dry weight of the raw material. Organic solvents (hexane, $CHCl_3$, and methanol) were used sequentially to extract Scoch Bonnet to afford a combined total of 26.7% extract, and contained only 0.5 and 0.09% capsaicin and dihydrocapsaicin, respectively [64].

As two of the most important spices, paprika and chili have been very well investigated as raw materials in the CO_2 extraction. The volatile oil content is very low in capsicum species. Usually it is below 0.2 wt.%, and is not considered in the specification. The pungency (for the chili) and the colour value (for the paprika) are the most important parameters when choosing the raw material. The pigment which is responsible for the red colour is capsanthin, which is a carotenoid present in all capsicums. Most of the pigments are in the pericarp: the seed of the paprika pod contains only very small amounts of pigment. Paprika pigments are oxidized very rapidly during storage, and so the industry is using large amounts of paprika oleoresin.

Before extracting the dried paprika, the fruits are crushed and the pericarp and seed separated.

After separation the following quantities are obtained:
- Pericarp: 40%
- Seed: 54%
- Stem: 6%

The pericarp is ground further to a particle size of 0.3 to 0.8 mm and put into the extractor basket.

Table 9.6-5
Colour value for different species of paprika

Species	Colour values of raw material	Colour values of oleoresins	Remarks
Sweet Paprika Hungary (whole pods)	50 ASTA	350 to 500 ASTA	With seeds and stems
Paprika sweet Hungary "Kalocsai"	120 ASTA	1,200 to 1,400 ASTA	Without seeds
Paprika sweet Hungary "Kalocsai"	140 ASTA	1,500 to 1,650 ASTA	Without seeds
Chilli sweet India	150 ASTA	1,800 to 2,000 ASTA	Without seeds
Paprika sweet Spain	200 ASTA	2,200 to 2,500 ASTA	Without seeds
Paprika sweet South Africa	240 ASTA	>2,500 ASTA	Without seeds

After extraction of the whole paprika, including seed and stems, the obtained colour-value of the extract is relative low, about 350 to 500 ASTA, but the extraction yield can reach 15 to 17 wt.%. High-quality Hungarian paprika species like Sweet Paprika Excelsior give an extract with a colour value between 1,500 and 1,800 ASTA, and a good flavour profile.

Paprika pericarp from Spain or South Africa have initial colour values of 200 to 240 ASTA, and extracts with 2,500 ASTA and more can be obtained.

560

All Indian sweet chilis, even those with a high pigment concentration, also contain the pungent component called capsaicin.

With CO_2 extraction it is possible to produce colour extracts up to 3,000 ASTA from Indian chili. These extracts still contain 2 to 2.5 wt.% capsaicin, which can be removed in a second CO_2 extraction step, if desired. The result is a high concentrated colour extract with a colour value up to 5,000 ASTA and more, and a highly pungent fraction containing capsaicin up to 20 wt.%. Such extracts are produced, for example, in a multi-purpose plant at Novo Agritech near Hyderabad, India.

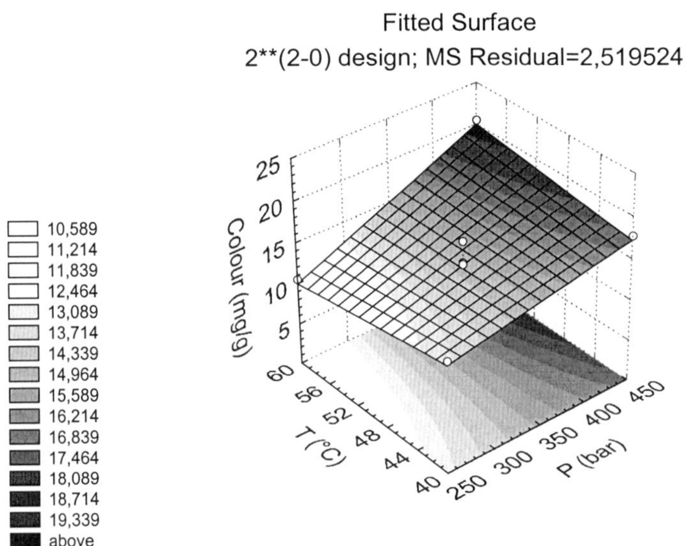

Figure 9.6-9. Effects of extraction parameters on the total pigment-content of the paprika oil ("Kalocsai Csemege", Hungary).

Designed experiments were carried out to determine the quantitative effects of extraction-pressure and -temperature on the pigment recovery. The 2^2 full-factorial-design was realized once, and tree-repeated experiments were performed in the centre of the design. The three-dimensional-response surface of the effects of extraction parameters on the pigment concentration in the oil is shown in Figure 9.6-9. It is seen that an increase of extraction pressure increases the pigment content significantly.

The optimal extraction pressure for producing paprika oleoresin should be 350 to 500 bar. Higher pressures extract too rapidly the fatty oil, which acts as an entrainer for the pigments. Below 300 bar, the solubility of the pigments is too small, and mainly the fatty oil is extracted.

As can be seen from Fig. 9.6-10 low pressure gives a poor yield, especially of capsanthin, the coloured fraction of paprika. The colour value is only about 1,370 ASTA (54,800 CU), or about 48% of initial capsanthin content. A higher extraction pressure like 500 bar allows a depleting yield of capsanthin.

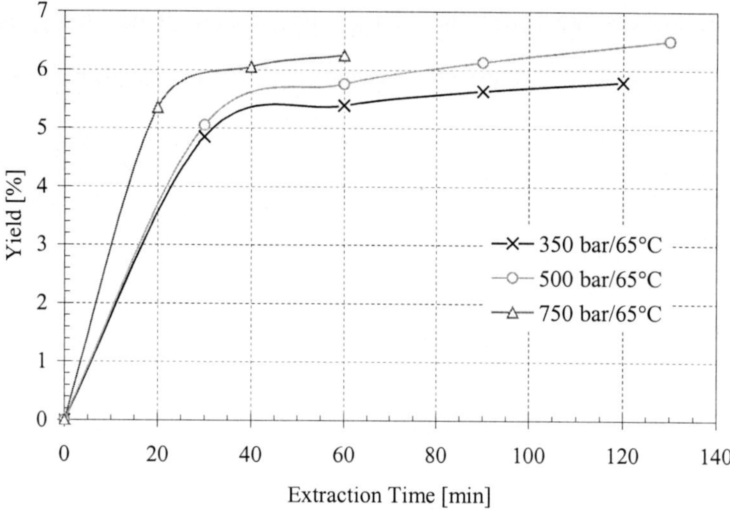

Figure 9.6-10. Yield of paprika colour

With a two-step separation it is possible to enrich the colour fraction in the first step, and to obtain most of the aroma fraction in the second separator, with a colour value of about 200 to 300 ASTA. The conditions in the first separator should be 220 to 230 bar and 40°C.

Turmeric

Turmeric is the rhizosomes of a plant which is a member of the Zingiberaceae family and native to southern and southeast Asia. The traditional name of this spice is *Curcuma longa.*

The dried and polished fingers of turmeric contain 5 to 10% fatty oils, up to 5% volatile oil, and 5 to 10% of the colouring matter, curcumin.

The pure oleoresin produced by solvents normally contains only pure curcumin, in a crystalline form. It is hardly soluble in liquid- and supercritical CO_2. Even at an extraction pressure of 450 bar, and with 2 hours' extraction time at 65°C, only 20% of the initial curcumin can be extracted. On the other hand, all the volatile oil and fatty oil is extracted, and a fat-free curcumin-starch mixture with a very low flavour-content can be produced. The total extraction yields are between 5 to 12%, with mostly fatty oil and volatile oil, and about 10% curcumin in the extract.

9.6.2.5 Production of natural antioxidants

Lipid oxidation is a degradative, free-radical-mediated process, responsible for the development of unpleasant odours and flavours in oils, fats, and foods containing them [90]. Moreover, chain oxidation processes initiated in the human body can cause carcinogenesis, atherosclerosis, infarction, and other pathological changes [91,92]. Antioxidants are used to minimize detrimental lipid peroxidation, because they neutralize and scavenge free radicals. In the early 1980s, consumers began looking for natural ingredients in their foods. Sources of natural antioxidant compounds are spices, herbs, tea, oils, cereals, grains, fruits and vegetables [93].

Many researchers have recently studied the antioxidant activity of plants and plant extracts [94,95].

The antioxidant activity of rosemary and sage (leaves and extracts) were most effectively investigated [96,97]. Traditional extracts of spices and herbs are obtained by steam distillation (essential oil) or by extracting the botanical with solvents such as alcohol, hexane, or acetone, and removing the solvents by evaporation. The SFE process for production of the inherent natural antioxidants is now the most gentle and effective method [70].

In the year 1952, Chipault *et al.* mentioned that rosemary and sage have the best antioxidant activities, followed by oregano, thyme, clove, allspice and black pepper [28]. By using the CO_2 extraction the carnosolic acid, the most effective substance in this respect, can be enriched to high concentrations.

Many undesired substances such as chlorophyll or other polar substances, which are extracted in the alcohol extraction, are only slightly soluble in CO_2. In case of necessity, the aroma components can be removed from the extract by a further purification step.

Table 9.6-6
Extraction of herbs with antioxidant activity

	750 bar/70°C Yield [wt.%]	Carnosolic acid content [wt.%]	500 bar/70°C Yield [wt.%]	Carnosolic acid content [wt.%]
SAGE *Salvia officinalis* Austria	8.5	8.01	5.3	10.5
OREGANO *Origanum vulgare* Austria	4.75	0.03	3.4	0.1
THYME *Thymus vulgaris* Austria	5.6	0.04	3.6	0.05
ROSEMARY *Rosemarinus officinalis*, Maroc	8.2	22.4	7.5	24.5

Rosemary is the raw material most used for the production of carnosolic acid. Medium qualities contain about 1.6 - 1.8 wt.%, and high-quality rosemary contains more than 2 wt.% carnosolic acid.

The optimized extraction pressure required in order to get a desired yield in a feasible time is about 500 bar. At 750 bar, carnosolic acid has a higher solubility, and the extraction efficiency of 90% can be reached in a shorter time, but other, undesired substances, are also in the extract.

At an extraction pressure lower than 400 bar the extraction efficiency for carnosolic acid is not sufficient.

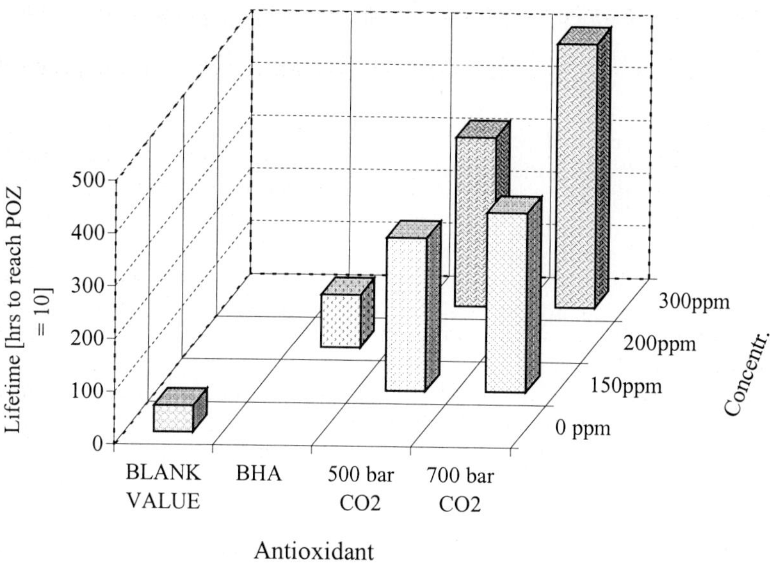

Figure 9.6-11. Antioxidant activity of CO_2-rosemary extract

Fig. 9.6-11 demonstrates the stability of lard stabilized with CO_2-rosemary extracts, expressed as the Peroxide Number (POZ), compared to BHA (Butylhydroxyanisole), and pure lard, as a blank test value.

9.6.2.6 Production of high-value fatty oils

The supercritical fluid extraction of oil seeds has been investigated extensively by several authors [34,98]. Possible applications of supercritical fluids in the edible-oil industry include deacidification, deodorization, and fractionation of crude oils and chemical conversion (like hydrogenation, and enzymatic reactions).

Removal of oil from commodity oil seeds (sunflower, soybeans, and rape) does not appear economic, owing to the expense of the high-pressure batch process. But there may be areas in which SFE can be useful in the extraction of high-value oils. From health food and speciality gourmet oils (almond, apricot, avocado, grapeseed, hazelnut, walnut), to pharmaceutical and cosmetic applications (corn germ, wheat germ, evening primrose, borage), there are speciality oils which cost more than commodity oils. Because of its high-capital costs, a commercial scale processing plant should be able to process various raw materials (multi-purpose plant). The operating pressure is also limited since the optimal conditions for oil extraction are much greater than for other raw materials, and high pressures in the order of 500 - 800 bar will entail high capital and operating costs.

Numerous studies have been conducted on SFE of corn germ. Christianson *et al.* [99] and List *et al.* [100], respectively, compared the composition and stability of oils obtained by SFE, or recovered by conventional hexane extraction. The quality and storage characteristics of the defatted germ flour obtained by SFE were superior to those of germ flours obtained by hexane

564

extraction. The oil yielded by SFE had a significantly lower refining loss with about 5 to 10 ppm of the phosphorus content, and was light-coloured but did not contain the native antioxidants. Therefore, it was much less stable to oxidation. Quirin *et al.* [101] reported a 100 ppm phospholipid content of CO_2-extracted-, and 1.3% phospholipid in hexane-extracted oils, while the free-fatty-acid and peroxide values were nearly the same. Wilp and Eggers [102] demonstrated that by using multi-stage high-pressure extraction combined with multi-stage separation it is possible to obtain oils nearly free from water and with a low content of free fatty acids, as well as oils with a high content of tocopherols and sterols. Furthermore, it is possible to fractionate the triglycerides.

The rate of extraction can be increased significantly by using ethyl alcohol as entrainer. The extraction curves, in which the total extraction yields are plotted against the specific CO_2-solvent-mass passed through the raw material, are shown in Figure 9.6-12. Increasing the alcohol concentration in CO_2 from 0% to 10% significantly improved the solubility of phospholipids (phospholipid content of oils increased from 0.026 wt.% to 0.756 wt.%). The use of alcohol as a co-solvent also improved the functionality of protein materials [103].

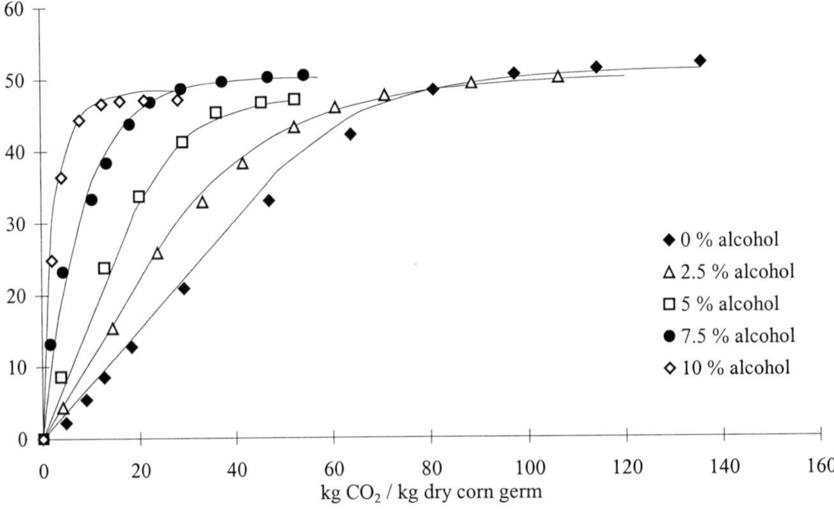

Figure 9.6-12. Effect of alcohol-entrainer content on the extraction curves

Extraction of γ-linolenic acid

High-grade oil with the essential polyunsaturated fatty acids can be extracted very successfully with CO_2. The physiological activity of these oils is determined by the ω-6,9-double bonds, and γ-linolenic acid, $C_{18:3}$ (*6c, 9c, 12c*) is of particular interest.

Carbon dioxide-extracted oils have the same fatty acid distribution as do hexane extracts, but they are clear, light yellow, high grade oils. It must be concluded that the phospholipids are not extracted with CO_2.

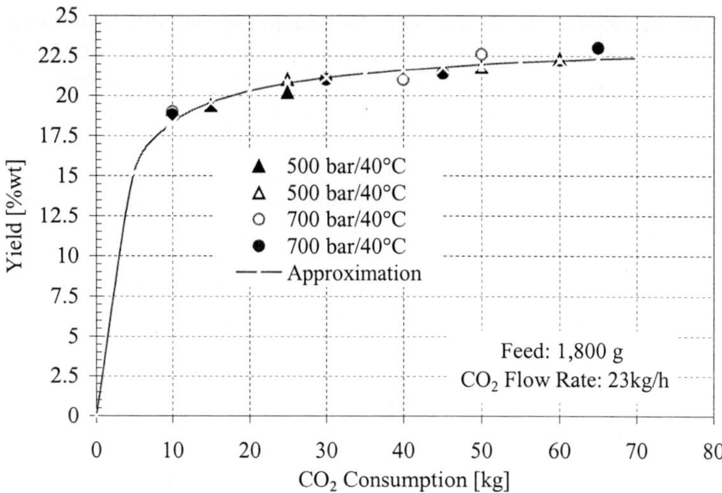

Figure 9.6-13. Extraction of evening primrose seeds

In contrast to coriander seeds, high extraction pressures do not have advantages: pressures of 500 bar are sufficient. As well as evening primrose, many other raw materials are very interesting for the production of γ-linolenic acid.

Table 9.6-7
List of raw materials for γ-linolenic acid:

Raw material	Yield CO_2 extraction [%]	Source	$C_{18:3}$ in the extract (γ-linolenic acid) [%]
Evening primrose	22	Poland	8 to 12
Borage	18	Canada	18 to 22
Hops seed	6	Czech	3 to 4
Hemp seed	35	Austria	3 to 6
Redcurrant seed	14	Austria	4 to 6
Blackcurrant seed	18	Austria	16 to 19
Rose hips	12	South Africa	24 to 31
Sea Buckthorn	12	Russia	26 to 30

9.6.3 Depestisation of vegetal raw materials

Many natural materials contain residues of plant-protection agents which are, understandably not desired owing to their possible effect on human or animal organisms.

Such residues are, despite reduced application to plants and the cultivation of parasite-resistant plants, unavoidable, and for food-use, are controlled in most countries or under legal regulations, respectively. For the reduction or removal of such residues only solvents such as CO_2 can be used, which are harmless and do not remain in the food.

Processes of such kinds can be found in the literature. E. Stahl, *et al.* [1] describe the decontamination of drugs, in particular, the reduction of pesticides from senna leaves by means of supercritical dry carbon dioxide [2]. The patents EP 0382116 A2 [4] and DE 4342874 A1 [5] specify processes to remove non-polar substances from ginseng with dense carbon dioxide, in which the ginseng is moisturized prior to extraction, and the second method also uses an entrainer. The European Patent Application EP 0925724 A2 from NATEX [3] explains a process for the removal of plant-protection agents and/or undesired substances from cereals.

9.6.3.1 Decontamination of the rice

Within this area of process development one of the first aims was to evaluate the solubility and extractability of selected representative carbamate pesticides, whose residual concentration should be as low as possible. Solubility data were obtained from experimental investigations of the related pesticides in SC - CO_2 from inert matrices (sea sand) first, and later on this experience was applied for the real matrix of rice.

Table 9.6-8
Instrument specification of the HP 7680 T SFE

Parameter	Hewlett Packard HP-7680T
Type of pump	Reciprocating, piston-type metering pump
Operating pressure range	77 to 383 bar
Flow range	0.5 to 4.0 ml/min (liquid CO_2 at 5°C) automatically settable per extraction step
Capacity	Continuous
Accuracy	+/-2% (340 bar, thermostat pump head)
Extraction fluid	CO_2
Extraction chamber	
Temperature range	40 to 150°C automatically settable per extraction step
Settable density range	0.15 to 0.95 g/ml CO_2; automatically settable per extraction
Extraction cartridge	
Description	Vertical 7.0 ml, stainless steel 1.1 cm id*8.3 cm length
Maximum pressure rating	540 bar
Back pressure device	
Dimensions of restrictor	Variable restrictor nozzle, with electronic control of dimensions. Automatically achieves/controls to pressure selected per extraction step.
Temperature range	5 to 150°C
Collection system	
Type of collection system	Adsorbent Trap (ODS, stainless steel etc.) With automatic presentation of extracted fractions to be used in standard autosampler vials.
Temperature	Variable heating/cooling –30°C to +120°C

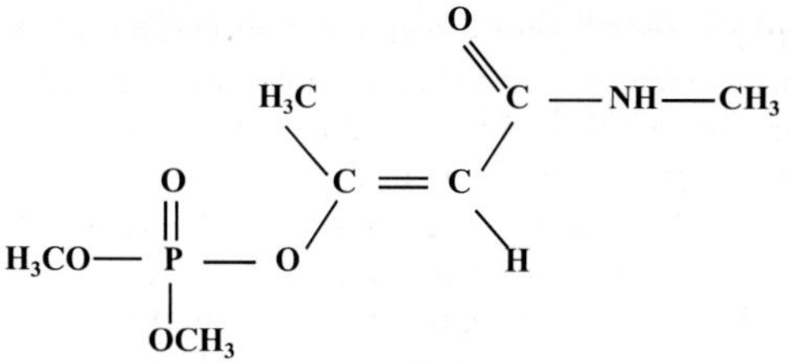

Monocrotophos

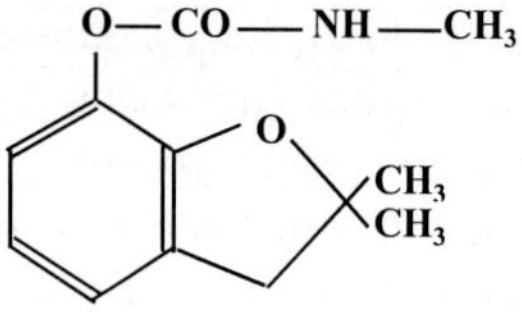

Carbofuran

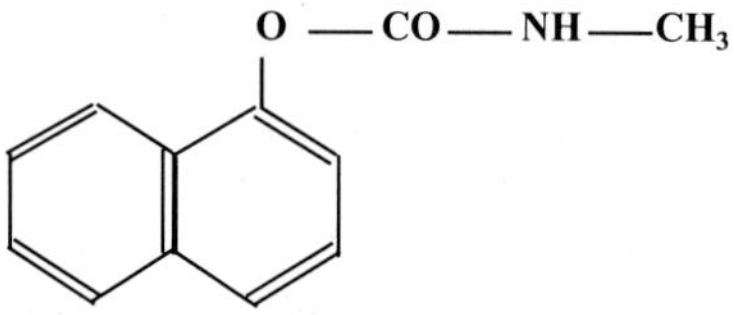

Carbaryl

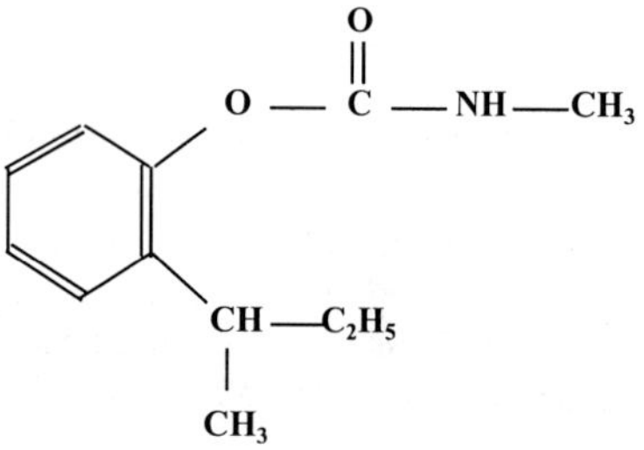

Fenobucarb

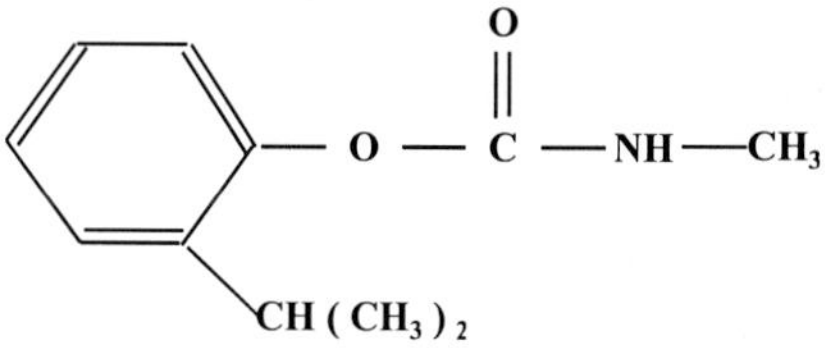

Isoprocarb

Figure 9.6-14. Chemical structure of carbamates

In order to have an overview of the possible chemical structures of all carbamates, the following pesticides were chosen (all of ultrapure analytical grade).

Pesticide analysis was done by using a Hewlett Packard-1050 HPLC coupled with fluorescence detection and diode-array detection. The system for obtaining supercritical conditions was an SFE Module produced by Hewlett-Packard, with a specification shown in Table 9.6-8.

After SF-treatment the rice was extracted three times, with 2 ml methanol per g rice, for 20 min on a rotary shaker. The extract was collected and evaporated to 1 ml with nitrogen and extracts analysed on HPLC. The results were calculated as μg/kg dry substance (ppb).

Because large amounts are to be treated on an industrial scale, and preboiling of rice must be avoided, the following conditions were selected:

- Extraction temperature below 60, preferably under 50°C.
- Extraction pressure below 400, preferably under 325 bar.
- CO_2 consumption as low as possible.

For trend-analysis a test series was performed with two fixed temperatures of 40°C (see Table 9.6-9) and 50°C (see Table 9.6-10) at different pressures but constant CO_2 consumption of 5.5 kg CO_2/kg sand. The relative recoveries are displayed in these tables.

Table 9.6-9
Set points: 40°C, 5.5 kg CO_2/kg sand. Relative recovery in percent of related pesticides from sea sand

	40°C/100 bar 20 min	40°C/170 bar 15 min	40°C/250 bar 14 min	40°C/360 bar 13 min	Rated value
Monocrotophos	70	62	75	70	100
Carbofuran	73	58	69	61	100
Carbaryl	52	39	50	43	100
Isoprocarb	80	66	74	59	100
Fenobucarb	92	76	83	77	100
Sum	73	60	70	62	100

Table 9.6-10
Set points: 50°C, 5.5 kg CO_2/kg sand. Relative recovery in percent of related pesticides from sea sand

	50°C/100 bar 31 min	50°C/170 bar 17 min	50°C/245 bar 15 min	50°C/350 bar 14 min	Rated value
Monocrotophos	26	81	79	16	100
Carbofuran	75	74	73	13	100
Carbaryl	34	50	47	10	100
Isoprocarb	94	82	81	11	100
Fenobucarb	101	81	99	20	100
Sum	66	76	76	14	100

The recovery appears more or less independent of pressure, and the temperature did not give a measurably higher recovery, but longer extraction times tend to give higher effectiveness.

From experiments with an inert matrix, further investigations focused on the humidity of the rice. The results with extraction conditions of 100 bar and 40°C and a CO_2 consumption of 5.5 kg/kg rice are shown in Fig. 9.6-15, for which the water content was set by spraying defined amounts of water on the surface of the rice. The best results are obtained with moisture contents between 10 and 15 wt.%. Significant lose of efficiency was observed with increased humidity.

Alternatively, continuous water addition into the CO_2 flow was tested, but problems in gaining uniform distribution were observed. The reduction from an initial total concentration of 2,524 ppb amounted to about 72.3% with 1% vol., and only about 64.3% with 2% vol. water addition.

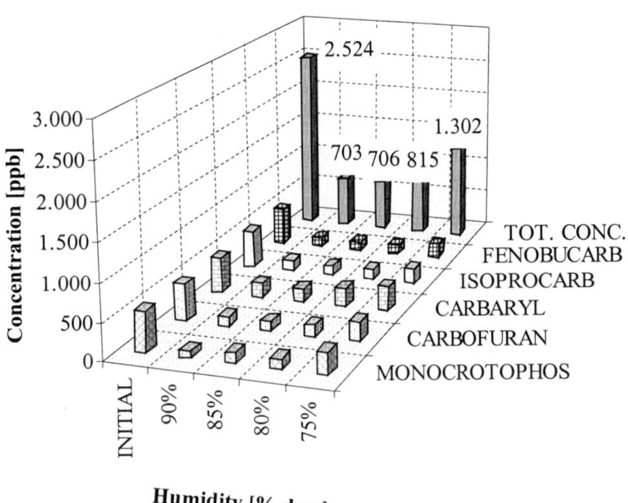

Figure 9.6-15. Influence of humidity

So far, the CO_2 consumption was kept constant and was the limiting factor for efficiency, and a step-wise increase of the quantity of CO_2 employed was investigated. To simulate natural conditions, contaminated rice samples at different levels were stored for 6 days at 20°C to support diffusion and adsorption of the diverse pesticides into and onto the rice. The concentrations of individual pesticides were adjusted to about 500/1,000 and 2,000 ppb. The results, as displayed in Fig. 9.6-16, show that the removal of pesticides above 90% is possible under given conditions (100 bar, 40°C).

In summary it can be said that supercritical CO_2 is a moderate solvent for the investigated group of carbamates, although the chemical structure would support more polar solvents for complete removal. The pressure and temperature influences are not really important, so gentle extraction conditions can be applied. Water used as a matrix- or CO_2-modifier is appropriate, with an optimum water content of around 10 - 15%. An amount of 7 - 15 kg CO_2 per kg rice provides an acceptable effect for pesticide reduction.

9.6.3.2 Depestisation of ginseng

Another application, as mentioned earlier in this Chapter and in Chapter 8.1.4, is in the decontamination, mainly from organochlorpesticides, of ginseng. The initial concentrations in the raw materials investigated were in the range of 3.1 to 4.4 ppm for hexachloro-benzene (HCB) and 45 - 64 ppm for quintozen. As both are similar in structure, in addition, another

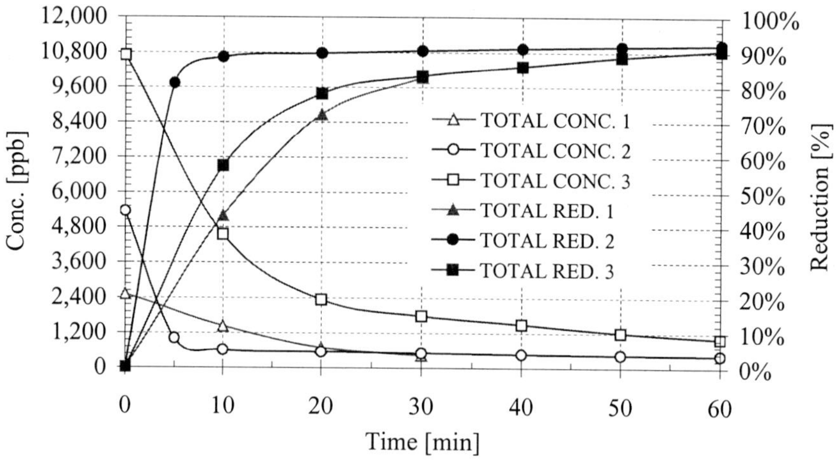

Figure 9.6-16. Pesticide reduction from different levels of contamination

pesticide - dieldrin - was spiked onto the ginseng in a range between 60 to 70 ppm prior to extraction.

In brief, the influence of the systematically investigated parameters mentioned in Chapter 8.1.4 are that for ground ginseng the temperature is highly significant, independent of the pressure (density) or whether a modifier is used or not, that higher densities result in better

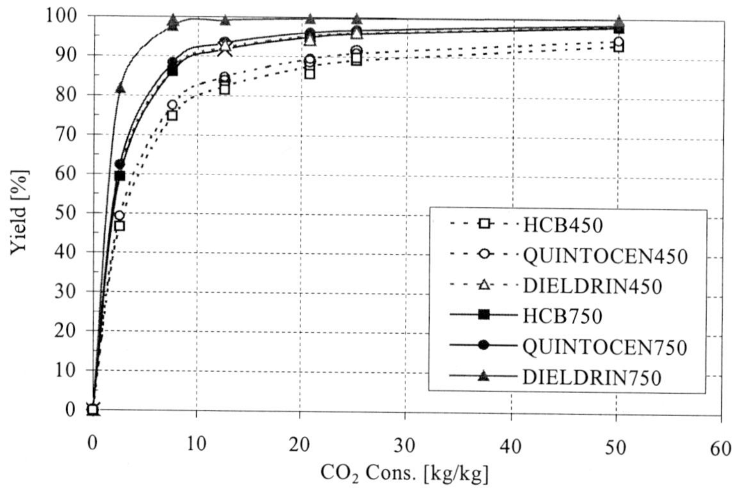

Figure 9.6-17. Reduction of diverse pesticides at different densities in relation to CO_2 consumption.

yields, and the influence of water as a modifier is negative, most likely owing to compacting of the ground material, which causes hindrance in diffusion. For larger ginseng pieces the yield is relatively poor, despite higher CO_2 consumption and matrix modification, most likely owing to the matrix/solute interaction and the slow desorption step.

The individual pesticide reductions under different extraction conditions, for two selected densities, in dependence on the CO_2 consumption, are given in Fig. 9.6-17. The illustrated densities are 750 kg/m³ and 450 kg/m³ obtained at temperatures of 70°C and 50°C with the corresponding pressures.

The influence of a modifier on ground ginseng and pieces, extracted under the same conditions (density of 750 kg/m³ with 70°C) is shown in Fig. 9.6-18. It is obvious that for pieces of ginseng the use of modifier is an advantage, unlike the situation with treated ground ginseng. The water addition of 30% to ginseng pieces might not be the optimum and requires further investigations.

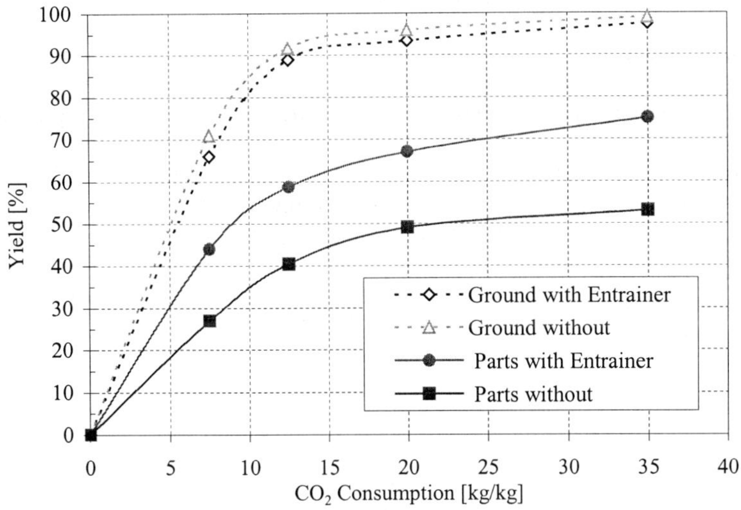

Figure 9.6-18. Influence of water as modifier on ground- and larger pieces of ginseng.

References

1. E. Stahl, K.W. Quirin, D. Gerard, Verdichtete Gase zur Extraktion und Raffination, Springer Verlag (1987) 231.
2. E. Stahl, G. Rau, Planta med., (1984) 71.
3. E. Lack, H. Seidlitz, F. Lang, E. Glanz, A. Trukses, Yi-Jen Liaw, European Patent Application EP 0 925 724 A2 (1998).
4. E. Schütz, H.R. Vollbrecht, European Patent Application EP 0382116 A2 (1990).
5. A. Forster, J. Schulmeyr, J. Sebald, Offenlegungsschrift DE 4342874 A1 (1993).
6. R.J. Clarke, R. Macrae, Coffee, Vol. 2: Technology, Elsevier Applied Science Publishers Ltd. (1987).

572

7. M.B. King and T.R. Bott, Extraction of natural products using near-critical solvents, Blackie Academic & Professional, Glasgow (1993).

8. K. Zosel, Separation with Supercritical Gases, Extraction with Supercritical Gases, VCH Verlagsgesellschaft mbH. (1980) 20.

9. E. Lack, H. Seidlitz and P. Toro, Proceedings of 13th Association Scientifique Internationale du Café, Paipa, Columbia, (1989) 236.

10. General Foods, European Patent Application 0331852A2 (1989).

11. O. Vitzthum and P. Hubert, German Patent Specification DE 2127642 (see also US Patent No. 4,167,589) (1973).

12. H. Klima, E. Schütz, and H.R. Vollbrecht, European Patent Specification 0200150 B1 (1990).

13. Hopfenextraktion HVG Barth, Raiser & Co., German Patent DE 3414767 C2 (1984).

14. P. Klüsters, Forum der Brauerei (1983) 17.

15. W. Sander and D. Deublein, Brauindustrie (1982) 16.

16. C.M. Grimmett, Chemistry and Industry (1980) 359.

17. L. Narziß, Brauwelt (1981) 1785.

18. H. Weyh, Brauwelt (1981) 1132.

19. L. Narziß, E. Reicheneder, Brauwelt (1983) 680.

20. R. Vollbrecht, Chem. Ind., (1982) 397.

21. M. Gehrig, Preprint of First International Symposium High Pressure Chemical Engineering, GVC-VDI, Düsseldorf, (1984). 281.

22. A.F. Prokopcuk, Izv. Piscev. Technol., 3 (1974) 7.

23. P. Hubert and O. Vitzthum, Angew. Chem. Ind., 17 (1978) 710.

24. H. Brogle, Chemistry and Industry, (1982) 19 June, 385.

25. G. Brunner and S. Peter, Chem.-Ing.-Tech., 53(7) (1981), 529.

26. K.T. Farrell, Spices Condiments and Seasonings, Second Edition, Van Nostrand, Reinhold, (1990).

27. D.R. Tainter, A.T. Grenis, Spices and Seasonings, A Food Technology Handbook, VCH Publishers, (1993).

28. J. Chipault, G. Mizuno, J. Hawkins and W. Lundberg, Food Res., 46 (1952) 46.

29. P. Hubert and O. Vitzthum, Angew. Chem., 90 (1978) 756.

30. C. Penny, Ingredients and Analysis International, (June/July), (1993) 13.

31. U. Paap, IFI Nr. 3, (1995) 43.

32. G. Hainrihar, IFI Nr. 4, (1991) 52.

33. D.A.J. Starmans and H.H. Nijhuis, Trends in Food Science and Technology, 7 (1996) 191.

34. E. Stahl, K.-W. Quirin and D. Gerard, Verdichtete Gase zur Extraktion und Raffination, Springer-Verlag, Heidelberg, (1987).

35. M.B. King and T.R. Bott, Extraction of natural products using near-critical solvents, Blackie Academic & Professional, Glasgow, (1993).

36. G. Brunner, Gas extraction, An introduction to fundamentals of supercritical fluids and the application to separation processes, Steinkopff, Darmstadt, (1994).

37. E. Reverchon, J. Supercrit. Fluids, 10 (1997) 1.

38. H. Knapp, R. Döring, L. Oellrich, U. Plöcker and J.M. Prausnitz, Vapor-liquid equilibria for mixtures of low-boiling substances, DECHEMA Chem. Data Series VI. (1981).

39. R. Fornari, P. Alessi and I. Kikic, Fluid Phase Equil., 57 (1990) 1.

40. R. Dohrn and G. Brunner, Fluid Phase Equil., 106 (1995) 213.
41. H. Kollmannsberger and S. Nitz, Chem. Mikrobiol. Technol. Lebensm., 15(3/4) (1993) 116.
42. E. Stahl and D. Gerard, Parfuem. Kosmet., 63 (1982) 117.
43. E. Reverchon, L. Sesti Osseo and D. Gorgoglione, J. Supercrit. Fluids, 7 (1994) 185.
44. B. Simándi, J. Sawinsky, A. Deák, S. Kemény, Á. Kéry, M. Then and É. Lemberkovics, Fractionated extraction of essential and fatty oils from spices with carbon dioxide, In: Solvent extraction in the process industries, Proceedings of ISEC '93 (ed: D.H. Logsdail and M.J. Slater), Elsevier, London, (1993) 676.
45. H. Sovova, R. Komers, J. Kucera and J. Jez, Chem. Eng. Sci., 49 (1994) 2499.
46. S. Nitz, H. Kollmannsberger and M. Punkert, Chem. Mikrobiol. Technol. Lebensm., 14 (1992) 108.
47. S.N. Naik, H. Lentz and R.C. Maheshwari, Fluid Phase Equil., 49 (1989) 115.
48. D.A. Moyler, CO_2 extraction of essential oils: Part III. pimento berry, coriander and celery seed oil, In Flavors and off-flavors, Developments in food science 24 (ed. G. Charalambous), Elsevier, Amsterdam, (1990) 263.
49. B. Simándi, M. Oszagyán, É. Lemberkovics, G. Petri, Á. Kéry and Sz. Fejes, J. Essent. Oil Res., 8 (1996) 305.
50. O. Vitzthum and P. Hubert, German Patent Specification DE 2127611 (see also US Patent 4,123,559) (1971).
51. N. Gopalakrishnan, P.P.V. Shanti and C.S. Narayanan, J. Sci. Food Agric., 50 (1990) 111.
52. K. Kerrola and H. Kallio, J. Agric. Food Chem., 41 (1993) 785.
53. B. Simándi, A. Deák, E. Rónyai, Y. Gao, T. Veress, É. Lemberkovics, M. Then, Á. Sass-Kiss and Zs. Vámos-Falusi, J. Agric. Food Chem., 47 (1999) 1635.
54. D.A. Moyler, Perf. Flavor., 9 (April/May) (1984) 109.
55. J.P. Bartley and P. Foley, J. Sci. Food Agric., 66 (1994) 365.
56. E. Reverchon, J. Supercrit. Fluids, 5 (1992) 256.
57. M. Taniguchi, R. Nomura, I. Kijima and T. Kobayashi, Agric. Biol. Chem., 51 (1987) 413.
58. M. Ondarza and A. Sanchez, Chromatographia, 30 (1990) 16.
59. B. Simándi, M. Oszagyán, É. Lemberkovics, Á. Kéry, J. Kaszács, F. Thyrion and T. Mátyás, Food Res. Int., 31 (1999) 723.
60. H. Coenen, R. Hagen and M. Knuth, German Patent Specification DE 3114593, (see also US Patent 4,400,398) (1982).
61. H. Coenen and R. Hagen, Gordian, 83(9) (1983) 164.
62. A. Mohri, K. Morikawa, T. Matsuya and S. Sato, German Patent Specification DE 4133047 (1992).
63. Z. Knez and R. Steiner, J. Supercrit. Fluids, 5 (1992) 251.
64. J. Yao, M.G. Nair and A. Chandra, J. Agric. Food Chem., 42 (1994) 1303.
65. N. Behr, H. van der Mei, W. Sirtl, H. Schnegelberger and O. Ettingshausen, European Patent Specification 0023680 (1980).
66. H. J. Gährs, Chem. Tech., 14(7) (1985) 51.
67. K.-W. Quirin, D. Gerard and J. Kraus, Gordian, 86(9) (1986) 156.
68. E. Reverchon and F. Senatore, J. Flavour Fragr., 7 (1992) 227.
69. B. Simándi, M. Oszagyán, E. Rónyai, J. Fekete, Á. Kéry, É. Lemberkovics, I. Máthé and É. Héthelyi, Supercritical fluid extraction of medicinal plants, Proceedings of the 3rd

International Symposium on High Pressure Chemical Engineering (eds: Ph. Rudolf von Rohr and Ch. Trepp), Elsevier, Zürich, (1996) 357.

70. D. Gerard, K.-W. Quirin and E. Schwarz, Food Marketing and Technology, 9(5) (1995) 45.

71. D. Semiond, S. Dautraix, M. Desage, R. Majdalani, H. Casabianca and J.L. Brazier, Anal. Lett., 29(6) (1996) 1027.

72. E. Reverchon, G. Della Porta and R. Taddeo, J. Supercrit. Fluids, 8 (1995) 302.

73. E. Rónyai, B. Simándi, É. Lemberkovics, T. Veress and D. Patiaka, J. Essent. Oil Res., 11 (1999) 499.

74. S.B. Hawthorne, M.L. Riekkola, K. Serenius, Y. Holm, R. Hiltunen and K. Hartonen, J. Chromatogr., 634 (1993).

75. H.J. Bestmann, J. Erler and O. Vostrowsky, Z. Lebensm. Unters.-Forsch., 180 (1985) 491.

76. M. Oszagyán, B. Simándi, J. Sawinsky, Á. Kéry, É. Lemberkovics and J. Fekete, J. Flavour Frag., 11 (1996) 157.

77. M. Oszagyán, B. Simándi, J. Sawinsky and Á. Kéry, J. Essent. Oil Res., 8 (1996) 333.

78. E. Stahl and D. Gerard, Perf. Flavor., 10 (April/May) (1985) 29.

79. A.J. Jay, T.W. Smith and P. Richmond, The extraction of food colours using supercritical carbon dioxide, in Proceedings of the International Symposium on Supercritical Fluids, Nice (ed. M. Perrut), (1988) 821.

80. Y. Gao, H.G. Daood, B. Simándi and P.A. Biacs, Supercritical carbon dioxide extraction and HPLC of pigments from fennel seeds, Papers presented at the 2nd International Symposium on Natural Colorants for Food, Nutraceuticals, Beverages, Confectionery & Cosmetics, S. I. C. Publishing Company, Puerto de Acapulco Mexico, (1996) 1.

81. M.L. Cygnarowicz, R.J. Maxwell and W.D. Seider, Fluid Phase Equil., 59 (1990) 57.

82. J. Jay and D. Steytler, J. Supercrit. Fluids, 5 (1992) 274.

83. K. Sakaki, J. Chem. Eng. Data, 37 (1992) 249.

84. M. Skerget, Z. Knez and M. Habulin, Fluid Phase Equil., 109 (1995) 131.

85. P. Subra, S. Castellani, H. Ksibi and Y. Garrabos, Fluid Phase Equil., 131 (1997) 269.

86. F. Favati, J.W. King, J.P. Friedrich and K. Eskins, J. Food Sci., 53 (1988) 1532.

87. P. Subra, S. Castellani and Y. Garrabos, Supercritical carbon dioxide extraction of carotenoids from carrots, in Proceedings of the 3rd International Symposium on Supercritical Fluids (eds.: G. Brunner, M. Perrut), Strasbourg, Tome 2, (1994) 447.

88. G.A. Spanos, H. Chen and S.J. Schwartz, J. Food Sci., 58 (1993) 817.

89. T. Lorenzo, S.J. Schwartz and P.K. Kilpatrick, Supercritical fluid extraction of carotenoids from Dunaliela algae, in Proceedings of 2nd International Symposium on Supercritical Fluids, (ed. M. A. McHugh), Boston, (1991) 297.

90. J.C. Allen and R.J. Hamilton, Rancidity in Foods, Applied Science Publishers, London, (1983) 85.

91. C.E. Cross, Ann. Internal Med. 107 (1987) 526.

92. M.L. Burr, J. Hum. Nutr. Diet. 7 (1994) 409.

93. P. Six, INFORM, 5 (1994) 679.

94. U. Bracco, J. Löliger and J.-L. Viret, J. Am. Oil Chem. Soc., 58 (1981) 686.

95. C. Banias, V. Oreopoulou and C.D. Thomopoulos, J. Am. Oil Chem. Soc., 69 (1992) 520.

96. K. Schwarz and W. Ternes, Z. Lebensm. Unters. Forsch. 195 (1992) 95.

97. M.-E. Cuvelier, H. Richard and C. Berset, J. Am. Oil Chem. Soc., 73 (1996) 645.
98. J.W. King and G.R. List, Supercritical fluid technology in oil and lipid chemistry, AOCS Press, Champaign, Illinois, (1996).
99. D.D. Christianson, J.P. Friedrich, G.R. List, K. Warner, E.B. Bagley, A.C. Stringfellow and G.E. Inglett, J. Food Sci., 49 (1984) 229.
100. G.R. List, J.P. Friedrich and D.D. Christianson, J. Am. Oil Chem. Soc., 61 (1984) 1849.
101. K.-W. Quirin, D. Gerard and J. Kraus, Fat. Sci. Technol., 89 (1987) 141.
102. Ch. Wilp and R. Eggers, Fat Sci. Technol., 93 (1991) 348.
103. E. Rónyai, B. Simándi, S. Tömösközi, A. Deák, L. Vígh and Zs. Weinbrenner, J. Supercrit. Fluids, 14 (1998) 75.
104. H. Seidlitz, E. Lack, European Patent EP 0 247 999 B1 (1987).

576

9.7 High pressure polymer processing

L. Kleintjens

DSM Research and Patents
Postbus 18, NL-6160 MD Geleen, The Netherlands

9.7.1 Introduction

Organic chemists and polymer scientists have made considerable progress in their efforts to copy nature and new spectacular results are reported daily. The complex structures and morphologies nature has developed for, for example, coatings (skin), information transport (nervous systems), transport- and separation systems, are already being imitated, although primitively, in a wide range of organic materials and polymer products.

Today's polymer industry, however, is mainly driven by economy of scale. High- pressure production units with a capacity up to 500,000 tonnes of polymer a year are being built.

Thermodynamic optimization of such processes has led to significant cost savings.

The thermodynamic models used in these industrial process optimizations are mostly of a semi-empirical nature because:

a) most of the process streams consist of at least five constituents [one of them being a (co)polymer with its intrinsic polydispersity].

b) almost all separation steps have to be carried out at an elevated pressure, and

c) polymer solutions are usually highly viscous and, thus, transport phenomena and the settlement of thermodynamic equilibria are affected by this viscosity.

There is room for improvement of the plant- specific models and processing recipes for such high-pressure plants, where the focus will be on polymer aspects such as the interplay between the complex kinetics in copolymer chemistry and the composition and chainstructure in copolymers and/or in the blend, or the polarity of monomer, (co)polymers and solvents.

9.7.1.1 State of the art in polymer thermodynamics

The miscibility behaviour of polymer systems has been studied extensively, and experimental data and thermodynamic models have been generated for (co)polymer solutions and for polymer blends.

The most striking phenomena in polymer systems are the occurrences of liquid liquid phase equilibria. Owing to the long-chain structure (high molar mass) of the polymer the combinatorial entropy of mixing in a solution is reduced to approximately half of that of a low molar mass mixture, and to nearly zero in a polymer blend. As a consequence of this low entropy of mixing the free-energy of mixing quite often has a positive value, because the interaction energy of mixing outweighs the entropy contribution.

The phase-behaviour of such a system thus may change completely with variations in pressure, temperature, molar mass, and chemical composition of the polymer(s). Thus, polymer systems may show upper critical, lower critical, or hourglass-type demixing behaviour.

The first data on polymer systems were collected via (laser-) light-scattering techniques [1] and turbidity measurements, further developed by Derham *et al.* [2,3]. Techniques based on the glass-transition of the polymer-blend constituents were also tested, such as DSC, Dynamic Mechanical Spectroscopy, and Dielectric relaxation [4]. Films made from solutions of polymer mixtures were studied via microscopic techniques such as optical microscopy, phase-contrast microscopy, TEM, SEM and, more recently, AFM [5], with temperature programmes via a hot stage. DSM developed a micro-extruder [6], which enabled the preparation of small amounts of blends (1 to 5 g) under processing conditions (pressure, temperature). Based on such samples, we started to study the effect of pressure on the phase-behaviour of polymer blends [7]. In blends of copolymers showing narrow miscibility windows [8], pressure has severe effects on the resulting state and morphology.

Suzuki *et al.* reported cloud-point temperatures as a function of pressure and composition in mixtures of poly(ethyl acrylate) and poly(vinylidene fluoride) [9]. Their data in terms of p(T) curves at constant composition show that miscibility in the same system may either improve or decline with rising pressure, depending on the blend's composition. Important consequences for blend-processing ensue. A planned two-phase extrusion may easily be jeopardized by the pressure building up in the extruder. Conversely, a homogeneous melt may be turned into a two-phase system when the pressure on the blend increases.

In production units, and during processing, polymer systems are subjected to considerable shear rates. Recently the effect of shear on the phase behaviour was clearly demonstrated by several groups, e.g. by Aelmans and Reid. [10].

Crystallization and the underlying kinetics are also essential for the final properties of polymer products. In blends in which one of the polymer constituents is (partially) crystallizable there is often interference between the liquid liquid phase behaviour and crystallization [11]. It has been shown in many papers that a given blend can exhibit considerable differences in microstructure and morphology depending on the processing route (pressure, cooling rate, annealing time) and crystallization kinetics. A similar behaviour is found when the liquid liquid phase behaviour of a given blend is affected by glass solidification in one of the phases. Within this phase the viscosity will increase. The resulting reduced mobility can be used to quench a blend into a given percolation structure [12].

9.7.1.2 Special polymer systems

In the last decade specialty polymers have been developed that make use of thermodynamic phenomena on a molecular scale. We here briefly describe the most important effects.

Thermoreversible Gelation

Upon cooling of solutions of long-chain polymers it may happen that only parts of the chain eventually crystallize together with parts of other polymer chains. When a number of such small crystallites are formed they operate as knots in a network of flexible polymer chains in solution and one obtains a gel. Processing of such gels into strong fibres and films is applied commercially and has been reviewed [13,14].

Interpenetrating Networks (IPN)

Polymer networks can be formed by chemical reactions between polymer chains (cross-linking) or by using trifunctional comonomers during the polymerisation. If such a network is dissolved in a second monomer and this second monomer is again polymerized into a second network, one obtains a structure in which both polymers are intertwined. These polymer chains only have very local mobility. In cases where both polymers are partially or completely immiscible the L_1/L_2 phase-separation is reduced to a very small scale. The properties of such an IPN are completely different from the uncross-linked polymer blend [15].

Liquid Crystalline Polymers (LCP)

By building - in combinations of aromatic rings into the polymer chains, chemists are able to produce polymer chains with very low chain flexibility. In the limit they reach rigid-rod-type op polymers. Such polymers show substantial temperature - pressure -concentration regions in which the stiff polymer chains arrange in some form of orientation. This phase behaviour gave them the name Liquid Crystalline Polymers (LCP) and LCP have unique properties.

Block copolymers

In block copolymers the sequence lengths of monomer-1 and of monomer-2 in a chain are usually long enough to allow for local ordering. In cases where polymer-segment, 1, and polymer-segment, 2, are (partially) incompatible the copolymer chains will form structures in the melt, or in solution. Sequences of monomer-1 will "look for" other segments of 1, and similarly for sequences of segment 2. Electron microscope studies of such polymer systems show the occurrence of several types of block-copolymer morphology that may vary with temperature and pressure.

It is well-known that one can reduce the droplet size of a demixed polymer blend by adding a well-chosen block-copolymer at given p and T, and achieve interesting improvements of the blend's mechanical properties.

9.7.1.3 Modelling polymer systems

Current thermodynamic theories for polymer systems are combinations of the Flory - Huggins, Guggenheim, and Equations-of-State approaches. All of these theories make use of empirical parameters and are based on assumptions about the underlying molecular model.

The Flory Huggins Staverman- type models

Thermodynamic descriptions of polymer systems are usually based on a rigid-lattice model published in 1941 independently by Staverman and Van Santen, Huggins and Flory where the symbol $\chi(T)$ is used to express the binary interaction function [16]. Once the interaction parameter is known we can calculate the liquid liquid phase behaviour.

Estimations of the value of χ can be obtained via the theory of the solubility parameter [17], or the group contribution approach [18].

An essential improvement in the description of the phase-behaviour can be obtained if the disparity in size and shape between the molecules is taken into account. The number of nearest-neighbour contacts may thus be different for each molecule or site, and set proportional to its contact surface area [19].

When the concentration of polymer, a, in a blend drops below a given value we have isolated coils of polymer a separated by regions of pure polymer b. This phenomenon will occur at both ends of the concentration axis in blends. Koningsveld *et al.* [20] expressed the interaction function of a polymer mixture, g, as the sum of three terms, one for each concentration range. The interaction term representing the uniform concentration regime is corrected with two terms expressing differences relevant for the regime dilute in one of the polymers as compared with the uniform segment density region.

Changes in the flexibility of polymer coils owing to concentration - variation may effect the entropy. Huggins [21] introduced two corrections for the athermal entropy of mixing, that take into account the influence a second polymer has on the 'stiffness' of the other polymer chain, and vice versa. From this model one can conclude that a stiff chain will be enhanced in its orientational possibilities when surrounded by an increasing amount of a flexible polymer while the reverse will occur when a polymer is embedded in stiffer polymer surroundings. This was demonstrated by Stroeks [22] for the system PIB (flexible)/PST (stiffer).

Equation-of-state models

In many production routes, and also during processing, polymer systems have to undergo pressure. Changes in the volume of a system by compression or expansion, however, cannot be dealt with in rigid-lattice-type models. Thus, non-combinatorial free volume ('equation of state') contributions to ΔG have been advanced [23 - 29]. Detailed interaction functions have been suggested (but all of them are based on adjustable parameters, for blends, *e.g.,* Mean-field lattice gas [30], SAFT [31], specific interactions [32]), and have been succesfully applied, for example, by Kennis *et al.* [33].

Polymer systems containing copolymers call for a further extension of the thermodynamic model. The interaction function for statistical copolymers was originally derived by Simha and Branson [34], discussed by Stockmayer [35] *et al.,* and experimentally verified by Glöckner and Lohmann [36].

The miscibility gap will be described more accurately when a meanfield lattice gas approach is choosen [30]. The mathematical form of the interaction function in all the above models may bring about a negative value for the effective interaction parameter, g, while all binary interactions by themselves are positive. The complexity of copolymer phase behaviour can be attributed to this peculiarity, like the 'miscibility-windows' in mixtures of a copolymer with another homopolymer [37], or with a second copolymer [38,39].

9.7.1.4 Experimental methods in modelling polymer systems

We now will present an overview of experimental methods that can be applied to polymer systems under pressure.

Window autoclave

Window autoclaves have been built to allow one to determine visually the appearance of turbidity in a homogeneous polymer solution, indicating the onset of phase separation, and to observe the resulting number of liquid and gaseous phases and optionally to determine their composition. Lentz [40] developed a mechanically driven stirrer fitted in a 2000-bar window autoclave. This apparatus allowed fast homogenation of concentrated polymer solutions and blends having low viscosity.

Laser-light-scattering autoclave

We developed an experimental procedure that can be applied to highly viscous polymer blends. In the 'DSM micro-extruder'[6], polymers are blended in the melt, at the desired temperature and pressure, and injected into a small capillary tube which is immediately sealed with a floating plug. This capillary cell is placed in a small window autoclave and a laser beam enters the capillary cell at the lens-shaped bottom end. The intensity of the light scattered by the polymer system is recorded at two scattering angles (as a function of pressure and temperature).

It has proved to be possible to measure liquid-liquid phase-separation boundaries and crystallization phenomena in high viscous molten polymer mixtures.

Pressure Pulse Induced Critical Scattering

The spinodal and the cloud point can be determined as a function of pressure and temperature (up to 150C, 1000 bar) via light-scattering measurements [41]. The intensity of the scattered light of the polymer solution is measured in a high-pressure optical cell during a pressure pulse in the polymer solution.

9.7.2 Phase behaviour of polymer blends under pressure

EOS models were derived for polymer blends that gave the first evidence of the severe pressure - dependence of the phase behaviour of such blends [41,42]. First, experimental data under pressure were presented for the mixture of poly(ethyl acetate) and poly(vinylidene fluoride) [9], and later for in several other systems [27,43,44,45]. However, the direction of the shift in cloud-point temperature with pressure proved to be system-dependent. In addition, the phase behaviour of mixtures containing random copolymers strongly depends on the exact chemical composition of both copolymers. In the production of reactor blends or copolymers a small variation of the reactor feed or process variables, such as temperature and pressure, may lead to demixing of the copolymer solution (or the blend) in the reactor. Fig. 9.7-1 shows some data collected in a laser–light-scattering autoclave on the blend PMMA/SAN [46].

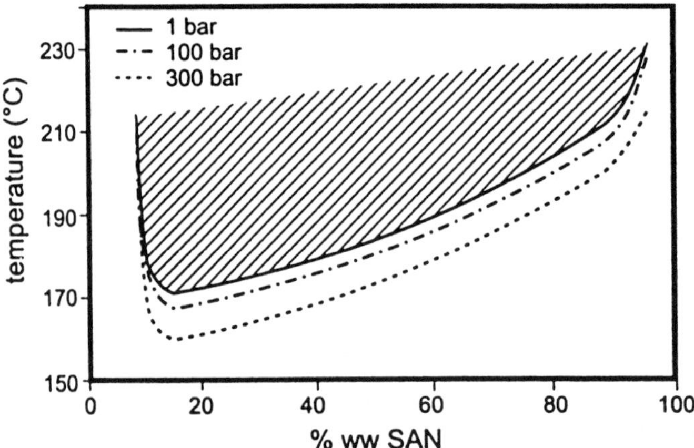

Figure 9.7-1 Experimental cloud-point curve of the polymer blend Poly(methyl methacrylate)/Poly(styrene-co-acrylonitrile (28%AN)) as a function of pressure.

9.7.3 High-pressure applications

We shall now highlight some successful developments in polymer systems under pressure. We will deal with some applications in production, processing, and the performance of polymers, and plastics recycling.

9.7.3.1 Process optimization

Downstream Separators

In many industrial polymerization processes, pressure is applied to control the thermodynamic state in the reactor as well as to effect downstream separations of products and reactants. Recent thermodynamic models, such as the Statistical Associating Fluid Theory (SAFT) [31] and Mean Field Lattice Gas (MFLG)-model [33], have been applied successfully to draw up adequate models for such unit operations even in systems involving many components.

On-line fractionation

Fast high-pressure fractionation of tails in the molar-mass-distribution and/or the chemical–composition-distribution of a (co)polymer [47] during the polymerisation process will indicate eventual drifts in (co)polymer composition. Several important polymer properties depend strongly on such tails in the (co)polymer distribution. On-
-line fractionation data are strongly needed.

Product purification

Impurities such as solvents and a variety of reactants can be removed from a polymer product to a very low (10 - 50 ppm) level via Supercritical Extraction (See chapter 9.6.) or Rapid Expansion Technology (See chapter 6.5.). In both methods, a compressed gas is used to extract undesired impurities out of the solid polymer (SCE), or as a compressed solvent for the polymer in a spray-drying operation. Several successes have been reported [48,49] and patented.

9.7.3.2 Enhanced processing of polymer blends

The influence of pressure on miscibility windows in (co)polymer blends is currently being studied. In polymer blends where the glass-transition temperature or the degradation temperature sometimes prevent the system from reaching a one-phase state, pressure variation may overcome such limitation in processing partially miscible copolymer blends, especially systems with lower critical demixing behaviour. Since extruders are built for operation at high pressures, sometimes up to 200 bar, pressure studies may lead to improved processing routes for such blends.

We further investigated the removal of polymeric binder material with compressed gases from a ceramic green product. We have shown that this extraction is well possible without any dimensional change in the ceramic body [50]. The time necessary for such a supercritical extraction is only a fraction of the time needed for conventional controlled soft pyrolysis of the polymeric binders.

Substantial knowledge and experience has been gained in spraying, spinning and surface-coating with polymers, starting from a compressed polymer solution. Composites where, besides the polymer, inorganic particles are dissolved or suspended in the compressed mixture, are very challeging systems for such treatments. After the initial spraying/spinning step one may optionally modify the resulting composite film or fibre, via, *e.g.,* mechanical deformation, UV or EB cross-linking or magnetic orientation. Thereafter the polymer can be removed, if necessary only partially, to preserve the coherence of the product.

Drugs and organic additives can be dissolved in supercritical solvents, while pigments and inorganic additives may be suspended in such systems. It proved to be possible [51] to deposit these components into polymer fibres and films by treating the polymer with such a high-pressure solution or suspension of the additive. The time necessary to deposit a pigment into a polymer fibre in this way proved to be of the order of 15-30 minutes, *e.g.*, for polyester fibres.

9.7.3.3 Polymer Particles

With the methods discussed in Chapter 6.5. we are able to produce small polymer particles from solutions of polymers in compressed gases [52,53]. First results indicate that, depending on the process conditions and the nozzle design, a wide variety of spherical particles, small fibrils, or longer fibres can be obtained. Sometimes the particle-size distribution obtained is very uniform, in other cases, a wide variety of sizes is obtained. In some cases, the monomer was used as solvent for the equivalent polymer. The particles, fibrils, etc., obtained may be used for direct spray-coating of surfaces but they also offer new opportunities as raw materials in their own right. New products are underway that are based on such technologies.

9.7.3.4 Plastics recycling

Most polymer products are offered for recycling in the form of mixed plastic waste (MPW). Compressed gases as solvents may offer advantages compared with liquid solvents in the necessary separations (*e.g.,* to collect families of polymers out of MPW) because such solvents are easy to recover and, furthermore, they make it possible to use pressure as an additional separation parameter or to remove the organic additives of the polymers. Finally, MPW may be added as a polymer solution in a compressed gas to a feed of a Naphtha cracker, and thus be recycled to base chemicals.

9.7.3.5 Reactive polymer blending

Much research effort has been focused on chemical modification of a blend during processing. In such systems, the chemistry interferes with the pressure-dependent liquid - liquid phase behaviour. Some successful applications are given below.

Reactor blends

A single copolymer with a bimodal distribution in chemical composition, made in a one- or two-autoclave reactor, might be regarded as a reactor blend. One frequently observes liquid-liquid phase behaviour in these (co)polymer blends, leading to very typical morphologies. Recent developments in this area are the Reactor Granule Technology (a multiphase process producing small multiphase polymer spheres) [54] and core-shell particles produced by encapsulation techniques. Reaction-induced phase separation in a thermoplastic-modified thermosetting polymer may generate complex morphologies in the resulting polymer material [55].

Reactive processing

Since the properties of a heterogeneous blend depend upon particle-size and interfacial adhesion, in-situ formation of a copolymer (during reactive processing) that acts as a "polymeric compatibilizer" represents a very attractive technology for producing and controlling the molecular architecture of the resulting two- or three-phase blends [56,57]. Grafting of monomers and polymers onto polymer chains, cross-linking, trans- -esterification, and imidization are commercially-used high-pressure technologies aimed at influencing miscibility (windows) and/or improving interfacial adhesion and reducing particle size in demixed polymer blends.

Reactive solvents

A quite flexible route for modifying the performance of thermoplastic polymers is based on the use of a compressed monomer as solvent for a polymer in a processing machine. Reactive solvents may offer an interesting alternative to blending. They circumvent the usual compromise between the need for low viscosity during processing, on the one hand, and the desire to give the final material high stiffness and impact-strength, on the other. In most studies the reactive solvent is polymerized after moulding (at low viscosity), whereafter liquid - liquid phase separation occurs. The first results on epoxy- and acrylic systems have already been published [58].

9.7.4 Future challenges

Polymer thermodynamics is a supporting science that has proved to be very successful owing to its intensive interactions with neighbouring disciplines. The future of polymer thermodynamics will thus strongly depend on developments made in these neighbouring research areas [59].
Let us thus first evaluate some current research activities and developments in organic- - polymer chemistry.

9.7.4.1 Controlled synthesis

Polymer chemists are nowadays capable of controlling chemical process steps to a very high degree, enabling them to produce well-designed macromolecules with a very specific molecular structure. To mention a few successes reported by polymer chemists, comb-shape-, star-shape-, and ladder-polymers, short-chain branched, long-chain branched, and hyper-branched polymers, catenanes and rotaxanes [60 to 63]. In the area of copolymerization, too, major advances have been made in controlling the block length of copolymers and the reproducibility of the chemistry, *e.g.,* via "living polymerizations" [64].
The market potential of such "high-tech" materials and engineering plastics is not quite clear. An essentially shorter time-to-market is to be expected with pharmaceutical (intermediate) products. Examples are dipeptides containing alpha-substituted amino acids [65]; and *d,l*-phenylalanine [66]).

9.7.4.2 Supramolecular structures

Partly as a result of the successful scale-up of dendrimer production by DSM (based on diamino-butane [67]), the field of nanotechnology is enjoying much attention worldwide. The use of cooperative weak interactions for the stepwise building of synthetic supramolecular 3-D structures is a very challenging research area. The self-organization of complex molecules into supramolecular arrangements is based on molecular-recognition phenomena (which are also relevant to replication processes occurring in living systems)

9.7.4.3 Morphology of polymer materials

At the recent European Symposium on Polymer Blends [59] about half of the contributions dealt with thermodynamic effects on molecular architecture, on polymer morphology, and on processing and performance of polymer blend materials. Although some attention has been focused mainly on the interface (material) in heterogeneous blends, in general most thermodynamic studies of such heterogeneous blends deal with two- or more bulk phases. Essential morphological features such as droplet size, cocontinuous phases, micellar or laminar morphology, interface composition, and crystal structure, seem to be beyond the reach of the present thermodynamic models. However, important material properties of structural polymer materials, such as impact strength, brittleness, gas diffusion, but also surface properties such as sliding friction, lubrication and surface wetting, or compatibility with bio-organisms, strongly depend on the macromorphology of the material.

Futurists predict the development of a new class of polymer-based materials [68 to 74] for application in sensor technology (ultra-fast response), opto-electronics, photovoltaics (artificial photosynthesis), electronics (single-electron tunnelling) and catalysis. However, the production of such functional materials will require new forms of process engineering, combining computerization with ultra-clean technologies which are essential for supplying products with the exceptionally high purity and precision required.

Polymers synthesized by living organisms (such as cellulose, resins, wood, silk, natural rubber, etc.) or made out of CO_2 or CO have a "green" image. Arguments used in favour of such "green" polymers are: recyclability, renewable resources, biodegradability and no waste. Although several of these arguments are irrelevant or even (partly) incorrect [75] several major projects are being carried out with the aim of copying nature by building biopolymers out of CO_2, CO and CH_4 [76-78]

References

1. R. Koningsveld, L.A. Kleintjens, *Macromolecules* 4, (1971) 637.
2. K. Derham, J. Goldsbrough, M. Gordon, *Pure Appl. Chem.* 38 (1974) 97.
3. K. Derham, J. Goldsbrough, M. Gordon, R. Koningsveld, L.A. Kleintjens, *Makromol. Chem. Suppl. 1* (1975) 401.
4. F. Karasz, *in Polymer Blends and Mixtures*, NATO ASI ser. 89 (1985) 25.
5. P. Viville, R. Lazzaroni, G. Lambin, J. Bredas, *in ref. 59* page 137.
6. M. Bulters, S. Martens, *DSM Internal Communication* (1989).
7. V. Zeuner, H. Lentz, L. Kleintjens, *Macromol. Symp.* 102 (1996) 337.
8. V. Reid, L.A. Kleintjens, J. Cowie, in *"Third European Symposium on Polymer blends"*, PRI press, C9(1990).
9. Y. Suzuki, Y. Miyamoto, H. Miyaji and K. Asai, : J. *Polymer Scie., Polymer Lett.* 20 (1982) 563.
10. N. Aelmans, V. Reid, *in ref 59*, page 153.
11. J. Runt, L. Jin, S. Talibuddin, C.R. Davis, *in ref 59*, pages 70 and 73.
12. S. Danchinov, Y. Shibanov, Y. Godovsky, *Macromol. Symp* 112 (1996) 69.
13. R. Kirschbaum, J. van Dingenen, in: *Integration of Polymer Science and Technology 3*, *Elsevier Appl. Sci. Publ.* (1989).
14. H. Berghmans, *Integration of Polymer Science and Technology 2, Elsevier Appl.sci. Publ.* (1988).
15. K.C. Frisch, *Integration of Polymer Science and Technology 13, Elsevier Appl.sci. Publ.* (1986).
16. P.J. Flory, *Principles of Polymer Chemistry,* Cornell Univ. Press, Ithaca, New York (1953).
17. J.H. Hildebrand, R.L. Scott, *The Solubility of Non-electrolytes*, Dover, NY (1964).
18. J. Holten-Andersen, P. Rasmussen, A. Fredenslund, *Ind. Eng. Chem., Process. Des. Dev.* (1986).

19. A.J. Staverman, *Rec. Trav. Chim.*, 56 (1937) 885.
20. R. Koningsveld, W.H. Stockmayer, J.W. Kennedy, L.A. Kleintjens, *Macromolecules* 7 (1974) 73.
21. M.L. Huggins, *J. Phys. Chem.* 74 (1970) 371, 75 (1971) 1255, 80 (1976) 1317.
22. A. Stroeks, *Master Thesis Chem. Eng.* Eindhoven Univ., NL (1987).
23. P.J. Flory, J. *Am. Chem. Soc.* 87 (1965) 1833.
24. P. McMaster, *Macromolecules,* 6 (1973) 760.
25. D. Patterson, Pure Appl. Chem., 31 (1972) 133.
26. R. Simha, Macromolecules, 10 (1977) 1025.
27. D.R. Paul and S. Newman, *Eds. Polymer Blends*, New York, Academic Press., (1978) .vol. I, p. 115.
28. O. Olabisi, Macromolecules 9 (1975) 316.
29. I.C. Sanchez, in *Polymer Compatibility and Incompatibility*, Harwood, N.Y. (1982).
30. R. Koningsveld, L.A. Kleintjens, *J. Pol. Sci*; Pol. Symp. 61 (1977) 221.
31. S. Chem, I.G. Economou, M. Radosz, *Macromolecules* 25 (1992) 4987.
32. M. Coleman, J. Graf, P. Painter, "Specific Interactions and the Miscibility of Polymer Blends" *Technomic Publ. Corp.,* Basel (1991).
33. H.A.J. Kennis, Th.W. de Loos, J. de Swaan Arons, R. van der Haegen, L.A. Kleintjens, *Chem. Eng. Sci.* 45 (1990) 1875.
34. R. Simha, and H. Branson, J. Chem. Phys. 12 (1944) 253.
35. W.H. Stockmayer,L.D. Moore Jr., M. Fixman and B.N. Epstein, J. Polym. Sci. 16 (1955) 517.
36. G. Glöckner and D. Lohmann, Faserforsch. Textiltechn., 24 (1973) 365, 25 (1974) 476.
37. V. Reid, *Ph.D. Thesis*, Stirling U.K. (1988).
38. G. Ten Brinke, F.E. Karasz and W.J. MacKnight, Macromolecules 16 (1983) 1827.
39. R. Koningsveld, L.A. Kleintjens and A.M. Leblans-Vinck, *Ber Bunsenges. Phys. Chem.* 89 (1985) 1234.
40. H. Lentz, Siegen Univ., *personal communication* (1982).
41. P.A. Wells, Th.W. De Loos, L.A. Kleintjens, *Fluid Phase Equil* 83 (1993) 3.
42. P. Zoller, P. Bolli, V. Pahud, H. Ackermann, *Rev. Sci. Instrum.* 47 (8), 984 (1976).
43. M. Schmidt, F. Maurer, in *Proceedings European Symp. Polymer Blends*, Maastricht (1996), page 1.
44. Y. Maeda, F. Karasz, W. MacKnight, J. Appl. Pol.Sci. 32 (1984) 4423
45. D. Schwahn, S. Janssen, H. Frielinghaus, K. Mortensen, in *Proceedings Europ. Symp. Polymer Blends, Maastricht (*1996), page 272.
46. U. Michel, *Diplom Thesis*, TU Siegen (1996).
47. M.A. McHugh, *Proceedings, 8th Rolduc Polymer Meeting*, (1992), page 40.
48. L.A. Kleintjens, *Integration of Pol.Sci. & Technology 3, Elsevier Appl. Sci.*, London (1989) page 91.
49. A.K. Lele, A.D. Shine, *AIChE J.*, 38 (1992) 742.
50. L.A. Kleintjens, *Proc. 2nd Int. Symp. High Pressure Chem. Eng.*, (Ed. Dechema) Erlangen (1990), page 201.
51. W. Saus, E. Schollmayer, H.J. Buschmann; *European Pat.* 92810343.1.

586

52.	D.W. Matson, J.L. Frulton, R.C. Petersen, R.D. Smith, *Ind. Eng. Chem. Res.* 26 (1987) 2298
53.	J.F. Brennecke, Notre Dame, *personal communication.*
54.	P. Galli, *Macromol. Symp* 112, (1996) 1.
55.	E. Girard-Reydet, J.P. Pascault, H. Sautereau, *in ref 59* pages 328, 528.
56.	J. Rösch, R. Mülhaupt, G. Michler, *Macromol. Symp. 112* (1996) 141.
57.	C. Koning, W. Bruls, F. Op den Buysch, L. v.d. Vondervoort, *Macromol. Symp 112,* (1996) 167.
58.	H. Meijer, R. Venderbosch, J. Goossens, P. Lemstra, *in ref 59* page 525.
59.	L.A.L. Kleintjens, *European Symposium on Polymer Blends Maastricht* 1996; Proceedings, Ed.
60.	G. Wenz, B. Keller. Angew. *Chem. Int. Ed. Engl.* 31 (1992) 325.
61.	D.B.Amabilino, I.W. Parsons, J.F. Stoddard, *Trends in Polym. Sci.* 2 (1994) 146.
62.	N.A. Platé, V.P. Shibaev, Comb-shaped Polymers and Liquid-Crystals, *Plenum*, New York 1985.
63.	D.A. Tomalia, A.M. Naylor, W.A. Goddard, *Angew. Chem.* 102 (1990) 119.
64.	M. Szwarc, M. van Beylen, *Ionic Polymerization and Living Polymers*, Chapman and Hall, New York 1993.
65.	B. Kaptein, V. Monaco, Q.B. Broxterman, H.E. Schoemaker, J. Kamphuis, *Recl.Trav.Chim. Pays-Bas*, 114 (1995) 231.
66.	J.G. de Vries, R.P. de Boer, M. Hogeweg, E. Gielens, *J.Org.Chem.* 61 (1996) 1842.
67.	E.M.M. de Brabander, E.W. Meijer, *Angew. Chem. Int .Ed. Engl.* 32 (1993) 1308.
68.	J. Huff, J.A. Preece, J.F. Stoddard, *Macromol.Symp.* 102 (1996) 1.
69.	R.P. Sybesma, F.H. Beijer, L. Brunsveld, B.J.B. Folmer, J.H.K.K. Hirschberg, R.F.M. Lange, J.K.L. Lowe, E.W. Meijer, *Science* 278 (1997) 1601.
70.	A.M. van de Graats, J.M. Warman, K. Müllen, Y. Geerts, J.D. Brand, *Adv.Matter* 10 (1998) 36.
71.	K. Katsuma, Y. Shirota, *Adv. Matter,* 10 (1998) 223.
72.	M.W.P.L. Baars, P.E. Froehling, E.W. Meijer, *Chem.Commun* (1997) 1959.
73.	V.V. Tsukruk, *Adv.Mater* 10 (1998) 253.
74.	*Chemistry, Europe & the Future*, report prepared by AllChemE (Alliance for Chemical Sciences and Technologies in Europe (1997) in a combined effort of CEFIC, CERC3, COST, ECCC/FECS and EFCE).
75.	J. Put, *Macromol. Symp.* 127 (1998) 1.
76.	M. Super, E. Berluche, C. Costello, E. Beckman, *Macromol.Symp.* 127 (1998) 89.
77.	BRITE/Euram Project *"Polymerization and Polymer Modification in Supercritical Fluids* (1997 to 2000).
78.	Japanese *"New Sunshine Programme 5.4.4."* (1990 to 1999).

9.8 Precipitation of solids with dense gases

Ž. Knez[a], E. Weidner[b]

[a] Faculty of Chemistry and Chemical Engineering, University of Maribor
P.O. Box 222 Smetanova 17, SI-2000 Maribor, Slovenia

[b] Lehrstuhl für Verfahrenstechnische Transportprozesse, University Bochum
Universitätsstr. 150, 44780 Bochum, Germany

9.8.1 Introduction

The particle-size and size-distribution of solid materials produced in industrial processes are not usually those desired for subsequent use of these materials and, as a result comminution and recrystallization operations are carried out. Well known processes for particle size redistribution are crushing and grinding (which for some compounds are carried out at cryogenic temperatures), air micronization, sublimation, and recrystallization from solution. There are several practical problems associated with the above-mentioned processes. Some substances are unstable under conventional milling conditions, and in recrystallization processes the product is contaminated with solvent, and waste solvent streams are produced. Applying supercritical fluids may overcome the drawbacks of conventional processes.

9.8.2 State of the art of material processing using supercritical fluids

The processes of high-pressure micronization which have been studied intensively by numerous groups are:
- High pressure crystallization;
- Rapid expansion of supercritical solutions (RESS); and
- Gas antisolvent recrystallization (GASR).

In the past ten years, modifications and additional processes have been proposed:
- Precipitation with a compressed Antisolvent (PCA) [1];
- Solution Enhanced Dispersion of Solids (SEDS); and
- Particles from Gas-Saturated Solutions (PGSS).

9.8.3 Crystallization from a supercritical solutions (CSS)

9.8.3.1 Fundamentals

The advantages over conventional crystallization of crystallization under supercritical conditions are particularly clear, when non-volatile, thermally labile pharmaceutical substances are to be crystallized. As non-contaminating solvents at close-to-ambient temperatures, supercritical fluids might be an attractive alternative to conventional organic solvents.

In conventional batch-cooling-crystallization a saturated solution is cooled from an initial temperature at which the solute has a high solubility to a final lower operating temperature (with lower solubility) along an optimal cooling curve. This cooling is used to maintain a reasonably constant level of supersaturation, and a constant crystal growth-rate.

When the solute-laden solution is a supercritical fluid, supersaturation may be induced not only by varying the temperature but also by pressure variation. An advantage is that the

588

pressure in the whole system might be changed almost instantaneously, while temperature changes are often slower, owing to limitations in heat transfer. Thus pressure and pressure-
-gradients would be additional means for generation of particles with the desired size, form and morphology. Pressure changes are, in general, not used during crystallization from supercritical solutions, but in other processes which are decribed below.

The formation of small particles is favoured when solids' formation is maintained via primary nucleation throughout the batch crystallization. A widely used relationship between the nucleation rate and supersaturation is usually given as:

$$B^0 \alpha \exp\left(- K_1 / \log^2 S\right) \qquad\qquad (9.8\text{-}1.)$$

where: B^0 is the nucleation rate; S, supersaturation as C/C^*; C, concentration of solute; C^*, saturation concentration of solute; and K_1, constant.

Applying supercritical fluids in CSS allows one to obtain supersaturation and to control nucleation- and growth rates by temperature-induced variation of the concentration of the solute in systems where no liquid solvents (e.g., VOC = volatile organic compounds) are present.

9.8.3.2 Design criteria

A high-pressure batch crystallization unit is presented in Figure 9.8-1. The substance is filled in a stirred autoclave, then the proper amount of supercritical fluid is pumped in, and the system is heated. When equilibrium is obtained, the temperature and/or pressure of the solution is varied until crystals are formed.

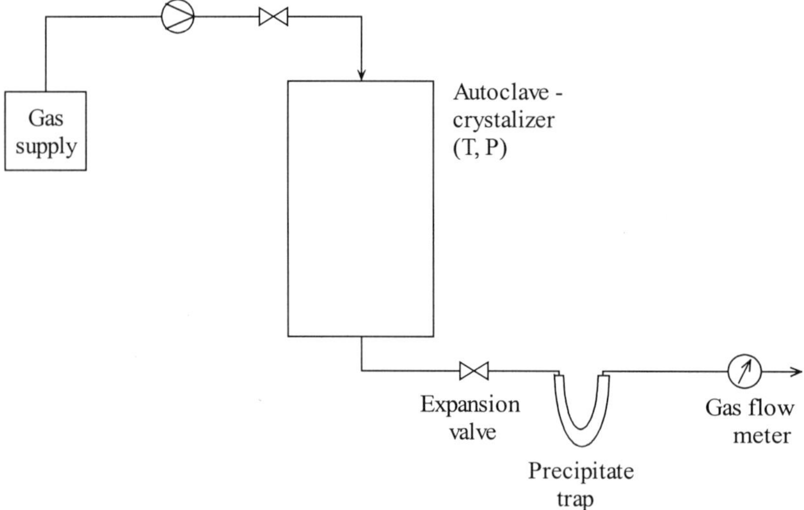

Figure 9.8-1. Batch crystallization apparatus.

9.8.3.3 Applications

Advantages of high pressure crystallization are: simple control of process parameters, and a solvent-free system.

Disadvantages of high pressure crystallization are: high volumes of solvents, owing to the low solubility of the substances to be crystallized, relatively high pressures, batchwise operation, and long cooling times.

Table 9.8-1
CSS (Crystallization from supercritical solutions) experiments for various products

Substance	SCF	Ref.
Benzoic Acid	CO_2	[2,3]
Benzoic Acid	CO_2	[4]
Para-/meta-Dichlorbenzene	N_2 or CO_2	[5]
Naphthalene/Biphenyl	N_2 or CO_2	[5]

9.8.4 Rapid expansion of supercritical solutions (RESS)

9.8.4.1 Fundamentals

By utilizing the rapid expansion of supercritical solutions, small-size particles can be produced from materials which are soluble in supercritical solvents. In this process, a solid is dissolved in a pressurized supercritical fluid and the solution is rapidly expanded to some lower pressure level which causes the solid to precipitate. This concept has been demonstrated for a wide variety of materials including polymers, dyes, pharmaceuticals and inorganic substances.

By varying the process parameters that influence supersaturation and the nucleation rate, particles can be obtained which are quite different in size and morphology from the primary material. Extremely high supersaturation can be obtained from cooling by depressurization and density reduction owing to the expansion of supercritical solutions.

The nucleation-rate expression is based on the following Equation, which describes the number of critical nuclei formed per unit time in a unit volume (9.8-2):

$$I = 2 \cdot N_{tot} \cdot \beta \cdot \sqrt{\sigma \cdot v_1^2 / k \cdot T} \cdot \exp\left[-16 \cdot \pi / 3 \cdot \left(\sigma \cdot v_1^{2/3}\right)^3 \cdot \left(1/\ln S - K \cdot y_1^e \cdot (S-1)\right)^2\right] \quad (9.8-2)$$

with: N_{tot}, total solute concentration in the bulk fluid phase; β, thermal flux of solute molecules; σ, interfacial tension; v_1, molecular volume of solid solute; k, Boltzmann's constant; S, supersaturation ratio; K, a function of temperature and pressure; and y_1^e, mole-fraction of solute in equilibrium [6].

9.8.4.2 Design criteria

For designing a RESS-experiment one is recommended to study the solubility of the substance to be crystallized in the supercritical gas phase near the upper critical end-point

590

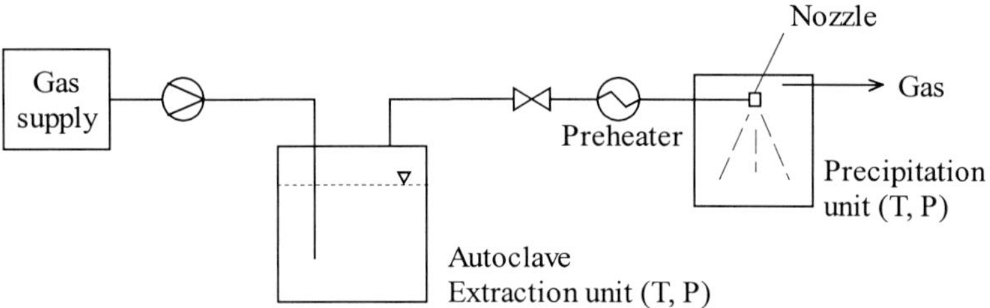

Figure 9.8-2. Experimental equipment for RESS.

(UCEP), in order to choose process parameters (P and T) which give the maximum amount of solute in the supercritical solution without appearance of a liquid phase.

For binary systems, a variety of phase behaviours possible when partial immiscibility of liquids and gases, and the occurrence of solid phases is considered. Phase-behaviours are classified into six basic types, and transitions between them are possible.

The systems which are often involved in RESS and also in PGSS are highly asymmetrical binary mixtures which contain two substances with large differences in molecular size, structure, and intermolecular interactions. Two main features of the phase-behaviours for such systems are:

- the triple-point temperature of the heavy component is much higher than the critical temperature of the light one, and
- the solubility of the light component in the liquid phase is quite limited.

A schematic representation of a typical RESS experimental apparatus is shown in Figure 9.8-2.

The S-L-V curve intersects the gas-liquid critical curve in two points: the lower critical end point (LCEP) and the upper critical end point (UCEP). At these two points, the liquid and gas phases merge into a single fluid-phase in the presence of excess solid. At temperatures between T_{LCEP} and T_{UCEP} a S-V equilibrium is observed. The solubility of the heavy component in the gas phase increases very rapidly with pressure near the LCEP and the UCEP. Near the LCEP the solubility of heavy component in the light one is limited by the low temperatures. In contrast, near the UCEP the solubility of heavy component in the light one is high, owing to the much higher temperatures [6].

The solute is solubilized in supercritical fluid at high pressure in an autoclave, and later the solution is decompressed through the nozzle. One key component of the apparatus is the nozzle, of which two types are used - a capillary of suitable diameter (< 100 μm) or laser-drilled nozzles of 20-60 μm diameter.

9.8.4.3. Applications

Several groups have studied the RESS process for substances which may be applied in various fields, such as pharmacy, the electronics industry, cosmetics, and the food industry.

Advantages of RESS are: very fine particles are given even of the size of some nanometres; controllable particle size; solvent-free, and rather well understood from the theoretical point of view.

Table 9.8-2
RESS experiments for different products

Substance	SCF	Pressure (bar)	Temperature (K)	Particle size (μm)	Ref.
Steroids	CO_2	130-250	313-333	1-10	[7]
Theophylline	CO_2	225	338	0.4	[8]
Salicylic acid	CO_2	223	318	< 4	[8]
Salicylic acid	CO_2				[9]
Anthracene	CO_2			1-10	[8]
Caffeine	CO_2			1-10	[9]
Caffeine	CO_2				[10]
Naproxen	CO_2	190-210	333	1-20	[11]
L-PLA (poly-L-lactic acid) + naproxen	CO_2	170-200	363-388	10-90	[11]
L-PLA (poly-L-lactic acid)	CO_2	210	373-393	1-20	[11]
Flavone	CO_2	250	308	10	[12]
L-PLA (poly-L-lactic acid)	CO_2	200	328-338	2-5	[13]
PGA (poly-glycolic acid)	CO_2	180-200	328	10-20	[14]
D-PLA (poly-D-lactic acid)	CO_2	200	328	10-20	[14]
Lovastatin-L-PLA	CO_2	200-250	328		[15]
Poly-methyl methacrylate	CCl_2F_2			0.2-0.6	[16]
Poly-ethyl methacrylate	CCl_2F_2			0.2-0.6	[16]
Poly-L-lactic acid	CCl_2F_2			0.2-0.6	[16]
Lovastatin	CO_2	125-400	328	0.04-0.3	[17]
PolyDL-lactic acid/lovastatin	CO_2				[18,19]
β-estradiol	CO_2	345	328		[20]
L-630.028 (steroid derivative)	CO_2				[21]
Mevinolin	CO_2			10-50	[21]
β - Carotene	C_2H_{4+} Toluene	300	343	20	[22]
Benzoic acid	CO_2	200-283	308-328		[2]
Benzoic acid	CO_2	160-280	308-338	2-10	[23]
Benzoic acid	CO_2				[24]
L-Leucin	CO_2	103			[25]
Stigmasterol	CO_2	100-150	373	0.05-2	[26]

Table 9.8-2 (continued)
RESS experiments for different products

Substance	SCF	Pressure (bar)	Temperature (K)	Particle size (μm)	Ref.
Carotenoids	CO_2, C_2H_6, C_2H_4	80-400	298-323		[27]
Salicylic acid	CO_2	200	333		[28]
Nifedipin	CO_2	600	343		[29]
Griseofulvin	CHF_3	180-220	333	0.9-2	[30]
Naphthalene	CO_2			1.5-3	[31]
Cholesterol	CO_2			< 0.35	[31]
Benzoic acid	CO_2			0.8-1.2	[31]
Anthracene +phenanthrene	CO_2	170	308		[32]
Ibuprofen	CO_2	130-190	308	< 2	[33]
Cholesterol	CO_2				[34]

Disadvantages of RESS are: high ratios of gas/substance owing to the low solubility of the substance, high pressures (SC conditions) and sometimes temperatures; difficult separation of (very) small particles from large volumes of expanded gas and supercritical solutions established discontinuously, and thus there is the requirement for large-volume pressurized equipment.

9.8.5. Gas anti-solvent processes (GASR, GASP, SAS, PCA, SEDS)

The gas anti-solvent processes overcome the limited solvent capacity of supercritical gases for substances with high molecular weight by using a classical liquid solvent.

Gas anti-solvent processes (GASR, gas anti-solvent recrystallization; GASP, gas anti-solvent precipitation; SAS, supercritical anti-solvent fractionation; PCA, precipitation with a compressed fluid anti-solvent; SEDS, solution-enhanced dispersion of solids) differ in the way the contact between solution and anti-solvent is achieved. This may be by spraying the solution in a supercritical gas, spraying the gas into the liquid solution.

9.8.5.1. Fundamentals

The application of supercritical fluids as anti-solvents is an alternative recrystallization technique for processing solids that are insoluble in SCF. This method exploits the ability of gases to dissolve in organic liquids and to lower the solvent power of the liquid for the compounds in solution, thus causing the solids to precipitate.

Depending upon the parameters that influence the supersaturation ratio, the nucleation rate and the rate of growth, particle size, size distribution, and shape, can be varied over a wide range.

The overall recrystallization process, consisting of the initial nucleation and the subsequent growth of nuclei, can be modelled by the basic equations describing these respective processes. The rate of formation of nuclei as developed is:

$$J = Z \exp\left(\Delta G_{max} / R \cdot T\right) \tag{9.8-3}$$

where: J is the production of nuclei, (number per unit time per unit volume); Z is the collision frequency, calculable from classical kinetic theory; and ΔG_{max} is the Gibbs free energy expression.

The difference in particle size and particle size distribution is regulated via the mode and the rate of addition of the anti-solvent.

9.8.5.2 Design criteria

The solvent is saturated with the substance to be powdered. This solution is diluted by contacting it with a supercritical gas. The solvent power of the classical solvent is reduced by the gas, initiating precipitation of the substance to be powdered.

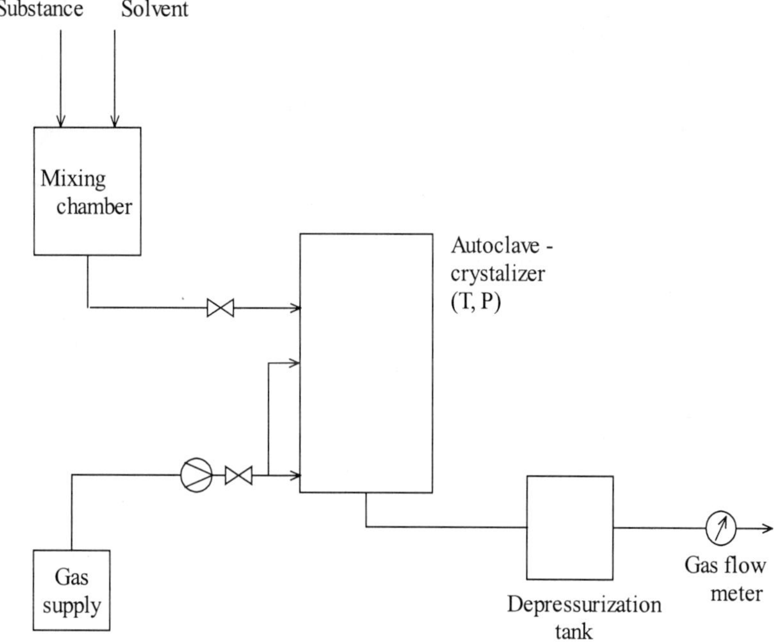

Figure 9.8-3. Scheme of the experimental apparatus for GASR.

9.8.5.3 Applications

Advantages of GASR are: (very) small particles may be produced; the particle sizes are easily controlled; and applicability shown for a wide variety of substances.

Disadvantages of GASR are: it is a batch process and uses organic solvents; scale-up of the mixing procedure is not yet investigated; separation of solid particles from solvent (organic) is required; a drying step is needed for the obtained powders; highly diluted product streams; separation of gas and solvent may be required in an industrial process.

Table 9.8-3
Gas anti-solvent experiments for different products using supercritical carbon dioxide (CO_2)

Process	Substance	Solvent	Pressure (bar)	Temperature (K)	Particle size (µm)	Ref.
GAS	Sodium cromoglycate (NaCrGlyc)	MeOH	95		0.1-2	[35]
	Hyaluronic acid ethyl ester (HYAFF) + Proteins	DMSO[a]	100		0-30	[36]
	Milk sugar	N,N-Dimethyl-formamide				[37]
	Acetaminophen	Ethanol	71	293		[38]
	Phenanthrene	Toluene				[39]
	Nitroguanidine	NMP[b], DMF[c]				[40]
	Polyamide	DMSO[a], DMF[c]				[41]
PCA	Hyaluronic acid ethyl ester (HYAFF)	DMSO[a]	104		50-500	[42]
	PLGA+ thymopentine (Thy)	CH_2Cl_2	85		40-60	[42]
	L-PLA 115 + 7000	CH_2Cl_2	65-125		7-50	[42]
	Hydrocortisone	DMSO[a]	104		0.2-1	[42]
	LPLA +Hyoscine-butyl-bromide (Hy)	CH_2Cl_2	90-200		10-15	[43]
	Indomethacin	CH_2Cl_2	200		8.2	[43]
	Piroxicam	CH_2Cl_2	90-200		6.0	[43]
	Thymopentine	CH_2Cl_2	90-200		6.6	[43]
	Methylprednisolone acetate (MPA)	THF	151		2.5-3	[44]
	Hydrocortisone acetate (HCA)	DMF	150		8.0	[44]
	Salmeterol xinafoate (SX)	Acetone	100-300		10	[45]

Table 9.8-3 (continued)
Gas anti-solvent experiments for different products using supercritical CO_2

Process	Substance	Solvent	Pressure (bar)	Temperature (K)	Particle size (μm)	Ref.
PCA	Chlorophenylamine maleate	CH_2Cl_2	69		1-5	[46]
	L-PLA 94000	CH_2Cl_2	81.6		1-5	[12]
	L-PLA 100000	CH_2Cl_2	76-97		0.5-5	[47]
	Paracetamol	Ethanol		313		[48]
	Ascorbic Acid	Ethanol		313		[48]
	Chloramphenicol	Ethanol		313		[48]
	Urea	Ethanol		313		[48]
	Soy Lecithin	Ethyl Alcohol	80-110	308	1-40	[49]
	Cu acetate $Cu(CH_3COO)_2 \cdot H_2O$	Ethanol			> 0.15	[50]
	Yttrium acetate $Y(CH_3COO)_3 \cdot xH_2O$	Ethanol			0.05-0.2	[50]
	Ba Acetate $Ba(CH_3COO)_2$	Ethanol			0.05-0.2	[50]
SAS	DL-PLA (poly-(DL-lactic acid)	$DMSO^{(a)}$	104		< 2	[51]
	Trypsin	$DMSO^{(a)}$	73-140		1-5	[52]
	Lysozyme	$DMSO^{(a)}$	73-110		1-5	[52]
	Insulin	$DMSO^{(a)}$	91-140		1-5	[52]
	Insulin	$DMSO^{(a)}$	86		2-4	[53]
	Eudragit E 100	Ethylacetate	52			[54]
	Eudragit E 100	Toluene	67			[54]
	Eudragit E 100	Acetone	46			[54]
	Eudragit E 100	1,4Dioxane	28			[54]
	HP-55 (Hydroxypropyl methylcellulose phthalate)	Acetone	52			[54]
	HP-55 (Hydroxypropyl methylcellulose phthalate)	1,4Dioxane	34			[54]
	Ethylcellulose	Acetone	43			[54]
	Ethylcellulose	Ethylacetate	53			[54]
	Yttrium acetate (AcY)	$DMSO^{(a)}$	120	323	0.1	[55]

Table 9.8-3 (continued)
Gas anti-solvent experiments for different products using supercritical CO_2

Process	Substance	Solvent	Pressure (bar)	Temperature (K)	Particle size (μm)	Ref.
SAS	Samarium acetate (AcSm)	DMSO[a]	150	313	0.14	[55]
	Dextran	DMSO[a]	150	313		[56]
	Inulin	DMSO[a]	150	313		[56]
	Poly-L-lactic acid (PLLA)	CH_2Cl_2	120	313		[56]
	HYAFF	DMSO[a]	150	313		[56]
		NMP	180	313	0.6	[57]
	Tetracycline	NMP	180	313	0.6	[57]
	Amoxicillin	NMP	150	298-323	0.2-10	[57]

(a) DMSO = Dimethyl sulfoxide
(b) (b) NMP = N-Methylpyrrolidone
(c) DMF = Dimethylformamide

9.8.6 Particles from gas-saturated solutions (PGSS)

In a relatively new process for production and fractionation of fine particles by the use of compressible media - the PGSS process (Particles from Gas-Saturated Solutions) - the compressible medium is solubilized in the substance which has to be micronized [58-61]. Then the gas-containing solution is rapidly expanded in an expansion unit (e.g., a nozzle) and the gas is evaporated. Owing to the Joule-Thomson effect and/or the evaporation and the volume-expansion of the gas, the solution cools down below the solidification temperature of the solute, and fine particles are formed. The solute is separated and fractionated from the gas stream by a cyclone and electro-filter. The PGSS process was tested in the pilot- and technical size on various classes of substances (polymers, resins, waxes, surface-active components, and pharmaceuticals). The powders produced show narrow particle-size distributions, and have improved properties compared to the conventional produced powders.

9.8.6.1 Fundamentals

By dissolving the compressible media in a liquid, a so-called gas-saturated solution may be formed. By expansion of such a solution in an expansion unit (e.g., a nozzle) the compressed medium is evaporated and the solution is cooled. Owing to the cooling caused by evaporation and/or the Joule-Thompson effect the temperature of the two-phase flow after the expansion nozzle is lowered. At a certain point, the crystallization temperature of the substance to be solidified is reached, and solid particles are formed and cooled further.

In Figure 9.8-4 a typical dependence between pressure, and gas-concentration in the liquid, and the temperature, is shown for the system glycerides-CO_2.

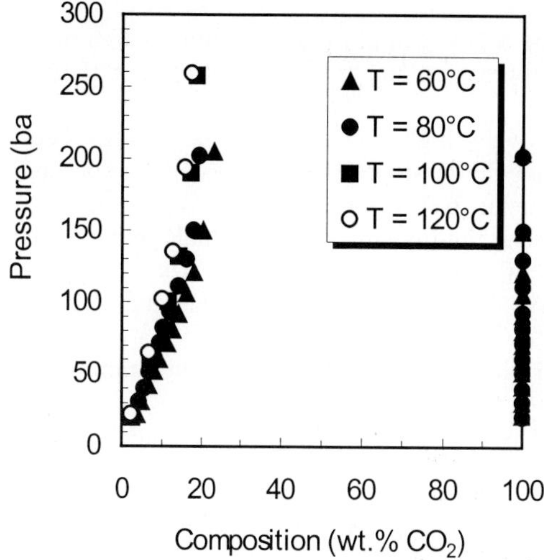

Figure 9.8-4. Pressure-concentration diagram.

The curves on the right-hand side represent the concentration of glyceride in the gas, while the curves on the left side represent the concentration of the gas in the liquid glycerides [62,63]. It can be seen that although the glycerides are not dissolved in the gas, a certain amount of gas is dissolved in the glycerides.

Considering an expansion of the gas-rich phase it is reasonable that the temperatures achieved after the expansion are in the range of those obtained after the expansion of pure gas. If the glyceride-rich phase is expanded, the temperature after the expansion is much higher. Nevertheless, it may be lower then the solidification temperature of the glycerides if a sufficient amount of gas is dissolved. Depending on the specific substance to be powdered, the minimum concentration of gas required for solidification may be as low as 5-8 wt.%.

The expansion phenomena (generation of low temperature by expansion) of compressible media is well described by the Joule-Thomson effect, which may be used as basic information on the applicability for a certain gas of the PGSS-process. If heat-transfer, changes in kinetic energy, and work-transfer for the flow process through an expansion valve are neglected, the process may be considered to be isenthalpic [62]. The effect of change in temperature for an isenthalpic change in pressure is represented by the Joule-Thomson coefficient, μ_{jt}, defined by:

$$\mu_{jt} = \left(\partial T / \partial P\right)_h \qquad (9.8\text{-}4)$$

The Joule-Thomson coefficient is the slope of the isenthalpic lines in the P-T projection. In the region where $\mu_{jt} < 0$, expansion through the valve (a decrease in pressure) results in an increase in temperature, whereas in the region where $\mu_{jt} > 0$, expansion results in a reduction in temperature. The latter area is recommendable for applying the PGSS process.

9.8.6.2. Design criteria

In PGSS, knowledge of the P-T trace of the S-L-V equilibrium gives information on the pressure needed to melt the substance to be micronized and form a liquid phase at a given temperature, and to calculate its composition.

When the supercritical fluid has a relatively high solubility in the molten heavy component, the S-L-V curve can have a negative dP/dT slope [64]. The second type of three-phase S-L-V curve shows a temperature minimum [65]. In the third type, where the S-L-V curve has a positive dP/dT slope, the supercritical fluid is only slightly soluble in the molten heavy component, and therefore the increase of hydrostatic pressure will raise the melting temperature and a new type of three-phase curve with a temperature minimum and maximum may occur [66].

In general, a system with a negative dP/dT slope and/or with a temperature-minimum in the S-L-V curve could be processed by PGSS.

The substance which has to be micronized is filled into a thermostatted feed vessel (A) (Figure 9.8-6). The solution from (A) is transferred into the thermostatted autoclave (C) after evacuation. The compressible medium is supplied by a high-pressure pump (B). The pressure in the autoclave is increased to a certain value, and the high-pressure circulation pump (D), connected with the autoclave, is installed to increase the efficiency of dissolving of the solute by circulating the liquid phase. The gas-saturated mixture is rapidly depressurized through a nozzle. In the spray-tower the solvent is set free, and the formed solid particles of the substance under consideration (>10 μm) are collected in the vessel at the bottom. The temperature in the spray can be measured and recorded via two thermoelements.

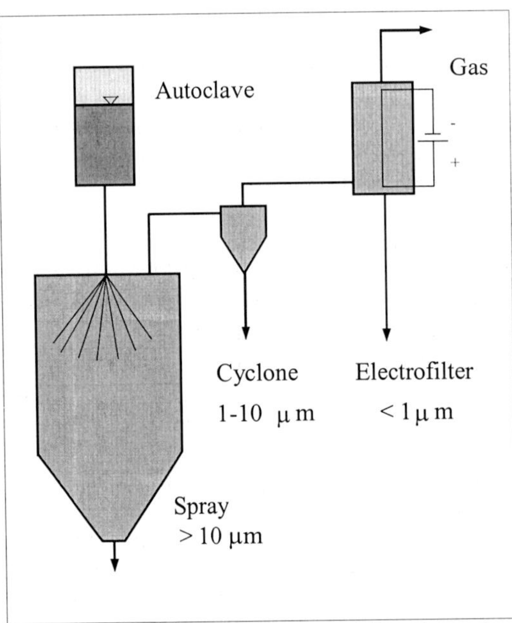

Figure 9.8-5. Scheme of experimental equipment.

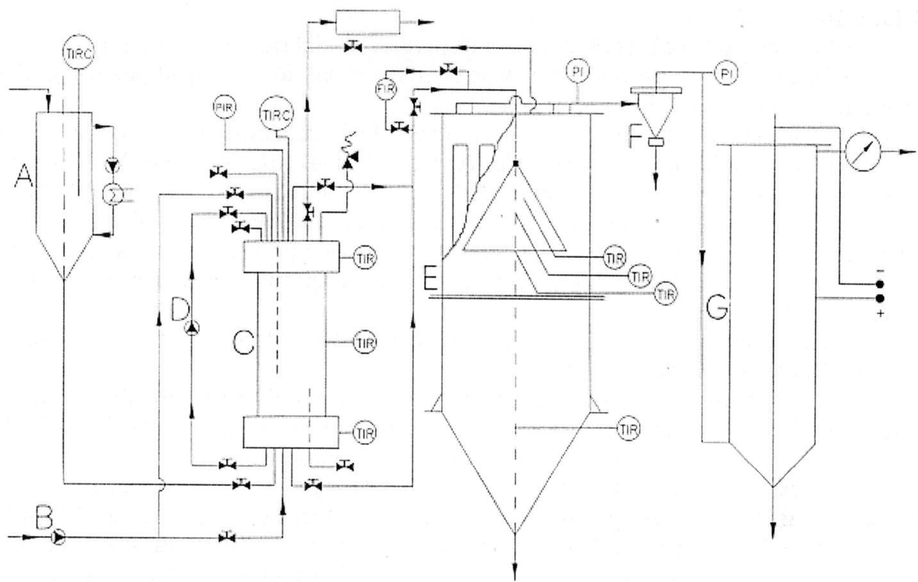

Figure 9.8-6. Basic scheme of experimental equipment for PGSS.

The behaviour of the spray inside the spray tower can be observed through windows. The depressurized solvent, which contains the smaller particles (1-10 μm) is led into the cyclone (F) where these are collected at the bottom of the vessel. For separation of smallest particles (<1μm) the gas stream is introduced into the electrofilter (G). Finally, the gaseous solvent passes through a gas-flow meter for determination of total flow.

9.8.7 Application of the PGSS process for micronization

One goal of RESS/CSS, the anti-solvent processes such GAS, and the PGSS process, is to obtain submicron- or micron-sized particles. Technological features of the various high-pressure micronization processes are summarised and compared in Table 9.8-4 [58].

The RESS and GAS-processes are suitable for obtaining submicron particles. Although several features concerning scale-up are not yet very well known, it is probable that these processes are, or may be, used for producing relatively small amounts of high-value-added substances.

Restrictions arising from the difficult product- and gas-recovery in the RESS and GASR, GASP, SAS/PCA/SEDS processes are avoided by the PGSS process.

The PGSS process has several advantages which favour its use for large scale applications. This process has promise for the processing of low-melting, highly viscous, waxy, and sticky compounds, even if the obtained particles are not of submicron size. The process already runs in plants with a capacity of some hundred kilograms per hour.

Table 9.8-4
Technological features of RESS, GASR and PGSS process

	RESS	GAS/SAS/PCA	PGSS
Establishing gas-containing solution	Discontinuous	Semicontinuous	Continuous
Gas demand	High	Medium	Low
Pressure	High	Low to medium	Low to medium
Solvent	None	Yes	None
Volume of pressurized equipment	Large	Medium to large	Small
Separation Gas/Solid	Difficult	Easy	Easy
Separation Gas/Solvent	Not required	Difficult	Not required

Up to the present time the application of the PGSS process has been investigated for following products:
➤ Polymers (polyethylene-glycols, polyetherurethanes, polymethyl-methacrylate (PMMA)),
➤ Waxes and Resins (Montanwaxes, polyethylene waxes),
➤ Natural Products (extracts from spices, phospholipids, menthol),
➤ Fat Derivatives (partial glycerides, fatty alcohols, fatty acids, alkylpolyglucosides) and
➤ Others (Nifedipine, synthetic antioxidants, UV-stabilizers).
The Highly Compressible Fluids which have been used are:
➤ Carbon Dioxide,
➤ Propane,
➤ Butane,
➤ Dimethyl ether,
➤ Freon, R23,
➤ Ethanol,
➤ Nitrogen, and
➤ Mixtures of above-mentioned gasses.

9.8.7.1 Glycerides

A systematic investigation of the influence of pre- and post-expansion conditions, and solute concentration (general process parameters) upon the crystallinity, particle size, and particle-size distribution, was performed for the model system glyceride-CO_2 [67-69]. A glyceride with the compositions, 50-55 wt.% of monoglycerides, 35-40 wt.% of diglycerides, 3-8 wt.% of triglycerides, and less than 1 wt.% of free fatty acids, was used.

Determination of melting point of monoglycerides under pressure

In the developed process it is important to know the influence of the pressure on the melting point of the substance in the presence of the gas "s-l-transition" and the phase-equilibrium "l-v-data" of the system.

The values of melting points were determined by a modified capillary method. As can be seen from Figure 9.8-7 the pressure influences the melting point of glycerides. The reduction of the melting point with increasing pressure is relatively high, especially at pressures up to 115 bar. At pressures over 200 bar the melting point of the substance increases again. This is probably due to an upper critical end point. The reduction of the melting temperature in the presence of a soluble gas allows one to establish liquid solutions without heating the substances to be micronized above their melting temperature at ambient pressure. This is of special interest if thermally sensitive substances have to be treated. The process of establishing the melt could be named as "gas-induced melting". Establishing liquid solutions is one boundary condition for designing spray processes. Using a compressed gas as solvent has several distinct advantages over the use of classical solvents. It is known that the viscosity of highly viscous substances can be reduced by several orders of magnitude if a gas is used as diluent. The viscosity reduction is much larger than given by classical solvents.

Additionally, the interfacial tension between the gas-saturated solution and the gas is much lower than that of the pure melt. Both properties are favourable either for establishing a spray – even from high viscous compounds - and for obtaining high interfacial areas, corresponding to small droplets and thus small particles. In addition to the fluid dynamic properties, the thermodynamic boundary conditions have to be considered. This can be illustrated in a pressure-concentration diagram of the system Monoglyceride-CO_2 (Figure 9.8-8).

Basically, the amount of gas dissolved in the liquid must be large enough to take up the energy which has to be removed from the liquid during particle formation. The lowest gas concentration at which a powdery product was obtained is 5 wt.%. At a melt-temperature of 60°C this corresponds in Figure 9.8-8 to a pressure of approx. 80 bar.

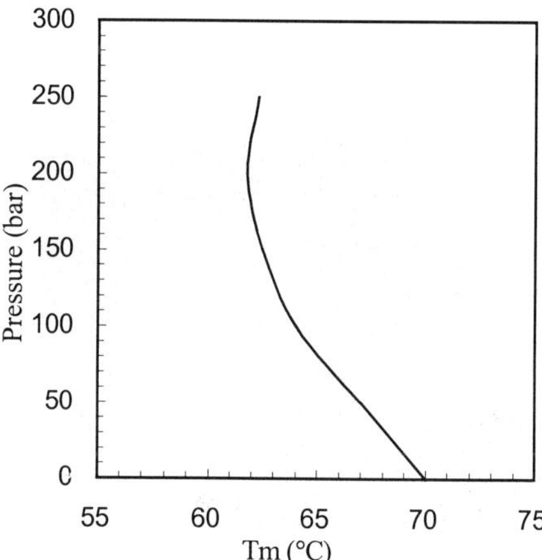

Figure 9.8-7. P-T diagram for solid-liquid transition of monoglyceride in presence of CO_2.

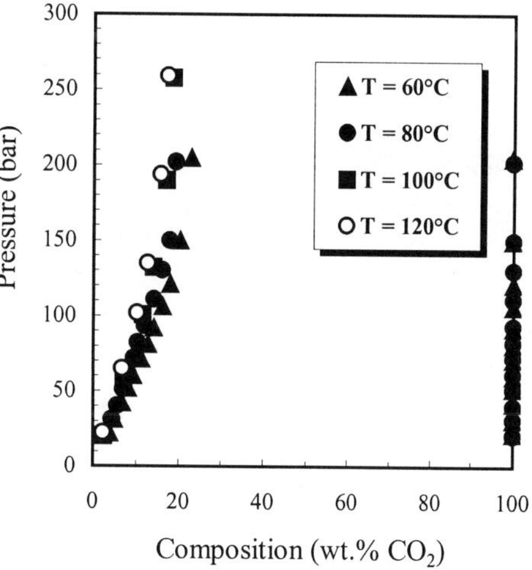

Figure 9.8-8. Pressure-concentration diagram.

602

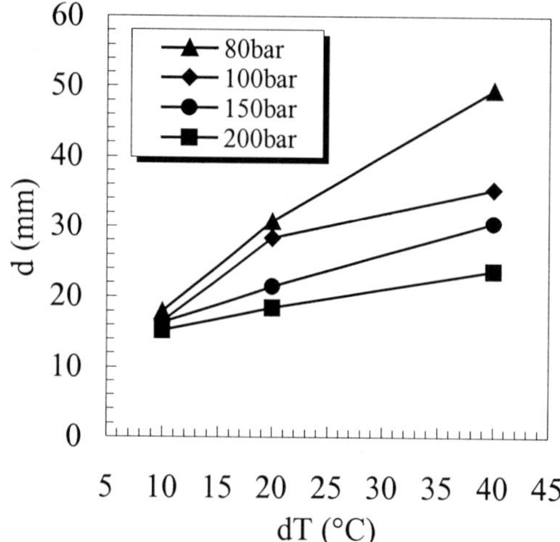

Figure 9.8-9. Influence of the pre-expansion temperature on the particle size in the system glyceride-CO$_2$.

At higher pressures, higher solubilities are observed, which is favourable for improved solidification. At pressures below 80 bar, instead of a powdery end-product, a suspension or a liquid is obtained.

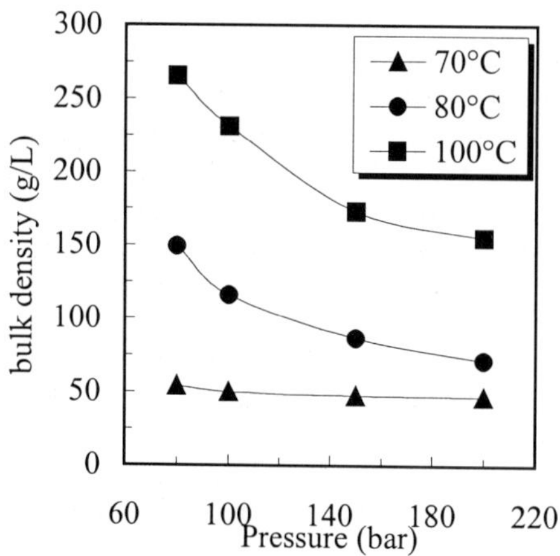

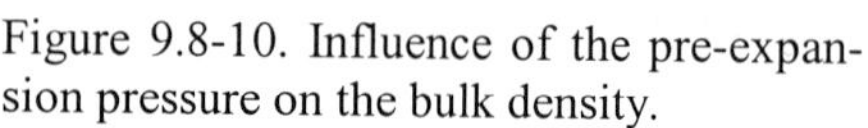

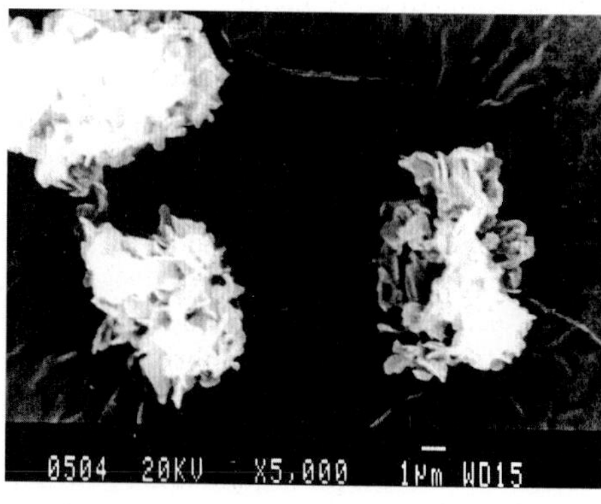

Figure 9.8-10. Influence of the pre-expansion pressure on the bulk density.

Figure 9.8-11. View of monoglyceride particle processed by PGSS.

The influence of the pre-expansion temperature of the equilibrated gas-saturated solution on the particle size was studied over a pressure range from 80-200 bar. The temperature is represented as the temperature-difference between the saturation temperature before expansion and the melting point of the substance. The results are shown in Figure 9.8-9.

The mean particle size is a function of the pre-expansion temperature: the particle size reduces with lower pre-expansion temperatures.The closer the saturation temperature is to the melting point of the substance, the smaller are the formed particles. The smallest particles were obtained at high pressure and low temperature, which could have been expected, owing to the higher solubility of CO_2 in the melted substances and the resulting higher level of supersaturation during the expansion.

Figure 9.8-10 shows the influence of the pre-expansion pressure on the bulk-density of the micronized glyceride at different temperatures. The PGSS process has changed the appearance of the glycerides. In the processed particles a sponge-like morphology with a rather large internal area can be observed. Differential scanning calorimetry has confirmed some lack of crystallinity in a freshly micronized sample, thus indicating the presence of the amorphous form (up to 40 %).

9.8.7.2 Cocoa butter

Cocoa butter (CB) is a vegetable fat, present in chocolate at levels up to 40%. Cocoa butter is a complex compound, containing various types of triglycerides. The characteristic physical properties of cocoa butter are related to the arrangement of the fatty acids in the triglycerides: there is a high content of symmetrical monounsaturated triglycerides, which have the unsaturated fatty acid in the 2-position and saturated fatty acids in the 1- and 3-positions. These triglycerides consist of 2-oleoyl-1-palmitoyl-3-stearoylglycerol (POS), 2-oleoyl-1,3-distearoylglycerol (SOS), and 2-oleoyl-1,3-dipalmitoylglycerol (POP), with POS being present in the largest amount. The melting behaviour of cocoa butter can be related to the melting points of these major triglycerides (POS, SOS, POP). Each of these triacylglycerols is polymorphic- they can solidify in various crystallographic forms. The process of solidification of cocoa butter is more complex than is the case for other fats. It can crystallize in five different polymorphic forms, each with its own melting point. The melting points of the major polymorphic forms (in order of increasing stability) are: γ (16-18 °C), α (21-24 °C), β_1 (27-29 °C), β (34-35 °C) and β_2 (36-37 °C). Phase β is always stable below its melting temperature, but its kinetics of nucleation and growth are very slow, so that under direct cooling a less stable phase is generally formed. The hardness of cocoa butter and the crystallinity have a major effect on chocolate's organoleptic characteristics.

The data for the phase equilibrium solid-liquid for the binary system cocoa butter-CO_2 and for the equilibrium solubility data of CO_2 in the liquid phase of cocoa butter have been presented [70].

Based on phase-equilibrium data in the Master diagram (Figure 9.8-12) (where S-l and l-V equilibrium data are presented) the experiments for cocoa butter micronization using the PGSS process were carried out. The pre-expansion pressure was in the range of 60 to 200 bar and at temperatures from 20 to 80°C. The micronization with the nozzle D = 0.25 mm resulted in fine solid particles with median particle sizes of about 62 μm. In Figure 9.8-13 the morphology of a cocoa-butter particle is presented.

604

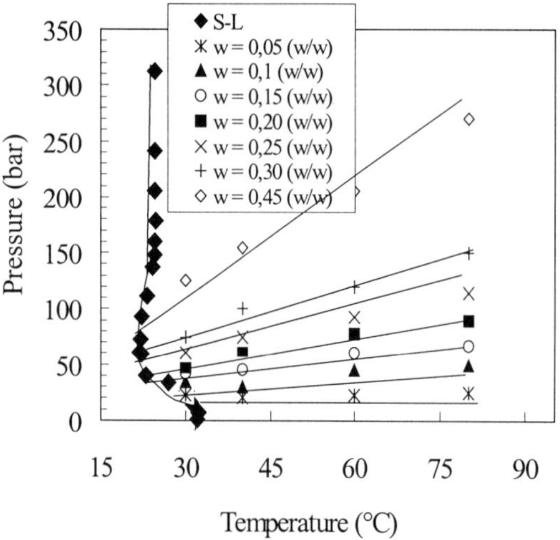

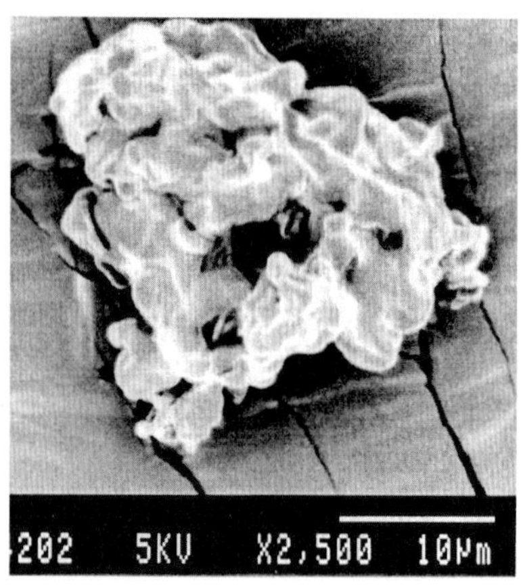

Figure 9.8-12. Master diagram for the binary system of cocoa butter-CO_2.

Figure 9.8-13. Morphology of a cocoa-butter particle.

From DSC measurements it was deduced that the crystallinity of a freshly micronized sample is about 80% and that it crystallizes in the stable polymorphic form β_2. Under the operating conditions mentioned, the PGSS process caused no degradation of cocoa butter and the product was a powder with a narrow and very controllable size-distribution.

9.8.7.3 Pharmaceuticals

The practically water-insoluble compounds nifedipine and felodipine, which are dihydropyridine calcium-channel-blockers were processed by the PGSS process with the aim of increasing their dissolution rate and hence bioavailability [71,72].

Table 9.8-5
Experimental conditions for micronization of drugs using the PGSS process

	Nifedipine	Felodipine
Melting point of drug (°C)	175-175	141-145
Pre-expansion pressure (bar)	100, 125, 150, 175, 190, 200	200
Pre-expansion temperature (°C)	165, 175, 185	150
Nozzle diameter (mm)	0.4, 0.25	0.4

Using the PGSS process, nifedipine was micronized at various pressures in the range from 100 to 200 bar and at temperatures 165, 175 and 185°C. The mean particle-size of the starting nifedipine was 50 μm, and it was decreased to 15-30μm, depending on the experimental conditions. The resulting particle size was a function of the process conditions. With increasing pre-expansion pressure the mean particle-size was reduced and, as a result, the dissolution rate was found to be higher for samples prepared at higher pre-expansion pressures (Figure 9.8-14).

The shape of the micronized particles was irregular and, according to SEM pictures, it was assumed that the particles were porous. With particle-size reduction and, therefore, increased specific surface area (external and internal) the dissolution rate increased to some extent, but the anticipated effective surface area was probably reduced by the drug's hydrophobicity and agglomeration of the particles during and after micronization.

In order to avoid agglomeration of micronized particles, and thermal degradation of nifedipine at high temperatures (175 and 185^{0}C), the hydrophilic polymer PEG 4000 was added to nifedipine to reduce its melting point. Micronization at pre-expansion temperatures between 50 and 70ºC was possible and fine powdered co-precipitates of nifedipine/PEG 4000 were obtained. Their dissolution rate was twice as high as that for pure micronized nifedipine.

The mean particle size of the starting felodipine was 60 μm, and reduced after micronization with the PGSS process to 42 μm. Specific surface areas measured using the BET method increased from 0.33 m^2/g for the starting felodipine to 1.33 m^2/g for micronized felodipine.

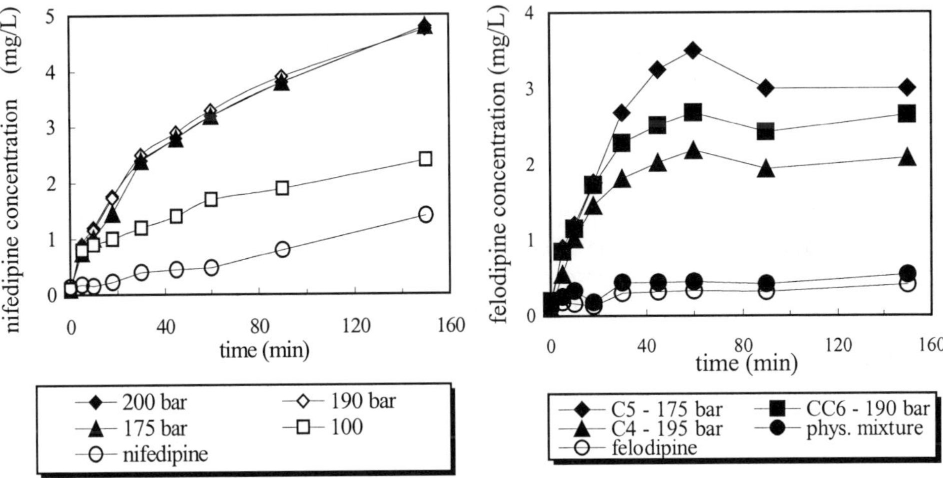

Figure 9.8-14. Dissolution profiles of nifedipine, micronized at 175°C and various pressures compared with original nifedipine.

Figure 9.8-15. Dissolution of micronized felodipine PEG 4000 samples prepared at different pre-expansion pressures, compared with original felodipine and a physical mixture of the same composition (1:4).

The starting felodipine, as well as micronized felodipine, practically do not dissolve in water, owing to the poor wettability of the felodipine particles. The amount of dissolved felodipine after 1 h was only 0.26 mg and 0.29 mg, for the starting felodipine and the micronized sample, respectively.

The dissolution profiles of PGSS-felodipine/PEG 4000 co-precipitates, along with starting felodipine and its physical mixture with PEG 4000 (1:4), are presented in Figure 9.8-15. The amount of felodipine dissolved in 1 h from felodipine/PEG 4000 co-precipitates micronized at pre-expansion pressures of 175, 190 and 195 bar is 13.5-, 10-, and 8-times higher than that of original felodipine.

With the PGSS process, micronized drugs and drug/PEG 4000 samples were prepared in a new way, which has some advantages over conventional methods for the micronization of pure drugs and for drug/carrier solid dispersion preparation, namely fusion methods and solvent processes.

In Figure 9.8-15 the impact of the pre-expansion pressure on the dissolution behaviour of felodipine from micronized felodipine/PEG 4000 co-precipitates can be seen. Similar observations were noted above for nifedipine/PEG 4000 samples.

Even at the pre-expansion pressure of 175 bar, enough CO_2 was dissolved in the melt to cause a rapid precipitation of the components. The lower dissolution rate of the sample micronized at 195 bar and 70^0C could be explained by the decreased solubility of CO_2 in the melt at a higher temperature. The removal of solvent is a problem of conventional co-precipitation or co-evaporation techniques where large amounts of organic solvents are needed and in which complete removal is often a long and difficult process. With the PGSS process, micronized drug or micronized drug/carrier can be obtained in one step without organic solvent.

9.8.7.4 PGSS of polyethyleneglycols

Polyethyleneglycols (PEGs) are water-soluble polymers which are, owing to their physiological acceptance, used in large quantities in the pharmaceutical-, cosmetic-, and food industries. The general formula of PEGs is $H(OCH_2CH_2)nOH$, where n is the number of ethylene oxide groups. PEG is, like other polymers, a mixture of homologues where n varies between certain numbers. PEGs up to molar mass of 600 g/mol are liquid, while those with higher molar masses are solid. PEG is usually manufactured as flakes, but in some applications it is used as a powder which is obtained by milling the flakes. Milling is possible for PEGs with molar masses between about 2000 g/mol and 10000 g/mol. PEGs of lower molar mass are too greasy, and higher PEGs are too hard. Not only with respect to flexibility for different feedstocks, the newly developed process - PGSS - is favourable for the generation of fine particles [73].

It was found that the solubility of CO_2 (expressed in weight percent) in PEGs above a certain chain-length is practically independent of the molecular mass of PEG, and is influenced only by the pressure and temperature of the system.

At low pressures, the liquefaction temperature of PEG in the presence of CO_2 increases slightly compared to the melting point of the pure PEG. This increase might be caused by a pressure effect or by a solid-transition of PEG induced by increasing pressure. Then a sharp reduction is observed. The liquefaction temperature passes through a minimum and increases again.

In Figure 9.8-16 the expansion for the system PEG/CO_2 according to the PGSS-process is represented by the isoplethes (the composition does not change during the expansion). The gas-saturated solution is established at pressures and temperatures which are on the right-hand side of the liquefaction curve (S-L-V line). On the left-hand side of the liquefaction curve, solid PEG coexists with gaseous CO_2. The starting conditions have to be chosen in such a way that the end-conditions after expansion are in the S-V-region of the binary system.

For the production of fine dispersed PEG by expanding CO_2-saturated solutions, a first assumption indicates that the starting conditions should be near the liquefaction curve in order to reach the solid-gas region after the expansion. An initial estimation can be performed by calculating an energy balance which takes into account the:

- enthalpies of CO_2 before and after expansion (e.g., from a T,S-diagram);
- enthalpies of PEG before and after expansion (e.g., from the heat capacity of the liquid- and solid PEG),
- enthalpy of a solution of CO_2 and PEG (e.g., derived from phase-equilibrium data) and
- the heat of crystallization of PEG (from literature data or DSC).

The mechanical energy for the formation of a new interface is in most cases, two to three orders of magnitude lower than the thermal energies, and can therefore be neglected.

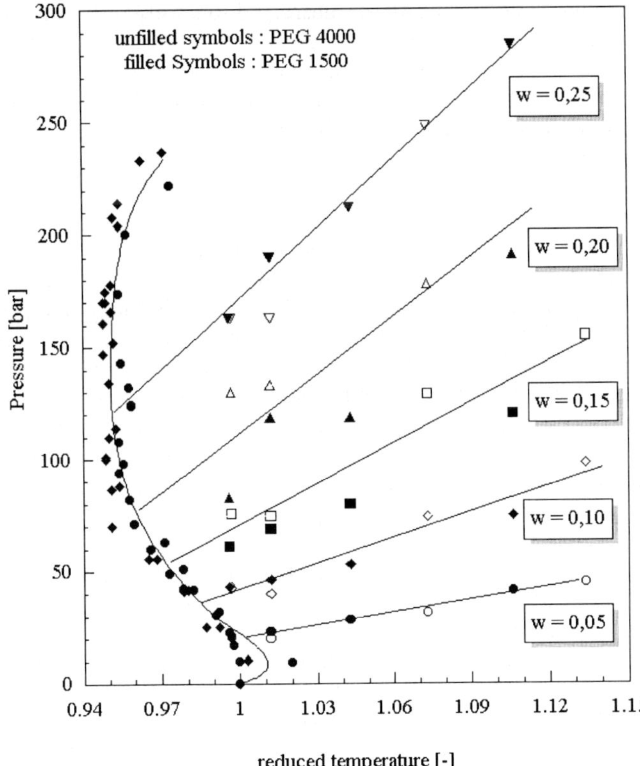

Figure 9.8-16. P-T_{red} diagram with melting-point line and constant composition lines.

608

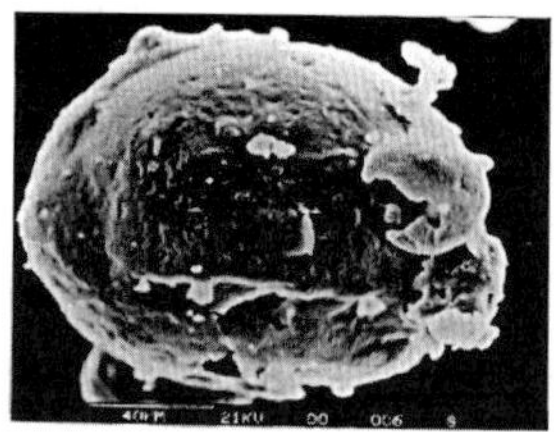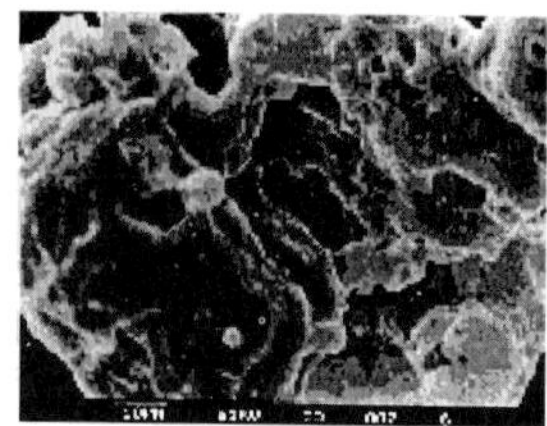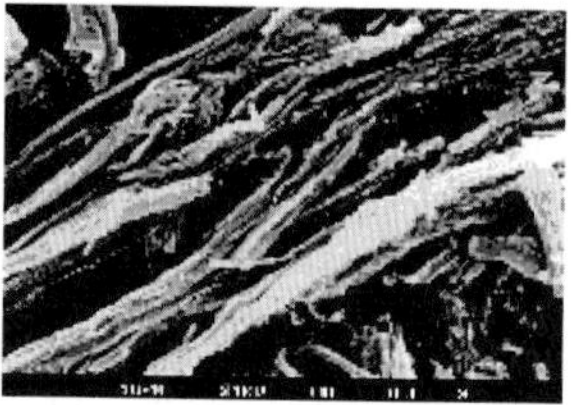

Figure 9.8.17. Shape of the particles of PEG obtained by the PGSS process.

The experiments on the PGSS of PEG were performed in a laboratory-scale plant with sample sizes of about 200-400 g powder, and in a small pilot plant with sample sizes of 1-3 kg powder. Depending on the nozzles (orifices 0.4 / 0.5 /1.0 mm, spraying angles 30° and 90°), the kind of PEG (MW 1500/4000/8000/35000) and on pressure (100-250 bar) and temperature (45-70°C) three classes of particles were obtained: fibres, spheres and sponges as presented in Figure 9.8-17.

Spheres are obtained when the initial temperature is so high that the product is not completely powdery (approx. 30 K above the melting point). Fibres are obtained preferentially, when the molar mass of the PEG increases. The "normal" configuration is the sponges.

From DSC measurements of fresh powders (some minutes old) it was deduced that the crystallinity is between 85 and 90%. Older powders (1-6 weeks) have a crystallinity which corresponds to the conventional PEGs.

In the pressure range investigated the saturation pressure has almost no influence on the mean particle-size, and no influence on the bulk densities.

The diameter of the nozzle influences the mean particle-size. A 1.0 mm nozzle gave particles between 350 and 500μm, while a 0.4 mm nozzle gave mean particle-sizes of 170-370 μm.

Overall CO_2 consumption is in the range between 0.17 and 0.7 kg / 1 kg of powdered PEG.

9.8.7.5 Economy of the process

Based on the results from laboratory-scale equipment and the first production plants, the basic economics of the process were calculated. The processing costs are between 0.15 and 0.60 Euro/kg (composition of costs: investment ~ 20%, personell ~ 37%, and operating ~ 43%) and vary depending on the substances to be micronized, the scale of the equipment, etc.

9.8.7.6 Advantages of PGSS

The advantages of the PGSS process over conventional methods of particle-size reduction are numerous.

- It is a versatile process, applicable for several substances;
- it uses moderate pressures;
- has low gas consumption;
- gives solvent-free powders;
- is suitable for highly viscous or sticky products;

- gives fine powders with narrow size-distribution,
- different morphologies,
- and is easy to scale-up.

Through the choice of the appropriate combination of solvent and operating conditions for a particular compound, PGSS can eliminate some of the disadvantages of traditional methods of particle-size redistribution in material processing. Solids formation by PGSS therefore shows potential for the production of crystalline and amorphous powders with a narrow and controllable size-distribution, thin films, and mixtures of amorphous materials.

9.8.8 Conclusions

Numerous processes for powder generation using supercritical fluids have been developed. The specific properties of dense gases allow obtaining fine dispersed solids, especially of substances with low melting point temperatures, high viscosities and very waxy or sticky properties. Economic evaluation of the process shows that these compounds cannot be efficiently and economically processed by conventioned mechanical processes and there is a big advantage of the use of supercritical fluids.

References

1. B. Bungert, G. Sadovski and W. Arlt, Chemie Ingeniur Technik, 69 (1997) 298-311.
2. A. Tavana and A.D. Randolph, AIChE J., 35 (10) (1989) 1625-1630.
3. A. Tavana and A.D. Randolph, AIChE Symp. Ser. 87 (284) (1991) 5.
4. C. Y. Tai and G.-S. You, Proceedings of the 5th Meeting on Supercritical Fluids, Nice (France), (1998) 297-299.
5. H. Freund and R. Steiner, High Pressure Chemical Engineering, P.R. von Rohr and C. Trepp (Eds.), Elsevier, Amsterdam, (1996) 211-216.
6. P.G. Debenedetti, AIChE Journal, 36 (9) (1990) 1289-1298.
7. P. Alessi, A. Cortesi, I. Kikic, N.R. Foster, S.J. Macnaughton and I. Colombo, Ind. Eng. Chem. Res., 35 (1996) 4718-4726.
8. P. Subra and P.G. Debenedetti, High Pressure Chemical Engineering, Edited by Ph. Rudolf von Rohr and Ch. Trepp, Elsevier, Amsterdam, (1996) 49-54.
9. E. Reverchon, G. Donsi and D. Gorgoglione, J. Supercrit. Fluids, 6 (4) (1993) 241-248.
10. H. Ksibi, P. Subra and Y. Garrabos, Adv. Powder Technol., 6 (1) (1995) 25-33.
11. J.-H. Kim, T.E. Paxton and D.L. Tomasko, Biotechnology Progress, 12 (1996) 650-661.
12. K. Mishima, K. Matsuyama, H. Uchiyama, M. Ide, J.-J. Shim and H.-K. Bae, The 4th International Symposium on Supercritical Fluids, Sendai, Japan, (1997) 267-270.
13. J. W. Tom, P.G. Debenedetti and R. Jerome, J. Supercrit. Fluids, 7 (1994) 9-29.
14. J.W. Tom and P.G. Denenedetti, Biotechnology Progress, 7 (1991) 403-411.
15. J.W. Tom, G.-B. Lim, P.G. Debenedetti, and Prud'homme, R.K., ACS Symp. Series 514; American Chemical Society: Washington, DC., 238-257.
16. A.K. Lele and A.D. Shine, Ind. Eng. Chem. Res., 33 (6) (1994) 1476-1485.
17. X. Kwauk and P.G. Debenedetti, J. Aerosol. Sci., 34 (1993) 445-469.
18. P.G. Debenedetti, J.W. Tom, X. Kwauk and S.D. Yeo, Fluid Phase Equilibria, 82 (1993) 311-321.
19. R.S. Mohamed, D.S. Halverson, P.G. Debenedetti and R.K. Prud'homme, ACS Symposium Series, 406 (1989) 355-378.

20. V. Krukonis, A.I.Ch.E. Meeting, San Francisco, November, (1984) 140f.
21. K.A. Larson and M.L. King, Biotechn. Prog., 2 (1986) 73-82.
22. C.J. Chang and A.D. Randolph, AIChE J., 35 (1989) 1876-1882.
23. E.M. Berends, O.S.L. Bruinsma and G.M. van Rosmalen, J. Crystal Growth, 128 (1993) 50-56.
24. M. Turk, B. Helgen, S. Cihlar and K. Schaber, High Pressure Chemical Engineering, (1999) 243-246.
25. S. Furuta, R.W. Rousseau and A.S. Teja, A.I.Ch.E. Annual Meeting, Miami, (1992) 706.
26. K. Ohgaki, H. Kobayashi, T. Katayama and N. Hirokawa, J. Supercrit. Fluids, 3 (1990) 103-107.
27. W. Best, F.J. Muller, K. Schmieder, R. Frank and J. Paust, J. Germ. Patent No. 29 43 267 (1979).
28. J.V. Tom, G.-B. Lim, P.G. Debenedetti and R.K. Prud'homme, ACS Symp. Series, Washington, 514 (1993) 238-257.
29. E. Stahl, K.-W. Quirin and D. Gerard. Dense Gases for Extraction and Refining, Springer Verlag, 1987.
30. E. Reverchon, G. Della Porta, R. Taddeo, P. Pallado and A. Stassi, Ind. Eng. Chem. Res., 34 (1995) 4087-4091.
31. M.Turk, J. Supercrit. Fluids, 15 (1) (1999) 79-89.
32. G.-T. Lin and K. Nagahama, J. Chem. Eng. Jpn., 30 (2) (1997) 293-301.
33. M. Charoenchaitrakool, F. Dehghani and N.R. Foster, Fifth Conference on Supercritical Fluids and their Applications, Garda (Verona), (1999) 485-492.
34. Kröber, Teipel: Please check in CIT from February (1999).
35. S. Jaarmo, M. Rantakyla and O. Aaltonen, The 4th International Symposium of Supercritical Fluids, Sendai, Japan, (1997) 263-266.
36. A. Bertucco, P. Pallado and L. Denedetti, High Pressure Chemical Engineering, Process Technology Proceedings, 12 (1996) 217-222.
37. J. Robertson, M.B. King and J.P.K. Seville, Fourth Italian Conference on Supercritical Fluids and their Applications, Capri (Italy), (1997) 365-372.
38. F.E. Wubbolts, C. Kersch and G.M. van Rosmalen, Proceedings of the 5th Meeting on Supercritical Fluids, Nice (France), (1998) 249-256.
39. E.M. Berends, Dissertation, TU Delft 1994.
40. P.M. Gallagher, M.P. Coffey, V.J. Krukonis and N. Klasutis, ACS Symp. Ser. 406 (1989) 335-354.
41. S.-D. Yeo, P.G. Debenedetti, M. Radosz and H.-W. Schmidt, Macromolecules 26 (1993) 23 6207-6210.
42. B. Subramaniam, R.A. Rajewski and K. Snavely, J. Pharm. Sci. 86 (1996) 885-890.
43. Bleich and B.W. Muller, Microencapsulation, 13 (1996) 131-139.
44. W.J. Schmitt, M.C. Salada, G.G. Shook and S.M. Speaker III, A.I.Ch.E. J., 41 (1995) 2476-2486.
45. M. H. Hanna, P. York and B. Yu. Shekunov, Proceedings of the 5th Meeting on Supercritical Fluids, Nice, (1998) 325-330.
46. R. Bodmeier, H. Wang, D.J. Dixon, S. Mawson and K.P. Johnston, Pharm. Res., 12 (1995) 1211-1217.
47. T.W. Randolph, A.D. Randolph, M. Mebes and S. Yeung, Biotechnol. Prog., 9 (1993) 429-435.

48. A. Weber, J. Tschernjaew and R. Kuemmel, Proceedings of the 5th Meeting on Supercritical Fluids, Nice (France), (1998) 243-248.
49. C. Magnan, E. Badens, N. Commenges and G. Charbit, Fifth Conference on Supercritical Fluids and their Applications, Garda (Verona), (1999) 479-484.
50. U. Ehrenstein, A. Weber, J. Tschernjaew and R. Kuemmel, Proceedings of the 6th Meeting on Supercritical Fluids, Nottingham (United Kingdom), (1999) 711-716.
51. B.L. Knutson, P.G. Debenedetti and J.W. Tom, Microparticulate Systems for the Delivery of Proteins and Vaccines, Drugs and the Pharmaceutical Sciences Series, Marcel Dekker, New York, 77 (1996) 89-125.
52. M.A. Winters, B.A. Knutson, P.G. Debenedetti, H.G. Sparks, T.M. Przybycien, C.L. Stevenson and S.J. Prestrelski, J. Pharm. Sci., 85 (1996) 586-594.
53. S.-D. Yeo, G.-B. Debenedetti and H. Bernstein, Biotechnol. And Bioeng., 41 (1993) 341.
54. A. Bertucco, F. Vaccaro and P. Pallado, Fourth Italian Conference on Supercritical Fluids and their Applications, Capri (Italy), (1997) 327-334.
55. E. Reverchon, G. Della Porta and A. Di Trolio, Fourth Italian Conference on Supercritical Fluids and their Applications, Capri (Italy), (1997) 335-342.
56. E. Reverchon, G. Della Porta, I. De Rosa, P. Subra and D. Letourneur, Fifth Conference on Supercritical Fluids and their Applications, Garda (Verona), (1999) 473-478.
57. E. Reverchon, G. Della Porta and M.G. Falivene, Proceedings of the 6th Meeting on Supercritical Fluids, Nottingham (United Kingdom), (1999) 157- 162.
58. E. Weidner, GVC-Fachausschuß High Pressure Chemical Engineering, Karlsruhe, Germany, (1999).
59. E. Weidner, Z. Knez, Z. Novak, Slovenian Patent, No. 9400079 (Feb.15.94); PCT WO 95/21688, 17.08.1995.
60. E. Weidner, Z. Knez, Z. Novak, Proceed. 3rd Int. Symp. on Supercritical Fluids, Strassbourg, Tome 3 (1994) 229-235.
61. E. Weidner, Z. Knez, R. Steiner, 3rd Int. Symp. on High Pressure Chem. Eng., Zürich, (1996) 223-228.
62. J.M. Prausnitz, R.N. Lichtenthaler, E.G. deAzevedo, Molecular Thermodynamics of Fluid Phase Equilibria, Prentice-Hall Inc., (1986) 371.
63. M.A. McHugh and V. Krukonis, Supercritical Fluid Extraction - Principles and Practice, Butterworths, Boston, (1986) 51.
64. J. De Swaan Arons and G.A.M. Diepen, Rec. Trav. Chim. Pays-Bas, 82 (1963) 249.
65. W.H. Tuminello, G.T. Dee and M.A. McHugh, Macromolecules, 28 (1995) 1506.
66. E. Weidner, V. Wiesmet, Z. Knez and M. Skerget, J. Supercrit. Fluids, 10 (1997) 139-147.
67. E. Weidner, Z. Knez, Z. Novak, VDI, GVC-Fachausschuss "Hochdruck Verfahrenstechnik", Regensburg, 1994.
68. E. Weidner, Z. Knez and Z. Novak, The 3rd International Symposium on Supercritical Fluids, Strasbourg (France), (1994) III/229-234.
69. Z. Novak, P. Sencar-Bozic, A. Rizner and Z. Knez, Confresso I Fluidi Supercritici e Le Loro Applicazioni, Grignano (Trieste), (1995) 231-238.
70. K. Kokot. Z. Knez and D. Bauman, Acta Alimentaria, 28 (1999) 197-208.
71. J. Kerc, S. Srcic, Z. Knez and P. Sencar-Bozic, Int. J. Pharm. 182 (1999) 33-39.
72. P. Sencar-Bozic, S. Srcic, Z. Knez and J. Kerc, International Journal of Pharmaceutics, 148 (1997) 123-130.
73. E. Weidner, Z. Knez, R. Steiner, 3rd Int. Symp. on High Pressure Chem. Eng., Zürich, (1996) 223-228.

9.9 Pharmaceutical processing with supercritical fluids

N. Elvassore[a], I. Kikic[b]

[a] Dipartimento di Principi e Impianti di Ingegneria Chimica (DIPIC) Università di Padova
Via Marzolo, 9 I-35131 PADOVA Italy

[b] Dipartimento di Ingegneria Chimica dell'Ambiente e delle Materie Prime
Università degli Studi di Trieste
Piazzale Europa, 1 I-34127 TRIESTE, Italy

9.9.1 Introduction

The utilization of supercritical fluids for the processing of pharmaceuticals has attracted considerable interest in recent years. The products' purity, the variety of the forms that can be produced, and the mild operating conditions are important features that make the supercritical techniques an important alternative to classical processes in the pharmaceutical field.

Supercritical fluids can be used to extract substances from natural products, as solvents or as anti-solvents to micronize drugs and biodegradable polymers, encapsulate drugs in polymeric matrices, resolve racemic mixtures of pharmacologically active compounds, fractionate mixtures of polymer and proteins, and sterilize bacterial organisms.

In this paragraph substances involved in pharmaceutical preparations, processed by supercritical fluids as drugs, polymers, vitamins, additives, preservatives, and nutraceuticals are considered.

The most important advantages of the supercritical fluid technique in the application to pharmaceuticals is the high quality of the products in terms of purity, their unique morphology, and the wide range of materials that can be processed. Furthermore, the mild operating conditions can be especially favourable to bio-labile molecules, such as protein- or genetic material involved in pharmaceutical applications.

The current regulation on the content of residual solvent is pushing versus higher degrees of purity, so that the higher cost of investment in a supercritical plant, relatively to a classical process where a larger amount of organics has to be used, could be justified. In fact, the concentration allowed for these solvents, controlled by international safety regulations [1], are generally restricted to a few ppm.

By far the most popular supercritical fluid in pharmaceutical applications is carbon dioxide; its inherent non-toxicity, low critical point, and low cost make this solvent a perfect substitute for the organic solvents.

The possibility of controlling the morphology of the product is relevant, especially in the forms of bio-polymer preparations and controlled delivery systems. Polymeric micro-particles, fibers, or three-dimensional networks can be produced by tuning the operating variables.

The purpose of this section is not to give an exhaustive list of pharmaceutical applications of supercritical fluids, but to review this topic and discuss some examples where these ideas have been applied successfully.

9.9.2 Separation

9.9.2.1 Fractionation/purification by precipitation

The potential of supercritical fractionation/purification through anti-solvent precipitation techniques has been exploited in few cases. Chang *et al.* [2] used CO_2 as a supercritical anti-solvent to precipitate β-carotene from carotene oxidation products, and to obtain the enrichment of *trans*-β-carotene from β-carotene isomers. In other applications, citric acid was separated selectively from fermentation broth containing organic acid, such as oxalic and malic acid. A continuous supercritical process for lecithin, called spray extraction, was proposed by Eggers and Wagner[3]. In this process lecithin was precipitated because is not soluble in CO_2. Catchpole *et al.* [4] also tested the precipitation of lecithin and soybean. They found that the presence of soybean oil in the solution modified the precipitation behavior of lecithin, depending on the percentages of the oil.

Selective fractionation of model proteins using vapour-phase carbon dioxide is another interesting application developed by Debenedetti and coworkers [5]. In the field of protein recovery and purification the precipitation remains an indispensable unit operation. Recently, an alternative process using CO_2 as anti-solvent for the precipitation of dry micro-particulate protein powders has been proposed by Yeo *et al.* [6]. In this technique, the protein precipitation from an organic solution is induced by the dissolution of the supercritical CO_2 in the liquid solution. Precipitates of model proteins exhibit minimal-, and only in high-molecular-weight protein, intermediate- losses of biological activity upon rehydration in aqueous buffer. In the work of Debenedetti and coworkers [5], the authors show that the gas techniques can represent very convenient tools for fractionating proteins using vapour-phase precipitants. Clearly, this process is not of general applicability because it is based on the resulting activity for redissolved protein precipitate.

Another important aspect of supercritical fluids application is in polymer fractionation, in order to obtain mono-dispersed molecular weights. The simulation of the fractionation of polyethylene from ethylene and hexane solutions into fractions of different molecular weights was proposed by Chen *et al.* [7].

9.9.2.2 Supercritical fluid chromatography

The first paper on supercritical fluid chromatography (SFC) by Klesper *et al.* in 1962 [8] considered analytical SCF applications. Only twenty years later, in 1982, a first patent on the use of SFC for production purposes was granted by Perrut [9]. During these years, various academic and industrial laboratories have demonstrated the feasibility and the applicability on a commercial scale of these chromatographic processes.

The combination of the high selectivity of chromatographic interactions and the unique properties of supercritical fluids leads to promising applications in the pharmaceutical field. A review on the topic has been presented by Perrut [10].

Recently, chiral separations have been studied as important steps in drug discovery. In particulars the US Food and Drug Administration requires investigators to evaluate the safety and effectiveness of therapeutic drugs if they can exist as racemic mixtures of two enantiomers. The two molecules may have different pharmacological or toxicological effects and, in these cases, the removal of the unwanted enantiomer is necessary. The recent introduction of the simulated moving bed (SMB) for enantiomer separation gives chromatographic techniques important application on the industrial scale. Furthermore, supercritical fluid chromatography (SFC) was shown to be useful and efficient tool for the

614

separation of chiral compounds. SFC offers some advantages over traditional high-performance liquid chromatography for both purity assessment and for larger, preparative scale, separation and isolation of the enantiomers from racemic mixtures. The organic solvent substitution becomes particularly important on a preparative scale, where grams of compound can be separated and collected in a few ml of solvents despite the litres of solvents used in the HPLC mobile phase. The SFC was found to have high separation efficiencies, and the overall economy of the process could be an advantage.

9.9.3 Extraction and Purification (SFE)

The use of near- or super-critical CO_2 for the extraction of a wide variety of natural products from complex matrices has been extensively reported and reviewed in section 9.6. However, most natural-product extraction, fractionation, and purification is related to the food, cosmetic, and nutraceuticals industries. There is no doubt that the fast-increasing application of SFE is opening new possibilities also in the pharmaceutical field where, owing to the high value of the drugs the problem of cost is much less relevant than in the food or cosmetic industries. Moreover, the potential benefits include faster analysis, protection from degradation by light, heat, or oxygen, high-load samples, the possibility of trace analysis, and the elimination of pesticides and residual solvents. Some examples can be cited: among them, solid-SFE of active molecules from natural products, liquid-SFE and fractionation of lipids and compounds from fermentation broths, SFE for the elimination of residual solvent from synthetic drugs, or monomers from polymeric structures or other toxic pollutants. Extraction of pharmaceuticals has been recently reviewed by Dean and Khundker [11].

The extraction of active principles and additives (pigments, preservatives) from natural products has been investigated for long, and probably most medicinal plants have now been evaluated. The polar nature of most pharmaceuticals often precludes the use of pure carbon dioxide, so it is common to find the addition of more polar solvents as modifier. A complete listing of applications under development would be difficult. However, the areas of major interest for pharmaceutical applications concern: alkaloids extraction and purification, steroids, lipids extraction or lipid-recovery and lipid-elimination from glycosides and proteins, polyunsaturated fatty acids, andanticancer drugs from vegetables and vitamins (mainly tocopherols).

The liquid-supercritical fluid extraction has not been investigated as intensively as the solid-fluid processes, although it can be very promising. Some examples of applications are given by the fatty-acid fractionation using supercritical CO_2. From fish oils, that have interesting pharmaceutical properties related to the cure or prevention of atherosclerosis, an EPA-rich phase can be obtained by CO_2 fractionation. The liquid-supercritical fluid extraction can be used for the elimination and purification of heavy and odorous compounds, or lipid fractionation including sterols, phospolipids or pesticide-elimination to obtain pharmaceutical-grade oils.

There is relatively little work covering the liquid-SFE from fermentation broths. Supercritical fluid extraction of aqueous fermentation materials have some advantages over conventional organic solvent extraction. The low toxicity, and the possibility of tuning the selectivity by altering the density, are real advantages, but the need of a high volume throughput and the handling an aqueous medium could be a problem. However, with the more restrictive legislation regarding organic-solvent disposal, SFE has become an effective alternative to solvent extraction in the area of antibiotics isolation [12].

In the same area, some authors [13,14] have shown that a new selective extraction from a fermentation broth in supercritical CO_2 can be driven by adding surfactants and molecular micelles. Promising preliminary results indicate new possibilities for treating water-soluble compounds with CO_2.

Most active principles and pharmaceutical forms are processed in the presence of organic solvents or reagents. The current regulations on products generally restrict to a few p.p.m. the amount of residual solvent. This very low concentration level could favour the CO_2 utilization when non-polar compounds have to be eliminated. On the other hand, the elimination of residual solvents from tablets, films or other pharmaceutical preparations in which organic solvent are involved has been addressed [15]. Another application is related to the removal of residues from medical materials such as monomers, additives or polymerization residues from polymers or elastomers. Purification of active principles includes elimination of other undesired molecules: pesticides from some vegetal extracts, and antibacterials suspected of toxic co-extracts from natural sources.

9.9.4 Particle formation

In the area of pharmaceuticals, particle formation is currently one of the most popular applications of supercritical fluids. The reasons can be found in the wide variety of particles obtained by the supercritical techniques. "Void free" particles or very "soft" particles, composed of polar or non-polar compounds, and with size ranging from 50 nanometr to 50 microns can be produced easily.

The conventional techniques for particle-size reduction include mechanical treatment such as crushing, grinding and milling, recrystallization of the solute particles by using liquid anti-solvent, freeze-drying and spray-drying. These techniques are often successful and simple to apply but they can lead to high local temperatures. The classical anti-solvent use and disposal of excessive amounts of organic solvent, exposes the pharmaceuticals to thermal and chemical degradation, gives high traces of residual solvent, and affects the crystallinity and the chemical stability of pharmaceuticals.

Moreover, the micro-encapsulation of pharmaceutical compounds in biodegradable polymer particles is of great interest for the development of new controlled drug-delivery systems. When conventional pharmaceutical methods for the production of drug-loaded micro-particles are considered, the obligatory use of organic solvents leads to high residual contents of toxic solvent in the final product, and low encapsulation efficiencies owing to the partitioning of the pharmaceuticals between the two immiscible liquid phases. In fact, these methods for particle encapsulation include emulsion- and double emulsion-solvent extraction, liquid anti-solvent, spray drying, and freeze drying.

Three research areas are under investigations.

- The production of powders of pharmaceutical, required to improve or modify their therapeutic action or to enhance their solubility.
- The production of polymers or bio-polymers which can be used as stationary phase or adsorbent, catalyst support, or as a matrix for drug impregnation.
- The simultaneous precipitation (co-precipitation) of drugs and polymers mainly focused on developing drug-delivery systems.

Depending on the particle morphology, several administration methods can be used: aerosol, inhalation, and systemic- and subcutaneous injections. Typically, particles in the range of 5 to 100 μm are subcutaneously injectable, 1-5 μm particles are suitable for aerosol delivery and

Table 9.9-1
RESS experiments for pharmaceuticals

Material	Supercritical Fluid	Particle size (μm)	Ref.
β -Estradiol	CO_2		[16]
Phenacetin	CO_2 - CHF_3		[17]
Mevinolin	CO_2	10-50	[18]
L-630.028 (steroid deriv.)	CO_2		[18]
Lovastatin	CO_2	0.04-0.3	[19]
β -carotene	C_2H_4 - Toluene	20	[20]
Benzoic Acid	CO_2		[21]
Benzoic Acid	CO_2	2-10	[22]
L-Leucina	CO_2		[23]
Stigmasterol	CO_2	0.05-2	[24]
Salicylic Acid	CO_2		[25]
Griseofulvin	CHF_3	0.9-2	[26]
Nifedipin	CO_2		[27]
Carotenoids	CO_2, C_2H_6, C_2H_4		[28]
Steroids	CO_2	1-10	[29]
Theophylline	CO_2	0.4	[30]
Salicylic Acid	CO_2	< 4	[30]
Naproxen	CO_2	1-20	[31]
Flavone	CO_2	10	[32]
Various polymers [*] - lysozime	CO_2 – organic solvents	10-50	[33]
Various polymers [*] - lipase	CO_2 – organic solvents	10-50	[33]

[*]Poly(lactide), PLA; Poly(lactide-co-glycolide), PLGA; poly(ethylene glicol), PEG; Poly(methyl methacrylate), PMMA.

inhalation therapy to the lungs, whereas nano-particles can be directly injected into the systemic circulation.

In the latest literature, the production by supercritical techniques of pharmaceuticals-loaded bio-polymer micro-particles is widely considered [34]. All of these applications take advantage of the solvent or anti-solvent power of CO_2. Various techniques have been proposed so far, such as the rapid expansion from supercritical solution (RESS) [35], the gas

Table 9.9-2
RESS experiments for polymers and pharmaceuticals

Material	Supercritical Fluid	Particle size (μm)	Ref.
L-PLA [a]	CO_2	2-5	[36]
D-PLA [b]	CO_2	10-20	[37]
PLGA [c]	CO_2	10-20	[37]
L-PLA	CO_2	1-20	[31]
Lovastatin - L-PLA	CO_2		[38]
L-PLA - naproxen	CO_2	10-90	[31]

[a] L-PLA, Poly(L-lactide); [b] D-PLA, Poly(D-lactide); [c] PLGA, Poly(lactide-co-glycolide).

anti-solvent (GAS) [39] and supercritical anti-solvent precipitation (SAS) [36], the precipitation with a compressed fluid anti-solvent (PCA) [40], and the aerosol solvent extraction system (ASES) [41]. In the case of polymers processing, owing to their low solubility in supercritical CO_2 (except for perfluorinated-polymers), major attention has been paid to SAS, GAS or PCA techniques [42-43].

9.9.4.1 Rapid expansion

The idea of dissolving the material at high pressure, exploiting the supercritical fluid's solvent power, and of precipitating it by decompression in the RESS process has been studied by a number of authors. The RESS process has been shown to produce drug micro-particles ranging in size from a few microns to several hundred microns. In order to produce fine powders, various pharmaceutical products have been processed; some examples are reported in Table 9.9-1. For a given drug, the supercritical fluid used, the morphology, and the particle size obtained are reported. As an alternative to pure pharmaceutical, bio-polymers have been used to produce fine polymeric powders which, in conjunction with the drug, are the basis for the preparation of controlled release systems. Results obtained with the polymers are reported in Table 9.9-2.

The properties required for the components to be processed using the RESS techniques obviously concern their solubility in supercritical fluids. This restriction becomes important for high-molecular-weight substances and, particularly for polymers. Very often the bio-polymers used in the preparation of controlled-release systems are practically insoluble in supercritical fluids; therefore, only low-molecular-weight polymers can be successfully processed by RESS. From a practical point of view, the order of magnitude of the solubility in the range of 0.01 wt. % or lower requires a very large amount of carbon dioxide for increased product throughput. A possible way of overcoming this problem is the addition of cosolvents as methanol. Recently, it was shown that small amounts of co-solvents can considerably improve the solubility because with a co-solvents the local density around the solute molecules is higher than the bulk one. However, even if a small percent of organic solvent is necessary, the process becomes more complex, and the environmentally favourable nature of the RESS is reduced.

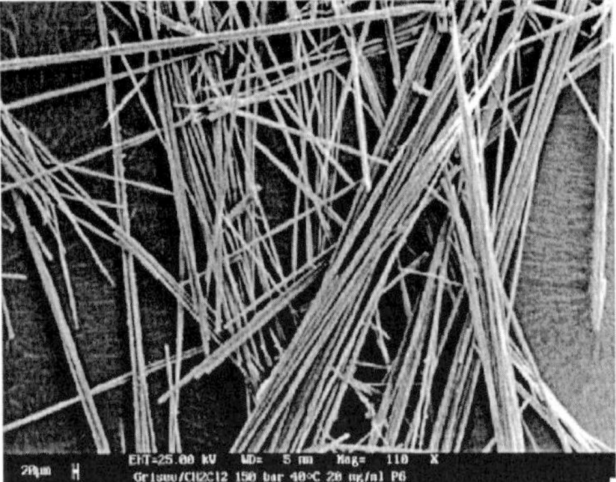

Figure 9.9-1. SEM micrographs of griseofulvin crystals precipitated from dichloromethane at 150 bar, 40°C [44].

Figure 9.9-2. SEM image of yttrium acetate nano-particles precipitated from DMSO at 120 bar, 40°C. Mean diameter about 150 nm [44].

9.9.4.2 Recrystallization by Supercritical anti-solvent

The main limitation of the RESS process is the very low solubility of many materials of pharmaceutical interest, such as biomolecules and polymers, in supercritical fluids. In the recrystallization process by supercritical anti-solvent (referred to as SAS, GAS, ASES, and PCA), the solid of interest is dissolved in a liquid, and a supercritical fluid having low solvent power with respect to the solute, but miscible with the liquid, is added to precipitate the solid. The low solubility in supercritical fluid is an advantage in recrystallization, whereas the solvent has to be completely miscible with the supercritical fluid at the temperature and pressure of the process. For the application to pharmaceuticals, recrystallization has been studied in both batch and continuous experimental set up. In the first case, a batch of solution is diluted by mixing either the compressed- or the supercritical fluid until the volumetric expansion and the lowering of the solvent strength forces the precipitation of the solute. On the other hand if the process is operated in a continuous mode, a solution of the pharmaceutical is sprayed through a nozzle into a compressed- or supercritical fluid environmental as fine droplets. The evaporation of solvent and the supercritical fluid influx when the liquid droplets contact an excess of gas phase give solute precipitation in the form of dry powders. Both of these processes have been applied for particles formation. Owing to its environmental safety and moderate critical temperature, compressed or supercritical CO_2 is widely used as the anti-solvent fluid in many pharmaceutical applications.

In Tables 9.9-3 to 9.9-5 a summary of the recrystallization process of pharmaceuticals and bio-polymers, and co-precipitation by supercritical anti-solvent is presented. In Figures 9.9-1 and 9.9-2 examples of very long needle crystal and nano-particles are reported.

An exhaustive review on particle formation by supercritical anti-solvent precipitation has recently been presented by Reverchon [44].

The use of therapeutic proteins is growing rapidly and it has been suggested that this class of drugs may soon represent a significant fraction of the pharmaceutical market [45]. There is an urgent need for the development of delivery systems for proteins because, so far, the use of therapeutic proteins is limited to their low oral and transdermal bioavailability.

Table 9.9-3
Anti-solvent experiments for pharmaceutical products

Material	Solvent	Process	Particle size (μm)	Ref
Insulin	DMSO [f]	SAS	2-4	[46]
Indomethacin	CH_2Cl_2	PCA	8.2	[47]
Piroxicam	CH_2Cl_2	PCA	6.0	[47]
Thymopentine	CH_2Cl_2	PCA	6.6	[47]
Hydrocortisone	DMSO	PCA	0.2-1	[48]
MPA[a]	THF [g]	PCA	2.5-3	[49]
HCA [b]	DMF [h]	PCA	8.0	[49]
SX [c]	Acetone	PCA	10	[50]
Insulin	DMSO	SAS	1-5	[51]
Lysozyme	DMSO	SAS	1-5	[51]
Trypsin	DMSO	SAS	1-5	[51]
CPM [d]	CH_2Cl_2	PCA	1-5	[52]
Indomethacin	CH_2Cl_2	PCA	1-5	[52]
NaCrGlyc [e]	MeOH	GAS	0.1-2	[53]

[a] MPA, Methylprednisolone acetate; [b] HCA, Hydrocortisone acetate; [c] SX, Salmeterol xinafoate; [d] CMP, Chlorophenylamine maleate; [e] NaCrGlyc, Sodium Cromoglycate; [f] DMSO, Dimethylsulfoxide; [g] THF, Tetrahydrofurane; [h] DMF, Dimethylformamide.

A first study on proteins processed by supercritical CO_2 was proposed by Debenedetti and his group [6], who used a supercritical anti-solvent precipitation to produce protein particles for aerosol delivery. In order to apply these methods successfully it is of fundamental importance that the biological activity of the labile compounds is maintained. In fact, the SAS process exposes proteins to organic and supercritical solvents, high pressure, and shearing stresses. Further studies by the same group [54] evidenced a conformational change of the secondary structure of the proteins. Surprisingly, despite the major conformation change induced during SAS precipitation, the precipitated proteins recovered their biological activity and native structure-content upon redissolution in aqueous media [55]. Owing to the amorphous nature of the micro-particles obtained, of diameter 1-5 μm, the evidence showed that it is possible to store the powder for long times under ambient conditions, without losses of activity.

In order to reduce the denaturation of possibly labile compounds an aqueous-based micronization can be used. In this case, the fine protein particles were precipitated from aqueous solutions and a modifier, such as ethanol, was used to increase the solubility of the

Table 9.9-4
Anti-solvent experiments for polymers

Material	Solvent	Process	Particle size (µm)	Ref
L-PLA [a] 100000	CH_2Cl_2	PCA	0.5-5	[56]
L-PLA 94000	CH_2Cl_2	PCA	1-5	[52]
L-PLA 115+7000	CH_2Cl_2	PCA	7-50	[48]
DL-PLA [b]	DMSO [e]	SAS	< 2	[57]
PLGA [c]	DMSO	PCA	15	[48]
HYAFF [d]	DMSO	PCA	50-500	[48]

[a] L-PLA, Poly(L-lactide); [b] DL-PLA, Poly(DL-lactide); [c] PLGA, Poly(lactide-co- glycolide); [d] HYAFF, Hyaluronic acid ethyl ether; [e] DMSO, Dimethylsulfoxide.

water in carbon dioxide. In the alternative technique proposed by York and his coworkers [58,59], named Solution Enhanced Dispersion by Supercritical fluids (SEDS), a coaxial nozzle was used to simultaneously spray an aqueous solution of ethanol and supercritical CO_2. Proteins precipitated from the aqueous solution by extracting water with a supercritical CO_2-ethanol mixture.

For the development of drug delivery systems it is of fundamental importance to achieve the co-precipitation of drugs and biodegradable polymers. The supersaturation working conditions of SAS allow, in many cases, fast and simultaneous precipitation of both polymer and drugs, so that the drugs can be trapped into the polymer matrix. A list of the co-precipitation experimental results is reported in Table 9.9-5.

Few works on proteins' encapsulation have been presented. A recent study by Young et al. [60] deals with the encapsulation of lysozyme in biodegradable polymer microspheres. A 1-10 µm lysozyme particle suspension in a polymer solution was sprayed into a CO_2 vapour phase through a capillary nozzle. The droplets solidified after falling into the liquid phase. By delaying the precipitation in the vapour phase, the larger microparticles obtained were able to encapsulate the suspended lysozyme. The final capsules were in the range of 5 - 70 µm. This work is a nice example of protein encapsulation for microparticle delivery systems.

Moreover, insulin-loaded PLA nanoparticles with high yields of encapsulation were produced by Elvassore and coworkers [61]. In this work, a homogeneous solution of protein and polymer is sprayed through a nozzle in a high-pressure vessel. In order to achieve nano-encapsulation, mixtures of dichloromethane (DCM) and dimethylsulfoxide (DMSO) were used to ensure the solubility of both the polymer and the protein. In Figure 9.9-3 a SEM image of the fine-particle powders produced is presented.

9.9.4.3 Impregnation with supercritical fluids
An alternative production of a controlled delivery system can be performed either by adding the filler to the matrix-formation mixture or by sorption of the guest molecule in the

Table 9.9-5
Anti-solvent experiments for pharmaceuticals and polymers

Material	Solvent	Process	Particle size (μm)	Ref
L-PLA [a] - Hy [b]	CH_2Cl_2	PCA	10-15	[47]
PLGA [c] - Thy [d]	CH_2Cl_2	PCA	40-60	[48]
HYAFF [e] - Prot.	DMSO [g]	GAS	0-30	[62]
L-PLA - gentamycin	CH_2Cl_2	PCA	1-5	[43]
PLGA - HC [f]	Various	SEDS	5-70	[63]
L-PLA - lysozime	CH_2Cl_2	PCA	5-70	[60]
PLGA - lysozime	CH_2Cl_2	PCA	10-50	[60]
L-PLA - insulin	DMSO-CH_2Cl_2	GAS	0.7-0.4	[61]

[a] L-PLA, Poly(L-lactide); [b] Hy, Hycosinebutylbromid; [c] PLGA, Poly(lactide-co-glycolide); [d] Thy, Thymopentine; [e] HYAFF, Hyaluronic acid ethyl ether; [f] HC, Hidrocortisone; [g] DMSO, Dimethylsulfoxide.

synthesized matrix (i.e., by diffusion from a gas or a liquid solution). In any case, careful removal of the solvent used in the impregnation process is essential in order to obtain an acceptable product characteristic on the basis of the regulation of the residual content [1]. The supercritical impregnation technique (SSI) is based on the simple idea of substituting the organic solvents in the traditional methods with a SF solvent used as carrier of the filler into a preformed matrix.

The basic understanding of the impregnation process deals with both the thermodynamic description of the ternary system (supercritical fluids, pharmaceutical, and polymer) and the kinetics (diffusion in the polymer) aspects. Studies have been made of the partition of organic compounds between a polymer and supercritical fluids [64, 65] and, surprisingly, despite their low solubility in supercritical fluids, high concentration of pharmaceutical in the polymer phase is easily reached. In the patent of Sand [66], partitioning of the solute between supercritical fluid and polymer, and the swelling phenomena of the polymeric phase were used with different solutes for polymer impregnation (SSI technique). In this patent, the technique is clearly limited to lipophilic solutes and to thermoplastic polymers (mainly ethylene copolymers). The depressurization time (from impregnation pressure to atmospheric pressure) has a strong influence on the physical aspects of the final product. Slow release of the pressure gives rise to micro-voids in some moulded polymers. These micro-voids contained in the interior of the polymeric structure can act as reservoirs for the impregnating substance, in the case of liquid-phase separation. Obviously this phenomenon is of importance for drugs impregnation purpouses.

622

Figure 9.9-3. SEM micrographs of insulin loaded PLA micro-particles produced from dichloromethane-dimethylsulfoxide at 150 bar, 20 ° C (adapted from [61]).

In another process, patented by Berens [67], the additive must again be solubilized in the supercritical solvent, as in Sand's process but, in this case, the patent is mainly concerned with reactive monomers that successively polymerize, modifying the host polymer. Nevertheless, in some examples (with a film of polyurethane), the problem of solid-drug impregnation is also considered.

In order to overcome the main limitations of the impregnation processes, connected to the limited solubility of the compounds in the supercritical fluids, Perman [68] proposed an alternative method. A supercritical impregnation process was coupled with a liquid solvent (preferentially water) to enhance the drug solubilization. The system composed of a liquid drug solution and the polymeric support was pressurized with the supercritical fluid. Consequently, the swelled polymer allows rapid diffusional transport of the solute into the polymeric substrate. In different examples, bovine serum albumin microspheres were impregnated with insulin, trypsin and gentamicin (see Table 9.9-5).

9.9.5 Future developments

This short presentation of the application of supercritical fluids in the pharmaceuticals field, gives an idea of the potential and the efficiency of this technique in processing pharmaceuticals and biological substances. Beyond "classical" supercritical fluids applications, such as the extraction of biologically active molecules and ultra-purification by chromatography, other types of processes are under investigation. Some of these are interesting for industrial developments: tablet coating, and new bio-compatible media of natural and synthetic materials (bones) can be obtained; cell disruption can be performed without heat degradation; sterilization and virus inactivation can easily be coupled with pharmaceutical processes; and reaction-fractionation can be used in down-stream processes.

The number of possible applications is very large and is growing rapidly, but the move of several plants from a laboratory scale to the industrial stage still requires additional work.

In fact, even if "clean chemistry" and "sustainable technology" concepts are pushing supercritical pharmaceutical processes under consideration up to the industrial scale, this development is still in its early stage. Future supercritical-fluids plants for the food- and pharmaceutical industries will need to address GMP requirements and FDA validation for pharmaceutical production. There are numbers of issues related to GMP validation, which range from the quality of the CO_2 in the process to the quality of the materials in the smallest individual components. Moreover, other issues include standardization of procedural software and validation of the hardware cleaning procedures, ASME certification, and so on. Of course, in the future, both the manufacturer of the equipment and the processor of the material have to consider and have a clear understanding of the GMP requirements and their associated cost. This will be a challenge for pharmaceutical industries as well as for supercritical fluids in the next ten years.

References

1. S. Kojima, Guidelines for Residual Solvents, Fourth International Conference on Harmonization (ICH4),Belgium, Bruxelles, 1997.
2. C.J. Chang A.D. Randolph, N.E. Craft, Biotechnol. Prog., 7 1991 275.
3. R. Eggers, H.J. Wagner, J. Supercrit. Fluids, 6 (1999) 31.
4. O.J. Catchpole, S. Hockman, S.R. Anderson, J. High Pres. Chem. Eng., 1 (1996) 309.
5. M.A. Winters, D.Z. Frankel, P.G. Debenedetti, J. Carey, M. Devaney, T.M. Przybycien, Biotec. Bioeng., 62 (1999) 247.
6. S.D. Yeo, G.B. Lim, P.G. Debenedetti, H. Bernstein, Biotechnol. Bioeng., 41 (1993) 341.
7. C.K. Chen, M.A. Duran, M. Radoz, Ind. Eng. Chem. Res., 33 (1994) 306.
8. Klesper e Corviwin a H Turner D A J. Org. Chem 27 1962 700
9. M. Perrut, French Patent No. 2 527 934 (1982).
10. M. Perrut, J. Chromatography, 658 (1994) 293.
11. J.R. Dean, S. Khundker, J Pharmac. Biomed. Analysis, 15 (1997) 875.
12. S. Cocks, S.K. Wrigley, Chicarelli M.I. Robinson, R.M. Smith, J. Chromatography 697 (1995) 115.
13. A.I. Cooper, J.D. Londono, G. Wignall, J.B. McClain, E.T. Samulski, J.S. Lin, A. Dobrynin, M. Rubinstein, A.L.C. Burke, J.M.J. Frechet, J.M. DeSimone, Nature 389 (1997) 368.
14. K.P. Johnston, K.L. Harrison, M.J. Clarke, S.M. Howdle, M.P. Heitz, F.V. Bright, C. Carlier, T.W. Randolph, Science, 271 (1996) 624.
15. R.F. Falk, T.W. Randolph, Pharmac. Res., 15 (1998) 1233.
16. V. Krukonis, A.I.Ch.E. Annual Meeting, San Francisco, USA, 1984.
17. H. Loth, E. Hemgesberg, Int. J. Pharm. 32 (1986) 265.
18. K.A. Larson, M.L. King, Biotech. Prog., 2 (1986) 73.
19. X. Kwauk and P.G. Debenedetti, J. Aerosol. Sci., 34 (1993) 445.
20. C.J. Chang and A.D. Randolph, A.I.Ch.E. J., 35 (1989) 1876.
21. A.Tavana and A.D. Randolph, A.I.Ch.E. J., 35 (1989) 1625.
22. E.M.Berends, O.S.L. Bruinsma and G.M. van Rosmalen, J. Crystal Growth, 128 (1993) 50.

624

23. S.Furuta, R.W. Rousseau and A.S. Teja, A.I.Ch.E.. Annual Meeting, USA, Miami, 1992.
24. K.Ohgaki, H. Kobayashi, T. Katayama and N. J. Hirokawa, J. Supercrit. Fluids, 3 (1990) 103.
25. E.Reverchon, G. Donsì and D. Gorgoglione, J. Supercrit. Fluids, 6 (1993) 241.
26. E. Reverchon, G. Della Porta, R. Taddeo, P. Pallado and A. Stassi, Ind. Eng. Chem. Res. 34 (1995) 4087.
27. E. Stahl, K.W. Quirin, D. Gerard (Eds), Dense Gases for Extraction and Refining, Springer-Verlag, Berlin, 1987.
28. E. Stahl, K.W. Quirin, D. Gerard (Eds), Dense Gases for Extraction and Refining, Springer-Verlag, Berlin, 1987.
29. P. Alessi, A. Cortesi, I. Kikic, N.R. Foster, S.J. Macnaughton and I. Colombo, Ind. Eng. Chem. Res., 35 (1996) 4718.
30. Ph. Rudolf von Rohr and Ch. Trepp (eds), High Pressure Chemical Engineering, Elsevier, Amsterdam 1996.
31. J.H. Kim, T.E. Paxton and D.L. Tomasko, Biotechnology Progress, 12 (1996) 650.
32. K. Mishima, K. Matsuyama, H. Uchiyama, M. Ide, J.J. Shim, and H.K. Bae, 4th International Symposium on Supercritical Fluids, Sendai, Japan, 1997.
33. K. Mishima, K. Matsuyama, D. Tanabe, S. Yamauchi, T.J. Young and K.P. Johnston, AIChE J., 46 (2000) 857.
34. B. Subramaniam, R. A. Rajewski, K. Snavely, J. Pharmac. Sci., 86 (1997) 885.
35. D.W. Matson, J.L. Fulton, R.C. Petersen, R.D. Smith, Ind. Eng. Chem. Res., 26 (1987) 2298.
36. K.P. Johnston and J.M.L. Penninger, (Eds) Supercritical Fluid Science and Technology. American Chemical Society, Washington DC, 1988.
37. J.W. Tom, P.G. Debenedetti, and R. Jerome, J. Supercrit. Fluids, 7 (1994) 9.
38. W.J. Tom and P.G. Debenedetti, Biotechnology Progress, 7 (1991) 403.
39. E. Kiran and J.F. Brennecke (Eds), American Chemical Society, Washington DC, 1993.
40. D.J. Dixon, K.P. Johnston; J. Applied Polymer Sci, 50 (1993) 1929.
41. J. Bleich, P. Kleinebudde and B.W. Mueller, Process. Int. J Pharm., 106 (1994) 77.
42. J. Bleich, B.W. Mueller, J Microencapsulation, 13 (1996) 131.
43. R. Falk, T. Randolph, J.D. Meyer, R. M. Kelly, M.C. Manning, J. Controlled Release, 44 (1997) 77.
44. E. Reverchon, J. Supercr. Fluids, 15 (1999) 1.
45. S.D. Putney, P.A. Burke, Nat. Biotechnol., 16 (1998) 153.
46. S.D. Yeo, G.B. Lim, P.G. Debenedetti and H. Bernstein, Biotechnol. and Bioeng. 41 (1993) 341.
47. J. Bleich and B.W. Muller, J. Microencapsulation 13 (1996) 31.
48. B. Subramaniam, R.A. Rajewski and K. Snavely, J. Pharm. Sci., 86 (1996) 885.
49. W.J. Schmitt, M.C. Salada, G.G. Shook, and S.M. Speaker III, A.I.Ch.E. J., 41 (1995) 2476.
50. M. Hanna, H. York, P. Shekunov, B. Yu. Proc. 5th Meeting on Supercritical Fluids, Nice, France, 1998.
51. M.A. Winters, B.A. Knutson, P.G. Debenedetti, H.G. Sparks, T.M. Przybycien, C.L. Stevenson, and S.J. Prestrelski, J. Pharm. Sci., 85 (1996) 586.
52. R. Bodmeier, H. Wang, D.J. Dixon, S. Mawson, and K.P. Johnston, Pharm. Res., 12 (1995) 1211.

53. S. Jaarmo, M. Rantakyla, and O. Aaltonen, 4th International Symposium on Supercritical Fluids, Sendai, Japan, 1997.
54. S. Yeo, P.G. Debenedetti, Y.P. Sugunakar, T.M. Przybycien, J. Pharmac. Sci., 83 (1994) 1651.
55. M.A. Winters, B.L. Knuston, P.G. Debenedetti, H.G. Sparks, T.M. Przybycien, C.L. Stevenson, S.J. Prestrelski, J. Pharmac. Sci., 85 (1996) 586.
56. T.W. Randolph, A.D. Randolph, M. Mebes, and S. Yeung, Biotechnol. Prog., 9 (1993) 429.
57. S. Cohen and H. Bernstein (Eds), Microparticulate Systems for the Delivery of Proteins and Vaccines, Drugs and the Pharmaceutical Sciences Series, Marcel Dekker, New York 1996.
58. R. M. Sloan, E. Hollowood, I. Kibria, G. O. Humphreys, W. Ashraf, P. York, 5th Meeting of Supercritical Fluids, Nice, France, 1998.
59. R. T. Forbes, R. Sloan, I. Kibria, M. E. Hollowood, G. O. Humphreys, P. York, World Congress on Particle Technology, Brighton, UK, 1998.
60. T. J. Young, K. M. Johnston, H. Tanaka, J. Pharmac. Sci., 88, (1999) 640.
61. N.Elvassore, A. Bertucco, P. Caliceti, Ind. Eng. Chem. Res., 4 (2001) 731.
62. Ph. Rudolf von Rohr and Ch. Trepp (eds), High Pressure Chemical Engineering, Elsevier, Amsterdam 1996.
63. R. Ghadieri, P. Artursson, J. Carlfors, Eur. J. Pharm. Sci., 10 (2000) 1.
64. J.J. Shim and K.P. Johnston, A.I.Ch.E. J. 35 (1989) 1097.
65. C.A. Eckert, S.G. Kazarian, B.L. West and N.N. Brantley, Proc. 5th Meeting on Supercritical Fluids, Nice, (1998).
66. M.L. Sand, U.S. Patent No. 4678684 (1987).
67. A.R. Berens, U.S. Patent No. 4820752 (1989).
68. C.A. Perman, U.S. Patent No. 5508060 (1996).

9.10 Treating micro-organisms with high pressure

A. Bertucco, S. Spilimbergo

Dipartimento di Principi e Impianti di Ingegneria Chimica (DIPIC) Università di Padova
Via Marzolo, 9 I-35131 Padova Italy

9.10.1 Introduction

Ultra-high-temperature treatment (UHT) is now the most widely exploited method in the food industry to stabilize microbiologically any foodstuff. It consists of heating at an ultra high-temperature for a short period of time: for example, a treatment at 145°C for 2 seconds is sufficient to assure a total microbial- and spore inactivation. The microbial death is principally due to irreversible cell damage (*e.g.*, of proteins, DNA, RNA, vitamins); enzymes are inactivated by heat which modifies their active sites.

Nevertheless this common technology has some disadvantages related to the high temperature employed; this may cause unwanted effects, such as the denaturation of temperature-sensitive substances (*e.g.*, proteins and vitamins) as well as the production of toxic compounds or the appearance of undesirable sensory features (odours and flavours).Thus, since the 1970s other alternative techniques have been studied to reduce the microbial activity in solid and liquid foods. Among these methods, in the last decade high-pressure treatments are becoming particularly relevant, in which high pressure instead of high temperature is employed as the stabilizing factor.

In high-pressure applications these are two main trends of research: high hydrostatic pressure treatment (2000–7000 bar) and supercritical CO_2 treatment. In both cases the aim is to inactivate the micro-organisms in order to protect and preserve foods, and so to prolong their shelf-life.

The former technique has occupied a larger number of researchers, as can be seen from the publications and from the number of congresses organized in the last decade (*e.g.*, [1]). The main limitation of this method is the difficulty of controlling and managing an operating pressure in an extreme range of values. Therefore its widespread diffusion in industry appears cumbersome.

The latter technique is less studied, and the mechanism of microbial reduction is not known yet. However, it could be exploited on the industrial scale in a way similar to classical chemical plants, and the safety requirements are exactly the same as those required in the chemical industry.

The results are not very different for the two techniques, as recent studies on *Escherichia coli* have demonstrated [2]: a treatment with CO_2 with a contact time of 15 min, an operating pressure of 150 bar, and constant temperature of 35°C, leads to the same microbial reductions as a treatment with a hydrostatic pressure of 3000 bar at the same temperature.

In the following pages we summarize the main operating features of each of these two methods, the equipment needed, and some applications.

9.10.2 Hydrostatic high pressure

9.10.2.1. State of the art

For many years researchers have been trying to understand the effect of high pressure on micro-organisms. Early studies, dated in the XIX and XX century, showed that short treatments with an operating pressure of a few thousand bars are able to reduce the microbial activity by many orders of magnitude [3].

In 1899, Hite [4] made known his studies on the sterilization of pressurized foods; he observed that pressurized milk kept fresh and unspoiled for a longer period of time than untreated milk, and this effect was seen as a consequence of the inactivation of the micro-organisms that turn the milk sour. Later, he showed that other microbes in vegetables and fruits can be inactivated if they undergo a high-pressure treatment for a few minutes [5].

In 1914, Bridgman described the coagulation of egg-white as a consequence of high hydrostatic-pressure treatment. In 1918 it seemed evident that all micro-organisms can be inactivated by a high-pressure treatment, just at room temperature, apart from those producing spores, and that each microbe requires individual and precise conditions to be inactivated.

In 1932, Basset and Machebouf published an article [6] about the sensitivity of various living organisms and biological compounds and showed that it was impossible to destroy completely bacterial spores, even with an operating pressure of 17000 bar, if the operation was carried out at ambient temperature.

After this work, no other studies about inactivation of spores were published until 1949 when ZoBell and Johnson [7] investigated the behaviour of sea- and earth- bacterial colonies under hydrostatic pressure.

Later, Timson and Short started systematic experiments to test the resistance of bacterial spores, and tried to inactivate them completely to obtain a total sterilization. These authors studied the behaviour of spores under a high-pressure treatment with a long residence time at constant pressure in a range of temperature between –25°C and 95°C. They noted the high insensitivity to pressure of the spores compared to vegetative forms [8].

In 1966, Lundgren [9] observed the existence of an optimum temperature for germination of spores, therefore Clouston and Willis [10, 11] deepened the process of germination and the inactivation of *Bacillus pumilius* under high-pressure treatment (HPT) and Wills [12] described the sensitivity of spores exposed to radiant energy.

Murrel and Wills [13] investigated the way in which temperature stimulates germination when the pressure is kept constant, to understand how inactivation takes place.

With the works of Gould and his co-workers [14, 15] the existence of optimal hydrostatic pressure to inactivate spores was demonstrated: after much research on various bacterial forms, these authors confirmed their hypothesis about the induction of germination by the action of pressure.

In 1990, Halbauer [16] demonstrated the existence of a minimum value of pressure for spore-inactivation and in the same year Ludwig's group [17, 18] stressed that, in order to kill the spores, it seemed necessary to germinate them first.

Ludwig investigated carefully the behaviour of spores under different conditions of HPT [19, 20] and introduced the cycle-type treatment that proved to be more efficient than the double level treatment [21–23]. Particularly studied was how the treatment time and pressure influenced the cycle processes. Furthermore, it was noted that the temperature, pressure, contact time under pressure, and average time of treatment were fundamental parameters in the optimization of germination.

The experiments led to the conclusion that if a higher temperature is kept, faster germination is obtained, as well as a wider is range of optimal pressure to induce germination. It was also found that dissolved salts, glucose, and amino acids [18] may influence the rate of germination.

All the above considerations led to the conclusion that the diffusive phenomenon during germination is fundamental to explain the results found, as too much high pressure could limit the diffusion of substances essential for germination, to the surroundings of the spore, while too low a pressure could be insufficient to activate germination at low operating temperatures.

The reason why micro-organisms are affected by high pressure is not completely clear. In 1994 Hayakawa developed a theory about the inactivation: he demonstrated that the inactivation of spores is caused by lysis of the internal membrane owing to the adiabatic expansion of water absorbed by the spore and that this causes the loss of substances fundamental for the survival of the cell [24, 25]. However, this theory was refuted by Ludwig and Heremans [26] who proved that water cannot enter through the spore and, furthermore, that the volumetric expansion is, in any case, negligible.

Particularly interesting seems to be the conclusion of Schreck and Ludwig [27], who hypothesized that the barometric resistance of micro-organisms is caused by a mechanical factor, but is also dependent upon the protein-structure of microbes, as there is a deep relationship between the effect of pressure and temperature on proteins and micro-organisms. In other words, pressure acts on proteins located in specific sites where they are particularly sensitive to mechanical stress.

Nowadays, researchers are trying to describe pressure-induced phenomena at the molecular level, to understand which limitations and future applications there could be in the food, pharmaceutical and cosmetic industries. They try to understand the influences of parameters such as temperature, pressure, space time, and the cycle of pressure and pre– and post–thermal treatment on the efficiency of inactivation.

In the meantime, a number of processes using high-pressure treatment (HPT) has already been put into the market. In 1990 the Meidi–ya Food Company, of Japan, introduced the first HPT-sterilized products–strawberry-, kiwi-, and apple jams, with a delicious taste. From 1993 the range of HPT-sterilized products has become even larger: other kinds of jams, sauces, milk-desserts, and fruit-jellies have been produced and commercialized. These products are packed in plastic bags and can be stored at 4°C for 2 months if sealed, or for one week after opening. Other companies, mainly Japanese, exploit HPT to stabilize and store foodstuffs: Wakayama produces 4 t/h of mandarin juice, operating with three autoclaves at 4000 bar, while Pokka produces 600 l/h of grapefruit-juice sterilized at 1500–2000 bar and 5°C for 10 minutes. Since 1994, Echigo–Seka has been producing and commercializing the "yomagi-mochi", a rice cake armoise-scented, whereas the QP Corporation exploits HPT to inactivate *Xantomonas campestris* without altering its property of inducing gelling–this microbe is used to cryo-concentrate wine, vinegar and fruit-juices.

It is noteworthy that HPT can be utilized on an industry scale for two purposes: to reduce the microbial activity, as well as to modify the consistency of foodstuffs, *e.g.*, to obtain pre-cooking induced by pressure.

9.10.2.2 Equipment and methods

Equipment

The basic elements of a prototype pressurization apparatus are: a high-pressure steel cell for the sample to be treated, a high-pressure generating system, a temperature controller, and a loading system for the material to be treated [28].

The sealing system of the steel container changes with the kind of application. If the step of loading and unloading is frequent the discontinuous interrupted thread is used, otherwise the continuous thread is preferred, as it is a cheapest choice when the time for loading and unloading is negligible compared to the operating time.

The system is sealed by a goods lift designed to centre the top and to screw it without damaging the threads. Alternatively it is possible to use a shoe-system which is, however, more cumbersome.

When the samples have been inserted into the container, it is filled with a fluid that transmits the pressure. Usually, water mixed with a small percent of soluble oil as lubricant is used, or a mixture of water and ethylene glycol (weight-ratio, 3:1).

The pressure is applied by a direct or indirect compression method. In the former, a piston coaxial with the container is required and the compressions are particularly fast; this method is employed only in laboratory-scale plant because of the sealing problem between the piston and the internal surface of the container (Fig. 9.10-1). The more widespread method is the indirect one, with a pressure booster to pump the liquid from the pressure-medium tank to the cell, until the desired pressure value is reached (Fig. 9.10-2).

There is a third method, in which the increase of pressure is provided by heating the medium. This technique is particularly useful when a double effect of high pressure and high temperature is required (this is not the case in the food industry, but is, for example, applied in quartz production).

The temperature controller consists of a simple electrical resistance if only heating is required. Otherwise a heating/cooling jacket, or a heat exchanger inside the cell is used (see Fig. 9.10-2). This is the most widespread method, as the metallic thermal inertia is very high, and does not permit a suitable temperature-control in the other cases.

As regards the sample, it is compulsory to use a treatment-resistant packaging, such as a plastic multi-layer envelope, or aluminium foil. When a rigid material (glass or metals) is preferred, a compressibility space must exist inside the packaging, otherwise the pressure is not transferred and the material could break. The container should be filled as much as possible, to reduce the empty space in order to minimize the stress imposed during the steps of pressurization and depressurization.

The form of the container is important for reaching a high volumetric efficiency, and therefore reducing the cost per operating unit. The form of the container is designed to minimize the dead space inside the autoclave. As the HTP cell is usually cylindrical, the containers are hexagonal.

The HTP is a batch process: the volume treated per unit time is a function of the cycle-time (that is the sum of the times of each step; sealing, pressurization, maintenance at the operating pressure, depressurization, opening, and unloading times), of the batch volume, and of the number of cells used simultaneously in parallel in the same cycle.

As regards the economics, in order to make the process economically feasible, the volumetric efficiency must be maximized. Also the ratio of temperature/pressure must be optimized to reduce the cost without damaging the sample. It is important to underline the fact that by increasing the temperature the pressure can be lowered, but the risk of thermal damage is increased.

Another parameter that can be minimized is the cycle-time: this is achieved, for example, by reducing the loading- and unloading times with an automation system, or the opening- and closing times with interrupted threads in the tap.

630

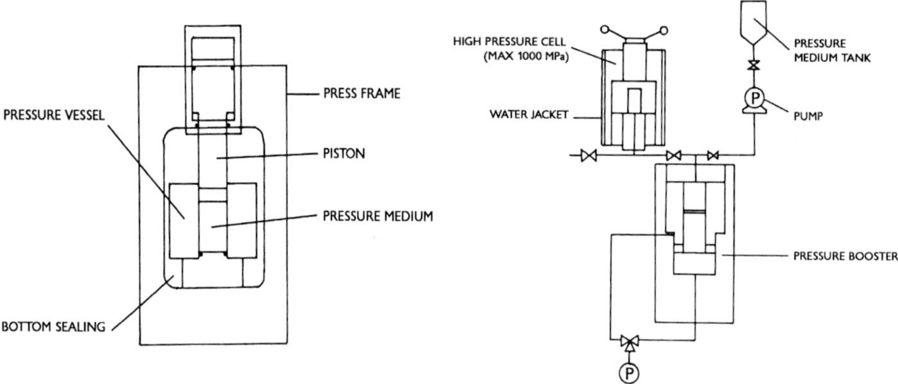

Fig. 9.10-1. Schematic diagram for the HPT using the direct method (adapted from [29]).

Fig. 9.10-2. Schematic diagram for the HPT using the indirect method (adapted from [24]).

Methods

In order to demonstrate the feasibility of HTP sterilization, we shall discuss an example taken from ref. 18. The pressure inactivation of *E.coli*, a gram-positive bacterium, is studied up to 5 kbar in the temperature range between 1– and 50°C; and the dependence on pressure and temperature of the germination of bacterial spores of *Bacillus stearothermophilus* is investigated.

The experimental device [30] consisted of ten small pressure vessels that were filled with the sample, simultaneously pressurized, and thermostatted. The single vessels could be opened independently in order to determine the number of viable organisms at any given time.

Figure 9.10-3 gives the results. The logarithm of colony-forming units, *i.e., log₁₀ (N)* $versus$ time, where N is the microbial count after treatment, per ml, is plotted on the ordinate and the time on the abscissa. Starting with 10^9 bacteria per ml, under an applied pressure of 2.5 kbar sterile solutions were obtained in 10 and 20 min at temperatures of 50 or 40 °C, respectively. Figure 9.10-4 the results under the same applied pressure of 2.5 kbar as before, but at a lower temperature, 25°C. This holds for all the lower temperatures down to 0°C. The shape of the inactivation curve does not depend on the initial concentration. For different initial concentrations it can shift vertically without changes. Figure 9.10-5 gives the temperature dependence of inactivation at 2 kbar.

The inactivation rate has a minimum at room temperature. The faster inactivation at lower temperature might be caused by the participation of hydrophobic forces and of membrane processes. The inactivation of *E.coli* is strongly dependent on pressure; this is shown in Fig. 9.10-6 which gives data for 9 min of pressure-treatment at 4°C. Figure 9.10-7 shows the kinetics of *Bacillus stearothermophilus* inactivation. A fast decrease of viable cells in the first ten minutes is followed by a much slower inactivation process; several hours are needed to obtain sterile solutions. Whereas in the fast step the vegetative *Bacilli* are killed, the slow step is caused by spores. In Figs. 8 and 9 the temperature- and pressure- dependence for vegetative forms can be seen; it looks similar to the diagrams of *E.coli* (Figs. 9.10-3 and 9.10-4) with the

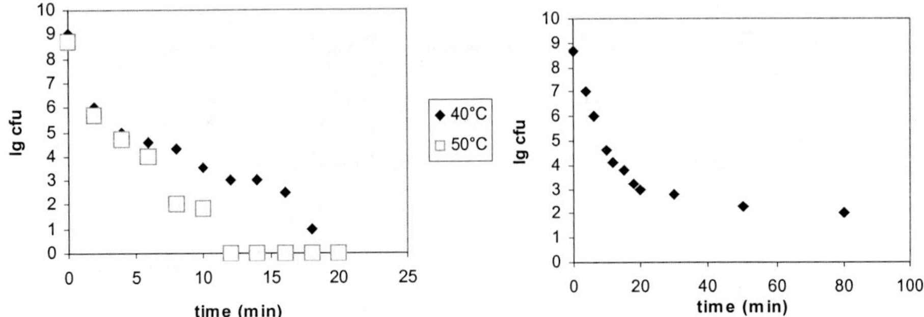

Fig. 9.10-3. Pressure inactivation of *E.coli* at 2.5 kbar at 40 and 50°C (adapted from[18])

Fig. 9.10-4. Pressure inactivation of *E.coli* at 2.5 kbar and 25°C (adapted from[18])

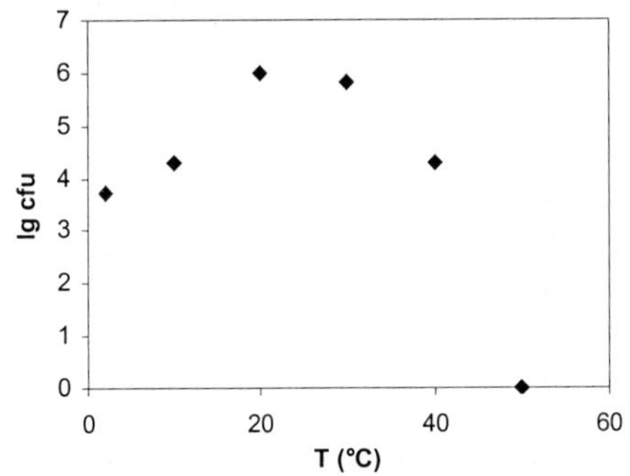

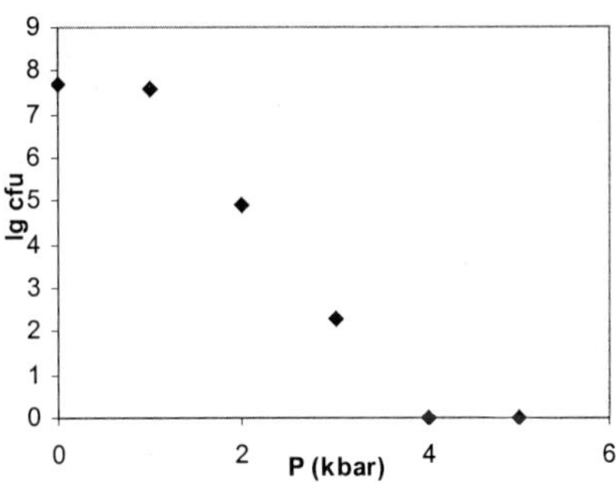

Fig. 9.10-5. Temperature dependence of *E.coli* inactivation, 15 min at 2 kbar (adapted from [18])

Fig. 9.10-6. Pressure dependence of *E.coli* inactivation, 9 min at 4°C (adapted from [18])

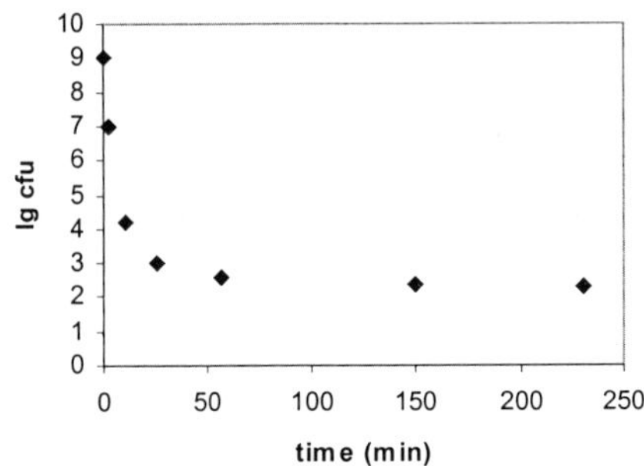

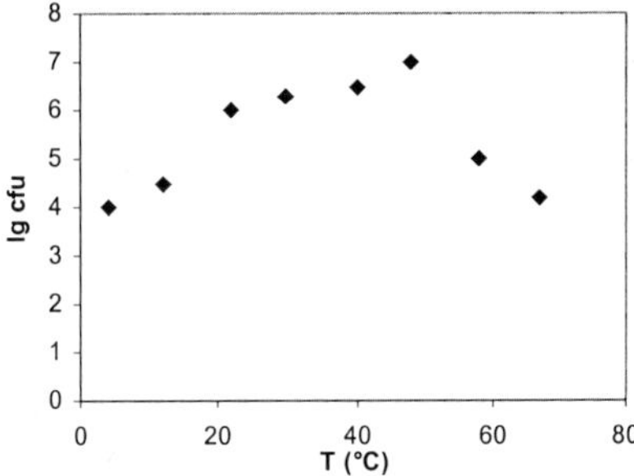

Fig. 9.10-7. Pressure inactivation of *B.Strarothermophilus* at 2.5 kbar and 60° (adapted from[18])

Fig. 9.10-8. Temperature dependence of *B.Stearothermophilus* inactivation, 15 min at 2 kbar (adapted from[18])

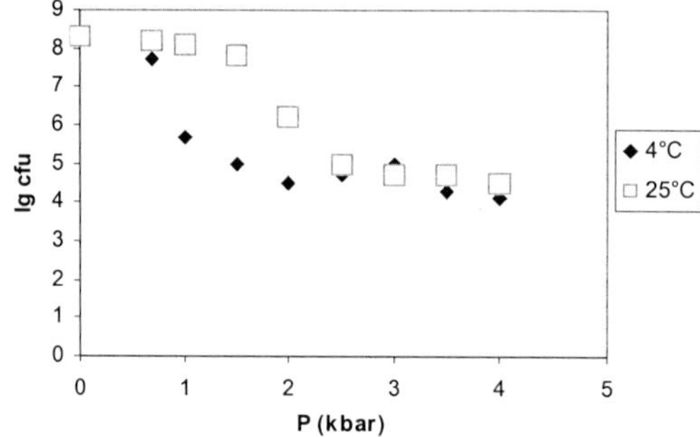

Fig. 9.10-9. Pressure dependence of *B.Stearothermophilus*
inactivation, 15 min, at 4°C and 25°C (adapted from [18])

only exception that plateaus are reached at high temperature and high pressure. These plateaus
again indicate the fractions of spores which are still alive.

9.10.3 Supercritical CO_2 treatment

9.10.3.1 State of the art

Nowadays, supercritical CO_2 is used in numerous technological processes for the
production or treatment of new materials. Among others, the food industry is seeking new
alternative techniques to protect and preserve foods. The use of CO_2, a non-inflammable,
atoxic, relatively inexpensive gas, under supercritical conditions, could become relevant in
this field. In fact, it has been demonstrated that the effect of microbial inactivation, assuring
healthy food preservation is already consistent at more moderate pressures (lower than 200
bar) than those employed by traditional hydrostatic-pressure HPT methods.

The open literature provides relatively few studies concerning bacterial inactivation by
CO_2, mainly published by American and Japanese authors. The first article is dated 1987 [31].
In this, a batch equipment was used, and samples of *E. coli*, *Staphylococcus aureus*, *Baker's
yeast*, and spores of *Aspergillus niger* were treated. An operating pressure of 200 bar at 35°C
with a residence time of 2 hours was sufficient to reduce the number of bacterial colonies by
many orders of magnitude ($8*10^4$ times for spores, $2*10^6$ times for yeasts). Wei *et al.*[32]
published interesting results on the microbial inactivation with CO_2: they treated pathogenic
microbes, *Listeria monocytogenes* and *Salmonella typhimurium* in pure water at different
operating pressures (54.4, 57.8, 61.2 atm) at 35°C for 2 hours. Only a treatment at the highest
pressure led to a totally sterile solution. The same treatment with N_2 did not lead to any
microbial reduction. Other experiments with different complex substrates were performed.

More recently [34] it was discovered that the CO_2 treatment is much more efficient if a
semi-continuos apparatus (called a "micro-bubble" method) is employed instead of the batch
apparatus: the contact surface between the gas and the medium where the microbes live is
much larger, as the liquid appears in the form of micro-bubbles created by the turbulence of
the gas-flow. In this way, the diffusion to the liquid phase is enhanced as well as the material

transport. Ishikawa *et al.* [33] performed their experiments with this new device: by treating *Lactobacillus brevis* and *Saccharomyces cerevisiae* they obtained a complete sterilization, reducing the permanent time compared to the batch system. Furthermore in 1997 [35] Ishikawa's group used the same method to inactivate spores of *Bacillus polymyxa*, as spores show a greater resistance to sterilization than do vegetative forms, and they tried to combine CO_2 treatment with a mild heating. For example a reduction of 100,000 times can be obtained at an operating pressure of 300 bar, a permanence time of 45 minutes and a temperature of 45°C; the same reduction could be performed with a permanence time of 15 min and a temperature of 60°C. Nowadays, no high-pressure hydrostatic treatments are able to completely kill spores.

Bacterial inactivation is achieved by CO_2 absorption in the liquid phase, even though the reason why it happens is still not clear. In this respect, batch- and semi-continuous operating modes are substantially different. In the batch system the residence time, *i.e.*, the time of contact between gas- and liquid phase, must be sufficient to allow the diffusion of CO_2 in the liquid, and is therefore a fundamental parameter to assure a desired efficiency. In the semi-continuous system the contact between the phases is localized in the surface of the moving micro-bubbles. In this second case, the efficiency of the process is influenced by temperature, pressure, gas flux, bubble diameter, and other parameters that modify the value of the mass-transfer coefficient. Therefore, it is not correct to use the residence time as a key parameter in the semi-continuous process. In fact, a remarkable microbial inactivation is reached even with an exposure time of 0 min (*i.e.*, pressurizing and immediately depressurizing the system): these two steps are sufficient to allow CO_2 to diffuse through the liquid phase.

Regarding the mechanism of the process, Haas in 1989 [36] suggested that in a batch system the operating pressure and the permanence time are fundamental parameters for bacterial inactivation, while the temperature must at least be higher than the critical value of CO_2 (31°C). It was observed that CO_2 in the liquid phase has no effect on microbes' inactivation and that the presence of water is fundamental, as experiments with dry samples lead to no useful results. It seems that the pH value plays a important role in the process. Anyway, neither the operating pressure nor the reduction in pH owing to CO_2-absorption can explain its strong stabilizing action. According to recent studies [37] it depends on the specific interactions between CO_2 and cell membranes, which cause the production of ladders that lead the loss of cytoplasmic material and the sudden reduction in intercellular pH.

Kumagai *et al.*[37] studied the relationship between the rate of CO_2 absorption inside the micro-organism and the value of the operating pressure and were successful in correlating the rate of absorption with a kinetic constant of inactivation. Recent work of Enamoto *et al.* [38] confuted the hypothesis that the inactivation is due to the explosion of the cell during the depressurization step.

Ishikawa and his group published his second work in 1998 [39]: the effect of CO_2 was investigated in a continuous co-current equipment. Pressure, temperature, and exposure time were kept constant during all experiments; only the CO_2 flow-rate varied. It was clearly proved that the effectiveness of microbial reduction was influenced by the CO_2 concentration in the liquid phase. The latest works were published in 1999: Debs-Louka and co-workers [40] succeeded in correlating pressure and exposure time with cell viability in a mathematical expression. They confirmed that the antimicrobial effect is gas-specificity dependent, and is increased by the water-content inside the cell. Dillow and co-workers [41] published an interesting paper regarding CO_2-treatment of a few bacteria and a colony of spores (*B. cereus*). The experiments were run in liquid medium as well as in biodegradable polymers

(PLA and PLGA, widely used in pharmaceutical applications). The most sensitive micro-organisms were *E. coli* and *P.vulgaris*, and the most resistant, *B. cereus*. As the SEM observations of *S.Aureus* and *P.Aeruginosa* showed intact cell walls, it was hypothesized that a pH reduction could be the main cause of cell death.

S.I. Hong *et al.* [42] confirmed that the inactivation rate increased with pressure, exposure-time and with decreasing pH of media. They stated that microbial inactivation was governed essentially by penetration of CO_2 into cells, and its effectiveness could be improved by enhancing the transfer rate. Microbial reduction of more than six powers of ten occurred within 30 min, under a CO_2 pressure of 2000 psi at 30°C. The authors hypothesized that cell death resulted from the lowered intracellular pH and damage to the cell membrane owing to penetration of CO_2.

Finally we quote a paper by Erkmen [43]. He investigated the inactivation of *E. faecalis* suspended in physiological and complex substrates (orange-, peach-, or carrot juice) and obtained a total inactivation under specific conditions of pressure, temperature and exposure time, apart from whole- or skimmed milk.

The dates of these references tell us about the current interest in this topic. However, despite the promising results obtained, and the great interest paid to this new alternative method, the mechanism of cell disruption is still unclear and the quantification of the technological process is not yet precise. Furthermore, it would be essential to understand the relationship between the degree of inactivation and the dimensions and the structure of cell membranes, and the shape of the micro-organism.

So far, no report of industrial applications is available, even though this new technique looks promising for protecting heat-sensitive foods. Also, in the cases where a chemical additive is needed for sterilization, this process uses lower pressures than other traditional sterilization methods and therefore its scale-up is promising and possibly economically convenient.

9.10.3.2 Equipment and Methods

Equipment

In Fig. 9.10-10 a diagram of typical batch equipment is shown, and in Fig. 9.10-11, a detail of the cell where samples can be located.

The apparatus consists of few elements. The CO_2 is supplied by a cylinder at the pressure of 55–60 bar. It passes through a filter (Fc) and then through a condenser, R, where it is cooled down (0–5°C) and liquefied.

The liquid arrives at the accumulator (A) and then is pumped by the volumetric pump (PCO_2). The CO_2 in the liquid phase evaporates in the heat-exchanger (W), and finally arrives inside the vessels (E1-E2). The temperature in the heat exchanger corresponds to the operating temperature of the experiment. In order to maintain the temperature in the vessel at a desired value, the temperature of the exchanger must be fixed slightly higher because of thermal dispersion through the line. The manual oulet valves must be heated by an electric resistance, otherwise they freeze during the depressurization step. At the end of the experiment, the gas is discharged into the atmosphere. The vessels are supplied with pressure- and temperature sensors and a temperature controller which acts on the flow-rate of the heating fluid of the jacket (Fig. 9.10-11). Figure 9.10-12 shows the diagram of the semi-continuous equipment used in our laboratory, and in Fig. 9.10-13, a detail of the cell where the liquid sample is located.

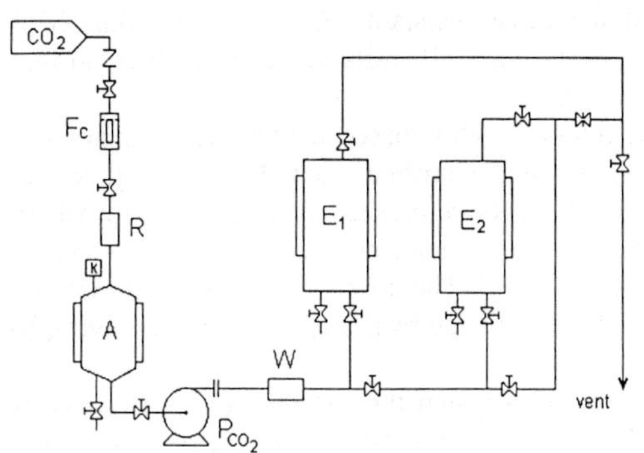

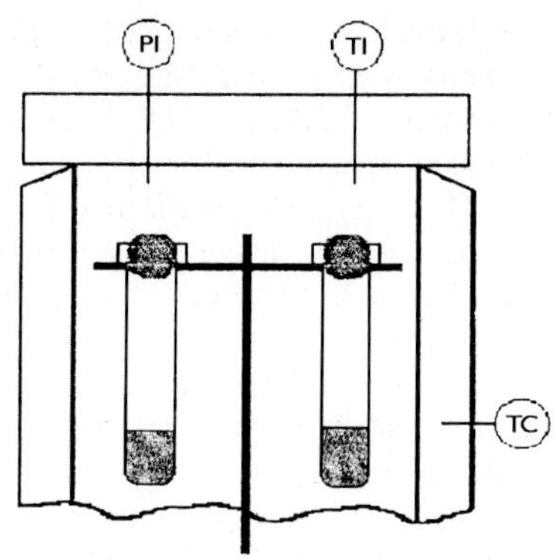

Fig. 9.10-10. Schematic diagram of prototype batch apparatus. Fc, filter; R, condenser; A, accumulator; Pco2, pump; W, heat exchanger; E1, E2, batch reactors (adapted from [29])

Fig. 9.10-11 Detail of a sample container. PI, pressure indicator; TI, temperature indicator; TC, temperature controller (adapted from [29])

As described above, CO_2 from the cylinder must be liquefied before being pumped, and heated after compressing. A HPLC pump with a maximum flow-rate of 25 ml/min is utilized.

The vessel (V) consists of a hollow cylinder and two taps which are screwed onto the vessel shell. They are supplied with one hollow, where a small quantity of bacteriological cotton is fitted in order to avoid losses of micro-organism to the surroundings, and with a porous metallic filter (average pore-diameter, 5 μm) acting as a CO_2-distributor. Such a filter (frit) allows the gas to flow and to get atomization in micro-bubbles, but it does not allow the liquid to leak. There are two seals: the first one is in Teflon, between the tap and the filter, the second is a neoprene O-ring between the vessel and the tap.

The vessel is supplied with a temperature-indicator (the probe, a Pt 100Ω, is located inside the vessel) and a temperature ON–OFF controller (see Fig. 13); it is heated by an electric resistance (R2), manually regulated. The vessel, as well as the resistance, is thermally insulated to avoid heat dispersion.

The outlet micrometric valve that is used to regulate the CO_2 pressure in the autoclave is heated by another resistance (R3) to prevent freezing, which could lead to a sudden and unsustainable pressure change.

Methods

In the following pages, we describe some experiments run in our High Pressure laboratory, in order to show how CO_2 treatment technology can be applied to micro-organisms of interest for the agro-alimentary industry.

A sourdough yeast strain (*S. cerevisiae*) and two bacterial species (*S. marcescens*, *B. subtilis*) were treated in a physiological buffer (PBS). The first experiments (with the yeast) were performed in a batch apparatus (Fig. 10) as well as in a semi-continuous apparatus (Fig. 12). As the latter device showed a higher efficiency in the microbial inactivation than the

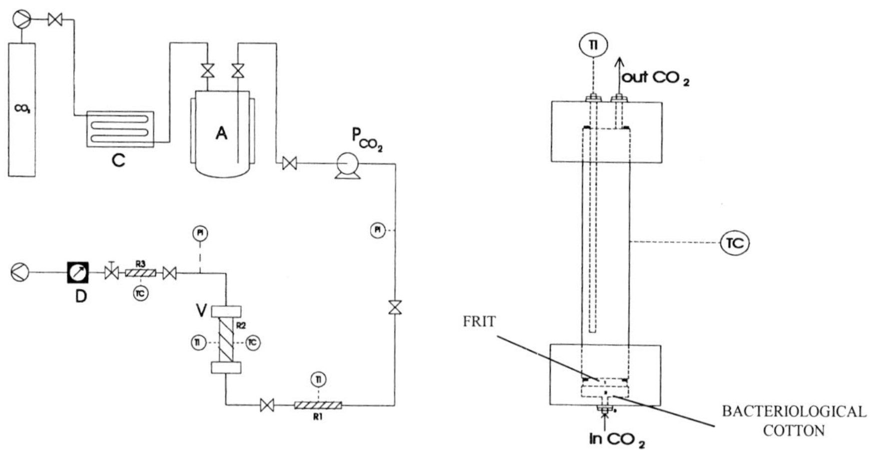

Fig. 9.10-12 Schematic diagram of prototype semi-continuos apparatus. C, Condenser; A, Accumulator; Pco$_2$, pump; R1, R2, R3, electrical resistances; V, pressurized vessel; D, rheostat

Fig. 9.10-13 Schematic diagram of vessel C. TC, temperature controller; TI, temperature indicator.

former, the other experiments, with the bacterial species, were performed in the semi-continuos apparatus. To perform the process, three steps are required: the loading of the vessel with CO_2 up to the selected P and T, the exposition at these conditions, and the discharge of CO_2 before collecting the tubes.

When the CO_2 treatment was finished, residual viable micro-organisms were counted according to plate standard procedures (for solid media and conditions see Table 9.10-1). The cultures were suspended in PBS buffer and diluted to reach about 10^7 colony-forming units, CFU/ml, for bacteria and 10^5 CFU/ml for yeast.

Table 9.10-1
Growth conditions in liquid medium for micro-organisms

Microorganism	Medium	Conditions
YEAST S.cerevisiae	YPD Broth	30°C, 48h
GRAM POSITIVE BACTERIA B.subtilis	Brain Heart Infusion	37°C, 24h
GRAM NEGATIVE BACTERIA S.marcescens	Brain Heart Infusion	37°C, 24h

In the test tubes, 10 ml of cell in PBS were employed and in the vessel, 25ml. Tables 9.10-2 and 9.10-3 give results for the first run of experiments with the batch system and the semi-continuous system, expressed as $log_{10}(N_o/N)$ *versus* time, where N_o is the initial microbial

Table 9.10-2
S.cerevisiae reduction at 40°C as a function of time and pressure in a batch process (adapted from [29])

Pressure (bar)	Time (min)	$Log_{10}(No/N)$
200	40	0.79
200	60	>5.00
200	80	>5.47
100	40	0.22
100	60	2.35
100	80	>4.44

Table 9.10-3
S.cerevisiae reduction at 38°C as a function of time and pressure in a semi-continuous process.

Pressure (bar)	Time (min)	$Log_{10}(No/N)$
55	30	>5
74	15	>5
74	10	>5
74	7.5	3.2
80	15	>5
80	10	>5

count and N is the microbial count after treatment. The values are the arithmetic means from three different runs. Data marked with the symbol ">" mean that no CFU was detected at those conditions.

From the data reported above, it is clear that the semi-continuous system is much more efficient than the batch one. If the main cause of microbial death is the CO_2 concentration in the liquid phase, in the flux process high gas concentrations are reached faster, as the bubbling enhances and favours the mass transport of CO_2 from the gas- to the liquid phase. In the semi-continuous process, a total sterilization is obtained with an operating pressure of 74 bar and an exposure time of 10 minutes, while in a batch process, 60 minutes are needed with an operating pressure of 200 bars. Tables 9.10-4 and 9.10-5 give results for other experiments run with the semi-continuous system.

The gram-positive bacteria, *B.subtilis*, as well as the gram-negative bacteria, *S.marcescens*, seem to be completely inactivated with an operating pressure of 74 bar and an exposure time of 15– or 10 minutes.

Table 9.10-4
B.subtilis reduction at 38°C as a function of time and pressure in a semi-continuos process.

Pressure (bar)	Time (min)	$Log_{10}(No/N)$
58	30	>7
74	30	>7
74	15	>7

Table. 9.10-5
S.marcescens reduction at 38°C as a function of time and pressure in a semi-continuos process.

Pressure (bar)	Time (min)	$Log_{10}(No/N)$
58	30	>7
74	0	4.7
74	7,5	>7
74	10	>7
74	15	>7

From the data above it is concluded that, in a continuous flow system, the pressure and the treatment time required for obtaining full inactivation are markedly lower than in a batch apparatus, so that the continuous process deserves to be investigated further.

9.10.4 Conclusions

Two relatively new methods for reducing the microbial activity of foods by means of high pressure have been presented. On the basis of the results obtained, each of them seems to be promising.

HTP has already become an alternative method to UHT and it is used nowadays for commercial production. It allows a total microbial inactivation, by operating with a relatively low temperature, and retaining unchanged the delightful tastes and smells of foodstuffs. Nevertheless, the cost of such a process is still high because of the extreme range of operating-pressure values required, and therefore it is not yet economically competitive with the traditional processes.

Results obtained for CO_2 treatment seem to be particularly attractive for the development of industrial applications; so far, mainly experiments on microbial suspensions have been carried out, and they are really successful. However, some questions remain as to the behaviour of micro-organisms in complex substrates, as it is well known that they are more resistant than in buffered solutions [32]. Another issue to be addressed in this case is the simultaneous effect of CO_2 as an extracting agent, as it could interfere with the quality of the final product.

References

1. International Conference on "High Pressure Bioscience & Biotechnology", August 30th, September 3rd, Heidelberg (D), 1998
2. J.P.P.M. Smelt, G.G.F. Rijke, High Pressure Biotechnology, C. Balny, R. Hayashi, K. Heremans, P. Masson Eds. (1992) 361
3. D. Farr, Trends Food Sci. Technol., 1 (1990) 14
4. B.H. Hite, West Va. Agr. Exp. Stat. Bull., (1899) 58
5. B.H. Hite, N.J. Giddings, J.R. Weakley, E. Chas, West Va. Agr. Exp. Stat. Bull., (1914) 146
6. J. Basset, M.A. Machebouf, Compt. Rend, Acad. Sci. Paris, 195 (1932) 1431
7. C.E. ZoBell, F.H. Johnson, J. Bacteriol., 57 (1949) 179
8. W.J. Timson, A.J. Short, Biotechnol. Bioeng., 7 (1965) 139

9. L. Lundgren, Physiologia Plantarum, 19 (1966) 403

10. J.G. Clouston, P.A. Willis, J. Bacteriol., 97 (1969) 684

11. J.G. Clouston, P.A. Willis, J.Bacteriol.,103 (1970) 140

12. P.A. Willis. Radiosterilisations of Parenteral Solutions, International Atomic Energy Agency, Vienna, 45

13. W.G. Murrel, P.A. Wills, J. Bacteriol., 129 (1977) 1272

14. G.W. Gould, A.J.H. Sale, J. Gen. Micribiol., 60 (1970) 335

15. G.W. Gould, A.J.H. Sale, W.A. Hamilton, J. Gen. Microbiol., 60 (1970) 323

16. K. Hallbauer, Diplomarbeit, University of Heidelberg, 1990

17. P. Butz, J. Ries, H. Traugott, H. Ludwig, Pharm. Ind., 52 (1990) 487

18. H. Ludwig, C. Bieler, K.E. Hallbauer, W. Scigalla, High Pressure Biotechnology, C. Balny, R. Hayashi, K. Heremans, P. Masson Eds., 224 (1992) 25

19. H. Ludwig, P. Gross, W. Scigalla, B. Sojka, High Pressure Research, 12 (1994) 193

20. I. Seyderhelm, D. Knorr, ZFL, 43 (1994) 17

21. H. Ludwig, B. Sokja, Pharm. Ind., 56 (1994) 660

22. H. Ludwig, B. Sokja, Pharm. Ind., 59 (1997) 335

23. H. Ludwig, B. Sokja, Pharm. Ind., 59 (1997) 436

24. I. Hayakawa, T. Kanno, M. Tomita, Y. Fujio, Journal of food science, 59 (1994) 159

25. I. Hayakawa, T. Kanno, M. Tomita, Y. Fujio, Journal of food science, 59 (1994) 164

26. K. Heremans, High Pressure Research in Biosciences and Biotechnology, K. Heremans (Ed.) Leuven University, Press. Leuven, Belgium

27. C. Schreck, H. Ludwig, High pressure Research in Biosciences and Biotechnology, K. Heremans (Ed.) Leuven University, Press. Leuven, Belgium

28. B. Mertens, G. Deplace, Food Technology, (1993) 164

29. A. Venturi, Thesis: "Effetti dell'anidride carbonica supercritica su microorganismi di interesse alimentare" University of Verona, Biothecnology-Agro Alimentary, 1998/1999

30. P. Butz, J. Ries, U. Traugott, H. Weber, H. Ludwig, Pharm. Ind., 52 (1990) 487

31. M. Kamihira, M. Taniguchi, T. Kobayashi, Agricoltural Biological Chemistry, 51 (1987) 407

32. C.I. Wei, M.O. Balaban, S.Y. Fernando, A.J. Peplow, Journal of Food Protection, 54 (1991) 189

33. H. Ishikawa, M. Shimoda, H. Shiratsuchi, Y. Osajima, Bioscience Biotechnology Biochemistry, 59 (1995) 1949

34. H. Ishikawa, M. Shimoda, T. Kawano, Y. Osajima, Bioscience Biotechnology Biochemistry, 59 (1995) 628

35. H. Ishikawa, M. Shimoda, K. Tamaya, A. Yonekura, T. Kawano, Y. Osajima, Bioscience Biotechnology Biochemistry, 61 (1997) 1022

36. G.J. Haas, H.E. Prescott, E. Dudley, R. Dik, C. Hintlian, L. Keane, Journal of Food Safety, 9 (1989) 253

37. H. Kumagai, C. Hata, K. Nakamura,. Bioscience Biotechnology Biochemistry, 61 (1997) 931

38. A. Enomoto, K. Nakamura, K. Nagai, T. Hashimoto, M. Hakoda, Bioscience Biotechnology Biochemistry, 61 (1997) 1133

39. M. Shimoda, Y. Yamamoto, J. Cocunubo-Castellanos, H. Tonoike, T. Kawano, H. Ishikawa, Y. Osaijima, Journal of Food Science, 63 (1998) 709

40. E. Debs-Louka, N. Louka, G. Abraham, V. Chabot, K. Allaf, Applied and Environmental Microbiology, 65 (1999) 626

41. A.K. Dillow, F. Dehghani, J.S. Hrkack , N.R. Foster, R. Langer, *PNAS*, 96 (1999) 10344
42. S. Hong, W-S. Park, Y-R. Pyun, International Journal of Food Science and Technology, 43 (1999) 125
43. O. Erkmen, Journal of the Science of Food and Agriculture, 80 (2000) 465

9.11 Dry-cleaning with liquid carbon dioxide

Paolo Pallado

via M. Ravel, 8   35132 Padova   Italy

9.11.1   Introduction

Carbon Dioxide (CO_2), in its liquid or supercritical state, is currently attracting much interest as an environmentally acceptable solvent, which is easy to handle, and available at low cost, even in high purity grade.

In the late 1960s, the potential of the non-toxic and environmentally benign of supercritical CO_2 (SC CO_2) for natural product extraction was recognized by several authors [1-3]. So far, the practical use of CO_2 has been addressed mainly as an extractive media in agricultural fields. The advantage of SC CO_2 over conventional extraction with liquid solvents is that the SC fluid has a very low viscosity and a high diffusion coefficient [4]. It penetrates solid substances efficiently and is removed after depressurization. The excess amount of adsorbed gas diffuses from the material under room conditions. Moreover, SC fluid's solvent power, which is related to solvent density, can be tuned by varying the pressure and temperature. Some substances can be selectively stripped away from other compounds from a raw material, and then the same substances are easily separated from the fluid by reducing the pressure and controlling the temperature.

Its application has been limited by the need for high pressures to dissolve substances having an affinity with CO_2, mainly non-polar compounds, but even small amounts of polar, amphiphilic, organometallic, or compounds with molecular weights up to 500 Da. Nevertheless, the poor solubility of many interesting target substances places a severe restriction on the widespread use of CO_2 as a solvent. Sometimes, unrealistically high pressures or expensive "CO_2-philic" compounds (substances with a high affinity for CO_2 solutions at lower pressures) are needed.

The specific properties of SC CO_2 make it an interesting "green" replacement for liquid organic solvents, which are often characterized by acute toxicity, ecological hazards, or drawbacks with disposal and recycling. The use of CO_2 has currently moved from the so-called conventional extraction process to several other areas as diverse as the dyeing and cleaning of fibre and textiles, polymerization and polymer processing [5], purification and crystallization of pharmaceuticals [6,7], comminution of difficult-to-handle solids [8], controlled drug delivery systems [9], and as a reaction medium for chemical synthesis [10].

Last, but not least, a great effort has been made by several companies worldwide for the development of the dry-cleaning of garments and textiles with liquid CO_2, in substitution for the organic solvents currently used, such as perchloroethylene, petroleum ether, and trichloroethane. The commercial target in this field is not limited to a few industrial plants, but to the huge number of industrial dry-cleaning companies, and to the dry-cleaning shops in commercial centers.

Especially thanks to these efforts, SC CO_2 can be handled safely in high-pressure equipment from the laboratory- to the industrial scale. Nevertheless, its limited power to dissolve polar or non-volatile compounds represents a major drawback in most of these applications, particularly in the garment-cleaning process, where the design of CO_2-philic solubilizers is of paramount importance.

9.11.2 Dry-cleaning processes

9.11.2.1 Conventional dry-cleaning

In the conventional dry-cleaning process, liquid organic solvents such as perchloroethylene and other petroleum-based substances are used. The machines have a washing chamber, inside which a rotating cylindrical basket is used for the motion and agitation of articles and garments. The removal of dirt, stains, and other contaminants on the articles to be washed is assured by the solvent's ability to dissolve the contaminants, and by the "fall and splash" effect caused by the rotation of the internal basket. The articles and the fluid are tumbled together by rotating the basket or "tumbler", so that the solvent is brought into intimate contact with the surfaces of the articles. As the solvent is absorbed by the absorbent portions, it dissolves these contaminants or maintains them in suspension. Solid particles and chemical- or organic contaminants are then forced from the internal part of the cylinder to the annular zone between the cylinder and the chamber's internal walls, and are removed by continuously re-circulating the liquid solvent from-and to-the washing chamber.

Dry-cleaning machines are usually provided with a centrifugal pump to force the liquid through a filter, and with a blower to move and filter the solvent vapour phase also. After the cleaning cycle, the solvent is drained from the cleaning chamber and may be reused, provided that the level of dissolved contaminants is not approaching saturation, otherwise it is regenerated by distillation or cleaned by mechanical filtration.

The conventional dry-cleaning process involves significant health and environmental risks associated with the toxic nature of conventional cleaning solvents. Percholoroethylene is suspected to be carcinogenic, while petroleum-based solvents are flammable and smog-producing. The search for alternative cleaning technologies, which are safe and environmentally acceptable, is still in progress, to substitute the solvents and methods of controlling exposure to dry-cleaning chemicals.

9.11.2.2 Dry-cleaning with liquid carbon dioxide

As CO_2 is an environmentally safe fluid, its efficient use as a dry-cleaning fluid avoids the risks and drawbacks associated with conventional cleaning solvents. Carbon dioxide has been identified as an inexpensive solvent, with an unlimited natural resource. It shows a high compatibility with garments, as it does not damage fabrics nor dissolve common dyes. In this way, CO_2 has been recognized as a good medium to be used in the dry-cleaning industry.

Also, the solvating properties of CO_2 are quite non-typical of those of more traditional solvents. As the fluid's solvent power is a function of its density, CO_2 must be compressed to enhance the cleaning efficiency. Keeping CO_2 at relative high densities requires that the vessels have to be designed for higher pressures than those of the process itself, increasing machine's cost and safety level. As a first step to contain the machine cost, the process is performed with liquid CO_2 to reduce the vessels' design-pressure values. The pressure ceiling can be fixed at 70 bar, but this constraint limits the maximum allowable working temperature to be lower than 18°C, reducing the mass-transfer and solute dissolution inside the fluid. The viscosity of fatty substances shows a large dependence on temperature, decreasing dramatically above 50°C. Heavy greasy compounds tend to segregate as a solid phase at temperatures lower then 10°C.

The pressure is the main cost-driver, since it affects all process-housing component costs, as well as operational costs. Standard petro-chemical components can be used, as the maximum design pressure value can be set close to the CO_2 critical value.

The physical characteristics of the fluid do not allow one to run the process at temperatures higher than 20°C using liquid CO_2, as the process-control ceases to be affordable when the pressure and temperature approach the near-critical zone. The use of SC CO_2, at temperatures higher than 35°C, will require the pressure above 200 bar to have a comparable value in density, raising the costs for the hardware and process control. The cleaning of the garments is thus assured, coupling the affinity of CO_2 towards dissolving greasy substances, with mechanical removal of fine particles and stains.

Agitation represents another major contribution to costs. In a pressure vessel, mechanical agitation produced with moving parts is cost-prohibitive in terms of capital and maintenance expenditures. It has to be provided without moving parts, that is, with no electrical or mechanical devices which require a sealing system, to avoid CO_2 leakage from the washing chamber.

9.11.3 The CO_2 dry-cleaning process

9.11.3.1 Fundamentals

Several dry-cleaning systems using CO_2 as a solvent have been proposed and are currently under development. As the topic is mainly concerned with the definition of a cleaning process for industrial or commercial applications, the literature refers to several patents which differ in type and performance of the washing machines.

The CO_2 dry-cleaning process was originally disclosed by Maffei [11], who designed a simple system formed by a chamber, a storage tank and an evaporator. Garments are placed inside a basket contained in the chamber, then CO_2 is gravity-fed thereto from the refrigerated tank, passes through the garments, and is finally transferred to the evaporator. In the evaporator, CO_2 is evaporated and separated from the soil, which is recovered from the bottom of the tank. The vapour CO_2 is compressed into a heat-exchanger to condense before being stored in the storage tank.

Many improvements have been made since Maffei's system. First of all, the washing chamber needs to be pressurized to maintain the CO_2 in the liquid state, otherwise the CO_2 must be very cold to remain liquid. To enhance the washing efficiency, garments have to be agitated to remove solid stains and particles which are not soluble in CO_2.

The simplified scheme depicted in Fig. 9.11-1 is useful to understand the machine units and process steps. The main representative vessel is the washing chamber, where garments are placed inside a basket. Normally, other two main vessels are used to handle the fluid during the process, a storage tank and an operative or service tank. The clean CO_2 is kept in the liquid state in the storage tank, which is used as an additional reservoir. The process liquid CO_2 is stored in the operative tank. Two small filtering vessels, a lint-trap with a filter train, and a button-trap, complete the number of vessels normally present in the machine. The button-filter is located at the bottom of the chamber; a wire filter avoids the dragging of fibres, buttons, and other macroscopic pieces which could seriously damage the liquid pump and the valves. The lint-filter is used as the main filtering device to capture the fine solid particles removed from the garments in the washing step, when the liquid is continuously pumped from and to the chamber. In some versions, active-carbon powder filtering-cartridges are used to

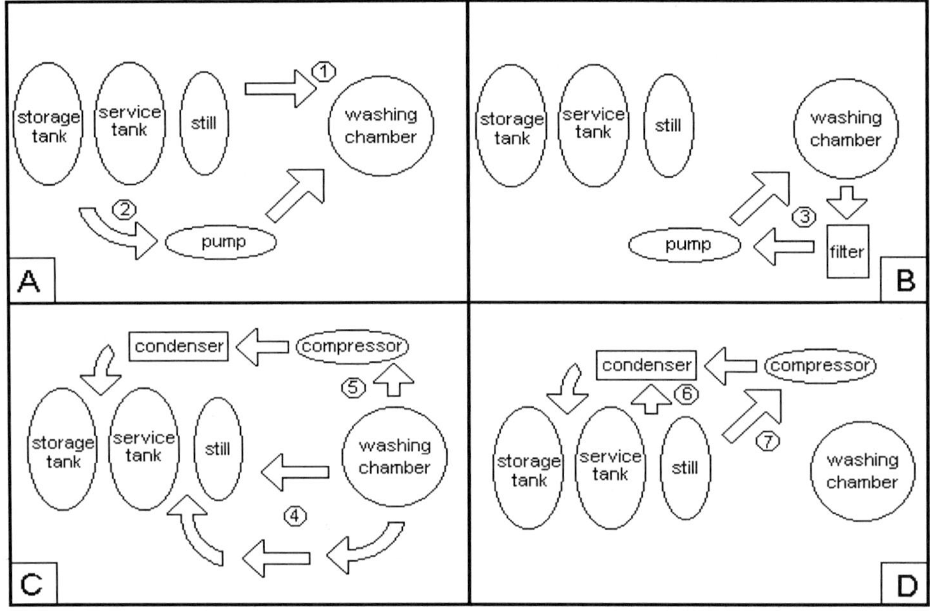

Figure 9.11-1: Schematic representation of the dry-cleaning process, displaying different steps from A to D.

trap unpleasant odours dissolved in the fluid.

The chamber is initially pressurized with vapour, fed from the still or the service tank (step 1), and then loaded with liquid from the service tank (step 2). A liquid piston pump assures the pumping of the liquid CO_2 in the system. Normally, the pump is used to displace the liquid CO_2 from the storage and operating vessels to the chamber, and vice versa, and through the chamber during the washing phase (step 3). Pressure-drops are limited to values lower than 20 bar, but very high flow-rates are needed to reduce the process time. At the end of the washing step, the liquid is drained from the chamber to the service tank (step 4).

A compressor is required to recover the vapour CO_2 from the chamber at the end of the washing cycle, to minimize the loss in CO_2 for each cycle (step 5). The vapour is condensed through a heat-exchanger and the liquid is recovered and returned to the storage vessel. The pressure-drop at the end of this step can be larger than 50 bar, with a temperature of the process-fluid higher than 80°C downstream the compressor.

The vapour phase is sometimes moved by heating in the distillation step, using the positive pressure-drop generated by the increase in temperature (step 6). Nevertheless, the compressor can be used to help the distillation process, sucking the vapour from the still to the heat exchanger and then to the storage or service tank (step 7).

The operative tank can be utilized as a full capacity still if it is provided with an internal heat exchanger, otherwise an additional still, smaller in size, is necessary to perform the distillation step and to clean the "dirty" liquid CO_2. As the percentage of lypophilic substances in the garments is very low, the "dirty" liquid CO_2 distillation by means of the little still is carried out after a series of washing cycles, depending on the amount of liquid loaded in the chamber. So, a distillation per dilution process is reached, as the operative tank remains partially dirty at its bottom. On the other hand, using a full capability still liquid CO_2 vaporizes and can be completely cleaned from stains and soluble substances.

Two main thermal fluxes characterize the system from a thermodynamic point of view. These come from the liquid-CO_2 distillation and, consequently, the vapour condensation to recover the liquid in the operative or storage tank, and the condensation of vapour CO_2 recovered from the chamber at the end of the washing step. Most frequently, the distillation is carried out in a vessel which is equipped with an internal heat exchanger, dip in the liquid to be cleaned. The heat exchanger can be of an electric resistance type, or a series of coils through which hot water flows. The heat exchange is very efficient at the beginning, when the exchanger is completely immersed in the liquid, then the efficiency reduces dramatically as the exchange-surface in contact with the liquid decreases.

As the overall heat coefficient of the vapour phase is very low, a plate heat-exchanger is used for the condensation. The machine is equipped with a refrigeration unit designed according to the thermal flux required for the condensation of the vapour CO_2. The refrigeration unit is then used to control the temperature in the chamber during the washing step or to prevent over-temperatures in the storage vessels. The system is finally optimized if the refrigerating fluid used to cool down the refrigeration unit process-fluid is fed to the still to supply the heat required for the distillation process.

A third heat flux is necessary for pre-heating the chamber before the gas recovery. To avoid the insertion of an additional plate heat-exchanger, it is preferable to use electric resistances which can supply high thermal fluxes per specific surface.

9.11.3.2 Garments agitation

Many systems have been proposed for the agitation of garments in CO_2 dry-cleaning systems. Deewes [12] proposed a rotating basket, used by the majority of the traditional dry-cleaning systems. The interior of the basket includes projecting vanes so that a tumbling motion is induced upon the garments when the basket is rotated by an electric motor, causing the garments to drop and splash into the liquid solvent.

Chao [13] disclosed a variety of agitation techniques, that is a "gas bubble and boiling agitation" where the liquid CO_2 in the basket is boiled, a "liquid agitation" where a nozzle spraying CO_2 tumbles the liquid and the garments, a "sonic agitation" where sonic nozzles induce vibrating waves, and a "stirring agitation" where an impeller creates the fluid displacement. In this patent, as in most of the developed systems, the basket in the washing chamber is completely drilled to allow the movement of solid particles and contaminants towards the annular free volume, from which they are removed inside the solvent stream.

A gas-jets agitation technique has been proposed by Purer [14] to remove particulate soils from fabric. The agitation occurs apart from the immersion of the garments in liquid CO_2; nevertheless CO_2 has been employed both as a gas and a solvent. Townsend and Purer [15] disclosed a rotating basket powered by a hydraulic flow generated by forcing the fluid through a number of nozzles. Finally, an improved nozzles agitation system has been applied in the Hughes Dry Wash™ machine [16].

Despite the several systems disclosed, garments' agitation is substantially operated in two ways, with a rotating basket or by fluid-jets. In the former, the mechanical force necessary to dislodge the soiling from the substrate is provided by the fall-and-splash effect of the solvent-loaded garment from the top of the rotating drum into the solvent pool below. The magnitude of the force is dependent on the height of fall, which depends on the solvent level in the chamber, and the solvent density. The relatively low density of liquid CO_2 discourages the application of this agitation mechanism.

In the fluid-jet system, the garments are submerged in liquid CO_2 within the perforated basket, set into motion, and are agitated by high-velocity fluid jets. The jets are discharged through nozzles, which are set in an appropriate configuration, and the garments agitated through a Venturi effect. As the garment is accelerated by the jet, its fibres are stretched, and the soil trapped into and between the fibres are expelled. As a summary, the power of the pump used to force the fluid through the nozzles is transformed into fluid velocity which is transferred to the garment as momentum allowing the expulsion of soiling. Four series of nozzles are fixed inside the chamber of the Hughes machine [16]. The chamber is completely filled with liquid CO_2 and pressurized, then the liquid is forced through two nozzles each time inside the chamber by opening and closing a valve located upstream of each pair of nozzle series. The nozzles are positioned so that they create an agitating vortex inside the basket as liquid CO_2 is forced through them. The soil-laden liquid CO_2 which leaves the basket is moved from the chamber to a lint-trap and a filter-train by means of a circulating pump.

Liquid CO_2 has a lower density than the liquid organic solvents normally used in conventional dry-cleaning processes, as shown in Table 9.11-1. Moreover, its viscosity is one order of magnitude lower than organic solvents. As a consequence, the jet-cleaning agitation seems to offer the more feasible approach for CO_2 dry-cleaning machines.

Table 9.11-1

Physical properties of solvents commonly used in dry-cleaning processes

solvent	Surface tension, Dynes/cm	Density, g/cm^3	Viscosity, cPoise
Perchloroethylene	32.0	1.60	0.88
1,1,1-Trichloroethane	25.6	1.40	1.20
Freon-113	19.6	1.60	0.68
Freon-114	12.0	1.50	0.38
Water	72.0	1.00	0.90
Liquid CO_2	3.0	0.85	0.08

This assumption is further supported when one considers that the rotating tumbler needs a special sealing system to prevent CO_2 escaping or leaking round the drive shaft which is connected to the tumbler through a wall of the pressure vessel. When a magnetically agitated shaft is used to avoid the constraint of rotating shaft seal, the thickness of stainless steel needed for pressure resistance dissipates the magnetic coupling, limiting the torque that can be transferred from the outer magnets to the internal shaft. These solutions have been found relatively ineffective, especially for commercial-size units, and have high maintenance costs.

However, in the jet-cleaning agitation process the chamber has to be completely filled with liquid to allow correct functioning of the system, and this feature reflects on the machine design, which requires larger tanks, more powerful pumping devices, and an increase in the thermal fluxes needed for the distillation and condensation steps. Operational times are also increased. Finally, the fall-and-splash effect requires a load of liquid which is sufficient just to submerge the garments.

9.11.3.3 Machine configurations

We refer to the simplified scheme depicted in Fig. 9.11-2, where the main vessels are the washing chamber C, the storage tank ST, the operative tank OT, and the evaporator or still ET. Of course, the chamber is provided with an agitation device and with a completely drilled basket, into which garments are placed.

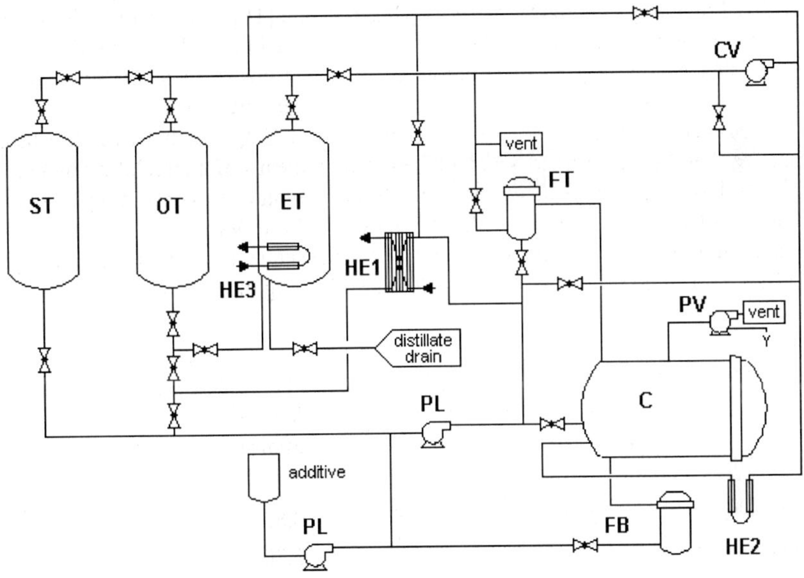

Figure 9.11-2: Simplified scheme of the high pressure lines for a dry-cleaning CO_2 unit. Legend: washing chamber, C; compressor, CV; evaporator or still, ET; button filter, FB; lint-trap, FT; heat-exchanger, HE#; operative tank, OT; additive pump, PA; liquid pump, PL; vacuum pump, PV; storage tank, ST.

The clean liquid CO_2 is stored in the storage tank, and dirty CO_2 in the operative tank. The storage and the operative tanks have the same internal capacity, while the still can be smaller.

In a steady-state configuration, the liquid CO_2 stored in the operative tank is used several times for washing, while the storage tank is half empty and is used as a reservoir, to fill up the chamber. After the garments are loaded in the basket, and the chamber closed, the air in the chamber is extracted by a vacuum pump, PV, then CO_2 vapour from the storage or the operative tank is fed thereto to avoid a sharp reduction in the temperature. To complete the chamber filling, liquid from the operative tank is fed by the liquid pump PL. At the end of the filling step, the chamber is over-pressurized to avoid vapour formation, and the washing step

648

can begin. The liquid is re-circulated continuously from and to the chamber, passes through the garments, and is filtered through the lint-trap, FT. A plate heat-exchanger, HE1, is used to keep the temperature lower than 20°C.

At the end of the washing step, the liquid is drained to the operative tank by the pump, and through the button filter, FB, then the gas-recovery step takes place to recover most of the vapour or gaseous CO_2 from the chamber. As the gas-recovery can be assumed to be like an adiabatic process, the fluid in the chamber is pre-heated by an electric resistance heat exchanger, HE2, to contain the reduction in temperature. The CO_2 vapour is compressed by the oil-free compressor CV and fed to HE1 to condense before being stored in the operative tank. Finally, the chamber is vented and the door can be opened to recover the cleaned garments.

9.11.4 Conclusions

The use of SC CO_2 for the cleaning of garments, as an alternative method to the liquid solvents, is very attractive from an environmental point of view, and for the safety and health of the operators involved.

Nevertheless, several issues have still to be addressed and improved, such as the ability of the solvent mixture to dissolve polar and heavy compounds, the entire washing and rinsing cycle to maximize stain removal, the machine's cost compared to the conventional machines.

As a final remark, the machine costs matched with the innovative character of the technology could confine the application just for industrial laundries, limiting the possibility to extend dry-cleaning machines to an end-user approach.

References

1. Brunner, G. and Peter, S. (1982), Ger. Chem. Eng., 5, pp. 181-195.
2. Stahl, E., Quirin, K.W. and Gerard, D. (1988). Dense Gases for Extraction and Refining, 1st Ed., Springer-Verlag, Berlin.
3. Williams, D.F. (1981), Chem. Eng. Sci., 36 (11), 1769-1788.
4. McHugh, M. and Krukonis, V. (1986). Supercritical Fluid Extraction: Principles and Practice, Butterworths, Boston, 217.
5. Gallagher, P., Proc. 3rd Int. Symp. on Supercritical Fluids, Strasbourg, France, 1994, 253-264.
6. Larson, K.A. and King, M.L. (1986), Biotech. Prog., 2 (2), 73-82.
7. Matson, D.W., Fulton, J.L., Petersen, R.C. and Smith, R.D. (1984), Ind. Eng. Chem. Res., 26, 2298-2306.
8. Gallagher, P.M., Coffey, M.P., Krukonis, V.J. and Klasutis, N. (1989), Supercritical Fluid Science and Technology, pp 334-354, ACS Symp. Ser., No. 406, Chap. 23, Ed. K.P. Johnston and J.M.L. Penninger.
9. Sang-Do Yeo, Gio-Bin Lim, Debenedetti, P.G. and Bernstein, H. (1993), Biotech. and Bioeng., 41, pp. 341-346.
10. Poliakoff, M., Howdle, S. M., Proc. 3rd Int. Symp. on Superc. Fluids, Strasbourg, France, 1994, 81-92.
11. Maffei, R. L., U.S. Patent No. 4.012.194, 1976.

12. Dewees, T. G., Knafele, F. M., Tayler R. G., Iliff, R. J., Carty, D. T., Latham, J. R., Lipton, T. M., U.S. Patent No. 5.267.455, 1990, The Clorox Company, Oakland, California.
13. Chao, S. C., Stanford, T. B., Purer, E. M., Wilkerson, A. Y., U.S. Patent No. 5.467.492, 1995, Hughes Aircraft Company, Los Angeles, California.
14. Purer, E. M., Wilkerson, A. Y., Townsend, C. W., Chao, S. C., U.S. Patent No. 5.651.276, 1997, Hughes Aircraft Company, Los Angeles, California.
15. Townsend, C. W., Purer, E. M., U.S. Patent No. 5.669.251, 1997, Hughes Aircraft Company, Los Angeles, California.
16. Chao, S. C., Purer, E. M., U.S. Patent No. 5.822.818, 1998, Hughes Engineering.

SUBJECT INDEX

654